# Nomenclature of Organic Compounds

# Nomenclature of Organic Compounds

## PRINCIPLES AND PRACTICE

**SECOND EDITION**

Robert B. Fox
Warren H. Powell

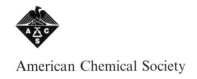

American Chemical Society

OXFORD
UNIVERSITY PRESS

2001

# OXFORD

UNIVERSITY PRESS

Oxford   New York
Athens  Auckland  Bangkok  Bogotá  Buenos Aires  Calcutta
Cape Town  Chennai  Dar es Salaam  Delhi  Florence  Hong Kong  Istanbul
Karachi  Kuala Lumpur  Madrid  Melbourne  Mexico city  Mumbai
Nairobi  Paris  São Paulo  Shanghai  Singapore  Taipei  Tokyo  Toronto  Warsaw

and associated companies in
Berlin   Ibadan

Developed and distributed in partnership by
American Chemical Society and Oxford University Press

Published by Oxford University Press, Inc.
198 Madison Avenue, New York, New York 10016

Library of Congress Cataloging-in-Publication Data

Fox, Robert B., 1922–
    Nomenclature of organic compounds : principles and practice.—2nd ed./
Robert B. Fox, Warren H. Powell.
       p. cm.
    Rev. ed. of: Nomenclature of organic compounds/edited by
John H. Fletcher, Otis C. Dermer [and] Robert B. Fox.
    Includes bibliographical references and index.
    ISBN 0-8412-3648-8
    1. Chemistry, Organic–Nomenclature.  I. Powell, Warren H., 1934–  II. Title.
QD291 .F6  2000
547'.0014—dc21      99-043801

9 8 7 6 5 4 3 2 1

Printed in the United States of America
on acid-free paper

This book is dedicated to the many who have devoted so much time, effort, and wisdom to the development of our chemical language. Some of you are mentioned in Chapter 1, but to all we say "Thank you".

# Preface

Chemical language, particularly the words and syllables used to describe structures and substances, has undergone considerable change since the mid 1970s. Much of this change is a result of the expanding scope of chemistry, especially organic chemistry, that has required refinements and extensions to existing language as well as the creation of new language for novel chemical structures. It is to those who wish to understand as well as those who simply have a need to use chemical language that this book is directed.

The problems of communication within organic chemistry have not changed, the computer notwithstanding. Organic chemists and those who are involved with the products of organic chemistry must still understand names for chemical structures and chemical substances in a variety of contexts in such a way that the information passed forward is the same as the information received. Contexts range from technical journals, formal and informal presentations, indexes, and catalogs to advertising, newspaper articles, and documents such as patents and tariff regulations; each with audiences that often overlap. Rules for systematic nomenclature of chemical substances adapt and change. Traditional names remain in use long after recommendations have been made to abandon them, and new names often seem to be created in a vacuum. Readers of this book will still have the seemingly simple need for an acceptable name for a given compound in a particular context, and there will always be questions about why one name is better than another. We hope that the pages that follow will respond to such needs and questions.

We wish to acknowledge the inspiration provided to us by the late Dr. Kurt Loening and express our appreciation to the reviewers of all or part of this book, Prof. H. A. Favre, Chairman of the International Union of Pure and Applied Chemistry (IUPAC) Commission on Nomenclature of Organic Chemistry (CNOC), Ms. J. E. Merritt, Senior Scientific Information Analyst at the Chemical Abstracts Service (CAS), Dr. Carlton Placeway, Senior Editor at the Chemical Abstracts Service (CAS); and to J. E. Blackwood and Dr. P. M. Giles, Jr. for many helpful discussions and insightful comments.

Finally, we wish to express our appreciation to Sandra Powell and (the late) Mattie Rae Fox whose support and encouragement over many years made this book possible.

# About This Book

This book begins with an introduction which talks about types of names, gives an overview of organic nomenclature, and compares nomenclature for general use with nomenclature for indexes. To provide a historical perspective, a discussion of the origin and evolution of organic chemical nomenclature is given in chapter 1. Chapters 2 and 3 discuss the conventions of organic nomenclature and explain the methods used in generating names for organic compounds. Chapter 4 notes some common errors and pitfalls on organic nomenclature.

Chapters 5–9 describe nomenclature for compounds that serve as parent hydrides for naming organic compounds, the basic structures on which organic names are based. The next fifteen chapters (10–24) are devoted to compound classes identified by means of various suffixes or class names used in naming them. Chapter 10 covers substituent groups that are not named by means of suffixes or class names. The next five chapters (25–29) and chapter 31 cover broad compound classes. Finally chapters 30, 32 and 33 deal with topics that apply generally to many of the other chapters, that is, stereoisomers, isotopically modified compounds, and radicals and ions.

The structures for the compounds named in this book are not all known; compounds not yet reported and hypothetical compounds need to be named, if for no other reason than discussion. It is recognized that there are many organic compounds whose precise structures are unknown; these also need names, but are beyond the scope of this book. Tautomerism is a particular problem for structure based nomenclature. Specific rules for choosing a preferred structure for tautomeric compounds have been established by the Chemical Abstracts Service (CAS). In this book, the structure drawn is the one named regardless of tautomerism; names for other tautomers of a given structure will be found in the chapter for its particular class. Most chapters includes a final section, called Additional Examples, that gives many examples illustrating a variety of structural types.

In this book alternative acceptable names are given for many structures. We encourage systematic names, but rigidity is not to be a hallmark of this volume, and we recognize that context may be quite important in choosing a name to be used. Acceptability of any name is often in the eye of the beholder, but an effort has been made to inform the reader of both advantages and disadvantages of particular names. Some names are clearly discouraged through the use of the word "not"; many of these are persistent traditional names, some of which are now obsolete and others that can be ambiguous or at least misleading. If such names can be pushed further from common use in modern chemical language, the writing of this book will have been worthwhile. Lack of a name that follows general principles given in the text of any chapter does not imply that it is not acceptable; the given names for a structure are meant to be illustrative, not exhaustive. In examples for which alternative names are given, the first name given corresponds to the numbering shown on the structure. Except for amino acids, their salts and esters, names used in biochemistry will not be given as alternatives; such names are discussed in chapter 31. Where alternative names differ in only a minor way, such as in the position of locants or the presence or absence of a vowel, only one will be cited. Some of these cases are noted below.

Frequent references are made to the recommendations of the International Union of Pure and Applied Chemistry (IUPAC) and to the nomenclature used by the Chemical Abstracts Service (CAS). CAS and IUPAC labels are attached to names only when they illustrate text in any chapter that compares the two systems. The format for names does not necessarily follow that of either IUPAC or CAS; however, if a name is identified as a CAS name, the format used by CAS is used and if identified as an IUPAC name, the format used by IUPAC is followed. CAS names are given in uninverted form, and it is understood that in the CAS indexes a name appears in inverted form and often uses periphrases or descriptive information.

Some of the more important principles and formats used in this book are as follows:

(1) a locant for a suffix is placed just before the suffix as recommended by IUPAC;

(2) enclosing marks follow the CAS nesting format, that is, parentheses are used first and then square brackets, as [[( )]]; braces (curly brackets) are not used as enclosing marks;

(3) all indicated hydrogen is cited except for that between two bivalent ring atoms; this follows CAS usage;

(4) names for organic acyl groups as substituent prefixes are derived from the name of the acid as recommended by IUPAC rather than formed additively, for example, cyclohexanecarbonyl and not cyclohexylcarbonyl;

(5) the numerical prefix for twenty is spelled eicosane as used by CAS rather than icosane as recommended by IUPAC;

(6) the suffix -ylidene attached to the name of a parent hydride will only mean that the substituent prefix is connected by a double bond, as recommended by IUPAC;

(7) brackets are used to enclose all component locants in fusion names as recommended by IUPAC;

(8) infixes rather than prefixes are used for functional replacement analogs of systematically named acids and related compounds, as in -sulfonothioic acid, not -thiosulfonic acid;

(9) in multiplicative names (see chapter 3), the CAS method will be followed when the total compound is symmetrical, while the IUPAC method will be used for unsymmetrically substituted parent structures that are multiplied;

(10) the multiplicative prefixes bis-, tris-, and so forth, followed by parentheses, are used when the multiplied parent structure is substituted or when there is the potential for ambiguity.

The ever changing nomenclature rules and recommendations and the perseverance of trivial and traditional names leads to the possibility of several names for a single compound, any of which may be perfectly acceptable, for example, the names propionic acid and propanoic acid. To avoid the presence of a large number of possible names for many examples, the following names are recognized as equivalent, but usually the first one given below will be used in this book:

Propanoic acid $\equiv$ Propionic acid

Butanoic acid $\equiv$ Butyric acid

Hydrazine $\equiv$ Diazine

hydrazinyl $\equiv$ hydrazino

thio(–S–) $\equiv$ sulfanyl (that is, methylthio $\equiv$ methylsulfanyl) or sulfanediyl (that is, 2,2'-thiodiethanol $\equiv$ 2,2'-sulfanediylidiethanol)

phosphanyl $\equiv$ phosphino

phosphoranyl $\equiv$ $\lambda^5$-phosphanyl

mercapto $\equiv$ sulfanyl

R-selenyl (R–Se–) $\equiv$ R-seleno (but not R-selanyl)

vinyl $\equiv$ ethenyl

The following general references were used in developing the material for this book. These references will not be cited in the various chapters unless there is a compelling reason for referring to a specific section or subsection in them.

1. International Union of Pure and Applied Chemistry, Organic Chemistry Division, Commission on Nomenclature of Organic Chemistry. *Nomenclature of Organic Chemistry, Sections A, b, C, D. E, F. and H*, 1979 edition; Rigaudy, J.; Klesney, S. P., Eds.; Pergamon Press: Oxford, U.K., 1979.
2. International Union of Pure and Applied Chemistry, Organic Chemistry Division, Commission on Nomenclature of Organic Chemistry. *A Guide to IUPAC Nomenclature of Organic Compounds, Recommendations 1993*; Panico, R.; Powell, W. H.; Richer, J-C., Eds.; Blackwell Scientific Publications: Oxford, U.K., 1993.
3. American Chemical Society, Chemical Abstracts Service. Chemical Substance Index Names Appendix IV in: *Chemical Abstracts Index Guide 1999*. Chemical Abstracts Service, Columbus, OH, 1999.

# Contents

# Introduction

The element carbon forms an enormous number and variety of chains and rings, with structures varying from very simple to incredibly complex. In addition to organic chemistry, organic compounds and their names are of concern in most other chemical fields, such as inorganic and coordination chemistry, polymer chemistry, and biochemistry. Although the nomenclature of organic compounds is complicated by the number and complexity of organic compounds, at the same time it is simplified by the characteristics of constant valence and definite bond order exhibited by carbon and its near neighbors.

## Types of Chemical Names

Chemical names have traditionally been divided into three general types, each with its own sphere of usefulness.

1. *Systematic name.* Nomenclature provides descriptions of chemical structures. Names formed in a systematic way define structures. The term *systematic*, however, is used in different ways and has often been misinterpreted. The 1979 Organic Rules define a systematic name on the basis of its components, that is "a name composed wholly of specially coined or selected syllables, with or without numerical prefixes, for example, pentane, oxazole". A more rigorous definition might require that a name fully describe the structure by the combination of single letters or groups of letters that have specific meanings in a particular context, as in 1,3-selen/az/ole and tri/az/ene. However, such a rigorous definition is too limiting; it would exclude all hydrocarbon names, since a name such as pentane does not specify that the chain element is carbon.

    Systematic chemical nomenclature may be defined as a set of principles, rules, and so on, arranged in an order that shows a logical plan for naming chemical compounds. Therefore, there can be many systematic names, and to be specific, the plan or method used must be identified; there are systematic IUPAC names, systematic CAS index names, systematic Beilstein names, systematic ACS names, and so on.

2. *Trivial or common name.* A name no part of which is used in a systematic manner (in the component sense noted above for systematic names); urea and xanthophyll are examples. Trade names are included in this category. Trivial names are applied to compounds and do not usually reveal composition or structure; this can be a real advantage for naming certain tautomeric compounds, for example, purine, bilin, and complex natural products. Trivial names are almost always shorter than systematic names, but the structure and composition of each compound known by a trivial name must be committed to memory. A very large number of trivial names is given in the *Chemical Abstracts Index Guide*,[1] with cross-references to systematic CAS names. There are other sources as well.[2,3]

A great many trivial names were invented for compounds whose structures were unknown, or, if structures were known, their systematic names were too long or complex for convenient use; this happens, for example, in enzymes and proteins. Often, history or a physical property, such as color, contribute to the formation of a trival name (see chapter 1).

3. *Semisystematic or semitrivial name.* Semisystematic names are names only a part of which is used in a systematic way, for example, meth*ane*, acet*ic acid*, phen*ol*. Many such names readily identify a characteristic group but other structural detail is revealed only when additional information is either given or remembered as with cholesterol. Other semisystematic names may identify one characteristic group, but not other characteristic groups or substituents, as exemplified by the names lactic acid, salicylic acid, and toluic acid. Still other semisystematic names may imply structure but give little or no information about composition, for example, cubane. Picric acid and barbituric acid are semisystematic names that suggest a chemical property and may erroneously imply a characteristic group. Such names really should be classified as trivial.

4. *Periphrases and descriptive names,* such as cyclohexane-1,2,4-triol monomethyl ether, naphthalene-1,3,5-tricarboxylic acid monoanhydride with benzoic acid, glycine ethyl ester or ethyl ester of glycine, are needed for compounds of unknown or partially known structure or for use in an inverted index format.

## Organic Nomenclature in a Nutshell

The basic concept of organic nomenclature is really rather simple. A structure is segmented into component parts that have names formed from a relatively small number of syllables and/or word fragments with defined meanings. These components are of two kinds: (1) parent hydrides, which are unbranched chains, rings, or ring systems consisting of skeletal atoms and hydrogen atoms; and (2) characteristic groups, which are individual atoms or heteroatomic groups of atoms sometimes combined with carbon atoms. Most organic names are formed from the names of these components by subsition, that is the conceptual replacement of hydrogen atoms by an atom, characteristic group, or other groups of atoms. However, other methods, usually combined with substitution, are used in naming many organic compounds. All of these methods are described in chapter 3.

The most important component of an organic name is that of the parent compound; it is the part of the structure on which a name is based. It can be the name of a parent hydride to which a characteristic group may or may not be attached as a suffix; or a structure having a trivial name, such as urea. It can also be a structure whose name implies or expresses a characteristic group or groups, for example, acetamide, phosphonic acid, and lactic acid. Parent compounds with implied or expressed characteristic groups are called functional parent compounds. Components attached to the parent compound may be chains, rings, ring systems, heteroatoms, or characteristic groups. Such components are considered to substitute for one or more hydrogen atoms of the parent compound. These typically are denoted by prefixes with traditional names or names derived from the names of parent hydrides.

## Conventional (or *Traditional*) Nomenclature versus Index Nomenclature

Nomenclature for general use need not conform to any one system nor should every substance necessarily have one and only one name. The only real requirement for conventional nomenclature is that it provide unambiguous and understandable names for the audience being addressed. The main goal of any systematic nomenclature is to convey the composition, and as far as possible, the structure of chemical compounds and substances. However, there are circumstances that need, even require, a unique name, that is, one and only one name for each substance. These include documents such as patents, tariff regulations, health and safety instructions, and the comprehensive alphabetical indexes provided by CAS and the Beilstein Institute. Except for the rigorous procedures used by these organizations, there are no officially recognized rules for generating unique names, nor do the procedures used by these organizations completely agree.

There are a few differences between conventional nomenclature and index nomenclature that should be mentioned. Conventional names are written from left to right. They sometimes begin with stereochemical descriptors, followed by substitutive prefixes, nondetachable prefixes (see chapter 2) describing modifications to the parent hydride, the parent hydride itself, and finally suffixes describing further modifications to the structure of the parent hydride and a particular substituent chosen to be highlighted as the principal substituent (the principal characteristic group). A principal characteristic group may be included, or implied, by a name for a single structure often called a functional parent compound, for example, acetic acid and phosphonic acid. Index names are inverted, that is the name of the parent compound or parent hydride, including prefixes that are considered nondetachable, and its suffixes is followed by a comma (the inversion comma) and the part of the name that precedes the parent compound in a conventional name, for example 1,4-naphthalenedicarboxylic acid, 2,3-dichloro-. 1,4-Naphthalenedicarboxylic acid is called the heading parent by CAS. Functional derivatives, such as esters, salts, anhydrides, hydrazones, oximes, and so on, identified by their class names follow the heading parent as in Benzoic acid, pentachloro-, ethyl ester. Finally, following any functional derivatives, for systematic names stereochemical information and other descriptive phrases that may be needed are cited. This procedure allows the collection of as many compounds as possible related to the parent structure at one point in an alphabetical arrangement.

## REFERENCES

1. American Chemical Society, Chemical Abstracts Service. *Chemical Abstracts Index Guide 1999*. Chemical Abstracts Service: Columbus, OH, 1999.
2. Giese, F. *Beilstein's Index: Trivial Names in Systematic Organic Chemistry*; Springer-Verlag: Berlin, 1986.
3. Nickon, A.; Silversmith, E. *Organic Chemistry: The Name Game (Modern Coined Terms and their Origins)*; Pergamon Press: Oxford, U.K., 1987.

Nomenclature of Organic Compounds

# 1

# Origin and Evolution of Organic Nomenclature

## History

The early history of chemical terminology and nomenclature has been documented comprehensively.[1] Other sources for this chapter include a history of the nomenclature of organic chemistry,[2] a history of the International Union of Pure and Applied Chemistry,[3] a history of nomenclature by P. A. S. Smith, Chairman of the International Union of Pure and Applied Chemistry Commission on Nomenclature of Organic Chemistry (1987 to 1995) and Chairman of the ACS Committee on Nomenclature (1990 to 1995),[4] and publications and presentations by K. L. Loening, Director of Nomenclature at Chemical Abstracts Service and Chairman of the ACS Committee on Nomenclature (1965 to 1990).[5-8]

Names of chemical substances usually were assigned by alchemists with the intention of concealing their work from colleagues. One of the major objectives of the alchemist was to make gold, and it was necessary to be obscure in reporting experiments for social and economic reasons. Alchemy was considered a divine science, and to reveal its secrets would bring on the wrath of the gods. Early terms for chemical compositions included "sugar of butter", "oil of vitriol", "cream of tartar", and "milk of lime", just to mention a few. "Spanish green" (basic copper acetate) was a name based on color; "flowers of sulfur", a term still used today, and "oil of tartar" (a concentrated solution of potassium carbonate) were names based on appearance. Crystalline form led to the names "horn silver" (fused silver chloride) and "cubic nitre" (sodium nitrate), while taste or smell gave names such as "sugar of lead" (lead acetate) and "stinking sulphureous air" (hydrogen sulfide). Names from medicinal properties are illustrated by "bitter cathartic salt" (magnesium sulfate). Names, especially those of metals, were often associated with planets; lead, silver, and gold have always been associated with Saturn, the moon, and the sun, respectively. Tin has been associated with Venus, Mercury, and Jupiter during various periods. Names have also been associated with places and personal names, for instance, vitriol of Goslar (zinc sulfate), Cyprian vitriol (copper sulfate), liquor of Boyle (ammonium polysulfide), Glauber's salt (sodium sulfate), and Zeise's salt, a name still used today, [potassium trichloro($\eta$-ethene) platinate(1-) monohydrate].

Even before the science of chemistry as we know it evolved, it was important to report discoveries and present theories either orally or through printed publications. To do so, it was recognized very early that a special, controlled language would be necessary. This language is called chemical nomenclature and today it is expected not only to reveal the atoms present, but also how these atoms are arranged in the molecule and chemical relationships with other chemical substances.

The earliest attempt at systematization appears to be a Sumerian cuneiform language especially suited for minerals.[1] It developed sometime around the seventh century B.C. The first part of the word described an outstanding property of a group of substances, and the second part indicated individual characteristics. Properties were indicated by means of suffixes, for example, ZA = rock or stone and GÌN = blue, so ZA.GÌN was a blue stone; AS = hard, so

ZA.GÌN.AS.AS was a very hard blue stone. This system was never carried into any other language.

It is generally agreed that modern chemistry began in the eighteenth century with the work of Lavoisier (1734–1794), who also played a key role in the beginning of modern chemical nomenclature.[9] Although the first attempt to provide systematization in chemical nomenclature was made by Bergman,[10] Guyton de Morveau is generally given credit for beginning the serious development of a systematic nomenclature,[11–12] but Lavoisier, Bertholet, and de Fourcroy all made important contributions.[13] These four have been called the first nomenclature committee.[6] Although the primary concern of this "committee" was the nomenclature of inorganic compounds, names such as "alcohol" and "ether" appeared in their recommendations along with names for a number of organic acids, such as succinic acid and malic acid, that are still in use today.

Although the beginning of organic chemistry is generally considered to be 1828 with the synthesis of urea by Wöhler,[14] many organic compounds were known much earlier. Vinegar was known to alchemists and glacial acetic acid was well known by the eighteenth century. An acid obtained from red ants was named formic acid (from the Latin for ant, *formica*). During the years 1780–1785, several new acids were prepared, including lactic acid, citric acid, mucic acid, oxalic acid, malic acid, and tartaric acid; their names were derived from the Latin or Greek names of their origins. Other acids named in the first half of the nineteenth century included oleic acid (from *oleum*, Latin for oil), butyric acid from butter (from *butyrum*, Latin for butter), capric and caproic acids from goat's milk (from *capra*, Latin for female goat), maleic acid and malonic acid (from *malum*, Latin for apple). In 1794, four Dutch chemists described a new gas that became known as olefiant gas in 1797 (a likely source for the class name olefin); it was also known by names such as etherene and hydride of vinyl before the name ethylene gained widespread use in spite of the pronouncement in 1892 by the Geneva Conference that its name should be ethene. A volatile liquid first prepared in the early seventeenth century was first known as pyro-acetic spirit, but became acetone in 1833. Methyl alcohol, prepared in 1812, was known as pyroligneous ether until classification as an alcohol from its properties some thirty years later.

One of the biggest problems in the early days of organic nomenclature was the use of different names for the same compound resulting from imperfect descriptions of properties, imprecise analyses, and ignorance of publications. In 1826, an oil that formed crystalline compounds with acids was called crystallin; in 1834, a basic compound from coal tar was called kyanol because it produced a violet-blue color with chloride of lime; in 1840, a compound from indigo was called anilin (from the Spanish word for indigo); and finally, in 1842, in Russia nitrobenzene was reduced to a product that was called benzidam. In the following year Hofmann noted that these four names described the same compound; he preferred the first name, crystallin. Berzelius, however, disapproved of this name because it depended on a property and endorsed the name anilin; the name aniline is almost universally used today. The evolution of the name propionic acid took a different course. The acid was first called met-acetic acid, then pseudoacetic acid, because it was similar, though not identical, to acetic acid. Still another name given to this acid was butyro-acetic acid because its properties were intermediate between the well-known acetic and butyric acids. In 1847, Dumas called it propionic acid, from the Greek word for first, because in the series of acids from acetic acid to the fatty acids it was the first one after acetic acid. This name became so well established that it was never considered as a candidate for systematization based on the number of carbon atoms.

Contractions of words have always been a part of the nomenclature of chemical compounds, especially organic nomenclature. The French chemist Chevreul introduced the practice in 1823 by stating in a publication that he formed the name ethal (now known as cetyl alcohol) by combining the first two syllables of the words ether and alcohol. Aldehyde was formed from the words alcohol dehydrogenatus, and aldol from the combination and contraction of aldehyde and alcohol.

The beginning of systematic organic nomenclature probably began when Guy-Lussac noted in a lecture that the reaction of chlorine with oils produced hydrochloric acid and a substance in which "a part of the chlorine...takes the place of the hydrogen which is removed",[15] an obvious observation of the principle of substitution. Radical names such as benzoyl,[16] ethyl[17] and methyl[18] appeared in the literature in the 1830s. Dumas and Laurent used a system of vowels to indicate degree of substitution in describing chloroacetic acids and chloronaphthalenes.

As more and more organic compounds were prepared, systematization of nomenclature became necessary. The need for an organization of names for organic compounds was noted very early by Berzelius[19] in connection with the so-called "pyro-acids", acids formed by pyrolyzing organic acids. The use of the term pyro for this purpose had already been suggested by de Morveau et al., in 1787. In 1840 Dumas thought that the development of a system of nomenclature for chemical compounds was possible,[20] and the question was discussed by Daubeny[21] in 1851 and Foster in 1857[22] before the British Association. In 1838 Berzelius suggested the use of Latin or Greek terms to indicate the number of atoms of an element, and in 1853 Gerhardt used such terms to indicate a position in an homologous series, for example, tritylene, tetryl, hexyl. In 1865, Hofmann arranged hydrocarbons in a series according to their empirical formulas and assigned names in which vowels were used to distinguish the degree of unsaturation;[23,24] he adapted the vowels used by Dumas[25] and promoted by Laurent[26,27] for differentiating chloro derivatives of acetic acid and napththalene. In 1868 Kekulé made the distinction between aliphatic and aromatic compounds.

The first international chemical congress was held in 1860 in Karlsruhe,[28-30] but it had little immediate effect on chemical nomenclature. Textbooks, however, were beginning to show tendencies of systematization. The numerical prefixes hexyl, heptyl, octyl, and nonyl to indicate chain length and the use of the "en" syllable, actually "ylen", for $C_nH_{2n}$ hydrocarbons appeared in a textbook by Fittig.[31] In 1879, a textbook appeared that divided hydrocarbons with an even number of carbon atoms into twelve series, the first of which was the homologous series of the alkanes, the second the alkenes, and the third alkynes.[32] The first six used the vowels adopted by Hofmann; the resulting endings were -ane, -ene, -ine, -one, and -une.

Journal editors now became concerned with nomenclature; editorial comments would appear as footnotes in papers. In England, H. E. Armstrong, a member of the Committee on Publications of the Journal of the Chemical Society, proposed guidelines for the nomenclature of chemical compounds, including recommendations for organic compounds,[34] leading to the publication of the "System of the English Chemical Society" in 1882.[35] An American Chemical Society Committee on Nomenclature and Notation, appointed in 1884,[36] issued a report in 1886[37] that endorsed the "English System" with only minor additions and alterations.

## International Nomenclature

The first effective international consideration of organic nomenclature began in the summer of 1889 when an International Commission of Chemical Nomenclature, later to be known as the International Commission for the Reform of the Chemical Nomenclature, was organized at an International Chemistry Conference in Paris.[38,39] A report of a Subcommission formed the basis for the 1892 International Conference of Geneva for the Reform of Chemical Nomenclature, a meeting of 34 of the leading chemists of the day from nine European countries. The main emphasis of the Geneva Conference was the need for names suitable for systematic indexing of organic compounds, and, although much more limited in scope than the Subcommission's report, the resulting "Geneva Nomenclature"[40-42] introduced principles still considered to be of primary importance in naming acyclic compounds, that is, the parent structure for the name should be the longest unbranched chain, and the presence of a functional or characteristic group should be expressed by a suffix. Very soon, proposals for extension and modifications appeared.[43-45] These and other proposals were discussed at the 21st Session of the l'Association française pour l'Advancement des Sciences in September, 1892.[46] The minutes of this meeting also contained a request by the Chemical Section that, although the advantage of systematic names for complex compounds was recognized, the

Subcommission should search for "short, concise, and euphonic names" for derivatives. Each compound in this category would have a systematic name formed according to the rules of nomenclature and a short, common name to indicate relationships with compounds having the same "molecular group". Clearly, the desire for trivial or common names existed even as the Geneva Nomenclature was being developed. In anticipation of a second international conference, proposals for extensions and modifications to the Geneva Nomenclature continued to appear,[47,48] but the conference failed to take place. Instead, with the assistance of the l'Association française pour l'Advancement des Sciences, the Subcommission that was responsible for the report for the Geneva meeting prepared a second report that was discussed by the Chemistry Section in 1897 at the meeting of the Association at Saint-Etienne.[49] This report contained the Geneva Nomenclature agreements only slightly modified and included a number of extensions. However, no international agreement on this report was recorded.

Although there were opportunities to continue discussions on the development of organic nomenclature and important proposals were put forward,[50] including that of von Baeyer for naming saturated bicyclic ring systems,[51] there were few further developments in organic nomenclature internationally until 1913 when the Council of the International Association of Chemical Societies, established in 1911,[52] created commissions on inorganic and organic nomenclature.[53] The work of these commissions was essentially halted by World War I. In 1921, three commissions for the reform of nomenclature in organic, inorganic, and biological chemistry were created by the International Union of Pure and Applied Chemistry (IUPAC)[54] that replaced the International Association of Chemical Societies in 1919. The Commission for the Reform of the Nomenclature of Organic Chemistry was to take the Geneva Nomenclature as the basis for its discussion, and apparently the Commission took this charge so literally that a large number of proposals were not seriously considered, including suggestions that were available at the Saint Etienne meeting,[49,50] von Baeyer's proposals for bicyclic ring systems,[51] publications such as Istrati's enormous book,[55] nomenclature for hetero-monocyclic rings,[56,57] Patterson's proposals,[58–60] ideas for carbocyclic ring systems,[61,62] replacement ("a") nomenclature for heterocyclic ring systems,[63,64] and others.

In 1930, the International Union of Chemistry (IUC) and its Organic Nomenclature Commission published a "Definitive Report" consisting of 68 rules.[65,66] This report is known as the Liege Rules; it was supplemented by less extensive reports of meetings in Lucerne in 1936[67] and Rome in 1938.[68] At the next meeting in London in 1947, the name of the Union was changed back to IUPAC and the word "reform" was removed from the names of the Commissions. The Commission on Nomenclature of Organic Chemistry (CNOC) began a program emphasizing codification of good nomenclature practices that already were in use rather than developing new ones. Thus, acceptable alternative methods were now to be recognized in IUPAC organic rules where, for a variety of reasons, limitation to a single method seemed to be undesirable or unfeasible. In 1949, CNOC published rules on organosilicon compounds, changes and additions to the definitive report with extended examples of radical names and a rule on "extra hydrogen";[69] these rules, known as the Amsterdam Rules, were officially accepted in 1951,[70] along with new rules on naming *cis*- and *trans*-isomers of olefinic hydrocarbons and skeletal replacement ("a") nomenclature.[70]

In 1951, the Organic Commission began an ambitious program of revision and extension of the Liege Rules with the goal of providing a general nomenclature for the field of organic chemistry other than the "biochemical" area. This effort resulted in the publication of two sections of "definitive rules" in 1958[71] and a third section in 1965.[72] A second edition of Sections A and B appeared in 1966,[73] and a third edition combined with a second edition of Section C appeared in 1971.[74] Another edition of the *Nomenclature of Organic Chemistry* appeared in 1979.[75] It included new editions of Sections A , B, and C and four additional sections D, E, F, and H. Section E on stereochemistry and Section H on isotopically modified compounds contain approved recommendations published in 1976[76] and 1979,[77] respectively. Section D, prepared jointly with the IUPAC Commission on Nomenclature of Inorganic Chemistry, is a second publication of provisional recommendations on nomenclature of organic compounds containing elements not covered by Section C; the first publication appeared as tentative recommendations in 1973.[78] Similarly, Section F is a second publication of provisional recommendations on the nomenclature of natural products; the first provisional recommendations were published in 1976.[79]

The latest recommendations published by CNOC (1993) is a guide to IUPAC nomenclature for organic compounds[80] that includes revisions, published[81–84] and unpublished, of the 1979 edition of the IUPAC *Nomenclature of Organic Chemistry*.[75] This guide is not intended to replace the 1979 edition, but to update and supplement it. Since the publication of the 1993 guide, the Commission has published recommendations for naming radicals and ions,[85] a glossary of class names of organic compounds (jointly with the IUPAC Commission on Physical Organic Chemistry),[86] basic terminology of stereochemistry (jointly with the IUPAC Commission on Physical Organic Chemistry),[87] recommendations for naming fused ring and bridged fused ring systems,[88] principles of "phane nomenclature",[89] and revised recommendations for natural products.[90]. The Commission is now working on revised recommendations for naming stereoisomers, recommendations for extending "phane nomenclature", for naming fullerenes, and the necessary hierarchical rules for generating a preferred IUPAC name.

From 1922 through 1949 the work of the Commission on Biological Nomenclature created in 1921 touched areas of interest to the Organic Commission, for example, carbohydrates, carotenoids, amino acids, and vitamins. These efforts eventually resulted in the recommendations important to organic nomenclature that are referenced in chapter 31. In 1949, IUPAC was reorganized to include Biological Chemistry in order to continue the work of the original biological commission. In 1953, the International Union of Biochemistry (IUB) was organized with an expressed interest in nomenclature, especially enzyme nomenclature, establishing a Committee on Enzymes in 1955. To avoid duplication of effort, in 1965 IUPAC and IUB established an IUPAC–IUB Combined Commission on Biochemical Nomenclature (CBN), the first joint nomenclature venture between major scientific unions (other examples of cooperation between national nomenclature committees are the joint American–British efforts that led to the 1952 rules for naming organophosphorus compounds[91] and the rules of carbohydrate nomenclature[92]). IUPAC disestablished its Commission on Nomenclature of Biological Chemistry and IUB dissolved its Committee on Enzymes at this point. CBN was associated with CNOC in IUPAC and its reports were published in *Pure and Applied Chemistry* as "tentative rules". IUB published CBN reports in several biochemical publications as recommendations as of a specific year. In 1977, CBN was dissolved and replaced with an IUB Nomenclature Committee (NC-IUB) and a Joint Commission on Biochemical Nomenclature (JCBN), which operate together under one Chairman and Secretary. JCBN is concerned primarily with compounds, such as carbohydrates, steroids, and porphyrins, and NC-IUB with topics such as enzymes, iron–sulfur proteins, human immunoglobins, and multienzyme proteins.

## Chemical Abstracts Index Nomenclature

*Chemical Abstracts* (*CA*), which began publication in 1907, and American nomenclature committees, especially the American Chemical Society Committee on Nomenclature, Spelling, and Pronunciation, established in 1911 (later to become known as the ACS Committee on Nomenclature) and ACS divisional nomenclature committees, have had a profound influence on the development of organic nomenclature. A. M. Patterson was editor of *CA* (1910–1914) and Chairman of the ACS Committee on Nomenclature until 1914 when E. J. Crane assumed these responsibilities until 1958. Both were also active participants in the activities of IUPAC nomenclature commissions (Patterson from 1924 until 1953 and Crane from 1922 until 1966) and thus were able to ensure that the same basic principles were followed for both the *CA* indexes and that recommended internationally. Of course, details had to vary somewhat because of special requirements that were needed for creating an effective index (see the Introduction to this book). In addition to Patterson's publications noted earlier, he was the architect of the system still used for naming rings and ring systems. With the assistance of L. T. Capell and under the direction of a joint committee of the

American Chemical Society and the National Research Council and with the cooperation of the International Union of Chemistry, Patterson produced the first edition of *The Ring Index*, a listing of the rings and ring systems in organic chemistry.[93] This work contained rules for naming rings and ring systems that were to appear without serious modification in the 1957 IUPAC Organic Nomenclature Rules.[71] In 1960, a second edition of *The Ring Index* was published[94] that had, as an appendix, the IUPAC rules for naming and numbering ring systems used in organic chemistry; supplements appeared in 1963, 1964, and 1965. This work contained over 14,000 rings and ring systems from the literature through 1963. Today, over 120,000 rings and ring systems are listed in the CAS *Ring Systems Handbook*. Under the chairmanship of E. J. Crane, the ACS Committee issued a number of reports, many based on the work of nomenclature committees of various ACS Divisions.[95–104]

The close association between CAS and international nomenclature bodies was continued by L. T. Capell from 1958, when E. J. Crane retired, until 1964 and by K. L. Loening from 1964 through 1989, both holding the positions of Director of Nomenclature at CAS and Chairman of the ACS Committee on Nomenclature. Although L. T. Capell had been involved with CNOC for a long time, he was an official member only from 1953 until 1964. K. L. Loening became a member of the Organic Commission in 1963 and served until 1981. Dr. Capell wrote much of the introduction to the 1945 *Chemical Abstracts* Subject Index.[105] Since IUPAC recommendations for nomenclature of organic compounds were rather incomplete at that time, this publication represented the only comprehensive manual for naming the whole spectrum of chemical compounds; it has been updated on a regular basis, the latest in 1999.[106]

ACS Divisional Committees prepared recommendations on various topics for consideration by the ACS Nomenclature Committee. Several of these were published;[107–110] all were forwarded to the appropriate IUPAC Nomenclature Commission. Some were later modified slightly and published as IUPAC recommendations.[110,111]

## Evolution of Nomenclature

The development of nomenclature recommendations from first proposals or ideas to full approval by IUPAC can be agonizingly slow. Proposals for nomenclature recommendations can come from a variety of sources, individuals, research groups, national committees, and international commissions. For a proposal originating in the United States, the procedure usually is as follows:

1. An individual or a group of workers in a particular field of chemistry suggests new or revised recommendations that seem to fit a particular need. Such an individual or group may consist of "experts" in that particular field appointed as an ad hoc subcommittee or it may not have any status at all. In either case, a report is prepared that presents the proposal and provides appropriate justification. Many nomenclature proposals originate in ACS divisional committees; three examples are the structure-based rules for linear polymers,[107] the rules for boron nomenclature,[108] and recommendations for naming highly fluorinated organic compounds.[109]
2. This report is then submitted for review, revision (if needed), and approval by an officially recognized nomenclature body, such as a nomenclature committee of one of the divisions of the American Chemical Society, and then by the ACS Committee on Nomenclature. If there is doubt as to an appropriate divisional committee, the report can be submitted to the ACS Committee which would then refer it to an appropriate divisional committee or establish an ad hoc subcommittee to consider it.
3. The ACS Nomenclature Committee can approve the report, thereby approving it for use in ACS publications (in 1972 one of the Committee's approved duties was to become the authority to approve nomenclature proposals on behalf of the ACS Council), and at the same time submit it via the National Academy of Sciences–National Research Council, the official adhering organization to IUPAC for the

U.S., to the appropriate IUPAC commission for consideration. The Committee can also seek comments from the IUPAC commissions before giving its approval.

4. The appropriate IUPAC commissions consider the report and its proposals along with any others on the same subject from other sources. Factors such as adaptability to other languages and overall usage are considered along with comparisons with past and current methods. They may even recommend further study by a subcommission or working group or refer it back to national committees for further study and/or development.

5. The IUPAC commissions, after approval by the Interdivisional Committee or Nomenclature and Symbols (IDCNS), submit the proposal to an international review process.[112] When this review is successfully completed, the recommendations are published as a book or in *Pure Appl. Chem.*, the official journal for publication of IUPAC reports and symposia. Today many IUPAC nomenclature publications appear on the Internet at the address http://www.chem.qmc.ac.uk/iupac/iupac.html.

6. The ACS Committee then reviews the IUPAC approved recommendations and either accepts the recommendations or recommends alternatives for use by the ACS. Each National Adhering Organization can adapt IUPAC recommendations to its own habits and language. This may be best illustrated by the publication of the 1957 IUPAC Rules for Nomenclature of Organic Chemistry, with comments, in *J. Am. Chem. Soc.*[71]

This process is not always followed rigorously. For instance, the structure-based rules for single-strand organic polymers[107] were introduced into CAS index nomenclature right after approval by Council in 1968, but IUPAC recommendations did not appear until 1975.[110] Furthermore, the ACS Committee on Nomenclature has not been diligent in reviewing recommendations approved by IUPAC. In addition, CAS has been inconsistent in adopting recommendations by the ACS Committee or by IUPAC. The inorganic boron rules, adopted by the ACS Council in 1968[108] and published by IUPAC in 1972[111] have been adopted only in part for CAS index nomenclature. Likewise, the $\lambda$-convention[82] is used only in a limited way in CAS index nomenclature and the $\delta$-convention[84] and revised recommendations for the Hantzsch-Widman nomenclature system[81] have not been adopted by CAS.

Clearly, the time span for this procedure can be significantly reduced and the quality of the final nomenclature recommendations can be improved if a proposal is considered concurrently by several groups. One way that this can be done is to invite experts in a particular field, regardless of nationality, members of nomenclature committees and/or commissions, journal editors, and members of professional staffs of secondary publishers, such as CAS and Beilstein, to review the proposals at appropriate stages in the development of the new recommendations. A problem with this method is that wide distribution or premature publication of new recommendations can not only complicate the process leading to official adoption of the new recommendations but can lead to considerable confusion among authors and journal editors as to which names are acceptable for use in their primary publications.

Ideally, each IUPAC Commission or ACS Committee should be the leader in its particular field of nomenclature and should be able to anticipate the needs of chemists throughout the world. Practically, however, this is quite impossible. New compounds of widely different types are being reported in the literature every day. Thus, the burden of providing acceptable (unambiguous) names for these compounds necessarily falls heavily on research workers, journal editors, and abstracting and indexing services. To maintain consistency in its index nomenclature, CAS has published a compilation of its own nomenclature practices, as noted above, not only for use by its own staff, but for the guidance of those using its indexes. At its own discretion, *CA* revises its index nomenclature rules as much as possible to be in agreement with new IUPAC recommendations. However, it often must name new compounds long before official recommendations can be established. This is done as far as possible to conform to the basic principles established by IUPAC even though some of the details may turn out to be different. IUPAC Commissions in turn rely heavily on abstracting and indexing services and on journal editors for guidance in the consideration of new nomenclature recommendations.

## Concluding Remarks

Although it is easy to point out imperfections and inconsistencies in the nomenclature rules of today, a few words must be said about their advantages and virtues. The language of chemistry is said to involve more terms and to be better ordered than the language of any other scientific discipline. All of this has come about despite the fact that until 1949 relatively few nomenclature recommendations had been officially approved internationally. The acceptance and codification of many new and revised recommendations by the IUPAC Commissions and the ACS Nomenclature Committees have done much to maintain clear and useful communication among chemists and other scientists. Continuation of these efforts will ensure that improved methods for describing and identifying composition and molecular structure of chemical substances will be available for use in our published scientific literature and indexes, in the classroom, and on the lecture platform.

### REFERENCES

1. Crosland, M. P. *Historical Studies in the Language of Chemistry*; Harvard University Press: Cambridge, MA, 1962; 2nd ed., Dover Publications: New York, 1978.
2. Verkade, P. E. *A History of the Nomenclature of Organic Chemistry*; D. Reidel: Dordrecht, The Netherlands, 1985 (a collection of articles, Études historiques sur la nomenclature de la chimie organique; published in *Bull. Soc. Chim. Fr.*, translated from the French by S. G. Davies, Oxford University).
3. Fennell, R. *History of IUPAC 1919–1987*; Blackwell Science: Oxford, U.K. 1994.
4. Smith, Peter A. S. Nomenclature. In *Kirk-Othmer Encyclopedia of Chemical Technology*, 4th ed.; John Wiley: New York, 1996; Vol. 17, pp 238–259.
5. Loening, K. L. Nomenclature. In *Kirk-Othmer Encyclopedia of Chemical Technology*, 3rd ed.; John Wiley: New York, 1981; Vol. 16, pp 28–46.
6. Loening, K. L. Historical Development in Chemical Nomenclature. Presented at the Symposium on History of Chemical Information Science, Part III, ACS National Meeting, Philadelphia, PA, August, 1984.
7. Loening, K. L. The ACS Committee on Nomenclature, 100 Years of Contributions to the Language of Chemistry. Presented at the Symposium on History of Chemical Nomenclature, ACS National Meeting, Miami Beach, FL, April, 1985.
8. Loening, K. L. Activities of the IUPAC/IUB Joint Commission on Biochemical Nomenclature (JCBN) and the Nomenclature Committee of IUB. In *The Terminology of Biotechnology: A MultiDisciplinary Problem*, Proceedings of the 1989 International Chemical Congress of Pacific Basin Societies, PACIFICHEM '89; Loening, K. L., Ed.; Springer-Verlag: Berlin, 1990; pp 1–8.
9. Lavoisier, A. L. *Traite Elementaire de Chemie*; Cuchet: Paris, 1789; 2 vol. (English translation: Kerr, R. *Elements of Chemistry*; William Creech: Edinburgh, 1790).
10. Bergman, T. O. *Meditations de Systemate Fossilium Naturali*; J. Tofani: Florence, 1784.
11. Guyton de Morveau, L. B. Sur les Denominations Chymiques, la necessite d'en perfectionner le system & les regles pour y parvenir. *Obs. phys. hist. nat. arts* [*J. Physique Chem. Hist. Nat. Arts*] **1782**, *19*, 370–382.
12. Guyton de Morveau, L. B. De quelches critiques de la nomenclature des chimistes francais. *Ann. Chim. Phys. (Paris)* **1798**, [1], *25*, 205–222.
13. Guyton de Morveau, L. B.; Lavoisier, A. L.; Bertholet, C. L.; de Fourcroy, A. F.; *Methode de Nomenclature Chimique*; Cuchet: Paris, 1787 (English translation with adaptations to the English language: St. John, J. *Method of Chemical Nomenclature*; G. Kearsley: London, 1788).
14. Wöhler, F. Über Künstliche Bildung des Harnstoffs. *Ann. Physik Chem. (Leipzig)* **1828**, [2], *12*, 253–256.
15. Gay-Lussac, J. L. In *Cours de Chemie: comprenant l'historie des sels, la chemie végétable et animale*; H. Tarlier: Brussels, 1829; Vol. 2, Leçon 28 (16 July, 1828), p 12.
16. Liebig, J. von; Wöhler, F. VI. Untersuchungen über das Radical der Benzoesäure. *Ann. Pharm.* **1832**, *3*, 249–282 [pp 261–262].
17. Liebig, J. von. Ueber die Constitution des Aethers und seiner Verbindungen. *Ann. Pharm.* **1834**, *9*, 1–39 [p 18].

18. Berzelius, J. J. *Jahres-Berichte über der Fortschritte der physischen Wissenschaften* **1836**, *15*, 380–381.
19. Berzelius, J. J. *Jahres-Berichte über der Fortschritte der physischen Wissenschaften* **1834**, *13*, 231–232.
20. Dumas, J. B. Mémoire sur la loi des substitutions et la théorie des types. In *C. R. Hebd. Séances Acad. Sci.* **1840**, *10*, 149–178; Nomenclature, pp 168–170.
21. Daubeny, C. On the nomenclature of organic compounds. In *Reports of the British Association for the Advancement of Science*, 1851; pp 124–138.
22. Foster, G. C. Suggestions towards a more systematic nomenclature for Organic bodies. In: *Reports of the British Association for the Advancement of Science*, 1857; pp 45–47.
23. Hofmann, A. W. von. Las über die Einwirkung des Phosphortrichlorids auf die Salze der Aromatischen Monamine, December 11, 1865. *Monatsberichte der königlichen Preuss. Akad. der Wissenschaften zu Berlin* **1866**, 649–660, footnote, pp 652–654.
24. Hofmann, A. W. von. On the action of trichloride of phosphorus on the salts of aromatic monoamines. *Proc. R. Soc. London* **1866/7**, *15*, 55–62, footnote, pp 57–58.
25. Dumas, J. B.; Péligot, E. Recherches de Chimie organique - Action der chlore sur l'huile de cannelle. *Ann. Chim. Phys.* **1834**, [2], *59*, 316–322.
26. Laurent, A. Sur la Nitronaphthalase, la Nitronapthtalése, et la Naphthalase. *Ann. Chim. Phys.* **1835**, [2], *59*, 376–397.
27. Laurent, A. *Chemical Method, notation, classification, and nomenclature* (translated by W. Odling), Cavendish Society: London, 1855; p 37.
28. Der Internationale Chemiker-Kongress in Karlsruhe am 3, 4, und 5 September 1860. In *August Kekulé*, R. Anschutz; Verlag Chemie: Berlin, 1929; Vol. I, pp 183–209.
29. Compte rendu des séances du Congrès international des chimistes rénui à Carlsruhe le 3,4, et 5 September 1860. In *August Kekulé*, R. Anschutz; Verlag Chemie: Berlin, 1929; Vol I, Appendix 8, pp 671–688.
30. Bogert, M. T. International Organization of Chemists. *Chem. Eng. News* **1949**, *27*, 1992–1995.
31. Fittig, R. *Wöhler's Grundriss der Organischen Chemie*, Siebente Umgebarte Auflage; Verlag von Duncker and Humblot: Leipzig, 1868; pp 48–50.
32. Semple, C. E. Armand. *Aids to Chemistry*; Bailliere, Trindall and Cox: London, 1879; Part III Organic, pp 21–23.
33. Armstrong, H. E. On the nomenclature of carbon compounds. Paper read before the Chemical Society, April 7, 1876 (discussion on April 20, 1876); reported in *Chemical News* **1876**, *33*, 156–157, 177–178.
34. Armstrong, H. E. On Systematic Nomenclature. *J. Chem. Soc.* **1876** Part I, 685.
35. Chemical Society (London) Publications Committee, Nomenclature and Notation, Appendix to Report of the Anniversary Meeting. *J. Chem. Soc., Transactions* **1882**, *41*, 247–252.
36. The American Chemical Society. Proceedings. *J. Am. Chem. Soc.* **1884**, *6*, 55.
37. The American Chemical Society, Committee on Nomenclature and Notation. Report. *J. Am. Chem. Soc.* **1886**, *8*, 116–125.
38. Congrés international de Chemie in Paris, am 30 Juli bis 3 August 1889. *Chem.-Ztg.* **1889**, *13*, 1371–1372; 1391–1393.
39. Internationaler Congress für Chemie in Paris. *Chem.-Ztg.* **1889**, *13*, 907–908.
40. Pictet, A. Le Congress International de Genève pour la Reforme de la Nomenclature Chimique. *Arch. Sci. Phys. Nat.* **1892**, [3], *27*, 485–520.
41. Tiemann, F. Ueber die Beschluesse des Internationalen in Genf von 19 bis 22 April 1892, versammelten Congresses zur Regelung der chemischen Nomenclatur. *Ber. Dtsch. Chem. Ges.* **1893**, *26*, 1595–1631.
42. Armstrong, H. E. The International Conference on Chemical Nomenclature. *Nature* **1892**, *46*, 56–58.
43. Armstrong, H. E. Contributions to an International System of Nomenclature, The Nomenclature of Cycloids. *Proc. Roy. Soc. London*, **1892**, *8*(127), 127–131.
44. Combes, A. La Nomenclature Chimique au Congrès de Pau. *Rev. Gen. Chim. Pure Appl.* **1892**, *3*, 597–601.
45. Siboni, G. Sulla reforma della nomenclatura chimica (I). *Boll. Chim. Farm.* **1892**, *31*, 390–394.
46. Association Francaise pour l'advancement des Sciences. *Compte Rendu. 21st Session*, Sept. 21 1892, Pau, **1892**, *1*, 189–190.

47. Siboni, G. Sulla reforma della nomenclatura chemica (II). *Boll. Chim. Farm.* **1893**, *32*, 101–110; (III) 354–359; IV 481–489.

48. Combes, A. Chimique (Nomenclature). In *Dictionnaire de Chemie pure et appliquée*; Wurtz, A., Ed.; Deuxième Suppl. Librairie Hachette: Paris, Priemiere Part, C, 1894; pp 1060–1075.

49. Association Francaise pour l'advancement des Sciences. *Compte Rendu, 26th Session*, Aug. 9, 1897, Saint-Etienne **1897**, *1*, 204–240.

50. Bouvealt, M. L. Nomenclature des composés a chaines fermées. In: Association Francaise pour l'advancement des Sciences. *Compte Rendu, 26th Session*, Aug 9, 897, Saint-Etienne **1897**, *1*, 243–264.

51. Baeyer, A. Systematik und Nomenclatur Bicyclischer Kohienwasserstoffe. *Ber. Dtsch. Chem. Ges.* **1900**, *33*, 3771–3775.

52. Ostwald, W. *Lebenslinien, Eine Selbstbiography*; Klasing: Berlin, 1927; Vol. 111, pp 262–286.

53. Proceedings of the Third Session of the Council of the International Association of Chemical Societies, Institute Solvay, Brussels, Sept 19–23, 1913.

54. International Union of Pure and Applied Chemistry. *Comptes Rendus de la Deuxieme Conference Internationale de la Chimie*, Brussels, June 27–30, 1921; pp 53–54.

55. Istrati, C. I. *Studiu relativ la o Nomenclatura generala in Chimica organica*; Librarille Socec: Bucharesti, 1913.

56. Hantzsch, A.; Weber, J. H. Ueber Verbindungn des Thiazol (Pyridins der Thiophenreihe). *Ber. Dtsch. Chem. Ges.* **1887**, *20*, 3118–3132.

57. Widman, O. Zur Nomenclatur der Verbindungen, welche Stickstoffkerne enhalten. *J. Prakt. Chem.* **1888**, *38*, 185–201.

58. Patterson, A. M.; Curran, C. E. A System of Organic Nomenclature. *J. Am. Chem. Soc.* **1917**, *39*, 1623–1638.

59. Patterson, A. M. Proposed International Rules for Numbering Organic Ring Systems. *J. Am. Chem. Soc.* **1925**, *47*, 543–561.

60. Patterson, A. M. The Nomenclature of Parent Ring Systems. *J. Am. Chem. Soc.* **1928**, *50*, 3074–3087.

61. Sudborough, J. J. I. The Systematic Nomenclature of Polycyclic Carbon Systems. *J. Indian Inst. Sci.* **1924**, *7*, 145–165.

62. Sudborough, J. J.; Ayyar, P. R. II. Polycyclic and Cage Systems of Carbon Compounds. *J. Indian Inst. Sci.* **1924**, *7*, 166–180.

63. Sudborough, J. J. III. The Systematic Nomenclature of Heterocyclic Compounds including Polycyclic Structures. *J. Indian Inst. Sci.* **1924**, *7*, 181–195.

64. Stelzner, R. Nomenklatur-Fragen II: Vorschläge zur Benennung und Bezifferung heterocyclisher Systems von beliebiger Ringgliederzahl. In *Literatur-Register der organischen Chemie*; Deutsche Chemische Gesellschaft: Braunschweig and Berlin, 1926; V, pp ix–xv.

65. International Union of Chemistry, Commission on the Reform of the Nomenclature of Organic Chemistry. Definitive Report. *Compt. Rend. de la Dixieme Conference*, Liege, Sept 14–20, 1930; pp 57–64 (translation with comments: Patterson, A. M. *J. Am. Chem. Soc.* **1933**, *55*, 3905–3925).

66. International Union of Chemistry, Definitive Report of the Committee for the Reform of Nomenclature in Organic Chemistry. *J. Chem. Soc.* **1931**, 1607–1616.

67. International Union of Chemistry, Commission on the Reform of the Nomenclature of Organic Chemistry. *Compt. Rend. de la Douzieme Conference*, Lucerne and Zurich, Sept 14–20, 1936; pp 39–42.

68. International Union of Chemistry, Commission on the Reform of the Nomenclature of Organic Chemistry. *Compt. Rend. de la Treizieme Conference*, Rome, May 15–21, 1938; pp 36–37.

69. International Union of Pure and Applied Chemistry, Commission on Nomenclature of Organic Chemistry. *Compt. Rend. de la Quinzieme Conference*, Amsterdam, Sept 5–10, 1949; pp 127–186.

70. International Union of Pure and Applied Chemistry, Commission on Nomenclature of Organic Chemistry. *Compt. Rend. de la Seizieme Conference*, New York and Washington DC, Sept 8–15, 1951; pp 100–102, 102–104.

71. International Union of Pure and Applied Chemistry, Commission on Nomenclature of Organic Chemistry. *Nomenclature of Organic Chemistry, Definitive Rules for Section A. Hydrocarbons and Section B. Fundamental Heterocyclic Systems*; Butterworths: London,

1958; pp 3–70 (also published with American comments in *J. Am. Chem. Soc.* **1960**, *82*, 5545–5574).

72. International Union of Pure and Applied Chemistry, Commission on Nomenclature of Organic Chemistry. *Nomenclature of Organic Chemistry, Definitive Rules for Section C. Characteristic Groups Containing Carbon, Hydrogen, Oxygen, Nitrogen, Halogen, Sulfur, Selenium, and/or Tellurium*; Butterworths: London, 1965 (also published in *Pure Appl. Chem.* **1965**, *11*, 1–260).

73. International Union of Pure and Applied Chemistry, Commission on Nomenclature of Organic Chemistry. *Nomenclature of Organic Chemistry, Definitive Rules for Section A. Hydrocarbons and Section B. Fundamental Heterocyclic Systems*, 2nd ed.; Butterworths: London, 1966; pp 3–70.

74. International Union of Pure and Applied Chemistry, Commission on Nomenclature of Organic Chemistry. *Nomenclature of Organic Chemistry, Definitive Rules for Section A. Hydrocarbons and Section B. Fundamental Heterocyclic Systems*, 3rd ed.; *Section C. Characteristic Groups Containing Carbon, Hydrogen, Oxygen, Nitrogen, Halogen, Sulfur, Selenium, and/or Tellurium*, 2nd ed.; Butterworths: London, 1971.

75. International Union of Pure and Applied Chemistry, Organic Chemistry Division, Commission on Nomenclature of Organic Chemistry. *Nomenclature of Organic Chemistry, Section A,B,C,D,E,F, and H*, 1979 edition; Rigaudy, J., Klesney, S. P. Eds.; Pergamon Press: Oxford, 1979.

76. International Union of Pure and Applied Chemistry, Organic Chemistry Division, Commission on Nomenclature of Organic Chemistry. Rules for the Nomenclature of Organic Chemistry, Section E. Stereochemistry (Recommendations 1974). *Pure Appl. Chem.* **1976**, *45*, 11–30.

77. International Union of Pure and Applied Chemistry, Organic Chemistry Division, Commission on Nomenclature of Organic Chemistry. Nomenclature of Organic Chemistry, Section H. Isotopically Modified Compounds (Approved Recommendations 1978). *Pure Appl. Chem.* **1979**, *51*, 352–380.

78. International Union of Pure and Applied Chemistry, Organic and Inorganic Chemistry Divisions, Commissions on Nomenclature of Organic and Inorganic Chemistry. Nomenclature of Organic Chemistry, Section D. Organic Compounds Containing Elements that are not Exclusively Carbon, Hydrogen, Oxygen, Nitrogen, Halogen, Sulfur, Selenium and Tellurium (Tentative). *IUPAC Inf. Bull. Append.* August, 1973, No. 31.

79. International Union of Pure and Applied Chemistry, Organic Chemistry Division, Commission on Nomenclature of Organic Chemistry. Nomenclature of Organic Chemistry, Section F. Natural Products and Related Compounds (Provisional). *IUPAC Inf. Bull. Append.* December, 1976, No. 53.

80. International Union of Pure and Applied Chemistry, Organic Chemistry Division, Commission on Nomenclature of Organic Chemistry. *A Guide to IUPAC Nomenclature of Organic Compounds*, Recommendations 1993, Panico, R., Richer, J-C., Powell, W. H., preparers; Blackwell Scientific: Oxford, 1993.

81. International Union of Pure and Applied Chemistry, Organic Chemistry Division, Commission on Nomenclature of Organic Chemistry. Revision of the Extended Hantzsch–Widman System of Nomenclature for Heteromonocycles (Recommendations 1982). *Pure Appl. Chem.* **1983**, *55*, 409–416.

82. International Union of Pure and Applied Chemistry, Organic Chemistry Division, Commission on Nomenclature of Organic Chemistry. Treatment of Variable Valence in Organic Nomenclature (Lambda Convention) (Recommendations 1983). *Pure Appl. Chem.* **1983**, *55*, 409–416.

83. International Union of Pure and Applied Chemistry, Organic Chemistry Division, Commission on Nomenclature of Organic Chemistry. Extension of Rules A-1.1 and A-2.5 Concerning Numerical Terms Used in Organic Chemical Nomenclature (Recommendations 1986). *Pure Appl. Chem.* **1986**, *58*, 1693–1696.

84. International Union of Pure and Applied Chemistry, Organic Chemistry Division, Commission on Nomenclature of Organic Chemistry. Nomenclature for Cyclic Organic Compounds with Contiguous Formal Double Bonds (The δ-Convention) (Recommendations 1988). *Pure Appl. Chem.* **1988**, *60*, 1395–1401.

85. International Union of Pure and Applied Chemistry, Organic Chemistry Division, Commission on Nomenclature of Organic Chemistry. Revised Nomenclature for

Radicals, Ions, Radical Ions and Related Species (Recommendations 1993). *Pure Appl. Chem.* **1993**, *65*, 1357–1455.

86. International Union of Pure and Applied Chemistry, Organic Chemistry Division, Commission on Nomenclature of Organic Chemistry and Commission on Physical Organic Chemistry. Glossary of Class Names of Organic Compounds and Reactive Intermediates Based on Structure (Recommendations 1995). *Pure Appl. Chem.* **1995**, *67*, 1307–1375.

87. International Union of Pure and Applied Chemistry, Organic Chemistry Division, Commission on Nomenclature of Organic Chemistry and Commission on Physical Organic Chemistry. Basic Terminology of Stereochemistry (Recommendations 1996). *Pure Appl. Chem.* **1996**, *68*, 2193–2222.

88. International Union of Pure and Applied Chemistry, Organic Chemistry Division, Commission on Nomenclature of Organic Chemistry. Nomenclature of Fused and Bridged Fused Ring Systems (Recommendations 1998). *Pure Appl. Chem.* **1998**, *70*, 143–216.

89. International Union of Pure and Applied Chemistry, Organic Chemistry Division, Commission on Nomenclature of Organic Chemistry. Phane Nomenclature, Part I (IUPAC Recommendations 1998). *Pure Appl. Chem.* **1998**, *70*, 1513–1545.

90. International Union of Pure and Applied Chemistry, Organic Chemistry Division, Commission on Nomenclature of Organic Chemistry. Revised Section F: Natural Products and Related Compounds (IUPAC Recommendations 1999). *Pure Appl. Chem.* **1999**, *71*, 587–643.

91. American Chemical Society. The Report of the ACS Nomenclature, Spelling, and Pronunciation Committee for the First Half of 1952. *Chem. Eng. News* **1952**, *30*, 4513–4526; E. Organic Compounds Containing Phosphorus, 4515–4522.

92. American Chemical Society. Rules of Carbohydrate Nomenclature. *J. Org. Chem.* **1963**, *28*, 281–291.

93. Patterson, A. M.; Capell, L. T., *The Ring Index*; Reinhold: New York, 1940.

94. Patterson, A. M.; Capell, L. T.; Walker, D. F. *The Ring Index*, 2nd ed.; American Chemical Society: Washington, DC, 1960.

95. American Chemical Society. Nomenclature of the Hydrogen Isotopes and Their Compounds. *Ind. Eng. Chem., News Ed.* **1935**, *13*, 200–201.

96. American Chemical Society. Report of the Committee for the Revision of the Nomenclature of Pectic Substances. *Chem. Eng. News* **1944**, *22*, 105–106.

97. American Chemical Society. Nomenclature of Carotenoid Pigments. *Chem. Eng. News* **1946**, *24*, 1235–1236.

98. American Chemical Society. Naming of *cis* and *trans* Isomers of Hydrocarbons Containing Olefin Double Bonds. *Chem. Eng. News* **1949**, *27*, 1203.

99. American Chemical Society. The Designation of 'Extra' Hydrogen in Naming Cyclic Compounds. *Chem. Eng. News* **1949**, *27*, 1303.

100. American Chemical Society. The Naming of Geometric Isomers of Polyalkyl Monocycloalkanes. *Chem. Eng. News* **1950**, *28*, 1842–1843.

101. American Chemical Society. The Report of the ACS Nomenclature, Spelling, and Prounciation Committee for the First Half of 1952. *Chem. Eng. News* **1952**, *30*, 4513–4545; A. Arene and Arylene, p 4513; B. Halogenated Derivatives of Hydrocarbons, pp 4513–4514; C. The Use of 'Per' in Naming Halogenated Organic Compounds, pp 4514–4515; D. The Use of 'H' to Designate the Positions of Hydrogen in Almost Completely Fluorinated Organic Compounds, p 4516; E. Organic Compounds Containing Phosphorus, pp 4515–4522; F. Organosilicon Compounds, pp 4517–4522; Nomenclature of Natural Amino Acids and Related Substances, pp 4522–4526.

102. American Chemical Society. A Proposed System of Nomenclature for Terpene Hydrocarbons. *Chem. Eng. News* **1954**, *32*, 1795–1797.

103. American Chemical Society. Recommendations of the Nomenclature Committee of the Organic Division of the ACS to the Nomenclature Committee of the National Society. *J. Am. Chem. Soc. Proc.* **1931**, 40–41.

104. American Chemical Society. Nomenclature of Terpene Hydrocarbons (a report of the Nomenclature Committee of the Division of Organic Chemistry of the American Chemical Society). In *Advances in Chemistry Series*, No. 14; American Chemical Society, 1955.

105. American Chemical Society, Chemical Abstracts Service. The Naming and Indexing of Chemical Compounds by Chemical Abstracts (a reprint of the Introduction to the *1945*

*Subject Index of Chemical Abstracts*, **1945**, *39* (Dec. 20)); Chemical Abstracts Service: Columbus, OH, 1945; pp 5867–5975.

106. American Chemical Society, Chemical Abstracts Service, Chemical Substances Index Names. Appendix IV in the *Chemical Abstracts* 1999 *Index Guide*. Chemical Abstracts Service: Columbus, Ohio, 1999.

107. American Chemical Society. A Structure-Based Nomenclature for Linear Polymers. *Macromolecules* **1968**, *1*, 193–198.

108. American Chemical Society. The Nomenclature of Boron Compounds. *Inorg. Chem.* **1968**, *7*, 1945–1964.

109. Young, J. A. Revised Nomenclature for Highly Fluorinated Organic Compounds. *J. Chem. Doc.* **1974**, *14*, 98–100.

110. International Union of Pure and Applied Chemistry, Macromolecular Division, Commission on Macromolecular Nomenclature. Nomenclature of Regular Single-Strand Organic Polymers (Rules Approved 1975). *Pure Appl. Chem.* **1976**, *48*, 373–385.

111. International Union of Pure and Applied Chemistry, Inorganic Chemistry Division, Commission on Nomenclature of Inorganic Chemistry. Nomenclature of Inorganic Boron Compounds. *Pure Appl. Chem.* **1972**, *30*, 681–710.

112. International Union of Pure and Applied Chemistry. Revised Procedure for Comment and Approval of IUPAC Recommendations on Nomenclature and Symbols. In IUPAC Recommendations on Nomenclature and Symbols: Revision of a Procedure. *Chem. Int.* **1983**, *5*, 51–53.

# 2

# Conventions in Organic Nomenclature

A wide variety of characters and symbols are used in describing the structure of organic compounds. Spelling and punctuation differ from system to system, from language to language, and often from country to country. Biochemical conventions sometimes differ from standard organic conventions. While the conventions used in this chapter reflect those used by CAS and IUPAC, deviations from the practices of each will occasionally be found.

## Spelling

The preferred spelling of element names and most other words in chemical nomenclature and terminology is that given in *Webster's Collegiate Dictionary* (10th ed.). Thus, *sulfur, aluminum*, and *cesium* are used rather than *sulphur, aluminium*, or *caesium* (the latter two spellings are recommended by IUPAC). Similarly, *center* and *stoichiometry* are used rather than *centre* and *stoicheiometry*.

There are two main kinds of spelling variations in chemical names, elision or addition of vowels, and elision of syllables resulting in contracted names. Elision and addition of vowels occur mainly for euphony. It is almost impossible to give all of the variations in the spelling of chemical names; the more common spelling conventions are mentioned in this chapter and others will be noted as they occur in various chapters.

The elision of vowels is not affected by the presence of interrupting locants or other characters. The following are important occasions where elision of vowels occurs in organic names:

- A terminal "e" of a parent hydride name is usually elided when followed by a suffix beginning with a vowel or the letter "y", as in pentanol (not pentaneol), cyclopentenyl (not cyclopenteneyl), and pyridinyl (not pyridineyl). Occasionally, there is a justifiable exception such as thiopheneol or pentapheneyl to avoid possible confusion with the name thiophenol or the prefix pentaphenyl.
- A terminal "e" of a parent hydride name is not elided when followed by a numerical prefix beginning with a vowel as in anthraceneoctol (not anthracenoctol).
- The terminal "a" of a replacement ("a") term (see chapter 3, table 3.8) is elided when followed by a vowel in a Hantzsch–Widman name (see chapter 9), as in azetane (not azaetane) and oxazole (not oxaazole), or in an alternating atom repeating unit, $a(ba)_n$ name (see chapter 9), for example, disiloxane (not disilaoxane).
- The terminal "a" of a numerical prefix (see table 2.1) is elided when followed by a suffix beginning with "a" or "o" as in tetramine (not tetraamine) and tetrol (not tetraol); when followed by a replacement prefix beginning with a vowel in a Hantzsch–Widman name (see chapter 9), for example, tetrazocane (not tetrazaocane); and when followed by "az" in names of polynuclear nitrogen parent hydrides, as in cyclopentazane (not cyclopentaazane). The last elision is no longer recognized by IUPAC.

- The terminal "o" of the hydrocarbon fusion prefixes acenaphtho, benzo, naphtho, perylo; and the terminal "a" of the fusion prefixes derived from monocyclic hydrocarbons are elided when followed by a vowel as in benzindole (not benzoindole) and cyclopentanthracene (not cyclopentaanthracene). These elisions are no longer recommended by IUPAC.[1]
- The terminal "o" of a functional replacement infix (see chapter 3, table 3.9) is elided when followed by a characteristic group suffix or part of a characteristic group suffix beginning with a vowel, except for chalcogen infixes when followed by -amide and -aldehyde, for example, -thiol (not -thiool) but -thioaldehyde (not -thialdehyde), and when followed by another functional replacement prefix beginning with a vowel as in -imidamidic acid (not -imidoamidic acid).
- The "o" of the oic acid suffix is elided when preceded by the functional replacement infixes -imid(o)- and -hydrazon(o)- as in -carboximidic acid (not -carboximidoic acid) and -hydrazonic acid (not -hydrazonoic acid).

Although it would be impossible to list all, or even most, occasions where vowels are not elided in organic names, a few of the most important instances are given here.

- The final "e" of the cyclic parent hydride component name of a conjunctive name is not elided before an acyclic component beginning with a vowel as in cyclopentane-acetic acid (not cyclopentanacetic acid).
- The final "a" of a numerical prefix (see table 2.1) or of a replacement prefix (see chapter 3, table 3.9) is not elided in a skeletal replacement name (see chapter 3) as in tetraazaoxacycloeicosane (not tetrazoxacycloeicosane).
- The final "a" of a numerical prefix (see table 2.1) is not elided when followed by the suffixes -ene or -yne, by a substituent prefix, or a class name that begins with a vowel as in -tetraene (not -tetrene), tetraalkyl (not tetralkyl ), and tetraoxide (not tetroxide).
- The final "i" of the numerical prefixes tri- and di- (or bi-), or the final "o" of mono are not elided as in, for example, triamine, triol, triiodide, trioxide, monoanhydride, and monooxide (not monoxide). Traditional names like carbon monoxide are exceptions.

The addition of vowels occurs very infrequently in organic names. The only important addition of a vowel is the insertion of an "o" between two consonants for euphony, as in butyrophenone (not butyrphenone) and -sulfonohydrazonic acid (not -sulfonhydrazonic acid).

In the evolution of organic nomenclature, many names of parent compounds, substituent suffixes, and prefixes have been shortened by the elision of syllables, presumably for euphony or to satisfy a natural desire for shorter names. Examples include methoxy from methyloxy, naphthyl from naphthalenyl, naphthol from naphthalenol, carbamic acid from carbonamidic acid, phosphinimyl from phosphinimidoyl, and sulfamide from sulfuric (or sulfonic) diamide. Such elisions can also result in simplifications elsewhere. For instance, the use of the contracted prefix silyl rather than silanyl avoids the need for the prefix bis to indicate two $H_3Si-$ groups. The prefix disilyl, not bis(silanyl), describes two silyl groups; the prefix disilanyl describes the $H_3SiSiH_2-$ group.

The elimination of syllables is sometimes deliberate. For example, the syllables "en" and "ur" are elided from the functional replacement infixes -selen(o)- and -tellur(o)-, respectively, in forming the parent hydride names selane and tellane and the suffixes -selone and -tellone; this avoids ambiguity with the Hantzsch–Widman names selenane and tellurane (see chapter 9) and the class names selenone and tellurone, analogs of sulfones (see chapter 24).

Elisions and slight modifications of names can promote understanding and avoid potential ambiguity. For example, the use of the suffix -carbaldehyde rather than -carboxaldehyde avoids the possible misconception that two oxygen atoms are present. However, care must be exercised in creating a contraction. For instance, the prefix phosphono, $(HO)_2P(O)-$, cannot be shortened to phospho, because the latter is the accepted prefix for $O_2P-$. On the other hand, avoiding a contracted name can simplify names; for example, the use of the prefix benzeno rather than benzo as a bridge prefix, avoids the need for the prefix epi or italicized *endo-* to differentiate a bridge prefix from a fusion prefix (see chapter 7).

The addition of syllables is often necessary to avoid ambiguity with well-established names. For example, the syllable "in" is added in front of established endings in the Hantzsch–Widman system (see chapter 9) to avoid ambiguity with mononuclear parent hydride names; for example, the Hantzsch–Widman name for a six-membered monocyclic ring with one boron atom is borinane, not the expected borane, which is the long-established name for $BH_3$.

## Italics

Italics are used in organic names to give special meaning to individual characters, abbreviations, prefixes, and words. They are also instrumental in alphabetizing names by identifying characters to be ignored in initial alphabetization processes.

Italicized element symbols are used as locants indicating substitution as in *N*-methylcarbamic acid; to distinguish isomers as in *O*-ethyl *S*-methyl carbonothioate; or to describe a structural feature, such as indicated or added hydrogen as in, for example, 2*H*-pyran and naphthalen-2(1*H*)-one, and isotopic modification as in methanol-$^{18}O$.

Lower case italic letters are used in a variety of ways:

- In fusion nomenclature, as in benz[*a*]anthracene (see chapters 7 and 9).
- *o*-, *m*-, and *p*- describe the three isomers of disubstituted benzenes; they are no longer used by CAS and IUPAC recommendations now prefer the numerical locants 1,2-, 1,3-, and 1,4-.
- *s*- and *as*- (or *uns*-) and *assym*- (or *unsym*-) distinguish between symmetrical and unsymmetrical features, such as *s*- and *as*-indacene and *s*- or *as*-triazine; they are no longer used in place of locants in CAS names nor recommended by IUPAC.
- *v*- or *vic*- (for vicinal) indicating adjoining positions, and *gem*- indicating two substituents at the same position; they are not used in CAS or IUPAC names.
- *sec*- (for secondary) and *tert*- (for tertiary) describe precisely two of the three branched butyl substituent prefixes; the prefix *tert*- has also been approved by IUPAC only for the unsubstituted substituent prefix *tert*-pentyl. Other than these specific meanings, use of these prefixes has not been precise and should be avoided. CAS discontinued use of *sec*- and *tert*- in 1972.
- *n*- (for normal) is used occasionally for unbranched hydrocarbon substituent prefixes; it is really not necessary since the unmodified alkyl name, by definition, is the *n*-isomer.
- *n*-, *s*-, and *t*- are used with the symbol Bu in formulas to distinguish among the three structural isomers, as *n*-Bu, *s*-Bu, and *t*-Bu.
- *aci*- (for "acid form") is used only in the prefix *aci*-nitro, but see chapter 10.
- The prefix *peri*- is used to indicate a relationship akin to that of positions 1 and 8 in naphthalene or the fusion of a ring to two adjacent rings of a polycyclic ring system, as *peri*-naphthindene (phenalene).
- The prefix *ar*- is used to indicate nonspecific substitution in an aromatic ring, not to be confused with the abbreviation Ar (see below) or the element argon.
- In stereodescriptors for describing configurational isomers or optical activity, such as *cis*- and *trans*-, *endo*- and *exo*-, *syn*- and *anti*-, *d*- (for *dextro*), *l*- (for *levo*), *m*- (for *meso*), *dl*- or *rac*- (for racemic), *ent*- (for enantiomer), *r*- (for reference or pseudoasymmetry), *c*- (for *cis*-) and *t* (for *trans*-), s- (for pseudoasymmetry), and *glycero*-, *erythro*-, *threo*-, *ribo*-, etc. (see chapters 30 and 31).
- The letters *d* and *t* are used to indicate deuterium and tritium isotopic substitution (see chapter 32).
- The descriptor *retro*- indicates a modification of the structure of a stereoparent (see chapter 31).

- Infixes, such as -*co*-, -*stat*-, -*ran*-, -*alt*-, -*per*-, -*block*-, -*graft*-, are used in naming organic polymers (see chapter 29).

Capital italic letters, such as *N*, *S*, and *As* are used as locants (see below). *R* and *S* describe absolute stereochemical configuration in chiral structures, and *Z* and *E* denote achiral configuration at a double bond (see chapter 31).

Italics are not used for the prefixes iso as in isopropyl and isooctane, cyclo as in cyclohexane and 3,5-cyclopregnane (see chapter 31), homo, nor, abeo, neo, friedo, and seco in natural product names (see chapter 31), or for poly and copoly in polymer names (see chapter 29).

## Punctuation

All punctuation in chemical names is significant. Misuse of punctuation can lead to names that are misleading and often ambiguous:

- Commas are used to separate individual locants in a series, for example, 1,4-dichlorobenzene, and to separate italic letters or sets of italic letters in fusion descriptors, as in dibenzo[*a, j*]anthracene and dibenzo[*def,mno*]chrysene (see chapter 7).
- Hyphens are used to separate a locant or a set of locants from an alphabetical part of a name, as in 2-chlorobenzene; the two parts of a primary fusion descriptor as in thieno[2,3-*b*]furan; different types of locants and locants that refer to different parts of a name, even if enclosing marks are also present as in, for example, *N*-2-naphthylacetamide and 2-(3-ethylphenyl)-1-naphthoic acid.
- The dash (or long hyphen) is used to separate the names of components of addition compounds as in borane—tetrahydrofuran and in the names of organic polymers to indicate connection between two blocks, for example, polystyrene—dimethylsilylene—polystyrene.
- Periods (full stops) are used to separate the arabic numbers of descriptors in von Baeyer and spiro names, for example, bicyclo[3.2.1]octane and spiro[4.5]decane.
- Colons are used to separate sets of locants already related by other punctuation as in 1,1′:4′,1″-terphenyl, 1,4:5,8-dimethanonaphthalene, and dinaphtho[2,1-*c*:1′,2′-*g*]phenanthrene.
- Semicolons are used to separate sets of locants in which a colon has already been used, as in benzo[1″,2″:3,4;4″,5″:3′,4′]dicyclobuta[1,2-*a*:1′,2′-*a*]diindene.
- Spaces are an important type of punctuation for organic names. Failure to use a space when one is required often results in ambiguous names. To use spaces where not required can be confusing. For example, the name ethyl phosphinate describes the ester $H_2P(O)-O-CH_2CH_3$ and the name ethylphosphinate describes the anion $(CH_3CH_2)HP(O)-O^-$. Generally, there is no space following punctuation or enclosing marks (see below). Spaces separate words in additive names such as phosphine oxide and functional class names such as benzoyl bromide, methylene dichloride, naphthoic acid and sodium hydrogen butanedioate.

Spaces are also important in class names, as in chloro alkylbenzoic acids versus chloroalkylbenzoic acid.

## Enclosing marks

Parentheses, brackets, and braces are the three types of enclosing marks used in organic names to set off portions of a name so as to convey the structure of a compound as clearly as possible. The "nesting order" of enclosing marks recommended by IUPAC is: ... {[({[( )]})]} ... CAS does not use braces but simply repeats square brackets following the use of parentheses; this is the method followed in this book.

*Parentheses* are used to enclose prefixes that consist of two or more parts, such as a substituted parent substituent like chloromethyl. There have always been two ways to apply this convention. One is to rigorously use parentheses to enclose all substituent prefixes consisting of two parts, as in (chloromethyl) and (dimethylamino); this is the method used by CAS and is the convention used in this book. The other method would not restrict the use of parentheses to two-part multipart substituent prefix names as long as there is no ambiguity, as in dimethylaminomethyl; this is the method used in the IUPAC Organic Rules before 1993.

Parentheses are often used simply to avoid ambiguity in names like hexane(dithioic) acid, 2-methyl(thiobenzoic acid), and 2-(thioformyl)pentanedioic acid. Names like chloromethylsilane and dichloromethylphosphane are potentially ambiguous. In such cases, the 1990 IUPAC Inorganic Rules[2] used parentheses to enclose one substituent prefix as in chloro(methyl)silane, which unambiguously describes the compound $CH_3$–$SiH_2$–Cl; and the name (chloromethyl)-silane, which describes the compound Cl–$CH_2$–$SiH_3$. In the first case parentheses would not be used by CAS. Differentiation between the two structures for CAS relies on the requirement that parentheses are *always* used to enclose prefixes that consist of two parts; if no parentheses appear then the structure must be $CH_3$–$SiH_2$–Cl because parentheses would always be used if the structure were Cl–$CH_2$–$SiH_3$. This attitude is fine provided one is fully familiar with the rules. However, the first method removes all possibility of ambiguity and is preferred; it is the method used in this book.

Parentheses are used around two-part prefixes even though square brackets are used within the name, for example, (4'-chloro[1,1'-biphenyl]-4-yl).

Parentheses are also used in the following instances:

- to enclose simple prefixes in order to separate locants that refer to different parts of a structure even though only one may be expressed, as in 4-(2-naphthyl)benzoic acid, 2-(pyrazinyl)propionic acid, and (2-naphthyl)anthracene (unknown anthracene locant) and 2-(naphthyl)anthracene (unknown naphthyl group locant);
- to enclose "added hydrogen" (see chapter 3) as in naphthalen-2(1*H*)-one, and to isolate a second locant of a double bond when it differs from the first by any number other than 1, as in bicyclo[3.3.2]dec-1(9)-ene;
- following the multiplying prefixes bis-, tris-, etc., as in 2,3-bis(2-bromoethyl)phenol and 2,4,6-tris(4-hydroxybutyl)cyclohexanecarboxylic acid; and sometimes poly-, as in poly(styrene), but not polyamine;
- to enclose the stereodescriptors *R, S, E, Z,* and related symbols (see chapter 30) as in (*E*)-but-2-enedioic acid, (*R,S*)-2,3-dihydroxybutanedioic acid, and (+)-tartaric acid.

*Square brackets* are almost always used to enclose substituent prefixes in which parentheses have already been used, as in 4-[2-(chlorobutyl)phenyl]pyridine and 2,2'-[ethylenebis(oxy)]dibenzoic acid; added hydrogen and the parenthetical number indicating the hydrogen population of borane parent hydrides (see chapter 27) are considered by CAS as exceptions leading to names such as (2-chloro-1(2*H*)-naphthyl) and (2-methyldiboran(6)-1-yl). However, CAS does use brackets when parentheses are used to enclose charge numbers in coordination names as in bis[hexafluorophosphate(1-)]. In this book, square brackets will be used whenever parentheses are used in a name for any reason, giving names like [2-chloro-1(2*H*)-naphthyl] and [2-methyl-diboran(6)-1-yl].

Square brackets also:

- enclose fusion descriptors as in dibenzo[*a,c*]cyclooctene; von Baeyer descriptors as in tricyclo[2.2.1.0$^{1,4}$]heptane; spiro descriptors as in dispiro[4.1.4.1]dodecane; and ring assembly names when followed by a suffix as in [2,2'-bipyridine]-6,6'-dicarboxylic acid;
- enclose locants that refer to a structural feature of a component in a ring system name as in dibenzo[*b,e*][1,4]dioxin, spiro[cyclobutane-1,9'-[1,4]methanonaphthalene, 7,16-[1,2]benzenoheptacene, and 1,3,5-propan[1]yl[3]ylidene-1*H*-benz[*e*]indene. In CAS names, such locants are enclosed in brackets only when they do not correspond to

the final numbering of the ring system, for example, naphtho[2,3-*b*]-1,4-dioxin but dibenzo[*b,e*][1,4]dioxin;
- are used in formulas to enclose a series of methylene units as in $CH_3[CH_2]_4$–OH and to enclose organic ions as in $[(CH_3)_4N^+]$ $Cl^-$ and $[C_4H_8]^+$.

## Locants

Locants are symbols used to indicate specific positions in a structure. In organic nomenclature italic Roman letters, Greek letters, and arabic numbers are the most common locants.

- Arabic numbers indicate positions in parent hydrides, in substituent prefixes derived from parent hydrides, and in some functional parent compounds, as in 4-(pyridin-2-yl)naphthalene-1-carboxylic acid and 2-bromoacetamide.
- Greek letters are used to indicate positions on an acyclic component of a conjunctive name as in $\alpha$-chloro-$\gamma$-phenyl-1-naphthalenebutanoic acid and on side chains of alkylated benzenes, such as toluene and mesitylene, as in $\alpha$-bromotoluene and $\alpha, \alpha', \alpha''$-trichloromesitylene.
- Greek letters also indicate end groups in polymer nomenclature, as in $\alpha$-chloro-$\omega$-(trichloromethyl)poly(1,2-dichloroethylene) (see chapter 29).
- Capital italic letters are used in element symbols that denote positions on heteroatoms of principal groups, such as amides, amines, and esters, and in functional parent compounds, such as urea, guanidine, and phosphonamidic acid, as in *N,N*-dimethy-cyclohexanecarboxamide and *O*-ethyl *N,P*-dimethylphosphonamidothioate.

Primes and superscripts are used to differentiate between multiple occurrences of the same locant in a name, as in *N*-ethyl-*N'*-methylbenzene-1,4-dicarboxamide, *N,N',N''*-trimethyl-guanidine, and $O^1$-methyl $O^3$-propyl 4-methylbenzene-1,3-dicarboxylate.

Traditionally, locants defining the position of unsaturation in an otherwise saturated parent hydride and locants defining the position of a principal group suffix or free valence have been cited in front of the name of the parent hydride unless other locants, such as those indicating the position of hetero atoms, were already cited there. In its 1993 recommendations, IUPAC placed such locants just before the part of the name to which they refer, except for traditional contracted names, for instance, hex-2-ene rather than 2-hexene, pyridin-4-ol rather than 4-pyridinol, and butane-1,4-diyl rather than 1,4-butanediyl; but 2-naphthyl *not* naphth-2-yl, and 2-pyridyl *not* pyrid-2-yl.

In generating names for complex organic structures, it is sometimes necessary to choose between different sets of locants. The following general priorities, given in decreasing order, are used by CAS for determining a preferred set of locants; they are, in general, followed by IUPAC.

- Capital italic Roman letters, in alphabetical order, that is, $As > N > P > S$. [*Note*: If the lower case italic letters *o*, *m*, and *p* are used, they take the order $o- > m- > p-$, which corresponds to the order of their numerical equivalents, $1, 2 > 1, 3 > 1, 4$. However, in the Beilstein Handbook[3], their normal alphabetical order $m- > o- > p-$, is used.]
- Greek letters, in Greek alphabetical order, that is, $\alpha > \beta > \gamma > \mu$.
- Arabic numbers, in numerical order, $1 > 2 > 3 > 4$.

Accordingly, $N > \alpha > 1 > 2$.

When there are two or more possibilities for a series of locants in the formation of an organic name, the series of locants that is lowest is preferred. The lower of two locant series is the one having the more preferred locant lower at the first difference when each series is written in ascending order and compared term by term. Thus, the locant series 1,2,5,6 is preferred to 1,3,4,5, because at the second term the locant 2 is preferred to 3. Other considerations in determining a lowest locant series include the following:

- Unprimed locants are preferred to the corresponding primed locant in an otherwise identical set of locants; primes are disregarded until the unprimed locant set is identical. Serially primed locants are ordered in the same way. Thus, $2 > 2'$; $1' > 2$; $\beta' > \beta''$; $\alpha' > \beta$; $N > N'$; $3' > 3''$; $3'' > 4'$; $N' > N'' > P$; $2',2'' > 2,3'$; and $2,3',4' > 2',3,4$.
- Arabic numbers followed by lower case Roman letters immediately follow the corresponding arabic number; thus, $8 > 8a > 8' > 8'a > 8b$.
- Superscript arabic numbers are considered only after all choices among parent locants have been made. Thus, the locant set $N^1,N^3,N^3,4$ is preferred to $N^3,N^1,N^1,6$ because the parent locant set $N,N,N,4$ is preferred to $N,N,N,6$; comparison of the superscript locant number sets 1,1,3 with 1,3,3 is not necessary. However the locant set $N^1,N^1,2$ is preferred to $N^4,N^4,2$; here the parent locant sets $N,N,2$ are identical; the superscript arabic number set must be compared, and 1,1 is preferred to 4,4. This consideration is needed in making choices among different von Baeyer names for polycyclic hydrocarbons (see chapter 6).

Locants are often omitted from organic names when their omission causes no ambiguity. For example, locants are not needed to describe substitution on the methyl group of acetic acid, but they are needed for acetamide because substitution is also possible on the amide nitrogen atom. Locants are also often omitted in the following circumstances:

- When all substitution sites on a parent compound have been substituted by the same substituent. This is common in fluorinated compounds, as in hexafluoroacetone and pentafluorophenol. In the latter example, substitution of the hydrogen atom of the hydroxyl group is not a normal operation in substitutive organic nomenclature; however, the name pentafluoroaniline would be ambiguous.
- From common indicated hydrogen isomers of rings and ring systems as in indene instead of $1H$-indene and fluorene instead of $9H$-fluorene. This omission is no longer followed by CAS.
- For hydro prefixes, locants are usually omitted in a fully saturated monocyclic or nonbridged fused ring systems in which all skeletal atoms are in their standard valency state as in octahydro-$1H$-indene.

The practice of omitting locants is certainly decreasing and will continue to do so as the generation and editing of organic names is done more and more by computer programs. CAS has adopted rigorous rules for citation of locants; such rigor is not needed in nomenclature for general purposes as long as ambiguity can be avoided.

## Other Symbols

- The Greek letters $\lambda, \delta, \eta, \kappa$, and $\mu$ have special meanings in organic (and coordination) nomenclature (see chapters 3, 27, and 28).
- The small capital Roman letters D and L are stereodescriptors in amino acid and carbohydrate nomenclature (see chapter 31), as in D-glucopyranose and L-alanine.
- The Greek letters $\alpha$ and $\beta$ can be used as stereodescriptors (see chapter 30).
- Italic Roman letters, such as $n$, $m$, $p$, $q$, as subscripts are used to indicate an unstated number of repeating units in a formula for a polymer or polymer segment.

## Numerical Terms (Multiplying Prefixes)

Numerical terms are used in organic nomenclature to indicate a multiplicity of identical structural features.

Table 2.1. Simple Numerical Terms

| Number | Prefix | Number | Prefix | Number | Prefix | Number | Prefix |
|--------|--------|--------|--------|--------|--------|--------|--------|
| 1 | mono[a] | 10 | deca | 100 | hecta | 1000 | kilia |
| 2 | di | 20 | eicosa or | 200 | dicta | 2000 | dilia |
| 3 | tri | | icosa[c] | 300 | tricta | 3000 | trilia |
| 4 | tetra | 30 | triaconta | 400 | tetracta | 4000 | tetralia |
| 5 | penta | 40 | tetraconta | 500 | pentacta | 5000 | pentalia |
| 6 | hexa | 50 | pentaconta | 600 | hexacta | 6000 | hexalia |
| 7 | hepta | 60 | hexaconta | 700 | heptacta | 7000 | heptalia |
| 8 | octa | 70 | heptaconta | 800 | octacta | 8000 | octalia |
| 9 | nona[b] | 80 | octaconta | 900 | nonacta | 9000 | nonalia |
| | | 90 | nonaconta | | | | |

[a] The prefix mono- is usually, but not always, omitted.

[b] Derived from Latin.

[c] The spelling eicosa- is used by CAS and the Beilstein Institute[3]; icosa is recommended by IUPAC. Except for the multiplying prefix heneicosa-, the initial "ei" or "i" of the numerical term eicosa or icosa- is elided when any other numerical term is prefixed to it; thus the multiplying prefix for 23 is tricosa- (not triicosa-) and for 26, hexacosa- (not hexicosa-).

The numerical terms used by IUPAC[4] and CAS are illustrated in table 2.1. With three exceptions, multiplying prefixes are formed systematically from these numerical terms by citing them in an order opposite to that of the constituent numerals in the number. For example, the multiplying prefix for 25 is pentacosa-; for 38, octatriaconta-; for 456, hexapentacontatetracta-; for 3598, octanonacontapentactatrilia-. In this way multiplying prefixes up through 9999 can be generated. The exceptions are as follows: (1) in association with the numerical term deca, the number "one" is represented by "un", resulting in the multiplying prefix undeca rather than hendeca-; (2) in association with numerical terms other than deca-, the number "one" is represented by "hen-"; thus, the multiplying prefix for 21 is heneicosa- (or henicosa-); for 31, hentriaconta-; and for 101, henhecta-; and (3) in association with any of the numerical terms in table 2.1, the numerical term for the number two is "do" (pronounced "dough", not "doo"). Thus, the multiplying prefix for 12 is dodeca-, for 22 is docosa-; for 32, dotriaconta-, and for 232, dotriacontadicta-.

The simple numerical prefixes given in table 2.1 are used to indicate a multiplicity of unsubstituted substituent prefixes, suffixes, replacement terms, conjunctive components, and unsubstituted functional modification terms, as in dichloro, dioxa, disulfide, dicarboxylic acid, dimethanesulfonate, and dimethanamine, provided that their use does not result in ambiguity. For instance, the numerical prefix tri- cannot be used to indicate the presence of three decyl groups or three phosphate anions because the name tridecyl- describes a thirteen-membered chain (see chapter 5), and triphosphate describes an anion derived from the condensed trinuclear acid, triphosphoric acid (see chapter 25). When this situation occurs, the multiplicative forms of the numerical prefixes described below are used.

Multiplicative forms of the simple numerical prefixes given in table 2.1 are used for substituted prefixes or functional modifiers and when the numerical prefixes given above cannot (or should not) be used because of ambiguity or potential misinterpretation; except for bis- and tris-, they are formed by adding -kis to the simple numerical prefixes given above. These prefixes are always followed by appropriate enclosing marks, for example bis(2-aminoethyl), tris[(4-chlorophenyl)methyl], bis(thioic) acid, tris(sulfate), and tetrakis(methylene). Bis(thioic) acid is needed to indicate two monothiocarboxylic acid groups because dithioic acid indicates one dithiocarboxylic acid group (see chapter 11). Tetrakis(methylene) indicates four distinct $-CH_2-$ groups, either as part of a multiplying prefix (see chapter 3) or, in CAS names, four $CH_2=$ substituents on a single parent structure, whereas tetramethylene is a traditional name for a chain of four methylene units $-CH_2CH_2CH_2CH_2-$ commonly found in polymer names (see chapter 29), but preferably named butane-1,4-diyl in substitutive nomenclature

Table 2.2. Numerical Terms from Latin Number Words

| Number | Name | Number | Name | Number | Name |
|--------|------|--------|------|--------|------|
| 2 | bi | 6 | sexi | 10 | deci |
| 3 | ter | 7 | septi | 11 | undeci |
| 4 | quater | 8 | octi | 12 | dodeci |
| 5 | quinque | 9 | novi | 13 | trideci |

(see chapter 5). In the name for the structure structure shown below, if di- were used the

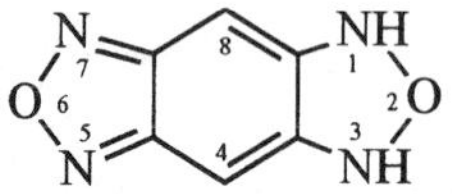

$1H,3H$-Benzo[1,2-$c$:4,5-$c'$]bis[1,2,5]oxadiazole

name would not actually be ambiguous because there would have to be four locants in the component ring system if the heteromonocyclic ring were dioxadiazole, but this use of bis does prevent confusion.

Numerical prefixes derived from Latin number names, given in table 2.2, are used in names for assemblies of identical units, such as biphenyl, biacetyl, and terpyridine.

Other numerical prefixes derived from Latin are also found in organic nomenclature, namely, semi- and hemi- for $\frac{1}{2}$ and sesqui- for $\frac{3}{2}$; another numerical prefix, sester- for $\frac{5}{2}$, has virtually disappeared from use.

## Alphabetization

An alphabetical order of citation of substituent prefixes in an organic name is very important for proper citation of detachable prefixes.

Simple substituent prefixes, that is, those describing atoms, unsubstituted heteroatomic groups, substituent prefixes derived from parent hydrides, and hydro prefixes, if treated as detachable (see below), are arranged in increasing alphabetical order of their unitalicized Roman letters. Numerical prefixes that are an integral part of the prefix name, such as di- in diazo or dimethylamino and tri- in trisilanyl, are included in the alphabetization process, but other multiplying prefixes as tri- in tributyl, as well as locants, are ignored. In the following examples, the letters that are alphabetized are underlined: 5-ethyl-2-methylcyclohexane-1-carboxylic acid, 1,2,3,8-tetrachlorooctahydro-5,8-dimethylnaphthalene, 8-(2-hydroxyethyl)-6-(1-hydroxypropyl)naphthalene-2-carboxylic acid, and 4,5-dichloro-1-diazooctane. If two or more substituent prefixes have identical Roman letters, the one with the lower locant at the first point of difference is cited first, for example,8-(1-chlorobutyl)-6-(2-chlorobutyl)tridecane and 3-$o$-chlorophenyl-1-$m$-chlorophenylnaphthalene.

Names of substituted heteroatomic groups and substituent prefixes derived from parent hydrides are first arranged within themselves according to the alphabetic ordering of the component simple prefixes as described above and then arranged in increasing alphabetical order of their initial unitalicized Roman letters. The mutiplicative prefixes, bis-, tris-, and so on, are considered for alphabetization in the same way as the simple numerical terms di-, tri-, and so on. In the following examples, the letters of the simple prefixes to be alphabetized are single underlined and the letters involved in the final alphabetization process are double underlined if not already single underlined: 6-(3-chloro-4-nitrophenyl)-1,3-bis(2,2,2-trichloro-1-fluoroethyl)isoquinoline, 3,4,5-trichloro-2,6-bis[2-(diethylamino)ethyl]-benzoic acid, and 4-[bis(2-chloroethyl)amino]-6,7-bis(2-bromopropyl)naphthalene-2-carboxylic acid. In the last example, the first bis- prefix is an integral part of the substituent name but the second bis- is only a multiplier and is ignored for alphabetization of the substituent prefixes.

## Abbreviations

In systematic organic nomenclature, the only abbreviations allowed in names are those described earlier in the section on the use of lower case italic letters.

In formulas, however, common abbreviations are Me (for methyl, not metal), Et (for ethyl), Pr (for propyl, not praseodymium), Bu (for butyl), Ph (for phenyl), Bz (for benzoyl), Ac (for acetyl, not actinium), Bzl (for benzyl), Py (for pyridine), and Cp (for cyclopentadienyl). Care must be taken in the use of abbreviations as many, many abbreviations are used in the literature; except for those given here, any abbreviation used in any publication or presentation should be precisely defined.

Although not truly abbreviations, acronyms such as THF (for tetrahydrofuran), MEK (for methyl ethyl ketone, known systematically as 2-butanone), and DMSO (for dimethyl sulfoxide) must be clearly defined especially when commonly used in a specialized field, such as polymer chemistry or biochemistry.

The symbol R can be used to denote a univalent hydrocarbon or heterocyclic group. Different R groups may be differentiated as $R'$, $R''$, $R'''$, and so on, or $R^1$, $R^2$, $R^3$, and so on. Multiple occurrences of the same R group are indicated by appropriate subscripts, for example, $R_2$, $R''_3$ or $R^2_2$. The symbol Ar is used to denote univalent groups derived from arenes.

## Bonding Number and the $\lambda$-Convention[5]

Organic chemical nomenclature is based on the concepts of constant valence of elements and of classical single, double, and triple covalent electron pair bonds between atoms. With these concepts, the hydrogen atoms required for exchange with other atoms or groups need not be expressed. Organic nomenclature was developed initially for compounds containing carbon atoms and a limited number of other atoms, in particular, oxygen, nitrogen, and the halogens; these elements have a constant valence in most organic compounds. As the kind of heteroatoms encountered in organic compounds increased, it became necessary to define a "standard valence" for each element, since some are commonly found in more than one valence. Because the term valence has come to mean different things to different people, the new term "bonding number", $n$, was introduced. This new term is the sum of the number of bonding equivalents, that is, classical valence bonds, that an atom of a parent hydride has to adjacent atoms. For example, whereas the bonding numbers of carbon, nitrogen, and oxygen are almost always 4, 3, and 2, respectively, the bonding number of the phosphorus atom in $PH_5$ is 5; of the sulfur atom in $SH_4$ is 4; and of each phosphorus atom in $H_2PPH_2$ is 3. In order to know the number of hydrogen atoms attached to each skeletal atom, the bonding number of each skeletal atom must be either implied or stated. Each element has been given a standard bonding number that need not be specified in names; these standard bonding numbers are given in chapter 3, table 3.8.

The $\lambda$-convention[5] is a method for describing nonstandard bonding numbers of skeletal atoms in parent hydrides. A neutral skeletal atom of a parent hydride having a nonstandard bonding number is denoted by the symbol $\lambda^n$, where $n$ is the bonding number of the skeletal atom; it is cited following an appropriate locant for the skeletal atom. The $\lambda$-convention is not used when a nonstandard bonding number is implied by the name itself, as in phosphorane.

The $\lambda$-convention is applied only to cyclic parent hydrides by CAS and even then not to all ring systems with nonstandard skeletal atoms. The format for CAS names using the $\lambda$-convention sometimes differs slightly from that used by IUPAC.

$SH_4$ $\qquad$ $\lambda^4$-Sulfane

$H_2PPH_3PH_2$ $\qquad$ $2\lambda^5$-Triphosphane
$\phantom{H_2P}3\ \ 2\ \ \ \ 1$

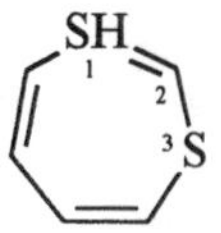

$1\lambda^4,3$-Dithiepin (IUPAC)
$1\lambda^4$-1,3-Dithiepin (CAS)

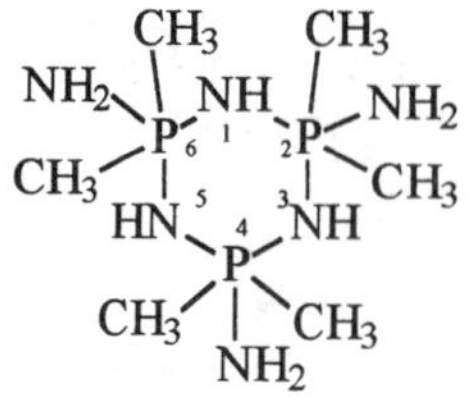

2,2,4,4,6,6-Hexamethyl-1,3,5,2$\lambda^5$,4$\lambda^5$,6$\lambda^5$-triazatriphosphinane-2,4,6-triamine (IUPAC)
2,2,4,4,6,6-Hexamethyl-2,2,4,4,6,6-hexahydro-1,3,5,2,4,6-tri-azatriphosphorinane-2,4,6-triamine (CAS)

## The δ-Convention[6]

The δ-convention[6] is a method for designating the presence of two or more formal double bonds attached to the same skeletal atom of a cyclic parent hydride. It is another method for ensuring that the number of hydrogen atoms at each skeletal atom can be precisely determined. A skeletal atom of a parent hydride that has two or more formal double bonds attached to adjacent skeletal atoms, and that otherwise has the maximum number of conjugated, that is, noncumulative, double bonds, is described by the symbol $\delta^c$, where "c" is an arabic number equal to the number of skeletal double bonds: it is cited after an appropriate locant and $\lambda^n$-symbol for that skeletal atom, if needed. CAS does not use the δ-convention.

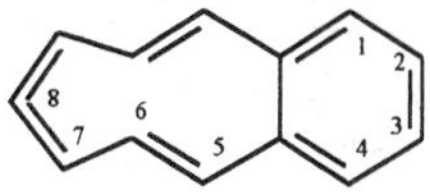

$8\delta^2$-Benzocyclononene
$8\delta^2$-Benzo[9]annulene

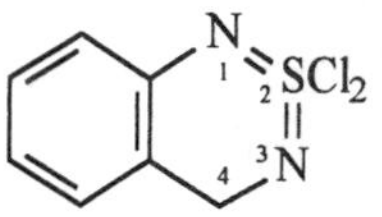

2,2-Dichloro-4$H$-2$\lambda^6\delta^2$,1,3-benzothiadiazine

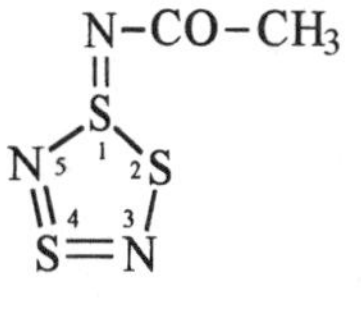

$N$-(1$H$-1$\lambda^4$,2,4$\lambda^4\delta^2$,3,5-Trithiadiazol-1-ylidene)acetamide

## Detachable and Nondetachable Prefixes

Prefixes that describe atoms or groups that have been exchanged for hydrogen atoms of a parent hydride and that are alphabetized in front of the name of the parent hydride are called detachable prefixes. This terminology doubtless arises from index nomenclature considerations where these prefixes are detached from the name of the parent compound and cited after it and a comma. For example, the prefixes 2-chloro and 3-methyl in the name 2-chloro-3-methylnaphthalene are detachable. Thus, in inverted index nomenclature, this name would be written as Naphthalene, 2-chloro-3-methyl-.

Nondetachable prefixes describe alterations to parent structures; for example, skeletal replacement ("a") prefixes indicate replacement of skeletal atoms of a hydrocarbon by other atoms, such as aza, azonia, thia, boranuida (see chapters 3, 9 and 33), and prefixes that modify structure, such as homo, nor, seco, and so on (see chapter 31). Each class of these prefixes has its own internal order of priorities; for example, replacement prefixes are ordered by the seniority of the element, and prefixes modifying structure are ordered alphabetically.

Other nondetachable prefixes such as cyclo, bicyclo and spiro describe ring formation; non-detachable prefixes such as benzo, pyrido, and ethano describe ring fusions and bridges; and still other nondetachable prefixes such as iso, *tert-*, neo, and *o-* (*ortho*) describe isomeric ring or chain positions. Nondetachable prefixes become an integral part of an unsubstituted parent compound; all other prefixes are detachable. In the name 3,6,10-trimethyl-2,5,8,11-tetraoxa-dodecane, the 2,5,8,11-tetraoxa prefix is nondetachable whereas the 3,6,10-trimethyl prefix is detachable. In an inverted alphabetical index, this name would appear inverted at tetraoxa and not dodecane, that is, 2,5,8,11-Tetraoxadodecane, 3,6,10-trimethyl-.

Hydro and dehydro prefixes are somewhat unique in that they can be detachable or non-detachable; CAS and Beilstein both treat them as detachable although not in the same way. IUPAC considered them as either detachable or nondetachable in the 1979 recommendations, but as only nondetachable in the 1993 recommendations. For example, 5,6,7,8-tetrahydro-4-methyl-1-naphthalenecarboxylic acid is the CAS name (inverted as 1-Naphthalenecarboxylic acid, 5,6,7,8-tetrahydro-4-methyl-) and an IUPAC name prior to the 1993 recommendations. The Beilstein name is 4-methyl-5,6,7,8-tetrahydro-1-naphthalenecarboxylic acid[3] (inverted as 1-Naphthalenecarboxylic acid, 4-methyl-5,6,7,8-tetrahydro-). The name in which the hydro prefixes are nondetachable is 8-methyl-1,2,3,4-tetrahydronaphthalene-5-carboxylic acid.

The epoxy prefix, and related prefixes, such as epithio, are also somewhat unique. As bridge prefixes in naming bridged fused ring systems (see chapter 9) they are nondetachable, as in 5,8-dihydro-5,8-dimethyl-5,8-epoxyquinoline; in the formation of a ring by connecting different atoms of a chain these prefixes are detachable, as in 1,3-epoxy-4-nitropentane.

## Indicated and Added Hydrogen

Heterocyclic ring systems and polycyclic hydrocarbons and their derivatives that have the maximum number of conjugated double bonds may also have isolated saturated ring atoms. As a result the same name can apply to isomers differing only in the position(s) of these saturated atoms. For parent rings and ring systems, the position of a saturated ring atom is indicated by an italic capital *H* preceded by the ring atom locant; this description is cited in front of the name of the ring parent. These "extra" hydrogen atoms are called "indicated hydrogens".

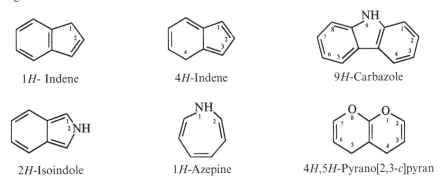

1*H*- Indene                    4*H*-Indene                    9*H*-Carbazole

2*H*-Isoindole                  1*H*-Azepine                   4*H*,5*H*-Pyrano[2,3-*c*]pyran

Indicated hydrogen atoms are assigned to positions of the parent ring system that are needed to accommodate other structural features of a compound, such as a bridge, a spiro atom, principal characteristic group, or free valence or radical.

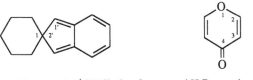

Spiro[cyclohexane-1,2'-[2*H*]indene]          4*H*-Pyran-4-one

5,7a-Methano-7a*H*-isoindole          2*H*-Imidazolidin-2-ylidene

Saturated positions may be created by a principal characteristic group, free valence or radical, spiro atom, or ring assembly (but not bridges) in ring systems that do not need indicated hydrogen or in which the indicated hydrogen of a parent ring cannot be used to accommodate such added structural features. The "extra" hydrogen at such positions is called "added hydrogen" and is described in the same way as indicated hydrogen, except that it is enclosed in parentheses and cited directly after the locant for the principal characteristic group, spiro atom, or ring assembly junction.

Naphthalen-1(2*H*)-one          Quinoline-1(4*H*)-carboxylic acid          1(2*H*)-Pyridyl
                                                                                                Pyridin-1(2*H*)-yl

1*H*-Benz[*e*]indene-1,2(3*H*)-dione          Spiro[imidazolidine-4,2'(1'*H*)-quinoxaline]

2(1*H*),4'-Biisoquinoline

Added hydrogen is not used if the structural features simply remove a double bond from the fully conjugated parent hydride structure.

Naphthalene-1,4-dione          Pyrazine-1,4-diyl

Indicated and added hydrogen should not be confused with the use of hydro prefixes to describe partially or fully saturated ring structures.

6,7-Dihydro-3a-methyl-3a*H*-indene

1,1,2,2,3,3,4,4-Octachloro-1,2,3,4-tetrahydro-
naphthalene

## REFERENCES

1. International Union of Pure and Applied Chemistry, Organic Chemistry Division, Commission on Nomenclature of Organic Chemistry. Nomenclature of Fused and Bridged Fused Ring Systems (Recommendations 1998). *Pure Appl. Chem.* **1998**, *70*, 143–216.
2. International Union of Pure and Applied Chemistry, Division of Inorganic Chemistry, Commission on Nomenclature of Inorganic Chemistry. *Nomenclature of Inorganic Chemistry Recommendations 1990*; Leigh, G. J., ed.; Blackwell Scientific: Oxford, U.K., 1993.
3. *Beilstein Handbook of Organic Chemistry*, 4th Ed.; Springer-Verlag: Berlin, 1972–1999.
4. International Union of Pure and Applied Chemistry, Organic Chemistry Division, Commission on Nomenclature of Organic Chemistry. Extension of Rules A-1.1 and A-2.5 Concerning Numerical Terms Used in Organic Chemical Nomenclature (Recommendations 1986). *Pure Appl. Chem.* **1986**, *58*, 1693–1696.
5. International Union of Pure and Applied Chemistry, Organic Chemistry Division, Commission on Nomenclature of Organic Chemistry. Treatment of Variable Valence in Organic Nomenclature (Lambda Convention)(Recommendations 1983). *Pure Appl. Chem.* **1984**, *56*, 769–778.
6. International Union of Pure and Applied Chemistry, Organic Chemistry Division, Commission on Nomenclature of Organic Chemistry. Nomenclature for Cyclic Organic Compounds with Contiguous Formal Double Bonds (δ-Convention)(Recommendations 1988). *Pure Appl. Chem.* **1988**, *60*, 1395–1401.

# 3

# Methods of Organic Nomenclature

The predominant method for naming organic compounds is substitutive, and organic nomenclature is often, but not completely correctly, referred to as substitutive nomenclature. Other methods are also used to generate names for organic compounds, such as replacement, multiplicative, functional class (including radicofunctional), conjunctive, additive, and subtractive nomenclature. One or more of these methods may be combined in the same name.

This chapter has two purposes. The first is to provide an overview of organic nomenclature by explaining its various methods without much discussion and only a few examples of names for specific compounds; elaboration is found in the chapters that follow. The second purpose is to collect together in tabular form some of the prefixes, suffixes, infixes, class names and other information essential to the application of these methods. IUPAC and CAS sometimes differ in this regard; where this occurs, either source will produce acceptable names unless otherwise noted.

Generally, organic molecules may be viewed as parent hydrides such as ethane and cyclohexane, parent compounds such as acetic acid and phosphonic acid, with or without attached substituents, or compounds belonging to one or more functional classes. For all compounds, the method or combination of methods most suitable for generating an acceptable unambiguous name is first determined. When one or more substituents of a parent hydride are characteristic groups, one, called the principal characteristic group, is cited as a suffix to the name of a parent hydride or used in a functional class name. Only one kind of characteristic group can be so used; all other characteristic groups must be cited as prefixes. Although, in principle, any characteristic group can be chosen, it is customary to select one on the basis of a seniority classification. A common list of compound classes in decreasing order of seniority is given in table 3.1; characteristic groups for these classes are enumerated in tables 3.2 and 3.10 and class names are found in tables 3.3 and 3.6. Finally, the name of the parent structure attached to the chosen characteristic group is generated and numbered in so far as possible and the remaining parts of the structure and any other characteristic groups are named and cited as prefixes.

## The Substitutive Method

The substitutive method is the system of choice for naming organic compounds and is also widely used for naming derivatives of inorganic molecular hydrides. It is based on the concept of exchanging hydrogen atoms of "parent hydrides" or "functional parent compounds" for other atoms or groups, a process called substitution. Parent hydrides are unbranched acyclic or cyclic skeletal structures to which only hydrogen atoms are attached, for example, ethane, hydrazine, trisilane, naphthalene, cyclohexane, toluene, and pyridine. Functional parent compounds are structures whose names imply the presence of at least one characteristic group in addition to hydrogen atoms attached to skeletal atoms, for example, acetic acid, aniline, and phosphonic acid. Hydrogen atoms attached to nitrogen atoms of amine, imine, imide, and amide characteristic groups may be "substituted", but such an exchange of hydrogen atoms

Table 3.1 Decreasing Seniority Order for Compound Classes

| | |
|---|---|
| 1. Radicals | 7. Amides |
| 2. Cations/Anions[a] | 8. Hydrazides[c] |
| 3. Acids (in the general order carboxylic, | 9. Imides[d] |
|    sulfonic, sulfinic, sulfenic, selenonic, | 10. Nitriles |
|    seleninic, selenenic, telluronic, | 11. Aldehydes |
|    tellurinic, tellurenic, carbon acids[b] | 12. Ketones |
|    (carbonic and formic), nitrogen acids, | 13. Alcohols (Phenols) |
|    phosphorus acids, arsenic acids, | 14. Hydroperoxides |
|    antimony acids, silicon acids, | 15. Amines |
|    boron acids) | 16. Imines |
| 4. Anhydrides[c] | 17. Ethers |
| 5. Esters[c] | 18. Peroxides |
| 6. Acid halides | |

[a] CAS ranks anions after cations; IUPAC ranks cations after anions.
[b] A CAS ranking; IUPAC has yet to decide on a ranking for these acids.
[c] These classes are treated by CAS as functional derivatives of acids and therefore
  are ranked according to the acid from which they are derived.
[d] Applies only when the compound is named as an imide.

attached to a chalcogen atom, including oxygen, of a characteristic group is not generally considered a substitution in organic nomenclature.

Names of the parent hydrides of systematic organic nomenclature, such as pentane, toluene, anthracene, 1,3-dioxole, hydrazine or diazane, bicyclo[2.2.1]heptane, and so on, are described in chapters 5–9; names for functional parent compounds, such as acetic acid, phosphonic acid, and carbamimidic acid, are found in other chapters of this book.

There are two types of substituents. Those of the first type are derived from a parent hydride (except for the mononuclear hydride of nitrogen and the chalcogen or halogen hydrides in their standard bonding states) by removal of one or more hydrogen atoms, as described in chapters 5–9; examples are methyl, cyclohexylidene, naphthalen-2-yl, hydrazinyl or diazanyl, piperidin-1-yl, and spiro[4.5]decan-1-yl. Substituents derived from parent hydrides are often called "radicals"; however, IUPAC has recommended that this term be restricted to what have been called "free radicals".[1–3] Substituents of the second type are called "characteristic groups" (formerly called functional groups); basically they are heteroatomic groups that are not of the first type. A more specific definition is given in the Glossary. Examples include single heteroatoms, such as –Cl (chloro) and =O (oxo); heteroatoms attached only to one or more hydrogen atoms or to heteroatoms that differ from it, such as –NH$_2$ (amino), –OH (hydroxy), and –ClO$_2$ (chloryl); heteroatomic groups consisting of a heteroatom or group attached to a single carbon atom, such as –CHO (formyl), –CO–NHNH$_2$ (hydrazinocarbonyl or hydrazinylcarbonyl), –CN (cyano), or that contain an isolated carbon atom, such as –NCO (isocyanato) and –NC (isocyano); and acyclic heteroatomic groups such as –OOH (hydroperoxy), –OO– (peroxy or dioxy), =N$_2$ (diazo), –N$_3$ (azido), and –S$_3$H (trithio).

Many characteristic groups can be expressed either as prefixes or suffixes attached to the name of a parent hydride. The fundamental characteristic groups for this type are given in table 3.2; chalcogen and nitrogen analogs derived by functional replacement nomenclature (see below) are illustrated in table 3.10. Trivial names often include a characteristic group and an implied parent hydride, such as benzoic acid, but a parent hydride is not always implied, for example, phenol and acetic acid. Such compounds are best considered as functional parent compounds.

Certain characteristic groups may only be expressed as prefixes, called compulsory prefixes, or by means of a class name; they are never expressed as suffixes. These prefixes are summarized in table 3.3 (see also chapter 10).

Mononuclear inorganic parent compounds having one or more hydrogen atoms attached to their central atoms are the basis for naming organic derivatives by substitution of these hydrogen atoms, thus providing for a very large number of characteristic groups. These parent compounds and substituent prefix names derived from them are illustrated in table 3.4.

Table 3.2. Fundamental Characteristic Group Suffixes and Prefixes[a]

| Group | Sux | Prefix |
|---|---|---|
| –COOH | -carboxylic acid | carboxy |
| –(C)OOH | -oic acid | — |
| –SO$_2$–OH | -sulfonic acid | sulfo |
| –SO–OH | -sulfinic acid | sulfino |
| –S–OH | -sulfenic acid | sulfeno |
| –SeO$_2$–OH | -selenonic acid | selenono |
| –SeO–OH | -seleninic acid | selenino |
| –Se–OH | -selenenic acid | seleneno |
| –TeO$_2$–OH | -telluronic acid | tellurono |
| –TeO–OH | -tellurinic acid | tellurino |
| –Te–OH | -tellurenic acid | tellureno |
| –CO–X[b] | -carbonyl halide[c] | halocarbonyl[d] |
| –(C)–OX[b] | -oyl halide[c] | — |
|  | -yl halide[c] | — |
| –CO–X′ (X′=NCO)[e] | -carbonyl isocyanate | isocyanatocarbonyl |
| –(C)O–X′(X′=NCO)[e] | -oyl isocyanate | — |
|  | -yl isocyanate | — |
| –CO–NH$_2$ | -carboxamide | carbamoyl (IUPAC) |
|  |  | aminocarbonyl (CAS) |
| –(C)O–NH$_2$ | -amide | — |
| –CN | -carbonitrile | cyano |
| –(C)N | -nitrile | — |
| –CHO | -carboxaldehyde (CAS) | formyl |
|  | -carbaldehyde (IUPAC) | formyl |
| –(C)HO | -al | — |
| >(C)=O | -one | oxo |
| –OH | -ol | hydroxy |
| –NH$_2$ | -amine | amino |
| =NH | -imine | imino |

[a] This table does not include characteristic groups derived by functional replacement nomenclature. For such characteristic groups, see table 3.10.

[b] X = –F, –Cl, –Br, –I, –At, or –N$_3$.

[c] Halide = fluoride, chloride, bromide, iodide, astatide, and the pseudohalide, azide.

[d] Halo = fluoro, chloro, bromo, iodo, astato, and the pseudohalide azido.

[e] X′ can be –NC as well as chalcogen analogs of –NCO, in which case the suffix names would be carbonyl isocyanide; carbonyl isothiocyanate; carbonyl isoselenocyanate; and carbonyl isotellurocyanate, respectively, and the prefix names would be isocyanocarbonyl or isothiocyanatocarbonyl, isoselenocyanatocarbonyl, and isotellurocyanatocarbonyl, respectively.

X′ could also be –CN, which leads to -carbonyl cyanide as a suffix and cyanocarbonyl as a prefix, although compounds with this characteristic are usually named as derivatives of nitriles (see chapter 20).

X′ could also be –CNO, which has been called isofulminato, and thus would be -carbonyl isofulminate as a suffix and isofulminatocarbonyl as a prefix, although compounds with this characteristic group are usually named as oxides of nitriles (see chapter 20).

Some mono and polynuclear oxo acids, such as sulfuric acid, phosphoric acid, diphosphoric acid, silicic acid, boric acid, and carbonic acid, do not have substitutable hydrogen atoms themselves and therefore cannot be directly involved in naming organic compounds with substitutive prefixes. However, they become functional parent compounds through the replacement of oxygen atoms or hydroxy groups by nitrogenous groups that do have substitutable hydrogen atoms. These oxo acids are also parent compounds for organic esters and anhydrides. Table 3.5 lists the common mononuclear oxo acids used in this way in organic nomenclature.

Still other characteristic groups may be described by class names (some of which may also be expressed by suffixes as given in table 3.2) or as prefixes. Some of these are given in table 3.6; however, characteristic groups whose suffixes given in table 3.2 include a class name, for example, acid halide (or halogenoid), are not included.

Table 3.3. Characteristic Groups Cited Only as Prefixes or Class Names

| Group | Prefix | Class name | Group | Prefix | Class name |
|-------|--------|-----------|-------|--------|-----------|
| Br– | bromo | bromide | $N_2=$ | diazo | — |
| Cl– | chloro | chloride | $N_3-$ | azido | azide |
| OCl– | chlorosyl[a] | — | CN– | isocyano | isocyanide |
| $O_2Cl-$ | chloryl[a] | — | ON– | nitroso[c] | — |
| $O_3Cl-$ | perchloryl[a] | — | $O_2N-$ | nitro[c] | — |
| F– | fluoro | fluoride | HO(O)N= | aci-nitro[d] | — |
| I– | iodo | iodide | NCO– | cyanato[e] | cyanate |
| OCN– | isocyanato | isocyanate | NCS– | thiocyanato[b,e] | thiocyanate |
| SCN– | isothiocyanato[b] | isothiocyanate | | | |

[a] The corresponding bromine, fluorine, and iodine analogs are named similarly: OBr– is bromosyl, OI– is iodosyl (also called iodoso), and $O_2I-$ is iodyl (also called iodoxy). Chalcogen analogs are named by functional replacement nomenclature: SCl– is thiochlorosyl, $S_2Br-$ is dithiobromyl, S(O)I– is thioiodyl, and Se(S)Cl– is selenothiochloryl.

[b] Selenium and tellurium analogs are named similarly: SeCN– is isoselenocyanato and TeCN– is tellurothiocyanato.

[c] Chalcogen analogs are named by functional replacement prefixes: thionitroso is SN– and thionitro is O(S)N–.

[d] IUPAC no longer recommends this prefix; compounds containing this characteristic group are named on the basis of the functional parent compound azinic acid (see table 3.4).

[e] Organic derivatives containing the NCO–, NCS–, and so on, groups may also be named as esters of cyanic acid, thiocyanic acid, and so on (see table 3.5); $C_6H_5-OCN$ can be named phenyl cyanate and $CH_3-SCN$ as methyl thiocyanate.

Table 3.4. Functional Parent Compounds and Derived Prefixes[a]

| Functional parent compound | Name | Group | Prefix name |
|---------------------------|------|-------|-------------|
| HB(OH)$_2$ | boronic acid | (HO)$_2$B– | borono<br>dihydroxyboryl |
| H$_2$B–OH | borinic acid | HO–B< or HO–B=<br>HO–B<<br>HO–B=<br>HO–HB– | hydroxyborylene<br>hydroxyboranediyl<br>hydroxyborylidene<br>hydroxyboryl |
| HN(O)(OH)$_2$ | azonic acid | (HO)$_2$N(O)–<br><br>HN(O)< or HN(O)=<br><br><br>(O$^-$)$_2$N(O)– | azono[b]<br>dihydroxynitroryl[b]<br>azonoyl[b]<br>hydronitroryl[b]<br>azinylidene<br>azonato[c] |
| H$_2$NO–OH | azinic acid | HO–N(O)<<br>HO–N(O)< or HO–NO=<br>H$_2$N(O)–<br><br>(O$^-$)N(O)< or (O$^-$)N(O)= | azinico[c]<br>hydroxynitroryl[b]<br>azinoyl[b]<br>dihydronitroryl[b]<br>azinato[c] |
| HP(O)(OH)$_2$ | phosphonic acid | (HO)$_2$P(O)–<br><br>HP(O)< or HP(O)=<br><br><br>(O$^-$)$_2$P(O)–<br>(O$^-$)(HO)P(O)– | phosphono<br>dihydroxyphosphoryl[b]<br>phosphonoyl[b,d]<br>hydrophosphoryl[b]<br>phosphinylidene[e]<br>phosphonato[b]<br>hydroxyoxidophosphoryl[b] |
| H$_2$PO–OH | phosphinic acid | HO–P(O)<<br>HO–P(O)< or HO–P(O)= | phosphinico<br>hydroxyphosphoryl[b]<br>hydroxyphosphinylidene[e] |

*(Continued)*

Table 3.4. Continued

| Functional parent compound | Name | Group | Prefix name |
|---|---|---|---|
| | | $H_2P(O)-$ | phosphinoyl[b,d] dihydrophosphoryl[b] phosphinyl[e] |
| | | $(O^-)P(O)<$  or  $(O^-)P(O)=$ | phosphinato[b] oxidophosphoryl |
| $HAs(O)(OH)_2$ | arsonic acid | $(HO)_2-As(O)-$ | arsono[d] dihydroxyarsoryl |
| | | $HAs(O)<$  or  $HAs(O)=$ | arsonoyl[d] hydroarsoryl arsinylidene[e] |
| | | $(O^-)_2As(O)-$ $(O^-)(HO)As(O)-$ | arsonato[d] hydroxyoxidoarsoryl |
| $H_2AsO-OH$ | arsinic acid | $HO-As(O)<$ | arsinico |
| | | $HO-As(O)<$  or  $HO-As(O)=$ | hydroxyarsoryl hydroxyarsinylidene |
| | | $H_2As(O)-$ | arsinoyl[d] dihydroarsoryl arsinyl[e] |
| | | $(O^-)As(O)<$  or  $(O^-)As(O)=$ | arsinato[d] oxidoarsoryl |
| $Sb(O)(OH)_2$ | stibonic acid | $(HO)_2Sb(O)-$ $(O^-)_2Sb(O)-$ $HSb(O)<$  or  $HSb(O)=$ | stibono[c] stibonato[c] stibonoyl[c] |
| $H_2SbO-OH$ | stibinic acid | $HO-Sb(O)<$ $H_2Sb(O)-$ $(O^-)Sb(O)<$  or  $(O^-)Sb(O)-$ | stibinico[c] stibinoyl[c] stibinato[c] |
| $HN(OH)_2$ | azonous acid[c] | $(HO)_2N-$ | dihydroxyazanyl dihydroxyamino |
| $H_2N-OH$ | azinous acid[c] hydroxylamine | $HO-N=$ $HO-N<$  or  $HO-N=$ $HO-N<$ | hydroxyazanylidene hydroxyimino hydroxyazanediyl |
| $HP(OH)_2$ | phosphonous acid | $(HO)_2P-$ | dihydroxyphosphanyl dihydroxyphosphino |
| $H_2P-OH$ | phosphinous acid | $HO-P=$ $HO-P<$  or  $HO-P=$ $HO-P<$ | hydroxyphosphanylidene hydroxyphosphinidene[e] hydroxyphosphanediyl |
| $HAs(OH)_2$ | arsonous acid | $(HO)_2As-$ | dihydroxyarsanyl dihydroxyarsino |
| $H_2As-OH$ | arsinous acid | $HO-As=$ $HO-As<$  or  $HO-As=$ $HO-As<$ | hydroxyarsanylidene hydroxyarsinidene[e] hydroxyarsanediyl |
| $HSb(OH)_2$ | stibonous acid[c] | $(HO)_2Sb-$ | dihydroxystibanyl dihydroxystibino |
| $H_2Sb-OH$ | stibinous acid[c] | $HO-Sb=$ $HO-Sb<$ | hydroxystibanylidene hydroxystibanediyl |

[a] This table does not include characteristic groups derived by functional replacement nomenclature; for such characteristic groups see table 3.10. Also it does not include mononuclear oxoacids from which functional parent compounds are derived by functional replacement nomenclature, for which see table 3.11.
[b] These names are given in the 1993 IUPAC organic nomenclature recommendations.
[c] These names are formed by analogy with phosphorus and arsenic prefixes.
[d] These names are given in the 1979 edition of the IUPAC Organic Nomenclature Rules.
[e] These names are given in the ACS rules for monophosphorus compounds.[4]

Table 3.5. Mononuclear Oxo Acid Parent Compounds and Derived Prefixes[a]

| Oxo acid | Name | Group | Prefix name |
|---|---|---|---|
| NC–OH | cyanic acid | NCO– | cyanato (see table 3.3) |
| $C(O)(OH)_2$ | carbonic acid | –CO–  or  OC< | carbonyl |
|  |  | OC< | oxomethanediyl |
|  |  | OC= | oxomethylidene |
| $B(OH)_3$ | boric acid | B≡ | borylidyne |
|  |  | –B< | boranetriyl |
| $Si(OH)_4$ | silicic acid | >Si< | silanetetrayl |
|  |  | =Si< | silanediylylidene |
|  |  | =Si= | silanediylidene |
| $N(O)_2OH$ | nitric acid | $O_2N$– | nitro (see table 3.3) |
| N(O)-OH | nitrous acid | ON– | nitroso (see table 3.3) |
| $P(O)(OH)_3$ | phosphoric acid | OP≡  or  –P(O)=  or  –P(O)< | phosphoryl[b] |
|  |  |  | phosphinylidyne[c] |
| $P(OH)_3$ | phosphorous acid | –P<  or  –P=  or  P≡ | phosphinidyne[c] |
|  |  | –P< | phospanetriyl[b] |
|  |  | P≡ | phosphanylidyne |
| $P(O)_2OH$ | phosphenic acid | $O_2P$– | phospho |
| P(O)-OH | phosphenous acid | OP– | phosphoroso |
| $As(O)(OH)_3$ | arsenic acid | OAs≡  or  –As(O)=  or  –As(O)< | arsoryl[b] |
|  | arsoric acid |  | arsinylidyne[c] |
| $As(OH)_3$ | arsenous acid | –As<  or  –As=  or  As≡ | arsinidyne[c] |
|  | arsorous acid | –As< | arsanetriyl[b] |
| $As(O)_2$–OH | arsenenic acid | $O_2As$– | arso |
| As(O)-OH | arsenenous acid | OAs– | arsenoso |
| $S(O)_2(OH)_2$ | sulfuric acid | $O_2S<$  or  $O_2S=$ | sulfonyl |
|  |  |  | sulfuryl[d] |
| $S(O)(OH)_2$ | sulfurous acid | OS<  or  OS= | sulfinyl |
|  |  |  | thionyl[d] |
| $S(OH)_2$ | sulfoxylic acid | –S– | thio |
|  |  |  | sulfanediyl |
|  |  | S= | thioxo |
|  |  |  | sulfanylidene |
| $Se(O)_2(OH)_2$ | selenic acid | $O_2Se<$  or  $O_2Se=$ | selenonyl |
| $Se(O)(OH)_2$ | selen(i)ous acid | OSe< | seleninyl |
| $Se(OH)_2$ | selenoxylic acid | –Se– | seleno |
|  |  |  | selanediyl |
|  |  | Se= | selenoxo |
|  |  |  | selanylidene |

[a] This table does not include characteristic groups derived by functional replacement nomenclature; for such characteristic groups see tables 3.10 and 3.11.
[b] These names are given in the 1979 edition of the IUPAC Organic Nomenclature Rules.
[c] These names are given in the ACS rules for monophosphorus compounds.[4]
[d] These names are given in the 1990 edition of the IUPAC Inorganic Rules.[5]
Recommendation I-8.4.2.2, p 113.

In substitutive nomenclature, most radicals and ions are named by suffixes, not all of which are substitutive suffixes. These suffixes are transformed into prefixes by the addition of -yl, -ylidene, -diyl, and so on (see chapter 33). Table 3.7 summarizes the suffixes used for radicals and ions derived from parent hydrides.

## Construction of Substitutive Names

Formation of names by substitutive nomenclature usually involves a series of criteria which are applied successively until no further choices are needed. The criteria used depend on the purpose of the name. Rigorous rules are required to generate a single preferred name for an

Table 3.6. Class Names and Prefixes

| Class | Class name | Prefix name |
|---|---|---|
| R–OH | alcohol | hydroxy |
| R–SH | mercaptan[a] | mercapto (CAS) |
| | thiol (IUPAC) | sulfanyl (IUPAC) |
| Acyl-O-Acyl' | anhydride | acyl(or acyl')oxy |
| Acyl-S-Acyl' | thioanhydride[b] | acyl(or acyl')thio[b] (CAS) |
| | anhydrosulfide[b,c] | acyl(or acyl')sulfanyl[b] (IUPAC) |
| R–O–R' | ether (oxide) | R(or R')oxy |
| R–OOH | hydroperoxide | hydroperoxy |
| R–OO–R' | peroxide | R(or R')peroxy |
| | | R(or R')dioxy |
| R–CO–R' | ketone | acyl(or acyl') |
| R–CS–R' | thioketone[b] | thioacyl(or thioacyl') |
| R–S–R' | sulfide (thioether) | R(or R')thio (CAS) |
| | | R(or R')sulfanyl (IUPAC) |
| R–SSH | hydrodisulfide[b] | disulfanyl (IUPAC) |
| | thiosulfenic acid[b] | thiosulfeno (CAS) |
| | dithiohydroperoxide | dithiohydroperoxy |
| R–SS–R' | disulfide[b] | R(or R')disulfanyl (IUPAC) |
| | thiosulfenate | R(or R')dithio (CAS) |
| RX (X = F, Cl, Br, I, At) | halide | halo |
| R–SO–R' | sulfoxide | R(or R')sulfinyl |
| R–SO$_2$–R' | sulfone | R(or R')sulfonyl |

[a] The class name mercaptan is no longer encouraged by IUPAC. The class name thioalcohols has been used and presumably selenoalcohols and telluroalcohols for the other chalcogen analogs.

[b] Selenium and tellurium analogs are named similarly, for example, hydrodiselenide and thio-selenenate.

[c] The class name anhydrosulfide means literally the removal of H$_2$S from two appropriate thioacids, whereas the class name thioanhydride is a functional replacement term indicating replacement of the anhydride oxygen atom by sulfur.

Table 3.7 Suffixes for Radicals and Ions from Parent Hydrides[3] (see also Chapter 33)

| Operation | Suffix | Suffix for substituent prefix |
|---|---|---|
| Loss of H• | -yl, -diyl, etc. | —[a] |
| | -ylidene | |
| | -ylidyne | |
| Loss of H$^-$ | -ylium | -yliumyl |
| Addition of H$^+$ | -onium | -oniumyl |
| | -ium | -iumyl |
| Loss of H$^+$ | -ide | -idyl |
| Addition of H$^-$ | -uide | -uidyl |

[a] There is no suffix creating a substituent prefix. The prefix "ylo" is added in front of the name of a substituent prefix, for example, yloethyl (See Chapter 33).

effective index. IUPAC rules and recommendations are less rigorous, leaving choices particularly among the types of nomenclature to be used. For general use, names may be formed according to criteria suitable for the purpose of the audience.

The first step in generating a substitutive name is to determine the principal characteristic group to be cited as a suffix or class name (see the section on functional class nomenclature below). Only one characteristic group can be cited as a suffix or class name; all other characteristic groups must be cited as prefixes. A widely accepted seniority order for compound classes is shown in table 3.1; it follows closely that used by CAS and IUPAC. Many compound classes have a seniority order within themselves that is not always the same for CAS, IUPAC, or Beilstein. For example, CAS places all peroxy acids at the head of the list of acids, whereas IUPAC prefers a normal acid to the corresponding peroxy acid. IUPAC also favors anions over cations whereas CAS prefers the reverse. Tables 3.2 and 3.10 (see p. 42) list characteristic groups paralleling the seniority of compound classes used as suffixes to parent hydride names.

In addition to a seniority order for characteristic groups, there must also be a seniority order for parent hydrides and parent compounds since not all compounds have characteristic groups, as in $H_2NNH-CH_2CH_2-P(CH_3)_2$ and $C_6H_5-CH_2[CH_2]_4CH_3$. Choices also need to be made between parent hydrides each attached to the same characteristic group, as in $HOOC-CH_2CH_2-SiH_2-COOH$. One useful order is parent hydrides having at least one of the heteroatoms earlier in the following list (carbon compounds fall at the end of the list): N, P, As, Sb, Bi, B, Si, Ge, Sn, Pb, O, S, Se, Te.

This seniority order applies to homogeneous parent hydrides (parent hydrides having only one kind of skeletal atom) and to heterogeneous parent hydrides and parent compounds (parent hydrides and compounds with different skeletal atoms), except when the choice is between two or more heterogeneous chains named by replacement ("a") nomenclature; for the latter see the section below on skeletal replacement nomenclature. For heterogeneous parent compounds, the seniority is that of the higher ranking element; for example, disiloxane is treated as a silicon compound, not an oxygen compound. In each class, a heterocycle is preferred to a heteroacyclic compound of the same heteroatom class. Further considerations are based on criteria such as size of the parent hydride and number of the preferred element which may be illustrated by the order: formazan > triazane > hydrazine (diazane) > hydroxylamine.

After the principal characteristic group is chosen, there may be alternative names that identify the rest of the structure. If needed, a set of hierarchical criteria is applied to determine a preferred name. The following criteria, patterned after those used by CAS, may be applied in order until a decision is reached (they are not used for choices between heteroacyclic compounds to which skeletal replacement ("a") nomenclature is applicable; for such choices see under the replacement nomenclature method below).

A preferred name should include the following:

1. The larger number of the preferred characteristic group denoted by a suffix or class name; this criterion includes principal characteristic groups expressed through multiplicative nomenclature (see below); CAS does not recognize this latter aspect.
2. At least one heteroatom cited earliest in the seniority of parent hydrides and parent compounds given above, that is, $N > P > As > Sb > Bi > B > Si > Ge > Sn > Pb > O > S > Se > Te$.
3. A ring or ring system; criteria for choosing between two or more ring systems, in decreasing order of priority are as follows (this priority order is different from the one used for choosing a preferred parent ring in fusion nomenclature, for which see chapters 7 and 9):
   (a) a nitrogeneous ring system;
   (b) a heterocyclic ring system not containing nitrogen;
   (c) a ring system with the larger number of rings;
   (d) a spiro ring system;
   (e) a bridged fused ring system; further choice goes to the bridged system with lowest locants for bridges;
   (f) a bridged nonfused (von Baeyer) ring system; further choice goes to the ring system having the larger number of ring atoms common to two or more rings followed by lowest locants for bridgehead atoms;
   (g) a fused ring system; further choice goes to the ring system with the larger individual ring and then the ring system with the most linear arrangment of rings;
   (h) a ring system with the larger number of ring atoms;
   (i) a ring system with the larger number of heteroatoms;
   (j) a ring system with the larger number of most preferred heteroatom occurring earlier in the following list: O, S, Se, Te, N, P, As, Sb, Bi, Si, Ge, Sn, Pb, B;

(k) a ring system with the lowest locants for heteroatoms;

(l) a ring system in the lowest level of hydrogenation;

(m) a ring system with the lowest locant for indicated hydrogen.

4. The longest chain of atoms.*

5. The larger number of multiple bonds.*

6. The larger number of double bonds.*

7. The lowest locants in the parent structure successively for hetero atoms, then principal characteristic groups, then multiple bonds regardless of type, and then double bonds.

There is a difference of opinion with regard to compounds consisting of a ring and a chain. One view has it that the ring should always be preferred no matter what the length of the chain may be; this is reflected in the criteria given above. The other viewpoint is that the largest parent hydride in terms of the number of skeletal atoms should be preferred, and only when the number of skeletal atoms is the same should the ring be preferred. Either opinion is acceptable.

If a compound contains two or more parent structures that satisfy the above criteria, the preferred name may be based on the one (or two) parent structures that is (or are) most centrally located with respect to all parent structures. If a further choice is needed, the preferred name is based on the parent structure with the most substituents cited as prefixes and then the one having the lowest locant set for the substituents cited as prefixes. Multiplicative nomenclature (see below), if applicable, is now applied. Finally, the preferred name is the one that is earliest alphabetically.

A few examples will illustrate the above selection criteria. Further details for the formation of these names will be found in later chapters of this book.

$$\underset{7\quad6\quad5\quad4\quad3\quad2\ 1}{\overset{\overset{\text{HO}}{|}\ \overset{\text{CH}_2\text{CH}_3}{|}\ \ \ \overset{\text{O}}{\|}}{\text{CH}_3\text{CHCHCHFCHFCCH}_3}}$$

The principal characteristic group is =O expressed as the suffix -one; it is attached to a seven carbon chain, chosen on the basis of the most substituents to be cited as prefixes. The parent compound is heptan-2-one. The substituent prefixes are ethyl, difluoro, and hydroxy, and the preferred substitutive name is 5-ethyl-3,4-difluoro-6-hydroxyheptan-2-one.

$$\underset{5\quad4\quad3\quad2\quad1}{\overset{\overset{\text{COOH}}{|}}{\text{CF}_3\text{CH}_2\text{CH}_2\text{CH}_2\text{CH}_2\text{CH}-\text{COOH}}}$$

The principal characteristic group is –COOH. Two carboxylic acid groups terminating a carbon chain are expressed by the suffix -dioic acid (see chapter 11). The parent compound is therefore propanedioic acid or, as an acceptable trivial name, malonic acid. The substituent prefix is 5,5,5-trifluoropentyl and accordingly the substitutive name is (5,5,5-trifluoropentyl)propanedioic acid or (5,5,5-trifluoropentyl)malonic acid.

Here, there is no principal characteristic group, but the pyridine ring is preferred to diazene. Both contain at least one nitrogen atom, but the ring is preferred to the chain. The substitutive name is 3-[4-[3-(ethyldiazenyl)propyl]phenyl]pyridine.

---

* IUPAC prefers unsaturation to chain length.

$$H_3Si-O-SiH_2-CH_2CH_2-Si(CH_3)_2-S-Si(CH_3)_2-S-Si(CH_3)_3$$

Here, again, there is no principal characteristic group, and the parent is disiloxane because oxygen is preferred to sulfur. The name is therefore 1-[2-(heptamethyl-trisilathian-1-yl)ethyl]disiloxane.

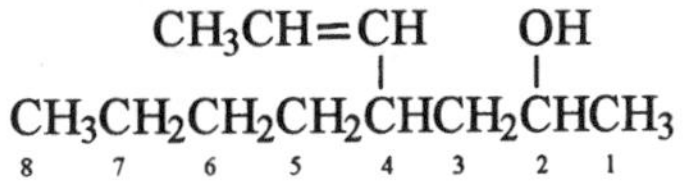

In this structure, the longest chain attached to the principal characteristic group, –OH, is an octane chain. For CAS, the parent compound is therefore 2-octanol, and the preferred substitutive name would be 4-(1-propenyl)-2-octanol. For IUPAC, the preferred chain would include the unsaturation and therefore 4-butylhept-5-en-2-ol is its preferred name.

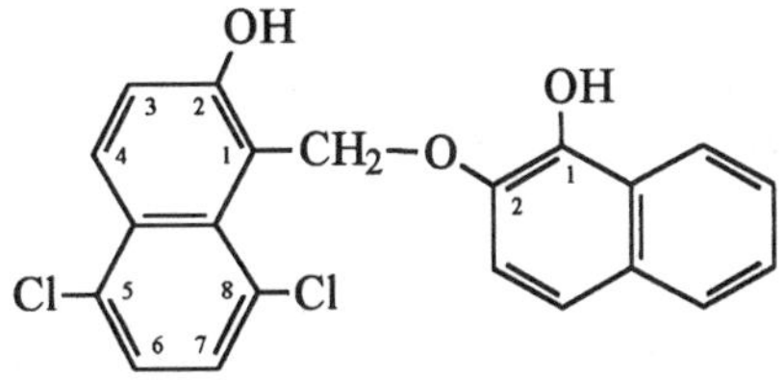

In this example, the principal characteristic group is –OH, and each naphthalene ring is attached to one –OH group. The parent compound is the naphthalenol with the lowest locant for the –OH group. The preferred substitutive name is therefore 2-[(5,8-dichloro-2-hydroxynaphthalen-1-yl)methoxy]naphthalen-1-ol.

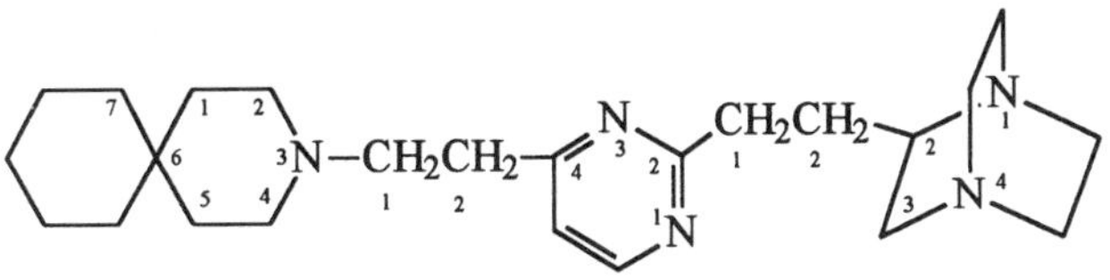

All three rings contain at least one nitrogen atom. The spiro system is preferred to the monoheterocyclic ring or the bridged von Baeyer ring system even though the latter contain more nitrogen atoms. Therefore the preferred name is 3-[2-[2-[2-(1,4-diazabicyclo[2.2.2]oct-2-yl)ethyl]pyrimidin-4-yl]ethyl]-3-azaspiro[5.5]undecane.

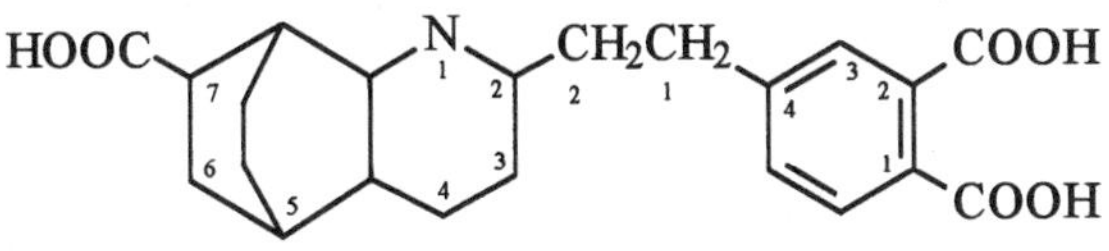

Based on the larger number of the principal characteristic group, the preferred name is 4-[2-(7-carboxydecahydro-5,8-ethanoquinolin-2-yl)ethyl]benzene-1,2-dicarboxylic acid or 4-[2-(7-carboxydecahydro-5,8-ethanoquinolin-2-yl)ethyl]-phthalic acid.

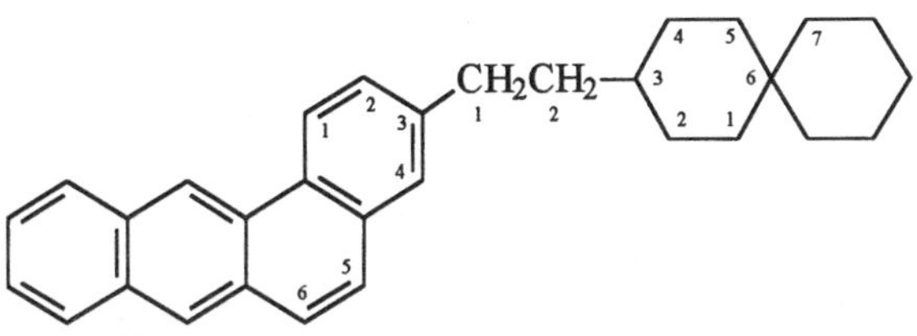

Based on the criterion of larger number of rings, the preferred name is, 3-(2-spiro[5.5]undec-3-ylethyl)benz[a]anthracene.

## The Replacement Method

The replacement method is a technique for changing one parent structure or characteristic group into another parent structure or characteristic group by replacing one or more atoms or groups by other atoms or groups. Names are formed by attaching prefixes to or inserting infixes into the name of a parent structure or characteristic group. There are two distinct types of replacement called skeletal replacement and functional replacement.

### Skeletal Replacement

In skeletal replacement a skeletal atom and any associated hydrogen atoms in a parent hydride are replaced by another atom with the appropriate number of hydrogen atoms. Replacement is indicated by nondetachable prefixes affixed to the name of the parent hydride. This type of replacement is often called "a" nomenclature because the terms adopted as prefixes to indicate replacement of carbon atoms by heteroatoms all end in "a". Skeletal replacement ("a") prefixes for the nonmetallic elements, the metallic elements of groups 14, 15, and 16 of the periodic table, and mercury with their standard bonding number (see chapter 2) are given in table 3.8 in order of decreasing priority. The standard bonding number is assumed when an unmodified prefix is used in an organic name. The $\lambda$-convention (see chapter 2) is used with these prefixes when an element is in a nonstandard valence state. Halogen elements with a standard valence of 1 are included because they are found in organic compounds in nonstandard valence states, for instance, dichloro(methyl)-$\lambda^3$-iodane and $2\lambda^3$,1,3-benziodadioxole, and names for cations, as in $1\lambda^3$-iodol-1-ylium (see chapter 33).

Replacement nomenclature has been extended to indicate replacement of noncarbon atoms by other atoms, including carbon. So far this extension has been limited to heteropolyboranes (see chapter 27) and modification of heterocyclic natural product names (see chapter 31). However, it has not supplanted the firmly established traditional method of using the prefixes thio, seleno, and telluro to replace skeletal oxygen atoms in heterocycles as in thiopyran and selenoxanthene (see chapter 9).

### Construction of Skeletal Replacement Names

Replacement names are formed by citing the appropriate replacement ("a") prefixes given in table 3.8 in front of the name of a parent hydride, usually a hydrocarbon. These prefixes are cited in the name in order of their occurrence in table 3.8. Lowest locants for replacing atoms are assigned first to all heteroatoms as a set and then to heteroatoms occurring earliest in table 3.8. Fixed numbering of a cyclic hydrocarbon, however, is retained.

Table 3.8. Skeletal Replacement ("a") Prefixes (in Decreasing Priority Order)[a]

| Element | "a" Prefix | Standard bonding number | Element | "a" Prefix | Standard bonding number |
|---------|-----------|------------------------|---------|-----------|------------------------|
| F | fluora | 1 | As | arsa | 3 |
| Cl | chlora | 1 | Sb | stiba | 3 |
| Br | broma | 1 | Bi | bisma | 3 |
| I | ioda | 1 | C | carba | 4 |
| O | oxa | 2 | Si | sila | 4 |
| S | thia | 2 | Ge | germa | 4 |
| Se | selena | 2 | Sn | stanna | 4 |
| Te | tellura | 2 | Pb | plumba | 4 |
| N | aza | 3 | B | bora | 3 |
| P | phospha | 3 | Hg | mercura | 2 |

[a] Replacement ("a") prefixes for all the elements have been given in the IUPAC Organic Recommendations.

In an acyclic compound, the longest chain of carbon and heteroatoms is named as if it were an acyclic hydrocarbon. For IUPAC, the chain must terminate with carbon atoms; for CAS, the chain can terminate with a heteroatom provided that a mononuclear parent hydride is available (see chapter 9). If the principal characteristic group is a monocarboxylic acid or a dicarboxylic acid, or a related characteristic group such as carboxamide or carboxaldehyde, the chain must terminate with a carbon atom because these characteristic groups include the carbon atom in their suffixes.

IUPAC does not provide definitive guidance as to when skeletal replacement should be used, saying only that it is intended to be used for acyclic structures when other methods produce cumbersome names, are difficult to use, or lead to unfamiliar names; in the last instance, characteristic groups such as amide, and functional derivative classes such as ester and anhydride can be obscured. CAS rigorously restricts skeletal replacement nomenclature to acyclic chains with four or more *hetero units*, none of which can be all or part of a principal characteristic group expressed as a suffix or class name; a hetero unit is defined as a single heteroatom or a series of contiguous heteroatoms, alike or different, that can be named with a simple substituent prefix. Thus, $-N=N-$ and $-SiH_2-O-SiH_2-$ are hetero units, but $-O-SiH_2-O-$ and $-O-S-$ are not.

If a choice is needed between two chains each qualifying for a skeletal replacement name, the preferred chain is chosen by the following criteria in descending order of priority:

1. The larger number of the preferred characteristic group named as a suffix or class name; this criterion includes principal characteristic groups expressed by means of multiplicative nomenclature (see below); CAS does not recognize this latter aspect;
2. the larger number of heteroatoms (units) of any kind;*
3. the longest chain of carbon and heteroatoms (units);**
4. the larger number of the more preferred heteroatoms (units);
5. the larger number of multiple bonds;
6. the larger number of double bonds;
7. the lowest locants for principal characteristic groups;
8. the lowest locants for heteroatoms regardless of type and then the lowest locants for the more preferred heteroatoms;
9. the lowest locants successively for principal characteristic groups, then multiple bonds regardless of type, and then double bonds.

The following examples illustrate the above criteria in the formation of skeletal replacement names. Further details for the formation of the names will be found in later chapters of this book.

$$\overset{\displaystyle CH_2CH_2CH_3}{\underset{\phantom{x}}{|}}$$
$$\underset{1}{CH_3}-\underset{2}{O}-\underset{3}{CH}-\underset{4}{O}-\underset{5}{CH_2}-\underset{6}{O}-\underset{7}{CH_2}-\underset{8}{O}-\underset{9}{CH_2}-\underset{10}{O}-\underset{11}{CH_3}$$

3-Propyl-2,4,6,8,10-pentaoxaundecane (larger number of heteroatoms)

$$\overset{\displaystyle O-CH_2CH_2-O-CH_3}{\underset{\phantom{x}}{|}}$$
$$\underset{1}{HO}-\underset{2}{CH_2CH_2}-\underset{3}{O}-\underset{4}{CH}\underset{5}{CH_2}-\underset{6}{O}-\underset{7}{CH_2}\underset{8}{CH_2}-\underset{9}{O}-\underset{10}{CH_2}\underset{11}{CH_2}-\underset{12}{O}-\underset{13}{CH_2}\underset{14}{CH_2}-OH$$

4-(2-Methoxyethoxy)-3,6,9,12-tetraoxatetradecane-1,14-diol (larger number of the principal characteristic group)

---

* CAS applies an additional criterion that the main chain must have at least one heteroatom occurring earlier in the seniority of parent hydrides and parent compounds given above in the section on substitutive nomenclature.

** IUPAC prefers a greater number of unsaturated sites, with double bonds preferred to triple bonds.

$$O-CH_2CH_3$$
$$CH_3-O-CH_2\overset{|}{C}H-O-CH_2-O-CH_2-O-CH_2CH_2-COOH$$

1   2   3   4   5   6   7   8   9   10   11   12

4-Ethoxy-2,5,7,9-tetraoxadodecan-12-oic acid
[*not* 4-(Methoxymethyl)-3,5,7,9-tetraoxadodecan-12-oic acid;
*not* 9-(Methoxymethyl)-4,6,8,10-tetraoxadodecanoic acid;
lower locants for heteroatoms are preferred over lower locants for principal
characteristic groups]

$$CH_2-NHNH-CH_2CH_2CH_3$$
$$CH_3-O-CH_2CH_2-O-CH_2CH_2-O-\overset{|}{C}H-O-CH_2CH_2-O-CH_2CH_3$$

1   2   3   4   5   6   7   8   9   10   11   12   13   14   15

9-[(2-Propylhydrazino)methyl]-2,5,8,10,13-pentaoxapentadecane (IUPAC)
[larger number of heteroatoms (units)]
9-(2-Ethoxyethoxy)-2,5,8-trioxa-11,12-diazapentadecane (CAS)
(preferred heteroatom, N, in the main chain)

$$CH_3CH_2-O-CH_2CH_2-O-CH_2CH_2-S-CH_2CH_2-O-CH_2CH_2$$

14  13  12  11  10   9   8   7   6   5   4   3   2   1

$$NH$$

$$CH_3CH_2-O-CH_2CH_2-O-CH_2CH_2-O-CH_2CH_2-O-CH_2CH_2$$

14  13  12  11  10   9   8   7   6   5   4   3   2   1

*N*-3,9,12-Trioxa-6-thiatetradec-1-yl-3,6,9,12-tetraoxatetradecan-1-amine
(larger number of most preferred heteroatom, O, in the parent chain)

$$CH_3CH=CH-O-CH=CH-O$$

15  14  13  12  11  10   9

$$O$$
$$CH-O-CH_2CH_2-O-CH_2CH_2-\overset{||}{C}-NH-CH_3$$

8   7   6   5   4   3   2   1   *N*

$$HO-CH_2CH_2CH_2-O-CH_2CH_2-O-CH_2$$

16  15  14  13  12  11  10   9

16-Hydroxy-*N*-methyl-8-[[2-(1-propenyloxy)ethenyl]oxy]-4,7,10,13-hexadecan-
amide (CAS) (longest chain of carbon and heteroatoms)
8-{[2-(3-Hydroxypropoxy)ethoxy]methyl}-*N*-methyl-4,7,9,12-tetraoxapentadeca-
10,13-dienamide (IUPAC) (larger number of multiple bonds)

## Functional Replacement

The replacement of oxygen atoms in characteristic groups by other atoms or groups denoted
by the infixes or prefixes given in table 3.9 is called functional replacement nomenclature. The
use of prefixes to indicate replacement is the older, more traditional, method. Prefixes are often
used to name inorganic acids, as in bromosulfuric acid, $Br-SO_2-OH$; selenocyanic acid,
$HS-CN$; and cyanosulfurous acid, $NC-SO-OH$, and they are particularly useful when the
structure of the acid is unknown, for example, dithiophosphoric acid for $H_3PS_2O_2$. The prefix
method, however, presents problems for an alphabetical index in that numerical prefixes
separate similar compounds, such as thiocarbonic acid, dithiocarbonic acid, and trithiocar-
bonic acid. Both prefixes and infixes are used in naming organic compounds. Infixes have a
distinct advantage for an index environment; for example, carbonothioic acid, carbonodithioic
acid, and carbonothiothioic acid all appear near each other in an alphabetical index. Whereas
the prefix method has never been fully codified, the infix method was highly developed
for phosphorus compounds.[4] Two or more infixes are cited in alphabetical order in a

Table 3.9. Prefixes and Infixes in Functional Replacement
Nomenclature

| Replacement operation | Prefix[a] | Infix[b,c] |
|---|---|---|
| –OH by –F | fluoro | fluorid(o) |
| –OH by –Cl | chloro | chlorid(o) |
| –OH by –Br | bromo | bromid(o) |
| –OH by –I | iodo | iodid(o) |
| –OH by –NHNH$_2$ | — | hydrazid(o) |
| –OH by –N$_3$ | azido | azid(o) |
| –OH by –NH$_2$ | amido | amid(o) |
| –OH by –OCN | cyanato | cyanatid(o) |
| –OH by –SCN | thiocyanato[d] | thiocyanatid(o)[d] |
| –OH by –NCO | isocyanato | isocyanatid(o) |
| –OH by –NCS | isothiocyanato[d] | isothiocyanatid(o)[d] |
| –OH by –CN | cyano | cyanid(o) |
| –O– by –OO– | peroxy | perox(o) |
| –O– by –SS– | dithioperoxy[d,e] | dithioperox(o)[d,e] |
| –O– by –OS– or –SO | thioperoxy[f] | thioperox(o)[f] |
| =O or –O– by =Se or –Se– | seleno | selen(o) |
| =O or –O– by =Te or –Te– | telluro | tellur(o) |
| =O or –O– by =S or –S– | thio | thi(o) |
| =O by =NNH$_2$ | hydrazono | hydrazon(o) |
| =O by =NH | imido | imid(o) |
| –OH and =O by ≡N | nitrido | nitrid(o) |

[a] Although prefixes have been long used, the only documentation appears briefly in the 1990 IUPAC Inorganic Nomenclature Recommendations[5] and in somewhat more detail in an ACS Inorganic Guidebook.[6]

[b] The use of infixes is documented in several places: first in the ACS rules for monophosphorus compounds;[4] the documentation of CAS names for chemical substances; and in the IUPAC Recommendations.

[c] When followed by an "a", "i", or "o", the final "o" of an infix is generally elided.

[d] Selenium and tellurium analogs use seleno or telluro in place of thio.

[e] Enclosing these affixes in parentheses greatly assists recognition of structure, for example -thioperoxoic acid is –C(S)–OOH and -(thioperoxoic) acid is –C(O)–OSH or –C(O)–SOH.

[f] This mixed chalcogen affix and other mixed affixes involving selenium and tellurium do not describe the chalcogen atom sequence.

name, as in phosphoramidothioic acid, $H_2N–P(OH)_2$, and phosphoramidochloridothioic acid, $H_2N–PCl(O)(SH)$. Two or more prefixes are usually cited in alphabetical order, as in chloroimidophosphoric acid, $Cl–P(O)(OH)–NH–P(O)(OH)$, and amidoimidothiosulfuric acid, $H_2N–S(=NH)(S)–OH$, but chalcogen prefixes often precede all other prefixes in names of polynuclear oxo acid analogs, as in thioimidodiphosphoric $O,O'$-acid, $(HO)_2P–NH–P(O)(OH)_2$.

Functional replacement nomenclature permits the generation of a large number of suffixes for characteristic groups, which, along with their corresponding prefixes, are illustrated in table 3.10. When the affixes in table 3.9 are used in combination, the names of thousands of mononuclear functional parent compounds can be derived; table 3.11 gives examples of these and their corresponding prefixes. CAS uses functional replacement infixes for mononuclear carbon, phosphorus, and arsenic acids and related compounds. Extension of infix functional replacement nomenclature to other mononuclear compounds, particularly those of boron, silicon, sulfur, selenium, and tellurium, is encouraged.

Although prefixes can be used to name functional replacement analogs of mononuclear oxo acids, they are more often used for polynuclear oxo acids. CAS uses a number of techniques to collect similar compounds closely together alphabetically. These include synonym line formulas and structure drawings (which are not recommended for general use); capital italic letter locants and arabic numbers are much preferred.

Table 3.10. Illustrative Characteristic Group Suffixes and Prefixes in Functional Replacement Nomenclature[a]

| Group | Suffix | Prefix |
|---|---|---|
| –C(O)–OOH | -carboperoxoic acid (CAS) | hydroperoxycarbonyl |
| | -peroxycarboxylic acid (IUPAC) | |
| –(C)(O)–OOH | -peroxoic acid | hydroperoxy...oxo |
| –CO–SH or –CS–OH | -carbothioic acid | thiocarboxy |
| –CO–SH | -carbothioic S-acid | mercaptocarbonyl |
| –CS–OH | -carbothioic O-acid | hydroxycarbonothioyl |
| –(C)O–SH or –(C)S–OH | -thioic acid | — |
| –(C)O–SH | -thioic S-acid | mercapto ... oxo |
| –(C)S–OH | -thioic O-acid | hydroxy ... thioxo |
| –C(S)-SH | -carbodithioic acid | dithiocarboxy |
| –(C)(S)-SH | -dithioic acid | mercapto ... thioxo |
| –C(S)-OSH or –C(S)-SOH | -carbothio(thioperoxoic) acid | thio(thioperoxy)carboxy |
| –C(S)-OSH | -carbothio(thioperoxoic) OS-acid | (mercaptooxy)carbonothioyl |
| | | (SO-thiohydroperoxy)carbonothioyl |
| –C(S)-SOH | -carbothio(thioperoxoic) SO-acid | (hydroxythio)carbonothioyl |
| | | (OS-thiohydroperoxy)carbonothioyl |
| –(C)(S)-OSH or –(C)(S)-SOH | -thio(thioperoxoic) acid | — |
| –(C)(S)-OSH | -thio(thioperoxoic) OS-acid | (mercaptooxy) ... thioxo |
| | | (SO-thiohydroperoxy) ... thioxo |
| –(C)(S)-SOH | -thio(thioperoxoic) SO-acid | (hydroxythio) ... thioxo |
| | | (OS-thiohydroperoxy) ... thioxo |
| –C(=NH)-OH | -carboximidic acid | imidocarboxy |
| | | hydroxycarbonimidoyl[b] |
| –(C)(=NH)-OH | -imidic acid | hydroxy ... imino |
| –C(=NH)-SH | -carboximidothioic acid | imidothiocarboxy |
| | | mercaptocarbonimidoyl[b] |
| –(C)(=NH)-SH | -thioimidic acid | imino ... mercapto |
| –C(S)-Cl | -carbothioyl chloride | carbonochloridothioyl |
| | | chloro(thiocarbonyl) |
| –(C)(S)Cl | -thioyl chloride | chloro ... thioxo |
| –C(=NH)-NH₂ | -carboximidamide | carbamimidoyl |
| | | aminocarbonimidoyl[b] |
| | -carboxamidine | amidino |
| –(C)(=NH)-NH₂ | -imidamide | amino ... imino |
| | -amidine | |
| –CHS | -carbothialdehyde (IUPAC) | thioformyl |
| | -carbothioaldehyde (CAS) | thioxomethyl |
| –(C)HS | -thial | thioxoalkyl |
| –CHSe | -carboselenaldehyde | selenoformyl |
| | | selenoxomethyl |
| –CHTe | -carbotelluraldehyde | telluroformyl |
| | | telluroxomethyl |
| >(C)=S | -thione | thioxo |
| >(C)=Se | -selone[c] | selenoxo |
| >(C)=Te | -tellone[c] | telluroxo |
| –S(O)₂-SH or –S(S)(O)-OH | -sulfonothioic acid | thiosulfo |
| | -thiosulfonic acid | |
| –S(O)₂-SH | -sulfonothioic S-acid | mercaptosulfonyl |
| | -thiosulfonic S-acid | |
| -S(S)(O)-OH | -sulfonothioic O-acid | hydroxysulfonothioyl |
| | -thiosulfonic O-acid | hydroxy(thiosulfonyl) |
| –S(Se)-OH or –S(O)-SeH | -sulfinoselenoic acid | selenosulfino |
| | -selenosulfinic acid | |
| –S(Se)-OH | -sulfinoselenoic O-acid | hydroxysulfinoselenoyl |
| | -selenosulfinic O-acid | hydroxy(selenosulfinyl) |

*(Continued)*

Table 3.10. Continued

| Group | Suffix | Prefix |
|---|---|---|
| –S(O)-SeH | -sulfinoselenoic $Se$-acid<br>-selenosulfinic $Se$-acid | selanylsulfinyl |
| –S-SeH | -sulfenoselenoic acid<br>-selenosulfenic acid | selenosulfeno<br>selanylthio |
| –SH | -thiol | mercapto (CAS)<br>sulfanyl (IUPAC) |
| –SeH | -selenol | selenyl (CAS)<br>hydroseleno (IUPAC/1979)<br>selanyl (IUPAC/1993)[d] |
| –TeH | -tellurol | telluryl[d] |

[a] The acid and related characteristic group suffixes and prefixes in this table are illustrated with the replacement prefix or infix thio; selenium and tellurium analogs are named using seleno or telluro in place of thio. The replacement prefix or infix imido is illustrated for carboxylic acids and amides; acyl halides as well as hydrazonic, hydroximic, and hydrazonic acids are named similarly. Sulfonic, sulfinic acids, and their selenium and tellurium analogs are named following the procedures for carboxylic acids.

[b] This prefix could be viewed as ambiguous since the nitrogen atom of the imido group has a substitutable hydrogen. In the latter case, however, the prefix would be a multiplicative prefix or a part of a multiplicative prefix. The possibility of ambiguity can be removed by attaching a letter locant, for example, $C$-hydroxycarbonimidoyl and $Se$-hydroxyselenonimidoyl.

[c] The suffixes -selenone and -tellurone cannot be used because they describe the $-SeO_2-$ and $-TeO_2-$ characteristic groups, respectively.

[d] The prefixes "selenanyl" and "telluranyl" cannot be used because they are prefix names for the Hantzsch–Widman names selenane and tellurane.

$$HS-SO_2-SH$$

Dithiosulfuric $S,S'$-acid
Thiosulfuric acid ($H_2S_3O_2$) (CAS)

$$\underset{\displaystyle Cl-S-OH}{\overset{\displaystyle NH}{\|}}$$

Chloroimidosulfurous acid
Sulfurochloridimidous acid

$$\underset{\displaystyle HO-C-OO-C-OH}{\overset{\displaystyle O \qquad O}{\| \qquad \|}}$$

Peroxydicarbonic acid

$$\underset{\displaystyle (HO)_2P-NH-P(OH)_2}{\overset{\displaystyle O \qquad O}{\| \qquad \|}}$$

Imidodiphosphoric acid

$$H_2N-\underset{NH_2}{\overset{O}{\underset{|{}^P}{\overset{\|}{P}}}}-O-\underset{NH_2}{\overset{O}{\underset{|{}^{P'}}{\overset{\|}{P}}}}-NH-\underset{NH_2}{\overset{O}{\underset{|{}^{P''}}{\overset{\|}{P}}}}-O-\underset{NH_2}{\overset{O}{\underset{|}{\overset{\|}{P}}}}-NH_2$$

$P',P''$-Imidotetraphosphoric hexaamide
(Numbering the P atoms in the chain would allow the name 2,3-Imidotetraphosphoric hexamide)

$$\underset{\displaystyle \underset{H_2N}{|}\quad \underset{SH}{|}}{\overset{\displaystyle S \quad S}{HP-S-PH}}$$

Amidotetrathiodiphosphonic acid

Infixes can sometimes be combined with prefixes, but the same group must be treated by the same method.

$$\underset{\displaystyle HO-C-SS-C-OH}{\overset{\displaystyle S \qquad S}{\| \qquad \|}}$$

(Dithioperoxy)dicarbonothioic $O,O'$-acid
Thioperoxydicarbonic acid ($[(HO)C(S)]_2S_2$) (CAS)

Table 3.11. Illustrative Functional Parent Compounds and Related Prefixes in Functional Replacement Nomenclature

| Parent compound | Parent compound name | Group | Prefix name |
|---|---|---|---|
| HO–C(S)–OH | carbonothioic O,O′-acid | SC< or SC= | carbonothioyl<br>thiocarbonyl |
| HO–C(=NH)–OH | carbonimidic acid | HN< or HN= | carbonimidoyl<br>imidocarbonyl |
| H₂N–C(O)–OH | carbamic acid (from carbonamidic acid) | H₂N–C(O)– | carbamoyl (from carbonamidoyl) (IUPAC)<br>aminocarbonyl (CAS) |
| H₂N–C(=NH)–OH | carbamimidic acid (from carbonamidimidic acid) | H₂N–C(=NH)– | carbamimidoyl (from carbonamidimidoyl) |
| H₂N–C(S)–NH₂ | carbonothioic diamide | H₂N–C(S)–NH– | carbamothioylamino |
| H₂N–C(=NH)–NH₂ | carbonimidic diamide | H₂N–C(=NH)–NH– | carbamimidoylamino |
| H₂N–S(O₂)–OH | sulfamic acid (from sulfuramidic acid) | H₂N–S(O₂)– | sulfamoyl (from sulfuramidoyl)<br>aminosulfonyl (CAS) |
| P(O)(NH₂)(OH)₂ | phosphoramidic acid | H₂N–P(O)< or H₂N–P(O)= | phosphoramidoyl |
| HP(S)(OH)₂ or HP(O)(OH)(SH) | phosphonothioic acid | HP(S)= or HP(S)= | phosphonothioylidene |
| P(S)(OH)₃ or P(O)(OH)₂(SH) | phosphorothioic acid | (HO)₂P(S)– or (HO)(HS)P(O)– | thiophosphono |
| P(S)(OH)₃ | phosphorothioic O,O′,O″-acid | (HO)₂P(S)– | dihydroxyphosphinothioyl<br>dihydroxyphosphorothioyl |
| HP(O)(NH₂)(OH) | phosphonamidic acid | (HO)(H₂N)P(O)– | aminohydroxyphosphinyl<br>amidohydroxyphosphoryl<br>phosphonamidoyl |
| | | (H₂N)HP(O)– | aminophosphinyl |
| HP(O)(SeH)₂ | phosphonodiselenoic Se,Se′-acid | (HSe)₂P(O)– | diselanylphosphinyl |
| HP(=NH)Cl(SH) | phosphonochloridimidothioic acid | (HS)ClP(=NH)– | chloromercaptophosphinimidoyl |
| ClHP(=NH)(OH) | phosphonochloridimidic acid | ClHP(=NH)– | phosphonochloridimidoyl |
| HP(S)(NH₂)₂ | phosphonothioic diamide | (H₂N)₂P(S)– | phosphorodiamidothioyl |
| HP(=NNH₂)(NH₂)Cl | phosphonamidohydrazonic chloride | (H₂N)ClP(=NNH₂)– | phosphoramidochloridohydrazonoyl |
| HP(=NH)(NHNH₂)(OH) | phosphonohydrazidimidic acid | (HO)(H₂NNH)P(=NH)– | hydrazinohydroxyphosphinimidoyl |
| | | (H₂NNH)HP(=NH)– | phosphonohydrazidimidoyl |
| HP(Se)(CN)(SH) | phosphonocyanidoselenothioic S-acid | (HS)(NC)P(Se)– | cyanomercaptophosphinoselenoyl |
| HP(S)(CN)(SeH) | phosphonocyanidoselenothioic Se-acid | (HSe)(NC)P(S)– | cyanoselanylphosphinothioyl |
| HP(Se)(CN)(OH) | phosphonocyanidoselenoic O-acid | (NC)HP(Se)– | phosphonocyanidoselenoyl |
| HP(≡N)Br | phosphononitridic bromide | BrP(≡N)– | phosphorobromidonitridoyl |
| HP(=NH)(OOH)₂ | phosphonimidodiperoxoic acid | (HOO)₂P(=NH)– | phosphorimidodiperoxoyl |

$$\underset{\displaystyle HOO-C-O-C-OOH}{\overset{\displaystyle O \quad\;\; O}{\phantom{H}\|\quad\;\;\|\phantom{H}}}$$

Dicarbonodiperoxoic acid

$$\underset{\displaystyle HO-C-O-C-O-C-OH}{\overset{\displaystyle NH \quad NH \quad NH}{\phantom{H}\|\qquad\|\qquad\|\phantom{H}}}$$

Tricarbonimidic acid

$$\underset{\displaystyle HO-C-NH-C-NH-C-OH}{\overset{\displaystyle NH \quad\;\; NH \quad\;\; NH}{\phantom{H}\|\qquad\;\;\|\qquad\;\;\|\phantom{H}}}$$

Diimidotricarbonimidic acid

$$\underset{\displaystyle \underset{\displaystyle SH \qquad\;\; SH}{|\qquad\qquad|}}{\underset{\displaystyle Cl-P-O-S-P-Cl}{\overset{\displaystyle S \qquad\quad S}{\|\qquad\quad\|}}}$$

(Thioperoxy)diphosphorodithioic dichloride

When the infix method is used, all substituents on nitrogen atoms, including those expressed as class names, are denoted by appropriate locants and cited in front of the name of the functional replacement compound. However, the method of citation of nitrogen substituents has not been specifically defined and the examples in the provisional Section D-5 of the IUPAC 1979 Organic Rules are not completely consistent. They would indicate, however, that substituents on nitrogen atoms are cited with the prefix (or class name) and alphabetized with other prefixes.

$C_6H_5-NH-SO_2-OH$

Phenylsulfamic acid
(Phenylamido)sulfuric acid

$$\underset{\displaystyle (CH_3)_2N-P-S-CH_3}{\overset{\displaystyle O}{\|}}$$

S-Methyl dimethylphosphoramidothioate
S-Methyl (dimethylamido)thiophosphate

$$\underset{\displaystyle (CH_3CH_2)_2P-Cl}{\overset{\displaystyle N-C_6H_5}{\|}}$$

Diethyl(phenylimido)phosphinic chloride
P,P-Diethyl-N-phenylphosphinimidic chloride

$$\underset{\displaystyle \underset{N'}{NH-CH_2CH_3}}{\underset{\displaystyle C_6H_5-P-NH-CH_2CH_3}{\overset{\displaystyle \overset{N}{O}}{\|}}}$$

Phenylphosphonic bis(ethylamide)
N,N'-Diethyl-P-phenylphosphonic diamide

The infix method seems to be superior and is favored in this book.

The large number of characteristic groups made possible by functional replacement nomenclature complicates the process of generation of a preferred name. The acid characteristic groups expressed as suffixes are all monobasic and therefore are easier to prioritize than polynuclear acids. The published order by IUPAC is quite incomplete. The following is an expansion of item 3 in table 3.1 based on the priority order used by CAS. The classes are in decreasing priority order, and in each case the preferred chalcogen order is S > Se > Te.

1. Peroxy acids expressed as a suffix in order of the corresponding nonperoxy acids that follow; mixed chalcogen replacement analogs are ranked according to the maximum number of the more preferred chalcogen atom.
2. Carboxylic acids, then chalcogen replacement analogs ranked according the more preferred chalcogen atom; mixed chalcogen analogs are ranked according to the maximum number of the more preferred chalcogen atom.
3. Carbohydrazonic acids, then chalcogen replacement analogs.
4. Carboximidic acids, then chalcogen replacement analogs.
5. Sulfonic acids, then chalcogen replacement analogs; mixed chalcogen analogs are ranked according to the maximum number of the more preferred chalcogen atom.

6. Sulfonohydrazonic acids, then chalcogen replacement analogs.
7. Sulfonodihydrazonic acids, then chalcogen analogs.
8. Sulfonohydrazonimidic acids, then chalcogen analogs.
9. Sulfonimidic acids, then chalcogen analogs.
10. Sulfonodiimidic acids, then chalcogen analogs.

Sulfinic, sulfenic, selenonic, seleninic, selenenic, telluronic, tellurinic, and tellurenic acids follow the pattern established above, where applicable.

Replacement analogs of functional parent acids require a more intricate set of criteria for ranking such characteristic groups. IUPAC offers some guidance, but it is contained in rules that have never been fully approved. The criteria used by CAS for carbon, phosphorus, and arsenic functional parent compounds is outlined below in decreasing priority order.

1. the larger number of acid groups;
2. the larger number of nuclear atoms;
3. the higher oxidation state of the nuclear atom(s);
4. the more preferred atom attached to the nuclear atom(s), according to the order of skeletal replacement prefixes given in table 3.8;
5. the larger number of the more preferred atom(s) attached to the nuclear atom;
6. the nature of the multiatomic groups attached to the central atom.

Accordingly, the CAS ordering for phosphorus acids would be illustrated by the following:

triphosphoric > peroxydiphosphoric > diphosphoric > imidodiphosphoric > diphosphorous > phosphoroperoxoic > phosphoric > phosphorothioic > phosphorodithioic > phosphorimidic > phosphorous > phosphorochloridic > phosphorohydrazidic > phosphoramidic > phosphonic > phosphinic > phosphinous.

However, many questions arise in the application of these criteria; much more detail is needed.

Other hierarchies have been discussed within IUPAC. As early as 1977, IUPAC's Commission on Nomenclature of Organic Chemistry (CNOC) was presented with a set of ordering criteria somewhat similar to the above except that it emphasized the organic principle of most substitutable hydrogen atoms attached to the nuclear atom(s). This problem is still under discussion by the Commission.

It is clear that more effort is needed in order to derive a fully workable seniority ordering hierarchy for functional parent compounds.


## The Additive Method

As the name implies, the additive method is is a technique by which two or more atoms or groups of atoms are combined together without loss of any atoms from any part. It finds use in organic nomenclature in a variety of ways.

Hydrogen atoms are "added" to rings or ring systems whose names imply the presence of double bonds, usually the maximum number of conjugated double bonds, by attaching the prefix hydro with an appropriate numerical prefix. The hydro prefixes can be either detachable or nondetachable (see chapter 2): in decahydro-1,2-dimethylnaphthalene they are detachable (naphthalene is the parent hydride) and in 1,2-dimethyldecahydronaphthalene they are nondetachable (decahydronaphthalene is the parent hydride).

Hydrogen cations are added to a parent hydride or to certain characteristic groups and hydrogen anions are added to a parent hydride to form cationic or anionic parent structures; this addition is denoted by the suffixes -ium and -uide, respectively (see table 3.7 and chapter 33), for example, pyridinium (addition of $H^+$ to pyridine) and boranuide, $BH_4$, (addition of $H^-$ to borane).

Many compound substituent prefixes and all multiplicative substituent prefixes (see below) are formed by the additive technique; examples include pentyloxy, butylthio, and ethyleneimino.

Ring assembly names formed by direct combination of a multiplicative prefix with a substituent prefix name of a ring or ring system are additive (see chapter 8), as in biphenyl, 2,2'-bipyridyl, and 1,1'-binaphthyl.

The insertion, that is, addition, of a methylene group into a ring or chain structure is indicated by the nondetachable prefix homo (see the section in skeletal modification, below); this prefix is used mainly in naming natural products (see chapter 31), as in 4a-homo-5$\alpha$-pregnane, but it has been used with trivial names, particularly polycyclic ring systems, for instance, homocubane.

Systematic structure-based polymer nomenclature (see chapter 29) is another example of the additive method.

## The Subtractive Method

The subtractive method is used to describe the removal of atoms or groups from a parent structure. The suffixes -ene, -yne, and so on, indicate the removal of pairs of hydrogen atoms from a saturated parent hydride (see chapters 5 and 6), as in hex-2-ene, cyclohepta-1,3-diene, oct-3-yne, and but-1-en-3-yne.

Functional class names such as anhydride and anhydrosulfide indicate the loss of water or hydrogen sulfide, respectively, from two acid groups, as in the names acetic anhydride, hexanedithioic anhydrosulfide, and acetic phosphoric monoanhydride (see chapter 12).

The detachable or nondetachable prefix anhydro denotes loss of a molecule of water. This prefix is commonly used in carbohydrate nomenclature (see chapter 31), for example, 4,6-di-O-methyl-2,3-anhydro-$\alpha$-D-glucopyranose; here anhydro is nondetachable.

The prefix "de" followed by the name of an atom (other than hydrogen) or a group indicates removal of that atom or group and its replacement by the appropriate number of hydrogen atoms; this aspect of subtractive nomenclature is commonly found the names of natural products (see chapter 31); examples include 2-deoxy-$\beta$-D-glucopyranose and demethylmorphine. The removal of pairs of hydrogen atoms is denoted by "de" followed by an appropriate numerical term and the prefix hydro, as in 7,8-didehydro-$\beta,\beta$-carotene and 1,2-didehydropyrene.

The nondetachable prefix "nor", when attached to the name of a natural product, describes the elimination of a methylene group from a chain or a ring of the structure described by that name, as in 19-nor-5$\alpha$-androstane (see the section on skeletal modification, below). The special convention in terpene nomenclature by which the prefix nor denotes the removal of methyl groups attached to a ring system and replacement by hydrogen atoms is not recommended for use elsewhere.

The removal of hydrogen atoms from parent hydrides and nitrogen suffix groups is described by suffixes such as -yl, -ylidene, and -diyl; examples include methyl (a contraction of methanyl), propan-2-yl, cyclohex-3-en-1-ylidene, pyridine-2,4-diyl, phosphanylidene, and butan-2-aminyl. Removal of hydrogen ions from parent hydrides and certain characteristic groups is shown by suffixes such as -ylium, -ide, and -ate (see chapter 33); examples include butylium, cyclobut-2-en-1-ylium, methanaminylium, ethanide, butan-1-olate, pentanoate, and cyclohexanedicarboxylate.

The prefix "cyclo" indicates removal of two hydrogen atoms from a chain with the formation of a ring, as in cyclohexane; it also is used to describe the formation of a new ring by the removal of two hydrogen atoms from different atoms of a natural product structure (see the section on skeletal modification, below).

## The Conjunctive Method

The conjunctive method describes a structure formed from two parent compounds joined together with the formal elimination of an appropriate number of hydrogen atoms from each component. A conjunctive name is a combination of the names of the parent compounds.

A ring assembly name is formed with multiplicative prefixes and the name of a cyclic parent hydride (see chapters 8 and 9); such names also illustrate the conjunctive method, as in 2,2'-bipyridine and 1,2'-binaphthalene.

Perhaps the best known examples of the conjunctive method are names formed by combining the name of a cyclic parent hydride with that of an acyclic chain containing a principal characteristic group; the elimination of one atom of hydrogen from each component is implied.* The ring or ring system and the principal characteristic group must terminate the ends of the chain; thus, the only locant to be cited, if needed, is the locant of the ring or ring system at which the acyclic chain is attached. The locant is placed before the name of the cyclic parent hydride unless locants for structural features, such as heteroatoms, are already there; in that case the locant precedes the name of the acyclic chain. A locant for the chain attachment is not needed since the chain must terminate at its junction with the ring. Substituents on the acyclic component are identified by Greek letter locants with $\alpha$ at the carbon atom next to the principal characteristic group. The advantage of this type of name is that it results in a larger parent structure.

2-Naphthaleneacetic acid

2$H$-1-Benzopyran-2-propanenitrile

$\alpha$-Methyl-4-pyridineethanol

$\beta,\beta'$-Diethyl-1,4-benzenedipropanoic acid

## The Multiplicative Method

The multiplicative usually provides shorter and simpler names by reflection of structural symmetry. Accordingly, identical structural fragments are cited only once in the name.** The method is most often applied to parent compounds that have a principal characteristic group, implied or expressed as a suffix or functional parent compound, and to heterocyclic ring systems. Since 1972, however, CAS has extended the method to include cyclic hydrocarbons.

The most important part of the structure for a multiplicative name is the part that links two or more identical parent compounds. The linking group must be expressed by a single bi- or multivalent prefix name, or a single bi- or multivalent prefix name must be the central group from which other identical (including substituents) bi- or multivalent prefix names radiate and terminate at the parent compound. The single (or central) bi- or multivalent group may be substituted, even unsymmetrically; all other component groups may only be substituted symmetrically. The component groups of a multipart linking group are cited additively. Substituents on the linking

---

* Before 1972 (Volume 76 of Chemical Abstracts), conjunctive names were used by CAS when there was a double bond between the components, when the acyclic component was unsaturated provided that it had a trivial name, with carbamic acid, sulfamic acid, and related compounds, and with acyclic components with two or more principal characteristic groups provided that there was no ambiguity.
** The current IUPAC Organic Rules allow unsymmetrical substitution on the functional parent compound but not on the linking group. Unsymmetrical linking groups are also not allowed.

groups are described by the usual substitutive methods The numerical prefixes di-, tri-, and so on are used when followed by only one unsubstituted group on each branch of the linking group; the multiplicative prefixes bis-, tris-, and so on, are used when followed by two or more groups or when a single group is symmetrically substituted. Linking groups that are not the central unit are numbered, when given a choice, so that the point of attachment nearer to the parent compound has the lowest number, which is cited last in the name regardless of its value.

Examples of multiplying substitutive prefixes are as follows:

—O—    Oxy

—CH—    Methanetriyl (IUPAC)
Methylidyne (CAS)

$\overset{CH_3}{\underset{}{-N-}}$    Methylimino

—CH$_2$—O—CH$_2$—    Oxydimethylene
Oxydimethanediyl

—O—CH$_2$—O—    Methylenebis(oxy)
Methanediylbis(oxy)

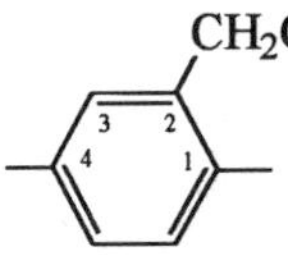

2-(Chloromethyl)-1,4-phenylene

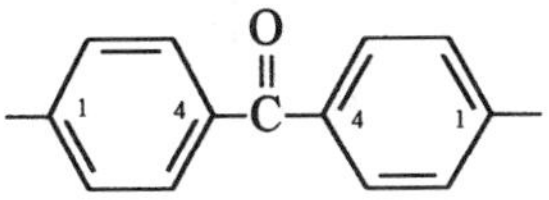

Carbonyldi-4,1-phenylene

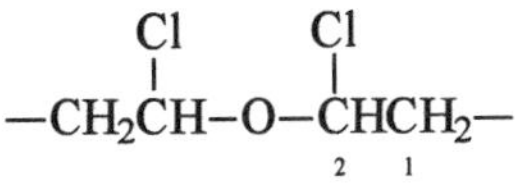

Oxybis(2-chloro-2,1-ethylene)
Oxybis(2-chloro-2,1-ethanediyl)

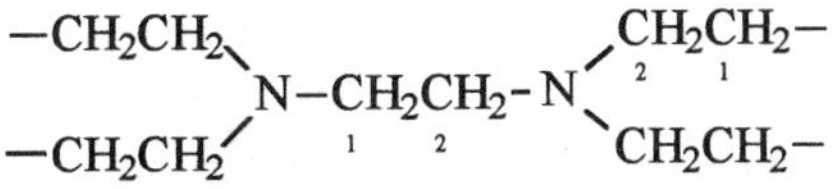

Ethylenebis(nitrilodiethylene)
Ethane-1,2-diylbis(nitrilodiethane-2,1-diyl)

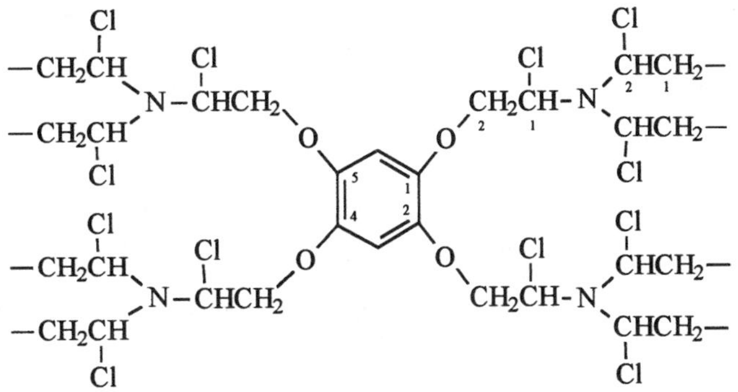

1,2,4,5-Benzenetetrayltetrakis[oxy(1-chloroethylene)nitrilo]bis(2-chloroethylene)]
1,2,4,5-Benzenetetrayltetrakis[oxy(1-chloroethan-2,1-diyl)nitrilo]-
   bis(2-chloroethane-1,2-diyl)]

Multiplicative names are illustrated by the examples below. The locants of the parent compound, primed serially according to its multiplicity, are cited first, followed by the name(s) of the linking groups, a numerical or multiplicative prefix indicating the number of parent compounds, and finally the name of the parent compound.

$$HOOC-CH_2-O-CH_2-COOH$$
$$\underset{2}{\phantom{xxx}}\underset{2'}{\phantom{xxxxxxx}}$$

2,2'-Oxybis(acetic acid)

$$\underset{3}{HOOC-CH_2CH_2}\diagdown \qquad \diagup \underset{3''}{CH_2CH_2-COOH}$$
$$N-CH_2CH_2-O-CH_2CH_2-N$$
$$\underset{3'}{HOOC-CH_2CH_2}\diagup \qquad \diagdown \underset{3'''}{CH_2CH_2-COOH}$$

3,3',3'',3'''-[Oxybis(ethylenenitrilo)]tetrapropanoic acid

There are two methods for naming identical parent compounds that contain substituents in addition to the principal group. One method, codified in the IUPAC Organic Rules, multiplies only the parent compound; other substituents are cited as prefixes in front of the name of the linking substituent group. The other method, used by CAS, requires that the parent compound be substituted symmetrically, including the principal characteristic group.

6,6'-Dichloro-3,3'-(methylenedioxy)dibenzoic acid (IUPAC)
3,3'-[Methylenebis(oxy)]bis[6-dichlorobenzoic acid] (CAS)*

## The Functional Class Method

The functional class method is a technique for naming compounds by citing a class term, such as chloride, oxime, imide, oxide, hydrazone, usually as a separate word,** following a term or terms that describes the rest of the compound. Acids and esters can be viewed as examples of this method. When the word or words preceding the class term are substituent prefixes derived from a parent hydride or an acid, the method has been called "radicofunctional", for example, dimethyl ether, ethyl methyl ketone, dibenzoyl peroxide, and acetyl chloride. Many "additive names" such as pyridine oxide (see above) are of this type. Subtractive names, such as anhydrides, and names for functional derivatives of mononuclear oxo acids and related compounds, such as hydrazides and amides, are also functional class names.

## Skeletal Modification

Skeletal modification is a technique usually applied to "stereoparent" parent compounds, such as natural products and related compounds (see chapter 31), for changing the structure of a well-known parent structure to avoid a more complicated systematic name. Skeletal modification is used to a limited extent with other trival names. Nondetachable prefixes are used to indicate a modification of the skeletal structure of a parent compound.

The prefix homo- describes the insertion of a methylene ($-CH_2-$) group between two skeletal atoms of a cyclic or cyclic/acyclic parent structure. The position of the new skeletal

---

* CAS uses "bis(oxy)" to avoid the possibility of misinterpretation of dioxy as meaning $-OO-$.
** A type of amine nomenclature considered in chapter 21 to be substitutive based on the parent name amine for $NH_3$ can be visualized as a functional class name without the space that usually precedes the class name; for example, methylamine instead of methyl amine.

atom in stereoparent compounds is given by a numerical locant followed by a lower case letter as described in chapter 31,* for example, 16a-homo-5α-pregnane and 23a-homo-5α-ergostane. This prefix is also used in naming the amino acids homoserine and homocysteine, but without the numbering convention just described. It is also found elsewhere, for instance with trivial names for polycyclic hydrocarbons, as in homoadamantane and homocubane, and with trivially named functional parent compounds, as in homophthalic acid and homoisovanillin; however, such usage is not encouraged.

The prefix nor- describes the elimination of one unsubstituted skeletal atom, saturated or unsaturated, of a stereoparent structure with its attached hydrogen atoms from a ring or side chain of a parent structure. The position occupied by the skeletal atom removed is indicated by the appropriate locant of the parent structure, for example, 18-nor-5α-pregnane, 20-nor-β,β-carotene, and 3-norlabdane. The special use of this prefix to remove multiple methylene groups from certain terpene parent hydrides, such as bornane, is no longer recommended.

The prefix nor- is also used to name two amino acids, norvaline and norleucine, but the meaning of the prefix in these names is not the same as that described above; the use of these trivial names is discouraged in IUPAC/IUB recommendations. The prefix nor- is sometimes found as a part of trivial names, but in very few cases does it have the systematic meaning given here, often having no discernible meaning at all; usage outside the meaning given above is strongly discouraged.

The prefix seco- describes the cleavage of a ring bond, saturated or unsaturated, with the addition of the appropriate number of hydrogen atoms at each new terminal group so created for example, 2,3-seco-5α-cholestane and 2,3-secoyohimban. It is often used in combination with cyclo- to indicate bond rearrangement.

The prefix cyclo- describes the formation of a ring, as in cyclohexane, or the creation of an additional ring in a polycyclic stereoparent structure by indicating a direct link between any two atoms of the structure, for example, 3α,5-cyclo-5α-pregnane and 5,19-cycloandrostane. It is often used in combination with seco- to indicate bond rearrangement, as in 6α,10-cyclo-5,10-secoandrostane.

The prefix x(y → z)-abeo- describes the migration of one end of a single bond between the positions "x" and "y" in a parent structure to the positions "x" and "z'; that is, the "y" end of the single bond moves to "z", for example, (3αH)-5(4 → 3)-abeopodocarpane. (The prefix "abeo-" has been in italic type in earlier versions of Section F of the 1979 IUPAC Organic Rules.)

The italicized prefix retro-, as used in the nomenclature of carotenes, preceded by a pair of locants, indicates a shift, by one position, of all double bonds of a conjugated polyene system delineated by the locants; the conjugated polyene system cannot include all or some of the double bonds that consititue the maximum number of conjugated double bonds of a ring or ring system, for example, 4′,11-retro-β,γ-carotene.

The prefix apo-, as used in the nomenclature of carotenes, preceded by a locant, indicates the removal of all atoms of a side chain of a carotene structure beyond the skeletal atom indicated by the locant. An example is 6′-apo-β-carotene.

The prefix neo-, as used in the nomenclature of terpenes, indicates a bond migration that converts a gem-dimethyl group directly attached to a ring carbon atom into an isopropyl group. This prefix is no longer recommended by IUPAC; the prefix abeo- is preferred. It is no longer used by CAS except for A′-neogammacerane terpenes. An example of its earlier use is A:B-neolupane.

The prefix friedo-, as used in the nomenclature of terpenes, indicates a migration of a methyl group from one position to another. This prefix is no longer recommended by IUPAC; the prefix

---

* Capital letters, associated with the locant(s) of the added skeletal atom(s) where needed, have been used to indicate enlargement of particular rings. This method is still used by CAS, but is not included in IUPAC recommendations. Also, the provisional IUPAC recommendations do not mention the lengthening of a side chain.

abeo- is preferred, and its use was discontinued by CAS in 1997. An example of its earlier use is D:A-friedolupane.

Combinations of the above prefixes are commonly used and are generally cited in alphabetical order. However, too many modifications can be as difficult as a complicated systematic name. Although it is an advantage to minimize the number of fundamental parent structures, the use of many structure-modifying prefixes can result in modified structures that are too drastically removed from the fundamental parent structure to be useful. Furthermore, combinations of structure-modifying prefixes can transform one fundamental parent structure into another.

## REFERENCES

1. International Union of Pure and Applied Chemistry, Organic Chemistry Division, Commission on Physical Organic Chemistry. Glossary of Terms Used in Physical Organic Chemistry. *Pure Appl. Chem.* **1994**, *66*, 1077–1184.
2. International Union of Pure and Applied Chemistry. *Compendium of Chemical Terminology (IUPAC Recommendations)*; 2nd Ed. McNaught, A. D., Wilkinson, A., Compilers; Blackwell Science: Oxford, U.K., 1997; p 336.
3. International Union of Pure and Applied Chemistry, Organic Chemistry Division, Commission on Nomenclature of Organic Chemistry. Revised Nomenclautre of Radicals, Ions, Radical Ions, and Related Species. *Pure Appl. Chem.* **1993**, *65*, 1357–1455.
4. American Chemical Society. The Report of the ACS Nomenclature, Spelling, and Pronunciation Committee for the First Half of 1952. E. Compounds Containing Phosphorus. *Chem. Eng. News* **1952**, *30*, 4517–4522.
5. International Union of Pure and Applied Chemistry, Inorganic Chemistry Division, Commission on Nomenclature of Inorganic Chemistry. *Nomenclature of Inorganic Chemistry Recommendations 1990*; Blackwell Scientific: Oxford, U.K., 1990.
6. Block, B. P.; Fernelius, W. C.; Powell, W. H., *Inorganic Chemical Nomenclature. Principles and Practice*; American Chemical Society: Washington, D.C., 1990.

# 4

# Common Errors, Pitfalls, and Misunderstandings

Since this book is designed to aid chemical communication through the use of reasonable and acceptable names for organic compounds, it seems only right to point out a few of the wrong turns that can be made in connecting names and structures. Some missteps are as simple as typing a typo. Many possible pitfalls are discussed in the various chapters. In this chapter, an additional number of easily avoided deviations and their consequences in nomenclature will be described.

It almost goes without saying that typos in names should be avoided. Some are obvious, but there are typos that produce perfectly acceptable names for compounds other than the one being described. Without a structure for comparison, these typos can be difficult to detect by another reader. "Ethyl" when "methyl" is meant is an example. "Sulfonyl" instead of "sulfi-nyl" can be very misleading, since two classes of closely related compounds may be involved; this kind of mistake is easy to make in the sulfur and phosphorus fields. "Amide" when "amine" is intended is less likely to cause trouble, but a train of thought by the reader is sure to be interrupted. Clearly, oxymorons are beyond the pale.

One misunderstanding is the idea that there must be a single "best acceptable" name for each compound. Nomenclature is language. As language is replete with synonyms, so too it will be seen that a large proportion of the compounds illustrated in this book come with more than one acceptable name. There are several useful systems of nomenclature (see chapter 3), and within a given system it may be desirable to emphasize a specific part of a structure. One name might be better than another for a particular purpose, but they all may be acceptable, and one should not be afraid to use them. On the other hand, indexes, chemical catalogs, and even tariff regulations must, with certain reservations, use a single name for a given com-pound. Their systems may differ from each other in various ways, and they may not corre-spond completely with IUPAC rules which enjoy widespread acceptance.

There is, however, a common thread. A parent structure is identified and named, and any structural modifications and substituents are incorporated into the name through the use of prefixes, infixes, or suffixes. Thus $(CH_3)_2CH–C_6H_5$ might be named isopropylbenzene, (1-methylethyl)benzene, or 2-phenylpropane, each of which is intelligible and unambiguous; the choice may be a matter of style. Rules change as nomenclature develops; the present IUPAC rules allow the name "cumene" for this compound as long as there are no substitu-ents. It would not be an error to use any of these names. At the same time, the name "cumol", by which this compound was once known, is no longer used, as is also the case with its cousins "benzol", "toluol", and "xylol". The suffix -ol has denoted an alcohol or phenol for over a century.

## Function

"Function" and its cousins "functional", "functional group", and "functionality" are terms often confused and misunderstood. "Functional group" is synonymous with, and in the IUPAC rules has been supplanted by, "characteristic group". The introduction of the term "characteristic group" was the result of difficulties in defining "function", one problem being whether or not unsaturation was a function. In this book, characteristic group, which is a defined (see chapter 3 and the glossary) atom or group of atoms (unsaturation and carbon-containing groups without associated heteroatoms are excluded by definition), is used while functionality will be used to refer to the ability of a characteristic group to affect the chemistry of the structure to which it is attached. Function and functional serve as class terms, as in "the ketone function" and "the ketone functional suffix is -one". The terms "functional class name", "functional replacement name", and "functional class nomenclature" are discussed in chapter 3.

## The Principal Characteristic Group

Functionality is the heart and soul of organic nomenclature, for it pertains to a center of potential reactivity in an organic molecule. In substitutive nomenclature, it is usually emphasized by means of a characteristic group cited as a suffix, a class name, or in the name of a functional parent compound. The name ethanol is always preferred to hydroxyethane. Phenylamine, benzenamine, and even aniline, which implies an amino group, are generally preferred to "aminobenzene". Functionality, not size of a parent hydride with or without substituents that are not characteristic groups, should take precedence in forming a name, as in the following:

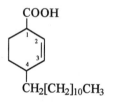

4-Dodecylcyclohex-2-ene-1-carboxylic acid
(*not* 3-Carboxy-6-dodecylcyclohex-1-ene or
4-carboxycyclohex-2-en-1-yldodecane)

It is unfortunate that some trade names, trivial names, and proprietary drug names have endings that imply the presence of a nonexistent or incorrect characteristic group. Examples are Telone (a mixture of chlorinated propenes) and silicone, neither of which are ketones; picric acid, which is a phenol; semicarbazide, which is not an azide; and theobromine, which has no bromine. Obsolete names such as carbolic acid or toluol should not be used in scientific communication.

A single center of functionality will normally appear as a characteristic group suffix or class name at the end of a name, as in hexanol, octanoic acid, or acetyl chloride. But what if there is more than one characteristic group? A choice, the "principal characteristic group", must be made. In substitutive nomenclature, this group is cited as a suffix, and the remaining characteristic groups and other substituents are denoted by prefixes. The choice for the principal characteristic group depends on the context or purpose of the name, or upon an order of precedence (see chapter 3). The latter ought to be followed unless there is good reason not to, such as a desire to emphasize a particular characteristic group or structure in a series of compounds. Even then, it should be recognized that indexes do follow an order of precedence. This order (see chapter 3) has come about through usage and tradition; gross violations often result in names that do not "sound right". The carboxy group is near the top of the preference list. Accordingly, most compounds containing this substituent will be named as acids, no matter what other substituents may be present. 4-Hydroxybutanoic acid, not 4-carboxybutan-1-ol, is the name for $HO–CH_2CH_2CH_2–COOH$. Ethyl 4-aminobenzoate would be preferred to 4-(ethoxycarbonyl)aniline or 4-carbethoxybenzenamine because esters are senior to amines in the seniority order. Almost any characteristic group will take precedence over the ether group. The substitutive name 2-ethoxyethanol is better than ethyl 2-hydroxyethyl ether, but both are better than 1-ethoxy-2- hydroxyethane. Sometimes, no decision is made and two suffixes are crowded into the end of a name; this is never a good idea. Traditional names like ethanolamine and 1-naphthol-2-sulfonic acid should not be used; the systematic names 2-aminoethanol and 1-hydroxynaphthalene-2-sulfonic acid are always preferred.

Occasionally one sees names based on obsolete or unofficial methods of nomenclature. Benzol and toluol are in this class. "Simple nucleus" names such as triphenylcarbinol for $(C_6H_5)_3C–OH$, tetramethylmethane for $(CH_3)_4C$, and diphenylcarbinyl for $(C_6H_5)_2CH–$ are no longer acceptable. Some names, such as "diacetic acid", once used for 3-oxobutanoic acid, look legitimate but actually have no meaning today. There is no need to perpetuate such names since they have been supplanted by readily understood systematic nomenclature.

As chemical nomenclature develops and rules change, many trivial names, at one time approved, are discouraged and will fall from use. There are several examples among the carboxylic acids: capric, caproic (both of these names lead to "caproyl" for the acyl group), and valproic acid are no longer acceptable for chemical purposes, although valproic acid is still in use in certain medical fields.

## Spaces

A common error is the deletion of spaces that belong in names. "Ethylacetate" (ethyl acetate) and "methylethylketone" (methyl ethyl ketone) are examples. Equally common is the unnecessary placing of spaces in names, as in "ethyl benzene" (ethylbenzene). The latter is an example of a substitutive name in which an ethyl group has replaced a hydrogen atom in a parent hydride, and the replacing group is named by a prefix that becomes part of the newly named compound. Similarly, spaces are not used in names with additive prefixes, such as 1,2,3,4-tetrahydronaphthalene. Parent hydrides are compounds, and in these two examples, benzene and naphthalene are names of compounds. "Acetate", "alcohol", "ketone", and so on, are not compounds, and a space should precede them in functional class names such as butyl alcohol; functional class names of acyl halides, ketones, ethers, sulfides, sulfoxides, acetals, anhydrides, and glycosides also contain spaces. "Oxide" is a separate word in names such as pyridine N-oxide. "Amine", on the other hand, is often considered to be derived from ammonia, which is a compound, and therefore no space appears in names such as methylamine. The compound methyl(octyl)amine also does not have spaces; the parentheses are not strictly required but do clearly indicate that this a secondary amine and not a methyl group substituted on the octyl group.

## Hyphens and Dashes

Hyphens in a name or a part of a name serve to maintain its status as a single word, as in butan-2-ol or 4-but-3-en-1-ylbenzoic acid. Dashes, however, are used to separate the components of addition compounds at an unknown position. Hyphens or dashes should not be used to indicate the components of a mixture of two or more chemicals. An example is naphthalene-acetic acid, which might be interpreted to be the compound naphthaleneacetic acid rather than an addition compound or a mixture of naphthalene and acetic acid. For a mixture a solidus could be used, as in "a naphthalene/acetic acid mixture", but the circumlocution "a mixture of naphalene and acetic acid" is clearly better.

## Locants and Parentheses

Omitting locants is much like omitting the house number in a street address. In the alkanoic acids, the locant 1 is assumed to be that of the carboxy group. However, locants can denote positions in a parent structure as well as its substituents; often the same numerals are involved. An example is 3-phenylpropanoic acid: the 3 refers to the location of the phenyl group on the propanoic acid molecule. Further substitution on the phenyl group can present a problem, since that group will also have locants. The answer is to enclose the substituted phenyl group name in parentheses; any locants within the parentheses apply only to the phenyl group:

3-Bromo-3-(3-chlorophenyl)propanoic acid

Without the bromine substituent, this example would be named 3-(3-chlorophenyl)propanoic acid, not 3-chlorophenylpropanoic acid, because there is a 3 position on both ring and chain; however, 3-*m*-chlorophenylpropanoic acid might be acceptable since *m* can apply only to the ring. Another example is 2,4-dinitrophenylhydrazine. Parentheses in the correct name (2,4-dinitrophenyl)hydrazine are often omitted despite the fact that there is a 2 position on the hydrazine part of the compound.

Potentially ambiguous names can arise from the lack of parentheses in names such as "chloromethylsilane", which for CAS means $CH_3–SiH_2–Cl$ while (chloromethyl)silane is $Cl–CH_2–SiH_2$; a better name for $CH_3–SiH_2–Cl$ is chloro(methyl)silane in which the parentheses clearly define the structure.

Primed locants are often used to differentiate among positions in various parts of a structure. In the example above, primed locants on the benzene ring were unnecessary because substituent names could be placed safely within the parentheses. However, where primed locants are used, they are all-important in a name: $N,N'$-dimethylethane-1,2-diamine and $N,N$-dimethylethane-1,2-diamine are quite different compounds. Trivial names of parent compounds frequently cover a multipart structure, and primed locants must be given to one part.

2'-Chloropropiophenone

Ring assemblies (see chapter 8) are other good examples of the need for primed locants.

2'-Methyl-1,1':4',1"-terphenyl

## Esters and Other Acid Derivatives

Spaces are necessary to separate the names of esterifying groups from each other and from the remainder of the name of the parent acid; other substituents remain with the parent. The term "hydrogen" should be used with partial esters of di- or polybasic acids, as in methyl dihydrogen phosphate; as a family, partial esters are acid esters, and this compound is a phosphoric acid ester. Other examples are:

Ethyl hydrogen phenylpropanedioate
Ethyl hydrogen phenylmalonate

$C_6H_5–O–CO–CH_2–CO–O–CH_2CH_3$

Ethyl phenyl propanedioate
Ethyl phenyl malonate

The above compounds are malonates or malonic acid esters, but not "malonate esters", since they are not esters of malonates. An analogous comment applies to the family names of esters of other acids: sulfonates or sulfuric acid esters, phosphonates or phosphonic acid esters, and so on. These compounds are esters of acids, not esters of esters. The same logic applies to salts (benzoic acid salts) and amides (cyclohexanecarboxylic acid amides); amides of acids is also acceptable, but specific names such as "octanoic acid amide" or "acetic acid anhydride" should not be used because such derivatives are not acids. The origin of these incorrect names may lie in their appearance in indexes, as in "octanoic acid, anhydride" and "acetic acid, ethyl ester".

In the past, substituent prefixes derived from acyclic hydrocarbons always placed the locant 1 at the free valence, and therefore this locant was omitted in names such as pentyl or butyl. This is still acceptable, but a name like "but-2-yl" is not. Current IUPAC recommendations allow the free valence in an acyclic parent hydride or a parent substituent prefix to be given the lowest locant consistent with the numbering of the longest chain. Accordingly, the 1 must be cited, as in butan-1-yl to distinguish it from butan-2-yl.

## Hybrid Names

A common error occurs in names derived by mixing systems of nomenclature. Isopropyl alcohol is a functional class name, and propan-2-ol is formed by substitutive nomenclature, but "isopropanol" mixes the systems and should not be used because therei s no parent hydrocarbon name to which the suffix -ol can be added. "Isobutene" (for 2-methylpropene), "*tert*-butanol" (for 2-methylpropan-2-ol or *tert*-butyl alcohol), and "pyrrole-2-aldehyde" (for pyrrole-2-carboxaldehyde) are additional examples of names that never should be used.

The principle in nomenclature of the inviolability of the longest chain of like atoms is somewhat sacred. "Simple nucleus" names such as trimethylmethanol for $(CH_3)_3C–OH$ are not preferred to 1,1-dimethylethanol or 2-methylpropan-2-ol even though they may seem to treat like things alike. Another example is *tert*-butyl ketene, for $(CH_3)_3CCH=CO$, better named as 3,3-dimethylbut-1-en-1-one. Similarly, $(CH_3)_2NNH_2$ is dimethylhydrazine, not amino(dimethyl)amine. Exceptions to this principle are seen where a characteristic group is bound to a chain of like atoms by the same kind of atom, as in nitrosohydrazine rather than 1-oxo-triazene for $H_2NNH–NO$ or trisulfane rather than disulfanethiol or sulfanedithiol for HSSSH.

## Detachable and Nondetachable Prefixes

One vexation in nomenclature is the detachability of prefixes (see chapter 3); the "error" can result in frustration in finding things in an index. Substitutive prefixes can be detached from the name of a compound and still leave a sensible name that in an index could serve as a heading. Remove 3-chloro- from 3-chlorooctanoic acid, and octanoic acid remains. A prefix is nondetachable when it modifies the structure of the parent compound. Examples are the "a" prefixes used in replacement nomenclature (see chapter 3), fusion prefixes (such as the "benz" in benz[*a*]anthracene), and the names of bridges such as ethano or epoxy. Such prefixes are part of the name of a new parent compound and stay with it through thick, thin, and inversion in an alphabetical index.

Unfortunately, some bridge names, and "epoxy" is one, are also detachable substitutive prefixes, as used in 3,4-epoxybutanoic acid, which can as easily be named oxiranylacetic acid. Names such as epoxy, hydrazi, and so on, can cause trouble in another sense: they should not be used to cite a group jointly held by the parent compound and a substituent. In the following example, 3-cyclohexylidenepropanamide is an acceptable name, but the corresponding epoxy derivative, in which the oxygen joins both a ring carbon atom and an acyclic carbon atom is best named as a spiro structure (see chapter 6).

3-Cyclohexylidenepropanamide

2-(1-Oxaspiro[2.5]octan-2-yl)acetamide

Bivalent prefixes in which the free valences of a group such as $-[CH_2]_n-$ are attached to a single atom are not acceptable in systematic nomenclature: "tetramethylenephosphane" is better named as a heterocycle.

## Italic Prefixes and Capitalization

When a sentence begins with a name that starts with an italicized prefix such as *para-* (or *p-*, *tert-*, *cis-*, and so on), the first letter of the prefix is not capitalized. On the other hand, italicized letter symbols denoting an element, such as *N*, *S*, *H*, and so on, are always capitalized no matter where they occur. Italicized prefixes are not used in alphabetization. Nondetachable prefixes such as iso and cyclo are capitalized at the beginning of a sentence and do take part in alphabetization. Examples of proper usage are as follows:

| | |
|---|---|
| *tert*-Butyl bromide is ... | not *Tert*-butyl bromide is ... |
| Isopropyl bromide is ... | not isoPropyl bromide is ... |
| *N*-Methylpiperidine is ... | not *N*-methylpiperidine is ... |

## Labor-Saving Devices

To save wear and tear on minds and keyboards, many compounds are known in the laboratory and in the written form by acronyms. Most acronyms are based on a name, often cumbersome, of the compound. In some cases, the name may no longer be acceptable, but the acronym lives on, and a reader may be forgiven for wondering where it came from. An example is

           1,4-Diazabicyclo[2.2.2]octane

known to most, but not all, as DABCO, which follows from the name. However, the compound is also known as TEA, from the older name triethylenediamine. The polymer field is full of acronyms. PTMO, PTMG, and PTHF are all used for the polymer whose structure-based name (see chapter 29) is poly(oxybutane-1,4-diyl). At the very least, an acronym should be defined at the first point it appears in a paper or presentation, no matter how common it may be in usage, and an acronym should never be used in the title of a paper.

Another labor-saving device that should be used with care is that of collecting locants into a single form. Stereochemical descriptors such as "*all-E*" (see chapter 30) and "perfluoro-" (see chapter 10) are examples.

## Limitations in Nomenclature Systems

Any nomenclature system has its limits. Stepping over the boundaries can often lead to ambiguity. One limit is imposed when contraction of a name is attempted, as in silaethene (a replacement name) to silene, which looks for all the world like an analog of carbene. Another lies in the overlap between Hantzsch–Widman names and systematically generated parent hydride names as is seen in selenane and tellurane, which are the names of six-membered rings in the Hantzsch–Widman system. The hydride $InH_3$ cannot be named indane because that is the name of an organic ring system. Phosphinyl would be the logical prefix name for $H_2P-$ derived from phosphine, but phosphino or phosphanyl must be used because phosphinyl is well-established for $H_2P(O)-$. Many instances of this kind are cited in the discussion sections of other chapters in this book.

## Conclusion

A key to good nomenclature is to avoid ambiguity. It is almost a given that if a name can be misunderstood, it will be. At the same time, an unambiguous name that describes the wrong structure is nothing but misinformation. An easy way out in a paper or an oral presentation, of course, is to give the structure and name it "I" or call it "that structure". However, this is not good communication. There is no substitute for the right word in the right place, and a bit of elegance might be achieved by correctly spelling that word as well.

# 5

# Acyclic Hydrocarbons

In this and the next three chapters, hydrocarbon nomenclature will be presented. Simplicity of structure undoubtedly was the basis for early international agreement on names for *acyclic hydrocarbons*, the subject of this chapter. The "Geneva" system adopted in 1892[1,2] has been universally accepted. As simple structures, hydrocarbons are readily viewed as parents or, as IUPAC now calls them, "parent hydrides", in which hydrogen atoms can be exchanged for other atoms or groups. Names of the resulting compounds are therefore logically based on the names of the hydrocarbons themselves; these names and those of the substituting groups are of fundamental importance in the language of organic chemistry.

## Acceptable Nomenclature

*Saturated branched or unbranched acyclic hydrocarbons* are classified as *alkanes*; their names are characterized by the ending -ane, preceded by terms that describe the hydrocarbon structure.

The first four unbranched alkanes of the series $H[CH_2]_nH$ are called methane, ethane, propane, and butane. For "$n$" greater than four, a numerical term is prefixed to -ane (see table 5.1). These numerical prefixes have been extended to 9999.[3]

Table 5.1. Unbranched Alkanes

| $n$ | | $n$ | | $n$ | |
|---|---|---|---|---|---|
| 5 | pentane | 14 | tetradecane | 31 | hentriacontane |
| 6 | hexane | 15 | pentadecane | 32 | dotriacontane |
| 7 | heptane | 20 | icosane (IUPAC) | 40 | tetracontane |
| 8 | octane | | eicosane (CAS) | 50 | pentacontane |
| 9 | nonane | 21 | henicosane (IUPAC) | 100 | hectane |
| 10 | decane | | heneicosane (CAS) | 101 | henhectane |
| 11 | undecane | 22 | docosane | 200 | dictane |
| 12 | dodecane | 23 | tricosane | 300 | trictane |
| 13 | tridecane | 30 | triacontane | 1000 | kiliane |

*Saturated branched alkanes* are named on the basis of the longest unbranched chain (the principal or main chain) in the structure, with side chains designated as alkyl groups whose names are derived from the names of alkanes (see below). The longest unbranched chain is numbered from one end to the other, the numbers being called "locants". Numbering of the chain begins at the end nearest to a side chain and follows the path that provides the "lowest locants" for the side chains. In a comparison of two sets of locant numbers, the one having "the lowest locants" is that which has the lowest number at the first point of difference; for example, 1,1,2,7,7 is lower than 1,1,3,6,7.

Side chains may be joined to the longest unbranched chain. These side chains are called "substituting groups" because they substitute for hydrogen atoms bound to a carbon atom. As entities, the side chains are called "substituents", and they are given names based on their structures. Substituent names take the form of prefixes attached to the name of the chain. In the names of the following examples, ethyl, methyl, and isopropyl are the names of substituents or substituting groups, each preceded by a locant to indicate position on the main chain. The names of side chains are cited in alphabetical order. To establish alphabetical order, prefixes such as di- and tri- are disregarded, as are prefixes defining structure, such as *sec-* and *tert-*, which are italicized and separated from the name by a hyphen.

If chains of equal length compete for selection as the parent chain, the choice goes, in descending order, to (a) the chain carrying the largest number of side chains; (b) the chain whose side chains have the lowest locants; (c) the chain with a side chain cited earlier in alphabetical order; and (d) the chain with the lowest locants for side chains earliest in alphabetical order.

$$
\underset{8\quad 7\quad 6\quad 5\quad 4\quad 3\quad 2\quad 1}{\overset{\displaystyle \overset{CH_3}{|}\qquad \overset{CH_3}{|}\ \ \overset{CH_3}{|}}{CH_3CHCH_2CH_2CHCH_2CHCH_3}}
$$

2,4,7-Trimethyloctane
(*not* 2,5,7-Trimethyloctane; 2,4,7 is a lower locant set than 2,5,7)

$$
\underset{12\ 11\ 10\ 9\quad 8\quad 7\quad 6\quad 5\quad 4\quad 3\ 2\ 1}{\overset{\displaystyle \overset{CH_3}{|}\qquad\quad \overset{CH(CH_3)_2}{|}\ \ \overset{CH_2CH_3}{|}\qquad \overset{CH_3}{|}}{CH_3CHCH_2CH_2CHCH_2CH_2CHCH_2CH_2CHCH_3}}
$$

5-Ethyl-8-isopropyl-2,11-dimethyldodecane
(*not* 8-Ethyl-5-isopropyl-2,11-dimethyldodecane;
first cited substituent takes the lower locant)

The names isobutane, $(CH_3)_3CH$, neopentane, $(CH_3)_4C$, and isopentane, $(CH_3)_2CHCH_2CH_3$, are acceptable, but they should not be modified by substitution because acceptable locants are not available.

*Hydrocarbon substituting groups* that join to a main chain by a single bond are called *alkyl groups*. Structurally, these groups are shown with a long hyphen or a dash at the position where the single bond junction to another chain would occur, as in $-CH_2CH_3$ or $-CH(CH_3)_2$. For the purpose of systematic nomenclature, the long hyphen or dash represents a "free valence", and such substituting groups are called "univalent groups". Bivalent groups, trivalent groups, and so on, are joined to other structures through more than one single bond or through double or triple bonds.

Alkyl groups are named by two methods. In the first, which is limited to saturated alkyl groups, the name is based on the structure having the free valence at a terminal carbon atom of the chain. The ending -ane in the name of the chain is replaced by -yl, and the locant 1, not cited in the name of the group, is assigned to the carbon atom with the free valence; accordingly, heptane becomes heptyl. If the free valence occurs at a point other than at a terminal carbon atom, the group is named on the basis of the longest chain terminating at the carbon with the free valence, and the remainder of the chain is cited as a 1-substituent. Branched alkyl groups are also named on the basis of the longest chain, with side chains named as substituents.

A second and more general method used to name alkyl groups is based on structures in which the free valence may occur at any point in an unbranched hydrocarbon. The numbering of the hydrocarbon is retained, and the free valence is indicated by replacing the final "e" of the hydrocarbon name with -yl preceded by its locant; the locant 1 must be included even if the free valence is at a terminal carbon atom, as in heptan-1-yl. The locant of the free valence should be as low as possible in the longest chain; its position is senior to that of any substituent.

$$
\underset{2\quad\ \ 1}{CH_3CH_2CH_2CH_2CH_2CH_2-}
$$
Hexyl
Hexan-1-yl

$$\overset{\text{CH}_3}{\underset{6\quad5\quad4\quad3\quad2\quad1}{\text{CH}_3\text{CH}_2\text{CHCH}_2\text{CH}_2\text{CH}_2-}}$$

4-Methylhexyl
4-Methylhexan-1-yl

$$\underset{5\quad4\quad3\quad2\quad1}{\text{CH}_3\text{CH}_2\text{CH}_2\text{CH}_2\overset{|}{\text{CH}}\text{CH}_2\text{CH}_3}$$

1-Ethylpentyl
Heptan-3-yl

$$\overset{\text{CH}_3}{\underset{8\quad7\quad6\quad5\quad4\quad3\quad2\quad1}{\text{CH}_3\text{CH}_2\text{CHCH}_2\text{CHCH}_2\text{CH}_2\text{CH}_3}}$$

6-Methyloctan-4-yl
   (*not* 3-Methyloctan-5-yl)
3-Methyl-1-propylpentyl

The alkyl group names in table 5.2 are acceptable if they are not substituted.

Table 5.2. Alkyl Groups

| | |
|---|---|
| $(\text{CH}_3)_2\text{CH}-$ | isopropyl |
| $(\text{CH}_3)_2\text{CHCH}_2-$ | isobutyl |
| $\overset{\text{CH}_3}{\underset{}{\text{CH}_3\text{CH}_2\text{CH}-}}$ | *sec*-butyl |
| $(\text{CH}_3)_3\text{C}-$ | *tert*-butyl |
| $(\text{CH}_3)_2\text{CHCH}_2\text{CH}_2-$ | isopentyl |
| $\overset{\text{CH}_3}{\underset{\text{CH}_3}{\text{CH}_3\text{CH}_2\text{C}-}}$ | *tert*-pentyl |
| $(\text{CH}_3)_3\text{CCH}_2-$ | neopentyl |

*Bivalent substituting groups* in which the two free valences form a double bond at one carbon atom are named in the manner of alkyl groups with the ending -ylidene. Where the free valences do not form a double bond, -diyl is used by IUPAC. CAS employs -ylidene for both types of bonding.

$$\underset{2\quad1}{\text{CH}_3\text{CH}_2\text{CH}_2\text{CH}_2\text{CH}=}$$

Pentylidene
Pentan-1-ylidene

$$\overset{\text{CH}_3}{\underset{7\quad6\quad5\quad4\quad3\;2\quad1}{\text{CH}_3\text{CHCH}_2\text{CH}_2\overset{\|}{\text{C}}\text{CH}_2\text{CH}_3}}$$

6-Methylheptan-3-ylidene
1-Ethyl-4-methylpentylidene
   (*not* 1-Isopentylpropylidene)

Isopropylidene, which should not be substituted, is an acceptable alternative for $(\text{CH}_3)_2\text{C}=$. Methylidene denotes only $\text{CH}_2=$ as a substituent, although CAS still uses methylene for both $\text{CH}_2=$ and $-\text{CH}_2-$. Bivalent groups in which the free valences are at different carbon atoms are preferably named through the use of the hydrocarbon name followed by appropriate lowest locants and the ending -diyl. Alternatively, but less preferred, they may be named by citing the number of $-\text{CH}_2-$ groups making up the chain between the free valences followed by the ending -methylene. Substituent alkyl groups are appended in the usual way. Methylene, $-\text{CH}_2-$, and ethylene, $-\text{CH}_2\text{CH}_2-$, are normally preferred over the corresponding diyl names; ethylene as a name for the hydrocarbon $\text{CH}_2=\text{CH}_2$ (ethene) is not encouraged.

$$-CH_2CH_2CH_2CH_2CH_2-$$
5  4  3  2  1

Pentane-1,5-diyl
Pentamethylene

$$CH_3\overset{|}{C}HCH_2CH_2CH_2-$$
5  4  3  2  1

Pentane-1,4-diyl
1-Methylbutane-1,4-diyl
1-Methyltetramethylene

$$CH_3CH_2\overset{|}{C}HCH_2-$$
2  1

Butane-1,2-diyl
Ethylethane-1,2-diyl
1-Ethylethylene
(*not* Butylene)

Bivalent groups are used in forming names such as 2,2'-(pentane-1,5-diyl)dinaphthalene, ethylidenecyclohexane, or in polymer names based on structure (see chapter 29).

*Multivalent substituting groups* may be named with the ending -ylidyne for a triply-bonded group or by combinations of -yl, -ylidene, and -ylidyne, cited in that order with appropriate locants; locants are not used with -ylidyne, since a free triple bond must be at the terminus of the chain. The lowest locants principle is applied to the entire set of free valences, then to free valences in the order -yl, -ylidene, -ylidyne.

$$CH_3CH_2C\equiv$$
3  2  1

Propanylidyne
Propylidyne

$$CH_3CH_2C\overset{\diagup}{\underset{\diagdown}{-}}$$
2  1

Propane-1,1,1-triyl (IUPAC)
Propylidyne (CAS)

$$CH_3CH_2\overset{|}{C}=$$
2  1

Propan-1-yl-1-ylidene (IUPAC)
Propylidyne (CAS)

$$-CH_2\overset{|}{C}(CH_3)CH_2\overset{|}{C}=$$
4   3   2    1

3-Methylbutane-1,3,4-triyl-1-ylidene (IUPAC)
   (1,1,3,4 is a lower locant set than 1,2,4,4)
2-Methyl-1,2-butanediyl-4-ylidyne (CAS) (1,2,4 is a lower
   locant set than 1,3,4)

$$-\overset{|}{C}H-$$

Methanetriyl (IUPAC)
   (*not* Methine)
Methylidyne (CAS)

$$-CH=$$

Methanylylidene (IUPAC)
Methylidyne (CAS)

Multivalent substituting group names are typically used in naming structures in which all the free valences are bound to identical parent structures.

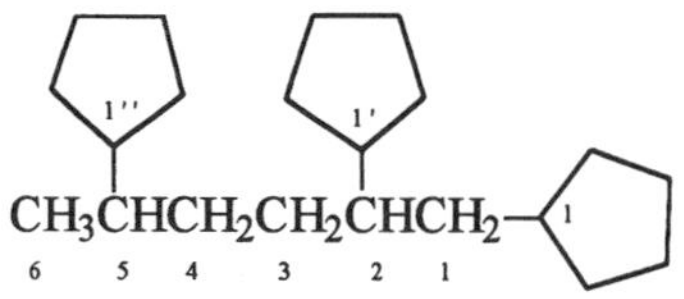

$$CH_3CHCH_2CH_2CHCH_2-$$
6    5    4    3    2    1

1,1',1''-(Hexane-1,2,5-triyl)tris(cyclopentane)

In this example, tris- is used to avoid confusion with von Baeyer names (see chapter 6) for alicyclic hydrocarbons.

*Acyclic unsaturated hydrocarbons* have the class names *alkene*, *alkadiene*, *alkyne*, and so on. They are named by replacing the ending -ane in the hydrocarbon name with an ending denoting unsaturation, preceded by appropriate positional locants. Only the lowest locant for a given position of unsaturation is cited, as in but-2-ene. Double bonds are indicated by the endings -ene, -adiene, -atriene, and so on, and triple bonds by -yne, -adiyne, -atriyne, and so on. Combinations such as -enyne, -adienyne, and so forth are also used with appropriate locants preceding each part of the combination. Locants as low as possible are given to unsaturated bonds, irrespective of their type, but if there is a choice, double bonds are preferred to triple bonds.

*Unbranched unsaturated hydrocarbon* names are based on the names of the corresponding saturated hydrocarbons.

$$CH_3CH_2CH_2CH{=}CHCH_2CH_3 \qquad \text{Hept-3-ene}$$
$$\phantom{xx}7\phantom{xx}6\phantom{xx}5\phantom{xx}4\phantom{xxx}3\phantom{xx}2\phantom{xx}1$$

$$CH_2{=}CHCH{=}CH_2 \qquad\qquad \text{Buta-1,3-diene}$$
$$\phantom{xx}4\phantom{xxxx}3\phantom{xx}2\phantom{xxx}1 \qquad\qquad\qquad (\textit{not}\ \text{But-1,3-adiene})$$

$$CH{\equiv}CCH_2CH{=}CH_2 \qquad \text{Pent-1-en-4-yne}$$
$$\phantom{xx}5\phantom{xxx}4\phantom{x}3\phantom{xx}2\phantom{xxx}1$$

Polyenes with successive (cumulative) double bonds in the chain have been called *cumulenes* as a class; they are named as above, as in penta-2,3-diene. The trivial name allene for $CH_2{=}C{=}CH_2$ is acceptable, but propadiene is preferred. Ethene is much preferred to ethylene for $CH_2{=}CH_2$ because of confusion with the use of ethylene for the bivalent group $-CH_2CH_2-$. Acetylene is usually preferred to ethyne for $HC{\equiv}CH$.

A *branched chain unsaturated hydrocarbon* name is based by CAS on the name of the longest unbranched chain; further choice is given to the longest chain with the most unsaturation, followed by that with the most double bonds. For IUPAC, names are based on the longest unbranched chain containing the maximum number of unsaturated bonds, followed in choice by the longest chain and then the chain with the most double bonds. Either method is acceptable. In other respects, the principles given above for saturated hydrocarbons are applied.

$$\begin{array}{l} CH_3 \\ | \\ CH_3C{=}CH_2 \\ \phantom{x}3\phantom{xx}2\phantom{xx}1 \end{array} \qquad \begin{array}{l}\text{2-Methylpropene} \\ (\textit{not}\ \text{Isobutene})\end{array}$$

$$\begin{array}{l} CH_3 \\ | \\ CH_2{=}CHC{=}CH_2 \\ \phantom{x}4\phantom{xxx}3\phantom{xx}2\phantom{xx}1 \end{array} \qquad \begin{array}{l}\text{2-Methylbuta-1,3-diene} \\ \text{Isoprene (unsubstituted only)}\end{array}$$

$$\begin{array}{l} CH_2\phantom{xxxx}C{\equiv}CH \\ \parallel\phantom{xxxxx}| \\ CH_3[CH_2]_3CCH_2CH_2CHCH{=}CH_2 \\ 10\phantom{x}9{\text -}7\phantom{xx}6\phantom{x}5\phantom{xx}4\phantom{xx}3\phantom{x}2\phantom{xx}1 \end{array} \qquad \begin{array}{l}\text{3-Ethynyl-6-methylene-1-decene (CAS)} \\ \text{2-Butyl-5-ethynylhepta-1,6-diene (IUPAC)} \\ \text{3-Ethynyl-6-methylidenedec-1-ene}\end{array}$$

*Unsaturated alkyl groups* are named by replacing the "e" ending of the name of the appropriate unsaturated hydrocarbon with -yl. Free valence positions, which need not be terminal, are preferred over unsaturation for the lowest locants; all locants, including 1, must be cited except where there is no ambiguity. Since CAS does not recognize nonterminal free valences for acyclic groups, the locant 1 is not cited in CAS names. Among trivial names, only vinyl, vinylidene, allyl, allylidene, and allylidyne should be used.

$$CH_2{=}CH- \qquad \begin{array}{l}\text{Ethenyl} \\ \text{Vinyl}\end{array}$$
$$\phantom{x}2\phantom{xxx}1$$

$$CH_2{=}CHCH_2- \qquad \begin{array}{l}\text{Prop-2-en-1-yl} \\ \text{Allyl}\end{array}$$
$$\phantom{x}3\phantom{xxx}2\phantom{xx}1$$

$$CH_3CH{=}CH- \qquad \text{Prop-1-en-1-yl}$$
$$\phantom{x}3\phantom{xxx}2\phantom{xx}1$$

$$CH_2=\underset{2}{\overset{\overset{\displaystyle CH_3}{|}}{C}}\underset{1}{-}$$

1-Methylethenyl (CAS)
Prop-1-en-2-yl (IUPAC)
Isopropenyl

$$CH_2=\underset{3}{\overset{\overset{\displaystyle CH_3}{|}}{C}}\underset{2}{C}H_2\underset{1}{-}$$

2-Methylprop-2-en-1-yl (IUPAC)
2-Methyl-2-propenyl (CAS)

$$CH_3CH_2CH_2\underset{3}{\overset{\overset{\displaystyle CH=CH_2}{|}}{C}}\underset{2}{=}CH\underset{1}{-}$$
$$\underset{5}{}\quad\underset{4}{}\quad\underset{3}{}\quad\underset{2}{}\quad\underset{1}{}$$

2-Ethenyl-1-pentenyl (CAS)
2-Propylbuta-1,3-dien-1-yl (IUPAC)

$$CH\equiv CCH_2\overset{\overset{\displaystyle CH_2}{||}}{C}CH=CH-$$
$$\underset{6}{}\quad\underset{5}{}\ \underset{4}{}\quad\underset{3}{}\ \underset{2}{}\quad\underset{1}{}$$

3-Methylene-1-hexen-5-ynyl (CAS)
3-(Prop-2-yn-1-yl)buta-1,3-dien-1-yl (IUPAC)
3-Methylidenehex-1-en-5-yn-1-yl

*Bi-* and *multivalent unsaturated substituting groups*, are denoted by adapting principles given earlier for saturated acyclic groups.

$$-CH_2CH=CH-$$
$$\underset{3}{}\quad\underset{2}{}\quad\underset{1}{}$$

Prop-1-ene-1,3-diyl
(*not* Propenylene)

$$-CH=CH-$$
$$\underset{2}{}\quad\underset{1}{}$$

Ethene-1,2-diyl
Vinylene

$$CH_2=C=$$
$$\underset{2}{}\quad\underset{1}{}$$

Ethenylidene
Vinylidene

$$=C=C=C=$$
$$\underset{3}{}\ \underset{2}{}\ \underset{1}{}$$

Propadienediylidene

$$CH_3CH_2\overset{|}{C}=CHCH=$$
$$\underset{5}{}\quad\underset{4}{}\ \underset{3}{}\quad\underset{2}{}\ \underset{1}{}$$

Pent-2-en-3-yl-1-ylidene (IUPAC)
1-Ethyl-1-propen-1-yl-3-ylidene (CAS)

## Discussion

The general term "aliphatic" should be used with care. By derivation, this adjective, with "hydrocarbon", does indeed describe the compounds of this chapter. The definition does not encompass cyclic compounds of any kind. Therefore, terms like "aliphatic cyclic hydrocarbons" and "cyclic aliphatic hydrocarbons" are unacceptable. Another general term that is seen increasingly often is "hydrocarbyl", to cover all monovalent hydrocarbon substituting groups; its value remains to be established.

"Free valence", a conceptual term used in describing substituting groups, should not be confused with a nonbonding electron in free radical structures (see chapter 33), even though the latter are known by the same names as the corresponding substituting hydrocarbon groups.

The symbol R has long been used for an alkyl group, as have the abbreviations Me, Et, Pr, i-Pr, Bu, s-Bu, and t-Bu for specific alkyl groups; their utility in structural formulas is undenied, but such symbols or abbreviations should not appear in names. Symbols such as $Bu^t$ should be avoided, as should "*t*-butyl" and so on. In naming saturated unbranched hydrocarbons, *n* (for normal) is often prefixed to the name of the hydrocarbon. This is an unnecessary appendage, since, for example, octane is by definition the saturated unbranched hydrocarbon with eight carbon atoms.

In the past, there has been some use of a "central nucleus" type of nomenclature, leading to names such as "trimethylmethane", "trimethylethylene", and so on. Choice of a parent "central nucleus" is arbitrary, and in extension, names can become cumbersome. The recommendations in this chapter are much preferred.

Hydrocarbons that can be viewed as a combination of two identical acyclic substituting groups have been given names such as "biallyl" for $CH_2=CHCH_2-CH_2CH=CH_2$. These names should not be used.

Names for unsaturated acyclic substituting groups related to fatty acids, such as "oleyl" for $CH_3[CH_2]_7CH=CH[CH_2]_8-$ also should not be used. Such names are too easily confused with acyl group names, such as oleoyl. Furthermore, there are no names for the corresponding unsaturated parent hydrocarbon from which such names would be derived.

## Additional Examples

1. 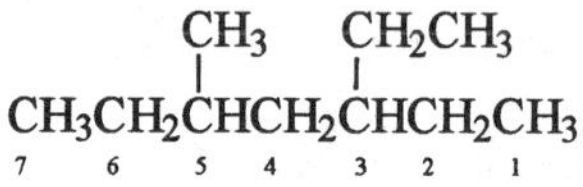  $\underset{7\ \ \ 6\ \ \ \ 5\ \ \ \ 4\ \ \ \ 3\ \ \ 2\ \ \ 1}{CH_3CH_2\overset{CH_3}{CH}CH_2\overset{CH_2CH_3}{CH}CH_2CH_3}$

    3-Ethyl-5-methylheptane

2.  $\underset{6\ \ \ \ 5\ 4\ \ \ \ 3\ \ 2\ \ \ 1}{CH\equiv CCH=CHCH=CH_2}$

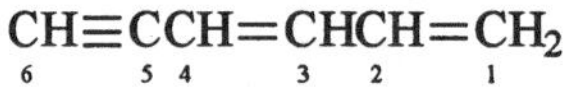

    Hexa-1,3-dien-5-yne

3.   $\underset{13\ \ \ 12\ \ \ 11\ \ \ \ 10\ \ \ 9\ \ \ 8\ 7\ \ 6\ \ \ 5\ \ \ 4\ 3\ \ 2\ \ \ \ \ 1}{CH_3CH_2CH=CHC\equiv C\overset{CH_3}{CH}CH_2C\equiv CCH_2CH=CH_2}$

    7-Methyltrideca-1,10-diene-4,8-diyne

4. 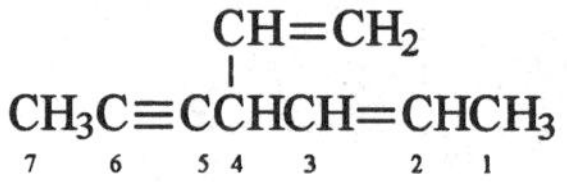  $\underset{7\ \ \ 6\ \ \ 5\ 4\ \ 3\ \ \ \ 2\ \ \ 1}{CH_3C\equiv C\overset{CH=CH_2}{CH}CH=CHCH_3}$

    4-Ethenyl-2-hepten-5-yne (CAS)
    4-Vinylhept-2-en-5-yne (IUPAC)

5. 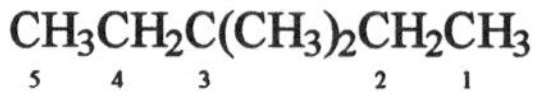  $\underset{5\ \ \ \ 4\ \ \ 3\ \ \ \ \ \ 2\ \ \ 1}{CH_3CH_2C(CH_3)_2CH_2CH_3}$

    3,3-Dimethylpentane

6. 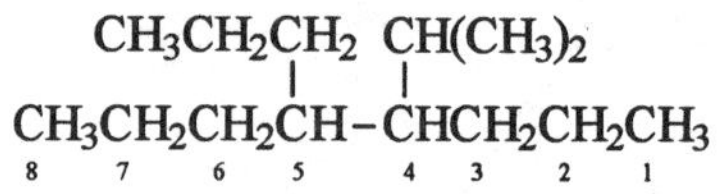  $\underset{8\ \ \ \ 7\ \ \ \ \ 6\ \ \ \ 5\ \ \ \ \ 4\ \ \ \ 3\ \ \ 2\ \ \ 1}{CH_3CH_2CH_2\overset{CH_3CH_2CH_2}{CH}-\overset{CH(CH_3)_2}{CH}CH_2CH_2CH_3}$

    4-Isopropyl-5-propyloctane
    4-(Propan-1-yl)-5-(propan-2-yl)-octane

7. 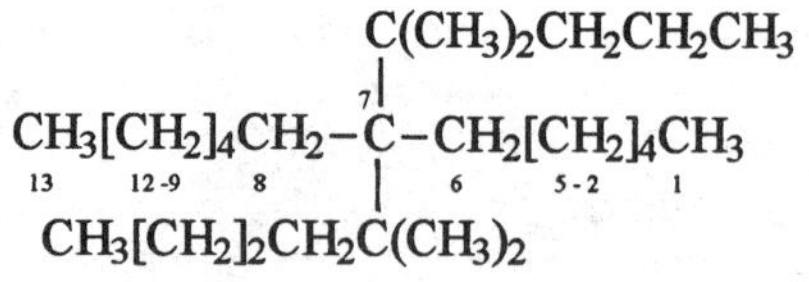  $\underset{\underset{13\ \ \ 12\text{-}9\ \ \ 8}{CH_3[CH_2]_4CH_2}}{}-\overset{7}{C}-\underset{6\ \ \ \ \ 5\text{-}2\ \ \ 1}{CH_2[CH_2]_4CH_3}$ with $C(CH_3)_2CH_2CH_2CH_3$ above and $CH_3[CH_2]_2CH_2C(CH_3)_2$ below

    7-(1,1-Dimethylbutyl)-7-(1,1-dimethylpentyl)tridecane (CAS)
    7-(2-Methylhexan-2-yl)-7-(2-methylpentan-2-yl)tridecane (IUPAC)

### References

1. Armstrong, H. E. The International Conference on Chemical Nomenclature. *Nature* **1892**, *46*, 56–58.
2. Pictet, A. Le Congress International de Genève pour la Reforme de la Nomenclature Chimique. *Arch. Sci. Phys. Nat.* **1892**, [3] *27*, 485–520.
3. International Union of Pure and Applied Chemistry, Organic Chemistry Division, Commission on Nomenclature of Organic Chemistry. Extension of Rules A-1.1 and A-2.5 Concerning Numerical Terms Used in Organic Nomenclature (Recommendations 1986). *Pure Appl. Chem.* **1986**, *58*, 1693–1696.

# 6

# Alicyclic Hydrocarbons

In chapter 5, nomenclature for hydrocarbons composed of chains of carbon atoms was considered. This chapter is concerned with saturated and unsaturated monocyclic and polycyclic hydrocarbon rings except for the six-membered ring, benzene, its polycyclic analogs, and related ring systems, which are taken up in chapter 7.

## Acceptable Nomenclature

*Saturated monocyclic hydrocarbons* are called *cycloalkanes* as a class. They are named by prefixing cyclo to the name of the saturated unbranched hydrocarbon (see chapter 5) with the same number of carbon atoms; names such as cyclopropane, cyclohexane, and cycloundecane are generated. Locant numbers are sequential around the ring. It is customary to draw the rings as polygons. For stereochemical aspects of these systems, see chapter 30.

Substitution for hydrogen in a parent cycloalkane is denoted in the usual way with attention to lowest locants for the substituent groups. Where a hydrocarbon substituent on a monocyclic ring is relatively large, or it is desired to emphasize such a substituent, the latter may be used as the parent with a cycloalkane substituting group as the substituent.

Pentylcyclobutane
1-Cyclobutylpentane

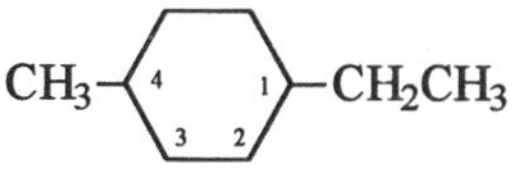

1-Ethyl-4-methylcyclohexane

Certain trivial names for substituted alicyclic hydrocarbons are acceptable in the field of terpenes (see chapter 31); examples are *p*-menthane (1-isopropyl-4-methylcyclohexane) and pinane (2,6,6-trimethylbicyclo[3.1.1]heptane).

*Unsaturated monocyclic hydrocarbons (cycloalkenes and cycloalkynes)* are named, like their acyclic counterparts (see chapter 5), by replacing the ending -ane of the corresponding cycloalkane name with endings such as -ene, -adiene, -yne, and so forth. The principle of lowest locants for the unsaturation is applied with preference for double bonds over triple bonds. A single point of unsaturation is given the locant 1, which does not need to be cited in the name. Substitution by hydrocarbon groups is denoted as above; locants are fixed by the ring unsaturation.

Cyclohexene

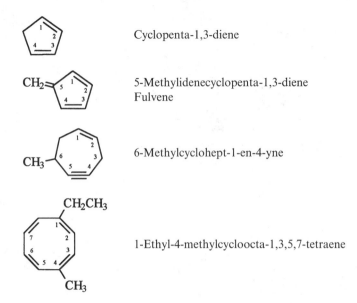

Cyclopenta-1,3-diene

5-Methylidenecyclopenta-1,3-diene
Fulvene

6-Methylcyclohept-1-en-4-yne

1-Ethyl-4-methylcycloocta-1,3,5,7-tetraene

Monocyclic hydrocarbon rings with seven or more ring atoms and maximum double-bond conjugation can be named as *annulenes*; the last example above would be 1-ethyl-4-methyl-[8]annulene, in which [8] indicates the number of carbon atoms in the ring. In rings with an odd number of carbon atoms, the extra hydrogen is cited as "indicated hydrogen", as in 1*H*-[9]annulene. This type of name is not used by CAS, although it is now allowed by IUPAC because of its value in fusion names (see chapter 7).

*Univalent substituting groups* derived from cycloalkanes and cycloalkenes are named by replacing -ane with -yl or the final "e" with -yl; the free valence is assigned the locant 1, although this locant is not expressed in the names of cycloalkyl (or cycloalkanyl) groups. In cycloalkenes, the free valence is cited by CAS, and it has precedence over locants for unsaturation or substituents. Names such as cyclohexyl or cyclohex-2-en-1-yl are generated.

5-Methylcyclohexa-2,4-dien-1-yl

*Bivalent substituting groups* in which both free valences are at the same ring carbon atom are given names with the ending -ylidene if the group is bound to a single atom of a parent hydrocarbon. If the carbon atom is bound to two different molecules, CAS also uses -ylidene, but IUPAC recommends the ending -diyl.

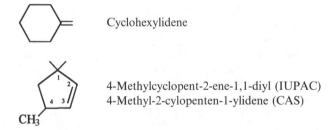

Cyclohexylidene

4-Methylcyclopent-2-ene-1,1-diyl (IUPAC)
4-Methyl-2-cylopenten-1-ylidene (CAS)

The ending -diyl is also used when the two free valences are at different ring atoms. The ending -ylene is now limited to methylene, ethylene, and phenylene; methylene refers to the structure –$CH_2$– where the two valences are attached to different atoms, although CAS uses methylene for both –$CH_2$– and $H_2C$= (see chapter 5). For example, cyclohex-1-ene-1,3-diyl and cyclohexane-1,4-diyl are used rather than "cyclohex-1-en-1,3-ylene" and "cyclohexan-1,4-ylene", respectively. Free valences are given lowest locants.

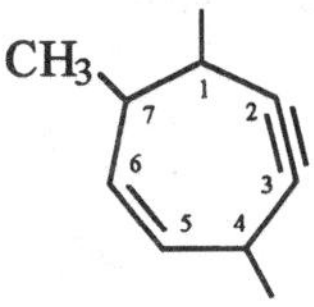

7-Methylcyclohept-5-en-2-yne-1,4-diyl

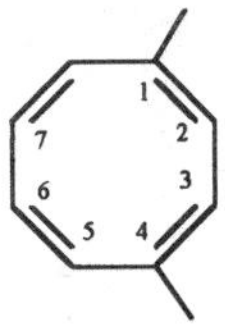

Cycloocta-1,3,5,7-tetraene-1,4-diyl
[8]Annulene-1,4-diyl

The endings -triyl, -tetrayl, and combinations of -yl and -ylidene, and so on, are used in other monocyclic multivalent groups in a manner similar to their acyclic analogs.

Cyclohexan-1-yl-4-ylidene

*Bridged alicyclic hydrocarbons*[1] consist of a "main ring" plus one or more bonds, atoms, or unbranched chains connecting nonadjacent atoms within the system. The points of connection are called "bridgeheads". Such systems are polycyclic structures; the number of rings is determined by the minimum number of bond scissions necessary to create an acyclic structure with the same number of atoms. If two scissions are required, the structure is termed "bicyclo", three gives a "tricyclo system", and so on. An example of a bicyclo system follows:

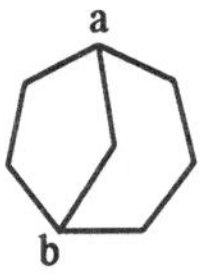

Junctions "a" and "b" are bridgeheads and are connected by as many atoms of the structure as possible to form the main ring and by a single atom, called the "main bridge". In this example, the entire system corresponds to an acyclic "octane". This system is named with the prefix bicyclo and a bracketed expression having, in descending order, the number of atoms in each segment of the main ring and the number of atoms in the bridge, followed by the name of the acyclic analog: the example above is bicyclo[3.2.1]octane.

Locants in a bicyclic system begin at a bridgehead atom and proceed around the main ring via the longest path to the second bridgehead and so on to the first, followed by numbering of the main bridge, beginning with the atom nearest the first bridgehead. Unsaturation and substituents are cited in the usual way.

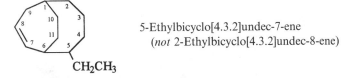

5-Ethylbicyclo[4.3.2]undec-7-ene
(*not* 2-Ethylbicyclo[4.3.2]undec-8-ene)

More complex systems have additional bridges other than a main bridge, which is the longest bridge; the main bridge should divide the main ring as symmmetrically as possible. The additional bridges are called "secondary bridges"; their lengths are also cited, usually in

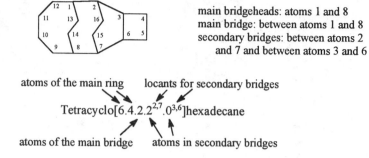

main bridgeheads: atoms 1 and 8
main bridge: between atoms 1 and 8
secondary bridges: between atoms 2
    and 7 and between atoms 3 and 6

**Figure 6.1.** A polyalicyclic ring system.

descending order, within the brackets along with pairs of superscript numbers denoting positions of their attachment to the rest of the structure.

For the purpose of locating secondary bridges, unsaturation, or substituents, after numbering of the system as described above, secondary bridges are numbered in order of decreasing value of their higher bridgehead locants, each beginning with the atom nearer the higher numbered bridgehead. If two secondary bridges are attached to the same higher numbered bridgehead, the bridge attached to the higher numbered bridgehead at the other end of the bridge is numbered first. As usual, the lowest set of locants is desired; the superscript locants for the secondary bridges should be as small as possible.

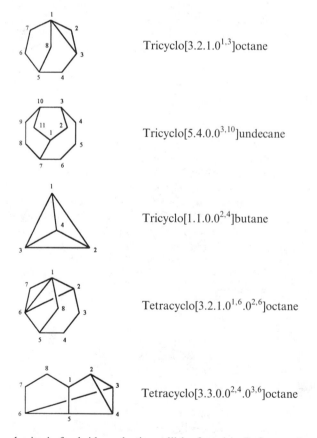

Tricyclo[3.2.1.0$^{1,3}$]octane

Tricyclo[5.4.0.0$^{3,10}$]undecane

Tricyclo[1.1.0.0$^{2,4}$]butane

Tetracyclo[3.2.1.0$^{1,6}$.0$^{2,6}$]octane

Tetracyclo[3.3.0.0$^{2,4}$.0$^{3,6}$]octane

Additional criteria for bridge selection will be found in Reference 1.

Locants for free valences and unsaturation are fixed by the numbering described above; free valences are preferred for low locants. IUPAC and CAS differ slightly in citing unsaturation when an arene is part of the system, as illustrated by example 10 in the additional examples at the end of this chapter.

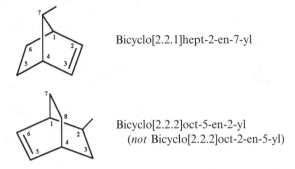

Bicyclo[2.2.1]hept-2-en-7-yl

Bicyclo[2.2.2]oct-5-en-2-yl
(*not* Bicyclo[2.2.2]oct-2-en-5-yl)

There are often a number of ways to draw the structures of polycyclic hydrocarbons, and in a few cases certain structures have generated obvious trivial names. The following are illustrative; in these specific instances, the trivial names are acceptable.

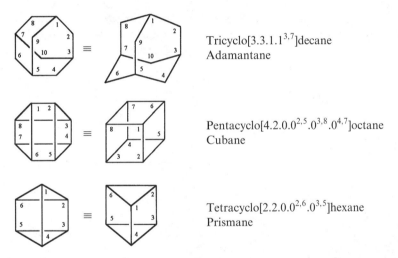

Tricyclo[3.3.1.1$^{3,7}$]decane
Adamantane

Pentacyclo[4.2.0.0$^{2,5}$.0$^{3,8}$.0$^{4,7}$]octane
Cubane

Tetracyclo[2.2.0.0$^{2,6}$.0$^{3,5}$]hexane
Prismane

Certain bicyclic hydrocarbons that can also be viewed as saturated analogs of fused hydrocarbons (see chapter 7) are best named as hydrogenated derivatives of the fused systems. Examples are decahydroazulene (not bicyclo[5.3.0]decane), dodecahydroheptalene (not bicyclo[5.5.0]dodecane), and decahydronaphthalene (not bicyclo[4.4.0]decane).

*Spiro alicyclic systems*[2] contain pairs of alicyclic rings with a single atom in common; that atom is termed a "spiro atom" or a "free spiro union". *Monospiro hydrocarbons* have two alicyclic rings joined in this manner. The name of the spiro system is based on the name of the acyclic hydrocarbon with the same number of carbon atoms, prefixed by spiro followed by (in brackets and in ascending order), the number of carbon atoms in each chain of atoms linking the spiro atom. Numbering of the system begins with the carbon atom adjacent to the spiro atom in the smaller ring, if one is smaller, and proceeds around the smaller ring, through the spiro atom, and around the second ring.

An alternative method of naming bicyclic spiro hydrocarbons, no longer recommended by IUPAC but still acceptable, cites the name of the larger ring followed by spiro and the name of the smaller ring. By this method, each ring retains its original numbering, including the spiro atom (assigned the lowest locant), with primes assigned to the locant numbers of the second-cited ring.

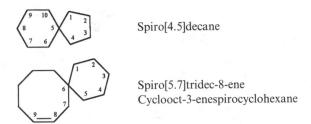

Spiro[4.5]decane

Spiro[5.7]tridec-8-ene
Cyclooct-3-enespirocyclohexane

*Polycyclic spiro hydrocarbons* consisting of three or more monocyclic rings joined through spiro atoms take the prefixes dispiro, tetraspiro, and so on, depending on the number of spiro atoms. Names are again based on the name of the corresponding acyclic hydrocarbon. Following the spiro prefix, the numbers of atoms joined to the spiro atoms are cited in brackets, beginning with the smallest terminal ring and proceeding by the shortest path through each of the spiro atoms to the other terminal ring and back to the origin. Locant numbering of the system follows the same path.

Dispiro[4.1.6.2]pentadecane

The appropriate spiro term may also be followed, in brackets, by the names of the components, cited in succession, beginning with the terminal ring lowest in alphabetical order. Each ring retains its own set of locants with unprimed numbers alloted to the first cited ring, and primed and doubly-primed locants, and so on, given to the successively cited rings. The example above would be named dispiro[cycloheptane-1,1'-cyclopentane-3',1"-cyclopentane]. This method is also applicable to branched spiro systems, but in complex systems, ambiguity can occur.

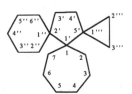

Trispiro[cycloheptane-1,1'-cyclopentane-2',1"-cyclohexane-5',1'''-cyclopropane]

If the individual rings are identical, a name may be formed by locants denoting the spiro atoms followed by the prefix spirobi or, as has been proposed,[2] dispiroter, and so on, depending on the number of rings, and, in brackets, the name of the component ring. Locants for each ring are unprimed, primed, doubly-primed, and so on, beginning at a terminal ring and proceeding across the system.

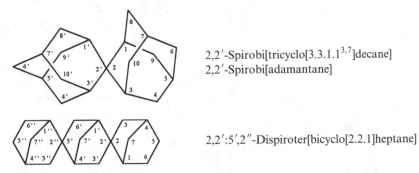

2,2'-Spirobi[tricyclo[3.3.1.1³,⁷]decane]
2,2'-Spirobi[adamantane]

2,2':5',2"-Dispiroter[bicyclo[2.2.1]heptane]

Where one or more rings in a spiro system is a bridged alicyclic ring system, the alternative method for naming polycyclic spiro hydrocarbons is used.

Dispiro[cycloheptane-1,2′-bicyclo[2.2.1]heptane-5′,1″-cyclopentane]

For spiro systems involving fused rings, see chapter 7. Assemblies of alicyclic rings joined by single or double bonds are considered in chapter 8.

## Discussion

The term "alicyclic", a combination of "aliphatic" and "cyclic", generally refers to both saturated and unsaturated cyclic compounds. However, when the unsaturation is such that the structure or parts of the structure can be viewed as an arene or a hydrogenated arene, special measures are taken to avoid overly complex names resulting from the usual methods of alicyclic nomenclature.

In naming bridged alicyclic hydrocarbons, the "polycyclo" method (also known as the "von Baeyer system") can be difficult to use where the structures are equivalent to saturated fused arenes (see chapter 7) with hydrocarbon bridges. It is best to employ a prefix such as perhydro (or *n*-hydro with locants if the system is not fully saturated) and to name the bridges with terms such as methano (–$CH_2$–), ethano (–$CH_2CH_2$–), but[2]eno (–$CH_2CH=CHCH_2$–), and so on.

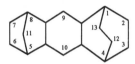

Perhydro-1,4-ethano-5,8-methanoanthracene
Tetradecahydro-1,4-ethano-5,8-methanoanthracene

A class of cyclic compounds, often called *cyclophanes*, may have saturated or unsaturated rings as members of a larger cyclic structure. "Polycyclo" names are applicable but can be cumbersome if the smaller ring members are arenes, since the unsaturation must be cited. Nomenclature for this class is discussed in chapter 7.

Under the latest IUPAC proposals, unsaturation in a spiro system may be cited by considering each ring as a separate parent hydrocarbon or by viewing the entire system as a parent and citing the unsaturation after the closing bracket (see example 12 in the following section). The latter method has the advantage of allowing unsaturation to be treated in the same way as a group named as a suffix.

## Additional Examples

1.

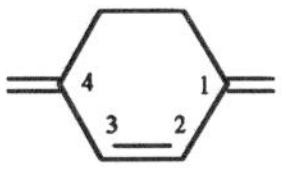

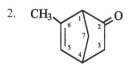

Cyclohex-2-ene-1,4-diylidene

2.   CH₃

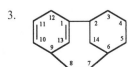

6-Methylbicyclo[2.2.1]hept-5-en-2-one

3.

Tricyclo[7.3.1.1²,⁶]tetradeca-1(13),10-diene

4.

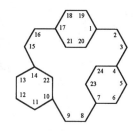

Tetracyclo[15.2.2.2$^{4,7}$.1$^{10,14}$]tetracosane

5.

5-Methyltricyclo[4.1.0.0$^{2,7}$]hept-2-ene

6.

Tricyclo[2.1.1.0$^{1,4}$]hexane

7.

Tricyclo[5.1.0.0$^{2,4}$]octane

8.

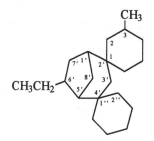

6′-Ethyl-3-methyldispiro[cyclohexane-1,2′-
   bicyclo[3.2.1]octane-4′,1″-cyclohexane]

9.

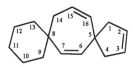

Dispiro[4.2.5.3]hexadeca-2,6,15-triene

10.

Bicyclo[6.2.2]dodeca-8,10,11-triene (CAS)
Bicyclo[6.2.2]dodeca-1(10),8,11-triene (IUPAC)
   (*not* Bicyclo[6.2.2]dodeca-1(11),8(12),9-triene)

11.

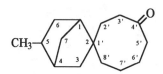

5-Methylspiro[bicyclo[2.2.1]heptane-2,1′-
   cyclooctan]-4′-one

12.

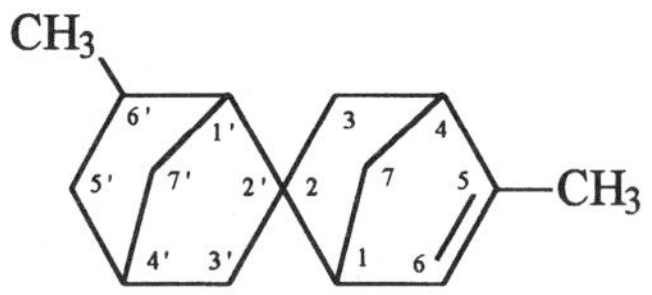

5,6′-Dimethylspiro[bicyclo[2.2.1]heptane-2,2′-bicyclo[2.2.1]hept-5-ene]

5,6′-Dimethyl-2,2′-spirobi[bicyclo[2.2.1]hept]-5-ene

## REFERENCES

1. International Union of Pure and Applied Chemistry, Organic Chemistry Division, Commission on the Nomenclature of Organic Chemistry. Extension and Revision of the von Baeyer System for Naming Polycyclic Compounds (Including Bicyclic Compounds) (IUPAC Recommendations 1999). *Pure Appl. Chem.* **1999**, *71*, 513–529.
2. International Union of Pure and Applied Chemistry, Organic Chemistry Division, Commission on the Nomenclature of Organic Chemistry. Extension and Revision of the Nomenclature of Spiro Compounds (IUPAC Recommendations 1999). *Pure Appl. Chem.* **1999**, *71*, 531–568.

# 7

# Arenes (Aromatic Hydrocarbons)

*Arenes* or *aromatic hydrocarbons* in this chapter include benzene and its derivatives and fused unsaturated polycyclic hydrocarbon ring systems. The latter consist of two or more rings fused to each other through two or more adjacent ring atoms. Arenes can be represented formally by cyclic structures containing the maximum number of alternating (conjugated) double bonds, called noncumulative double bonds by IUPAC.

Included in this chapter are partially or fully hydrogenated polycyclic arenes, bridged arenes, and spiro ring systems containing polycyclic arenes as components.

Polycyclic ring systems in which the rings are joined directly but do not share atoms are known as ring assemblies (see chapter 8). The nomenclature of monocyclic rings other than benzene that have the maximum number of alternating (conjugated) double bonds is taken up in chapter 6 on alicyclic compounds.

## Acceptable Nomenclature

### Monocyclic Arenes

The most common monocyclic arene has the trivial name benzene, a name much preferred to cyclohexa-1,3,5-triene (see chapter 6). Alkyl derivatives of benzene are named systematically with lowest number locants or by using *o*- (ortho = 1, 2-), *m*- (meta = 1, 3-), or *p*- (para = 1, 4-) where there are only two identical substituents. Names such as *p*-chlorotoluene are acceptable, but not encouraged. In general, substitutive nomenclature is used where there are substituents on the side chains of alkylbenzenes. If there is a principal characteristic group on the side chain, conjunctive nomenclature (see chapter 3) may be used in which Greek letter locants are used in the side chains, with α adjacent to the principal group.

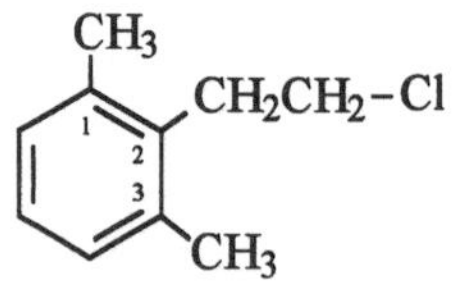

2-(2-Chloroethyl)-1,3-dimethylbenzene

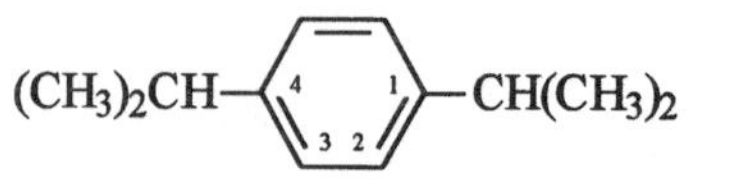

1,4-Diisopropylbenzene
*p*-Diisopropylbenzene

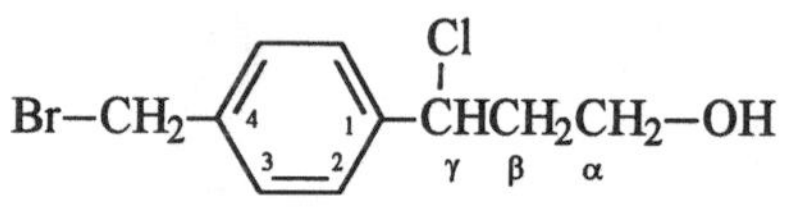

4-(Bromomethyl)-γ-chlorobenzenepropanol
3-[4-(Bromomethyl)phenyl]-3-chloro-
    propan-1-ol

Some alkyl and alkenyl derivatives of benzene have well-established trivial names. With further substitution limited by IUPAC to the ring, they include toluene, $C_6H_5$–$CH_3$; xylene (three isomers), $C_6H_4(CH_3)_2$; mesitylene, $1,3,5$-$C_6H_3(CH_3)_3$; and styrene, $C_6H_5$–$CH=CH_2$. However, substitution on the methyl groups of toluene, the xylenes, and mesitylene can be indicated by the locants $\alpha$, $\alpha'$, and $\alpha''$. The vinyl group of styrene should not be substituted; $\alpha$ and $\beta$ have been used, but inconsistently.

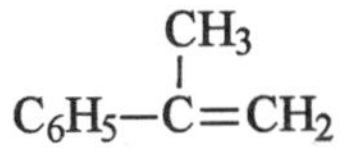

$$C_6H_5-\overset{\overset{\displaystyle CH_3}{|}}{C}=CH_2$$

Isopropenylbenzene
Prop-1-en-2-ylbenzene (IUPAC)
(1-Methylvinyl)benzene
(1-Methylethenyl)benzene (CAS)
(*not* $\alpha$-Methylstyrene)

Other derivatives of benzene, which should not be further substituted, are

$C_6H_5$–$CH(CH_3)_2$          Cumene

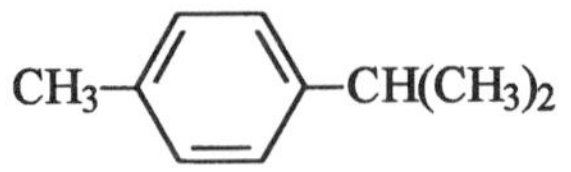

$CH_3$—⟨ring⟩—$CH(CH_3)_2$          Cymene (*p*-isomer shown)

$C_6H_5$–$CH=CH$–$C_6H_5$          Stilbene

Where the acyclic part of the structure is larger than the arene portion or it is desired to place emphasis on the acyclic part, the structure may be named as an aryl derivative of the acyclic hydrocarbon. This method is also advantageous where more than one aryl group is attached to a hydrocarbon chain.

$C_6H_5$–$CH_2$–$C_6H_5$          Diphenylmethane

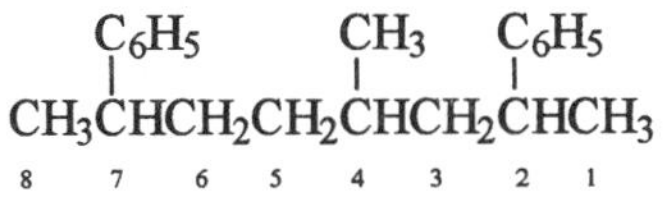

$$\underset{8\quad7\quad6\quad5\quad4\quad3\quad2\quad1}{CH_3\overset{\overset{\displaystyle C_6H_5}{|}}{C}HCH_2CH_2\overset{\overset{\displaystyle CH_3}{|}}{C}HCH_2\overset{\overset{\displaystyle C_6H_5}{|}}{C}HCH_3}$$          4-Methyl-2,7-diphenyloctane

*Substituting groups* derived from monocyclic arenes are called *aryl groups*. They are generally named by the principles used for alicyclic ring systems (see chapter 6). The trivial names phenyl, $C_6H_5$–, (*o*-, *m*-, *p*-)tolyl, $CH_3$–$C_6H_4$–, and (*o*-, *m*-, *p*-)phenylene, –$C_6H_4$–, are acceptable, but trivial names such as "xylyl" or "mesityl" for other aryl groups should not be used.

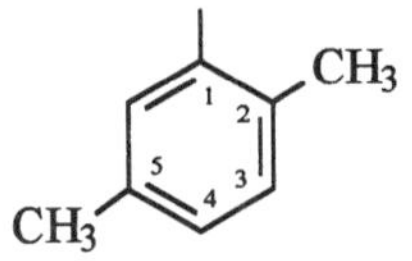

2,5-Dimethylphenyl
(*not p*-Xylyl)

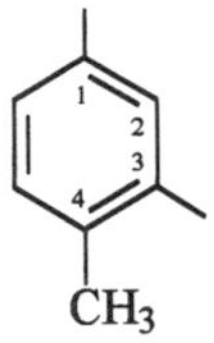

4-Methyl-1,3-phenylene
4-Methyl-*m*-phenylene
(*not* 3,4-Tolylene)

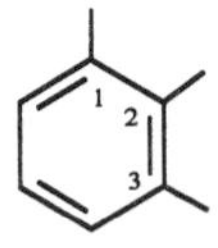

Benzene-1,2,3-triyl

Where a free valence resides on a side chain, the group is best named as an aryl derivative of the acyclic group. Certain trivial names are acceptable: benzyl, $C_6H_5-CH_2-$; phenethyl (a contraction of 2-phenylethyl), $C_6H_5-CH_2CH_2-$; styryl, $C_6H_5-CH=CH-$; cinnamyl, $C_6H_5-CH=CHCH_2-$. Although substitution of side chains, except for styryl, may be indicated by $\alpha$, $\beta$, and so on (with $\alpha$ adjacent to the free valence), this practice is declining and not encouraged. The names benzhydryl, $(C_6H_5)_2CH-$, and trityl, $(C_6H_5)_3C-$, are acceptable only in unsubstituted form. Closely related multivalent group names derived from these trivial names, such as benzylidene, $C_6H_5-CH=$ or $C_6H_5-CH<$; phenethylidene, $C_6H_5-CH_2CH=$ or $C_6H_5-CH_2CH<$; benzhydrylidene, $(C_6H_5)_2-C=$ or $(C_6H_5)_2-C<$; benzylidyne, $C_6H_5-C\equiv$,

$C_6H_5-\overset{|}{C}=$, or $C_6H_5-C\overset{/}{\underset{\backslash}{}}$; and phenethylidyne, $C_6H_5-CH_2-C\equiv$, $C_6H_5-CH_2-\overset{|}{C}=$, or

$C_6H_5-CH_2-C\overset{/}{\underset{\backslash}{}}$, are also allowed.

| | |
|---|---|
| $CH_3-$⟨ring⟩$-CH_2CH_2-$ | 2-(4-Methylphenyl)ethyl |
| | 2-(p-Tolyl)ethyl |
| | 4-Methylphenethyl |

$$\underset{3\quad 2\quad 1}{CH_3\overset{\overset{\displaystyle C_6H_5}{|}}{C}HCH_2-}$$  2-Phenylpropyl

$$\underset{3\quad 2\quad 1}{CH_3\overset{\overset{\displaystyle C_6H_5}{|}}{C}=CH-}$$  2-Phenylprop-1-en-1-yl

$$\underset{5\quad 4\quad\;\; 3\; 2\quad 1}{=CHCH=\overset{\overset{\displaystyle C_6H_5}{|}}{C}CH_2CH_2-}$$  3-Phenylpent-3-en-1-yl-5-ylidene

Multivalent compound group names are formed by the additive method (see chapter 3).

| | |
|---|---|
| $-$⟨ring⟩$-CH_2-$⟨ring⟩$-$ | Methylenedi-4,1-phenylene |
| | Methylenedi-p-phenylene |

| | |
|---|---|
| ⟨ring⟩ with $CH_2-$ and $CH_2-$ | 1,2-Phenylenebis(methylene) |

## Polycyclic Arenes

The polycyclic arenes form an exceedingly large class of hydrocarbons. There are, for example, nearly 1500 ways to fuse eight benzene rings together.[1] The nomenclature of *polycyclic arenes* is built on a foundation of parent polycyclic arenes having well-established systematic and trivial names. Approved polycyclic hydrocarbon names are shown in table 7.1. In table 7.2, these names are listed in ascending order of seniority for use as parent arenes (or base components) in the generation of fused hydrocarbon names. These hydrocarbons can be substituted in the same way as benzene.

Some of the arene names in table 7.1 are formed in a systematic way. For example, a linear series of benzene rings fused through ortho carbon atoms is named by attaching a numerical prefix to the ending -acene, as hexacene. Two linear series of two or more benzene rings fused to a single benzene ring at the 1,2 and 3,4 positions, respectively, are named by a numerical prefix and the ending -aphene (pentaphene). Arenes formed by an ortho fusion of two identical monocyclic rings are named by attaching a numerical prefix for the number of atoms in each

Table 7.1. Parent Polycyclic Arenes

| | |
|---|---|
| 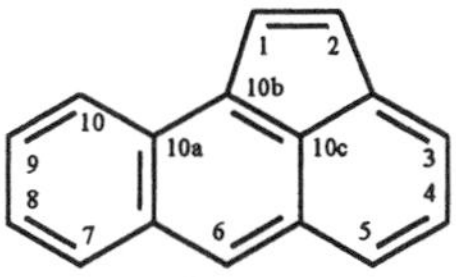 | aceanthrylene |
| 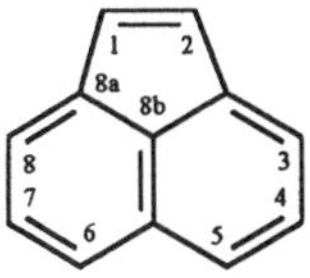 | acenaphthylene |
| 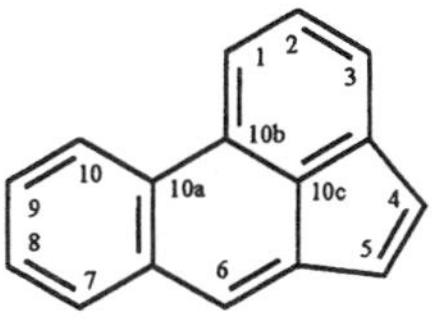 | acephenanthrylene |
| 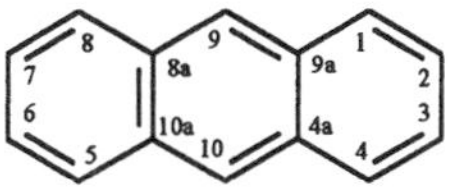 | anthracene |
| 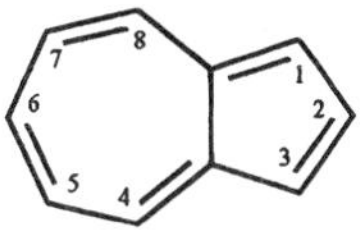 | azulene |
| 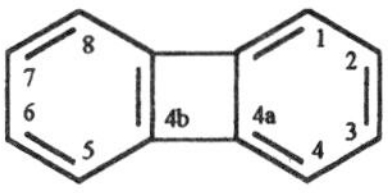 | biphenylene |
| 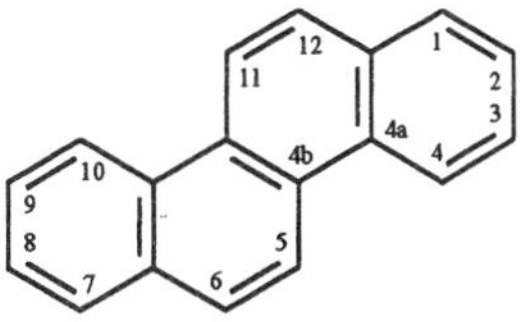 | chrysene |
| 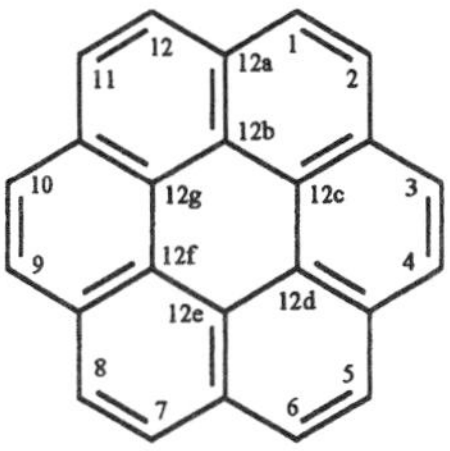 | coronene |
| 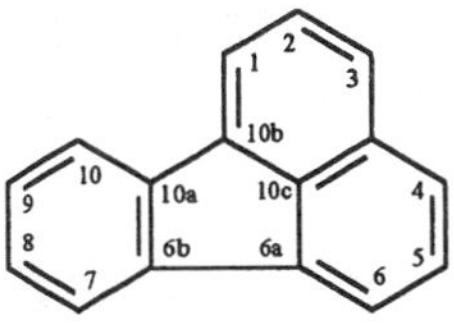 | fluoranthene |
| 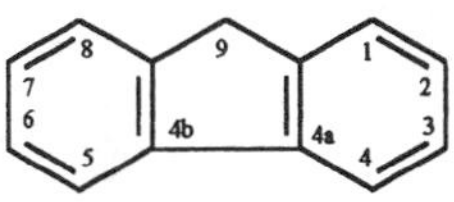 | fluorene (9H- isomer shown) |

*(Continued)*

Table 7.1. Continued

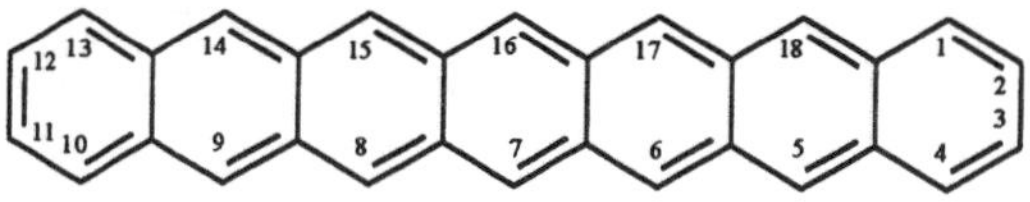

heptacene

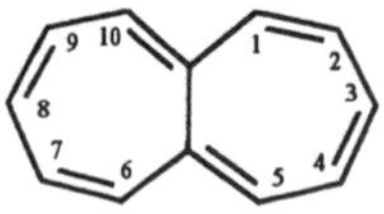

heptalene

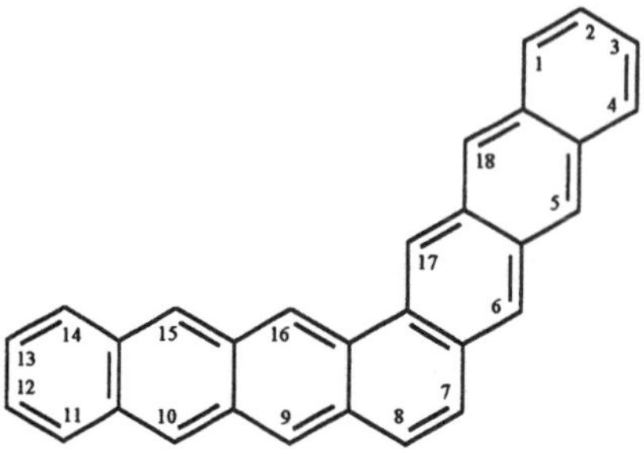

heptaphene

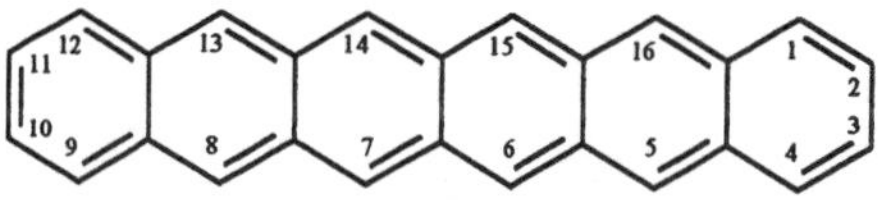

hexacene

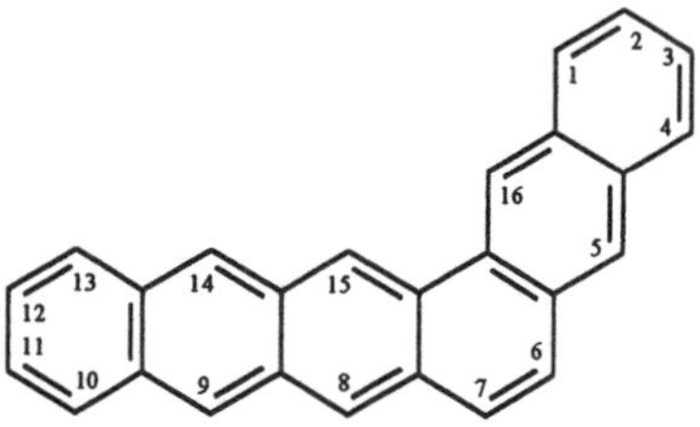

hexaphene

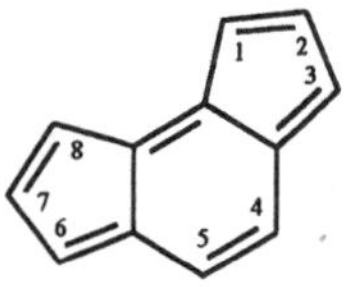

as-indacene

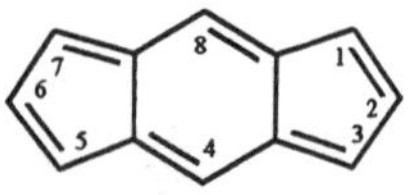

s-indacene

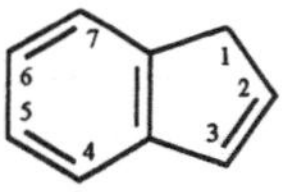

indene (1H- isomer shown)

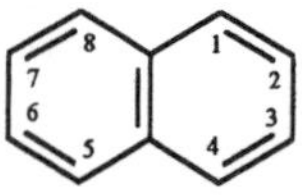

naphthalene

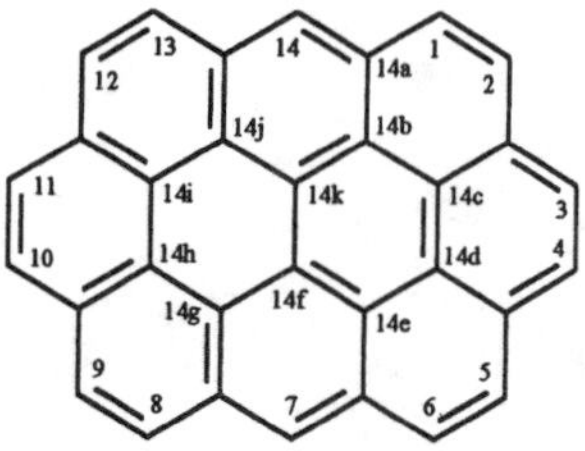

ovalene

(Continued)

Table 7.1. Continued

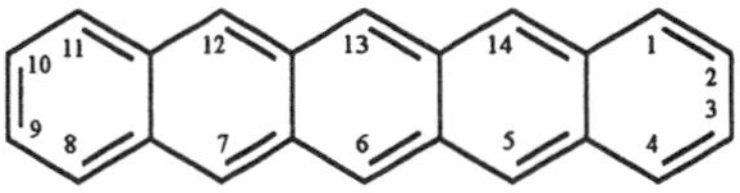

pentacene

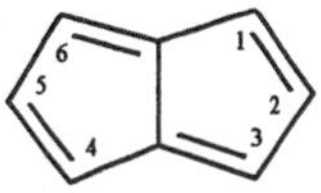

pentalene

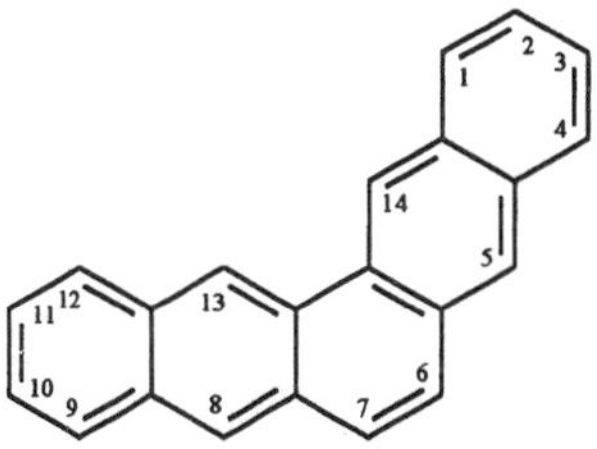

pentaphene

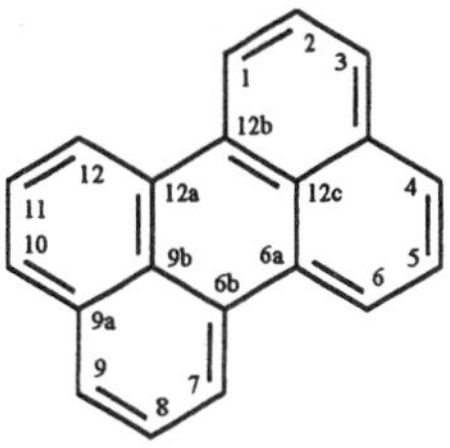

perylene

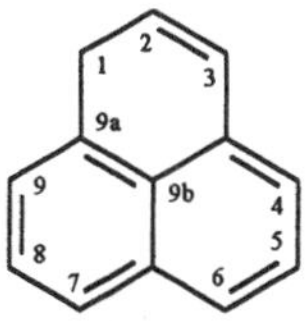

phenalene (1*H*- isomer shown)

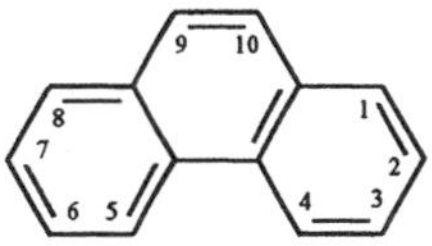

phenanthrene

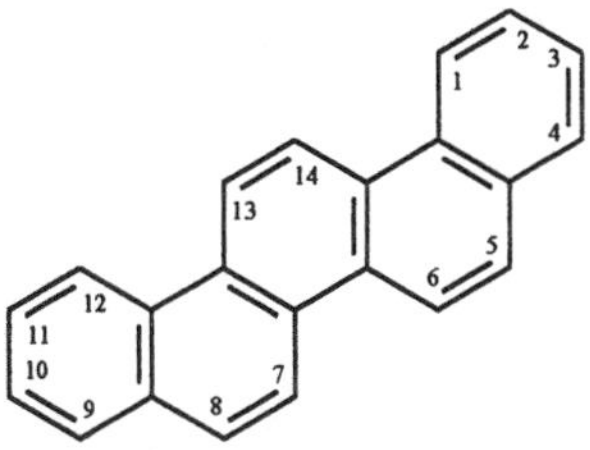

picene

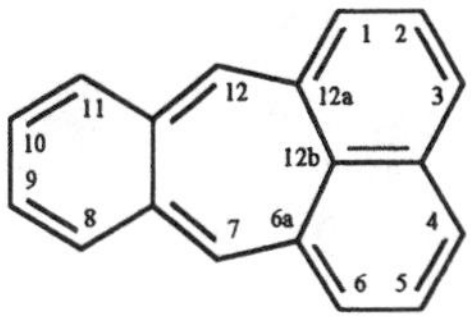

pleiadene

*(Continued)*

Table 7.1. Continued

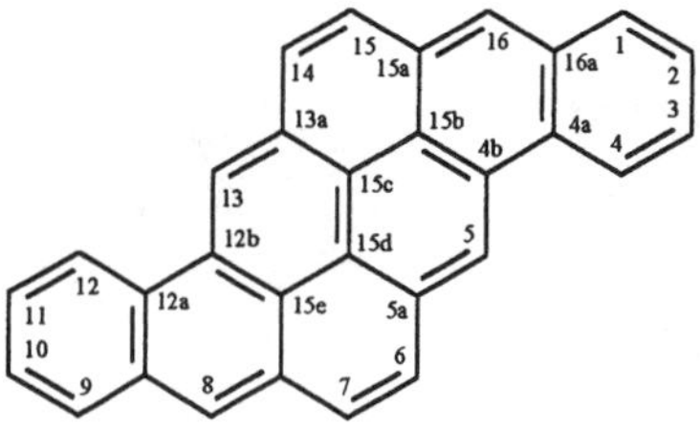

pyranthrene

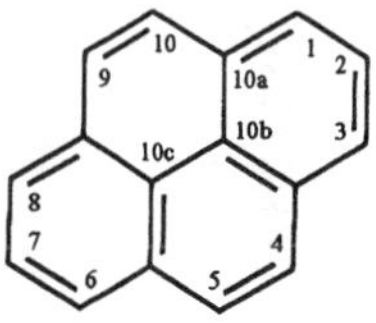

pyrene

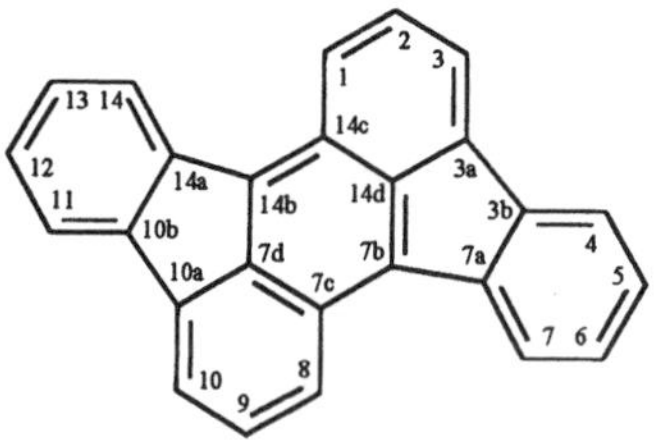

rubicene

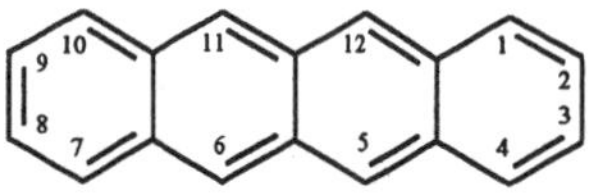

tetracene (IUPAC)
naphthacene (CAS)

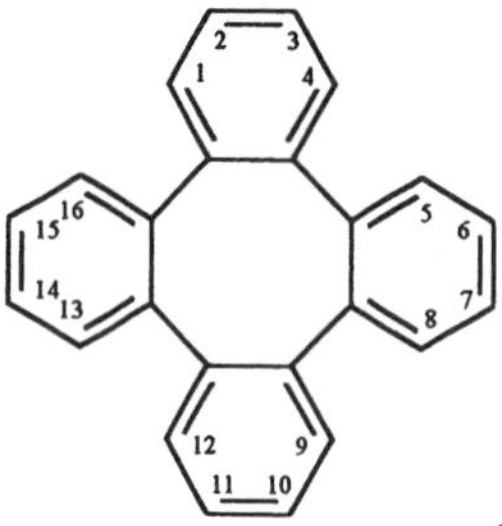

tetraphenylene

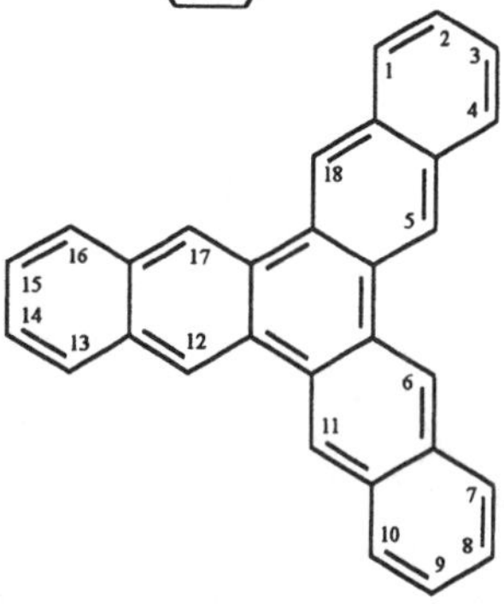

trinaphthylene

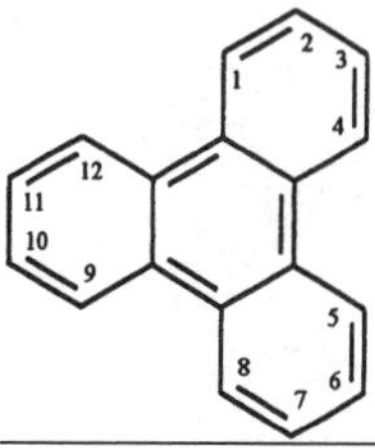

triphenylene

Table 7.2. Parent Arenes in Ascending Order of
Precedence in Fusion Names

Columns are to be read newspaper style

| | | |
|---|---|---|
| pentalene | anthracene | pentaphene |
| indene | fluoranthene | pentacene |
| naphthalene | acephenanthrylene | tetraphenylene |
| azulene | aceanthrylene | hexaphene |
| heptalene | triphenylene | hexacene |
| biphenylene | pyrene | rubicene |
| *as*-indacene | chrysene | coronene |
| *s*-indacene | tetracene (IUPAC) | trinaphtylene |
| acenaphthylene | naphthacene (CAS) | heptaphene |
| fluorene | pleiadene | heptacene |
| phenalene | picene | pyranthrene |
| phenanthrene | perylene | ovalene |

ring to the ending -alene (heptalene); the trivial name naphthalene is an exception. A mono-cyclic hydrocarbon with an even number of carbon atoms fused on alternate sides to benzene rings results in a polycyclic hydrocarbon that is named by adding a numerical prefix denoting the number of benzene rings to the ending -phenylene, as in biphenylene. Naphthalene rings similarly fused only at the 2,3 positions results in hydrocarbons that can be named with the ending -naphthylene, as in trinaphthylene. An additional method, recently adopted by IUPAC but not represented in table 7.1, uses a numerical prefix and the ending -helicene to name a helical arrangement of more than five benzene rings fused through ortho carbon atoms. CAS does not use helicene names, and until 1993 recognized the additional trivial name trindene (cyclopent[*e*]-*as*-indacene or tricyclopenta[*a,c,e*]benzene), placing it between anthracene and fluoranthene in table 7.2.

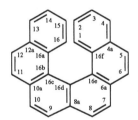

Hexahelicene (IUPAC)
Phenanthro[3,4-*c*]phenanthrene (CAS)

These arenes and other monocyclic structures are used as components which are fused together to form more complex polycyclic arenes. Their names are called *fusion names*. They take the form "fused-component[*x*]parent-component", where "*x*", in brackets, represents the manner and location of the shared ring atoms of the fusion. An example of a fusion name is benz[*a*]anthracene, in which a benzene ring is fused to the 1,2 side of anthracene.

To form a fusion name for an arene more complex than those given in table 7.1, a parent component must first be selected. This will be the portion of the structure that is most senior, as illustrated in table 7.2. To designate the location of attached components, the sides of the parent component rings are lettered "a" for the 1,2 side, "b" for the 2,3 side, continuing clockwise around the periphery of the parent component. This procedure follows the number-ing of the ring, including fusion atoms, except where the numbering is not systematic, as in the case of anthracene and phenanthrene.

Benzene and other monocyclic hydrocarbons are also used as parent components. The mono-cyclic hydrocarbon components take the ending -ene, which signifies the maximum number of alternating double bonds in fusion names. Potential confusion between the monounsaturated hydrocarbon cyclooctene, and "cyclooctene", the parent fusion component, can be alleviated by the use of annulene names (see chapter 6) as monocyclic parent fusion components.

The remaining portion(s) of the complex arene are denoted by prefixes derived from the names in table 7.1. They are cited alphabetically, with the first alphabetically matched to the lowest possible letter in the parent component. Monocyclic components with the maximum number of alternating double bonds have prefix names formed by combining "cyclo" with a numerical term for the number of carbon atoms in the ring: cyclopenta, cycloocta, and so on, except that the six-carbon ring is named benzo. Annuleno is not yet approved for use as a component prefix. Prefix names for polycyclic components are formed by replacing the ending -ene in the hydrocarbon name with -eno. A few abbreviated forms are used: benzo for benzeno, anthra for anthraceno, naphtho for naphthaleno, and phenanthro for phenanthreno; acenaphtho and perylo are also acceptable. In practice, the "a" or "o" at the end of these prefixes is usually deleted if the parent arene name begins with a vowel. A lowest locant number pair in the attached component is used to locate the position of its fusion to the parent structure, although citation of these locants is unnecssary if there is no ambiguity, as with benzo and other monocyclic hydrocarbon fusion prefixes. Once the fused structure has been named, it is renumbered (see below), ignoring the original numbering of the components, for the purpose of providing locants for substituents.

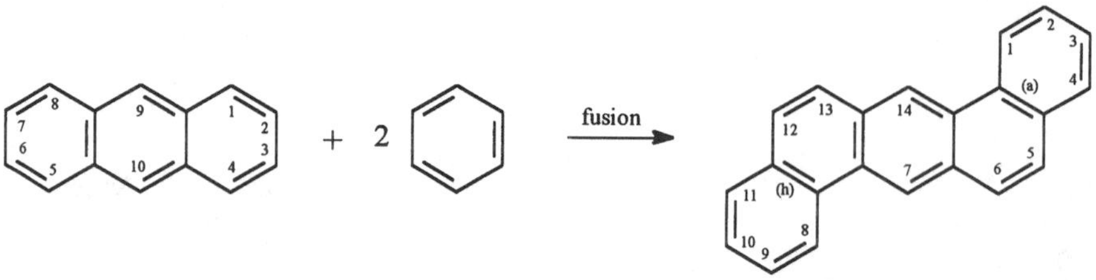

Anthracene                                          Dibenz[a,h]anthracene
(parent component)                                  (final numbering)

For numbering, the complete structure should be oriented in such a way that the maximum number of rings lie in a horizontal row, and a maximum number of rings lie above and to the right of this horizontal row. Numbering begins at an atom not involved in fusion at the most counterclockwise position in the top ring furthest to the right and proceeds clockwise around the periphery of the system. Angular atoms take the preceding locant followed by "a", "b", and so on. CAS numbers interior atoms similarly, taking the highest available primary locant from the periphery (see table 7.1 for examples). IUPAC[2] recommends that the locant for an interior atom be that of the nearest peripheral atom with a superscript numeral to denote the number of single or double bonds between them.

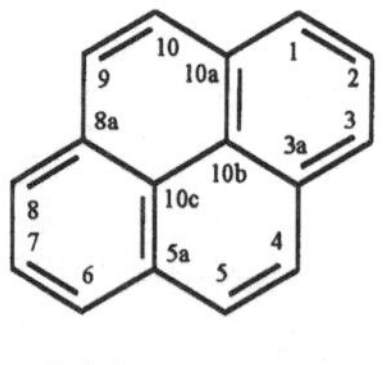

CAS Numbering 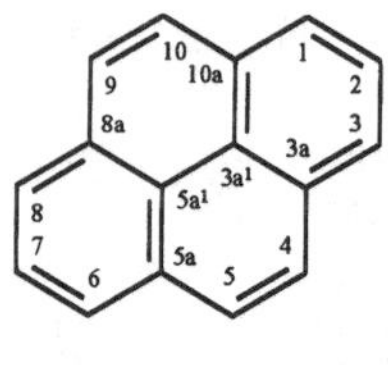                                       IUPAC Numbering

For complicated systems, CAS has published additional rules for the orientation and numbering of interior atoms.[3]

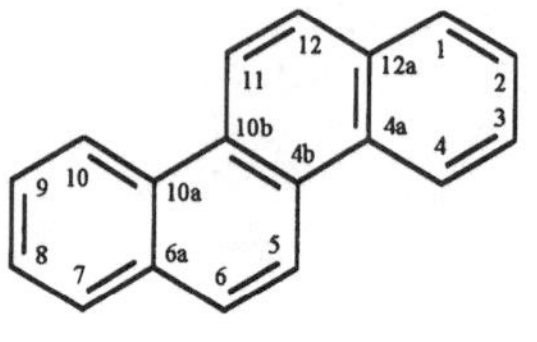

Chrysene

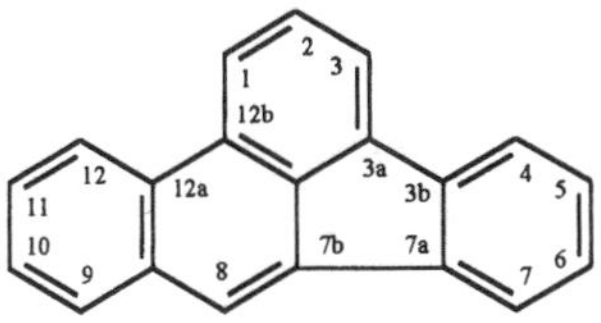

Benz[*e*]acephenanthrylene

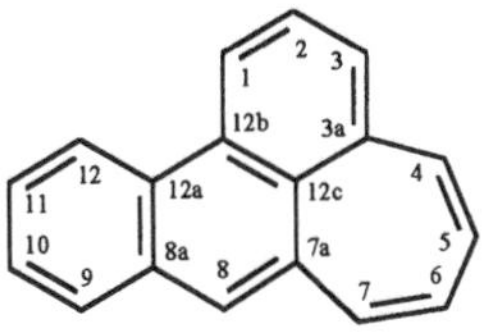

Cyclohepta[*jk*]phenanthrene

Some polycyclic arenes have isomeric forms, depending on the location of an "extra" hydrogen atom (see chapter 2) that usually appears as a $-CH_2-$ group in the ring structure. Fluorene and phenalene are examples in table 7.1. In a name, this "extra" or indicated hydrogen is cited ahead of the name by a prefix consisting of the appropriate locant and an italic *H*; this device becomes part of the name and is retained in place even when the extra hydrogen is substituted.

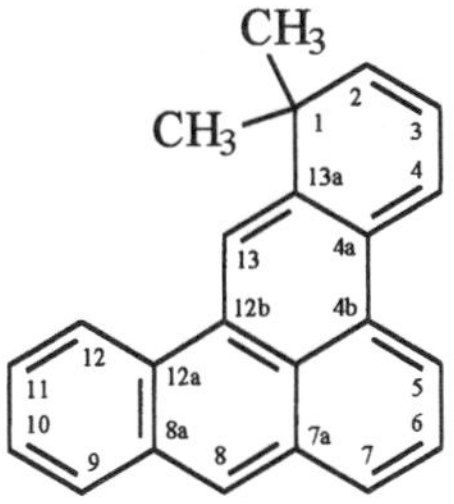

1,1-Dimethyl-1*H*-dibenz[*a,de*]anthracene

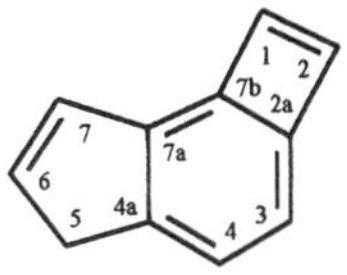

5*H*-Cyclobut[*e*]indene

Polycyclic arenes with fewer than the maximum number of alternating double bonds are named as derivatives of the corresponding polycylic arene with prefixes such as dihydro, tetrahydro, and so on. Perhydro is acceptable for complete hydrogenation, although it is not used by CAS. In the absence of substituents, lowest locants are given to these prefixes, but if indicated hydrogen is needed, it takes precedence for numbering.

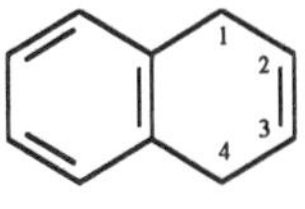

1,4-Dihydronaphthalene

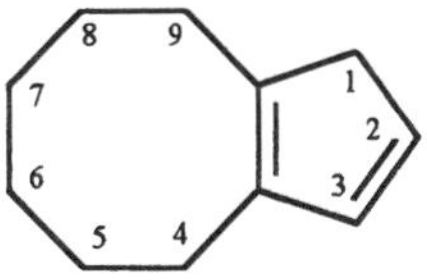

4,5,6,7,8,9-Hexahydro-1*H*-cyclopentacyclooctene

The names tetralin for 1,2,3,4-tetrahydronaphthalene and decalin for perhydronaphthalene are commonly used alternatives. In addition, although the alternative "hydro" names are preferred, trivial names for the following compounds are recognized:

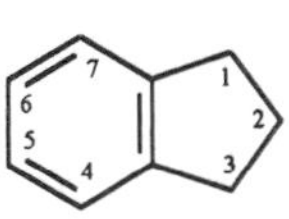

2,3-Dihydro-1*H*-indene
Indan

1,2-Dihydroacenaphthylene
Acenaphthene

1,2-Dihydrobenz[*j*]aceanthrylene
Cholanthrene

1,2-Dihydroaceanthrylene
Aceanthrene

4,5-Dihydroacephenanthrylene
Acephenanthrene

Dehydrogenation to form a triple bond in an arene is denoted by use of the prefix didehydro with appropriate locants to generate names such as 1,2-didehydrobenzene, which is preferred to cyclohexa-1,3-dien-5-yne; the trivial name benzyne is also used for this compound.

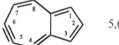

5,6-Didehydroazulene

*Bridged polycyclic arenes* are named by methods that differ from those used for bridged alicyclic hydrocarbons (see chapter 6). Bridges in polycyclic arenes are cited as prefixes derived from acyclic hydrocarbon names (see chapter 5) or from arene names by replacing the ending -ane or -ene with -ano or -eno. Locants in the names of bridges are enclosed in brackets since they refer to the bridge and not to the locants of the parent arene. Examples of bridge names are as follows:

| | |
|---|---|
| Methano | $-CH_2-$ |
| Metheno | $-CH=$   or   $-\overset{\vert}{\underset{\vert}{C}}H$ |
| Pentano | $-CH_2CH_2CH_2CH_2CH_2-$ |
| Etheno | $-CH=CH-$ |
| But[2]eno | $-\underset{4}{CH_2}\underset{3}{CH}=\underset{2}{CH}\underset{1}{CH_2}-$ |
| [1,4]Benzeno | |
| [2,3]Naphthaleno | |

[1,4]Benzenomethano

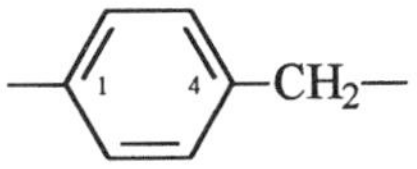

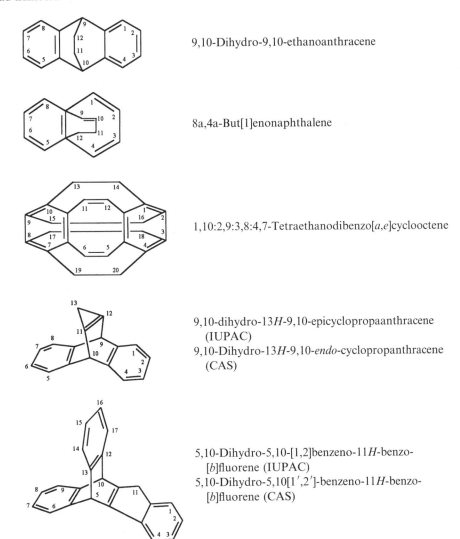

It seems reasonable to extend the alkano bridge names to their cyclic counterparts, but they are not used by CAS nor have they been adopted by IUPAC. Locants for the two points of the bridge that are attached to the arene, with lowest numbers if there is a choice, just precede the bridge prefix name. If the bridge is cyclic, these locants are followed, in brackets, by those of the bridge itself (CAS uses primes for these locants and a slightly different format, as in the fifth example below). If the name of a cyclic bridge is the same as its name as a fusion component, it is prefixed by the term epi- (IUPAC) or endo- (CAS). Atoms in the bridge are numbered consecutively, starting with the atom next to the bridgehead with the higher locant; for cyclic bridges, low numbers are given to the shorter path in the bridge ring. Where there are two or more bridges, they are numbered in decreasing order of their higher bridge-head numbers.

9,10-Dihydro-9,10-ethanoanthracene

8a,4a-But[1]enonaphthalene

1,10:2,9:3,8:4,7-Tetraethanodibenzo[a,e]cyclooctene

9,10-dihydro-13H-9,10-epicyclopropaanthracene
   (IUPAC)
9,10-Dihydro-13H-9,10-endo-cyclopropanthracene
   (CAS)

5,10-Dihydro-5,10-[1,2]benzeno-11H-benzo-
   [b]fluorene (IUPAC)
5,10-Dihydro-5,10[1′,2′]-benzeno-11H-benzo-
   [b]fluorene (CAS)

Where there is a choice, the largest arene is selected as the parent hydrocarbon, and a saturated bridge is preferred to the corresponding unsaturated bridge.

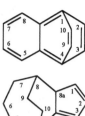

1,4-Ethenonaphthalene

4,5,6,7,8,8a-Hexahydro-4,8-ethanoazulene
(*not* 4,5,6,7-Tetrahydro-4,7-propano-1*H*-indene)

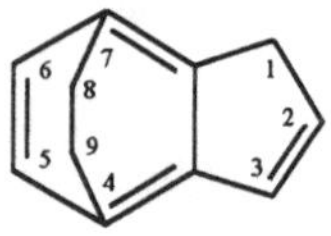

4,7-Ethano-1*H*-indene
(*not* 5,6-Dihydro-4,7-etheno-1*H*-indene)

*Spiro hydrocarbons involving polycyclic arenes* are named in the same way as those involving polycyclic alicyclic rings (see chapter 6).

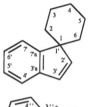

Spiro[cyclohexane-1,1′-[1*H*]indene]

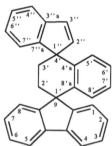

2′,3′-Dihydrodispiro[[9*H*]fluorene-9,1′-naphthalene-4′,1″-[1*H*]indene]

The names for *substituting groups* derived from polycyclic arenes are generated in much the same way as for monocyclic arenes. Carbon atoms with free bonds are given lowest locants consistent with the fixed numbering of the arene. Abbreviated names such as 2-naphthyl, 9-anthryl, and 2-phenanthryl are acceptable if they are not part of a fusion name.

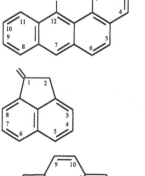

Benz[*a*]anthracen-12-yl

Acenaphthen-1-ylidene
Acenaphthylen-1(2*H*)-ylidene

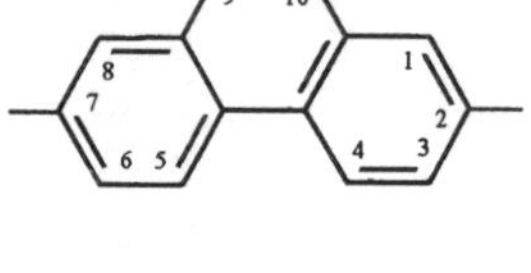

Phenanthrene-2,7-diyl

## Cyclophanes

Cyclophanes are a class of cyclic compounds in which an arene is connected at two or more positions by a closed system of bonds, atoms, acyclic chains, and/or other cyclic structures. The simplest case has one arene ring connected at two positions to an acyclic chain. Many are readily named as fused arenes or as unsaturated bridged acyclic hydrocarbons (see chapter 6).

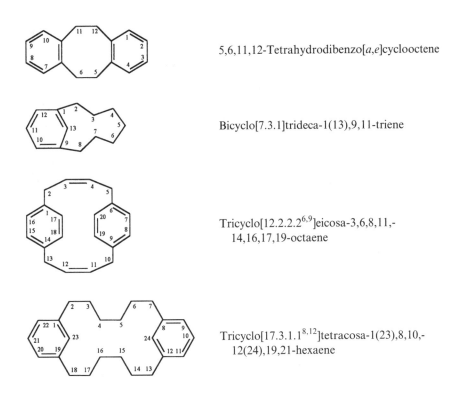

5,6,11,12-Tetrahydrodibenzo[*a,e*]cyclooctene

Bicyclo[7.3.1]trideca-1(13),9,11-triene

Tricyclo[12.2.2.2$^{6,9}$]eicosa-3,6,8,11,-14,16,17,19-octaene

Tricyclo[17.3.1.1$^{8,12}$]tetracosa-1(23),8,10,-12(24),19,21-hexaene

As complexity increases, as in the "calixarenes", these methods rapidly become unwieldy.

IUPAC is developing a method for naming cyclophanes[4] that can be summarized only briefly here. The basic idea is simple. Each arene ring is collapsed to a "superatom" that, along with the acyclic atoms, is named and numbered as an alicyclic system (see chapter 6). Where there is a choice, superatoms are preferred for low locants. The presence of a cyclophane system is indicated by changing the -ne ending of the alicyclic hydrocarbon to -phane.

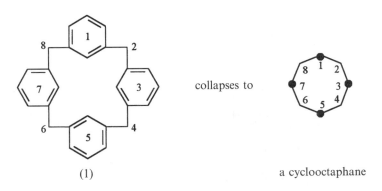

collapses to

(1)                                                           a cyclooctaphane

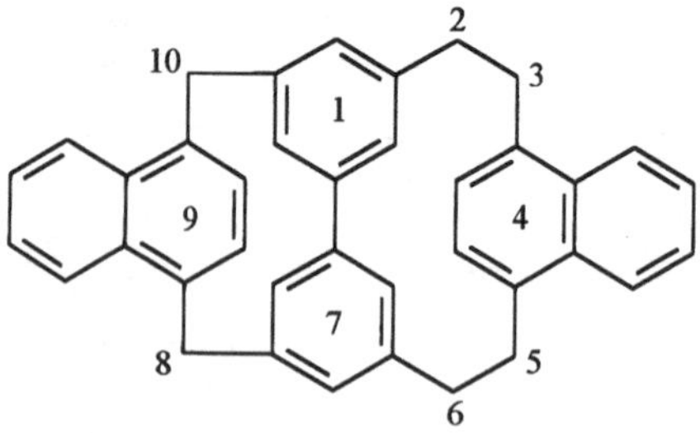

(2)                          collapses to                    a bicyclo[5.3.0]decaphane

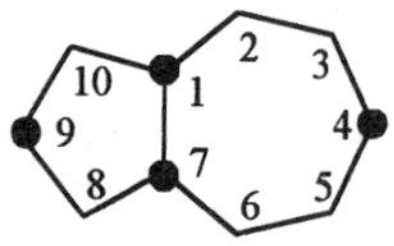

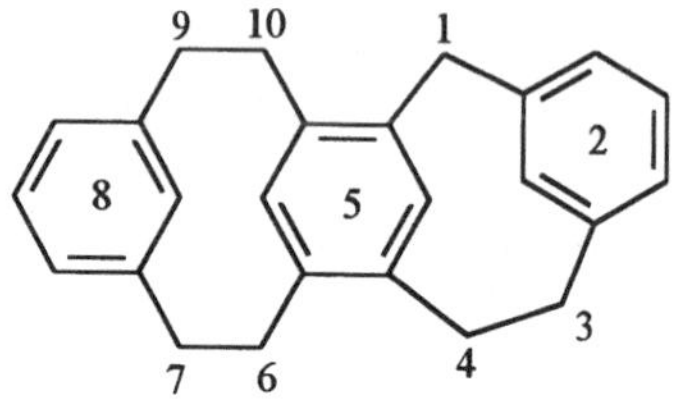

(3)                          collapses to                    a spiro[4.5]decaphane

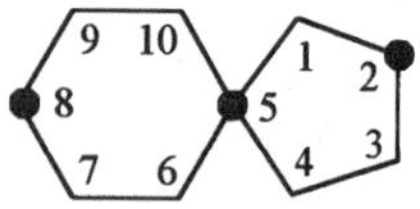

The arene rings represented in the "collapsed structure" by superatoms are described by prefixes derived from the name of the arene by replacing the final "e" of its name by "a", as in benzena and phenanthrena. These prefixes are cited in order of their ring seniority (see table 7.2) in front of the fundamental phane name as given above. The atoms of each arene that are attached to the normal atoms of the alicyclic chains are cited within parentheses immediately after the locant for the respective superatom. When there is a choice, the locant of the arene ring attached to the atom of the alicyclic system with the lower locant is cited first. Two or more identical locant pairs may be combined and cited after all superatom locants. Phane names and the corresponding systematic names for the examples above are as follows:

(1) 1,3,5,7(1,3)-Tetrabenzenacyclooctaphane
    Pentacyclo[19.3.1.1$^{3,7}$.1$^{9,13}$.1$^{15,19}$]octacosa-1(25),3,5,7(28),9,11,13(27)-
    15,17,19(26),21,23-dodecaene
(2) 4,9(1,4)-Dinaphthalena-1,7(1,3,5)-dibenzenabicyclo[5.3.0]decaphane
    6,11,12,19,20,25-Hexahydro-5,26:13,18-dietheno-7,10:21,24-dimethenobenzo[a]benzo-
    [7,8]cyclododeca[1,2-h]cyclotetradecene
(3) 2,8(1,3),5(1,3,4,6)-Tribenzenaspiro[4.5]decaphane
    Pentacyclo[10.9.1.1$^{2,11}$.1$^{4,8}$.1$^{15,19}$]pentacosa-1,4,6,8(25),11(24),12(22),15,17,19(23)-
    nonaene

In cyclophane names, locants on the arene ring are cited as superscripts to the superatom locant, as in $2^3$, $4^2$.

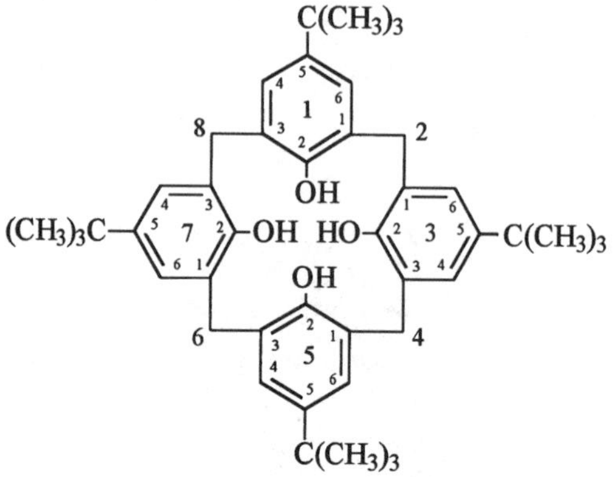

$1^5,3^5,5^5,7^5$-Tetra-*tert*-butyl-
1,3,5,7(1,3)-tetrabenzenacyclo-
octaphane-$1^2,3^2,5^2,7^2$-tetrol

Although IUPAC has not completed its recommendations, indicated *H* in an arene component could be cited with the appropriate arene prefix; it can also be cited in front of the phane name. Unsaturation in the acyclic portion of the phane name can be given by the usual endings and locants. For example, a phane name for example (3) above would be 1,6(1,4)-dibenzenacyclodecaphane-3,8-diene.

The principles of cyclophane nomenclature can be extended to chains of arenes interconnected by bonds, atoms, or hydrocarbon chains, by omitting "cyclo".

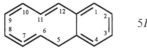

1,9(1),3,4,7(1,4)-Pentabenzenanonaphane

## Discussion

The family name "aromatic hydrocarbons" had its origin in the odor of certain compounds related to these hydrocarbons. "Aromaticity" and structure is the subject of some debate, and the smell of one's relatives seems a poor basis for a family name. Therefore, in this chapter "arene" has been adopted as the more structurally related term. For the present purpose, "arene" has been arbitrarily restricted to the six-membered ring, benzene, among monocyclic structures, although there are many others with a maximum number of alternating double bonds, such as cycloocta-1,3,5,7-tetraene. Such compounds can be named by the methods outlined in chapter 6.

Completely hydrogenated polycyclic arenes can be named with the perhydro prefix or hydro with appropriate numerical prefixes.

With the exception of biphenylene, pentalene, tetracene, pentaphene, and their analogs, the names in table 7.1 are the only trivial names retained by IUPAC. This number has been steadily decreasing in favor of longer and more complex systematic names.

In table 7.1, there are two important traditional exceptions to systematic numbering: anthracene and phenanthrene. This numbering is maintained for the corresponding perhydro hydrocarbons, and the positioning of substituents must be handled with care. Fused polycyclic hydrocarbons based on these parent arenes are, of course, renumbered, and as a consequence, the numbering of the parent is lost. Also, by tradition, a special numbering is used for the parent polycyclic structure cyclopenta[*a*]phenanthrene in steroid nomenclature (see chapter 31).

IUPAC has proposed a method, called the "δ-convention",[5] for indicating contiguous double bonds in a polycyclic arene name. Prefixed to the name of the arene is $x\delta^2$, where "x" is the locant for the carbon atom to which two double bonds are attached (see chapter 2).

5*H*-7$\delta^2$-Benzocyclodecene

The convention of dropping the "a" or "o" from the prefixes used in forming fusion names if the parent arene name begins with a vowel is no longer followed by IUPAC.

"Benzol", "toluol", and "xylol" are obsolete names for benzene, toluene, and xylene. They should not be used. The abbreviation *ar-* is sometimes used to indicate "aromatic" substitution when position is general or unknown. Abbreviations such as "ang" (angular) and "lin" (linear) to denote arrangements of rings in arenes should be avoided.

The nomenclature of *fullerenes*, which may be viewed as a class of arenes, is not yet fully developed. IUPAC is considering recommendations for naming and numbering fullerenes that appear to differ in some respects from the CAS methods. For CAS,[6] fullerenes are defined as carbon clusters with an even number of 20 or more atoms, each having a connectivity of three, that close upon themselves to form a spherical or distorted spherical structure. The sizes of the rings that make up the structure are indicated by bracketed numbers, cited in increasing numerical value, prefixed to the name "fullerene". The number of carbon atoms is added as a subscript to the capital letter "C" followed by the point group symmetry symbol[7] to differentiate isomeric structures, as in [5,6]fullerene-$C_{60}$-$I_h$ and [5,5]fullerene-$C_{20}$-$T_d$.

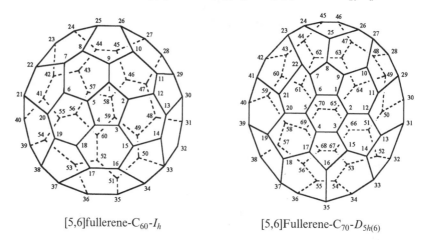

[5,6]fullerene-$C_{60}$-$I_h$               [5,6]Fullerene-$C_{70}$-$D_{5h(6)}$

There are two $C_{70}$-fullerenes with the same $D_{5h}$ point group symbol. They are distinguished by CAS by adding an italicized parenthetical subscript (5) or (6) to the point group symbol, indicating the types of rings that surround the five-membered ring on the principal symmetry axis. Fullerenes are numbered by CAS in a continuous spiral numbering, if possible. A preferred numbering starts with a ring, bond, or atom on the highest proper rotational symmetry axis. If no continuous spiral numbering is found with the highest preferred symmetry axis, other proper rotational axes are tried in order from highest to lowest.

Fullerenes do not have hydrogen atoms for substitution by atoms and groups; therefore hydro prefixes must be used to add hydrogen atoms that are replaced, for example, 1-fluoro-1,9-dihydro[5,6]fullerene-$C_{60}$-$I_h$. Fully saturated fullerenes are called fulleranes, which are named and numbered the same as the corresponding fullerenes, as in pentacosafluoro[5,6]fullerane-$C_{60}$-$I_h$. Fullerenes can be modified with some of the skeletal prefixes used for natural products (see chapters 3 and 31) such as nor (removal of a carbon atom), 1,9-dinor[5,6]fullerene-$C_{60}$-$I_h$; homo (addition of a carbon atom), 7,8(8a)-homo[5,6]fullerene-$C_{70}$-$D_{5h(6)}$; and seco (cleavage of a carbon–carbon bond), 1,2:5,6:11,12-triseco[5,6]fullerene-$C_{70}$-$D_{5h(6)}$. Arenes and heteroarenes can be fused to fullerenes, giving names such as 3′H,3″H-dicyclopropa[1,9:52,60][5,6]fullerene-$C_{60}$-$I_h$ and 2′H,5′H-[5,6]fullereno-$C_{60}$-$I_h$-[1,9-c]furan. Bridges can connect nonadjacent atoms of fullerenes, producing compounds with names such as 1,4-ethano[5,6]fullerene-$C_{70}$-$D_{5h(6)}$, and carbon atoms can be replaced by heteroatoms, generating names such as 1,6-diaza[5,6]fullerene-$C_{60}$-$I_h$.

## Additional Examples

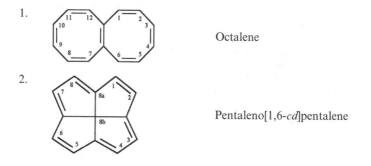

1.                                                 Octalene

2.                                                 Pentaleno[1,6-cd]pentalene

3.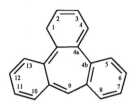

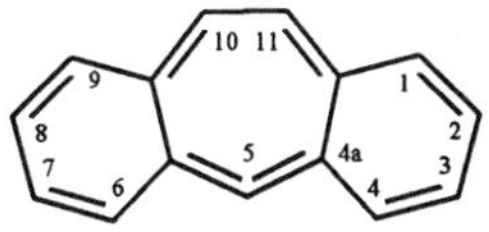

4a,5-Didehydro-4a*H*-dibenzo[*a,d*]cycloheptene
5δ²-Dibenzo[*a,d*]cycloheptene
4a,5-Didehydro-4a*H*-dibenzo[*a,d*][7]annulene

4.

1*H*-Tribenzo[*a,c,e*]cycloheptene
1*H*-Tribenzo[*a,c,e*][7]annulene

5.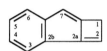

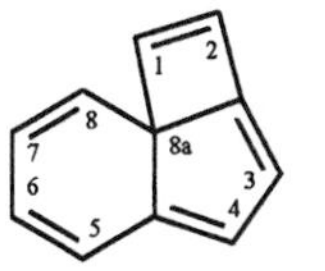

Cyclobut[*c*]indene

6.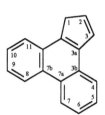

1*H*-Cyclobut[*a*]indene

7.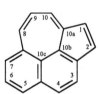

1*H*-Cyclopenta[*l*]phenanthrene

8.

Naphth[2,1,8-*cde*]azulene

9.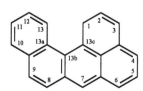

1,12-Ethenobenzo[4,5]cyclohepta[1,2,3-*de*]-
    naphthalene

10.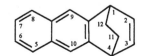

1H-Dibenz[*a,kl*]anthracene

11.

1,4-Dihydro-1,4-ethanoanthracene
    (*not* 1,2,3,4-Tetrahydro-1,4-etheno-
    anthracene)

12.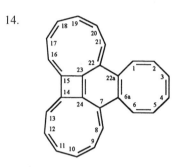

1,4:5,8-Dimethanonaphthalene

13.

1,4-Ethano-5,10:6,9-dimethanobenzo[3,4]-
    cyclobuta[1,2]cyclooctene
1,4-Ethano-5,10:6,9-dimethanobenzo[3,4]-
    cyclobuta[1,2][8]annulene

14.

7,14,15,22-Ethene[1,2,3,4]tetrayl-
    cyclooctacyclooctadecene

## REFERENCES

1. Balaban, A. T. Enumeration of Cyclic Graphs. In *Chemical Applications of Graph Theory*; Academic Press: London, 1976.
2. International Union of Pure and Applied Chemistry, Organic Chemistry Divisioin, Commission on the Nomenclature of Organic Chemistry. Nomenclature of Fused and Bridged Fused Ring Systems (Recommendations 1998). *Pure Appl. Chem.* **1998**, *70*, 143–216.
3. Gladys, C. L.; Goodson, A. L. Numbering of Interior Atoms in Fused Ring Systems. *J. Chem. Inf. Comput. Sci.* **1991**, *31*, 523–526.
4. International Union of Pure and Applied Chemistry, Organic Chemistry Division, Commission on the Nomenclature of Organic Chemistry. Phane Nomenclature. Part 1: Parent Phane Names (Recommendations 1998). *Pure Appl. Chem.* **1998**, *70*, 1513–1545.
5. International Union of Pure and Applied Chemistry, Organic Chemistry Division, Commission on the Nomenclature of Organic Chemistry. Nomenclature for Cyclic Organic Compounds with Contiguous Formal Double Bonds (δ-Convention) (Recommendations 1988). *Pure Appl. Chem.* **1988**, *60*, 1395–1401.
6. Goodson, A. L.; Gladys, C. L.; Worst, D. E. Numbering and Naming of Fullerenes by Chemical Abstracts Service. *J. Chem. Inf. Comput. Sci.* **1995**, *35*, 969–978.
7. Colthup, N. B.; Daly, L. K.; Wiberly, S. E. *Introduction to Infrared and Raman Spectroscopy*, 3rd ed.; Academic Press: San Diego, 1990; pp 115–119.

# 8

# Hydrocarbon Ring Assemblies

Two or more hydrocarbon rings or ring systems joined to each other directly by single or double bonds are termed *ring assemblies* if the structure has one more ring or ring systems than it has connecting bonds. The following are examples of ring assemblies:

but the following is a fused ring system:

By this definition, ring-containing macrocycles, including cyclophanes (see chapter 7), are not ring assemblies.

Heterocyclic ring assemblies are discussed in chapter 9, but the methods described below are generally followed.

## Acceptable Nomenclature

Ring assemblies consisting of identical rings or ring systems joined by single bonds have names based on that of the component ring or ring system. The name of the component hydrocarbon or, with two-component assemblies of monocycles, the corresponding substituting group name is used. A few names such as binaphthyl and bianthryl are acceptable, but for assemblies of benzene rings, only the substituting group name phenyl is employed. The ring name is preceded by locants denoting the junction points followed by a prefix such as bi-, ter-, quater-, and so on (see table 2.2, chapter 2), to show the number of components in the system; pairs of junction locants are separated by colons. The terminal component ring (or ring system) with the lowest junction locant is given unprimed locants, and the succeeding components are primed in sequence. The locants 1,1′ are usually not cited in the name of a two-component system of monocyclic rings. Locant numbers are preferred to *o*-, *m*-, or *p*- for assemblies containing benzene rings; in any case, numbers are required for placing substituents.

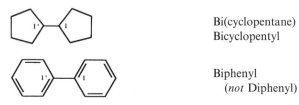

Bi(cyclopentane)
Bicyclopentyl

Biphenyl
  (*not* Diphenyl)

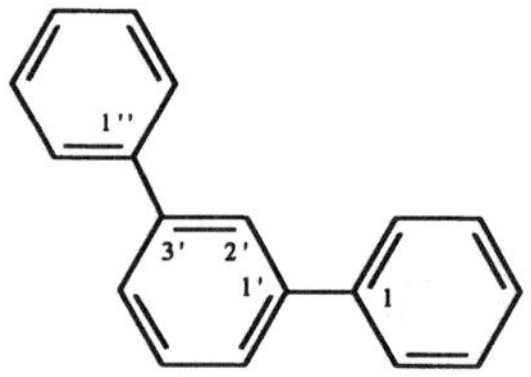

1,1':3',1''-Terphenyl
(*not m*-Terphenyl)

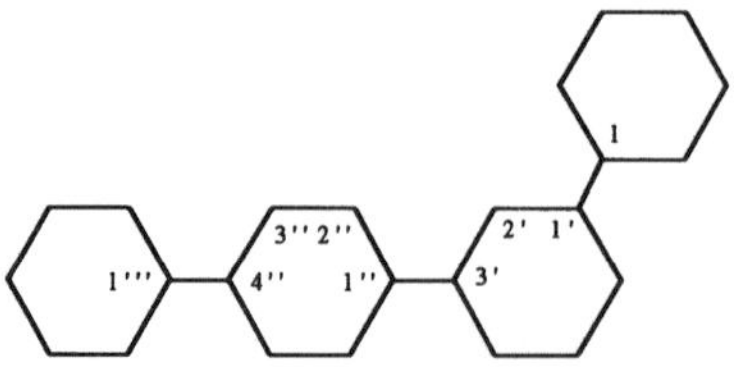

1,1':3',1'':4'',1'''-Quatercyclohexane
(*not* 1,1':4',1'':3'',1'''-Quatercyclohexane)

Parentheses are used in names like bi(cyclopentane) to avoid confusion with "bicyclo" names of bridged hydrocarbons (see chapter 6). In the last example above, the name given is preferred on the basis of lowest locants.

Where the individual rings have fixed locants, the junction locants for the assembly may differ from each other.

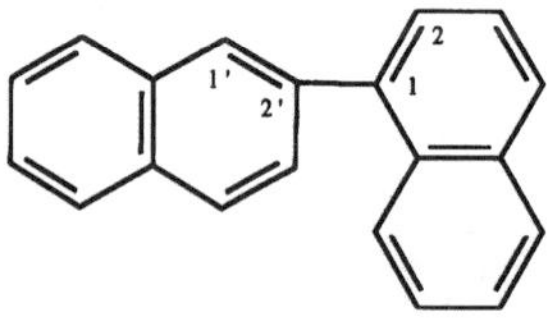

1,2'-Binaphthalene
1,2'-Binaphthyl

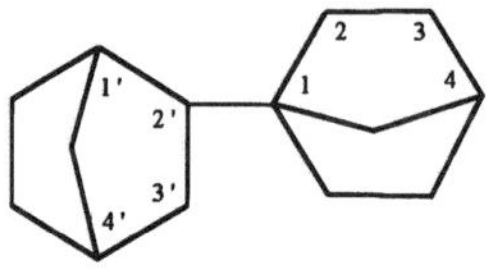

1,2'-Bibicyclo[2.2.1]heptane

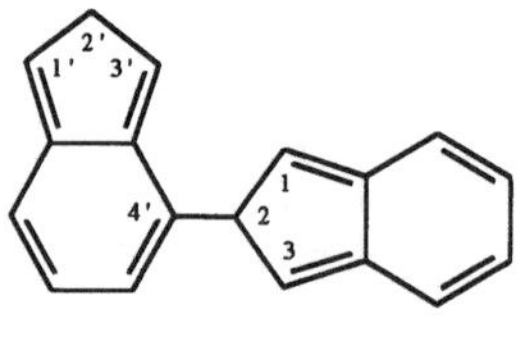

2,4'-Bi-2*H*-indene

In the latter example, the extra hydrogen indicated by "2*H*" must be at the same carbon in each ring. A possible modification allowing for different locations for the extra hydrogen would lead to names like 1*H*,2'*H*-2,4'-biindene and 1'*H*,2*H*-2,4'-biindene.

In symmetrical two-component assemblies derived from unsaturated hydrocarbons, locants for a name based on the parent hydrocarbon name are fixed by the unsaturation, whereas in a name based on the substituting group, the junction locants are those of the free valences of that group. Accordingly, a name based on the senior substituting group (see chapter 3) is probably the best choice. If the unsaturated assembly is unsymmetric, substitutive nomenclature with primed locants on the substituent is preferred.

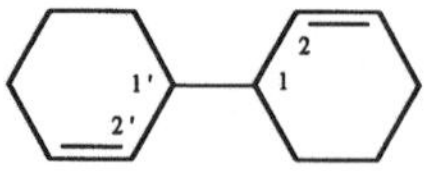

Bicyclohex-2-en-1-yl
3,3'-Bi(cyclohex-1-ene)

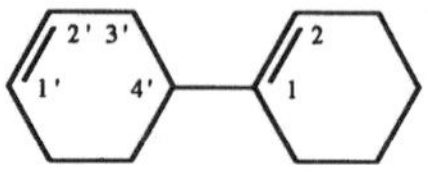

1-(Cyclohex-3-en-1-yl)cyclohex-1-ene
1,4'-Bi(cyclohex-1-ene)

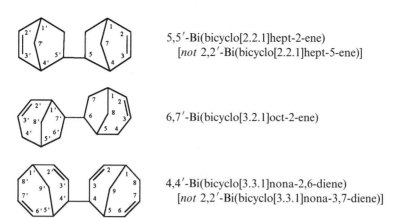

5,5'-Bi(bicyclo[2.2.1]hept-2-ene)
[*not* 2,2'-Bi(bicyclo[2.2.1]hept-5-ene)]

6,7'-Bi(bicyclo[3.2.1]oct-2-ene)

4,4'-Bi(bicyclo[3.3.1]nona-2,6-diene)
[*not* 2,2'-Bi(bicyclo[3.3.1]nona-3,7-diene)]

Partially hydrogenated polycyclic arene ring assemblies are named as hydro derivatives of the assembly.

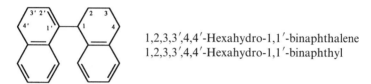

1,2,3,3',4,4'-Hexahydro-1,1'-binaphthalene
1,2,3,3',4,4'-Hexahydro-1,1'-binaphthyl

Two-component assemblies joined by a double bond are named by prefixing bi- to the name of the group with the ending -ylidene; these and assemblies with more than two components may also be named as substituted derivatives of one of the components. The method by which a double bond between components is indicated by a device such as $\Delta^{1,1'}$- prefixed to an assembly name based on parent components is no longer recommended.

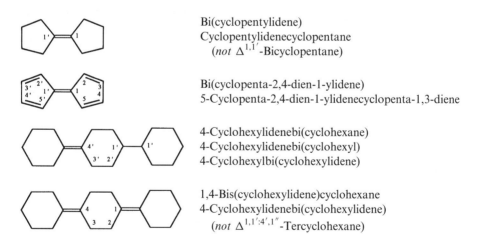

Bi(cyclopentylidene)
Cyclopentylidenecyclopentane
  (*not* $\Delta^{1,1'}$-Bicyclopentane)

Bi(cyclopenta-2,4-dien-1-ylidene)
5-Cyclopenta-2,4-dien-1-ylidenecyclopenta-1,3-diene

4-Cyclohexylidenebi(cyclohexane)
4-Cyclohexylidenebi(cyclohexyl)
4-Cyclohexylbi(cyclohexylidene)

1,4-Bis(cyclohexylidene)cyclohexane
4-Cyclohexylidenebi(cyclohexylidene)
  (*not* $\Delta^{1,1';4',1''}$-Tercyclohexane)

Substituents are preferentially given unprimed or lowest primed locants unless priming is fixed by the assembly itself. An unprimed locant is senior to the corresponding primed locant.

CH₃CH₂CH₂     CH₃

2-Methyl-3'-propylbiphenyl
  (*not* 3-Methyl-2'-propylbiphenyl;
    the locant set 2,3' is preferred to 2',3)

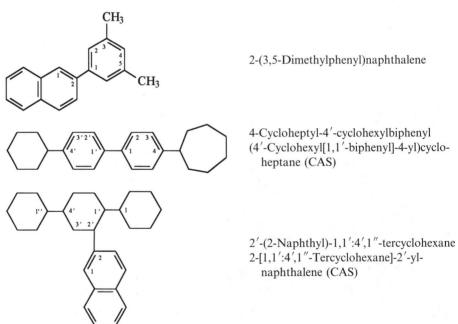

1′-Ethyl-4,4′-dimethyl-1,2′-binaphthalene
1′-Ethyl-4,4′-dimethyl-1,2′-binaphthyl

2,3′,4,4′,5-Pentamethylbiphenyl
   (The locant set shown is preferred to 2′,3,4,4′,5)

5′-Phenyl-1,1′:3′,1″-terphenyl

Ring assemblies with unlike rings or ring systems as components are named by selecting one component as the parent and naming the others as substituents. Selection of the parent may be based on, in descending order, (a) the component with the larger number of rings; (b) the component with the larger ring; and (c) the component with the lowest state of hydrogenation. A more complete list of criteria for ring seniority is given in chapter 3. CAS uses an alternative seniority order based on the following components: biphenyl > benzene > bicyclohexyl.

2-(3,5-Dimethylphenyl)naphthalene

4-Cycloheptyl-4′-cyclohexylbiphenyl
   (4′-Cyclohexyl[1,1′-biphenyl]-4-yl)cyclo-
      heptane (CAS)

2′-(2-Naphthyl)-1,1′:4′,1″-tercyclohexane
2-[1,1′:4′,1″-Tercyclohexane]-2′-yl-
      naphthalene (CAS)

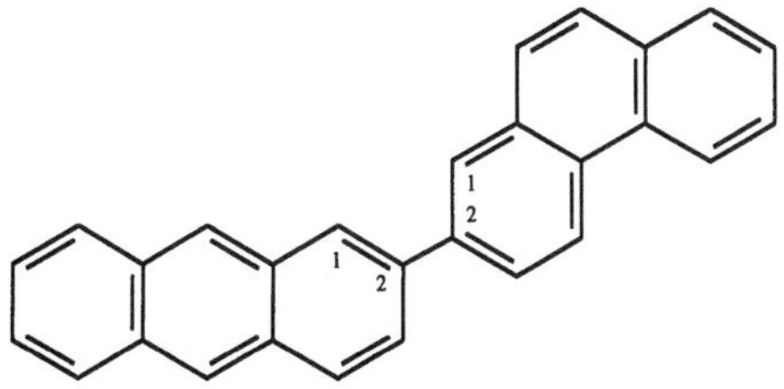

2-(Phenanthren-2-yl)anthracene
2-(2-Anthryl)phenanthrene

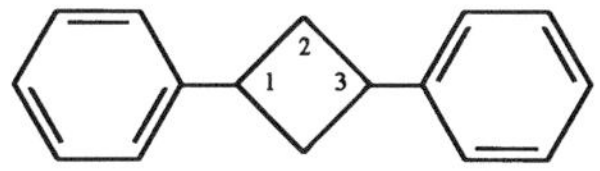

(3-Phenylcyclobutyl)benzene
1,3-Diphenylcyclobutane
Cyclobutane-1,4-diyldibenzene

Substituting group names derived from ring assemblies are formed in the usual way by adding, with its locant(s), -yl, -ylidene, -diyl, and so on to the name of the assembly; numbering of the assembly is retained. The use of parentheses or brackets around the parent assembly name often adds clarity.

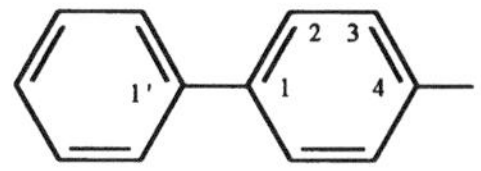

[1,1'-Biphenyl]-4-yl

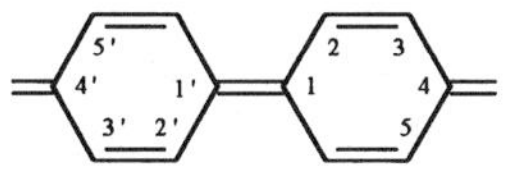

[Bicyclohexa-2,5-dien-1-ylidene]-4,4'-diylidene

[1,1':4',1"-Terphenyl]-4,4"-diyl

## Discussion

While both hydrocarbon and hydrocarbon substituting group names are acceptable as a basis for two-component ring assembly names, it is probably better to use the substituting group method for assemblies of monocycles. This is because, although the -yl endings indicate the presence of a free valence, numbering for unsaturation can be a problem.

Where the prefix bi- precedes a hydrocarbon name beginning with cyclo, the hydrocarbon name could be enclosed in parentheses to avoid possible confusion with bridged system names (see chapter 6) that begin with bicyclo. Parentheses are unnecessary when the assembly is named by the substituting group method.

CAS no longer names assemblies joined by double bonds as ring assemblies.

"Polyphenyls" has been used as a class name for ring assemblies composed of benzene rings joined by single bonds, and it constitutes a subject heading in the CAS General Subject Index. This practice should be discouraged because of the polymer implication. In structure-based polymer nomenclature (see chapter 29) the class would be properly named "polyphenylenes".

Names such as biphenylene or binaphthylene should not be used as ring assembly substituting groups; these names describe polycyclic hydrocarbons (see chapter 7) and not ring assemblies.

## Additional Examples

1.

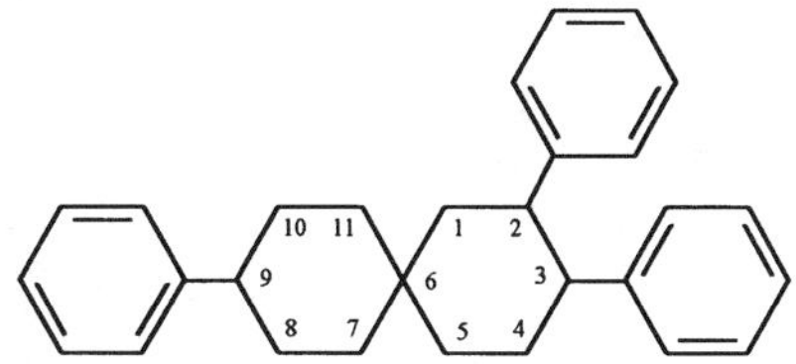

2,3,9-Triphenylspiro[5.5]undecane
[*not* 2',3',5',6'-Tetrahydro-4-
phenylspiro[cyclohexane-1,4'(1'*H*)-
1,1':2',1"-terphenyl]]

2.

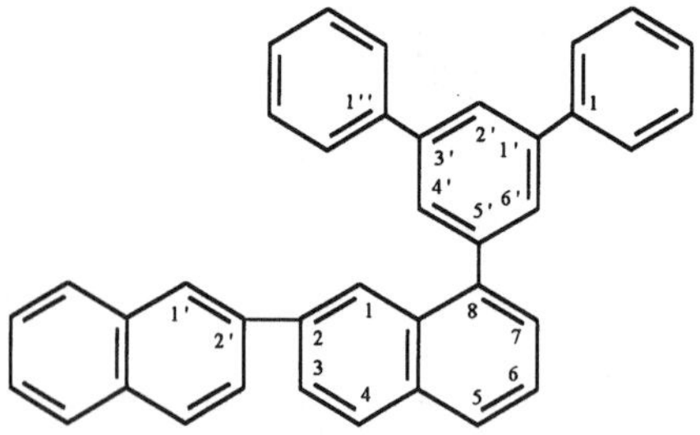

5′-[2,2′-Binaphthalen]-8-yl-
1,1′:3′,1″-terphenyl
8-[1,1′:3′,1″-Terphenyl]-5′-yl-2,2′-
binaphthalene

3.

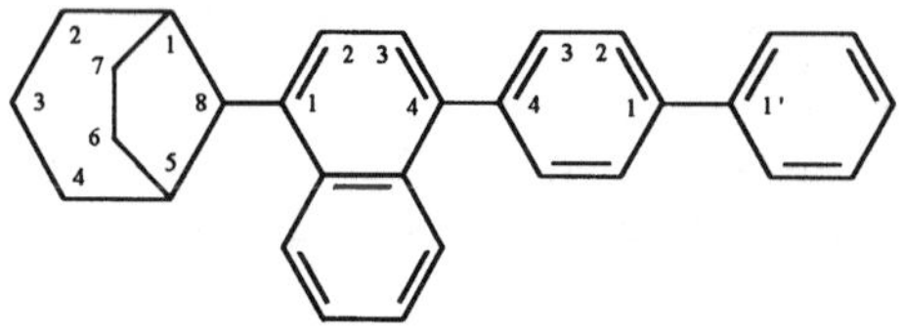

1-(Bicyclo[3.2.1]oct-8-yl)-4-[1,1′-
biphenyl]-4-ylnaphthalene
8-(4-[1,1′-Biphenyl]-4-ylnaphthalen-
1-yl)bicyclo[3.2.1]octane

4.

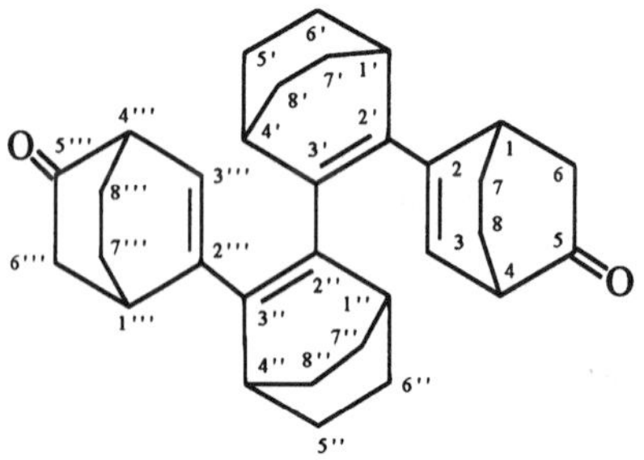

[2,2′:3′,2″:3″,2‴-Quaterbicyclo-
[2.2.2]oct-2-ene]-5,5‴-dione

5.

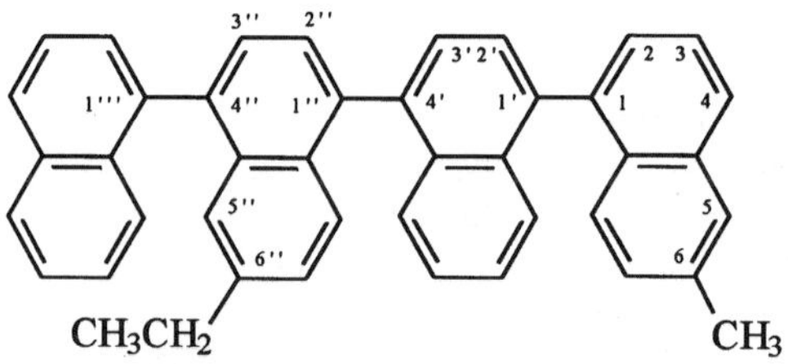

6″-Ethyl-6-methyl-1,1′:4′,1″:4″,1‴-
quaternaphthalene

# 9

# Heteroacyclic and Heterocyclic Compounds

In the previous four chapters, nomenclature for acyclic and cyclic hydrocarbons was presented. Structural analogs of hydrocarbons having one or more atoms of elements other than carbon, called heteroatoms, are considered in this chapter. Many heteroorganic compounds can be named, at least in part, by the principles described in the hydrocarbon chapters. Heterocyclic structures are frequently encountered in natural products (see chapter 31). Rings and chains not having carbon atoms, but which are often found as organic derivatives, are also included to a limited extent; examples in other chapters are the siloxanes (chapter 26), phosphazenes (chapter 25), and boron compounds (chapter 27).

## Acceptable Nomenclature

### Parent hydrides

The names of mononuclear heteroatomic hydrides used in substitutive nomenclature are given in table 9.1. A numerical term can be used with these names to form the name of an acyclic polynuclear parent hydride.

$$H_2PPH_2 \qquad \begin{array}{l} \text{diphosphane} \\ \text{diphosphine} \end{array}$$

$$H_3SSSH_3 \qquad 1\lambda^4,3\lambda^4\text{-trisulfane}$$

See chapter 2 for the use of $\lambda$. Names in parentheses in table 9.1 are equivalents, not always cited with examples.

### Replacement Nomenclature

The replacement nomenclature method, sometimes called "a" nomenclature, (see chapter 3) is generally applicable to both heteroacyclic chains and rings containing heteroatoms. However, as discussed later in this chapter, there are classes of compounds, such as monocyclic rings with 10 members or less, in which another nomenclature method is usually preferred. In replacement nomenclature a chain or ring is treated as a hydrocarbon structure in which specified carbon atoms have been replaced by heteroatoms. The name of the structure is based on that of the hydrocarbon. Heteroatoms and their positions are denoted by locants and prefixes ending in "a", such as aza for N, and oxa for O. Common prefixes, in descending order of seniority, are given in table 9.2 (see chapter 3, table 3.8, for a more complete list); these prefixes are cited in the order given in the table. Bonding numbers (also called "valences") other than those shown are indicated by the $\lambda$-convention"[1] (see chapter 2), in which the position locant is followed by the Greek letter $\lambda$ and a superscript arabic number

Table 9.1. Mononuclear Heteroatom Hydrides

| | | | |
|---|---|---|---|
| $BH_3$ | borane | $SH_4$ | $\lambda^4$-sulfane |
| $SiH_4$ | silane | $SH_6$ | $\lambda^6$-sulfane |
| $GeH_4$ | germane | $SeH_2$ | selane |
| $SnH_4$ | stannane | $SeH_4$ | $\lambda^6$-selane |
| $PbH_4$ | plumbane | $SeH_6$ | $\lambda^6$-selane |
| $NH_3$ | (azane)[a] | $TeH_2$ | tellane |
| $PH_3$ | phosphane (phosphine) | $TeH_4$ | $\lambda^4$-tellane |
| $PH_5$ | phosphorane ($\lambda^5$-phosphane) | $TeH_6$ | $\lambda^6$-tellane |
| $AsH_3$ | arsane (arsine) | $FH_3$ | $\lambda^3$-fluorane |
| $AsH_5$ | arsorane ($\lambda^5$-arsane) | $ClH_3$ | $\lambda^3$-chlorane |
| $SbH_3$ | stibane (stibine) | $ClH_5$ | $\lambda^5$-chlorane |
| $SbH_5$ | stiborane ($\lambda^5$-stibane) | $BrH_3$ | $\lambda^3$-bromane |
| $BiH_3$ | bismuthane | $BrH_5$ | $\lambda^5$-bromane |
| $BiH_5$ | $\lambda^5$-bismuthane | $IH_3$ | $\lambda^3$-iodane |
| $OH_2$ | (oxidane)[a] | $IH_5$ | $\lambda^5$-iodane |
| $SH_2$ | sulfane | $IH_7$ | $\lambda^7$-iodane |

[a] Seldom used in substitutive nomenclature.

Table 9.2. Common Replacement ("a") Prefixes

| Element | Bonding number | Prefix |
|---|---|---|
| oxygen | 2 | oxa |
| sulfur | 2 | thia |
| selenium | 2 | selena |
| nitrogen | 3 | aza |
| phosphorus | 3 | phospha |
| arsenic | 3 | arsa |
| silicon | 4 | sila |
| tin | 4 | stanna |
| boron | 3 | bora |

that gives the bonding number: the locant for a pentavalent phosphorus atom at a 2 position, for example, would be $2\lambda^5$. Cationic heteroatoms (see chapter 33) are described by the prefixes oxonia, azonia, and so on; these prefixes would be placed in table 9.2 immediately after the prefix for the corresponding uncharged atom. Locants for heteroatoms take precedence over those for unsaturation.

*Chains containing heteroatoms* are best named by replacement nomenclature when substitutive nomenclature, for example, would lead to undesirably complex names. This method can also be used to emphasize a particular structural aspect. Under the IUPAC rules, unbranched chains must terminate with carbon atoms, while CAS also allows termination on P, As, Sb, Bi, Si, Ge, Sn, Pb, and B atoms. In the absence of a principal characteristic group, the chain is numbered from one end to the other, giving lowest locants to the heteroatoms.

$$CH_3CH_2-O-CH_2-SH_2-CH_2CH_2-O-CH_2CH_2-O-CH_3$$

$$\phantom{}12\quad 11\quad\ 10\ \ 9\quad\ 8\quad 7\ \ 6\quad\ 5\ \ 4\quad 3\ \ \ 2\ \ \ 1$$

2,5,10-Trioxa-8$\lambda^4$-thiadodecane
(*not* 3,8,11-Trioxa-5$\lambda^4$-thiadodecane)

When functionality must be considered, the IUPAC rules give a principal characteristic group preference for lowest locants. In CAS nomenclature heteroatoms are given lowest locants in preference to principal characteristic groups, which permits the same numbering principles to be used for heteroatoms in both acyclic and cyclic structures.

$$CH_3-O-CH_2CH_2-O-CH_2CH_2-O-CH_2CH_2-O-CH_2CH_2-NH-CH_2CH_3$$
$$\phantom{C}1\phantom{CCC}2\phantom{CC}3\phantom{CCCC}4\phantom{CC}5\phantom{CCCC}6\phantom{CC}7\phantom{CCCC}8\phantom{CC}9\phantom{CCCC}10\phantom{CCC}11\phantom{CC}12\phantom{CC}13\phantom{CCC}N$$

N-Ethyl-2,5,8,11-tetraoxatridecan-13-amine (CAS)
N-Ethyl-3,6,9,12-tetraoxatridecan-1-amine (IUPAC)
2,5,8,11-Tetraoxa-14-azahexadecane

This example would be named N-[2-[2-[2-(2-methoxyethoxy)ethoxy]ethoxy]ethyl]ethanamine or N-ethyl-2-[2-[2-(2-methoxyethoxy)ethoxy]ethoxy]ethanamine by substitutive nomenclature. Oligoethers, oligoamines, oligosulfides, and related structures, especially when they have more than one kind of heteroatom, provide examples where replacement names are clearly the best alternative.

$$CH_3CH_2-NH-CH_2-O-CH_2CH_2-O-CH_2CH_2-O-CH_2CH_2-COOH$$
$$\phantom{C}1\phantom{CC}2\phantom{CCC}3\phantom{CC}4\phantom{CC}5\phantom{CCCC}6\phantom{CC}7\phantom{CCC}8\phantom{CC}9\phantom{CCC}10\phantom{CC}11\phantom{CC}12\phantom{CC}13\phantom{CC}14$$

5,8,11-Trioxa-3-azatetradecan-14-oic acid (CAS)
4,7,10-Trioxa-12-azatetradecanoic acid (IUPAC)

3-[2-[2-[(Ethylamino)methoxy]ethoxy]ethoxy]propanoic acid (substitutive name)

In the IUPAC name above, the locant 1 is not cited since alkanoic acid names do not require the locant 1 (see chapter 11).

Substitutive nomenclature is used for substituents other than the principal characteristic group.

$$CH_2CH_3$$
$$|$$
$$CH_3-S-CH_2-O-CH_2CH_2-O-CH_2-O-CH_2CHCH_2-COOH$$
$$\phantom{C}1\phantom{CC}2\phantom{CC}3\phantom{CCC}4\phantom{CC}5\phantom{CCC}6\phantom{CCC}7\phantom{CC}8\phantom{CCC}9\phantom{C}10\phantom{CC}11\phantom{C}12\phantom{CCC}13$$

11-Ethyl-4,7,9-trioxa-2-thiatridecan-13-oic acid (CAS)
3-Ethyl-5,7,10-trioxa-12-thiatridecanoic acid (IUPAC)

3-[[[2-[(Methylthio)methoxy]ethoxy]methoxy]methyl]pentanoic acid;
3-(2,4,7-Trioxa-9-thiadecyl)pentanoic acid (substitutive names)

$$O-CH_2CH_2CH_3$$
$$|$$
$$HOOC-CH_2-O-CH-S-CH_2CH_2CH_2-O-CH_2CH_2-NH-CH_2-O-CH_2-COOH$$
$$\phantom{CC}16\phantom{CC}15\phantom{CC}14\phantom{C}13\phantom{CC}12\phantom{C}11\phantom{CCC}10\phantom{CC}9\phantom{CCCC}8\phantom{CC}7\phantom{CCC}6\phantom{CCC}5\phantom{CC}4\phantom{CC}3\phantom{CC}2\phantom{CCC}1$$

13-Propoxy-3,8,14-trioxa-12-thia-5-azahexadecanedioic acid
(the set of locants 3,5,8,12,14 for the heteroatoms is
lower than 3,5,9,12,14)

Replacement nomenclature is sometimes used for emphasis even when the substitutive name is not particularly complex.

$$CH_3-S-CH_2CH_2-S-CH_3$$
$$\phantom{CC}6\phantom{CCC}5\phantom{CC}4\phantom{CC}3\phantom{CCC}2\phantom{CC}1$$

2,5-Dithiahexane

1,2-Bis(methylthio)ethane;
1,2-Bis(methylsulfanyl)ethane
(substitutive names)

$$CH_2=P-N=CH_2$$
$$\quad 4 \quad\; 3 \;\; 2 \quad\; 1$$

2-Aza-3-phosphabuta-1,3-diene

*N,P*-Bis(methylidene)phosphanamine;
*N,P*-Bis(methylidene)phosphinous amide
(substitutive names; see chapter 25)

The rule that chains must terminate with specific atoms such as carbon need not be absolute. If it were desirable to compare the last example with $CH_2=CH–CH=NH$, the latter might well be called 1-azabuta-1,3-diene.

In naming branched chains, the longest chain with the larger number of heteroatoms that can be used to express the greatest number of the principal characteristic group is selected as the main chain. Among chains with the same number of heteroatoms, the larger chain is preferred, followed by that with the greater number of most senior heteroatoms (tables 3.8 and 9.2); for additional criteria, see chapter 3. Side chains are usually named substitutively, but where the side chains are complex, they also may be named by replacement nomenclature as described above. In such cases, the free valence of the substituting group is numbered 1, even by CAS.

$$NH-CH_2-O-CH_2-O-CH_3 \qquad\qquad OH$$
$$\qquad\qquad\qquad | \qquad\qquad\qquad\qquad\qquad\qquad\qquad\qquad |$$
$$HO-CH_2CH-O-CH_2CH_2-NH-CH_2-OO-CH_2CH-O-CH_3$$
$$\quad\; 13 \;\; 12 \qquad 11 \quad\; 10 \quad\;\; 9 \qquad\; 8 \qquad\; 7 \qquad\; 6\;5 \qquad 4 \qquad 3 \quad\; 2 \quad 1$$

12-[[(Methoxymethoxy)methyl]amino]-2,5,6,11-tetraoxa-8-azatridecane-3,13-diol (CAS)
2-[(2,4-Dioxapentyl)amino]-3,8,9,12-tetraoxa-6-azatridecane-1,11-diol (IUPAC)

$$O-CH_3$$
$$\qquad\qquad\qquad\qquad\qquad |$$
$$CH_3-NH-CH_2-NH-CHCH_2CH_2-S-CH_2-NH-CH_2CH_2-OH$$
$$\;1 \qquad 2 \qquad\; 3 \qquad\; 4 \quad\; 5 \;\; 6 \quad\; 7 \qquad 8 \quad\; 9 \qquad 10 \qquad 11 \;\; 12$$

5-Methoxy-8-thia-2,4,10-triazadodecan-12-ol (CAS)
8-Methoxy-5-thia-3,9,11-triazadodecan-1-ol (IUPAC)

Heteroacyclic chains having two alternating atoms may be named with the prefixes of table 9.2 (or 3.8), cited in reverse order to that in the tables, followed by the ending -ane.

$$(CH_3)_3Si-NH-Si(CH_3)_3 \qquad 1,1,1,3,3,3\text{-Hexamethyldisilazane}$$
$$\qquad\qquad 3 \quad\; 2 \quad\;\; 1$$

Structures of this kind are considered in greater detail in chapters 25–27.

With *heterocyclic systems* replacement nomenclature is a must for (a) heteromonocycles with more than 10 ring atoms; (b) heteroalicyclic bridged nonfused ring systems; (c) spiro compounds with only monocyclic components; and (d) fused ring systems where fusion names cannot be properly formed. Replacement nomenclature is used by CAS for many rings containing silicon.

For heteromonocycles with more than 10 members, the name is based on that of the corresponding monocyclic ring (see chapter 6). Heteroatoms that replace carbon atoms are described by the prefixes of table 9.2 (or table 3.8). The most senior heteroatom is assigned the locant 1. Then lowest locants are given to the heteroatoms, first as a set (disregarding the kinds of heteroatoms), and second, in order of seniority. These prefixes are cited in a name in order of seniority.

1,6-Dioxa-10-azacyclotetradecane

*Heteroalicyclic spiro and bridged nonfused ring systems* are named on the basis of the corresponding saturated spiro or bridged hydrocarbon (see chapter 6). Appropriate replacement terms are prefixed to the name of the hydrocarbon; these terms are given lowest locants consistent with the fixed numbering of the spiro or bridged hydrocarbon. If there is no fixed numbering or the fixed numbering of the ring still leaves choices for numbering heteroatoms, lowest locants are given to the heteroatoms as a set without regard to the kind of heteroatoms. If there are still further alternatives, lowest locants are assigned to the heteroatoms in the order given in table 9.2 (or table 3.8). Locants for the heteroatoms take precedence over those for unsaturation. Where the heteroatom(s) occur only in bridges, replacement nomenclature is an alternative to the use of bridge names such as epoxy.

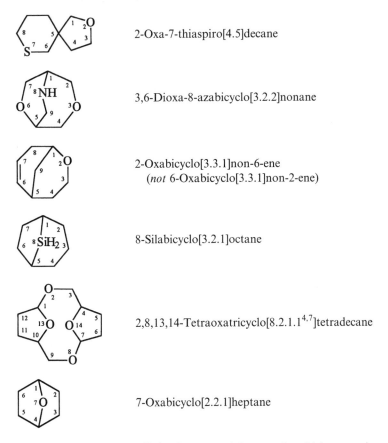

2-Oxa-7-thiaspiro[4.5]decane

3,6-Dioxa-8-azabicyclo[3.2.2]nonane

2-Oxabicyclo[3.3.1]non-6-ene
(*not* 6-Oxabicyclo[3.3.1]non-2-ene)

8-Silabicyclo[3.2.1]octane

2,8,13,14-Tetraoxatricyclo[8.2.1.1$^{4,7}$]tetradecane

7-Oxabicyclo[2.2.1]heptane

The latter example is sometimes called 1,4-epoxycyclohexane, in which epoxy is used as a substitutive prefix; the replacement name is preferred.

Where at least one component of a spiro heterocycle is a bridged or fused system, the prefixes spiro, dispiro, and so on are followed in brackets, in succession beginning with the ring earliest in alphabetical order, by the names of the components separated by locants denoting the points of the spiro unions. Fixed numbering of the components is retained, and those of the second and succeeding cited components are given primes in order. Heteroatoms in a ring component can be described by replacement nomenclature if they are not expressed otherwise within the component name.

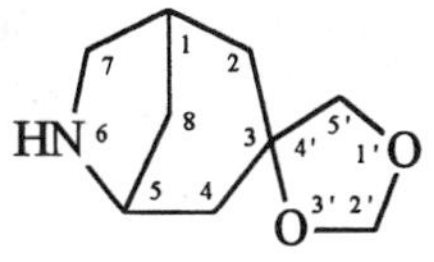

Spiro[6-azabicyclo[3.2.1]octane-3,4′-[1,3]dioxolane)]

Spiro[piperidine-4,9′-[9H]xanthene]

## Hantzsch–Widman Nomenclature[2]

Heteromonocycles with 10 or fewer ring atoms are named by the Hantzsch–Widman system as the much preferred alternative to replacement nomenclature. This system does not require reference to a hydrocarbon ring name and is applicable to rings not having carbon atoms. The prefixes of table 9.2 (or table 3.8), with elision of the final "a" if followed by a vowel, are utilized in combination with one of the endings shown in table 9.3.

Table 9.3. Hantzsch–Widman Endings[2]

| Ring size | Maximum conjugated double bonds | Saturated |
|---|---|---|
| 3 | irene[a] | irane[b] |
| 4 | ete | etane[b] |
| 5 | ole | olane[b] |
| 6A (O, S, Se, Te, Bi) | ine[c] | ane |
| 6B (N, Si, Ge, Sn, Pb) | ine[c] | inane[e,f] |
| 6C (B[g], P, As, Sb) | inine[c,d] | inane[d,f] |
| 7 | epine[c] | epane[f] |
| 8 | ocine[c] | ocane[f] |
| 9 | onine[c] | onane[f] |
| 10 | ecine[c] | ecane[f] |

[a] -Irine may be used for rings containing nitrogen as the only heteroatom; CAS uses -irine for any three-membered ring containing nitrogen.
[b] -Iridine, -etidine, and -olidine are preferred for rings containing nitrogen.
[c] For CAS, the final "e" in these endings is dropped for rings not containing nitrogen.
[d] The prefixes phosphora, arsena, and stibina are used by CAS rather than phospha, arsa, and stiba.
[e] CAS does not use this ending for saturated rings based on Si, Ge, Sn, or Pb; saturation is expressed by hydro prefixes.
[f] For nitrogen-containing rings, CAS expresses saturation with hydro prefixes; "perhydro" is not used.
[g] For CAS, B is in Group 6B.

The endings denote ring size and saturation. Numbering of the ring begins with the most senior atom in table 9.2 (or table 3.8) and proceeds in the direction that gives lowest locants to the remaining heteroatoms, first as a set, and then to the more senior heteroatoms. Locants for the heteroatoms are placed together in front of the replacement prefixes; both locants and their prefixes are cited in the decreasing order of seniority shown in table 9.2 (or table 3.8). The final "e" of the endings in table 9.3 is considered optional by IUPAC. Partial saturation is described by dihydro, tetrahydro, and so on, with names having endings denoting the maximum number of conjugated double bonds. Pyridine and pyran must be used rather than the corresponding Hantzsch–Widman names azine and oxine.

In the following examples, the Hantzsch–Widman name is given first. For comparison, the replacement name (see above) is also cited; these names are not used in fusion nomenclature.

1,3-Diazetidine
1,3-Diazacyclobutane

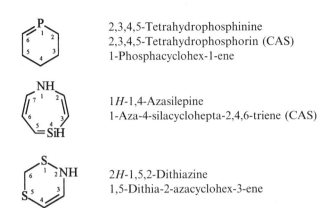

2,3,4,5-Tetrahydrophosphinine
2,3,4,5-Tetrahydrophosphorin (CAS)
1-Phosphacyclohex-1-ene

1*H*-1,4-Azasilepine
1-Aza-4-silacyclohepta-2,4,6-triene (CAS)

2*H*-1,5,2-Dithiazine
1,5-Dithia-2-azacyclohex-3-ene

Table 9.4 contains a list of well-established trivial names for saturated and partially hydrogenated heterocyclic rings. These names are acceptable but are not to be used as components in the formation of fusion names (see below). In table 9.4, names marked with an asterisk are not used by CAS.

## Fusion Nomenclature

*Heteropolycyclic systems* are preferably named by the principles used for fused polycyclic hydrocarbons (see chapter 7); replacement nomenclature is an alternative used only when fusion nomenclature cannot be applied. Both Hantzsch–Widman and, with a few exceptions, the trivial names listed in table 9.5 serve as component names in the fusion method.[3] When these names denote an attached component, any final "e" is changed to "o". The following contracted prefixes are commonly used: furo, imidazo, isoquino, pyrido, pyrimido, quino, and thieno. A heterocyclic ring is preferred to a hydrocarbon ring as the base component. Then, to select the parent component (see table 9.6), the following criteria are followed in descending order: a component containing (a) nitrogen; (b) a heteroatom most senior in table 9.2 (or table 3.8); (c) the greatest number of rings; (d) the largest ring; (e) the most heteroatoms. Additional criteria are available in the IUPAC rules.[3] Heteroatoms at fusion sites are expressed in both component names; in the fused polycycle, they are given whole number locants rather than locants such as 3a, as in the first example below. Locants referring to a component only are enclosed in brackets (CAS does not use brackets if the locants are the same as the final numbering of the ring system). Indicated hydrogen is denoted in the same way as it is for fused hydrocarbons.

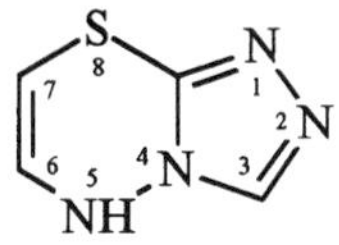

5*H*-[1,2,4]Triazolo[3,4-*b*][1,3,4]thiadiazine

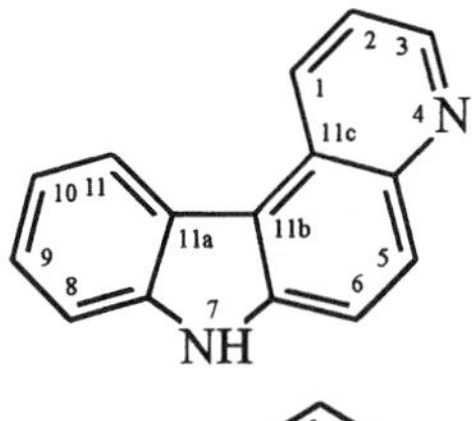

7*H*-Pyrido[2,3-*c*]carbazole

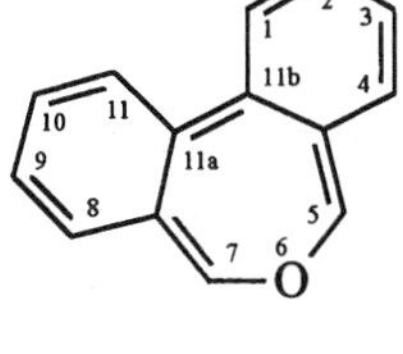

Dibenzo[*c,e*]oxepine

Table 9.4. Trivial Names for Hydrogenated Heterocycles[a]

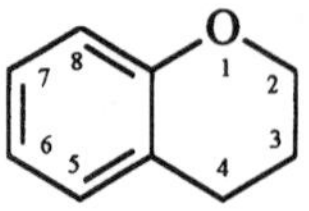

chroman(e)*
(chalogen analogs are named
"thiochroman(e)" and so on)

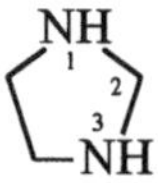

imidazolidine

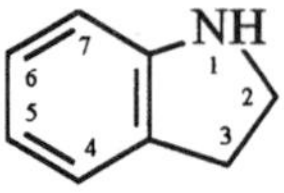

indoline*

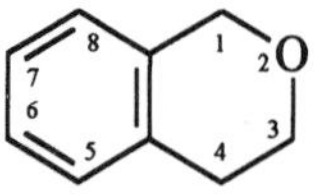

isochroman(e)*
(chalcogen analogs are
"isothiochroman(e)" and so on)

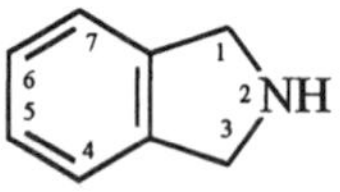

isoindoline*

morpholine

piperazine

piperidine

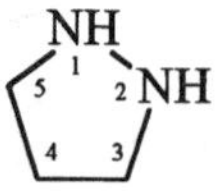

pyrazolidine

pyrrolidine

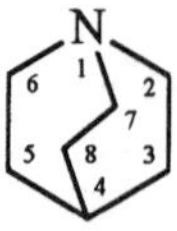

quinuclidine*

[a] This table contains hydrogenated heterocyclic rings with accepted trivial
names; these names are not used as components in fusion names. Names for
partially hydrogenated heteromonocycles such as pyrroline, imidazoline, and
pyrazoline are no longer acceptable. Names not used by CAS are marked with
an asterisk

Table 9.5. Trivial Names for Heterocyclic Parent Hydrides[a]

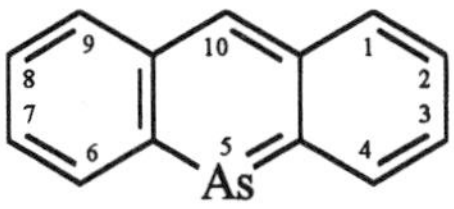

acridarsine

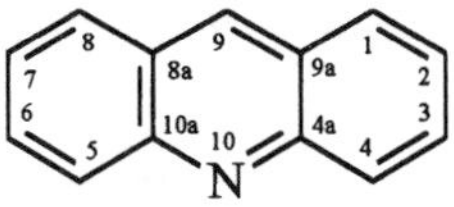

acridine

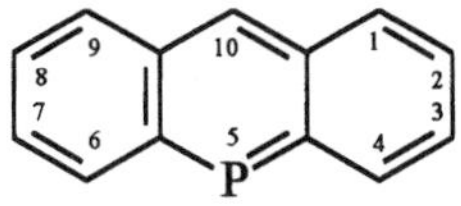

acridophosphine

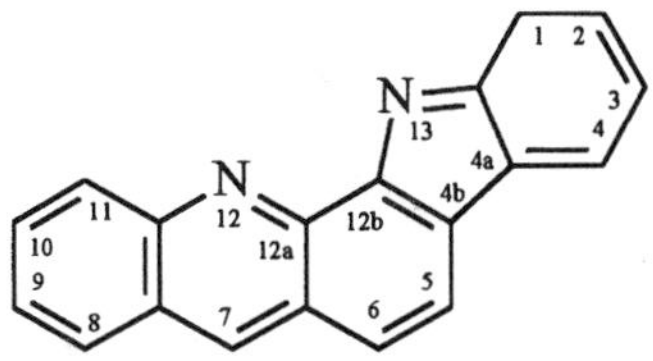

1H-acrindoline†

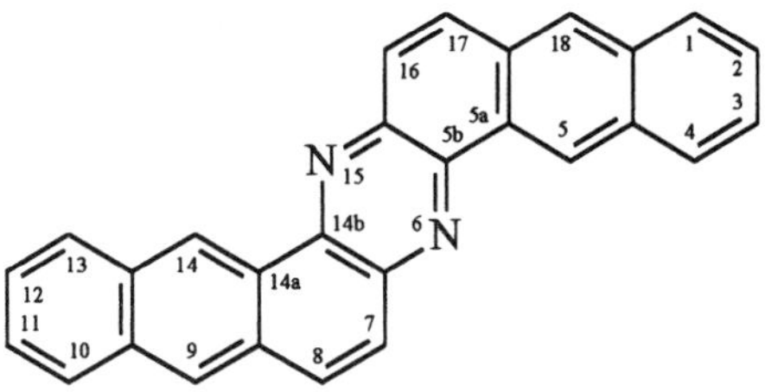

anthrazine†

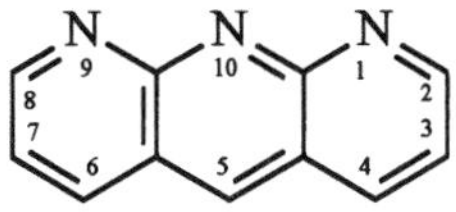

anthyridine†

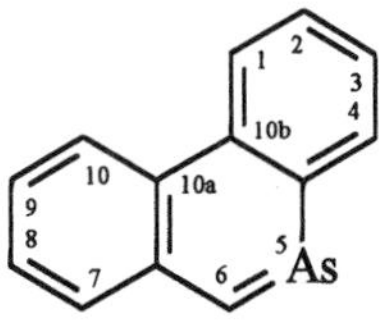

arsanthridine

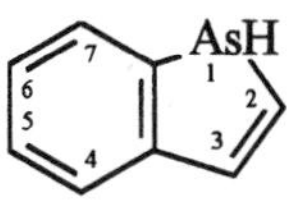

arsindole (1H- shown)
1H-1-benzarsole

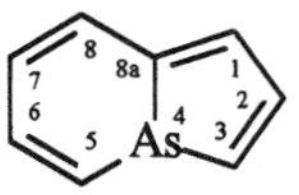

arsindolizine

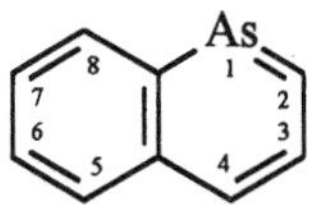

arsinoline
1-benzarsinine

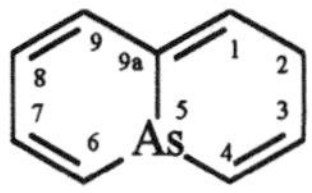

arsinolizine (2H- shown)

(Continued)

Table 9.5. Continued

| | |
|---|---|
| 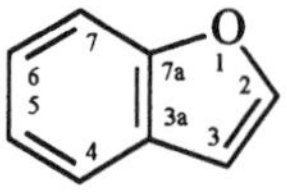 | benzofuran<br>1-benzofuran<br>1-benzoxole |
| 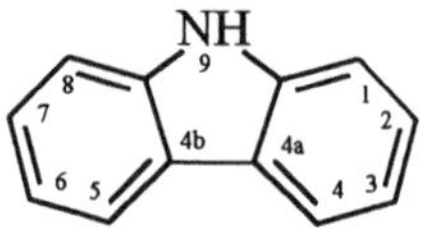 | carbazole (9H- shown) |
| 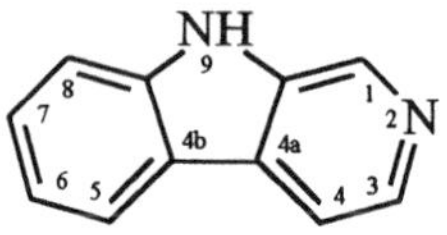 | β-carboline (9H- shown)*<br>9H-pyrido[3,4-b]indole (CAS) |
| 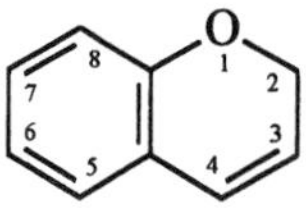 | chromene* (2H- shown)<br>   (chalcogen analogs are named<br>   "thiochromene" and so on)<br>2H-1-benzopyran (CAS) |
| 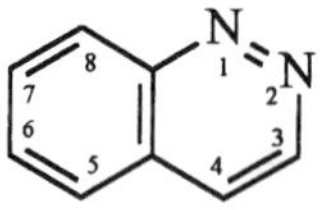 | cinnoline<br>1,2-benzodiazine |
|  | furan<br>oxole |
| 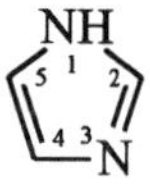 | imidazole (1H- shown)<br>1H-1,3-diazole |
| 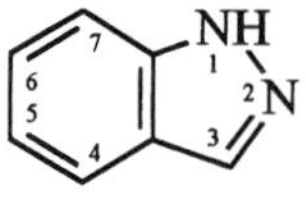 | indazole (1H- shown)<br>1H-1,2-benzodiazole |
| 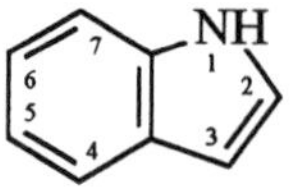 | indole (1H- shown)<br>1H-1-benzazole |
| 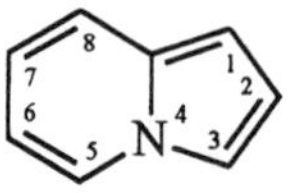 | indolizine |
| 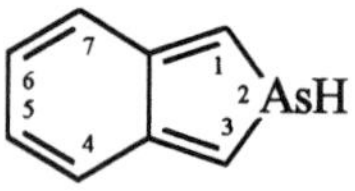 | isoarsindole (2H- shown)<br>2H-2-benzarsole |
| 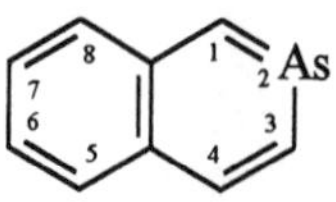 | isoarsinoline<br>2-benzarsinine |
| 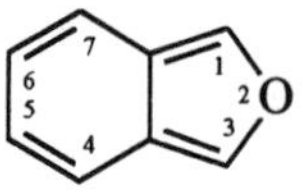 | isobenzofuran<br>2-benzoxole<br>2-benzofuran |

*(Continued)*

Table 9.5. Continued

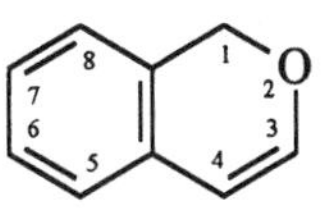

isochromene* (1*H*- shown)
  (chalcogen analogs are
    "isothiochromene" and so on)
1*H*-2-benzopyran (CAS)
  (chalogen analogs are "1*H*-2-
    benzothiopyran" and so on)

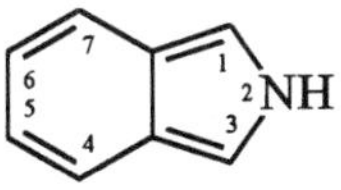

isoindole (2*H*- shown)
2*H*-2-benzazole

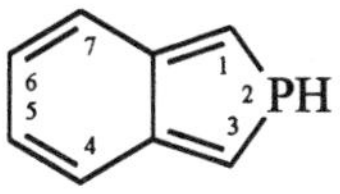

isophosphindole (2*H*- shown)
2*H*-2-benzophosphole

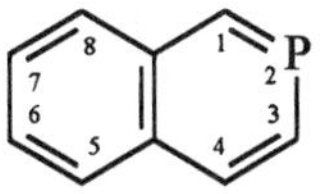

isophosphinoline
2-benzophosphinine

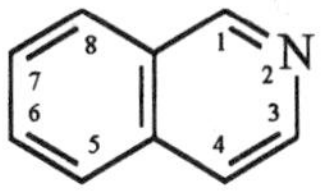

isoquinoline
2-benzazine

isothiazole
1,2-thiazole

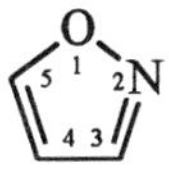

isoxazole
1,2-oxazole

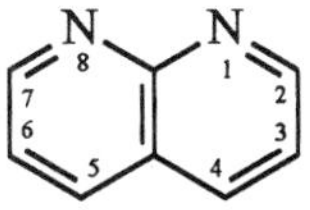

naphthyridine (1,8- isomer shown)

oxazole
1,3-oxazole

perimidine (1*H*- shown)

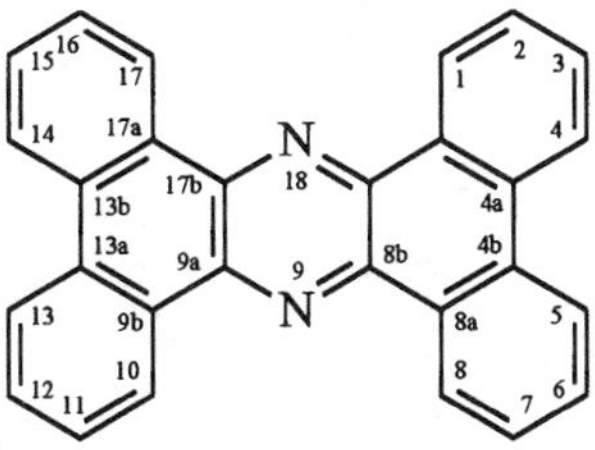

phenanthrazine†

*(Continued)*

Table 9.5. Continued

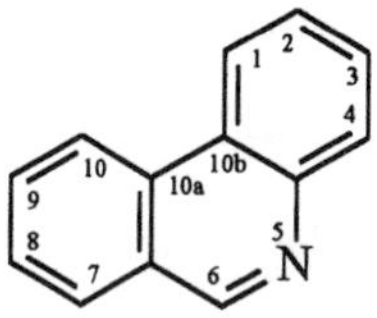

phenanthridine

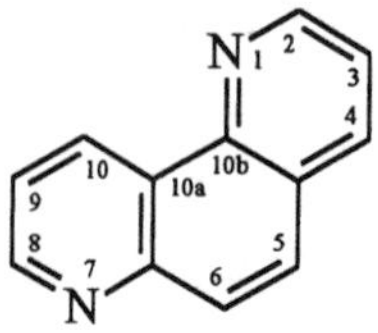

phenanthroline (1,7- isomer shown)

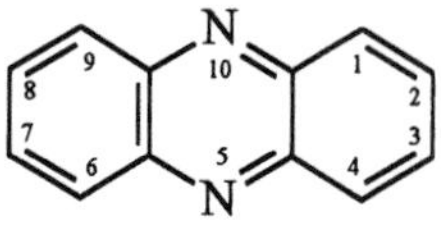

phenazine
azanthrene

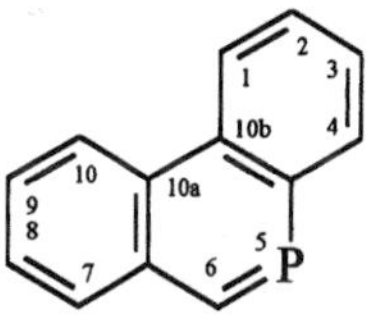

phosphanthridine

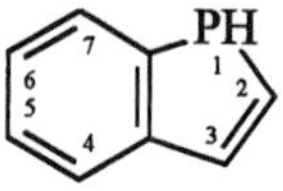

phosphindole (1*H*- shown)
1*H*-1-benzophosphole

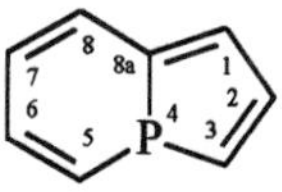

phosphindolizine

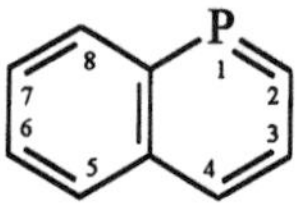

phosphinoline
1-benzophosphinine

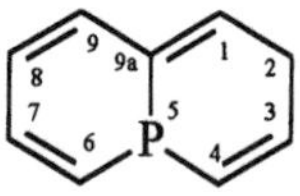

phosphinolizine

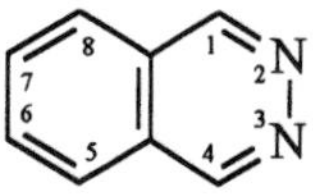

phthalazine
2,3-benzodiazine

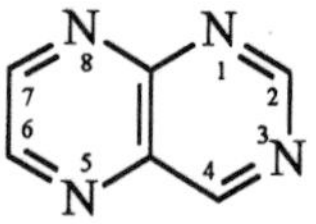

pteridine

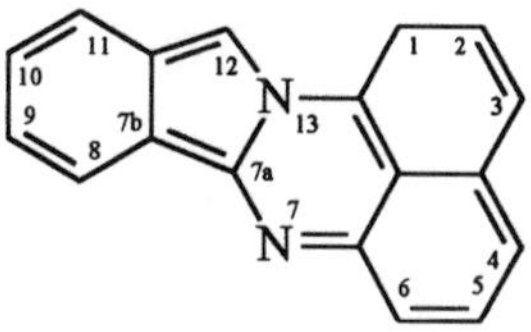

phthaloperine†

*(Continued)*

Table 9.5. Continued

| | |
|---|---|
| 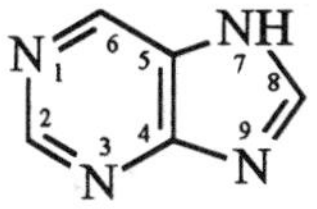 | purine |
| 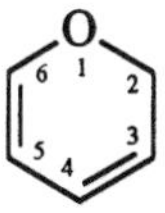 | pyran (2H- shown)<br>(*not* 2H-oxine)<br>(chalcogen analogs are named "thiopyran"<br>and so on, and 2H-thiin and so on, are<br>acceptable) |
| 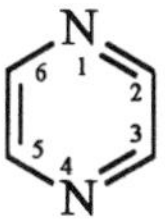 | pyrazine<br>1,4-diazine |
| 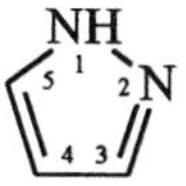 | pyrazole (1H- shown)<br>1H-1,2-diazole |
|  | pyridazine<br>1,2-diazine |
| 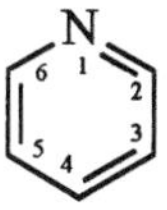 | pyridine<br>(*not* azine) |
| 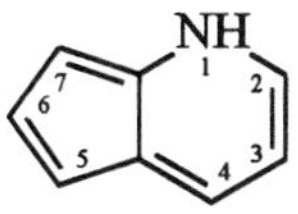 | pyrindine* (1H-1- shown)<br>(a useful parent name once used<br>by CAS, which now names it<br>cyclopenta[b]pyridine) |
| 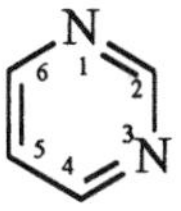 | pyrimidine<br>1,3-diazine |
|  | pyrrole (1H- shown)<br>1H-azole |
| 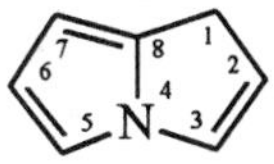 | pyrrolizine (1H- shown) |
| 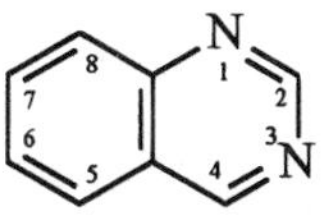 | quinazoline<br>1,3-benzodiazine |
| 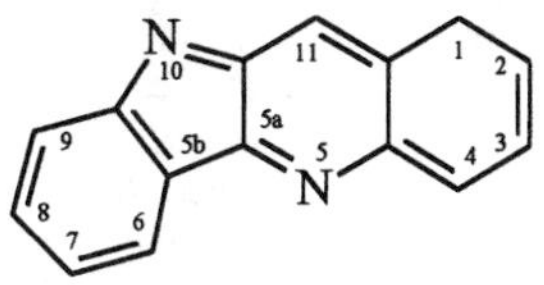 | 1H-quindoline† |

*(Continued)*

Table 9.5. Continued

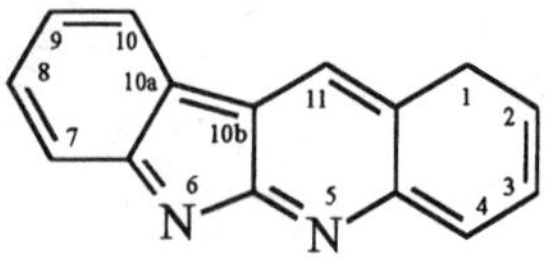

1H-quinindoline†

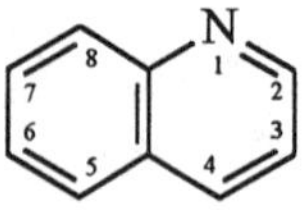

quinoline
1-benzazine

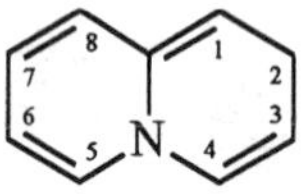

quinolizine (2H- shown)

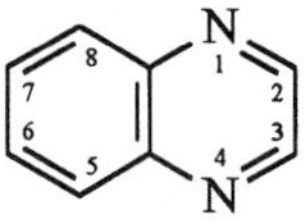

quinoxaline
1,4-benzodiazine

selenophene
1H-selenole

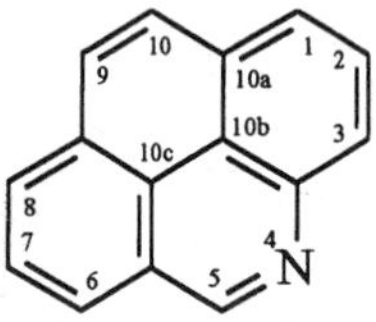

thebenidine†

thiazole
1,3-thiazole

thiophene
1H-thiole

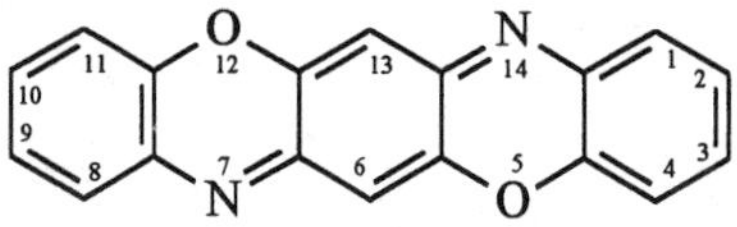

triphenodioxazine†

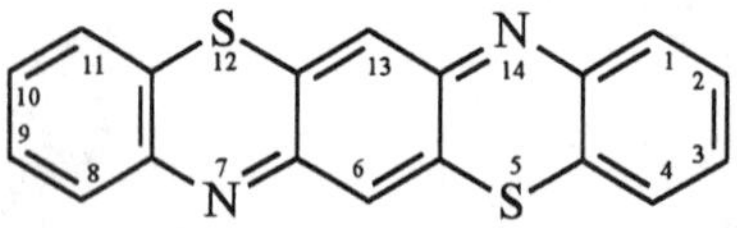

triphenodithiazine†

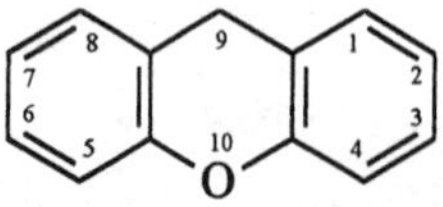

xanthene (9H- shown)
   (chalcogen analogs are named
   "thioxanthene" and so on)

---

[a] This table has a selection of heterocyclic rings with accepted trivial names that can be used in forming fusion names. Names not used by CAS are marked with an asterisk; those used by CAS but not listed by IUPAC are marked with a dagger. β-Carboline and pyrindine are no longer used by either CAS or IUPAC. As appropriate, the corresponding Hantzsch–Widman names and "benzo" names are also given.

Table 9.6. Selected Trivial Names for Heterocycles in Ascending
Order of Precedence in Fusion Names

Columns are to be read newspaper style. A more complete list will be
found in ref. 5.

| | | |
|---|---|---|
| arsindolizine | benzofuran | phthalazine |
| isoarsindole | xanthene | naphthyridine |
| arsindole | phenoxarsinine | quinoxaline |
| arsinolizine | phenoxaphosphinine | quinazoline |
| isoarsinoline | phenoxaselinine | cinnoline |
| arsinoline | phenoxathiin | pteridine |
| arsanthridine | pyrrole | carbazole |
| acridarsine | imidazole | β-carboline |
| arsanthrene | pyrazole | phenanthridine |
| phosphindolizine | isothiazole | acridine |
| isophosphindole | thiazole | perimidine |
| phosphindole | isoxazole | phenanthroline |
| phosphinolizine | oxazole | phenazine |
| isophosphinoline | pyridine | anthyridine |
| phosphinoline | pyrazine | phenarsazinine |
| phosphanthridine | pyrimidine | phenophosphazinine |
| acridophosphine | pyridazine | phenoselenazine |
| selenophene | pyrrolizine | phenothiazine |
| selenanthrene | indolizine | phenoxazine |
| thiophene | 1-pyrindine | thebenidine |
| thiopyran | isoindole | quindoline |
| thianthrene | 2-pyrindine | quinindoline |
| furan | indole | phthaloperine |
| pyran | indazole | acrindoline |
| isochromene | purine | triphenodithiazine |
|   (1H-2-benzopyran) | quinolizine[a] | triphenodioxazine |
| chromene | isoquinoline | phenanthrazine |
|   (2H-1-benzopyran) | quinoline | anthrazine |
| isobenzofuran | | |

[a] Follows quinoline in the CAS list, which, except for 1- and 2-pyrindine, is the same
as the order above.

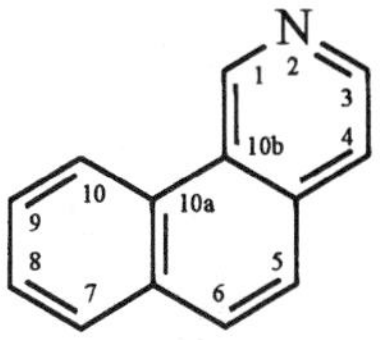

Benzo[h]isoquinoline

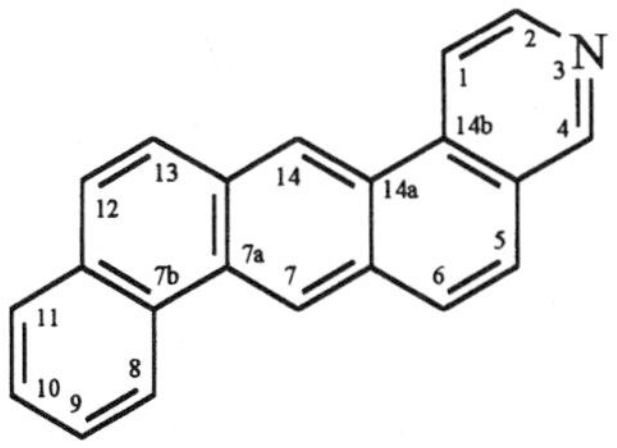

Phenanthro[2,3-f]isoquinoline

While interior heteroatoms are given whole number locants, CAS and IUPAC number
interior carbon atoms differently, as shown in the following example:

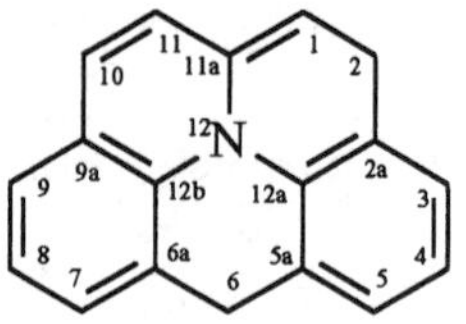

CAS 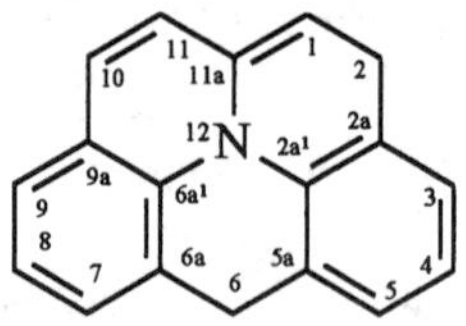                                    IUPAC

2*H*,6*H*-Quinolizino[3,4,5,6,7-*defg*]acridine

Heterocycles consisting of a benzene ring fused to a heteromonocycle need not be named by fusion principles. Rather, benz or benzo is prefixed to the Hantzsch–Widman or trivial name of a monocycle with maximum conjugated double bonds. The resulting composite name is prefixed by locants giving the positions of the heteroatoms in the final system. Saturated positions are denoted by hydro or indicated hydrogen prefixes. Trivial names (see table 9.4) for monocyclic components are preferred to Hantzsch–Widman names. These "benzo" names are sometimes employed as parent components in fusion nomenclature.

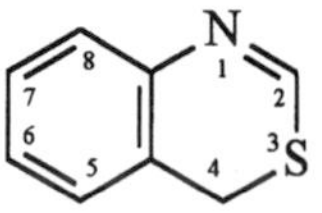

4*H*-3,1-Benzothiazine
4*H*-Benzo[*d*][1,3]thiazine (fusion name)

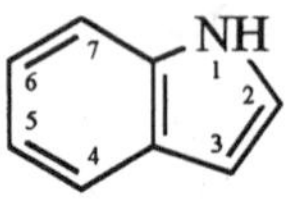

1*H*-Indole (preferred trivial name)
1*H*-Benzopyrrole
1*H*-1-Benzo[*b*]pyrrole (fusion name)

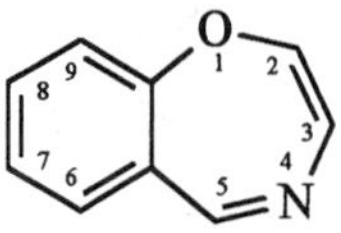

1,4-Benzoxazepine
Benzo[*f*][1,4]oxazepine (fusion name)

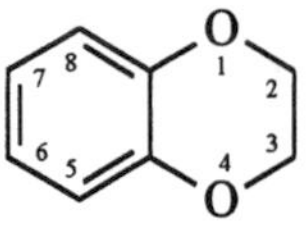

2,3-Dihydro-1,4-benzodioxin
2,3-Dihydrobenzo[*b*][1,4]dioxin (fusion name)

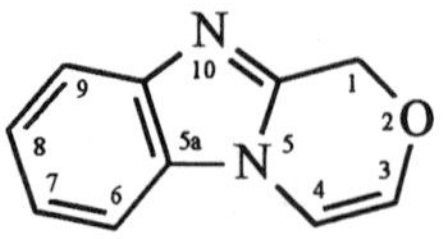

1*H*-[1,4]Oxazino[4,3-*a*]benzimidazole
1*H*-Benzo[4,5]imidazo[2,1-*c*][1,4]oxazine
    (full fusion name)

The replacement name for a saturated heteromonocycle with more than 10 ring atoms can also be used as a fusion component. As a parent component, the ending -ane is changed to -in (-ine if the ring contains nitrogen) to indicate a maximum number of conjugated double bonds. As a prefix, the -in or -ine ending becomes -ino.

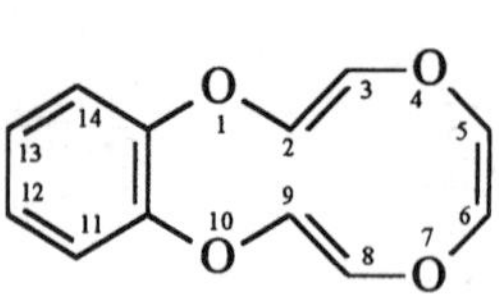

1,4,7,10-Benzotetraoxacyclododecin
Benzo[*b*][1,4,7,10]tetraoxacyclododecin (fusion name)

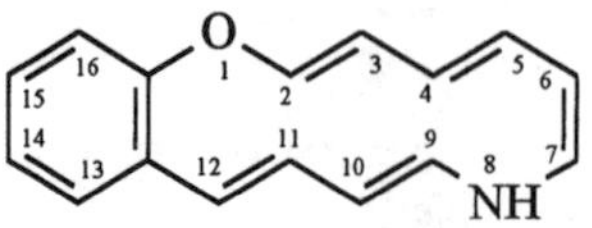

8*H*-1,8-Benzoxaazacyclotetradecine
Benzo[*b*][1,8]oxaazacyclotetradecine (fusion name)

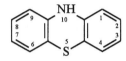

[1,4,7,10]Tetraoxacyclododecino[2,3-*g*][1,3]benzoxazole
[1,4,7,10]Tetraoxacyclododecino[2′,3′:3,4]benzo-
[1,2-*d*][1,3]oxazole (full fusion name)

In the example above "tetraoxacyclododecin" is not an approved name for the 12-member monocycle with maximum conjugation, which has the name 1,4,7,10-tetraoxacyclododeca-tetraene.

While many heterocyclic systems have acceptable trivial names (see tables 9.4 and 9.5), other names can be derived from some of them in a systematic way. For example, three linearly fused six-membered rings with the same heteroatom (other than nitrogen) at the two unfused central positions take the ending -anthrene, preceded by the "a" prefix for the heteroatom, as in silanthrene.

Arsanthrene

When the heteroatoms are different, the system is named with the prefix "phen" or "pheno" followed by the "a" terms for the heteroatoms and the ending -in (-ine if one of the atoms is N, P, or As), as in phenoxathiin. Usually, the central ring has its Hantzsch–Widman name.

Phenothiazine

Phosphorus and arsenic analogs of indole and isoindole take phosph- or ars- as prefixes to the trivial name, as in phosphindole, or as infixes between "iso" and the trivial name, as in isophos-phindole. With quinoline and isoquinoline, the "qu" is omitted, leading to names like arsinoline.

*Bridged fused heterocyclic systems* are named and numbered in slightly different ways by IUPAC and CAS. Under the IUPAC rules[3] the preferred fused ring system is selected by the following criteria taken in order: (a) the system, before bridging, having the most rings; (b) the system that includes the most ring atoms; (c) the system with the fewest number of hetero-atoms in the ring (CAS prefers more heteroatoms in the unbridged ring system); and (d) the most preferred fused ring system as described above in the section on fusion nomenclature. Additional criteria will be found in the IUPAC rules.[3] For example, the size and complexity of the bridges are minimized. After selection, the preferred system is named and numbered in the usual way.

Acyclic bridges are given "alkano" names in the same manner as they were with bridged polycyclic arenes (see chapter 7). Acyclic heteroatomic bridges usually take names beginning with "ep" (which is omitted when not the initial component of a complex bridge), such as epoxy, –O–; epithio, –S–; epimino (CAS uses imino), –NH–; episilano (CAS uses silano), –SiH$_2$–; epidiazano (CAS uses biimino), –NHNH–; epoxythiooxy, –O–S–O–; and epoxy-ethano, –O–CH$_2$CH$_2$–. If the bridge name is the same as the corresponding fusion prefix, IUPAC precedes it with epi-, while CAS uses *endo*-. The bridged system is then named by prefixing, in alphabetical order, the locants and names of the bridges to the name of the fused ring system; the first ring locant cited for bridge attachment is that for connection to the lowest numbered bridge atom.

8a,4a-(Epiminomethano)quinoline
(*not* 4a,8a-(Epiminomethano)quinoline)

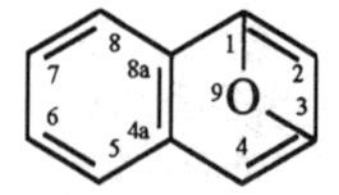

1,3-Epoxynaphthalene (IUPAC)
1,3-Metheno-1$H$-2-benzopyran (CAS)

Names for heterocycles as bridges are derived from the name of the heterocycle itself with locants in square brackets and the ending "o", as in [3,4]furano (CAS uses [3′,4′]-furano).

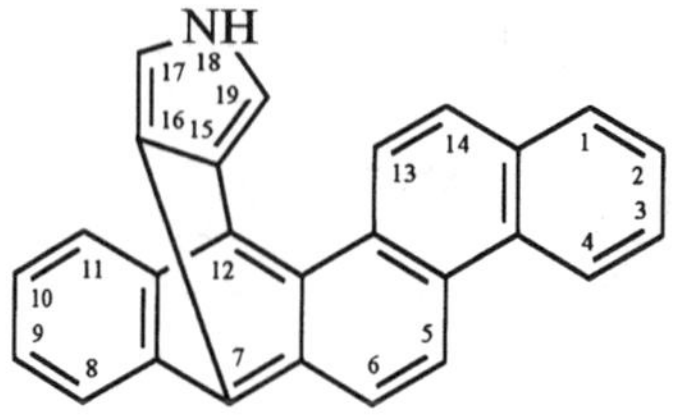

18$H$-7,12-[3,4]Epipyrrolobenzo[$b$]chrysene
   (IUPAC)

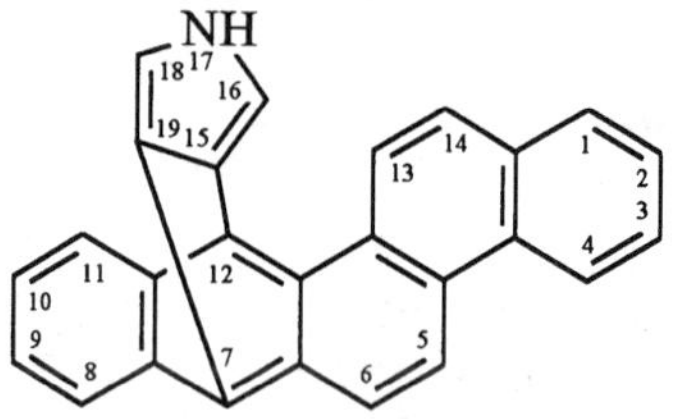

17$H$-7,12[3′,4′]-*endo*-Pyrrolobenzo[$b$]chrysene
   (CAS)

## Heterocyclic Ring Assemblies

Assemblies of identical heterocyclic rings are named in the same way as hydrocarbon ring assemblies (see chapter 8). Numbering of the individual heterocyclic rings is retained, and indicated hydrogen is cited just ahead of the name of the specific ring involved.

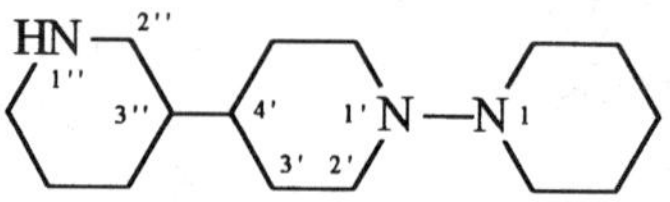

1,1′:4′,3″-Terpiperidine

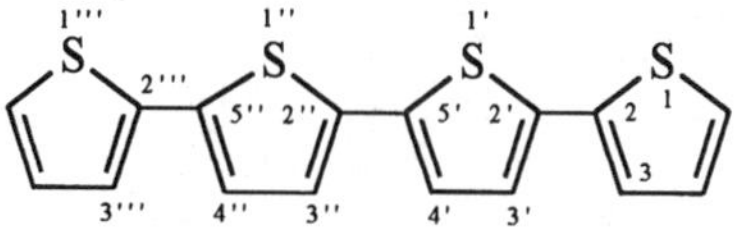

2,2′:5′,2″:5″,2‴-Quaterthiophene

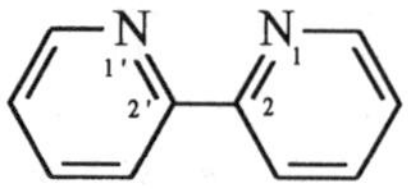

2,2′-Bipyridine
2,2′-Bipyridyl

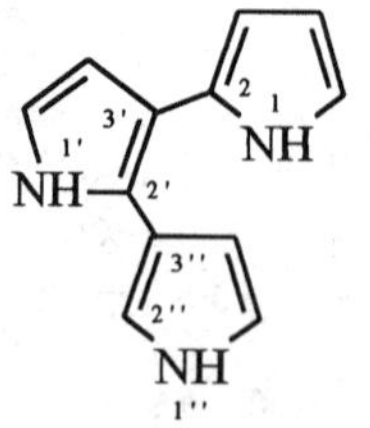

2,3′:2′,3″-Ter-1$H$-pyrrole
   (*not* 3,2′:3′,2″-Ter-1$H$-pyrrole)

## Heterocycles Without Carbon Atoms

Heteromonocyclic systems without carbon as a ring atom but containing fewer than 11 ring atoms can be named by the Hantzsch–Widman system (see above). If the ring atoms are identical, names analogous to those of the cycloalkanes may be derived.

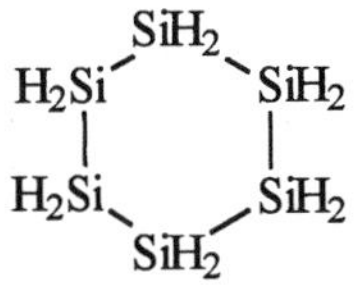

Hexasilinane
Cyclohexasilane

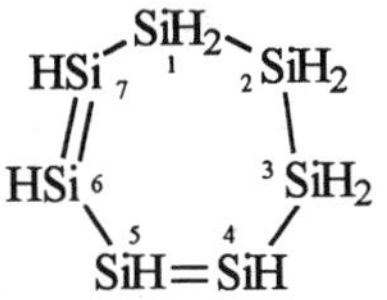

2,3-Dihydro-1*H*-heptasilepin
Cycloheptasila-1,3-diene

3*H*-1,2,4,3,5-Triazadiphosphole

Saturated heteromonocyclic systems consisting only of repeating units of alternating hetero-atoms may be named with the prefix cyclo followed by a multiplying term, the "a" prefixes for the heteroatoms cited in reverse order of table 9.2, and the ending -ane. Hantzsch–Widman names may also be used.

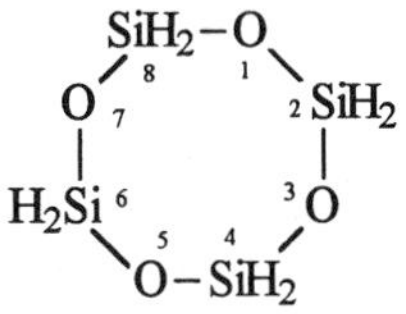

Cyclotetrasiloxane
1,3,5,7,2,4,6,8-Tetraoxatetrasilocane

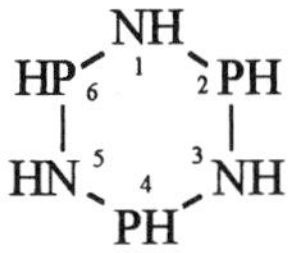

Cyclotriphosphazane
1,3,5,2,4,6-Triazatriphosphinane (IUPAC)
1,3,5,2,4,6-Triazatriphosphorinane (CAS)

Since recommendations for naming unsaturated systems of this type have not been established, Hantzsch–Widman names are preferred. However, in the P–N series, the $-P^5=N-$ structure has the well-established trivial name phosphazene.

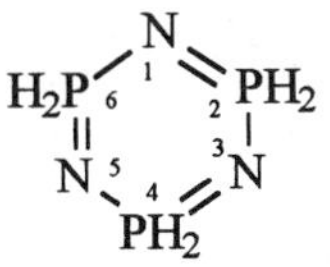

2,2,4,4,6,6-Hexahydro-1,3,5,2,4,6-triazatriphosphorine (CAS)
1,3,5,2$\lambda^5$,4$\lambda^5$,6$\lambda^5$-Triazatriphosphinine (IUPAC)
Cyclotriphosphazene

Substituting group names for heteroatom-containing systems are formed in the usual way, with locants for free valences in cyclic structures consistent with the fixed numbering of the system. In acyclic chains, lowest locants are given to free valences rather than heteroatoms.

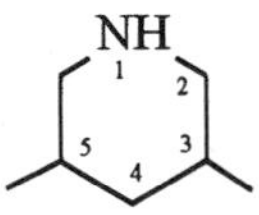

Piperidine-3,5-diyl

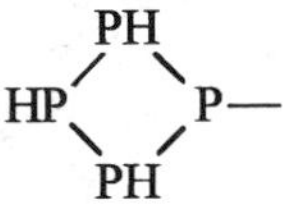

Cyclotetraphosphanyl
Tetraphosphetanyl

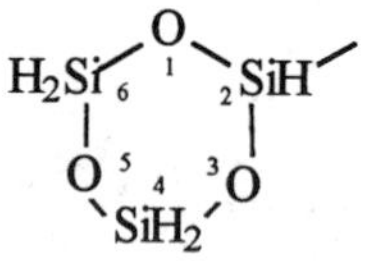

$$\underset{\substack{\text{SiH}_2}}{\overset{\text{O}}{\underset{\text{O}\,5\,4\,3\,\text{O}}{\text{H}_2\text{Si}\,6\,1\,2\,\text{SiH}}}}$$

Cyclotrisiloxanyl

$$\underset{8\quad7\quad6\quad\;5\quad4\quad3\quad\;2\quad\;1}{\text{CH}_3-\text{O}-\text{CH}_2-\text{O}-\text{CH}_2\text{CH}_2-\text{NH}-\text{CH}_2-}$$

5,7-Dioxa-2-azaoctyl

## Discussion

Given a knowledge of hydrocarbon nomenclature, the reader will surely find replacement nomenclature an easy method to use in naming the structures discussed in this chapter. The system, however, is not without problems, including the important one of familiarity. The concept of functionality, so essential to organic chemical communication, can be completely lost when a heteroatom is in the interior of a chain or in a ring named by replacement. There has been no resolution of the problem of end groups that involve principal characteristic groups: is the –NH– in the –NH–CH$_3$ group at the end of a chain described by "aza" or by "amine" or "amino"? The principal characteristic group is recognized in replacement nomenclature. Thus, other things being equal, an –NH$_2$ group at the end of a heteroacyclic chain will be expressed as "amine" or amino. Is it right to use a hydrocarbon name as the basis for the name of a ring that has only one or no carbon atoms as members? Despite these supposed defects, replacement nomenclature can go far toward the elimination of the multitude of meaningless trivial names for known heterocyclic structures: three-quarters of the two-component structures with trivial names in table 9.5 can be given replacement names based on indene or naphthalene, and two-thirds could be given "benzo" names. However, there are no provisions for using replacement names in the formation of other fusion names.

Replacement nomenclature should not be used with heterocycle parent names as a basis. Names such as "5-azaquinoline" or "7-azaindole" are not acceptable.

As will be seen in many of the chapters in this book that deal with characteristic groups, there are a variety of heterocyclic compounds that, in reality, are cyclic derivatives of a characteristic group expressed as a suffix; lactams, lactones, and cyclic anhydrides are examples. In the interest of simplifying nomenclature, it is often recommended that such structures be named as heterocycles. Today, cyclic amines, ethers, and sulfides are almost always named as heterocycles. Among the cyclic ethers, ethylene oxide is still acceptable, but other "oxide" names such as "tetramethylene oxide" or "styrene oxide" are obsolete.

In rare instances, heterocycles with contiguous formal double bonds are encountered (see also chapters 2 and 7). To avoid cumbersome prefixes such as "didehydrodihydro", the "$\delta$-convention" has been devised.[4] It employs the symbol $x\delta^c$, where "x" is the locant of the atom to which "c" double bonds are attached. This symbol can be used with both carbon and heteroatoms, as shown below:

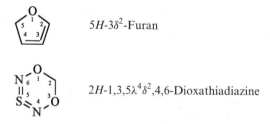

$5H$-$3\delta^2$-Furan

$2H$-$1,3,5\lambda^4\delta^2,4,6$-Dioxathiadiazine

Generalization of some of the names in table 9.5 should be done with caution. A name such as "naphthyridine (1,8- isomer shown)" or "phenanthroline (1,7- isomer shown)" suggests that other isomers with corresponding names having different locants are available. While this may be correct, limits to which these names may be taken have not been specified. Some of these isomers themselves have acceptable trivial names such as quinoxaline or phenanthridine, and locants for the heteroatoms are not cited in such names.

## Additional Examples

1.

$$H_2P-CH_2CH_2-O-CH_2-O-CH=CH-S-CH_2-O-CH_2-PH_2$$

13  12  11  10  9  8  7  6  5  4  3  2  1

3,8,10-Trioxa-5-thia-1,13-diphosphatridec-6-ene (CAS)
1,11-Bis(phosphanyl)-2,7,9-trioxa-4-thiaundec-5-ene (IUPAC)

[[[[2-[(2-Phosphanylethoxy)methoxy]ethenyl]thio]methoxy]methyl]phosphane (substitutive name)

2.

$$H_3C-O-[CH_2]_2-O-Si(CH_3)_2-O-[CH_2]_2-O-CO-NH-[CH_2]_6-NH-CO-O-CH_2CH_3$$

1  2  3-4  5  6  7  8-9  10  11  12  13-18  19  20

Ethyl 6,6-dimethyl-11-oxo-2,5,7,10-tetraoxa-12,19-diaza-6-silaeicosan-20-oate (CAS)
Ethyl 4,4-dimethyl-3,5,8-trioxa-4-silanonyl hexane-1,6-diyldicarbamate (IUPAC)

Ethyl 2-[(2-methoxyethoxy)dimethylsiloxy]ethyl hexane-1,6-diyldicarbamate (substitutive name)

3.

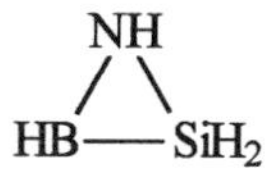

Azasilaboriridine
Azasilaboracyclopropane

4.

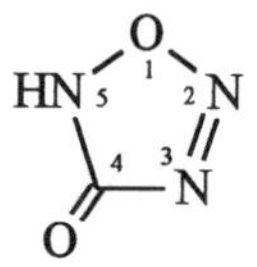

1,2,3,5-Oxatriazol-4(5H)-one
4,5-Dihydro-4-oxo-1,2,3,5-oxatriazole
1-Oxa-2,3,5-triazacyclopent-2-en-4-one

5.

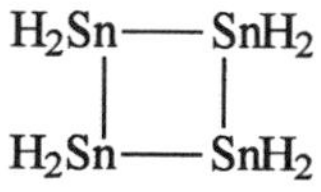

Cyclotetrastannane
Tetrastannetane

6.

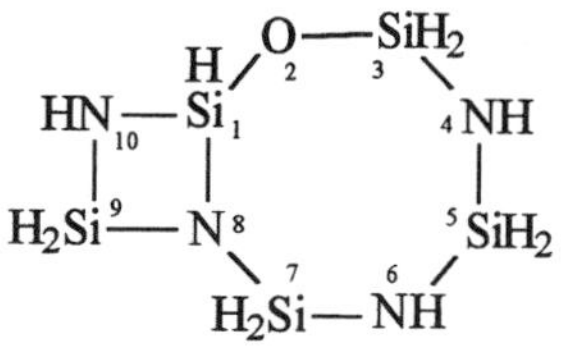

2-Oxa-4,6,8,10-tetraaza-1,3,5,7,9-pentasila-bicyclo[6.2.0]decane

7.

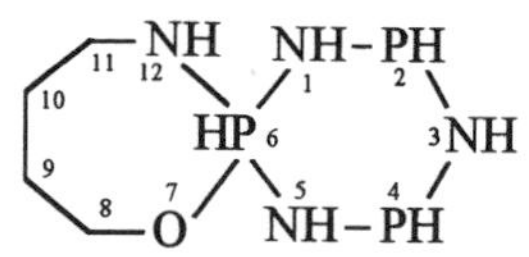

7-Oxa-1,3,5,12-tetraaza-2,4,6$\lambda^5$-triphospha-spiro[5.6]dodecane

8.

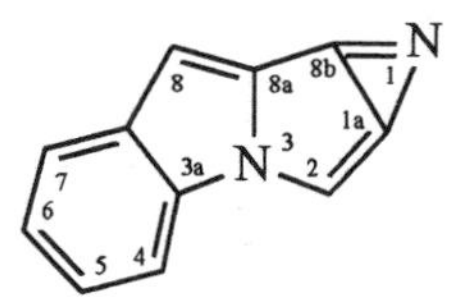

Azirino[2',3':3,4]pyrrolo[1,2-a]indole

9.

[1,2,4]Triazolo[3,4-*b*][1,3,4]thiadiazepine

10.

Tribenzo[*c*, *g*, *k*][1,2,5,10]tetraazacyclododecine

11.

1,3:7,10-Diepoxycyclohepta[*de*]naphthalene (IUPAC)
7,10-Epoxy-1,3-metheno-1*H*-cyclohepta[*de*]-2-
benzopyran (CAS)

12.

Pyrido[2,1,6-*de*]quinolizine

13.

5,8-Epoxy-3a,9-(epoxymethano)-3a*H*-benz[*f*]indene

14.

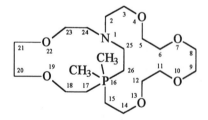

16,16-Dimethyl-4,7,10,13,19,22-hexaoxa-1-aza-16$\lambda^5$-
phosphabicyclo[14.8.2]hexaeicosane

## REFERENCES

1. International Union of Pure and Applied Chemistry, Organic Chemistry Division, Commission on Nomenclature of Organic Chemistry. Treatment of Variable Valence in Organic Chemistry (Lambda Convention) (Recommendations 1983). *Pure Appl. Chem.* **1984**, *56*, 769–778.
2. International Union of Pure and Applied Chemistry, Organic Chemistry Division, Commission on Nomenclature of Organic Chemistry. Revision of the Extended Hantzsch–Widman System of Nomenclature for Heteromonocycles (Recommendations 1982). *Pure Appl. Chem.* **1983**, *55*, 409–416.

3. International Union of Pure and Applied Chemistry, Organic Chemistry Division, Commission on Nomenclature of Organic Chemistry. Nomenclature of Fused Ring and Bridged Fused Ring Systems (Recommendations 1998). *Pure Appl. Chem.* **1998**, *70*, 143–216.
4. International Union of Pure and Applied Chemistry, Organic Chemistry Division, Commission on Nomenclature of Organic Chemistry. Nomenclature for Cyclic Organic Compounds with Contiguous Formal Double Bonds ($\delta$-Convention) (Recommendations 1988). *Pure Appl. Chem.* **1988**, *60*, 1395–1401.

# Groups Cited Only by Prefixes in Substitutive Nomenclature

In substitutive nomenclature, certain characteristic groups do not have suffix names and thus cannot be cited as a principal characteristic group (see chapter 3). They have no seniority and are therefore cited only as prefixes; the majority of these groups contain halogen, oxygen, or nitrogen. In this chapter, the nomenclature of compounds containing these characteristic groups and their chalcogen analogs are covered. Acid halides, acid halogenoids, and acid azides are discussed in chapter 11.

## Acceptable Nomenclature

Compounds containing one or more characteristic groups of the type shown in table 10.1 (see also chapter 3, table 3.3), each bonded to a nonacyl carbon atom, are named substitutively by combining the prefixes in alphabetical order with the name of the parent structure. Lowest locants are used only after assignment of locants for unsaturation, heteroatoms, principal characteristic groups, and any fixed numbering of the parent compound.

$CH_3-NO_2$                    Nitromethane

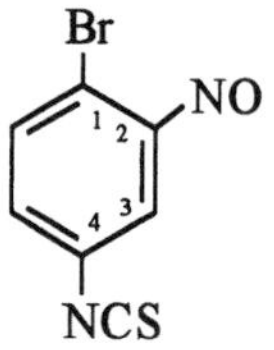

1-Bromo-4-isothiocyanato-2-nitrosobenzene

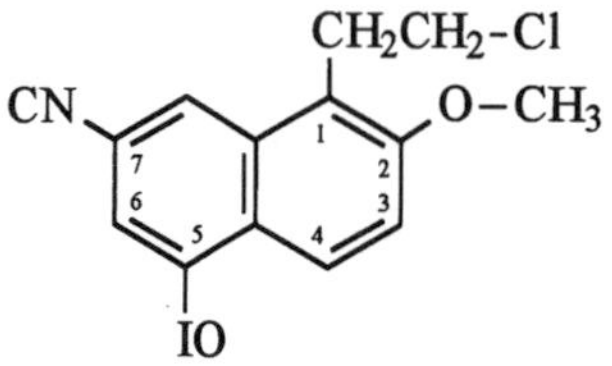

1-(2-Chloroethyl)-5-iodosyl-7-isocyano-2-methoxynaphthalene

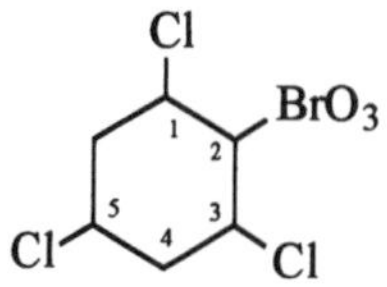

1,3,5-Trichloro-2-perbromylcyclohexane

Table 10.1. Some Prefixes Without Corresponding Suffix Names

| Group | Prefix | Group | Prefix |
|-------|--------|-------|--------|
| Br– | bromo | R–O– | R-oxy |
| Cl– | chloro | $N_2=$ | diazo |
| I– | iodo | $N_3–$ | azido |
| F– | fluoro | ON– | nitroso |
| OCl– | chlorosyl | $O_2N–$ | nitro |
| OBr– | bromosyl | CN– | isocyano |
| $O_2Cl–$ | chloryl | NCO–[a] | cyanato |
| $O_3Cl–$ | perchloryl | NCS–[a] | thiocyanato |
| HOO– | hydroperoxy | OCN– | isocyanato |
| R–OO– | R-peroxy | SCN– | isothiocyanato |
| HSO– | mercaptooxy[b] | | |

[a] Usually named as esters of cyanic or thiocyanic acid.
[b] Since 1993, IUPAC has recommended sulfanyloxy.

1-Bromo-2-chloro-4-hydroperoxybenzene

7-Bromo-2-(mercaptooxy)-1-nitronaphthalene
7-Bromo-2-(sulfanyloxy)-1-nitronaphthalene
(IUPAC)

1,1′-Methylenebis(4-isocyanatobenzene)
Bis(4-isocyanatophenyl)methane
Methylenedi-4,1-phenylene diisocyanate

The list in table 10.1 is by no means exhaustive; chalcogen analogs are treated similarly. Many of these groups appear in functional class nomenclature (see chapter 3). Some characteristic groups, such as ether and sulfide, are expressed by prefixes, but with different meanings depending on context. In substitutive nomenclature, the prefixes oxy and thio describe –O– and –S– not only as part of a substituting group, as in alkoxy or alkylthio, but also as a multiplying prefix in names such as 1,1′-oxydinaphthalene or as part of a multiplying prefix such as thiobis(methylene), $–CH_2–S–CH_2–$. They also appear in bridging prefixes, such as epoxy and epithio, which can be substituting prefixes that are not part of a parent name or they may be bridge components in a heterocyclic parent name, as in 1,4-epoxynaphthalene.

Some of these kinds of compounds can be alternatively, but usually less desirably, named by functional class nomenclature (see chapter 3); for example, ethyl bromide and 1,4-phenylene diisocyanate. Carbon tetrachloride is usually preferred over tetrachloromethane or tetrachlorocarbon, and the trivial names chloroform, fluoroform, bromoform, iodoform, phosgene (or the acyl halide names carbonyl (or carbonic) dichloride), and thiophosgene [or carbonothioyl (or carbonothioic) dichloride] are acceptable.

Compounds or organic groups in which *all* hydrogen atoms of the compound or group have been replaced by one kind of halogen atom can be named with "per" combined with the name of the compound or group, as in perfluoro, perchloro, perbromo, or periodo. This labor-saving device should be used with care (see Discussion); normal numerical prefixes with locants as needed are preferred. Parentheses should be used to indicate clearly the portion of the name to which "per" refers, as in (perfluorooctyl) acetate and perfluoro(octyl acetate).

A fully fluorinated parent hydride can be named with the prefix perfluoro, but the perfluoro prefix cannot apply if another substituent is present, since this would imply substitution for fluorine rather than hydrogen. If all but one of the hydrogen atoms in a compound or group have been substituted by the same kind of halogen atom, and the remaining hydrogen atom has been substituted by another kind of halogen atom, only the locant for that atom needs to be cited; CAS, however, cites all the locants.

$C_6F_5-CF_2CF_2CF_2CF_3$ 

Pentafluoro(perfluorobutyl)benzene
Perfluoro(butylbenzene)

$C_6F_5-CH_2CH_2CH_2CH_3$ 

Butylpentafluorobenzene
    (*not* Butylperfluorobenzene)

$C_6H_5-CF_2CF_2CF_2CF_3$ 

(Perfluorobutyl)benzene
(Nonafluorobutyl)benzene

$CF_3[CF_2]_4CF_2-I$
6        5-2      1

Tridecafluoro-1-iodohexane
    (*not* Perfluoro-1-iodohexane)
1,1,1,2,2,3,3,4,4,5,5,6,6-Tridecafluoro-6-iodohexane (CAS)

$CF_3[CF_2]_6CF_2-OH$
8        7-2      1

Heptadecafluorooctan-1-ol
    (*not* Perfluorooctan-1-ol)

$CF_3[CF_2]_6CH_2-OH$
8        7-2      1

2,2,3,3,4,4,5,5,6,6,7,7,8,8,8-Pentadecafluorooctan-1-ol
[*not* (Pentadecafluoroheptyl)methanol;
    *not* (Perfluoroheptyl)methanol]

## Discussion

Along with nitroso and nitro in table 10.1, one might expect to find *aci*-nitro ("*aci*" means "acid form") for the characteristic group HO–N(O)=. This prefix name is used by CAS and was recommended in the 1979 IUPAC rules but not included in the 1993 recommendations. In the latter, the group is viewed as a derivative of the parent compound azinic acid, $H_2N(O)$–OH, leading to names such as methylideneazinic acid for $CH_2 = N(O)$–OH. In 1993, IUPAC recommended hydroxynitroryl for HO–N(O)=, an additive name based on N(O)≡, which was called nitroryl by analogy to phosphoryl for P(O)≡. Other prefixes derived from nitroryl were also prescribed, such as hydronitroryl for HN(O)= (see chapter 22). For the present, it is hard to go wrong with the prefix hydroxynitroryl or the traditional *aci*-nitro as a compulsory prefix.

The seemingly simple "per device" is fraught with complications, most of which have been introduced to save wear and tear on keyboard keys and eyeballs when a long string of locants would otherwise be used. Most common is the problem of a "partially per" structure, exemplified by $CF_3[CF_2]_6CH_2$–OH, which has been named $1H,1H$-perfluorooctanol.[1] In itself, this method is reasonable, but it should not be extended beyond the point where it produces a name more cumbersome than the normal substitutive name. It should never be used for atoms other than hydrogen, as in $1Cl,1Cl$-perfluorooctanol. The "H device" has also been suggested, but is not recommended, for names not involving "per", as in $2H,2H$-decafluoropentane for $CF_3[CF_2]_2CH_2CF_3$. "Per" should never be used for atoms or groups other than simple halogen, as in permethylpentane or in perchlorylpentane, which might be interpreted as either $(O_2Cl)_3C[C(ClO_2)_2]_3C(ClO_2)_3$ or $CH_3[CH_2]_3CH_2$–$ClO_3$. Another extension seen from time to time is the use of "F" as a replacement for "perfluoro", as in "F-butane"; this, too, is not recommended.

Substituent prefix names derived from trivially named halogen compounds must be used with care. For example, the trivial name chloroform might be thought to lead to chloroformyl, but in fact this is one of the names for Cl–CO–, derived from the obsolete name chloroformic acid; the group $Cl_3C–$ is named trichloromethyl. The trivial names "nitroglycerin" and "chloropicrin" should be abandoned.

## Additional Examples

1. $CH_3–I$

   Iodomethane
   Methyl iodide

2. $CH_2=CH–Cl$

   Chloroethene
   Vinyl chloride

3. $Cl_3C–CHO$

   Trichloroacetaldehyde
   Trichloroethanal

4. $Cl_3C–NO_2$

   Trichloronitromethane

5. 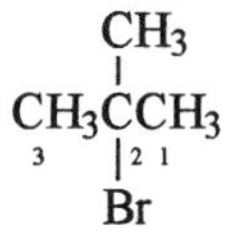 $\overset{\displaystyle CH_3}{\underset{\underset{\displaystyle Br}{|}}{\underset{3\ \ \ |2\ 1}{CH_3CCH_3}}}$

   2-Bromo-2-methylpropane
   *tert*-Butyl bromide
   2-Methylpropan-2-yl bromide

6. $\underset{4\quad\ 3\quad\ 2\quad\ \ 1}{O_2N–CH_2CH_2CH=CH_2}$

   4-Nitrobut-1-ene

7. $C_6H_5–CH_2–Cl$

   (Chloromethyl)benzene
   $\alpha$-Chlorotoluene
   Benzyl chloride

8. 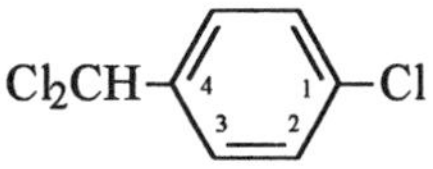 $Cl_2CH–\text{(ring)}–Cl$

   1-Chloro-4-(dichloromethyl)benzene
   $a,\alpha,4$-Trichlorotoluene
   4-Chlorobenzylidene dichloride

9. 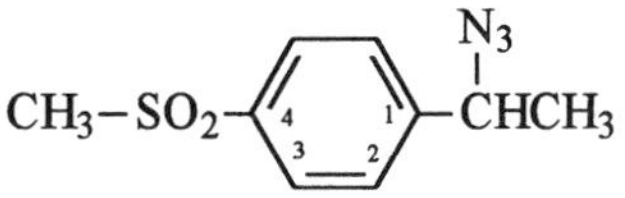 $CH_3–SO_2–\text{(ring)}–\overset{\displaystyle N_3}{\underset{|}{CHCH_3}}$

   1-(1-Azidoethyl)-4-(methylsulfonyl)benzene
   4-(1-Azidoethyl)phenyl methyl sulfone
   1-[4-(Methylsulfonyl)phenyl]ethyl azide

10. $C_6F_5–CF_3$

    Pentafluoro(trifluoromethyl)benzene
    Octafluorotoluene
    Perfluorotoluene

11.

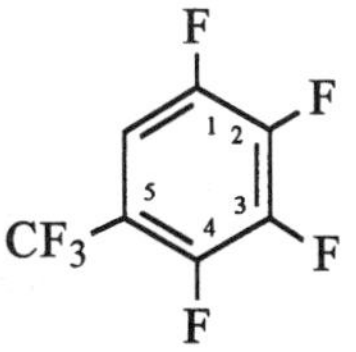

    1,2,3,4-Tetrafluoro-5-(trifluoromethyl)benzene
    $\alpha,\alpha,\alpha,2,3,4,5$-Heptafluorotoluene

12. $\underset{4\quad\ 3\quad\ 2\quad\ 1}{CH_3CH_2CH_2CH_2–O–CH_3}$

    1-Methoxybutane
    Butyl methyl ether

13.   Cl
      |                                          3-Chloro-1-(trifluoromethoxy)butane
      $CH_3CHCH_2CH_2-O-CF_3$                    3-Chlorobutyl trifluoromethyl ether
      4   3   2   1

14.   $CF_3CF_2CF_2CF_2-O-CFCl_2$               1-(Dichlorofluoromethoxy)nonafluorobutane
      4   3   2   1

15.   $CF_3CF_2CF_2CF_2-O-CF_3$                  Nonafluoro-1-(trifluoromethoxy)butane
      4   3   2   1                              Perfluorobutyl perfluoromethyl ether
                                                 Nonafluorobutyl trifluoromethyl ether

16.                                    OCN       1-Isocyanato-2-[(4-isocyanatophenyl)methoxy]-
                                                     benzene
      OCN⟨4  1⟩—$CH_2$—O—⟨2⟩          4-Isocyanatobenzyl 2-isocyanatophenyl ether
           3  2                                  1-Isocyanato-4-[(2-isocyanatophenoxy)methyl]-
                                                     benzene

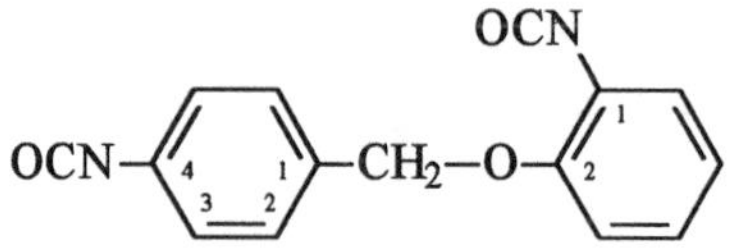

17.                                              1-Methyl-4-thiocyanatobenzene
      NCS—⟨4  1⟩—$CH_3$                          4-Methylphenyl thiocyanate
           3  2

18.   Cl  $C_6H_5$
      |    |                                     (3-Chlorobutan-2-yl)benzene
      $CH_3CHCHCH_3$                             (2-Chloro-1-methylpropyl)benzene
      4   3   2   1

## REFERENCE

1. Young, J. A. Revised Nomenclature for Highly Fluorinated Organic Compounds. *J. Chem. Doc.* **1974**, *14*, 98–100.

# 11

# Carboxylic Acids, Acid Halides, and Replacement Analogs

In this chapter, nomenclature for *carboxylic acids* (R–COOH) and their replacement analogs is considered. Chalcogen analogs of carboxylic acids contain groups such as –CS–OH, –CO–SeH, or –CS–SH; the sulfur analogs are called *thiocarboxylic acids*. Other analogs include *peroxycarboxylic acids*, with the –CO–OOH group, and nitrogen analogs such as *imido-carboxylic acids*, having the –C(=NH)–OH group. Also covered in this chapter are mono-carboxylic acids in which another part of the structure has been substituted with, for example, an amide group (*amic* and *anilic acids*) or an aldehyde group (*aldehydic acids*). Hydroxy, oxo, and certain amino derivatives will be found here, but *α-amino acids*, although they may be named by the principles described here, constitute a subclass of sufficient importance to justify a specialized nomenclature (see chapter 31). *Acid halides* and *pseudohalides* in which acidic –OH groups have been replaced by halogen or a pseudohalogen are discussed in this chapter; other acid derivatives such as esters, salts, and anhydrides are covered in chapter 12.

## Acceptable Nomenclature

*Unsubstituted acyclic monocarboxylic and dicarboxylic acids*, in which a –COOH group conceptually replaces one or both terminal –CH$_3$ groups of an acyclic hydrocarbon chain, are named by adding -oic acid or -dioic acid to the name of the hydrocarbon (see chapter 5) from which the final "e" has been elided if followed by a vowel. For the location of substituents and unsaturation, numbering of the chain begins with the carbon of a carboxy group.

$$\underset{6\quad5\quad4\quad3\quad2\quad1}{CH_3CH_2CH_2CH_2CH_2-COOH}$$
Hexanoic acid

$$\underset{4\quad3\quad2\quad1}{CH_3CH=CH-COOH}$$
But-2-enoic acid

$$\underset{7\quad6\text{-}2\quad1}{HOOC-[CH_2]_5-COOH}$$
Heptanedioic acid

$$\underset{7\quad6\text{-}4\quad3\quad2\quad1}{HOOC-[CH_2]_3\overset{\displaystyle C_6H_5}{\overset{|}{C}HCH_2-COOH}}$$
3-Phenylheptanedioic acid

Systematic and traditional trivial names for a number of common carboxylic acids are given in table 11.1. All are acceptable, but, except for methanoic and ethanoic aid, the systematic names are preferred.

*Polycarboxylic acids* with more than two –COOH groups attached to the same unbranched chain are named with suffixes such as tricarboxylic acid appended to the name of the parent hydrocarbon; carboxy groups need not terminate the parent chain. In keeping with

129

Table 11.1. Carboxylic Acid Names

| Systematic | Trivial | Formula |
|---|---|---|
| methanoic | formic | $HCOOH$ |
| ethanoic | acetic | $CH_3-COOH$ (2, 1) |
| propanoic | propionic | $CH_3CH_2-COOH$ (3, 2, 1) |
| butanoic | butyric | $CH_3[CH_2]_2-COOH$ (4, 3-2, 1) |
| 2-methylpropanoic | isobutyric[a] | $\begin{array}{c} CH_3 \\ \| \\ CH_3CH-COOH \end{array}$ (3, 2, 1) |
| pentanoic | valeric | $CH_3[CH_2]_3-COOH$ (5, 4-2, 1) |
| 2,2-dimethylpropanoic | pivalic[a] | $CH_3C(CH_3)_2-COOH$ (3, 2, 1) |
| dodecanoic | lauric | $CH_3[CH_2]_{10}-COOH$ (12, 11-2, 1) |
| tetradecanoic | myristic | $CH_3[CH_2]_{12}-COOH$ (14, 13-2, 1) |
| hexadecanoic | palmitic | $CH_3[CH_2]_{14}-COOH$ (16, 15-2, 1) |
| octadecanoic | stearic | $CH_3[CH_2]_{16}-COOH$ (18, 17-2, 1) |
| propenoic | acrylic | $CH_2=CH-COOH$ (3, 2, 1) |
| 2-methylpropenoic | methacrylic[a] | $\begin{array}{c} CH_3 \\ \| \\ CH_2=C-COOH \end{array}$ (3, 2, 1) |
| propynoic | propiolic | $CH\equiv C-COOH$ (3, 2, 1) |
| cis-octadec-9-enoic | oleic | $CH_3[CH_2]_7 \underset{H}{\overset{}{C}} = \underset{H}{\overset{}{C}} [CH_2]_7-COOH$ (18, 17-11, 10, 9, 8-2, 1) |
| ethanedioic | oxalic | $HOOC-COOH$ |
| propanedioic | malonic | $HOOC-CH_2-COOH$ (3, 2, 1) |
| butanedioic | succinic | $HOOC-[CH_2]_2-COOH$ (4, 3-2, 1) |
| pentanedioic | glutaric | $HOOC-[CH_2]_3-COOH$ (5, 4-3, 1) |
| hexanedioic | adipic | $HOOC-[CH_2]_4-COOH$ (6, 5-2, 1) |
| trans-butenedioic | fumaric | $\underset{HOOC}{\overset{H}{\diagdown}} C = C \overset{COOH}{\underset{H}{\diagup}}$ (4, 3, 2, 1) |
| cis-butenedioic | maleic | $\underset{H}{\overset{HOOC}{\diagdown}} C = C \overset{COOH}{\underset{H}{\diagup}}$ (4, 3, 2, 1) |
| benzenecarboxylic | benzoic | $C_6H_5-COOH$ |

*(Continued)*

Table 11.1. Continued

| Systematic | Trivial | Formula |
|---|---|---|
| benzene-1,2-dicarboxylic | phthalic | |
| benzene-1,3-dicarboxylic | isophthalic | |
| benzene-1,4-dicarboxylic | terephthalic | |
| naphthalenecarboxylic (2-isomer shown) | naphthoic | |
| pyridine-2-carboxylic | picolinic | |
| pyridine-3-carboxylic | nicotinic | |
| pyridine-4-carboxylic | isonicotinic | |

[a] These acids should not be substituted, since acceptable locants are unavailable.

the principle of maximizing the number of principal characteristic groups in the name of a parent compound, none of the –COOH groups should be denoted by the prefix carboxy (see the first example below). However, carboxy is used for a –COOH group as part of a substituent or in a structure with a more senior group such as a cationic center (see chapter 33). Since the –COOH group as a substituent is not part of the parent hydrocarbon, it is not included in the numbering of the chain.

$$\underset{8\;\;7\;\;\;\;6\;\;\;\;5\;\;\;4\;\;\;\;3\;\;\;2\;\;1}{CH_3CHCH_2CH_2CHCH_2CHCH_3}$$ with COOH groups at positions 2, 4, 7

Octane-2,4,7-tricarboxylic acid
   (*not* 4-Carboxy-2,7-dimethyloctanedioic acid)

$$\underset{5\;\;\;\;4\;\;\;\;3\;\;\;\;2\;\;\;\;\;1}{HOOC-CH_2CHCH_2-COOH}$$ with CH$_2$–COOH branch at position 3

3-(Carboxymethyl)pentanedioic acid

$$\left[\underset{\;\;\;\;\;\;\;\;\;\;2\;\;\;\;1\;\;\;\;\;N}{HOOC-CH_2CH_2-\overset{+}{N}(CH_3)_3}\right] Cl^-$$

(2-Carboxyethyl)trimethylammonium chloride
2-Carboxy-*N,N,N*-trimethylethanaminium chloride

*Cyclic and heterocyclic carboxylic acids* and other structures having a –COOH group as the principal characteristic group are named with suffixes such as -carboxylic acid, -dicarboxylic acid, and so on.

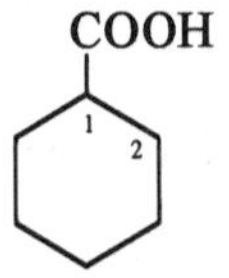

Cyclohexanecarboxylic acid

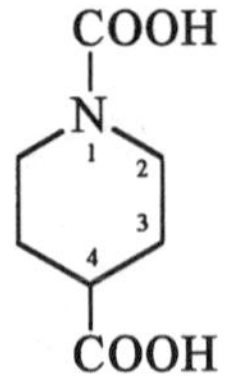

Piperidine-1,4-dicarboxylic acid

$$H_3SiSiH_2-COOH$$

Disilanecarboxylic acid

Trivial names abound in the field of carboxylic acids; some of the more common, and acceptable, names are given in table 11.1. However, in a desirable evolution, many are fading from use. Under the 1993 IUPAC recommendations, valeric, lauric, and myristic acids would be deleted from this table, and propionic, butyric, isobutyric, methacrylic, palmitic, stearic, oleic, glutaric, and adipic acids would be retained but not used with substituents attached to the carbon atoms.

Analogs of carboxylic acids in which one or more oxygen atoms are replaced by other atoms or groups are systematically named by functional replacement nomenclature (see chapter 3). In this method, prefixes or infixes denoting the replacing atoms or groups are added or inserted into the systematic or trivial name of the parent acid. Accordingly, replacement of oxygen in a –COOH group by sulfur, selenium, or tellurium atoms provides, as class names, *thiocarboxylic acids* and their analogs *selenocarboxylic acids* and *tellurocarboxylic acids*. Replacement of one oxygen atom by a sulfur atom in the –COOH group of a monocarboxylic acid without specifying the position or tautomeric form is denoted by the ending -thioic acid or -carbothioic acid in systematically named acids and by the prefix thio with acceptable trivial names of acids. Selenium and tellurium analogs are named with endings such as -selenoic acid or -carbotelluric acid or the prefixes seleno and telluro.

$$CH_3[CH_2]_6-C\{O,S\}H$$

Octanethioic acid

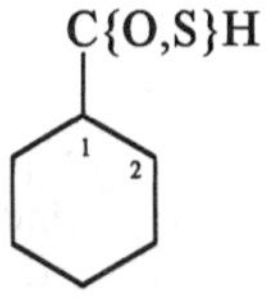

$$C\{O,S\}H$$

Cyclohexanecarbothioic acid

$$CH_3-C\{O,S\}H$$

Thioacetic acid
Ethanethioic acid

$$C_6H_5-C\{O,Se\}H$$

Selenobenzoic acid
Benzenecarboselenoic acid

Where two oxygen atoms are replaced, dithio, selenothio, and so on are used in a similar fashion.

$$CH_3[CH_2]_4-CSSH$$

Hexanedithioic acid

$$CH_3CH_2CH_2-C\{S,Se\}H$$

Selenothiobutyric acid
Butaneselenothioic acid

In polycarboxylic acids, oxygen atoms in the –COOH groups can be replaced by one or more sulfur, selenium, or tellurium atoms. When the replacement is symmetrical, suffixes such as -bis(thioic) acid, -dicarbothioic acid, -bis(selenothioic) acid, or -tricarboselenothioic acid can be used.

$$H\{O,S\}C-[CH_2]_4-C\{O,S\}H$$

6   5-2   1

Hexanebis(thioic) acid

$$HSeSeC-CH_2CH_2-CSeSeH$$

4  3  2  1

Butanebis(diselenoic) acid

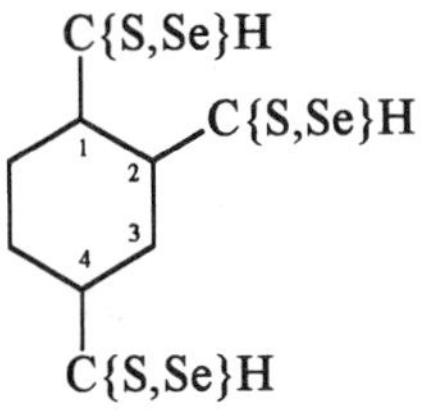

C{S,Se}H

Cyclohexane-1,2,4-tricarboselenothioic acid

For unsymmetrical replacement, or where all of the acid groups cannot be included in the name of the parent structure, prefixes such as thiocarboxy (H{O,S}C–), selenothiocarboxy (H{S,Se}C–), mercaptocarbonyl or sulfanylcarbonyl (HS–CO–), are employed. The senior acid group is the one having the greatest number of the most preferred chalcogen atom in the order O > S > Se > Te.

$$H\{O,S\}C-\overset{Cl}{\underset{}{CH}}[CH_2]_3-COOH$$

5   4-2   1

5-Chloro-5-(thiocarboxy)pentanoic acid

$$HO-\overset{Cl}{\underset{}{CSe-CH}}[CH_2]_3-C\{O,S\}H$$

5   4-2   1

5-Chloro-5-(hydroxycarbonoselenoyl)-
pentanethioic acid

An acceptable, but not encouraged, alternative for dibasic acids with trivial names in which one, three, or four (but not two, because of ambiguity) oxygen atoms have been replaced by other identical chalcogen atoms is to use prefixes such as thio or triseleno with the trivial name of the acid. Numbering begins with the senior acid group.

$$H\{O,S\}C-[CH_2]_3-COOH$$

5   4-2   1

Thioglutaric acid

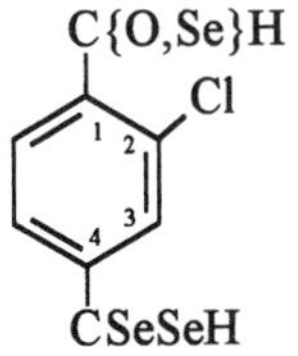

2-Chlorotriselenoterephthalic acid

Specificity between tautomeric forms is denoted by adding the terms O-acid or S-acid, for –OH or –SH, respectively, to the name of the acid. For numbering, an O-acid is preferred to an S-acid and an S-acid of a thioic acid is preferred to an Se-acid of a selenothioic acid.

$$CH_3-CO-SH$$

2   1

Thioacetic S-acid

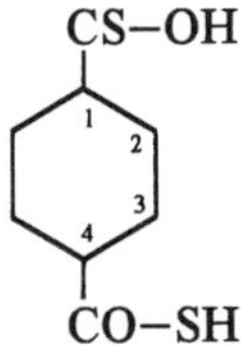

Cyclohexane-1,4-dicarbothioic
O,S-acid

$$CH_3CH_2CH_2-CS-SeH$$
$$\underset{4}{\phantom{C}}\underset{3}{\phantom{H}}\underset{2}{\phantom{C}}\underset{1}{\phantom{S}}$$    Butaneselenothioic *Se*-acid

*Imidic*, *hydrazonic*, and *hydroximic acids*, in which the carbonyl oxygen of a –COOH group is replaced by =NH, =NNH$_2$, or =N–OH, respectively, are named by replacing the endings of the corresponding carboxylic acid names with -imidic acid or -carboximidic acid for –C(=NH)–OH and -hydrazonic acid or -carbohydrazonic acid for –C(=NNH$_2$)–OH. For –C(=N–OH)–OH, IUPAC recommends -hydroximic acid or -carbohydroximic acid, while CAS uses *N*-hydroxy … imidic acid or *N*-hydroxy … carboximidic acid names; both are acceptable. These names are specific for these structures. The tautomeric amide, hydrazide, and hydroxyamide forms are preferred by CAS; they are discussed in chapters 18 and 19. Where nonspecificity of tautomeric forms is intended, the amide or hydrazide forms and names are generally preferred.

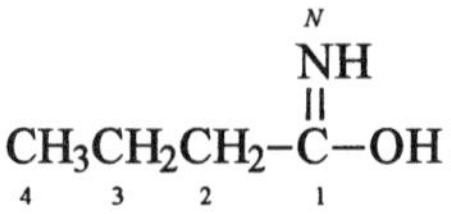

Butanimidic acid

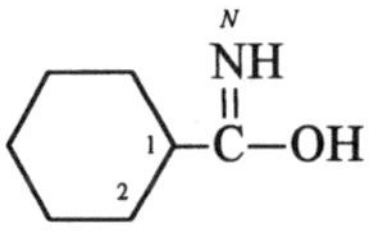

Cyclohexanecarboximidic acid

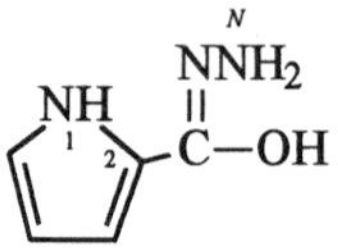

1*H*-Pyrrole-2-carbohydrazonic acid

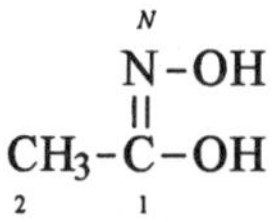

*N*-Hydroxyacetimidic acid
Acetohydroximic acid

The corresponding prefixes, with subunits cited alphabetically, are hydroxy(imino)methyl, hydrazono(hydroxy)methyl, and hydroxy(hydroxyimino)methyl. This treatment of prefixes means that these groups do not retain the identity they had as suffixes (compare -carboxy and -carboxylic acid) and that the "methyl" portion, when it is part of an acyclic chain, is included as a chain atom. If identity is to be preserved, a prefix name such as *C*-hydroxy-carbonimidoyl would have to be used for HO–C(=NH)–; the locant "*C*" is needed to avoid confusion with "*N*" substitution.

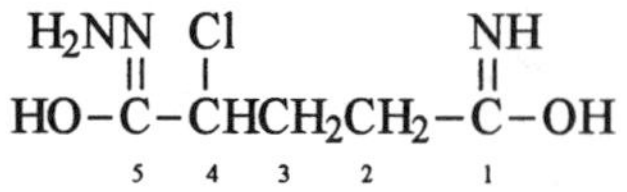

4-Chloro-5-hydroxy-5-hydrazonopentaneimidic acid
2-Chloro-5-hydroxy-5-iminopentanehydrazonic acid
4-Chloro-4-(*C*-hydroxycarbonohydrazonoyl)butaneimidic acid
2-Chloro-4-(*C*-hydroxycarbonimidoyl)butanehydrazonic acid

*Peroxycarboxylic acids* have the characteristic group –CO–OOH replacing the –COOH group in carboxylic acids. They are named with the prefixes peroxy, diperoxy, and so on, placed before an accepted trivial acid name or by means of suffixes such as -peroxoic acid, -carboperoxoic acid, -diperoxoic acid, and so on. Alternatively, prefixes such as peroxy, monoperoxy, diperoxy, and so on are placed before the name of an "oic" acid or inserted ahead of the suffix -carboxylic acid. With dibasic acids in which one acid function is peroxy, one or the other function must be named by a prefix to avoid ambiguity in locating other substituents; the prefix monoperoxy can be used with no substituents, but this technique should not be extended to polybasic acids with three or more acid groups. Under the IUPAC rules, the –COOH group is senior to the –CO–OOH group, while this order is reversed by CAS. The latter choice is recommended, since carboxy is a simpler prefix than hydroperoxycarbonyl. Performic, peracetic, and perbenzoic acid are commonly used trivial names, but their use is discouraged.

$$CH_3-CO-OOH$$
2    1

Peroxyacetic acid
Ethaneperoxoic acid
Peroxyethanoic acid

$$CH_3[CH_2]_6-CO-OOH$$
8       7-2      1

Octaneperoxoic acid
Peroxyoctanoic acid

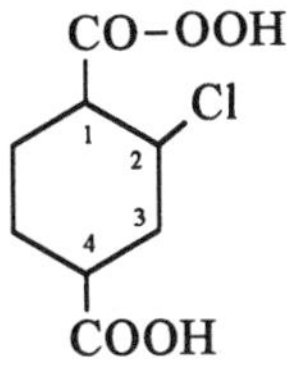

CO–OOH
Cl
1 2 3 4
COOH

4-Carboxy-2-chlorocyclohexanecarboperoxoic acid (CAS)
3-Chloro-4-(hydroperoxycarbonyl)cyclohexane-carboxylic acid (IUPAC)

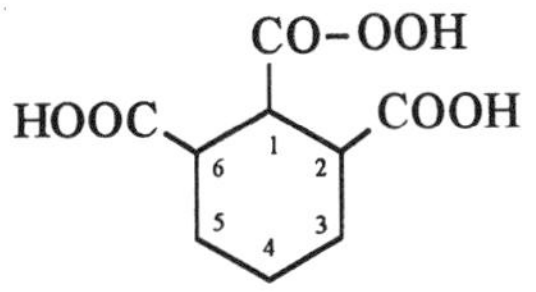

CO–OOH
HOOC    COOH
6 1 2 5 4 3

2,6-Dicarboxycyclohexanecarboperoxoic acid (CAS)
2-(Hydroperoxycarbonyl)cyclohexane-1,3-dicarboxylic acid (IUPAC)
2,6-Dicarboxycyclohexaneperoxycarboxylic acid

$$HOOC-CH_2CH_2-CO-OOH$$
3       2        1

3-Carboxypropaneperoxoic acid (CAS)
3-(Hydroperoxycarbonyl)propanoic acid (IUPAC)
Monoperoxysuccinic acid
4-Hydroperoxy-4-oxobutanoic acid

Cl
$$HOOC-CHCH_2CH_2-CO-OOH$$
4      3      2       1

4-Carboxy-4-chlorobutaneperoxoic acid (CAS)
2-Chloro-4-(hydroperoxycarbonyl)butanoic acid (IUPAC)
2-Chloro-5-hydroperoxy-5-oxopentanoic acid

Structures are readily envisioned in which the =O and –OH portions of the –COOH group are replaced by various combinations of the groups discussed above, leading to analogs such as –CS–OOH, –C(=NH)–SH, and so on. Such groups are named by the methods already described. In the case of the thioperoxy group, which can take the form –OSH or –SOH, the class name "acid" is preceded by OS- or SO-, respectively. Suffix forms take endings such as -carbo(thioperoxoic) acid. Examples of prefix names are given in table 11.2.

Illustrative examples of replacement in carboxylic acids are

$$CH_3-CS-OOH$$
2      1

(Thioperoxy)acetic OO-acid
Ethaneperoxothioic OO-acid

$$CH_3[CH_2]_6-CS-OSH$$
8       7-2      1

Octanethio(thioperoxoic) OS-acid

Table 11.2. Prefixes for Thioperoxy Acids

| | |
|---|---|
| H{S,O}–CO— | carbono(thioperoxoyl) |
| | (thiohydroperoxy)carbonyl |
| H{S,O}–CS— | carbono(thioperoxo)thioyl |
| | (thiohydroperoxy)carbonothioyl |
| HSO–CO— | (mercaptooxy)carbonyl |
| | (sulfanyloxy)carbonyl |
| | carbono(SO-thioperoxoyl)$^a$ |
| HOS–CO— | sulfenocarbonyl |
| | (hydroxythio)carbonyl |
| | (hydroxysulfanyl)carbonyl |
| | carbono(OS-thioperoxoyl)$^a$ |
| HSS–CO— | (thiosulfeno)carbonyl |
| | carbono(dithioperoxoyl) |
| | disulfanylcarbonyl |
| | (mercaptothio)carbonyl |
| HSS–CS— | carbono(dithioperoxo)thioyl |
| | (thiosulfeno)carbonothioyl |
| | (dithiohydroperoxy)carbonothioyl |
| | disulfanylcarbonothioyl |

$^a$ SO- and OS- specify attachment of O and S, respectively, to the carbonyl group.

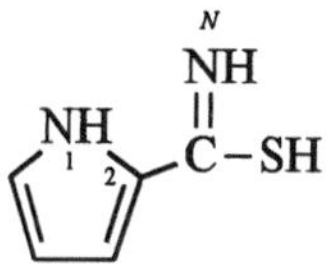

1H-Pyrrole-2-carboximidothioic acid

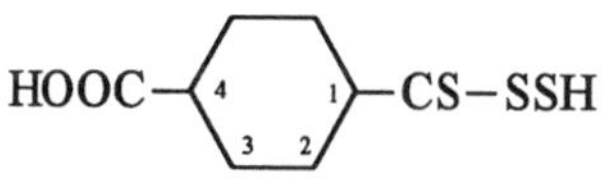

4-Carboxycyclohexanecarbo(dithioperoxo)thioic acid (CAS)
4-[(Dithiohydroperoxy)carbonothioyl]-cyclohexanecarboxylic acid (IUPAC)
4-[(Carbono(dithioperoxo)thioyl]-cyclohexanecarboxylic acid

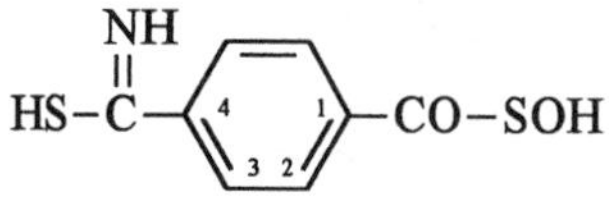

4-[Imino(mercapto)methyl](thioperoxy)benzoic SO-acid
4-(Imidothiocarboxy)benzenecarbo(thioperoxoic) SO-acid
4-(C-Mercaptocarbonimidoyl)-benzenecarbo(thioperoxoic) SO-acid

Although substituted carboxylic acids can be named systematically, many trivial names are retained, and in a few cases specialized nomenclature is acceptable; indeed, with *amino acids*, the specialized nomenclature (see chapter 31) is preferred. The names of *aldehydic acids*, which conceptually result when one –COOH group of a dicarboxylic acid has been replaced by a –CHO group, are derived from trivial names of dicarboxylic acids (see table 11.1) by changing the -ic acid ending to -aldehydic acid; numbering begins with the –COOH group. These compounds can also be named with the prefixes formyl, oxo, or oxomethyl. Chalcogen analogs are named systematically.

$$OHC–CH_2CH_2–COOH$$
$$\quad\quad 3\quad\quad 2\quad\quad 1$$

Succinaldehydic acid
3-Formylpropanoic acid
4-Oxobutanoic acid

Similarly, when one –COOH of a dicarboxylic acid is replaced by a carboxamide group, –CO–NH$_2$, the resulting *amic acid* may be named by changing -ic acid in the trivial name to -amic acid. An *N*-phenyl amic acid is named in a similar way using the ending *anilic acid*; numbering begins with the carboxy group, and primed locants are used for positions on the *N*-phenyl group.

5-Chlorophthalamic acid
2-Carbamoyl-5-chlorobenzoic acid

4′-Bromomalonanilic acid
[(4-Bromophenyl)carbamoyl]acetic acid
3-[(4-Bromophenyl)amino]-3-oxopropanoic acid

A number of hydroxy, oxo, and amino acids (other than $\alpha$-amino acids) have acceptable trivial names, to be used without further substitution. Among them are glycolic acid, HO–CH$_2$–COOH; lactic acid, CH$_3$CH(OH)–COOH; glyceric acid, HO–CH$_2$CH(OH)–COOH; tartaric acid, HOOC–[CH(OH)]$_2$–COOH; benzilic acid, (C$_6$H$_5$)$_2$C(OH)–COOH; and glyoxylic acid, OHC–COOH. Others, such as salicylic acid, 2–HO–C$_6$H$_4$–COOH; pyruvic acid, CH$_3$–CO–COOH; acetoacetic acid, CH$_3$–CO–CH$_2$–COOH; and anthranilic acid, 2-H$_2$N–C$_6$H$_4$–COOH, may be substituted, but systematic names are preferred. Chalcogen and nitrogen analogs are named systematically; chalcogen prefixes are not used.

The class name *acyl* is applied to substituting groups formed by the removal of the –OH group from carboxylic acids and their analogs. When acid names have the ending -oic acid or -dioic acid (see table 11.1) and –OH is removed from all acid groups, those endings are replaced with -oyl or -dioyl to form the names of the corresponding acyl groups. Acids with names ending in carboxylic acid form acyl group names ending in carbonyl.

Octanoyl

Octanedioyl

Cyclohexanecarbonyl (IUPAC)
Cyclohexylcarbonyl (CAS)

In the last example, cyclohexanecarbonyl illustrates the type of name used in functional class nomenclature to give names such as cyclohexanecarbonyl chloride (see below). Names such as cyclohexylcarbonyl may be employed when the acyl group is a substituent. The latter is followed by CAS, but IUPAC now recommends that cyclohexanecarbonyl names also be used as substitutive prefixes. Removal of the –OH group from one or more, but not all, carboxy groups in a polybasic acid generates acyl groups in which –COOH groups are denoted by carboxy and the carbonyl groups indicated by -oyl or carbonyl.

4-Carboxycyclohexanecarbonyl (IUPAC)
(4-Carboxycyclohexyl)carbonyl (CAS)

Acids with trivial names having the ending -ic acid (see table 11.1) form acyl group names ending in -oyl or, in the following cases, -yl: formyl, HCO–; acetyl, CH$_3$–CO–; propionyl, CH$_3$CH$_2$–CO–; butyryl, CH$_3$CH$_2$CH$_2$–CO–; oxalyl, –CO–CO–; malonyl, –CO–CH$_2$–CO–; and succinyl, –CO–CH$_2$CH$_2$–CO–. Acyl groups also use -oyl when they are derived from hydroxy, aldehydic, amic, and anilic acids with trivial names as in succinaldehydoyl or malon-aniloyl; for acyl groups derived from $\alpha$-amino acids, see chapter 31.

The foregoing methods also apply to acyl groups derived from replacement analogs of carboxylic acids. Infixes such as -thio-, -imid-, and so on, are inserted into acid names to produce endings such as -thioyl, -carbothioyl, -carbohydrazonoyl, -imidoyl, and -carbox-imidoyl. For the nitrogen analogs, acyl group names such as benzimidoyl, derived from the corresponding acids, are much preferred over additive prefixes such as *C*-phenylcarbonimid-oyl, in which the "*C*" locant is needed to avoid ambiguity.

$\underset{4\quad3\quad2\quad1}{CH_3CH_2CH_2-CS-}$          Butanethioyl

$-CS-\overset{4\quad1}{\underset{3\quad2}{\langle\quad\rangle}}N-CS-$          Piperidine-1,4-dicarbothioyl (IUPAC)
          Piperidine-1,4-diylbis(carbonothioyl) (CAS)

$\underset{6\quad5\text{-}2\quad1}{CH_3[CH_2]_4-}\overset{\overset{N}{NH}}{\underset{}{\overset{\|}{C}}}-$          Hexanimidoyl

$\overset{\overset{N}{NNH_2}}{\underset{}{\overset{\|}{C}}}-$          Cyclopentanecarbohydrazonoyl
          *C*-Cyclopentylcarbonohydrazonoyl
              (Cyclopentylcarbonohydrazonoyl is ambiguous)

$\overset{\overset{N}{N-OH}}{\underset{}{\overset{\|}{C}}}-$          Benzohydroximoyl
          Benzenecarbohydroximoyl
          *N*-Hydroxybenzimidoyl
          *N*-Hydroxy(phenyl)carbonimidoyl

When used as multiplying groups in symmetrical compounds or where both free bonds are attached to a single atom, the acyl group takes the forms carbonothioyl, carbonimidoyl, and so on.

$CH_3-N=C=$   or   $-\overset{\overset{N-CH_3}{\|}}{C}-$          *N*-Methylcarbonimidoyl

$-\overset{\overset{NH}{\|}}{C}\overset{4\quad1}{\underset{3\quad2}{\langle\quad\rangle}}\overset{\overset{NH}{\|}}{C}-$          1,4-Benzenedicarboximidoyl
          *C,C'*-1,4-Phenylenebis(carbonimidoyl)
          [1,4-Phenylenebis(carbonimidoyl) is ambiguous]

It is often useful to supplant acyl names with additive group names, particularly for groups derived from carbonic acid and its analogs (see below). For example, while chloroformyl is still acceptable for Cl–CO–, chlorocarbonyl is now recommended, and chlorooxomethyl can be used; the functional replacement name is carbonochloridoyl. CAS prefers aminocarbonyl over carbamoyl or carbonamidoyl, and hydroxy(imino)methyl over *C*-hydroxycarbonimidoyl.

*Acyl halides* (also called *acid halides*) and *acyl pseudohalides* are classes of compounds in which acyl group names are used. Such compounds are named by citing the name of the acyl group followed, as separate words, by the name(s) of a halide or pseudohalide such as brom-ide, cyanide, isocyanate, azide, and so on, in alphabetical order. In CAS, only the most senior acyl halide group is named in this way as the principal characteristic group, and the remaining groups are named substitutionally. Seniority among halides and pseudohalides is in the order –F > –Cl > –Br > –I > –N$_3$ > –NCO > –NCS > –NC > –CN; note that –OCN is not included because its acyl derivatives are usually viewed as anhydrides and named as such (see chapter 12). Compounds of the type RCO–CN are named as 2-oxonitriles or acyl cyanides (see chapter 20). Where different types of acyl groups (carbonyl, sulfonyl, and so on) are present, the principal characteristic group is that derived from the most senior acid (see chapter 3), and the remaining acyl groups are named substitutionally or by means of prefixes such as chlorocarbonyl or carbonochloridoyl.

$$CH_3-CO-Br$$
$$\phantom{CH_3-}2\phantom{-CO-}1$$

Acetyl bromide

$$CH_3[CH_2]_3-CO-NCO$$
$$\phantom{CH_}5\phantom{[CH_2]}4\text{-}2\phantom{-CO-}1$$

Pentanoyl isocyanate
Valeryl isocyanate

$$Cl-CO-CH_2-CO-Br$$

Malonyl bromide chloride
(Bromocarbonyl)acetyl chloride
3-Bromo-3-oxopropanoyl chloride
Carbonobromidoylacetyl chloride

$$\overset{\displaystyle Br}{\underset{\displaystyle |}{}}$$
$$OCN-CO-CH[CH_2]_5-CO-Cl$$
$$\phantom{OCN-C}8\phantom{-CH}7\phantom{[CH_2]}6\text{-}2\phantom{-CO-}1$$

7-Bromo-8-isocyanato-8-oxooctanoyl chloride
7-Bromooctanedioyl 1-chloride 8-isocyanate
7-Bromo-7-carbonoisocyanatidoylheptanoyl chloride

Cyclohexanecarbothioyl azide

$$Cl-CO-\text{(ring)}-CO-Cl$$

Terephthaloyl dichloride
1,4-Benzenedicarbonyl dichloride

$$NC-CO-\text{(ring)}-CO-O-CH_3$$

Methyl 4-(cyanocarbonyl)benzoate
Methyl 4-carbonocyanidoylbenzoate

Carbonic acid, often considered to be an inorganic acid, is, in its derivatives and analogs, clearly related to organic carboxylic acids. Many of the analogs are named by the methods described above, as in dithiocarbonic acid (which is ambiguous and only states that two O have been replaced by S) or carbonodithioic acid, $C\{O, S, S\}H_2$; carbonimidic acid, $HO-C(=NH)-OH$, and its tautomer, carbamic acid, $H_2N-COOH$; and monoperoxy-carbonic acid or carbonoperoxoic acid, $HOO-C(O)-OH$. Carbamic acid and oxamic acid, $H_2N-CO-COOH$, are retained as contractions of the names carbonamidic acid and oxalamic acid. A compound such as Cl–COOH is commonly named chloroformic acid, but the functional replacement method (see chapter 3), which leads to the name carbonochloridic acid, is used by CAS for compounds of the type X–COOH and its analogs, in which X may be –Cl, –CN, –NCO, and so on.

A variety of polycarbonic acids and the corresponding analogs is possible. Symmetrical polycarbonic acids are named dicarbonic acid, tricarbonic acid, and so on; analogs are similarly named with appropriate infix and prefix terms. To differentiate among isomers, CAS uses a name plus a line and/or structural formula to indicate number and position of replacement atoms.

$$\overset{\displaystyle NH\phantom{xx}NH}{\underset{\displaystyle HO-C-O-C-OH}{\| \phantom{xxx} \|}}$$

Dicarbonimidic acid

$$HS-CO-O-CO-O-CO-SH$$

Dithiotricarbonic $S,S'$-acid

$$HO-CS-NH-CS-OH$$

Imidodicarbonothioic $O,O'$-acid

$$HOOC-OO-COOH$$

Peroxydicarbonic acid

Many polycarbonic acid analogs may be viewed as anhydrides or thioanhydrides and can be named as such (see chapter 12). This method is particularly applicable to unsymmetrical structures. It cannot be applied to nitrogen analogs of anhydrides.

$$\underset{HO-C-S-C-OH}{\overset{\displaystyle NH \quad\; NH}{\overset{\displaystyle \|\quad\;\; \|}{}}}$$        Carbonimidic thiomonoanhydride

$$\underset{HOOC-O-C-OH}{\overset{\displaystyle NH}{\overset{\displaystyle \|}{}}}$$        Carbonic carbonimidic anhydride

*Ortho acids*, R–C(OH)$_3$, the hypothetical hydrated forms of carboxylic acids, are best named as triols (see chapter 15). Orthocarbonic acid, C(OH)$_4$, is better named methanetetrol.

## Discussion

Since -oic acid refers to =O and –OH replacing the three hydrogen atoms of a –CH$_3$ group terminating an unbranched carbon chain, the name and numbering of the chain still includes the terminal carbon atoms. Specific locants for the acid group(s) are thus unnecessary. The suffix -carboxylic acid, on the other hand, denotes a –COOH group, including the carbon atom. That atom is therefore not part of the parent structure to which it is attached and is not included in its numbering; locants defining the position of –COOH groups are usually required.

In the older literature, Greek letter locants were traditionally used to designate substituent positions in trivial names of carboxylic acids; these locants began with the carbon atom adjacent to the –COOH group. Thus, $\alpha$-chloropropionic acid was the name for 2-chloroprop-anoic acid. Today, Greek letter locants are used mainly in conjunctive nomenclature (see chapter 3), which is frequently seen in the names of carboxylic acids, notably in indexes such as those of CAS. In this system, 3-(naphthalen-1-yl)propanoic acid is the same as 1-naphthalenepropanoic acid, and Greek letter locants are used for positions on the side chain, beginning with the carbon atom next to the –COOH group.

In the seniority order for compound classes (see chapter 3, table 3.1), acids rank just above anhydrides and below anionic groups. Among acids themselves, the –COOH group is senior in the IUPAC list, followed by the –CO–OOH group, while the reverse order is followed by CAS. The recommended full descending order of seniority among carbon acids is peroxy acids > carboxylic acids > carbohydrazonic acids > carboximidic acids > carbohydroximic acids; within each group of acids, the decreasing order for the chalcogen atoms of the suffixes is O > S > Se > Te, and within each type of chalcogen analog, the order is –CS–OH > –CO–SH > –CS–SH. Observance of this seniority order is necessary to prevent ambiguity when another substituent is present, as in

$$\underset{\underset{\scriptstyle 4 \quad\; 3 \quad\;\; 2 \qquad 1}{HO-C-CHCH_2CH_2-CO-OSH}}{\overset{\displaystyle HN \quad COOH}{\overset{\displaystyle \|\quad\;\; |}{}}}$$

4-Carboxy-4-[hydroxy(imino)methyl]butane(thioperoxoic) *OS*-acid
2-[Hydroxy(imino)methyl]-4-[(mercaptooxy)carbonyl]butanoic acid
2-[Hydroxy(imino)methyl]-5-[(mercaptooxy)-5-oxopentanoic acid
4-Carboxy-5-hydroxy-5-iminopentane(thioperoxoic) *OS*-acid

Many of the traditional trivial names for acids are easily confused, as with the stereo-isomers angelic acid and tiglic acid; some trivial names, such as valproic acid (2-propylpenta-noic acid), are used in sciences other than chemistry. These names are not encouraged: a systematic name tells it as it is. "Diacetic acid" is ambiguous and unacceptable; at one time CAS used "diacetic acid" as a name for CH$_3$–CO–CH$_2$–COOH. Pseudosystematic names like "undecylenic acid", "heptylic acid", and "nonoic acid" should never be used. The trivial names "*o*-, *m*-, or *p*-phthalic acid" are not acceptable in place of phthalic, isophthalic, or terephthalic acid. The terms "thiolo" and "thiono" have been proposed[1] but never officially adopted as a means of distinguishing the positions of sulfur in thio acids; "*S*-acid" and "*O*-acid" are preferred for this purpose. Thio, seleno, and so on, should not be used as a prefix to trivial names of substituted acids: "thiosalicylic acid" has been incorrectly used for 2-mercaptobenzoic acid, as has "thioglycolic acid" for mercaptoacetic acid. Under the IUPAC rules, the latter would be named sulfanylacetic acid.


## Additional Examples

1. $Cl-CH_2CH_2-COOH$
   $\phantom{Cl-}{}_3\phantom{-CH_2}{}_2\phantom{CH_2-}{}_1$

   3-Chloropropanoic acid

2. 

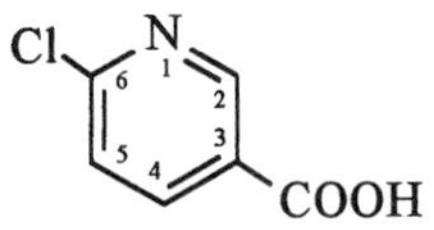

   6-Chloropyridine-3-carboxylic acid
   6-Chloronicotinic acid

3. $(CH_3)_3Si-O-SiH_2-COOH$
   $\phantom{(CH_3)}{}_3\phantom{Si-O-}{}_2\phantom{SiH_2-}{}_1$

   3,3,3-Trimethyldisiloxane-1-carboxylic acid

4. 
   $\phantom{HOOC-[CH_2]_4}\overset{\textstyle Cl}{\underset{\textstyle|}{\phantom{C}}}$
   $HOOC-[CH_2]_4CHCH_2-COOH$
   $\phantom{HOOC-}{}_8\phantom{[}{}_{7\text{-}4}\phantom{CH_2]_4}{}_3\phantom{CH}{}_2\phantom{CH_2-}{}_1$

   3-Chlorooctanedioic acid

5. 

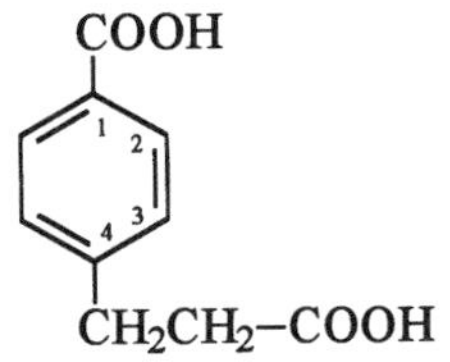

   4-(2-Carboxyethyl)benzoic acid
   3-(4-Carboxyphenyl)propanoic acid
   4-Carboxybenzenepropanoic acid

6. 

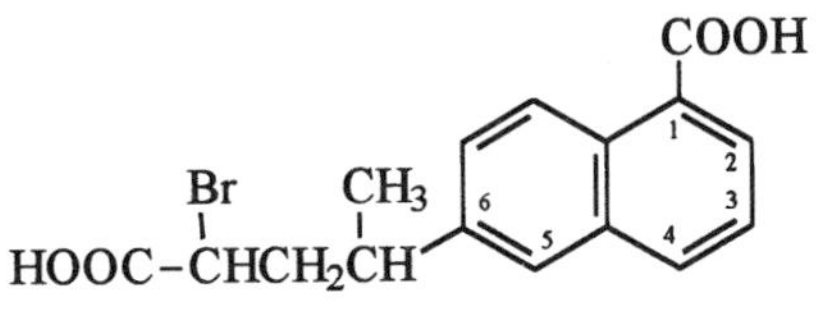

   6-(3-Bromo-3-carboxy-1-methylpropyl)-naphthalene-1-carboxylic acid
   6-(3-Bromo-3-carboxy-1-methylpropyl)-1-naphthoic acid
   6-(4-Bromo-4-carboxybutan-2-yl)-naphthalene-1-carboxylic acid
   2-Bromo-4-(5-carboxy-2-naphthyl)-pentanoic acid
   α-Bromo-5-carboxy-γ-methyl-2-naphthalenebutanoic acid

7. 

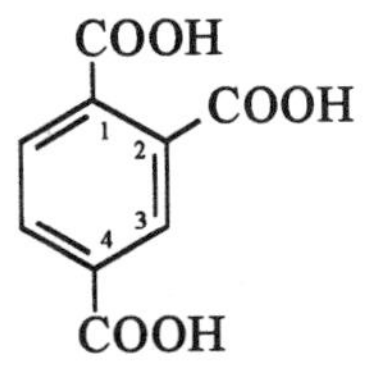

   Benzene-1,2,4-tricarboxylic acid

8. $HOOC-CH_2CH_2-O-CH_2CH_2-COOH$
   $\phantom{HOOC-}{}_{3'}\phantom{CH_2CH_2-O-}{}_3\phantom{CH}{}_2\phantom{CH_2-}{}_1$

   3,3'-Oxydipropanoic acid

9.

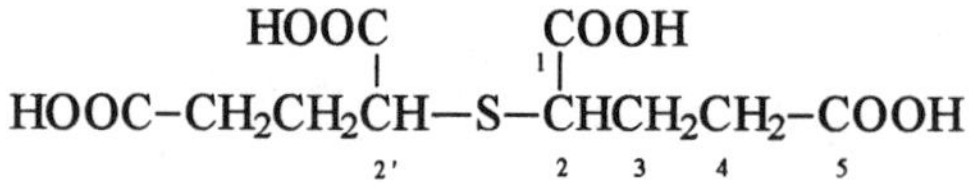

$$\underset{2'}{HOOC-CH_2CH_2}\overset{\overset{\displaystyle HOOC}{|}}{\underset{2}{CH}}-S-\overset{\overset{\displaystyle COOH}{|}}{\underset{3}{CH}}\underset{4}{CH_2CH_2}\underset{5}{-COOH}$$

2,2′-Thiodiglutaric acid
2,2′-Thiodipentanedioic acid

10.

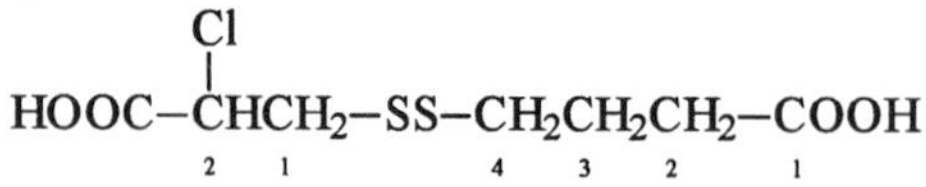

$$\underset{2}{HOOC-}\overset{\overset{\displaystyle Cl}{|}}{\underset{1}{CH}}CH_2-SS-\underset{4}{CH_2}\underset{3}{CH_2}\underset{2}{CH_2}\underset{1}{-COOH}$$

4-[(2-Carboxy-2-chloroethyl)-
  dithio]butanoic acid
4-[(2-Carboxy-2-chloroethyl)-
  disulfanyl]butanoic acid
2-Chloro-4,5-dithianonanedioic
  acid

11.

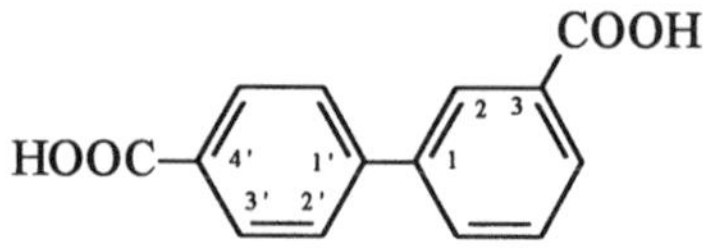

Biphenyl-3,4′-dicarboxylic acid

12.

$$\underset{3}{CH_3}\underset{2}{CH_2}\underset{1}{-CSSH}$$

Dithiopropionic acid
Propanedithioic acid

13.

$$H\{S,O\}C-CH_2CH_2-C\{O,S\}H$$

Butanebis(thioic) acid

14.

$$\underset{2'}{HOOC-CH_2}-SS-\underset{2}{CH_2}\underset{1}{-COOH}$$

2,2′-Dithiobis(acetic acid)
2,2′-Disulfanediylbis(acetic acid)

15.

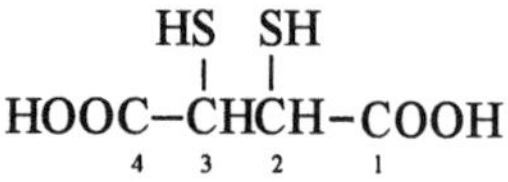

$$HOOC-\overset{\overset{\displaystyle HS}{|}}{\underset{4}{}}\underset{3}{CH}\overset{\overset{\displaystyle SH}{|}}{\underset{}{CH}}\underset{2}{-}\underset{1}{COOH}$$

2,3-Dimercaptosuccinic acid
2,3-Dimercaptobutanedioic acid
2,3-Disulfanylbutanedioic acid

16.

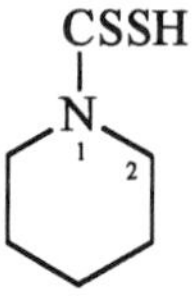

Piperidine-1-carbodithioic acid

17.

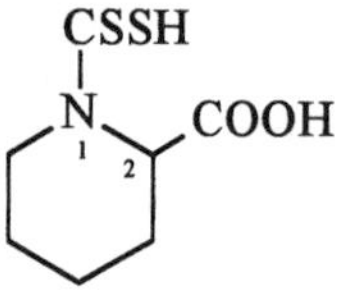

1-(Dithiocarboxy)piperidine-2-carboxylic
  acid

18.

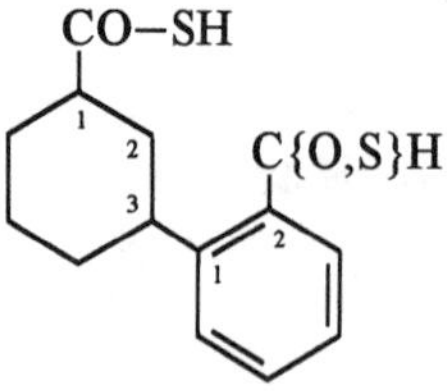

3-[2-(Thiocarboxy)phenyl]cyclo-
  hexanecarbothioic *S*-acid
2-[3-(Mercaptocarbonyl)cyclohexyl]-
  thiobenzoic acid
2-[3-(Sulfanylcarbonyl)cyclohexyl]-
  benzenecarbothioic acid

19.

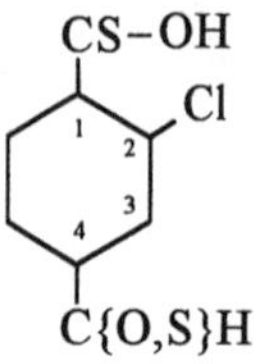

2-Chloro-4-(thiocarboxy)cyclo-
  hexanecarbothioic *O*-acid
    (the known tautomer is preferred to
    the unknown tautomer)

20. 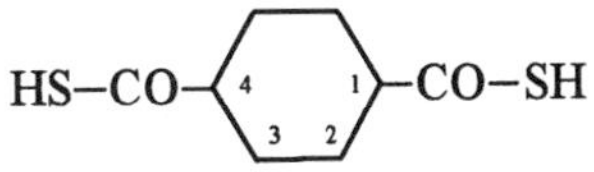   HS−CO−⟨4  1⟩−CO−SH          Cyclohexane-1,4-dicarbothioic $S,S'$-acid
      (3  2)

21. 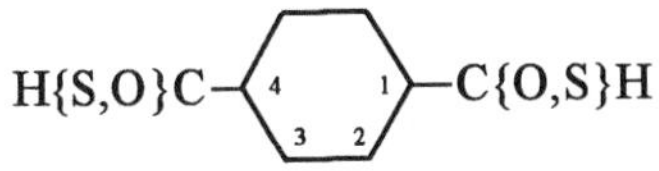   H{S,O}C−⟨4  1⟩−C{O,S}H          Cyclohexane-1,4-dicarbothioic acid
      (3  2)

22.   HS−CS−SH          Carbonotrithioic acid
                        Trithiocarbonic acid

23. 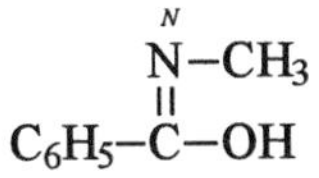   $\overset{N}{\underset{\parallel}{N}}$−CH₃          $N$-Methylbenzimidic acid
      C₆H₅−C−OH          $N$-Methylbenzenecarboximidic acid

24. 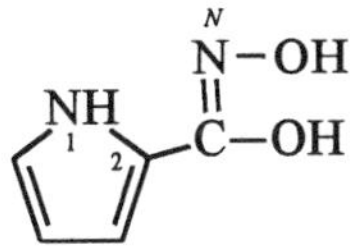   $\overset{N}{N}$−OH          1$H$-Pyrrole-2-carbohydroximic acid
      NH   C−OH          $N$-Hydroxy-1$H$-pyrrole-2-carboximidic
      ⟨1  2⟩                acid

25. 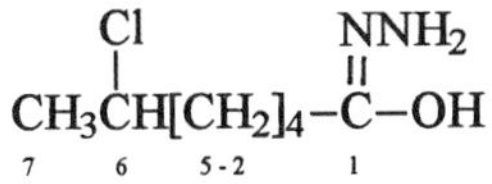   Cl        NNH₂          6-Chloroheptanehydrazonic acid
      CH₃CH[CH₂]₄−C−OH
      7   6   5-2    1

26. 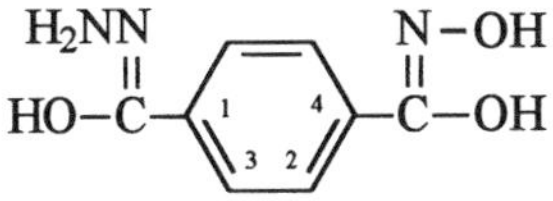   H₂NN        N−OH          4-[Hydroxy(hydroxyimino)methyl]-
      HO−C−⟨1  4⟩−C−OH             benzohydrazonic acid
          (3  2)             4-(Dihydroxycarbonimidoyl)-
                                benzenecarbohydrazonic acid

27. 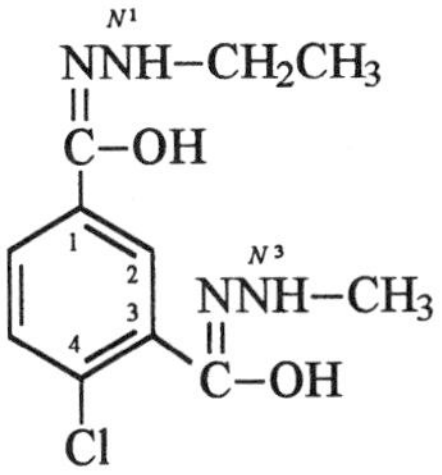   $\overset{N^1}{}$          4-Chloro-$N^1$-ethyl-$N^3$-methylbenzene-
      NNH−CH₂CH₃             1,3-dicarbohydrazonic acid
      C−OH
      
      ⟨1 2⟩
      ⟨  3⟩ $\overset{N^3}{}$ NNH−CH₃
      ⟨4  ⟩ C−OH
      Cl

28.   H₂N−CO−CH₂CH₂−COOH          Succinamic acid
                                 3-Carbamoylpropanoic acid
                                 4-Amino-4-oxobutanoic acid

29.   HOO−CO−OOH          Carbonodiperoxoic acid
                          Diperoxycarbonic acid

30. 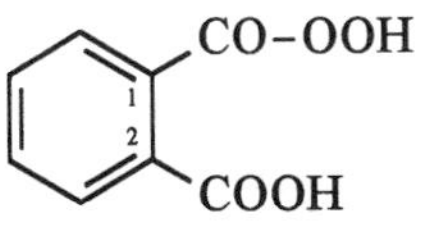   ⟨ ⟩CO−OOH          Monoperoxyphthalic acid
      ⟨1⟩                2-Carboxybenzenecarboperoxoic acid
      ⟨2⟩COOH               (CAS)
                         2-(Hydroperoxycarbonyl)benzoic acid
                            (IUPAC)

31. 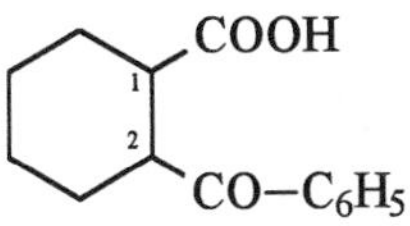   ⟨ ⟩COOH          2-Benzoylcyclohexanecarboxylic acid
      ⟨1⟩
      ⟨2⟩CO−C₆H₅

32.

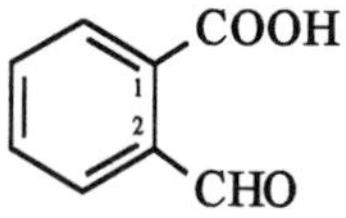

Phthaldehydic acid
2-Formylbenzoic acid

33.

$$OHC-CH_2CH_2-CO-CH_2-COOH$$
$$\phantom{OHC}6\phantom{HC-C}5\phantom{H_2CH_2}4\phantom{-CO}3\phantom{-CH_2}2\phantom{-COO}1$$

3-Oxoadipaldehydic acid
3,6-Dioxohexanoic acid
5-Formyl-3-oxopentanoic acid

34.

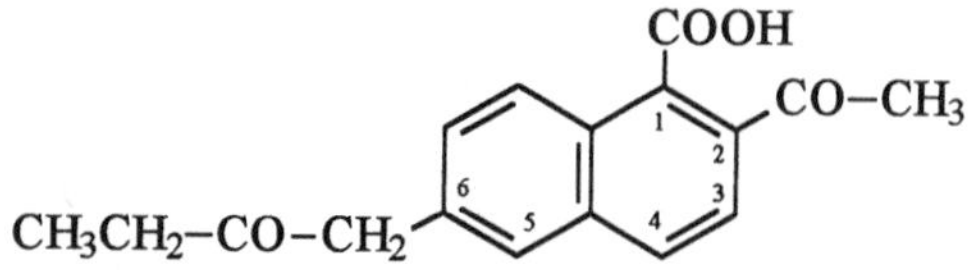

$$CH_3CH_2-CO-CH_2$$

2-Acetyl-6-(2-oxobutyl)-1-
    naphthoic acid
2-Acetyl-6-(2-oxobutyl)-
    naphthalene-1-carboxylic acid

35.

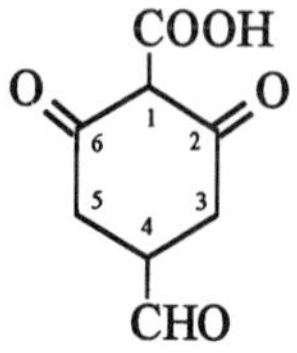

4-Formyl-2,6-dioxocyclohexane-1-
    carboxylic acid

36.

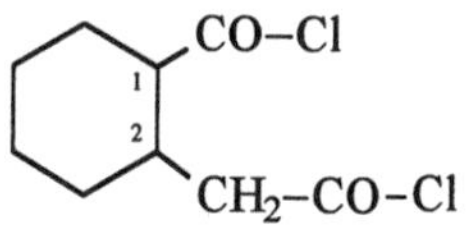

2-[(Chlorocarbonyl)methyl]cyclo-
    hexanecarbonyl chloride
2-(2-Carbonochloridoyl)cyclo-
    hexaneacetyl chloride
2-(2-Chloro-2-oxoethyl)cyclo-
    hexanecarbonyl chloride

37.

$$N_3-\overset{\overset{NH}{\|}}{C}-CH_2CH_2-\overset{\overset{NH}{\|}}{C}-N_3$$

Butanediimidoyl diazide
Succinimidoyl diazide

38.

$$NC-\overset{\overset{NH}{\|}}{C}-CH_2CH_2-CO-Cl$$
$$\phantom{NC-C}4\phantom{-CH}3\phantom{_2CH_2}2\phantom{-CO-}1$$

4-Cyano-4-iminobutanoyl chloride
3-(Carbonocyanidimidoyl)propanoyl chloride
4-Chloro-4-oxobutanimidoyl cyanide

39.

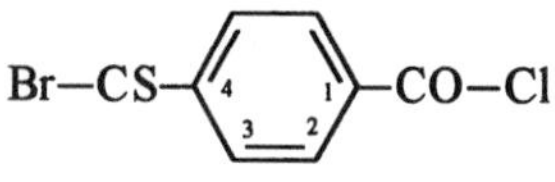

$$Br-CS-\!\!\!\underset{3\quad 2}{\overset{4\quad 1}{\bigcirc}}\!\!\!-CO-Cl$$

4-(Bromocarbonothioyl)benzoyl chloride
4-Carbonobromidothioylbenzoyl chloride

40.

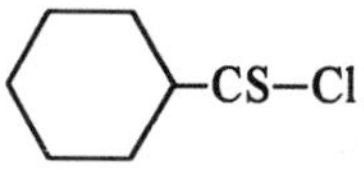

$$CS-Cl$$

Cyclohexanecarbothioyl chloride

41.

$$Cl-CO-CH_2-COOH$$

(Chlorocarbonyl)acetic acid
3-Chloro-3-oxopropanoic acid
Carbonochloridoylacetic acid

42.

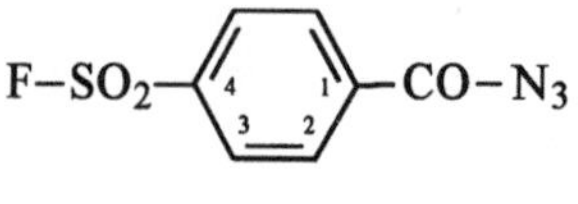

$$F-SO_2-\!\!\!\underset{3\quad 2}{\overset{4\quad 1}{\bigcirc}}\!\!\!-CO-N_3$$

4-(Fluorosulfonyl)benzoyl azide
4-(Fluorosulfonyl)benzenecarbonyl azide
4-(Sulfonofluoridoyl)benzoyl azide

**REFERENCE**

1. American Chemical Society. The Report of the ACS Committee on Nomenclature, Spelling, and Pronounciation for the First Half of 1952. E. Organic Compounds Containing Phosphorus. *Chem. Eng. News* **1952**, *30*, 4517–4522.

# 12

# Carboxylic Esters, Salts, and Anhydrides

A wide variety of derivatives of the carboxylic acid characteristic group (see chapter 11) is possible through the replacement of the hydrogen atom of the carboxy group. In this chapter, *salts*, *esters*, including *lactones* and *lactides*, and acyclic and cyclic *anhydrides* will be considered. The corresponding classes in which sulfur replaces oxygen are also covered. Additional examples of salts will be found in chapter 33. Amides, amidines, nitriles and other nitrogenous derivatives of carboxylic acids are discussed in chapters 18, 19, 20, and 22, respectively. Acid halides will be found in chapter 11 along with carboxylic acids.

The general order of seniority of compound classes (see chapter 3, table 3.1) in substitutive nomenclature is cations > acids > anhydrides > esters. It will be seen below that the acid group is often emphasized in this order.

## Acceptable Nomenclature

Replacement of hydrogen in a carboxy group is considered in the 1993 IUPAC recommendations to be functionalization rather than substitution. Therefore, within those recommendations, the trivial names retained by IUPAC for unsubstituted acids (see chapter 11) may be used in the formation of names discussed in this chapter.

*Neutral salts* of carboxylic acids are simply named by citing the name(s) of the cation(s) as separate words in alphabetical order followed by the name of the anion derived from the acid. The anions are named by replacing -ic acid with -ate in the name of the carboxylic acid. Salts of $\alpha$-amino acids (see chapter 31), in which the carboxy group is not indicated by a suffix, are named by a descriptive phrase, such as "sodium salt of methionine" or "methionine sodium salt". Descriptive names like naphthalene-1-carboxylic acid sodium salt may be used but are not recommended unless mandated by circumstance.

$Na^+ \left[ CH_3-CO-O^- \right]$      Sodium acetate

$Ca^{++} \left[ \text{cyclohexane with } CO-O^- \text{ at position 1 and } CO-O^- \text{ at position 2} \right]$      Calcium cyclohexane-1,2-dicarboxylate

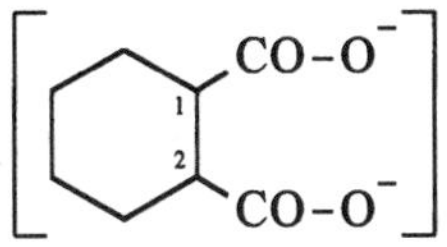

$Li^+ Na^+ \left[ {}^-O-CO-[CH_2]_6-CO-O^- \right]$      Lithium sodium octanedioate

$Ba^{++} \left[ CH_3CH_2-CO-O^- \right]_2$      Barium dipropanoate

$2Na^+ \left[ {}^-O-CO-O^- \right]$      Disodium carbonate

145

With structures having more than one anionic center, the senior center is designated by an anionic suffix and the other(s) by anion prefix names ending in -ato or -ido (see chapters 15 and 33). For example, $^-$O–CO– is carboxylato, $^-${O, S}C– is thiocarboxylato, and $^-$O– is oxido.

$$2\,Na^+ \left[ \,^-O-\overset{\overset{NH}{\|}}{C}-\underset{\substack{3\ 2}}{\overset{4\qquad 1}{\bigcirc}}-CO-O^- \right]$$

Disodium 4-carboximidatobenzoate
Disodium 4-(iminooxidomethyl)benzoate

$$Mg^{++} \left[ \underset{2}{\overset{1}{\bigcirc}} \begin{array}{l} CO-O^- \\ CS-S^- \end{array} \right]$$

Magnesium 2-dithiocarboxylatocyclohexane-
carboxylate

$$Na^+\ K^+ \left[ Cl \underset{\substack{6\ 5\ 4\ 3}}{\overset{\substack{CH_2CH_2-CO-O^-}}{\bigcirc}} \begin{array}{l} \\ CH_2-CO-O^- \end{array} \right]$$

Potassium sodium 3-[2-(carboxylatomethyl)-
6-chlorophenyl]propanoate
Potassium sodium 2-(carboxylatomethyl)-
6-chlorobenzenepropanoate

When both positive and negative centers are in the same structure, the cation is usually cited as a prefix (see chapter 33).

$$\bigcirc \overset{+}{N}H-\underset{3}{CH_2}\underset{2}{CH_2}-\underset{1}{CO-O^-}$$     3-(1-Piperidinio)propanoate

*Acid salts* are named by the same methods as neutral salts with hydrogen, dihydrogen, and so on inserted between the names of the cations and that of the anion.

$$Na^+ \left[ HOOC-CH_2CH_2-CO-O \right]^-$$

Sodium hydrogen succinate
Sodium hydrogen butanedioate

$$K^+\ Na^+\ H^+ \left[ \,^-OOC-[CH_2]_4\overset{\overset{COO^-}{|}}{\underset{\substack{6\text{-}3\qquad 2\quad 1}}{C}}HCH_2-CO-O^- \right]$$

Potassium sodium hydrogen 1,2,6-hexanetricarboxylate

*Neutral esters* are generally named by the methods used in salt nomenclature, with the name of the cation replaced by that of the esterifying group. In mixed esters of polybasic acids, the numbering of the original acid is retained to specify positions of substituents and the esterifying groups. For chalcogen analogs, *S*, *O*, and so on, are used to indicate the location of ester groups. Esters derived from polyols, such as *acylals* and *glycerides*, are named as esters. Prefix forms are alkoxycarbonyl for R–O–CO– and acyloxy for R–CO–O–; acetoxy is an acceptable contraction for $CH_3$–CO–O–. As the terminal group in an acyclic chain, R–O–CO– is named substitutively, as in 5-methoxy-5-oxopentanoyl, or additively, as in 4-(methoxycarbonyl)butanoyl.

$$CH_3CH_2-CO-O-CH_3$$          Methyl propanoate

$$\underset{\substack{4\qquad 3\qquad 2\qquad 1}}{C_6H_5-O-CS-CH_2CH_2-CO-S-CH_3}$$

*S*-Methyl *O*-phenyl butanebis(thioate)
*O*-Phenyl 4-(methylsulfanyl)-4-oxo-
butanethioate
*O*-Phenyl 3-[(methylthio)carbonyl]-
propanethioate

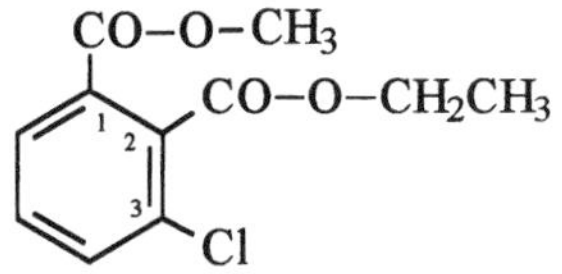

$$CO-O-CH_3$$
$$CO-O-CH_2CH_3$$

2-Ethyl 1-methyl 3-chlorophthalate

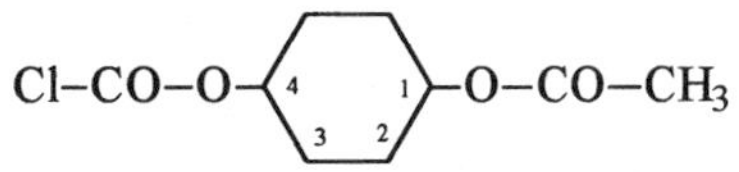

$$Cl-CO-O-\langle 4\ 1 \rangle-O-CO-CH_3$$

Cyclohexane-1,4-diyl acetate
  chlorocarbonate
Cyclohexane-1,4-diyl acetate
  carbonochloridate

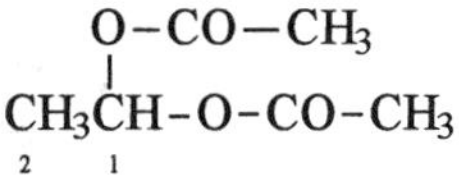

$$O-CO-CH_3$$
$$CH_3CH-O-CO-CH_3$$

Ethane-1,1-diyl diacetate

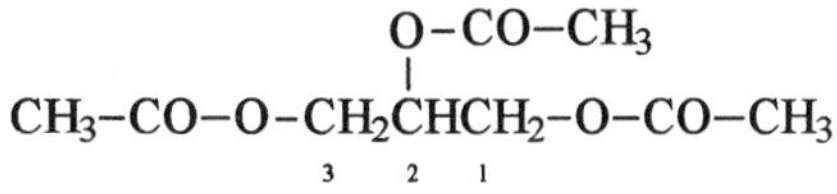

$$O-CO-CH_3$$
$$CH_3-CO-O-CH_2CHCH_2-O-CO-CH_3$$

Propane-1,2,3-triyl triacetate

$$CH_3-O-CO-CH_2CH_2-O-CO-C_6H_5$$

2-(Methoxycarbonyl)ethyl benzoate
3-Methoxy-3-oxopropyl benzoate
Methyl 3-(benzoyloxy)propanoate

Functional class nomenclature is often used with natural products (see chapter 31), as in cyclohexane-1,4-diol diacetate. This method also finds utility in naming esters of unknown structure, such as butane-1,2,4-diol monoacetate.

In the names of *acid esters* and their *salts*, the cations are cited first, followed by the esterifying groups, hydrogen, and anion in that order, as needed. Generally, acid esters are named as esters, but they may be named as acids substituted with an ester group; numbering will depend on the alternative chosen.

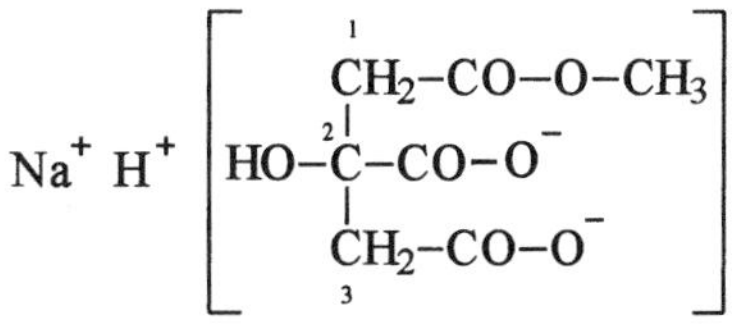

$$Na^+ H^+ \left[ \begin{array}{c} CH_2-CO-O-CH_3 \\ HO-C-CO-O^- \\ CH_2-CO-O^- \end{array} \right]$$

Sodium 1-methyl hydrogen citrate
Sodium hydrogen 2-hydroxy-3-(methoxy
  carbonyl)propane-1,2-dicarboxylate
Sodium methyl hydrogen 3,4-dicarboxylato-
  3-hydroxybutanoate
Sodium hydrogen 2-hydroxy-4-methoxy-
  4-oxobutane-1,2-dicarboxylate

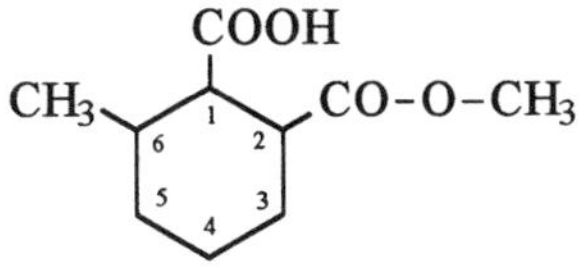

$$COOH$$
$$CH_3 \quad CO-O-CH_3$$

2-(Methoxycarbonyl)-6-methylcyclo-
  hexanecarboxylic acid
1-Methyl hydrogen 3-methylcyclohexane-
  1,2-dicarboxylate

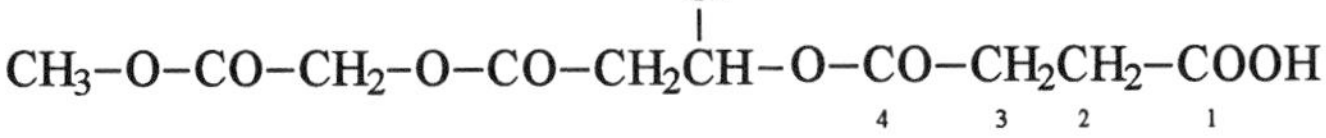

$$Cl$$
$$CH_3-O-CO-CH_2-O-CO-CH_2CH-O-CO-CH_2CH_2-COOH$$

4-[1-Chloro-3-(2-methoxy-2-oxoethoxy)-3-oxopropoxy]-4-oxobutanoic acid
1-Chloro-3-(2-methoxy-2-oxoethoxy)-3-oxopropyl hydrogen butanedioate

Compounds of the type R–C(OR′)$_3$ may be named as ethers, as in 1,1,1-trimethoxyethane or as esters of hypothetical ortho acids, R–C(OH)$_3$, as in triethyl orthoacetate for CH$_3$–C(OCH$_2$CH$_3$)$_3$ and tetramethyl orthocarbonate for C(OCH$_3$)$_4$. The related *acetals*, R$_2$C(OR′)$_2$, are named as acetals or as ethers (see chapter 16).

*Lactones* and *lactides* are cyclic esters with the following general structures:

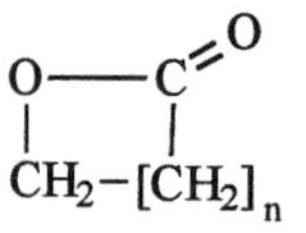

Lactones 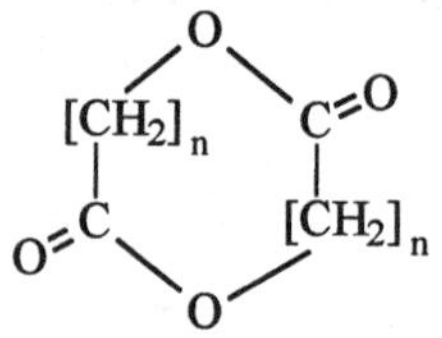                          Lactides

Members of these classes are better named as heterocyclic compounds (see chapter 9). Lactones may also be described by substituting -olactone for -ic acid of an acceptable trivial name for a hydroxy acid or -lactone may replace -ic acid in the -oic acid name of the corresponding nonhydroxylated parent acid; a locant inserted between "o" and "lactone" indicates the position of the ester –CO–O– bond. The ending -olide and names such as butyrolactone have been used in the past but are no longer recommended. In cyclic structures in which one or more (but not all) rings are lactone rings, the –O–CO– bridge can be denoted by placing -carbolactone after the name of the structure that existed before two hydrogen atoms were replaced by the lactone ring; locants for the position of the lactone ring are cited with the locant for the carbonyl group given first.

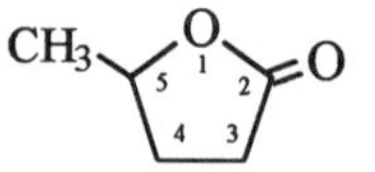

Dihydro-5-methylfuran-2(3$H$)-one
Pentano-4-lactone
Valero-4-lactone

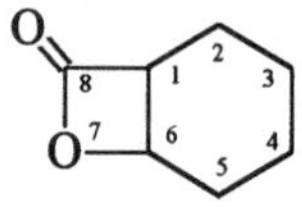

7-Oxabicyclo[4.2.0]octan-8-one
Cyclohexane-1,2-carbolactone

Self-esterification of two or more molecules of a hydroxy acid can produce a lactide, again better named as a heterocycle. However, if the hydroxy acid has an acceptable trivial name (see chapter 11), the ending -ic acid is replaced by -ide and the name is prefixed by di-, and so on, to indicate the number of molecules involved. For example, two molecules of lactic acid can form

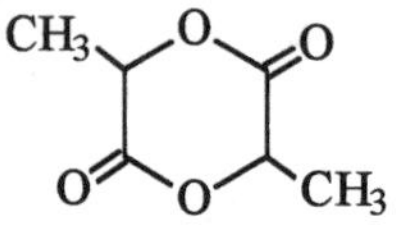

Dilactide
3,6-Dimethyl-1,4-dioxane-2,5-dione

Lactides can be given descriptive names such as 3-hydroxybutanoic bimol. cyclic ester.

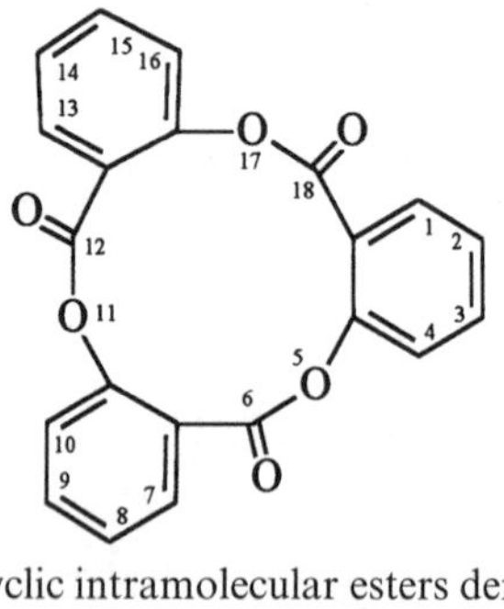

6$H$,12$H$,18$H$-Tribenzo[$b,f,j$][1,5,9]trioxacyclododecin-
    6,12,18-trione
Trisalicylide
2-Hydroxybenzoic trimol. cyclic ester

Cyclic intramolecular esters derived from unlike hydroxyacids are difficult to name as lactides; it is preferable to name them as heterocycles (see chapter 9).

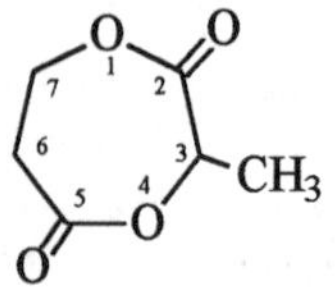

3-Methyl-1,4-dioxepane-2,5-dione

*Symmetric anhydrides*, derived by the loss of a molecule of water from two molecules of a substituted or unsubstituted monocarboxylic acid, are named by replacing the term "acid" with "anhydride".

$CH_3-CO-O-CO-CH_3$                                      Acetic anhydride

$CH_3CH_2CH_2-CO-O-CO-CH_2CH_2CH_3$                      Butanoic anhydride

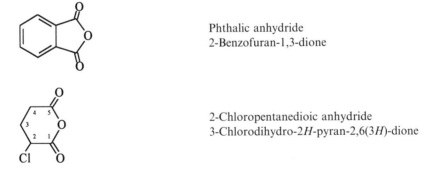

4-Chlorocyclohexanecarboxylic anhydride

$C_6H_5-CS-O-CS-C_6H_5$                                  (Thiobenzoic) anhydride

In the last example, parentheses are used to reduce the possibility of confusion with the corresponding thioanhydride (see below).

*Unsymmetric* or *mixed anhydrides* are named with the acid constituents (without the word acid) cited in alphabetical order as separate words followed by the word anhydride. This type of nomenclature also applies to organic anhydrides involving noncarboxylic acids. Mono-, di-, and so on, are prefixed to anhydride as needed. In CAS, a descriptive phrase is used with mixed anhydrides, as in benzoic acid anhydride with acetic acid.

$CH_3CH_2CH_2-CO-O-CO-CH_2-Cl$      Butyric chloroacetic anhydride
                                    (*not* chloroacetic butyric anhydride)

⬡—$CO-O-CO-CH_3$                    Acetic cyclohexanecarboxylic anhydride

$CH_3[CH_2]_6-CO-O-SO_2-OH$          Octanoic sulfuric monoanhydride

Dicarboxylic acids can form several anhydrides. Intramolecular anhydrides derived by the loss of water from two carboxyl groups attached to the same parent are named either as heterocycles or as above on the assumption that the anhydride has been formed from a single molecule of the dibasic acid.

Phthalic anhydride
2-Benzofuran-1,3-dione

2-Chloropentanedioic anhydride
3-Chlorodihydro-2*H*-pyran-2,6(3*H*)-dione

Both cyclic and acyclic anhydrides may be formed from two molecules of a dicarboxylic acid. For anhydrides from an unsubstituted acid such as pentanedioic acid, CAS uses the descriptive phrase "bimol. monoanhydride" to describe the acyclic monoanhydride. The cyclic dianhydride is named as a heterocycle, but pentanedioic bimol. cyclic anhydride is acceptable. These methods may also be used for symmetrical anhydrides derived from substituted dicarboxylic acids. However, with unsymmetric substitution, as in 2-chloropentanedioic acid, two distinct symmetric and one unsymmetric acyclic monoanhydrides, plus two cyclic dianhydrides are possible. In such cases, it is recommended that cyclic monoanhydrides be named as heterocycles and that the acyclic monoanhydrides be named as mixed anhydrides or as derivatives of a monobasic acid.

$$\text{HOOC-[CH}_2\text{]}_2\overset{\overset{\displaystyle Cl}{|}}{\underset{2}{\text{CH}}}\text{-}\underset{1}{\text{CO}}\text{-O-CO-}\overset{\overset{\displaystyle Cl}{|}}{\text{CH}}\text{[CH}_2\text{]}_2\text{-COOH}$$

4-3    2    1

4-Carboxy-2-chlorobutanoic anhydride
2-Chloropentanedioic bimol. monoanhydride

$$\text{HOOC-[CH}_2\text{]}_2\overset{\overset{\displaystyle Cl}{|}}{\text{CH}}\text{-CO-O-CO-CH}_2\overset{\overset{\displaystyle Cl}{|}}{\text{CH}}\text{CH}_2\text{-COOH}$$

4-3   2    2      1   2   3   4

4-Carboxy-2-chlorobutanoic 4-carboxy-3-chlorobutanoic anhydride
5-[(4-Carboxy-2-chlorobutanoyl)oxy]-3-chloro-5-oxopentanoic acid

*Sulfur analogs of anhydrides* have the class name *thioanhydride*; the prefix thio means replacement of the anhydride oxygen only. Other sulfur atoms are indicated by thio affixes in the names of the acid groups. The class name *anhydrosulfide* used by CAS indicates removal of $H_2S$ from two thioacids; it can be used if the appropriate sulfur atoms are expressed in the parent acid name.

$CH_3\text{-CO}-\text{S}-\text{CO-CH}_3$        Acetic thioanhydride
                                 Thioacetic anhydrosulfide

$CH_3\text{-CS}-\text{S}-\text{CS-CH}_3$        Thioacetic thioanhydride
                                 Dithioacetic anhydrosulfide

$CH_3CH_2\text{-CO-O-CS-CH}_3$       Propanoic thioacetic anhydride

$C_6H_5\text{-CO}-\text{S}-\text{CS}-\text{[CH}_2\text{]}_5CH_3$    Benzoic heptanethioic thioanhydride
                                 Thiobenzoic heptanedithioic anhydrosulfide

Dihydro-4-thioxothiophene-2(3*H*)-one
2-Oxo-4-thioxothiolane

Nitrogen analogs of anhydrides are imides; they are discussed in chapter 18.

## Discussion

Partial esters of polybasic acids have been described in a variety of ways, none of which are an improvement over a name like ethyl hydrogen adipate. Ethyl adipate might be either the mono- or the di- ester, and ethyl adipic acid is a half-acid name easily confused with adipic acid substituted with an ethyl group; ethyl hydrogen adipate or its systematic alternative, ethyl hydrogen hexanedioate, says it all.

Esters of polyols present a different problem: ethylene diacetate and 2-hydroxyethyl acetate are straightforward names for esters derived from ethylene glycol, but there are a number of "systems" in use for *glycerides*, or esters of glycerol. The group name glyceryl has often been used for 1,2,3-propanetriyl, and either can be used for triesters from a single acid; names such as olein or triolein (for the trioleic ester) are in use but are not recommended. Names ending in "in" have also been used for mixed esters, with locants $\alpha, \beta, \alpha'$ or 1, 2, 3 on the glycerol backbone, leading, for example to $\beta$-stearo-$\alpha$, $\alpha'$-oleopalmatin and 2-stearo-1,3-oleopalmitin for the ester with the preferred name 2-(stearoyloxy)-1,3-propanediyl dioleate.

While *urethane* is used as a class name for the products of alcohol-isocyanate reactions, individual compounds in this class should be named as esters of carbamic acid, as in ethyl methylcarbamate for $CH_3$–NH–CO–O–$CH_2CH_3$. Similarly, *xanthate* is a class name for RO–CS–SH, but these compounds are best named as *O*-alkyl carbonodithioates. Compounds of the type R–O–CN, R–S–CN, and R–O–$ClO_3$ are cyanates, thiocyanates, and perchlorates, respectively.

In the 1979 IUPAC Organic Rules, a recommended method for lactones involved appending -olide to the name of the (nonhydroxylated) hydrocarbon with the same number of carbon atoms. Names such as 4-butanolide are generated, but the corresponding butano-4-lactone is more descriptive and is preferred for this compound if heterocyclic nomenclature is not used. "Glycolide" and "lactide", derived from glycolic and lactic acid, respectively, are discouraged.

Some anhydride names come complete with redundancies, as in acetic acid anhydride and phthalic acid cyclic anhydride, in which the word "acid" can be excess baggage. Phthalic anhydride is an accepted trivial name for a heterocyclic compound, while the corresponding acyclic anhydride is bis(2-carboxybenzoic) monoanhydride or phthalic bimol. monoanhydride. The "bis" prefix is frequently omitted; occasionally this prefix or at least the parentheses can reduce ambiguity.

## Additional Examples

1.  Cl–CO–O–$CH_2CH_3$

    Ethyl carbonochloridate
    Ethyl chlorocarbonate

2.  $Na^+$ $\left[ CH_3-O-CO-\underset{\underset{2\ \ \ 3}{\overset{1\ \ \ 4}{}}{\bigcirc}\!\!\overset{Cl}{} -CH_2-CO-O^- \right]$

    Sodium methyl 4-carboxylatomethyl-
      3-chlorobenzoate
    Sodium 2-chloro-4-(methoxycarbonyl)-
      phenylacetate
    Sodium 2-chloro-4-(methoxycarbonyl)-
      benzeneacetate

3.  $C_6H_5$–CO–O–$\underset{1}{CH_2}\underset{2}{CH_2}$–COOH

    2-Carboxyethyl benzoate
    3-(Benzoyloxy)propanoic acid

4.  $CH_3$–CO–O–$\underset{\underset{3\ \ \ 2}{\overset{4\ \ \ 1}{}}{\bigcirc}$–O–CO–$CH_2$–Cl

    4-Acetoxyphenyl chloroacetate
    1,4-Phenylene acetate chloroacetate

5.  $CH_3CH_2$–O–CO–$\underset{\underset{4\ \ \ \ \ 3}{}}{CH_2}\underset{2}{\overset{Cl}{\underset{|}{CH}}}$–$\underset{1}{CO}$–O–$CH_3$

    4-Ethyl 1-methyl 2-chloro-
      butanedioate
    Methyl 2-chloro-3-(ethoxycarbonyl)-
      propanoate
    Methyl 2-chloro-4-ethoxy-4-oxo-
      butanoate

6.  $CH_3$–CO–O–$\underset{4}{CH_2}\underset{3}{\overset{OH}{\underset{|}{CH}}}\underset{2}{CH_2}$–$\underset{1}{CO}$–O–$CH_2CH_3$

    Ethyl 4-acetoxy-3-hydroxybutanoate

7.  $CH_3$–CO–O–$CH_2CH_2$–O–CO–$CH_3$

    Ethane-1,2-diyl diacetate
    Ethylene diacetate

8.

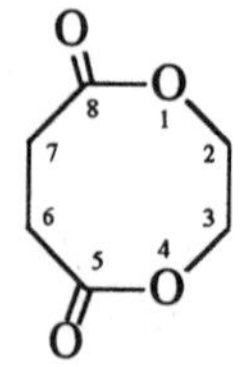

1,4-Dioxocane-5,8-dione
Ethylene succinate
Ethane-1,2-diyl butanedioate

9.

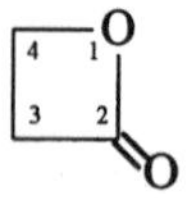

Oxetan-2-one
Propano-3-lactone

10.

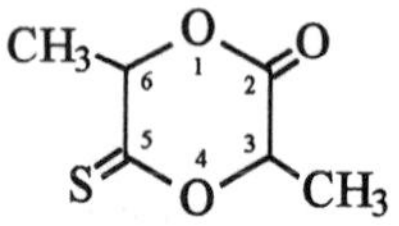

3,6-Dimethyl-5-thioxo-1,4-
   dioxan-2-one

11.

$$\underset{CH_3-C-O-C_6H_5}{\overset{N-O-CH_2CH_3}{\overset{\parallel}{}}}$$

Phenyl $O$-ethylacetohydroximate
Phenyl $N$-ethoxyacetimidate

12.

$$Cl-CO-[CH_2]_3-CO-O-CH_3$$
$$\phantom{Cl-CO-}{\scriptstyle 4-2}\phantom{--}{\scriptstyle 1}$$

Methyl 4-(chlorocarbonyl)butanoate
Methyl 5-chloro-5-oxopentanoate

13.

$$CH_3-NH-CO-NH-CO-O-CH_2CH_3$$

Ethyl (methylcarbamoyl)carbamate
Ethyl [(methylamino)carbonyl]-
   carbamate

14.

$$CH_3CH_2-CS-O-CS-CH_2CH_3$$

Propanethioic anhydride
(Thiopropanoic) anhydride

15.

$$\underset{CH_3-C-S-C-CH_3}{\overset{NH\quad\ NH}{\overset{\parallel\qquad\parallel}{}}}$$

Acetimidic thioanhydride
Ethanimidothioic anhydrosulfide

16.  $C_6H_5-CS-S-CHO$

Formic thiobenzoic thioanhydride
Dithiobenzoic thioformic
   anhydrosulfide
Benzenecarbodithioic methane-
   thioic anhydrosulfide

17.

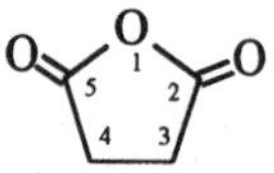

Dihydrofuran-2,5-dione
Succinic anhydride

18.

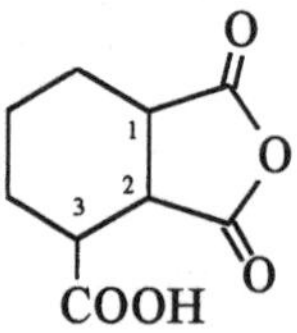

Cyclohexane-1,2,3-tricarboxylic
   1,2-anhydride
Octahydro-1,3-dioxo-2-benzo-
   furan-4-carboxylic acid

19.

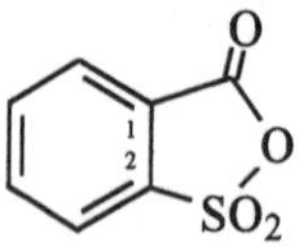

2-Sulfobenzoic cyclic anhydride
3$H$-2,1-Benzoxathiol-3-one
   1,1-dioxide

20.  $$HOOC-[CH_2]_4-CS-O-CO-[CH_2]_4-C\{O,S\}H$$
$$\phantom{HOOC-}{\scriptstyle 5-2}\phantom{--[CH_2]_4-}{\scriptstyle 1}\phantom{-O-}{\scriptstyle 1}\phantom{-[CH_2]_4-}{\scriptstyle 2-5}$$

5-Carboxypentanethioic 5-(thiocarboxy)pentanoic anhydride
6-[[5-(thiocarboxy)]pentanoyl]oxy]-6-thioxohexanoic acid

# Aldehydes and Their Chalcogen Analogs

Compounds containing a carbonyl group, $>C=O$, attached only to carbon or hydrogen have been divided traditionally into classes called *aldehydes*, R–CHO, and *ketones*, R–CO–R'. The appellative distinction between these structurally similiar categories has its origins in synthesis. "Aldehyde" is a contraction of *alcohol dehydrogenatus* (1835), while the dry distillation of salts of carboxylic acids produced compounds called (1833) acetone, stearone, and so on, from the names of the acids plus the suffix -one; later, the general name "ketone" was devised for the class.

The nomenclature of compounds having an aldehyde group, usually written –CHO, or a chalcogen analog, such as a thioaldehyde group, –CHS, is discussed in this chapter. Also included are certain derivatives of aldehydes such as acetals and hemiacetals, containing the groups –CH(OR)$_2$, –CH(OR)(OR'), and –CH(OH)(OR), respectively; these classes are now generally named as ethers (see chapter 16) or alcohols (see chapter 15). Chalcogen analogs of acetals and hemiacetals are likewise generally named as alcohols, thiols, ethers, or sulfides, depending on their structure. Cyclic acetals and their chalcogen analogs are now preferably named as heterocycles. Acylals, with the group –CH(OCOR)$_2$, are best named as esters (see chapter 12). Nitrogenous derivatives of aldehydes, such as imines ("aldimines"), oximes and hydrazones, are found in chapter 22.

## Acceptable Nomenclature

*Acyclic monoaldehydes* and *dialdehydes* are named substitutively by adding the suffix -al or -dial to the name of the corresponding hydrocarbon; the final "e" in the name of the hydrocarbon is elided when it precedes -al. Locant numbering includes the –CHO group(s), and in the case of dials, the parent hydrocarbon chain is the longest one terminated by the –CHO groups. Conjunctive nomenclature (see chapter 3) is applied by CAS, generating names such as benzenebutanal for C$_6$H$_5$–CH$_2$CH$_2$CH$_2$–CHO.

$$\underset{6\quad\ 5\quad\ 4\quad\ 3\quad\ 2\quad\ 1}{CH_3CH_2CH_2CH_2CH_2\text{-}CHO}$$  Hexanal

$$\underset{5\quad\ 4\quad\ 3\quad\ 2\quad\ 1}{CH_2\text{=}CHCH_2CH_2\text{–}CHO}$$  Pent-4-enal

$$\underset{6\quad\ 5\quad\ 4\quad\ 3\quad\ 2\quad\ 1}{OHC\text{—}CH_2\overset{\overset{\displaystyle CH_2CH_2CH_2CH_3}{|}}{C}HCH\text{=}CH\text{—}CHO}$$  4-Butylhex-2-enedial

*Acyclic polyaldehydes* with more than two –CHO groups attached to the same straight chain are named substitutively by IUPAC with suffixes such as -tricarbaldehyde or -tetracarbaldehyde (CAS uses -carboxaldehyde) appended to the name of the chain to which the –CHO groups are attached; locant numbering does not include the carbon atoms of the –CHO groups.

$$\underset{4\ \ 3\ \ 2\ \ 1}{\overset{\overset{\displaystyle CHO}{\displaystyle |}}{OHC-CH_2CH_2CHCH_2-CHO}}$$

Butane-1,2,4-tricarbaldehyde (IUPAC)
Butane-1,2,4-tricarboxaldehyde (CAS)

The prefix formyl is used for the –CHO group. Formyl may be added to the name of a dial that incorporates the principal chain to produce a name like 3-formylhexanedial for the example above, but this is not encouraged. Where –CHO groups are attached to more than one branch of a multi-branched chain, the name of the longest chain with the greatest number of –CHO groups is used with the suffix -dial, -tricarbaldehyde, and so on; the remaining –CHO groups are expressed by the prefix formyl.

$$\underset{8\ \ 7\ \ 6\ \ 5\ \ 4\ \ 3\ \ 2\ \ 1}{OHC-CH_2CH_2\overset{\overset{\displaystyle CH_2-CHO}{\displaystyle |}}{CH}CH_2CH_2CH_2\overset{\overset{\displaystyle CHO}{\displaystyle |}}{CH}CH_2-CHO}$$

6-(Formylmethyl)octane-
1,2,8-tricarbaldehyde

Alternatively, a –CHO group at the end of a chain may be named by the prefix oxo; in the example above, formylmethyl is equivalent to 2-oxoethyl. Formyl, however, is much preferred to oxomethyl.

Aldehydes in which one or more –CHO groups are attached to a cyclic structure are likewise named by adding suffixes such as -carbaldehyde, -dicarbaldehyde, or -tricarbaldehyde (CAS uses -carboxaldehyde and so on) to the name of the ring system. These suffixes are also applied when the –CHO group is bound to a heteroatom such as nitrogen or silicon, even though such compounds are not aldehydes according to the traditional definition.

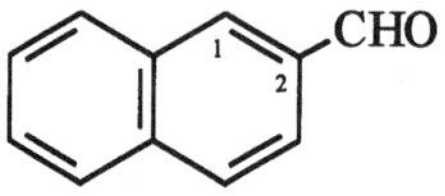

Naphthalene-2-carbaldehyde

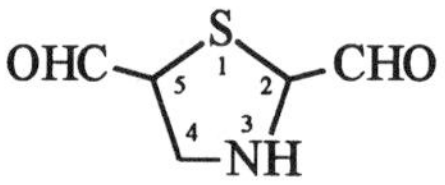

Thiazolidine-2,5-dicarbaldehyde

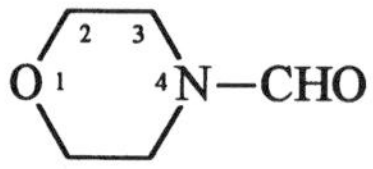

Morpholine-4-carbaldehyde

$H_3Si-CHO$          Silanecarbaldehyde

The prefix formyl is also used to express the aldehyde group in the presence of a group having priority for citation as the principal characteristic group.

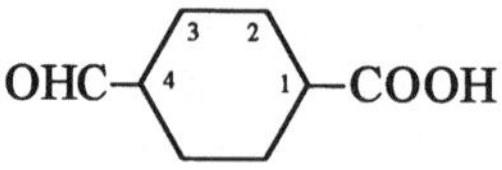

4-Formylcyclohexanecarboxylic acid

Trivial names for some common aldehydes are formed by changing the ending of the corresponding acid trivial name (see table 11.1 in chapter 11) from -ic acid or -oic acid to -aldehyde. When polyaldehydes are named in this manner, all –COOH groups implied by the trivial name must have been converted to –CHO groups.

$$\underset{3\ \ \ \ 2\ \ \ \ 1}{CH_2=CH-CHO}$$          Acrylaldehyde

$$\underset{4\ \ \ 3\ \ \ 2\ \ \ 1}{OHC-CH_2CH_2-CHO}$$          Succinaldehyde

A few trivial names for aldehydes such as vanillin, piperonal and glyoxal are acceptable although not encouraged in systematic nomenclature. Amino aldehydes are named as amino derivatives of the appropriate aldehyde rather than from the trivial names of amino acids unless the latter end in -ic acid (see Discussion).

$$\underset{\overset{|}{OHC-CH_2CH-CHO}}{NH_2}$$

Aspartaldehyde
(*not* Aspartal)

$$\underset{2 \quad \ 1}{H_2N-CH_2-CHO}$$

Aminoacetaldehyde
(*not* Glycinaldehyde or glycinal)

The aldehyde characteristic group in *aldehydic acids* is expressed by the prefix formyl, except when one carboxyl group of a dicarboxylic acid with a trivial name (table 11.1) has been replaced by an aldehyde group. Such aldehydic acids are described by changing -ic acid in the acid name to -aldehydic acid (see chapter 11).

$$\underset{8-2 \qquad \quad 1}{OHC-[CH_2]_7-COOH}$$

8-Formyloctanoic acid
9-Oxononanoic acid

$$\underset{4 \quad \ 3 \quad \ 2 \quad \ 1}{OHC-CH_2CH_2-COOH}$$

Succinaldehydic acid

*Chalcogen analogs of aldehydes* are named analogously to aldehydes by IUPAC with suffixes such as -thial, -selenal, and -carbothialdehyde (CAS uses -carbothioaldehyde) in place of -al and -carbaldehyde; thioformyl is the prefix name for the –CHS group, although thioxo, for S=, can also be used when the –CHS group is terminal to a carbon chain. Chalcogen analogs of aldehydes with trivial names in which all of the oxygens in the aldehyde groups are replaced by the same chalcogen are often named by prefixing thio, seleno, dithio, and so on, to the name of the aldehyde.

$$\underset{3 \quad \ 2 \quad \ 1}{CH_3CH_2-CHS}$$

Thiopropionaldehyde
Propanethial

$$\underset{\qquad 3 \quad 2}{SHC-CH_2-\langle \overset{4 \quad 1}{\phantom{xx}} \rangle -CHS}$$

4-[(Thioformyl)methyl]cyclohexanecarbothialdehyde
4-(Thioformyl)cyclohexaneethanethial
2-[4-(Thioformyl)cyclohexyl]ethanethial
4-(2-Thioxoethyl)cyclohexanecarbothialdehyde

$$\underset{10 \ \ 9 \quad 8 \quad \ 7 \quad 6 \quad \ 5 \quad \ 4 \quad \ 3 \quad 2 \quad 1}{OHC-\overset{\overset{\displaystyle CHS}{|}}{C}HCH_2CH_2\overset{\overset{\displaystyle CH_2-CHS}{|}}{C}HCH_2CH_2CH_2\overset{\overset{\displaystyle CH_2CH_2CH_3}{|}}{C}H-CHO}$$

2-Propyl-9-(thioformyl)-6-[(thioformyl)methyl]decanedial
2-Propyl-9-(thioformyl)-6-(2-thioxoethyl)decanedial

Compounds of the type R–CH(OR′)–OR″ are called *acetals*. They are named substitutively as alkoxy, and so on, derivatives of the parent hydrocarbon. Alternatively, they can be named by functional class nomenclature from the appropriate aldehyde name followed by the R′ and R″ group names in alphabetical order and, finally, the class name "acetal".

$$\underset{3 \quad \ 2 \quad \ 1}{CH_3CH_2\overset{\overset{\displaystyle O-CH_3}{|}}{C}H-O-CH_2CH_3}$$

1-Ethoxy-1-methoxypropane
Propanal ethyl methyl acetal

Sulfur analogs are named substitutively or as monothio- or dithioacetals with italic letter indicators used as necessary.

$$S-CH_2CH_2CH_3$$
$$|$$
$$CH_3\overset{}{C}H-S-CH_2CH_2CH_3$$
$$\phantom{CH_3}{}_{2}\phantom{CH-S-CH_2CH}{}_{1}$$

1,1-Bis(propylthio)ethane
Acetaldehyde dipropyl dithioacetal
1,1'-[Ethane-1,1-diylbis(thio)]dipropane

$$S-CH_2CH_2CH_3$$
$$|$$
$$CH_3CH_2CH_2\overset{}{C}H-O-CH_2CH_3$$
$$\phantom{CH_3CH_2CH}{}_{4}\phantom{CH}{}_{3}\phantom{CH}{}_{2}\phantom{H-O-CH}{}_{1}$$

1-Ethoxy-1-(propylthio)butane
Butanal O-ethyl S-propyl monothioacetal

"Mercaptal" in names for R–CH(SR')$_2$ is no longer acceptable; it was used by CAS prior to 1972.

*Hemiacetals*, R–CH(OR')–OH, are named substitutively as alkoxy, and so on, derivatives of a parent alcohol (see chapter 15), thiol (see chapter 24), and so forth. They may also be named in the same way as acetals with the class names hemiacetal, monothiohemiacetal, or dithiohemiacetal and appropriate italic letter locants.

$$O-CH_2CH_3$$
$$|$$
$$CH_3CH_2CH_2\overset{}{C}H-OH$$
$$\phantom{CH_3CH_2CH}{}_{4}\phantom{CH}{}_{3}\phantom{CH}{}_{2}\phantom{H-O}{}_{1}$$

1-Ethoxybutan-1-ol
Butanal ethyl hemiacetal

$$O-CH_2CH_3$$
$$|$$
$$CH_3CH_2\overset{}{C}H-SH$$
$$\phantom{CH_3CH}{}_{3}\phantom{CH}{}_{2}\phantom{H-S}{}_{1}$$

1-Ethoxypropane-1-thiol
Propanal O-ethyl monothiohemiacetal

*Acylals* have the structure R–CH(OCOR')$_2$, R$_2$C(OCOR')$_2$, and so on, and are named as esters (see chapter 12).

## Discussion

In biochemical nomenclature,[1] aspartaldehyde and aminoacetaldehyde may be named aspartal and glycinal, respectively, but this usage is not recommended for systematic organic nomenclature.

Because of their structural similarity, it would be useful to name aldehydes and ketones by the same system. With aldehydes, this might be done through use of the suffix -1-one, leading to names like hexan-1-one, or the prefix oxo, providing prefix names such as 6-oxohexyl and oxomethyl. With carboxylic acids substituted by oxo groups (see chapter 11), names such as 5-oxopentanoic acid and 3,5-dioxopentanoic acid are allowed under the IUPAC recommendations. However, it seems best to express the aldehyde function at the end of a chain in the same way as carboxy, cyano, or carbamoyl. Accordingly, these characteristic groups are recognizable by both prefixes and suffixes.

The suffix -carbaldehyde is recommended for –CHO; nonetheless, the older form -carboxaldehyde is frequently encountered, and it is used by CAS. It has been argued that in -carboxaldehyde, both the "ox" and the "aldehyde" could be interpreted as representing the carbonyl group and therefore constitute a possibly misleading redundancy.

A variety of trivial and semisystematic names for mono- and polyaldehydes are still in use but should be abandoned. Examples are acrolein (for acrylaldehyde), furfural (for 2-furaldehyde or furan-2-carbaldehyde), acetal (for acetaldehyde diethyl acetal), chloral (for trichloroacetaldehyde), and hexaldehyde (for hexanal). Chalcogen analogs of monoaldehydes having trivial names such as formaldehyde and benzaldehyde may be named with the prefixes thio, seleno, and so on, and the class name thioaldehyde is acceptable, but in general the recommendations given above for chalcogen analogs of aldehydes are preferred. Functional class names for acetals and hemiacetals of polyaldehydes can become unwieldy with structural complexity, as illustrated by example 17 below.

## Additional Examples

1.
$$CH_3$$
$$CH_3\overset{|}{CH}-CHO$$
<sub>3   2   1</sub>

2-Methylpropanal

2. $CCl_3-CHO$
<sub>2   1</sub>

Trichloroacetaldehyde
Trichloroethanal

3. $OHC-CHO$

Ethanedial
Oxalaldehyde
Glyoxal

4. $OHC-CH_2-CHO$
<sub>3   2   1</sub>

Propanedial
Malonaldehyde

5.
$$O$$
$$OHC-CH_2\overset{\|}{C}CH_2-COOH$$
<sub>4   3 2     1</sub>

3,5-Dioxopentanoic acid
3-Oxoglutaraldehydic acid
4-Formyl-1,3-oxobutanoic acid

6. $OHC-[CH_2]_4-CHO$
<sub>6   5-2   1</sub>

Hexanedial
Adipaldehyde

7. 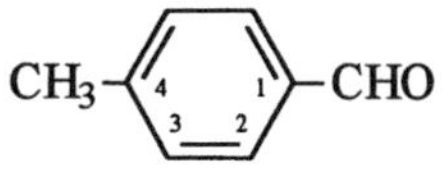 $CH_3-\langle\!\!\!\bigcirc\!\!\!\rangle-CHO$
<sub>4   1</sub>
<sub>3   2</sub>

4-Methylbenzenecarbaldehyde
4-Methylbenzaldehyde

8. 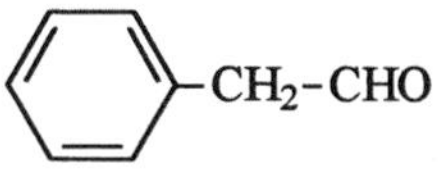 $\langle\!\!\!\bigcirc\!\!\!\rangle-CH_2-CHO$

Phenylacetaldehyde
Benzeneacetaldehyde

9. 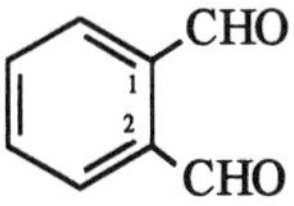 $\langle\!\!\!\bigcirc\!\!\!\rangle\overset{CHO}{\underset{CHO}{}}$
<sub>1</sub>
<sub>2</sub>

Benzene-1,2-dicarbaldehyde
Phthalaldehyde

10. $HC(CHO)_3$

Methanetricarbaldehyde

11.  $OHC-CH_2-\langle\!\!\!\bigcirc\!\!\!\rangle-CH_2-CHO$
<sub>4   1</sub>
<sub>3 2</sub>

1,4-Phenylenediacetaldehyde
1,4-Benzenediacetaldehyde

12. 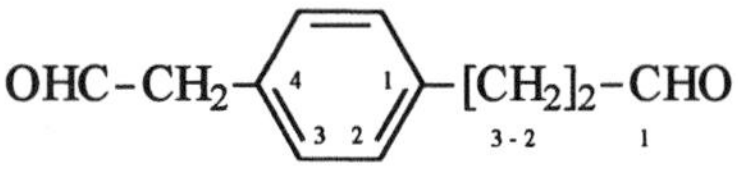 $OHC-CH_2-\langle\!\!\!\bigcirc\!\!\!\rangle-[CH_2]_2-CHO$
<sub>4   1</sub>
<sub>3 2</sub>   <sub>3-2   1</sub>

3-[4-(Formylmethyl)phenyl]propanal
4-(2-Oxoethyl)benzenepropanal

13. $CH_3CH_2-CHS$
<sub>3   2   1</sub>

Propanethial
Thiopropionaldehyde

14. 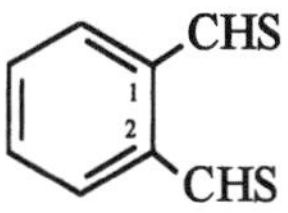 $\langle\!\!\!\bigcirc\!\!\!\rangle\overset{CHS}{\underset{CHS}{}}$
<sub>1</sub>
<sub>2</sub>

Benzene-1,2-dicarbothialdehyde
Dithiophthalaldehyde

15.

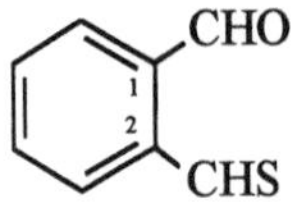

2-(Thioformyl)benzaldehyde

16.

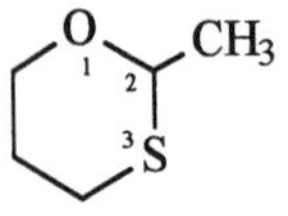

2-Methyl-1,3-oxathiane

17.

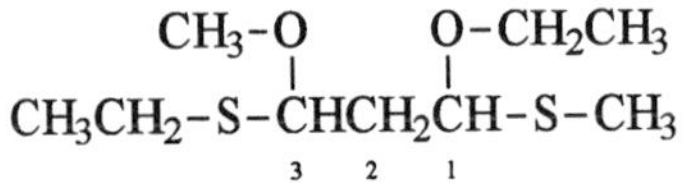

$$CH_3-O \qquad O-CH_2CH_3$$
$$CH_3CH_2-S-CHCH_2CH-S-CH_3$$
$$\phantom{CH_3CH_2-S-CHCH_2CH}3 \quad\quad 2 \quad\quad 1$$

1-Ethoxy-3-(ethylthio)-3-methoxy-1-(methylthio)propane
Propanedial $O,S'$-diethyl $O',S$-dimethyl bis(monothioacetal)
Propanedial $O^1,S^3$-diethyl $O^3,S^1$-dimethyl bis(monothioacetal)

**REFERENCE**

1. International Union of Pure and Applied Chemistry and International Union of Biochemistry, Joint Commission on Biochemical Nomenclature (JCBN). Nomenclature and Symbolism for Amino Acids and Peptides (Recommendations 1983). *Pure Appl. Chem.* **1984**, *56*, 595–624; International Union of Biochemistry and Molecular Biology. *Biochemical Nomenclature and Related documents*, 2nd ed.; Liébecq, C., Ed.; Portland Press: London, 1992; pp 39–69 (includes additions and corrections).

# 14

# Ketones and Their Chalcogen Analogs

*Ketones*, $R_2C=O$, are compounds in which the carbon atom of a carbonyl group is a non-terminal part of a carbon chain or part of a ring structure. In the chapter on aldehyde nomenclature (see chapter 13), it was noted that the etymology of ketones has been different from that of the structurally similar aldehydes. Acetone itself was discovered in 1669 and named "burning spirit of Saturn" in the belief that it was a lead compound derived from the lead salt of vinegar. By the late eighteenth century, the compound had been labelled "pyroacetic spirit" from its flammability and its acid source. That source in 1833 prompted the combination of the prefix acet (from the Latin for vinegar) and the suffix -one (from the Greek for descendent). The class name "ketone" was adopted by Gmelin in 1848.

The nomenclature of ketones and their chalcogen analogs is covered in this chapter. It includes heterocyclic "ketones" that can also be viewed as cyclic esters or anhydrides (see chapter 12) or amides (see chapter 18). In addition, the nomenclature of quinones, ketenes, ketals, hemiketals, acyloins, and certain oxo acids is included.

## Acceptable Nomenclature

*Monoketones* in which the carbon of the carbonyl group, –CO–, is part of a carbon chain or carbocyclic ring are named substitutively by adding the suffix -one to the name of the corresponding hydrocarbon after elision of the final "e". Alternatively, simple acyclic ketones may be named through functional class nomenclature by citing as separate words in alphabetical order the names of the groups attached to the –CO– group, followed by the class name "ketone". The trivial name acetone, for $CH_3$–CO–$CH_3$, is acceptable.

$$\underset{5\quad 4\qquad 3\qquad 2\quad 1}{CH_3CH_2-CO-CH_2CH_3}$$

Pentan-3-one
Diethyl ketone

$$\underset{6\quad 5\quad 4\qquad 3\qquad 2\quad 1}{CH_3CH_2CH_2-CO-CH=CH_2}$$

Hex-1-en-3-one
Propyl vinyl ketone

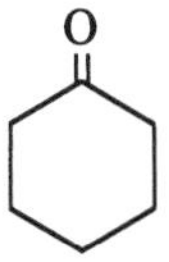

Cyclohexanone

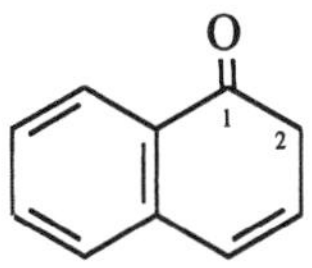

Naphthalen-1(2*H*)-one

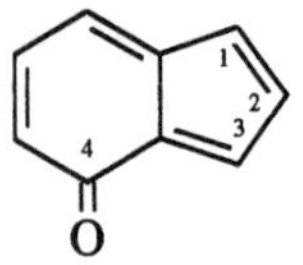

4*H*-Inden-4-one

Monoketones in which the –CO– group is part of a heterocyclic ring (see chapter 9) are named with the -one suffix even when the –CO– is bound to only one or to two heteroatoms of a ring. This nomenclature is a convenient alternative way of naming intramolecular esters (lactones, see chapter 12), amides (lactams, see chapter 18), and related compounds. Contracted names such as 3-pyridone and 9-acridone are not recommended.

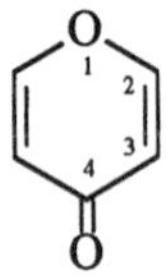

4*H*-Pyran-4-one

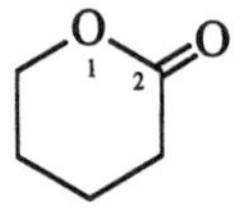

Tetrahydro-2*H*-pyran-2-one
Oxan-2-one

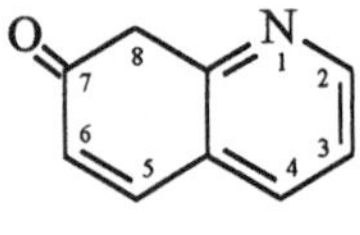

Quinolin-7(8*H*)-one

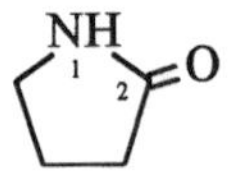

Pyrrolidin-2-one
Azolidin-2-one

*Monoacyl derivatives of cyclic compounds* are named as ketones with the ending -one, although functional class names may also be used. Acceptable trivial names for $C_6H_5$–CO–R or $C_{10}H_7$–CO–R are formed with the ending -phenone or -naphthone and a prefix derived from the name of the acid R–COOH; examples are acetophenone, $C_6H_5$–CO–$CH_3$; propano-phenone, $C_6H_5$–CO–$CH_2CH_3$; and benzophenone, $(C_6H_5)_2CO$. Locants on the acid residue are unprimed while those on the benzene or naphthalene ring are primed.

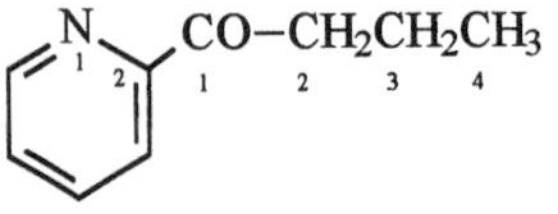

1-(2-Pyridyl)butan-1-one
Propyl pyridin-2-yl ketone

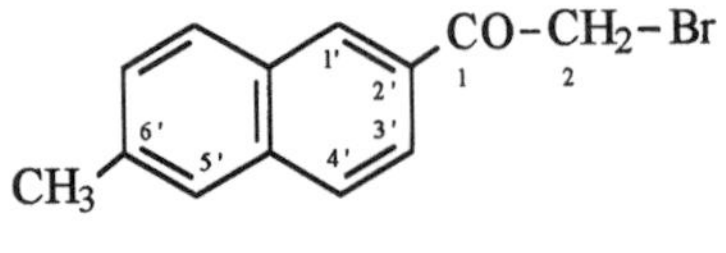

2-Bromo-6'-methyl-2'-acetonaphthone
2-Bromo-1-(6-methylnaphthalen-2-yl)ethan-1-one
Bromomethyl 6-methyl-2-naphthyl ketone

Structures with cyclic groups bound to both sides of a carbonyl group are named by functional class nomenclature or substitutively as derivatives of methanone.

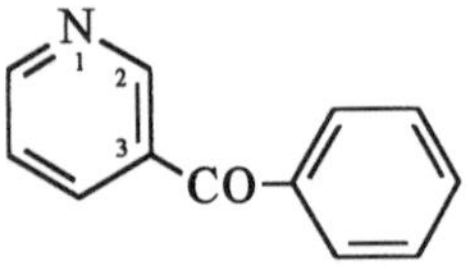

Phenyl(pyridin-3-yl)methanone
Phenyl 3-pyridyl ketone

*Polyketones* are named substitutively by adding the appropriate suffix -dione, -trione, and so on, to the name of the corresponding hydrocarbon or ring system. The functional class method may also be used for compounds with contiguous carbonyl groups. The names biacetyl for $CH_3$–CO–CO–$CH_3$ (butane-2,3-dione) and benzil for $C_6H_5$–CO–CO–$C_6H_5$ (diphenylethanedione) are not encouraged, although the 1993 IUPAC recommendations retain them in unsubstituted form.

$$CH_3CH_2-CO-[CH_2]_2-CO-CH_3$$

Heptane-2,5-dione

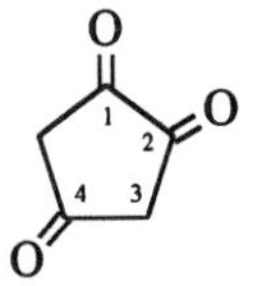

Cyclopentane-1,2,4-trione

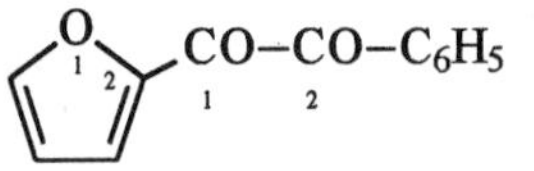

2-Furyl(phenyl)ethanedione
2-Furyl phenyl diketone

Diketones derived from aromatic hydrocarbons by conversion of two –CH= groups into –CO– groups with rearrangement of the ring double bonds as needed may also be named by adding the suffix -quinone to the name (sometimes with contraction) of the parent hydrocarbon.

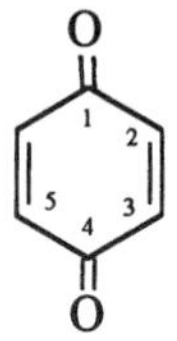

Benzo-1,4-quinone
Cyclohexa-2,5-diene-1,4-dione

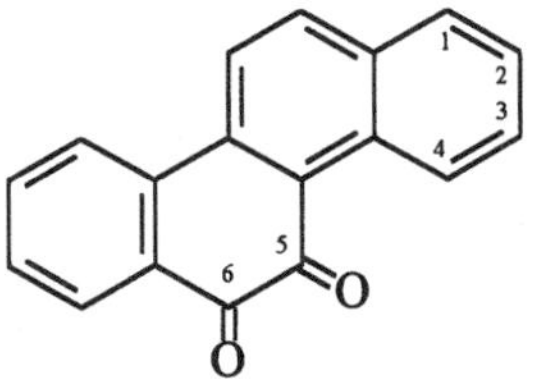

Chrysene-5,6-quinone
Chrysene-5,6-dione

As a substituent, ketonic O= is named with the prefix oxo. The –CO– group also appears as part of a substituent in the names of *acyl groups*, R–CO– (see chapter 11). These groups are named on the basis of the names of the acids from which they are derived by loss of OH. Acid names ending in -oic acid become -oyl in the acyl group, as in octanoyl from octanoic acid. Likewise, trivial acid names ending in -ic acid become -yl or -oyl as in acetyl or adipoyl, and acid names ending in carboxylic acid become carbonyl as in cyclohexanecarbonyl (CAS uses cyclohexylcarbonyl). Trivial names such as acetonyl for $CH_3$–CO–$CH_2$–, acetonylidene for $CH_3$–CO–CH=, acetoacetyl for $CH_3$–CO–$CH_2$–CO–, phenacyl for $C_6H_5$–CO–$CH_2$–, and phenacylidene for $C_6H_5$–CO–CH= are acceptable.

These prefixes are used to denote the presence of ketonic oxygen in names of *oxo acids*, since the acid characteristic group is senior to the carbonyl group. In substitutive nomenclature, oxo is typically used when the carbonyl group is not terminal to an acyclic chain, although it is used by CAS for terminal carbonyl as well.

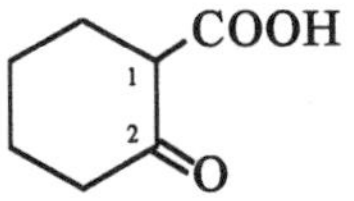

2-Oxocyclohexanecarboxylic acid

$$OHC-CH_2-CO-CH_2-COOH$$

4-Formyl-3-oxobutyric acid
3,5-Dioxopentanoic acid
3-Oxoglutaraldehydic acid

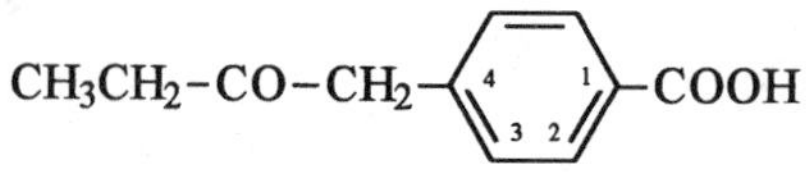

$CH_3CH_2-CO-CH_2$—[benzene ring, positions 4,3,2,1]—$COOH$     4-(2-Oxobutyl)benzoic acid

When the carbonyl group terminates a side chain, the acyl group form is applicable; CAS uses acetyl and benzoyl only when the carbonyl group is not part of a chain.

$$CO-C_6H_5$$
$$|$$
$$CH_3CHCH_2-COOH$$
$$4\quad3\quad2\quad\quad1$$

3-Benzoylbutanoic acid
β-Methyl-γ-oxobenzenebutanoic acid

*Thioketones* are named systematically in a manner analogous to that described for ketones by using the substitutive suffixes -thione, -dithione, -trithione, and so on, and the prefix names thioxo for S= and -carbonothioyl- for –CS–; if (thiocarbonyl) is used for –CS– parentheses are needed to avoid confusion with -thiocarbonyl-, –S–CO–. The functional class name is *thioketone*. Thio as a prefix in trivial names such as thioacetone or thioacetyl is acceptable. The corresponding terms for *selenium analogs* are -selone, -selenoxo, and (selenocarbonyl) or -carbonoselenoyl-, and for *tellurium analogs*, tellone, telluroxo, and (tellurocarbonyl) or -carbonotelluroyl-.

$CH_3CH_2-CS-CH_2CH_3$
$\quad5\quad\;4\quad\;\;3\quad2\quad\;1$

Pentane-3-thione
Diethyl thioketone

$SHC-CH_2-CS-CH_2CH_2-COOH$
$\quad5\quad\quad4\quad\quad3\quad\;\;2\quad\quad1$

5-(Thioformyl)-4-thioxopentanoic acid
4,6-Dithioxohexanoic acid

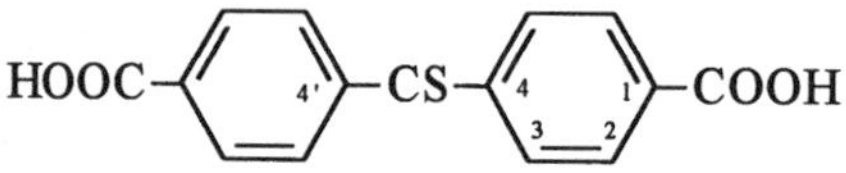

$HOOC$—[benzene ring, position 4']—$CS$—[benzene ring, positions 4,3,2,1]—$COOH$

4,4'-(Thiocarbonyl)dibenzoic acid
4,4'-Carbonothioyldibenzoic acid

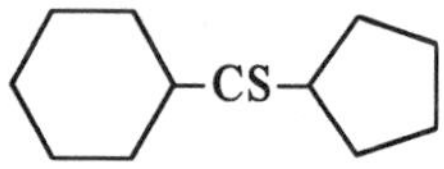

[cyclohexyl]—$CS$—[cyclopentyl]

Cyclohexyl(cyclopentyl)methanethione
Cyclohexyl cyclopentyl thioketone

$C_6H_5-CSe-CH_3$

Methyl(phenyl)methaneselone
Methyl phenyl selenoketone
Selenoacetophenone

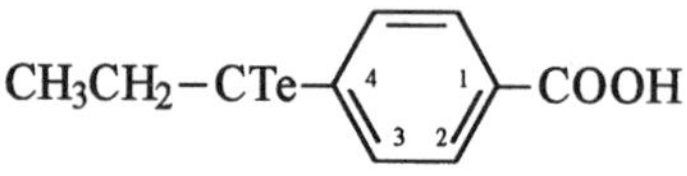

$CH_3CH_2-CTe$—[benzene ring, positions 4,3,2,1]—$COOH$

4-(1-Telluroxopropyl)benzoic acid
4-(Telluropropionyl)benzoic acid
4-Propanetelluroylbenzoic acid

*Ketene* is a name for the parent compound $CH_2$=C=O as well as a class name for this type of compound. Derivatives are named either on the basis of the parent compound or systematically as ketones; CAS names them only as ketones. The sulfur analogs are named as thioketenes or thiones. A name such as phenyl(thioketene) requires parentheses to avoid ambiguity.

$C_6H_5-CH=C=O$
$\qquad\quad2\quad\;\;1$

Phenylethen-1-one
Phenylketene

$(C_6H_5)_2C=C=S$
$\qquad\quad2\quad\;1$

Diphenylethene-1-thione
Diphenyl(thioketene)

$$CH_3CH_2CH_2CH_2$$
$$|$$
$$CH_3CH_2CH_2CH_2-C=C=O$$
$$6\quad\;5\quad\;\;4\quad\;\;3\quad\;\;2\quad1$$

2-Butylhex-1-en-1-one
Dibutylketene

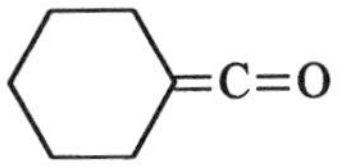

Cyclohexylidenemethanone

*α-Hydroxy ketones*, known by the class name *acyloins*, are best named by substitutive nomenclature. A few names derived from the trivial names of the R–COOH corresponding to $R_2C=O$, such as acetoin for 3-hydroxybutan-2-one and benzoin for 2-hydroxy-1,2-diphenylethanone are still seen, but the substitutive names are preferred.

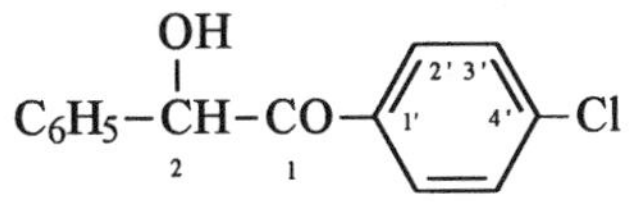

4'-Chloro-2-hydroxy-2-phenylacetophenone
1-(4-Chlorophenyl)-2-hydroxy-2-phenylethan-1-one

Compounds of the type $RR'C(OR'')(OR''')$ and $RR'C(OR'')(OH)$ are also best named substitutively on the basis of the parent hydrocarbon or alcohol. The functional class names are *ketal* and *hemiketal*, respectively.

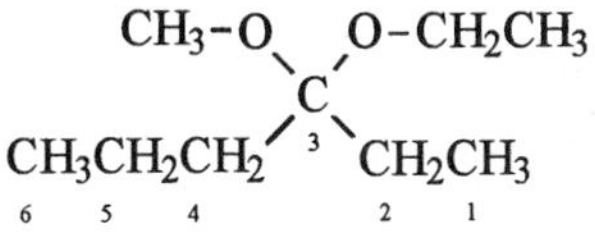

3-Ethoxy-3-methoxyhexane
Hexan-3-one ethyl methyl ketal

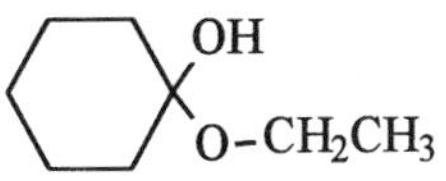

1-Ethoxycyclohexan-1-ol
Cyclohexanone ethyl hemiketal

Sulfur analogs of these compounds are named substitutively or by means of the class names *monothioketal* and *dithioketal*.

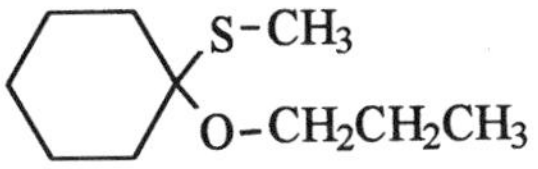

1-(Methylthio)-1-propoxycyclohexane
Cyclohexanone *S*-methyl *O*-propyl monothioketal

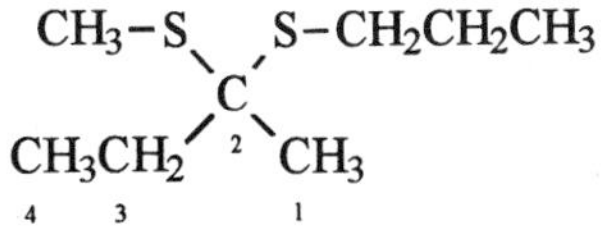

2-(Methylthio)-2-(propylthio)butane
2-Butanone methyl propyl dithioketal

## Discussion

Nomenclature for both ketones and aldehydes (see chapter 13) appears to be in a transition state. Familiar trivial names like acetone and benzil, and semisystematic names such as acetophenone and benzophenone abound and can hardly be ignored. The same may be said for most of the functional class names: they are mainly useful for the simpler and usually unsubstituted compounds such as ethyl methyl ketone (alphabetical order is not always followed, and, in this case, it leads to the acronym "MEK"). In general, systematic substitutive nomenclature is preferred, even though the names may be a few letters longer.

On the other hand, the older well-established "trivial" class names seem unlikely to be abandoned. Ketals and hemiketals (such as 1-hydroxy-1-methoxycyclohexane) were named as acetals in the 1979 IUPAC Organic Rules but "ketal" and "hemiketal" were reinstated in 1993; acetal and ketal names are no longer used by CAS. The class name "acylion" for α-hydroxyketones seems to be on the way out: old-timers have no problem with an acylion name, but the systematic name makes more sense to newcomers. Ketenes and quinones are here to stay, and, indeed, they are the basis for a systematic nomenclature; however, quinone is not a suitable synonym for 1,4-benzoquinone.

Contracted names such as 2-pyridone were limited to nitrogenous rings in the 1979 IUPAC recommendations, and in the 1993 recommendations were further restricted to quinolone, isoquinolone, and pyrrolidone. Other than elimination of two letters ("in" before -one), they offer no advantage and should no longer be used.

Keto should not be used as a prefix for O= in place of oxo in substitutive names. Similarly, although acyl prefixes such as hexanoyl are used for substituting groups with O= at the point of attachment, there is increasing pressure for 1-oxohexyl, since it relates readily with names like 3-oxohexyl and 2,5-dioxohexyl.

## Additional Examples

1.
$$CH_3$$
$$CH_3\overset{|}{C}HCH_2-CO-CH_3$$
5   4   3   2   1

4-Methylpentan-2-one
Isobutyl methyl ketone

2.
$$CH_3$$
$$CH_2{=}\overset{|}{C}-CO-CH_3$$
4   3   2   1

3-Methylbut-3-en-2-one
Isopropenyl methyl ketone

3.
$$CH_3CH_2-CO-CH{=}CH-C_6H_5$$
5   4   3   2   1

1-Phenylpent-1-en-3-one
Ethyl styryl ketone

4.
$$CH_3-CO-CH_2-SH$$
3   2   1

1-Mercaptopropan-2-one
1-Sulfanylpropan-2-one
Mercaptomethyl methyl ketone

5.
$$C_6H_5$$
$$CH_3-CO-\overset{|}{C}H-OH$$
3   2   1

1-Hydroxy-1-phenylpropan-2-one
$\alpha$-Hydroxybenzyl methyl ketone
1-Hydroxy-1-phenylacetone

6.
$$CH_3CH_2-CS-CS-CH_3$$
5   4   3   2   1

Pentane-2,3-dithione
Ethyl methyl dithiodiketone

7.
$$CH_3-CO-CH_2-CO-CH_3$$
5   4   3   2   1

Pentane-2,4-dione
Acetylacetone

8.
$$CH_3-CS-CH_2CH_2-CO-CH_3$$
6   5   4   3   2   1

5-Thioxohexan-2-one

9. 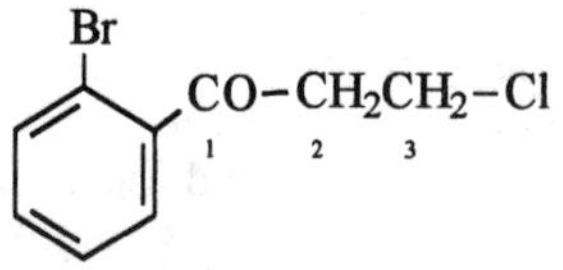
Br
$$CO-CH_2CH_2-Cl$$
1   2   3

1-(2-Bromophenyl)-3-chloropropan-1-one
2'-Bromo-3-chloropropanophenone

10.

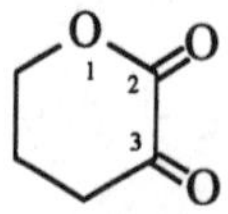

Dihydro-2H-pyran-2,3(4H)-dione
Oxane-2,3-dione

11.

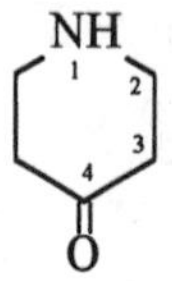

Piperidin-4-one
4-Piperidone

12.

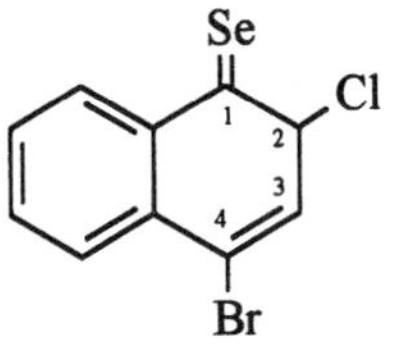

4-Bromo-2-chloronaphthalene-1(2*H*)-selone

13.

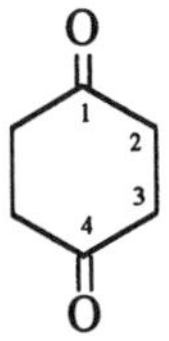

Cyclohexane-1,4-dione

14.

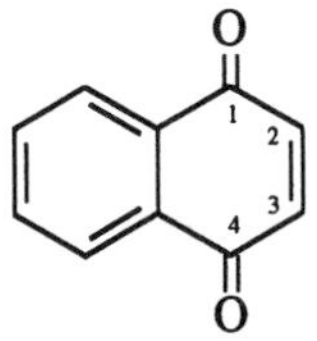

Naphtho-1,4-quinone
Naphthalene-1,4-dione

15.

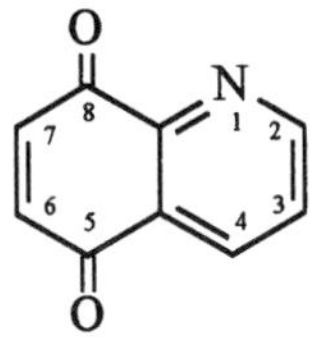

Quinoline-5,8-dione

16.

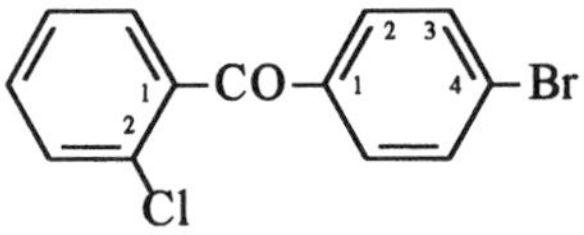

(4-Bromophenyl)(2-chlorophenyl)methanone
4-Bromophenyl 2-chlorophenyl ketone

17. $CH_3-CS-CO-C_6H_5$
       3    2   1

1-Phenyl-2-thioxopropan-1-one

18.

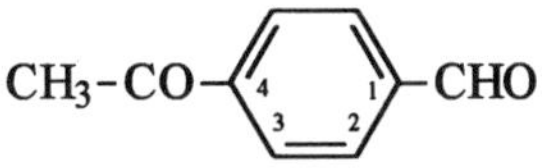

4-Acetylbenzaldehyde

19.

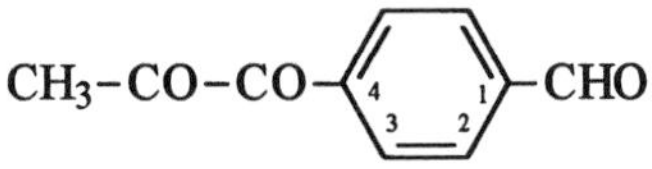

4-(1,2-Dioxopropyl)benzaldehyde
4-(2-Oxopropanoyl)benzaldehyde

20.

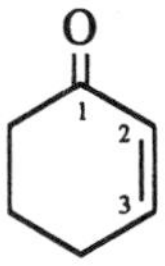

Cyclohex-2-en-1-one

21. $C_6H_5-CO-CH=C=O$
           3    2   1

3-Phenylprop-1-ene-1,3-dione
Benzoylketene

22.
$$CH_3$$
$$CH_3-\overset{|}{C}=C=C=O$$
   4    3   2   1

3-Methylbuta-1,2-dien-1-one
Isopropylideneketene

23. 

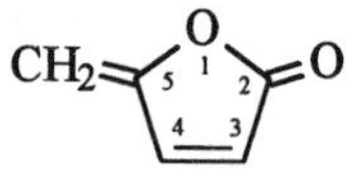

5-Methylidenefuran-2(5$H$)-one

24.

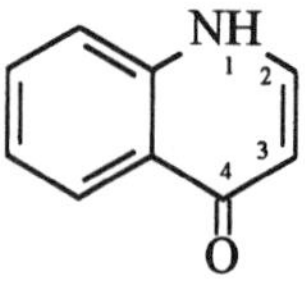

Quinolin-4(1$H$)-one

25.

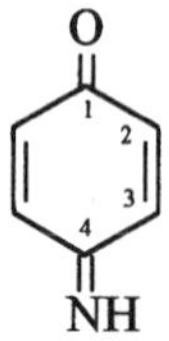

4-Iminocyclohexa-2,5-dien-1-one

26.

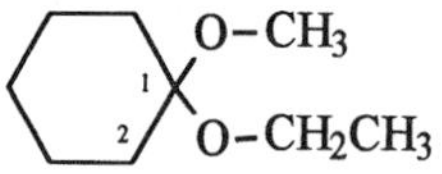

1-Ethoxy-1-methoxycyclohexane
Cyclohexanone ethyl methyl ketal

27.

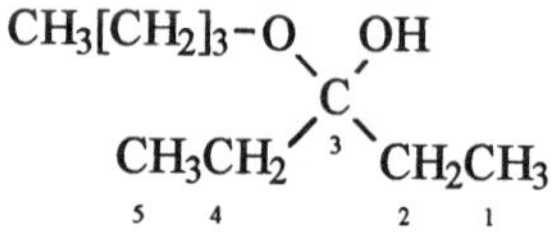

3-Butoxypentan-3-ol
Pentan-3-one butyl hemiacetal

28. $CH_3CH_2-CS-CH_3$

Butane-2-thione
Ethyl methyl thioketone

29. 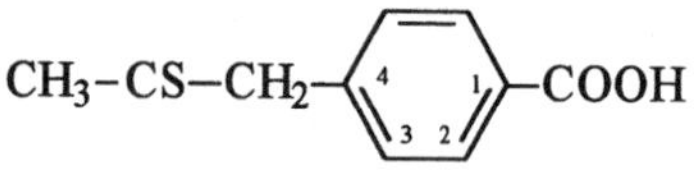 $CH_3-CS-CH_2$—〈 〉—COOH

4-(2-Thioxopropyl)benzoic acid

30.

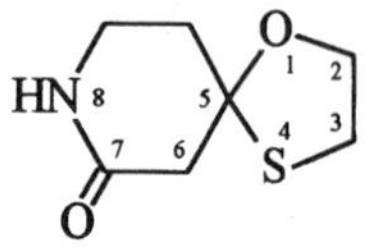

1-Oxa-4-thia-8-azaspiro[4.5]decan-7-one

31. 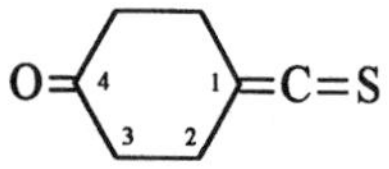 O=〈 〉=C=S

4-(Oxocyclohexylidene)methanethione
4-(Thioxomethylidene)cyclohexan-1-one
4-(Carbonothioyl)cyclohexan-1-one

32. 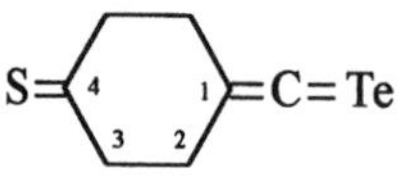 S=〈 〉=C=Te

4-(Telluroxomethylidene)cyclohexane-1-thione
4-(Carbonotelluroyl)cyclohexane-1-thione
4-(Thioxocyclohexylidene)methanetellone

33. $CH_3-CS-CO-CH_3$

Dimethyl thiodiketone

# Alcohols and Phenols

The title terms of this chapter have respectable beginnings. "*Al kohl*", with ancient origins, was first a descriptive name for any fine black powder, especially antimony sulfide. Only in the sixteenth century did Paracelsus apply it to the "spirit of wine". Phenol was first labeled "carbolic acid" when it was obtained from coal tar, but soon thereafter Laurent related it to "phene" and called the compound "*hydrate de phenyle*" ("phene" itself came from the Greek for "I light", and it was applied to the benzene obtained from illuminating gas); this name was converted by Gerhardt to phenol, and the term stuck.

The nomenclature of compounds having one or more hydroxy groups, –OH, attached to carbon atoms of parent hydrides is considered in this chapter. *Alcohols*, in which the –OH group is bound to a carbon atom in an acyclic or alicyclic parent hydride, and *phenols*, with the –OH group attached to a carbon atom of an arene or a heteroarene ring, constitute the major classes. Other classes of compounds with –OH groups bound to carbon are, for example, *enols*, *orthoacids*, *hemiacetals*, *acyloins*, and *halohydrins*. *Glycols* and other polyols are also covered, but hydroxy derivatives related to carbohydrates and cyclitols are found in chapter 31. Methods for naming alcohols and phenols have been extended to compounds with –OH groups bound to heteroatoms.

Nomenclature for chalcogen analogs of alcohols and phenols discussed here will be found in chapter 24.

## Acceptable Nomenclature

*Monohydric alcohols* and *phenols* in which the hydroxy group is the principal characteristic group are named substitutively by adding the suffix -ol to the name of the parent hydrocarbon or heterocycle after removal of the terminal "e", if present. The hydroxy group can be at any position in a chain of carbon atoms; the lowest locant from the end of the chain is used. Enols, R–CH=CH–OH, hemiacetals, R–CH(OR)–OH, and halohydrins, R–CH(X)CH$_2$–OH, are examples of classes of compounds also named by this method. Functional class names for simple alcohols are formed by citing the name of the alkyl or aryl group to which the –OH is attached, followed by the separate word "alcohol". Generally, substitutive names are preferred, although CAS uses conjunctive names such as benzenemethanol.

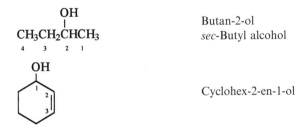

Butan-2-ol
*sec*-Butyl alcohol

Cyclohex-2-en-1-ol

$(C_6H_5)_3C-OH$

Triphenylmethanol
α,α-Diphenylbenzenemethanol
Trityl alcohol

$CH_2=CHCH_2-OH$
3      2    1

Prop-2-en-1-ol
Allyl alcohol

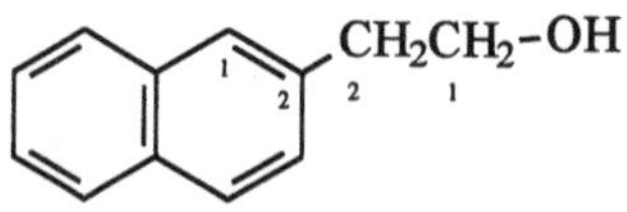

$CH_2CH_2-OH$

2-(2-Naphthyl)ethanol
2-Naphthaleneethanol

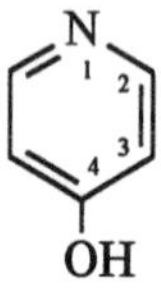

Pyridin-4-ol

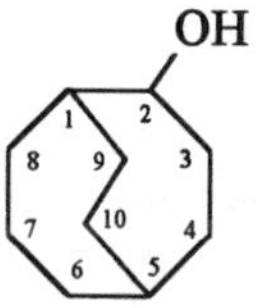

Bicyclo[3.3.2]decan-2-ol

Among the phenols, some trivial and contracted names are acceptable. Among them are phenol itself, 1-naphthol, 1-anthrol, 1-phenanthrol, and their positional isomers. Others, such as cresol, thymol, and picric acid are still recommended by IUPAC when unsubstituted, but it is best to name them by systematic substitutive nomenclature.

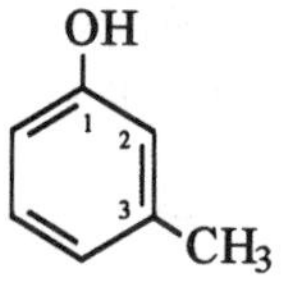

3-Methylphenol
*m*-Cresol

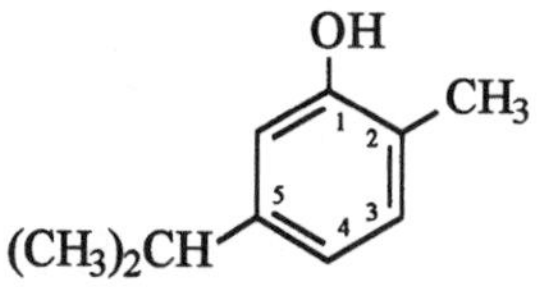

5-Isopropyl-2-methylphenol
    (*not* Carvacrol)
2-Methyl-5-(1-methylethyl)phenol
2-Methyl-5-(propan-2-yl)phenol

*Polyhydric alcohols* and *phenols* are named substitutively with the suffixes -diol, -triol, and so on, without elision of the final "e" in the hydrocarbon name.

$HO-CH_2[CH_2]_3CH_2-OH$
5     4-2    1

Pentane-1,5-diol

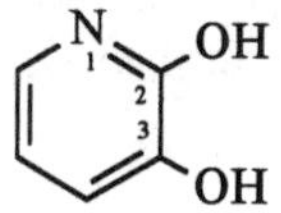

Pyridine-2,3-diol

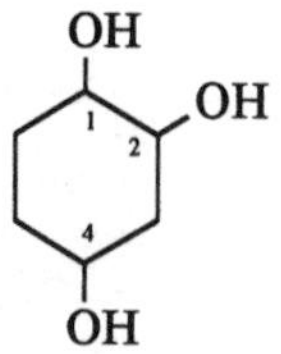

Cyclohexane-1,2,4-triol

For the series H–[O–CH$_2$CH$_2$]$_n$–OH, ethylene ($n = 1$), diethylene ($n = 2$), triethylene ($n = 3$), and tetraethylene ($n = 4$) glycol are acceptable, although IUPAC recommends only the first member. Longer chains should be named by replacement nomenclature (see chapters 3 and 9). "Propylene" analogs are named substitutively.

$$HO-CH_2CH_2-O-CH_2CH_2-O-CH_2CH_2-OH$$

2,2'-[Ethane-1,2-diylbis(oxy)]diethanol
Triethylene glycol

$$HO-CH_2CH_2-O-CH_2CH_2-O-CH_2CH_2-O-CH_2CH_2-O-CH_2CH_2-OH$$

3,6,9,12-Tetraoxatetradecane-1,14-diol

Glycerol, HO–CH$_2$CH(OH)CH$_2$–OH, pentaerythritol, C(CH$_2$OH)$_4$ and pinacol, (CH$_3$)$_2$C(OH)C(OH)C(CH$_3$)$_2$, are still in use, as are the following, but systematic names are preferred for substituted compounds.

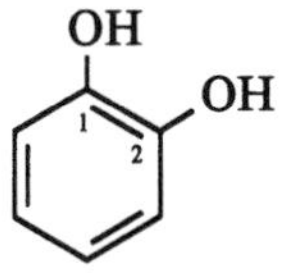

Pyrocatechol
Benzene-1,2-diol

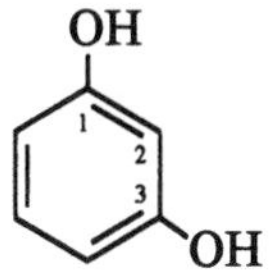

Resorcinol
Benzene-1,3-diol

Hydroquinone
Benzene-1,4-diol

*Hydroxy derivatives of ring assemblies* (see chapter 8) are usually named on the basis of the corresponding hydrocarbon ring assembly, but they can be named as derivatives of the component having the largest number of –OH groups.

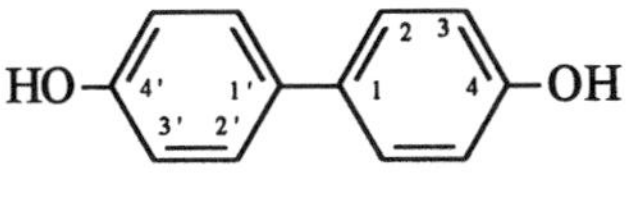

[1,1'-Biphenyl]-4,4'-diol
4,4'-Biphenol

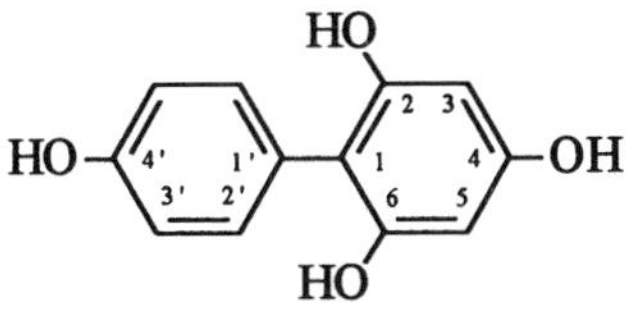

[1,1'-Biphenyl]-2,4,4',6-tetrol
2-(4-Hydroxyphenyl)benzene-1,3,5-triol

In the presence of a group of higher seniority (see chapter 3), the –OH group is cited with the prefix hydroxy; this prefix is also used when the –OH group is on a side chain. This method is used to name *acyloins*, R–CH(OH)–CO–R, and it may be applied to *hydroxycarboxylic acids*.

$$\underset{7}{CH_3}\underset{6}{\overset{\overset{\displaystyle OH}{|}}{CH}}\underset{5\text{-}2}{[CH_2]_4}-\underset{1}{CHO}$$

6-Hydroxyheptanal

6-Hydroxy-2-naphthoic acid
6-Hydroxynaphthalene-2-carboxylic acid

$$HO-\underset{4}{CH_2}\underset{3}{CH_2}\underset{2}{\overset{\overset{\displaystyle CH_2-OH}{|}}{CH}}\underset{1}{CH_2}-OH$$

2-(Hydroxymethyl)butane-1,4-diol

2-(Hydroxymethyl)-6-methylphenol
2-Hydroxy-3-methylbenzyl alcohol
2-Hydroxy-3-methylbenzenemethanol

The trivial names of several hydroxycarboxylic acids such as lactic and glycolic acid (see chapter 11) are acceptable.

Although the suffix -ol recommended here is normally used with compounds in which the –OH group is bound to carbon, it is also used with compounds in which an –OH group is bound to silicon, as in trimethylsilanol (see chapter 26), or when attached to other heteroatoms, as in piperidin-1-ol. The prefix hydroxy, however, is usually used to denote substitution by an –OH group at a noncarbon atom other than silicon in a heterocyclic ring system, as in 1-hydroxypiperidine, and in certain hydrazine and hydroxylamine derivatives (see chapter 22).

*Salts* and *anions* of alcohols (see also chapter 33) are named by changing the ending -ol to -olate in substitutive or trivial names; cations are cited first as separate words. Alternatively, salts may be named as oxides or, if the anionic group R–O$^-$ has a name ending in -oxy, as alkoxides or as alkyl oxides.

$$Na^+\left[CH_3CH_2-O^-\right]$$

Sodium ethanolate
Sodium ethoxide

$$K^+\left[C_6H_5-CH_2-O^-\right]$$

Potassium phenylmethanolate
Potassium benzyl oxide
Potassium benzenemethanolate

## Discussion

Today, substitutive nomenclature is the most broadly useful system for naming the compounds of this chapter. The trend is definitely away from names ending in "alcohol" and memory-taxing trivial names for di- and polyhydroxy derivatives of benzene and its alkylated analogs. A few trivially named classes, such as the *acyloins*, better named as α-hydroxy ketones (see chapter 14), are still with us. Trivial names are decried as a basis for further substitution.

There are many instances of pseudo names for alcohols, and they should be avoided. "Isopropanol", "*sec*-butanol", and "*tert*-butanol" have no corresponding hydrocarbon names (there is no name "*sec*-butane"), and therefore the use of suffixes like -ol is inappropriate. Isopropyl is a legitimate group name, and isopropyl alcohol is thus an acceptable name, but "stearyl alcohol" and "phthalyl alcohol" are incorrectly formed since there are no hydrocarbon names from which "stearyl" or "phthalyl" can be derived; however, one such name, vinyl alcohol, is firmly established in the polymer field in the form of poly(vinyl alcohol). "Carbinol" nomenclature (as in "triphenylcarbinol", in which "carbinol", for $CH_3$–OH, is the parent) is no longer used; the same can be said for the name "methylol", for the –$CH_2$–OH group. Terms such as "toluenol" and "xylenol" are ambiguous, since there is no way to locate the –OH group. Halohydrin and cyanohydrin names should be abandoned in favor of systematic names. Likewise, ortho acids and hemiacetals should be named as alcohols.

In some languages, but not in English, names of salts are formed by changing the ending "-yl alcoholate" to "-ylate", generating names such as "sodium benzylate".

Occasionally, two functional group names are combined in the ending of the compound name. While ethanolamine is widely used, it is inconsistent with systematic organic nomenclature; 2-aminoethanol is the correct name.

## Additional Examples

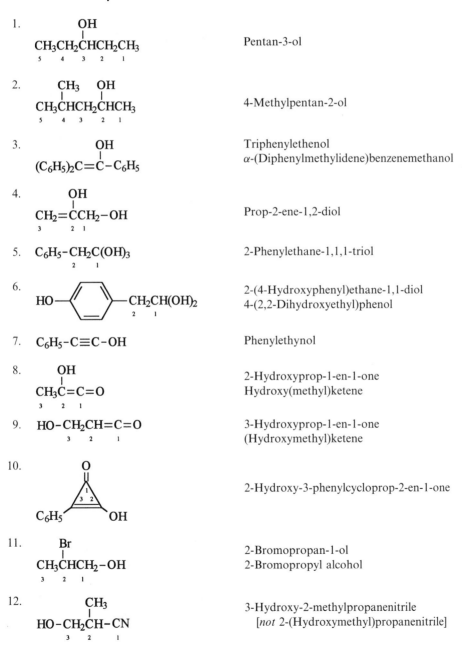

1.  $\underset{\substack{5 \quad 4 \quad 3 \quad 2 \quad 1}}{CH_3CH_2\overset{\displaystyle OH}{\overset{|}{C}}HCH_2CH_3}$    Pentan-3-ol

2.  $\underset{\substack{5 \quad 4 \quad 3 \quad 2 \quad 1}}{CH_3\overset{\displaystyle CH_3}{\overset{|}{C}}HCH_2\overset{\displaystyle OH}{\overset{|}{C}}HCH_3}$    4-Methylpentan-2-ol

3.  $(C_6H_5)_2C{=}\overset{\displaystyle OH}{\overset{|}{C}}{-}C_6H_5$    Triphenylethenol
    α-(Diphenylmethylidene)benzenemethanol

4.  $\underset{\substack{3 \quad\quad 2 \quad 1}}{CH_2{=}\overset{\displaystyle OH}{\overset{|}{C}}CH_2{-}OH}$    Prop-2-ene-1,2-diol

5.  $\underset{\substack{2 \quad 1}}{C_6H_5{-}CH_2C(OH)_3}$    2-Phenylethane-1,1,1-triol

6.  $HO{-}\langle\!\!\!\bigcirc\!\!\!\rangle{-}\underset{\substack{2 \quad 1}}{CH_2CH(OH)_2}$    2-(4-Hydroxyphenyl)ethane-1,1-diol
    4-(2,2-Dihydroxyethyl)phenol

7.  $C_6H_5{-}C{\equiv}C{-}OH$    Phenylethynol

8.  $\underset{\substack{3 \quad 2 \quad 1}}{CH_3\overset{\displaystyle OH}{\overset{|}{C}}{=}C{=}O}$    2-Hydroxyprop-1-en-1-one
    Hydroxy(methyl)ketene

9.  $\underset{\substack{3 \quad 2 \quad\quad 1}}{HO{-}CH_2CH{=}C{=}O}$    3-Hydroxyprop-1-en-1-one
    (Hydroxymethyl)ketene

10.  (cyclopropenone structure with $C_6H_5$ and OH)    2-Hydroxy-3-phenylcycloprop-2-en-1-one

11.  $\underset{\substack{3 \quad 2 \quad 1}}{CH_3\overset{\displaystyle Br}{\overset{|}{C}}HCH_2{-}OH}$    2-Bromopropan-1-ol
    2-Bromopropyl alcohol

12.  $\underset{\substack{3 \quad 2 \quad 1}}{HO{-}CH_2\overset{\displaystyle CH_3}{\overset{|}{C}}H{-}CN}$    3-Hydroxy-2-methylpropanenitrile
    [*not* 2-(Hydroxymethyl)propanenitrile]

13.

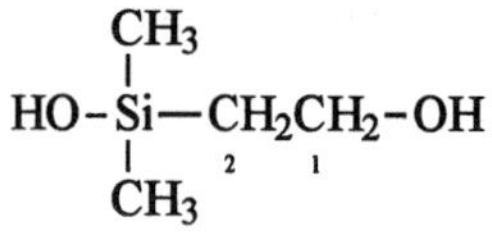

$$CH_3$$
$$HO-Si-CH_2CH_2-OH$$
$$CH_3$$

2-[Dimethyl(hydroxy)silyl]ethanol
Dimethyl(2-hydroxyethyl)silanol

14.

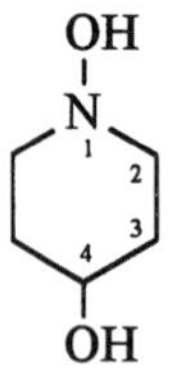

OH
N

OH

1-Hydroxypiperidin-4-ol
1,4-Dihydroxypiperidine
Piperidine-1,4-diol

15.     $CH_3-NH-CH_2CH_2-OH$

2-(Methylamino)ethanol
   (*not* *N*-Methylethanolamine)

16.

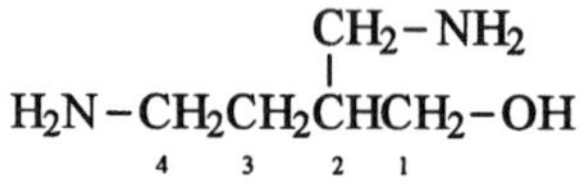

$$CH_2-NH_2$$
$$H_2N-CH_2CH_2CHCH_2-OH$$

4-Amino-2-(aminomethyl)butan-1-ol

17.

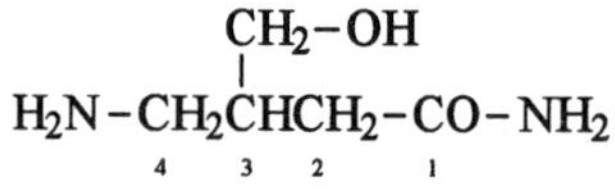

$$CH_2-OH$$
$$H_2N-CH_2CHCH_2-CO-NH_2$$

4-Amino-3-(hydroxymethyl)butanamide

18.

$$CH_3-CO-CH_2-OH$$

1-Hydroxypropan-2-one
Hydroxyacetone
   (*not* Acetol)

19. 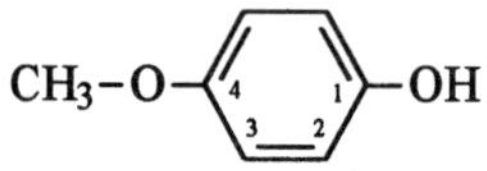     $CH_3-O-\langle\rangle-OH$

4-Methoxyphenol
   (*not* 4-Hydroxyanisole)

20. 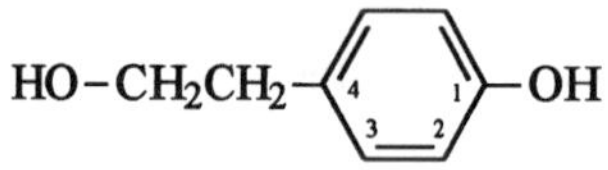     $HO-CH_2CH_2-\langle\rangle-OH$

4-(2-Hydroxyethyl)phenol
2-(4-Hydroxyphenyl)ethanol
4-Hydroxybenzeneethanol

21. 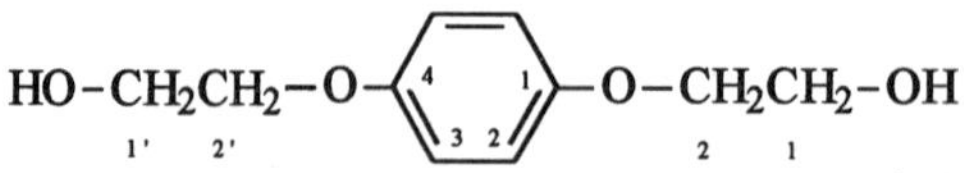     $HO-CH_2CH_2-O-\langle\rangle-O-CH_2CH_2-OH$

2,2'-[1,4-Phenylenebis(oxy)]diethanol

22. 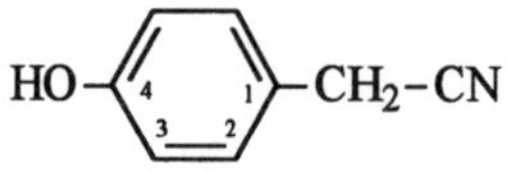     $HO-\langle\rangle-CH_2-CN$

(4-Hydroxyphenyl)acetonitrile
4-Hydroxybenzeneacetonitrile

23.

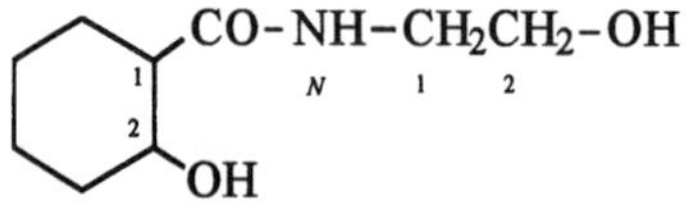

CO-NH-CH_2CH_2-OH

OH

2-Hydroxy-*N*-(2-hydroxyethyl)cyclohexane-
   carboxamide

24.

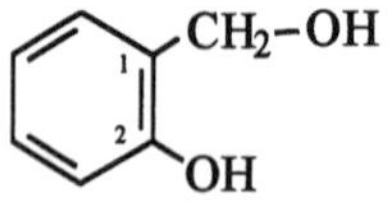

CH_2-OH

OH

(2-Hydroxyphenyl)methanol
2-Hydroxybenzenemethanol
2-Hydroxybenzyl alcohol
2-(Hydroxymethyl)phenol

25.

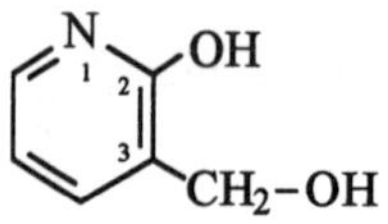

N     OH

CH_2-OH

3-(Hydroxymethyl)pyridin-2-ol
2-Hydroxy-3-pyridinemethanol
2-Hydroxy-3-(hydroxymethyl)pyridine

26.

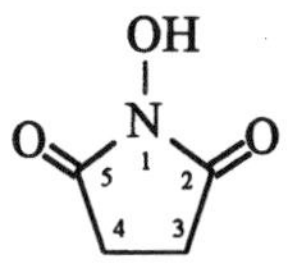

1-Hydroxypyrrolidine-2,5-dione
N-Hydroxysuccinimide

27.

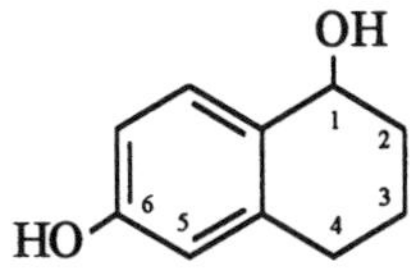

1,2,3,4-Tetrahydronaphthalene-1,6-diol

28.

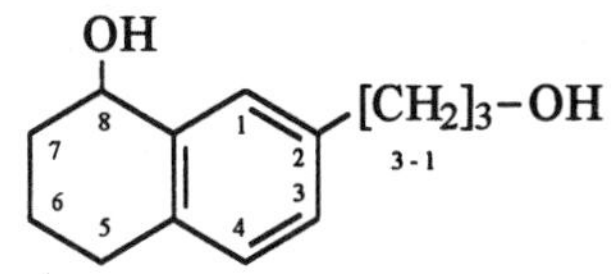

3-(5,6,7,8-Tetrahydro-8-hydroxy-2-
   naphthyl)propan-1-ol
5,6,7,8-Tetrahydro-8-hydroxy-2-
   naphthalenepropanol

29.

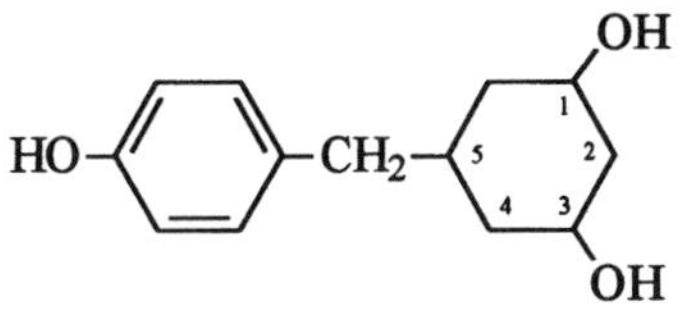

5-(4-Hydroxybenzyl)cyclohexane-1,3-diol
5-[(4-Hydroxyphenyl)methyl]cyclohexane-
   1,3-diol

# 16

# Ethers

In the laboratories of 1540, oil of vitriol plus spirit of wine gave "sweet oil of vitriol", called *spiritus vini aethereus* by Frobenius in 1729 and *ether sulfurique* in the *Methode* (1787). These names all described the same compound, diethyl ether, but in the early nineteenth century, it was *etherial spirit*, and the class name "ether" meant any neutral volatile liquid. History, but not the dictionary, teaches us to avoid calling diethyl ether "ether"; the latter can be used for the general class of R–O–R' compounds or for substances that once helped us sleep during our operations.

As a class, compounds with the structure R–O–R, where the R groups may be alike or different, are called *ethers, alkoxy compounds* or, in some languages, *oxides*. The oxygen atom may form part of a chain or a ring; in the latter case, the compound is usually, but not always, viewed as a heterocycle and is named as such (see chapter 9). Ethers stand very low in the seniority list of compound classes (see chapter 3, table 3.1). Therefore, in the presence of other characteristic groups, the R–O– group is almost always named as a substituent. Acetals (see chapter 13), ketals (see chapter 14) and ortho esters (see chapter 11) may also be considered as subclasses of ethers. Chalcogen analogs of ethers, that is, sulfides, selenides, and tellurides (see chapter 24), are named in the same manner as ethers. Thioacetals, thioketals, and thio(ortho esters) (see chapters 13, 14, and 11, respectively), can be named by principles closely following those for ethers.

## Acceptable Nomenclature

Names for ethers may be generated by substitutive, functional class, or, for polyethers, replacement nomenclature. An unsymmetrical ether, R–O–R', is named substitutively as an R'–O– derivative of the most senior parent hydrocarbon, (see chapter 3). Names of R'–O– groups take the form alkyloxy, aryloxy, or, in simple cases, contracted names, limited to methoxy, ethoxy, propoxy, butoxy, and phenoxy. Functional class names of ethers are formed by citing as separate words the names of R and R' in alphabetical order, followed by the class name *ether*. When R and R' are identical, the prefix di- or bis- should be used, although di- is often incorrectly omitted in the names of common ethers. Functional class names are usually preferred for simple symmetrical ethers; CAS uses substitutive nomenclature for these compounds.

$$CH_3CH_2-O-CH_2CH_3$$

Diethyl ether
  (*not* Ethyl ether)
Ethoxyethane

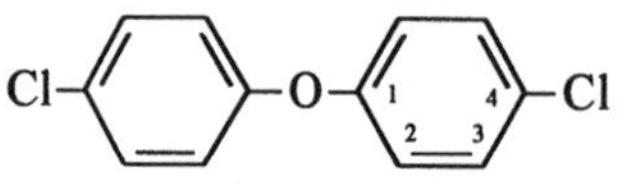

Bis(4-chlorophenyl) ether
1-Chloro-4-(4-chlorophenoxy)benzene
1,1'-Oxybis(4-chlorobenzene)

$CH_3CH_2CH_2CH_2-O-CH_3$
4   3   2   1

1-Methoxybutane
Butyl methyl ether

Cl
|
$CH_3CH-O-CH=CH_2$
2   1

(1-Chloroethoxy)ethene
1-Chloroethyl vinyl ether
1-Chloroethyl ethenyl ether

$Cl-CH_2-O-[CH_2]_2-O-[CH_2]_3-Br$
2-1     3-1

1-Bromo-3-[2-(chloromethoxy)ethoxy]propane
2-(3-Bromopropoxy)ethyl chloromethyl ether
3-Bromopropyl 2-(chloromethoxy)ethyl ether

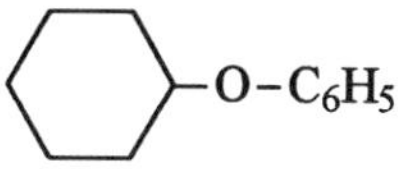

$\bigcirc-O-C_6H_5$

(Cyclohexyloxy)benzene
Cyclohexyl phenyl ether

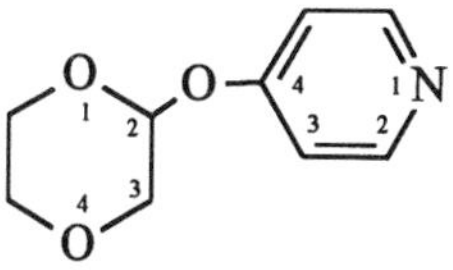

4-(1,4-Dioxan-2-yloxy)pyridine
1,4-Dioxan-2-yl 4-pyridyl ether

Partial ethers of polyhydroxy compounds are preferably named substitutively on the basis of the preferred characteristic group, alcohol, or by citing the name of the polyhydroxy compound followed by a numerical prefix (including mono) plus the etherifying group and the class name *ether*.

$CH_3-O-CH_2CH_2-OH$
2    1

2-Methoxyethanol
Ethane-1,2-diol monomethyl ether

Symmetrical ethers having a group with priority for citation as a suffix are named in the same way as assemblies of identical units (see chapter 3) with prefixes such as oxydi, oxybis, diylbis(oxy), or ylenebis(oxy) followed by the name of the units being multiplied.

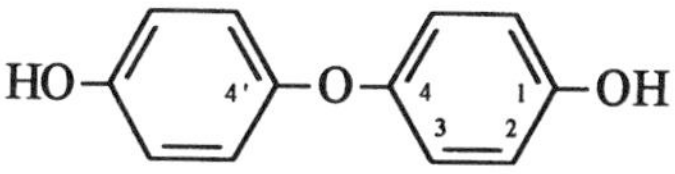

$HO-\bigcirc-O-\bigcirc-OH$

4,4′-Oxydiphenol

$H_2N-CH_2CH_2-O-CH_2CH_2-NH_2$
2′     2   1

2,2′-Oxydiethanamine
2,2′-Oxybis(ethylamine)
(Oxydiethane-2,1-diyl)diamine

$HOOC-CH_2-O-CH_2CH_2-O-CH_2-COOH$

[Ethane-1,2-diylbis(oxy)]bis(acetic acid)
[Ethylenebis(oxy)]bis(acetic acid)

Replacement nomenclature (see chapter 3), with the prefix oxa for an "O" atom taking the place of a "C" atom in a hydrocarbon chain, is a method used to simplify names for linear polyethers having multiple ether oxygen atoms; CAS and IUPAC differ in the criteria for forming these names in the presence of a principal characteristic group (see chapter 3). Polymeric polyethers are discussed in chapter 29. The replacement method may also be convenient in some situations even for diethers: the example immediately above would be named 3,6-dioxaoctanedioic acid by this method.

$CH_3-O-CH_2CH_2-O-CH_2-O-CH_2-NH-CH_2CH_2-OH$
11   10   9    8     7    6    5    4    3     2    1

5,7,10-Trioxa-3-azaundecan-1-ol (IUPAC)
2,5,7-Trioxa-9-azaundecan-11-ol (CAS)

2-[[[(2-Methoxyethoxy)methoxy]methyl]amino]ethanol (substitutive name)

Symmetric polyether diols are named by replacement or substitutive nomenclature; "glycol" names such as diethylene glycol are acceptable (see chapter 15).

Structures with one or more oxygen atoms as part of a ring system are usually named as heterocyclic compounds (see chapter 9).

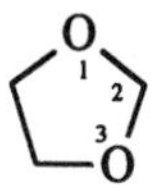

1,3-Dioxolane

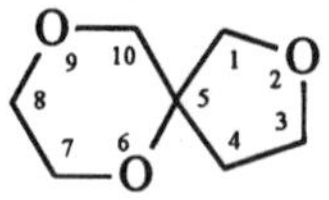

2,6,9-Trioxaspiro[4.5]decane

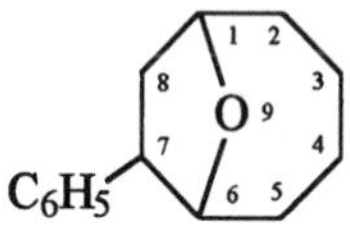

7-Phenyl-9-oxabicyclo[4.2.1]nonane

Replacement nomenclature is the method of choice for heteromonocycles having more than 10 ring atoms, such as *crown ethers*. The "cyclo(ab)$_n$" method (see chapter 9) or a modification of structure-based polymer nomenclature (see chapter 29) may also be used.

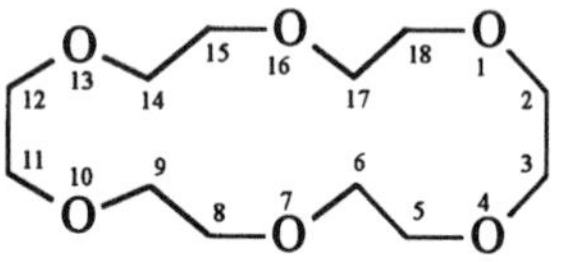

1,4,7,10,13,16-Hexaoxacyclooctadecane
Cyclohexa(oxyethylene)

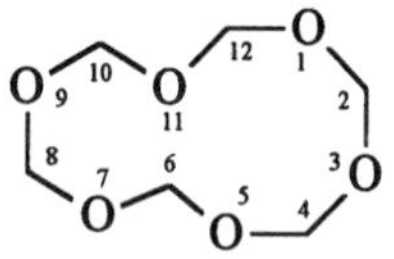

1,3,5,7,9,11-Hexaoxacyclododecane
Cyclohexacarboxane
Cyclohexa(oxymethylene)

The term epoxy is often used for an oxygen atom connecting two carbon atoms in a chain; however, a heterocyclic name is preferred.

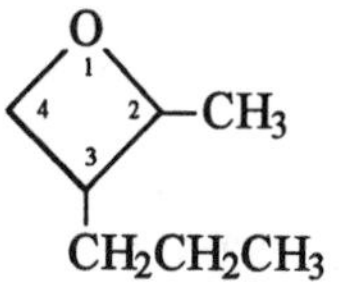

2-Methyl-3-propyloxetane
1,3-Epoxy-2-propylbutane

An alternative, but less preferable, method of naming simple epoxy compounds is to cite the name of the corresponding olefin, followed by the separate word oxide; names such as styrene oxide [(1,2-epoxyethyl)benzene or 2-phenyloxirane] result. Ethylene oxide (1,2-epoxyethane or oxirane) is also acceptable as a common name, despite the incorrect use of "ethylene" (see chapter 5).

Acetals (see chapter 13), with the structure R–CH(OR′)(OR″), ketals, RR′C(OR″)(OR‴) (see chapter 14), and ortho esters, R–C(OR′)$_3$ (see chapter 11) may be named substitutively as shown above for ethers. Cyclic acetals, ketals, and ortho esters are heterocyclic compounds and are usually named as such.

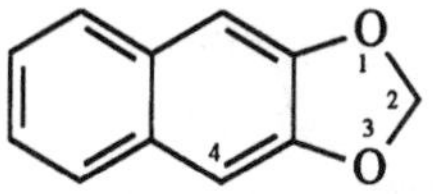

Naphtho[2,3-*d*][1,3]dioxole

A few trivial names for arene ethers are still widely used and acceptable but should not be substituted. Examples include the following:

$$C_6H_5-O-CH_3 \qquad \text{Anisole}$$

$$C_6H_5-O-CH_2CH_3 \qquad \text{Phenetole}$$

## Discussion

The ether class only exceeds its chalcogen analogs and peroxide in priority for citation as the compound class (see chapter 3, table 3.1). Thus, an ether is almost always expressed in prefix form (alkoxy and so on) in substitutive nomenclature of compounds with mixed functions. In functional class nomenclature, a somewhat related order of seniority of class names does exist, but the main reason for using this system is that it emphasizes a specific class. A compound that is both an ether and an alcohol might well be named as a hydroxy-substituted ether, despite the higher seniority of alcohol, if it is desired to stress the ether class.

The term epoxy is used in two different ways to indicate –O– connected to two carbon atoms that are already part of a ring or a chain. In the first method –O– is considered to form a "bridge" between the two carbon atoms that are already part of a ring system, and in this way it becomes part of the parent system. The prefix epoxy is part of the name of the parent and is therefore not alphabetized like the names of substituent prefixes.

 1,2,3,4-Tetrahydro-1,4-epoxynaphthalene

In the second method –O– is viewed as substituting for two hydrogens at two carbon atoms in the same chain; as a substituent, the prefix epoxy is alphabetized just as any other substituent would be. In this way, the name of the original hydrocarbon is preserved.

 2,4-Epoxy-5-methylhexane
2-Isopropyl-4-methyloxetane

The term "oxy", meaning –O–, occurs not only in prefixes such as ethoxy, $CH_3CH_2-O-$, but also in prefixes for groups such as $-O-CH_2CH_2-O-$, ethylenedioxy. Bis(oxy) is used by CAS in prefix names such as ethane-1,2-diylbis(oxy) to avoid confusion with prefixes involving –OO–, dioxy, which was recommended by IUPAC in the 1979 rules and might be seen in a name like ethyldioxy, $CH_3CH_2-OO-$. There are many heterocycles that involve an "epoxy" prefix (see chapter 9). "Oxy" also turns up in other guises such as ethoxycarbonyl, $CH_3CH_2-O-CO-$, ethoxyimino, $CH_3CH_2-O-N=$, or azoxy, $-N(O)=N-$.

Names based on "glyme", for 1,2-dimethoxyethane, are not acceptable.

The general principles of ether nomenclature have occasionally been applied to structures in which –O– is bound to a carbon atom and a hetero atom. An example is found in silicon compounds (see chapter 26) such as $H_3Si-O-R'$, alkoxysilane or alkyl silyl ether; care must be taken with this kind of usage, since, after all, $(RO)_4Si$ is an ester of silicic acid.

## Additional Examples

1.  $CH_3CH_2-O-CH_2CH_2-Br$      1-Bromo-2-ethoxyethane
2-Bromoethyl ethyl ether

2. $Cl-CH_2CH_2-O-CH_2CH_2-Cl$
        2        1

1-Chloro-2-(2-chloroethoxy)ethane
Bis(2-chloroethyl) ether
1,1'-Oxybis(2-chloroethane)

3. $H_2N-CH_2-O-CH_2-NH_2$
       1'       1     N

1,1'-Oxybis(methylamine)
[Oxybis(methylene)]diamine
Bis(aminomethyl) ether
1,1'-Oxydimethanamine

4. $H_2N-CH_2CH_2-O-CH_2CH_2-OH$
              2        1

2-(2-Aminoethoxy)ethanol
2-Aminoethyl 2-hydroxyethyl ether

5. $CH_3-O-CH_2CH_2CH_2-O-CH_3$
               3      2      1

1,3-Dimethoxypropane
3-Methoxypropyl methyl ether

6.
$$CH_3-[CH_2]_4-O-\underset{\underset{CH_3}{|}}{CH}-O-[CH_2]_4-CH_3$$
      4'-1'                          1-4    5

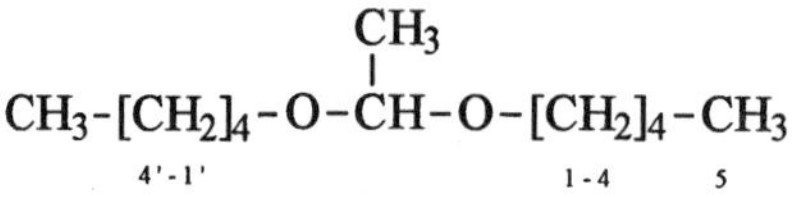

1,1'-[Ethane-1,1-diylbis(oxy)]dipentane
1-[1-(Pentyloxy)ethoxy]pentane
Acetaldehyde dipentyl acetal

7. $HO-CH_2CH_2-O-CH_2CH_2-O-CH_2CH_2-OH$
          2'                2        1

2,2'-[Ethylenebis(oxy)]diethanol
2,2'-[Ethane-1,2-diylbis(oxy)]diethanol

8.
$-CH_2-O-C_6H_5$

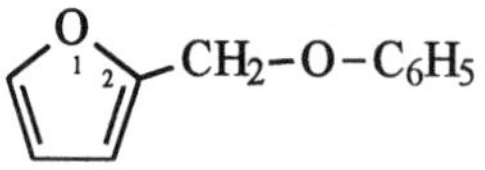

2-(Phenoxymethyl)furan
2-Furylmethyl phenyl ether

9.

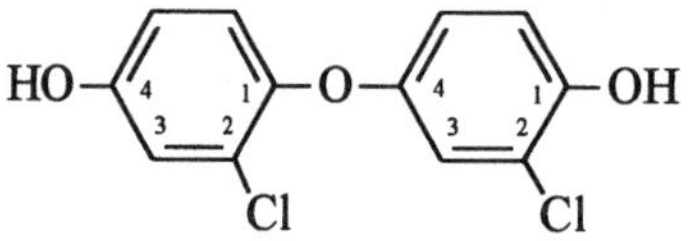

2-Chloro-4-(2-chloro-4-
   hydroxyphenoxy)phenol
2,3'-Dichloro-4,4'-oxydiphenol
2-Chloro-4-hydroxyphenyl 3-chloro-4-
   hydroxyphenyl ether

10.
$HOOC-$$-O-CH_2CH_2-O-$$-COOH$

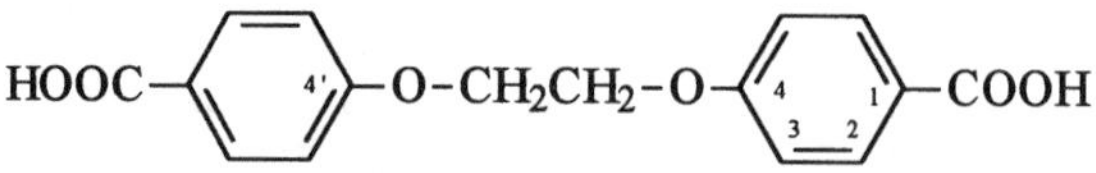

4,4'-[Ethylenebis(oxy)]dibenzoic acid

11.
$-O-CH_2CH_2-O-CH_2CH_2-O-CH_3$

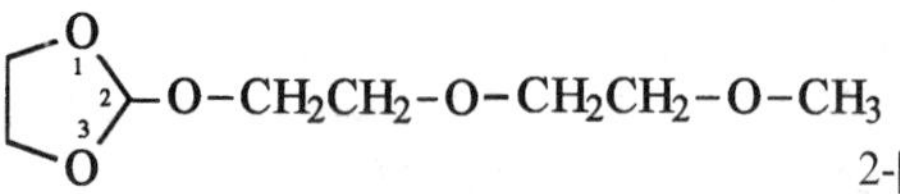

2-[2-(2-Methoxyethoxy)ethoxy]-1,3-
   dioxolane
2-(1,3-Dioxolan-2-yloxy)ethyl 2-methoxy-
   ethyl ether

12.
$$CH_3CH_2-O \overset{O-CH_2CH_3}{\underset{CH_3 \quad O-CH_2CH_3}{C}}$$

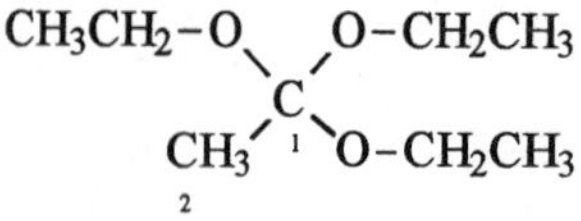

1,1,1-Triethoxyethane
Triethyl orthoacetate
[Ethane-1,1,1-triyltris(oxy)]triethane

13. $CH_3CH_2-O-CH=CH-CN$
            3      2    1

3-Ethoxyacrylonitrile
3-Ethoxyprop-2-enenitrile
2-Cyanovinyl ethyl ether

14.

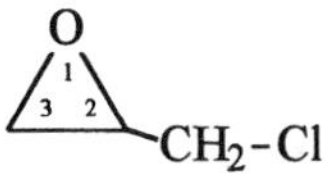

(Chloromethyl)oxirane
1-Chloro-2,3-epoxypropane
 (*not* Epichlorohydrin)

15.

2,3-Oxiranedicarboxylic acid
Epoxysuccinic acid

16.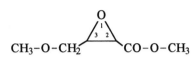

Methyl 3-(methoxymethyl)oxirane-2-
 carboxylate
Methyl 2,3-epoxy-4-methoxybutanoate

17.

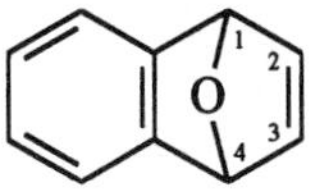

1,4-Dihydro-1,4-epoxynaphthalene

18.

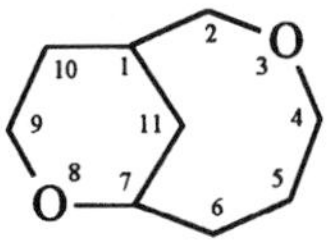

3,8-Dioxabicyclo[5.3.1]undecane

# 17

# Peroxides and Hydroperoxides

Organic compounds with the general structure R–OO–R′ are classified as *peroxides*; R–OOH are called *hydroperoxides*. The nomenclature of these and their sulfur, selenium, and tellurium analogs are discussed in this chapter. Structures in which R or R′ is an acyl group are also covered under *peroxy acids* (see chapter 11) and their esters (see chapter 12). Where both R and R′ are acyl groups, the compounds could be treated as mixed anhydrides but are usually named as peroxides.

The nomenclature of peroxides tends to parallel that of ethers (see chapter 16). In practice, however, functional class principles are used more than substitutive nomenclature where the peroxide characteristic group is likely to be expressed by a prefix because peroxides are low in the seniority list of compound classes (see chapter 3, table 3.1). Although hydroperoxides are also named like peroxides, they follow alcohols in order of seniority, possibly reflecting the sense that they are "peroxy alcohols".

## Acceptable Nomenclature

*Hydroperoxides*, R–OOH, may be named with the class name hydroperoxide or substitutively with the prefix hydroperoxy for the HOO– group.

$$CH_3CH_2{-}OOH$$
2  1

Ethyl hydroperoxide
Hydroperoxyethane

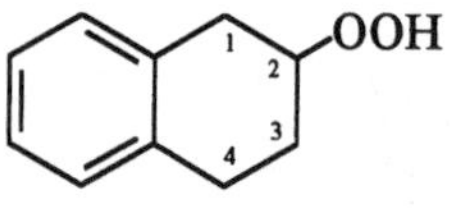

1,2,3,4-Tetrahydro-2-naphthyl hydroperoxide
1,2,3,4-Tetrahydro-2-hydroperoxynaphthalene

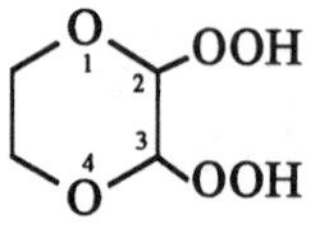

1,4-Dioxane-2,3-diyl bis(hydroperoxide)
2,3-Bis(hydroperoxy)-1,4-dioxane

Salts of hydroperoxides are given functional class names ending in peroxide in the same manner as one of the methods for naming salts of alcohols (see chapter 15).

$$Na^+ \left[ C_6H_5{-}CH_2{-}OO^- \right]$$
Sodium benzyl peroxide

*Peroxides*, R–OO–R′, have names derived from the names of R and R′, which can be chains or rings, followed by the word peroxide. In substitutive nomenclature, the bivalent group –OO– was named dioxy in the 1979 IUPAC rules, but peroxy is now recommended;

CAS uses dioxy. Both forms are acceptable, and dioxy is preferred in names such as dioxy-bis(acetic acid), where peroxy would be potentially ambiguous. These group names may appear in prefixes such as alkylperoxy or alkyldioxy for R–OO–. As a bridge connecting two carbon atoms in a ring or a chain, –OO– is called epidioxy.

$CH_3CH_2-OO-CH_3$
 ₂   ₁

Ethyl methyl peroxide
(Methylperoxy)ethane
(Methyldioxy)ethane

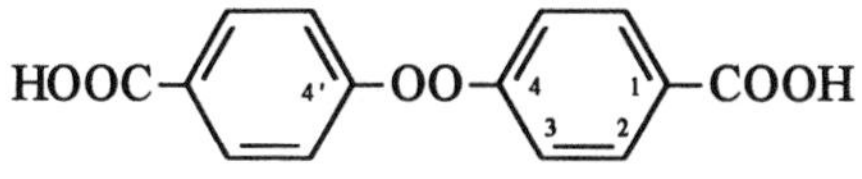

$HOOC$——OO——$COOH$   (4', 4, 3, 2, 1)

4,4'-Dioxydibenzoic acid

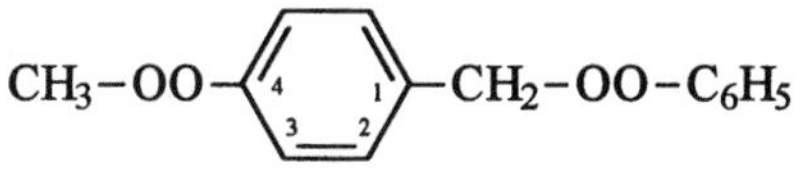

$CH_3-OO$——$CH_2-OO-C_6H_5$   (4, 1, 3, 2)

4-(Methylperoxy)benzyl phenyl peroxide
4-(Methylperoxy)-1-[(phenylperoxy)-
    methyl]benzene

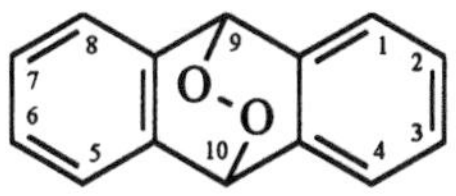

9,10-Dihydro-9,10-epidioxyanthracene

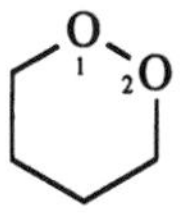

1,2-Dioxane
1,4-Epidioxybutane

Peroxides in which both R and R′ are acyl groups are named by the same methods; the name benzoyl peroxide is commonly used, but dibenzoyl peroxide is preferred. Compounds of the type R–CO–OO–R′ are named as esters of peroxy acids (see chapter 12).

Replacement of oxygen by another chalcogen in a hydroperoxide or a peroxide leads to compounds that can be named acceptably in a variety of ways: (a) as hydroperoxide or peroxide analogs with class names such as thiohydroperoxide or selenothioperoxide with letter locants to indicate the position of the R groups and prefixes such as thiooxy or sulfanyloxy, oxythio or oxysulfanyl, selenothio or selenosulfanyl, and thioseleno or sulfanylseleno; (b) as sulfenic acids, R–S–OH, and their chalcogen analogs with endings such as sulfenothioic acid (see chapter 23); or (c) as disulfides or disulfanes (see chapter 24).

$C_6H_5-S-OH$

*S*-Phenyl thiohydroperoxide
Benzenesulfenic acid

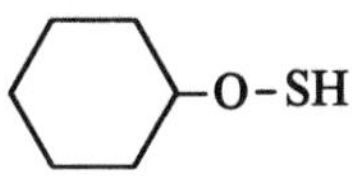

——$O-SH$

*O*-Cyclohexyl thiohydroperoxide

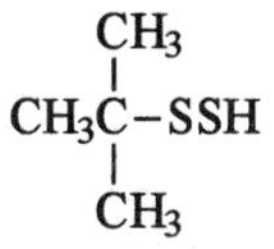

$CH_3$
|
$CH_3C-SSH$
|
$CH_3$

*tert*-Butyl hydrodisulfide
2-Methylpropane-2-sulfenothioic acid
(2-Methylpropan-2-yl)disulfane

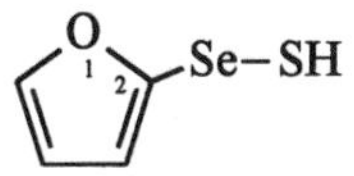

$Se-SH$   (1, 2)

*Se*-2-Furyl selenothiohydroperoxide
Furan-2-selenenothioic acid

$CH_3CH_2-O-S-C_6H_5$

*O*-Ethyl *S*-phenyl thioperoxide
Ethyl benzenesulfenate

$CH_3CH_2-SS-C_6H_5$

Ethyl phenyl disulfide
Phenyl ethanesulfenothioate
Ethyl(phenyl)disulfane

$CH_3CH_2-Te-S-CH_3$

*Te*-Ethyl *S*-methyl tellurothioperoxide
Ethyl methanesulfenotellurenate
Methyl ethanetellurenothioate

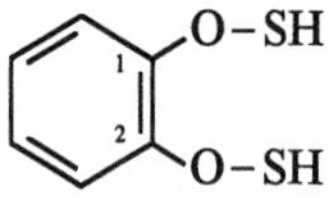

*O,O'*-1,2-Phenylene bis(thiohydroperoxide)
1,2-Bis(mercaptooxy)benzene
1,2-Bis(sulfanyloxy)benzene

When two or more dissimilar chalcogen analogs of –OOH or –OO–R groups are present, method (b) may be used if one of the groups is an acid or ester, such as –S–OH or –SS–R, that can serve as principal characteristic group; remaining groups are cited as prefixes, such as thiohydroperoxy, H{O,S}–; mercaptooxy or sulfanyloxy, HS–O–; or hydroxythio or hydroxysulfanyl, HO–S–. Under method (a), the group selected for the class name is the one with the most oxygen atoms or that with the atom having the lowest atomic number nearest to the hydrogen atom, that is, –OOH > –O–SH > –S–OH > –O–SeH, and so on.

4-Chloro-5-(mercaptooxy)pentane-1-sulfenic acid
*S*-[4-Chloro-5-(mercaptooxy)pentyl thiohydro-
    peroxide

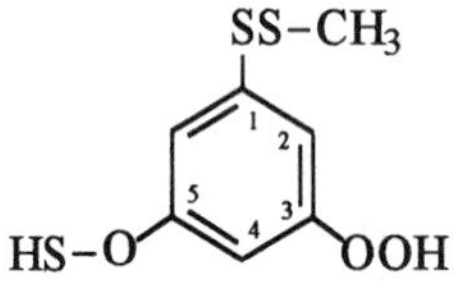

Methyl 3-hydroperoxy-5-(mercaptooxy)-
    benzenesulfenothioate
Methyl 3-hydroperoxy-5-(sulfanyloxy)-
    benzenesulfenothioate

$$HO-S-CH_2CH_2CH_2-COOH$$

4-(Hydroxythio)butanoic acid
4-Sulfenobutanoic acid

## Discussion

The compound types covered in this chapter provide an example of areas in which functional class and substitutive nomenclature are equally acceptable, no matter how desirable it would be to institute the latter throughout organic nomenclature. Perhaps it would be useful in substitutive nomenclature to declare the –OO– group in peroxides to be a "nonfunction", just as it is in ethers. The functional class route has been chosen by CAS for both peroxides and hydroperoxides. "Dioxide" has not been used as an analog of disulfide.

It might be debated whether the multiplier di- or bis- should be used before hydroperoxy, which has the look of a complex group that would need the recommended bis-. Care must be taken, however, where chalcogen replacement is involved. Names such as methylene thiobis(hydroperoxide) for HOO–CH$_2$–{O,S}H and methylene bis(thiohydroperoxide) for CH$_2$[{O,S}H]$_2$ are unambiguous but undesirable, and methylene dithiobis(hydroperoxide) is ambiguous. Italic letters as locants can remove this ambiguity for known structures, as in *O,S*-methylene bis(thiohydroperoxide) for HS–O–CH$_2$–S–OH.

## Additional Examples

1. $C_6H_5-CH_2-OOH$

Benzyl hydroperoxide
$\alpha$-(Hydroperoxy)toluene

2.

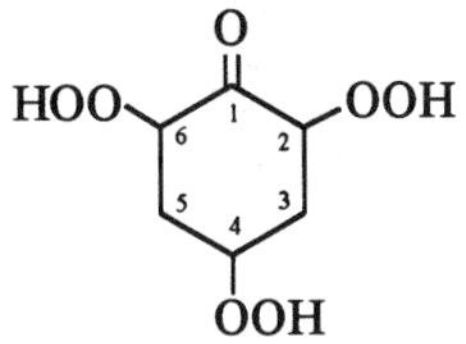

2,4,6-Tris(hydroperoxy)cyclohexan-1-one

3. $HOO-CH_2CH_2-OOH$
   $\quad\quad\;\; 2 \quad\;\; 1$

Ethylene bis(hydroperoxide)
1,2-Bis(hydroperoxy)ethane

4. $Na^+ \left[ HOO-CH_2CH_2-OO^- \right]$
   $\quad\quad\quad\quad\quad 2 \quad\;\; 1$

Sodium 2-(hydroperoxy)ethyl peroxide

5.

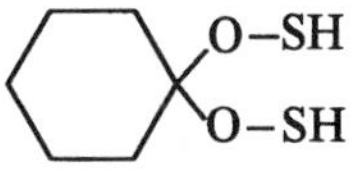

$O,O'$-Cyclohexane-1,1-diyl bis(thiohydroperoxide)

6.

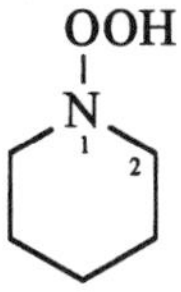

1-Hydroperoxypiperidine
Piperidin-1-yl hydroperoxide

7.

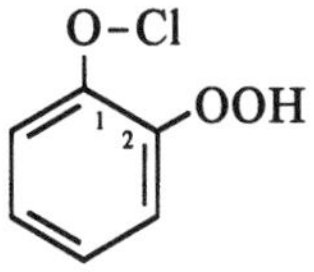

2-(Hydroperoxy)phenyl hypochlorite
2-(Chlorooxy)phenyl hydroperoxide

8. $HOOC-CH_2-OO-CH_2-COOH$

Dioxybis(acetic acid)

9. $CH_3C(CH_3)_2-OO-C(CH_3)_2CH_3$
   $\quad\quad 2' \quad\quad\quad\quad 2 \quad\quad\;\; 1$

2,2'-Peroxybis(2-methylpropane)
Bis(1,1-dimethylethyl) peroxide
Di-*tert*-butyl peroxide

10. $Cl-CH_2CH_2-OO-CH_3$
    $\quad\quad\;\; 2 \quad\; 1$

2-Chloroethyl methyl peroxide
1-Chloro-2-(methylperoxy)ethane

11. $C_6H_5-CH_2-O-Se-CH_2-C_6H_5$

Dibenzyl selenoperoxide
$\alpha,\alpha'$-(Selenoperoxy)ditoluene

12. 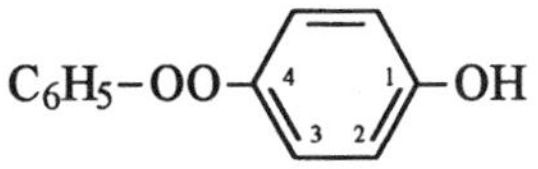 $C_6H_5-OO-$

$-OH$

4-(Phenyldioxy)phenol
4-(Phenylperoxy)phenol

13.

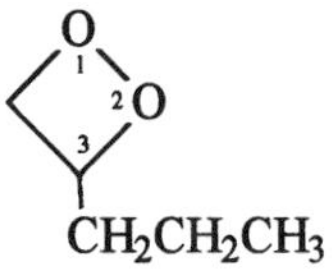

$CH_2CH_2CH_3$

3-Propyl-1,2-oxetane
1,2-Epidioxypentane

14.

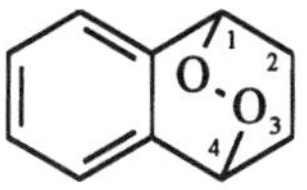

1,2,3,4-Tetrahydro-1,4-epidioxynaphthalene

15.

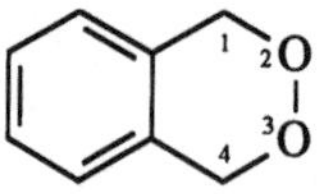

1,4-Dihydro-2,3-benzodioxin

16. $CH_3CH_2-CO-OO-CO-CH_3$

Acetyl propionyl peroxide

17. $C_6H_5-CO-OO-CO-[CH_2]_3CH_3$

Benzoyl pentanoyl peroxide

18.

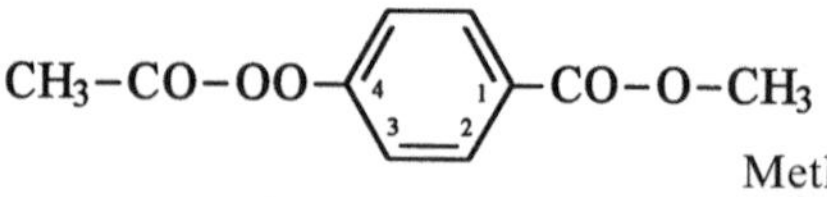

$CH_3-CO-OO-$ $-CO-O-CH_3$

Methyl 4-(acetylperoxy)benzoate (IUPAC)
4-(Methoxycarbonyl)phenyl peroxyacetate
4-(Methoxycarbonyl)phenyl ethaneperoxoate
    (CAS)

19. $CH_3-CO-OO-CH_2CH_3$

Acetyl ethyl peroxide
Ethyl ethaneperoxoate
Ethyl peroxyacetate

20.

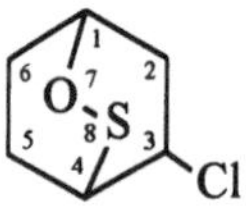

3-Chloro-7-oxa-8-thiabicyclo[2.2.2]octane

21. $C_6H_5-Se-OH$

*Se*-Phenyl selenohydroperoxide
Benzeneselenenic acid

22.

$$CH_3-O-S-CH_2\overset{OOH}{\underset{}{CH}}-COOH$$
$$\quad\quad\quad\quad_3\quad_2\quad_1$$

2-Hydroperoxy-3-(methoxythio)propanoic acid
2-Hydroperoxy-3-(methoxysulfanyl)propanoic
    acid

23. $CH_3-CO-OO-\underset{1}{CH_2}\underset{2}{CH_2}-OOH$

2-(Hydroperoxy)ethyl peroxyacetate
2-(Hydroperoxy)ethyl ethaneperoxoate
2-(Acetylperoxy)ethyl hydroperoxide

24.

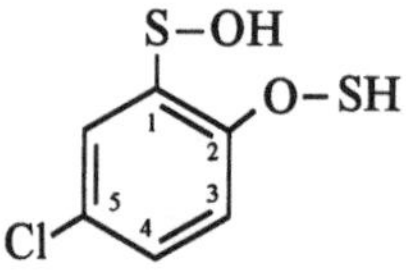

5-Chloro-2-(mercaptooxy)cyclohexanesulfenic
    acid
5-Chloro-2-(sulfanyloxy)cyclohexanesulfenic acid
*O*-(4-Chloro-2-sulfenophenyl) thiohydroperoxide

25.

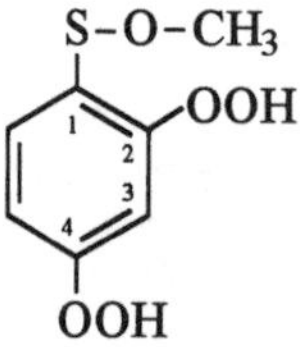

Methyl 2,4-bis(hydroperoxy)benzenesulfenic acid
4-(Methoxythio)-1,3-phenylene bis(hydroperoxide)
4-(Methoxysulfanyl)-1,3-phenylene
    bis(hydroperoxide)

26. $CH_3-CO-S-O-CO-C_6H_5$

*S*-Acetyl *O*-benzoyl thioperoxide

# 18

# Carboxylic Amides, Hydrazides, and Imides

*Amides of carboxylic acids* are compounds in which the –OH of the –COOH group has been replaced by an unsubstituted or substituted –$NH_2$ group; as a class, these compounds are called *amides* or *carboxamides*. They are typified by structures such as R–CO–$NH_2$, R–CO–NHR′, or R–CO–NR′R″, that is, as acyl derivatives of $NH_3$ or amines (see chapter 21). Analogous acyl derivatives of hydrazine are called *hydrazides*. *Imides* (or *carboximides*) are compounds containing the –CO–NH–CO– or –CO–NR–CO– group, especially when the group is part of a ring or ring system. In addition, nomenclature is discussed for amides of carbonic acid, such as urea, $H_2N$–CO–$NH_2$, and its derivatives, as well as carboxylic amides and imides in which sulfur has replaced oxygen.

Analogs of amides in which the oxygen atom of the carbonyl group is replaced by =NH are generally known as amidines; they are considered in chapter 19. The naming of peptides, a subclass of amides derived from $\alpha$-amino carboxylic acids, is covered briefly in chapter 31. Amides and imides derived from noncarboxylic acids, such as those with sulfur or phosphorus as the central atom, are taken up in chapters 23 and 25, respectively.

## Acceptable Nomenclature

*Carboxamides* of the type R–CO–$NH_2$ are named by changing the ending of the corresponding acid name from -oic acid or -ic acid to -amide or the suffix -carboxylic acid to -carboxamide.

$$\underset{2\quad 1\quad N}{CH_3\text{-}CO\text{-}NH_2}$$ Acetamide

$$\underset{5\quad 4\quad 3\quad 2\quad 1\quad N}{CH_3CH_2CH_2CH_2\text{-}CO\text{-}NH_2}$$ Pentanamide

–CO–$NH_2$    Cyclohexanecarboxamide

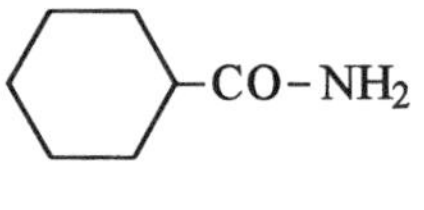

Amides derived from polycarboxylic acids in which all of the acid groups have been replaced by amide groups are also named as above.

$$\underset{N'\quad 2\quad 1\quad N}{H_2N\text{-}CO\text{-}CO\text{-}NH_2}$$ Oxamide (a contraction of oxalamide)
Ethanediamide

$$\underset{N'\quad 3\quad 2\quad 1\quad N}{H_2N\text{-}CO\text{-}CH_2\text{-}CO\text{-}NH_2}$$ Malonamide
Propanediamide

$$\underset{N'\quad 8\quad 7\text{-}2\quad 1\quad N}{H_2N\text{-}CO\text{-}[CH_2]_6\text{-}CO\text{-}NH_2}$$ Octanediamide

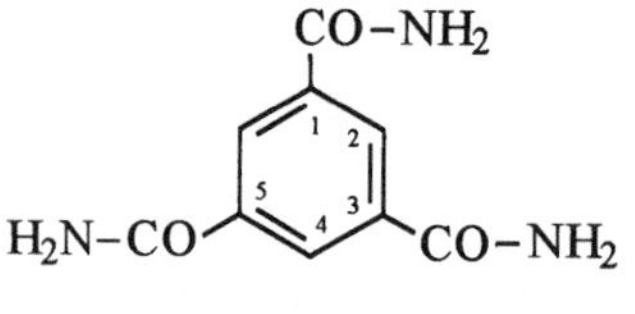

$$CO-NH_2$$

Benzene-1,3,5-tricarboxamide

$$H_2N-CO \quad CO-NH_2$$

Substituents on the amide nitrogen atom are indicated by an italic *N*; in symmetrical poly-amides, the locants *N*, *N′*, *N″*, and so on, are used. For unsymmetrical polyamides, *N* with a superscript corresponding to the locant of the carboxamide group is employed.

$$C_6H_5-CO-N(CH_3)_2$$

*N,N*-Dimethylbenzamide

$$(CH_3)_2N-CO \quad N \quad CO-NH-CH_2CH_3$$

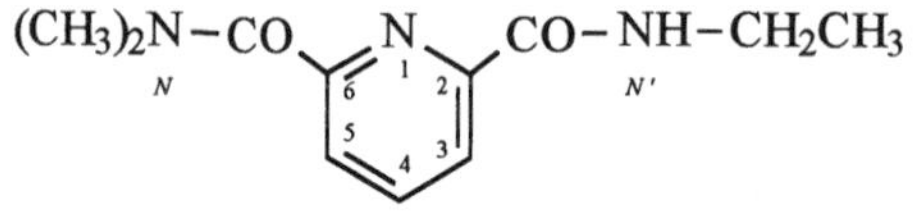

*N′*-Ethyl-*N,N*-dimethylpyridine-2,6-dicarboxamide

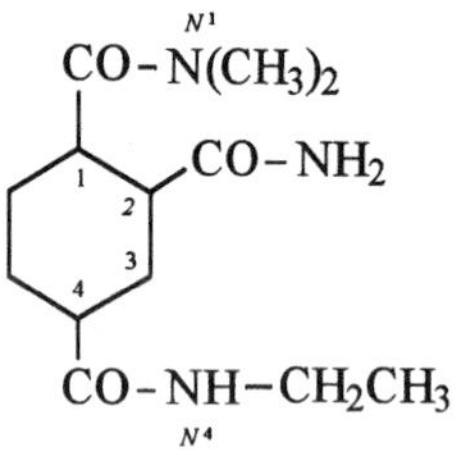

$$CO-N(CH_3)_2$$
$$CO-NH_2$$
$$CO-NH-CH_2CH_3$$

$N^4$-Ethyl-$N^1,N^1$-dimethylcyclohexane-1,2,4-tricarboxamide

$$CH_3$$
$$CH_3CH_2-NH-CO-CH_2CH_2CH-CO-NH-CH_3$$

$N^5$-Ethyl-$N^1$,2-dimethylpentanediamide

*N-Hydroxyamides*, R–CO–NH–OH, are named as such, as hydroxamic acids, or as *N*-acyl derivatives of hydroxylamine. *N-(Acyloxy)amides* may also be named as such or as *O*-acyl derivatives of hydroxylamine (see chapter 22).

$$CH_3CH_2-CO-NH-OH$$

*N*-Hydroxypropanamide
Propanohydroxamic acid
*N*-Propanoylhydroxylamine

$$C_6H_5-CO-NH-O-CH_3$$

*N*-Methoxybenzamide
*N*-Benzoyl-*O*-methylhydroxylamine
Methyl benzohydroxamate

*Mono- and polyamides of polybasic acids* in which one or more –COOH groups remain are named as derivatives of the acid with the most –COOH groups; the prefix for $H_2N-CO-$ is carbamoyl (CAS uses aminocarbonyl or amino and oxo prefixes), used in the names of structures having a more senior function such as –COOH. Monoamides of dibasic acids with trivial names are called *amic acids* and are named by replacing -ic in the acid name with -amic; *N*-phenyl derivatives are given the ending -anilic acid (see chapter 11). For $H_2N-COOH$ and $H_2N-CO-COOH$, the names carbamic and oxamic acid, respectively, are in common use; their *N*-phenyl derivatives are carbanilic acid and oxanilic acid.

$$H_2N-CO-[CH_2]_4-COOH$$

5-Carbamoylpentanoic acid
Adipamic acid
6-Amino-6-oxohexanoic acid

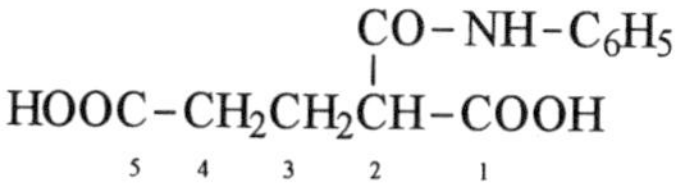

$$\underset{5}{CH_3} \underset{4}{CH_2CH_2}\underset{3}{CH}-\underset{2}{\overset{\overset{\displaystyle CO-NH-C_6H_5}{|}}{C}}H-\underset{1}{COOH}$$

HOOC–CH₂CH₂CH–COOH

2-(Phenylcarbamoyl)glutaric acid
(*not* 2-Carboxyglutaranilic acid)
2-[(Phenylamino)carbonyl]pentanedioic acid

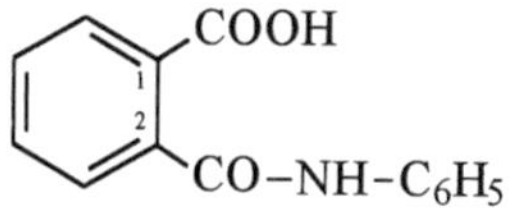

COOH
CO–NH–C₆H₅

Phthalanilic acid
2-(Phenylcarbamoyl)benzoic acid
2-[(Phenylamino)carbonyl]benzoic acid

Compounds of the type R–CO–NHR′ and R–CO–NR′R″, where R′ and R″ are alkyl, aryl, or acyl groups, are named as *N*-substituted amides. When R′ and R″ are acyl groups, the largest or most preferred acyl group is chosen for the parent amide name. However, where two or three identical acyl groups with trivial names are bound to a nitrogen atom, names such as dipropanamide for HN(CO–CH₂CH₃)₂ or tribenzamide for N(CO–C₆H₅)₃ may be formed; these names are not recommended for substituted acyl groups. An *N*-phenyl monoamide may be named by changing -amide to -anilide, but extension of this type of name to substituted *N*-phenyl monoamides ("toluidides" and so on) is discouraged. Primed numbers are used as locants for substituents on the benzene ring of anilides. Amides in which the amide nitrogen atom is part of a ring structure are named as acyl derivatives of heterocyclic compounds. In the presence of a more senior characteristic group the amide group, R–CO–NH–, may be expressed as a prefix by (1-oxoalkyl)amino or acylamino or by changing the ending -amide to -amido.

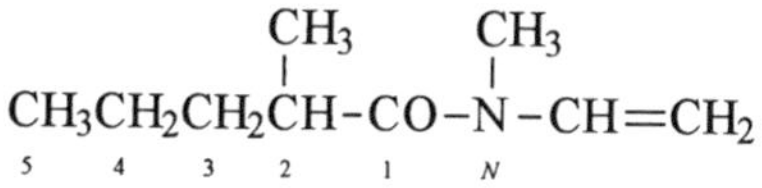

CH₃CH₂CH₂CH–CO–N–CH=CH₂

*N*,2-Dimethyl-*N*-vinylpentanamide
*N*-Ethenyl-*N*,2-dimethylpentanamide

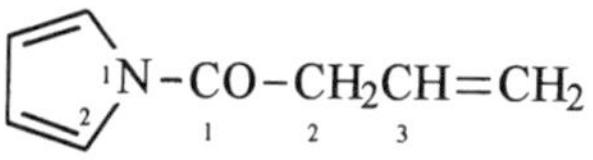

N–CO–CH₂CH=CH₂

1-(But-3-enoyl)-1*H*-pyrrole
1-(1-Oxobut-3-en-1-yl)-1*H*-pyrrole

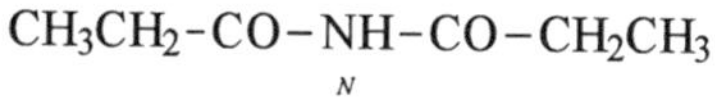

CH₃CH₂–CO–NH–CO–CH₂CH₃

Dipropanamide
*N*-Propanoylpropanamide
*N*-(1-Oxopropyl)propanamide

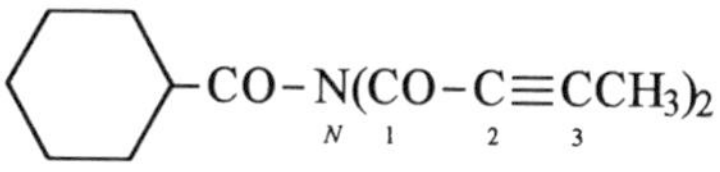

CO–N(CO–C≡CCH₃)₂

*N*,*N*-Dibut-2-ynoylcyclohexanecarboxamide
*N*,*N*-Bis(1-oxobut-2-yn-1-yl)cyclo-
hexanecarboxamide

CH₃CH₂–CO–N–CO–CH₂CH₃

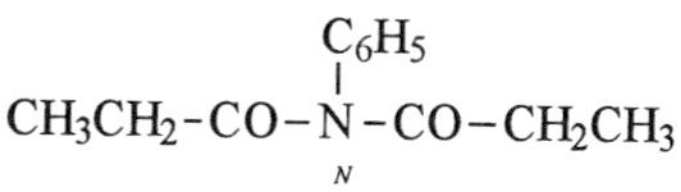

*N*-Phenyldipropanamide
*N*-Phenyl-*N*-propanoylpropanamide
*N*-(1-oxopropyl)-*N*-phenylpropanamide

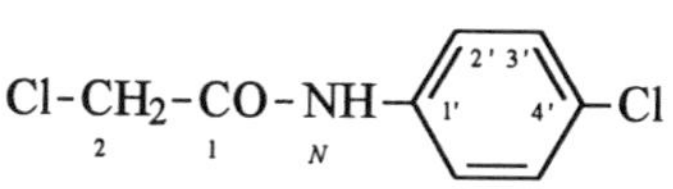

Cl–CH₂–CO–NH——Cl

2,4′-Dichloroacetanilide
2-Chloro-*N*-(4-chlorophenyl)acetamide

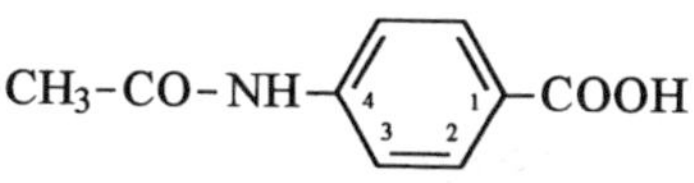

CH₃–CO–NH——COOH

4-Acetamidobenzoic acid
4-(Acetylamino)benzoic acid
4-[(1-Oxoethyl)amino]benzoic acid

Structures containing the groups –CO–NH– or –C(OH)=N– as part of a ring are called *lactams* or *lactims*, respectively, and can be named analogously to lactones (see chapter 12), but they are best named as heterocyclic compounds (see chapter 9). However, ε-caprolactam is an acceptable trivial name for hexahydro-2*H*-azepin-2-one or hexano-6-lactam.

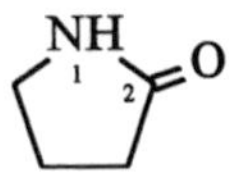

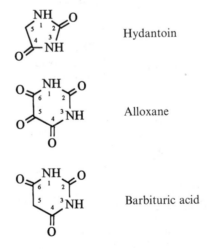

Pyrrolidin-2-one
2-Pyrrolidone
Butano-4-lactam

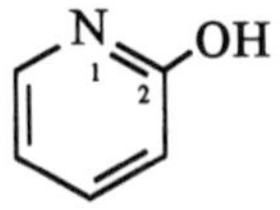

Pyridin-2-ol
2-Hydroxypyridine
2,4-Pentadieno-5-lactim

*Imides*, for the purpose of nomenclature, are usually considered to be compounds in which a –CO–NH–CO– group is part of a cyclic structure; they are best named as heterocycles. However, imides may also be derived from dibasic acids. If the acids have systematic names, the imides may be named by changing the endings -dioic acid and -dicarboxylic acid to -imide and -dicarboximide, respectively; the final "e" of a parent hydride name is elided before "imide". Imides formed from dibasic acids with trivial names are named by changing the -ic acid ending of the acid name to -imide, as in succinimide and phthalimide; in prefix form, these names become succinimido and phthalimido. Although systematically named cyclic imides can be named as prefixes in this way, heterocyclic names are preferred.

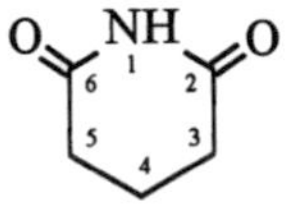

Piperidine-2,6-dione
Pentanimide

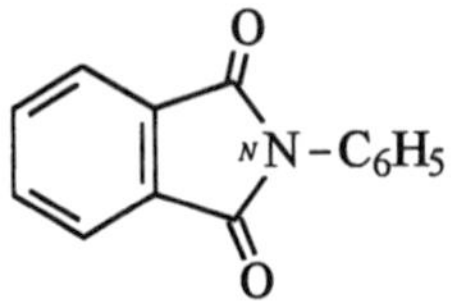

N-Phenylphthalimide
2-Phenyl-1H-isoindole-1,3(2H)-dione
N-Phenylbenzene-1,2-dicarboximide

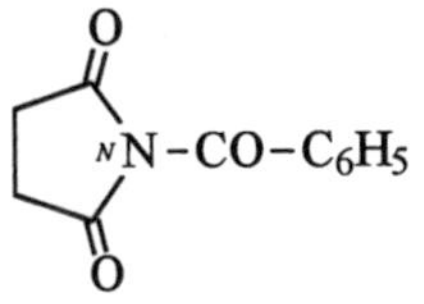

N-Benzoylsuccinimide
1-Benzoyl-2,5-dioxopyrrolidine
1-Benzoylpyrrolidine-2,5-dione

The following trivial names are also acceptable:

Hydantoin

Alloxane

Barbituric acid

The –CO–NH–CO– group is expressed as a multiplying term by iminodicarbonyl (CAS) or by imidodicarbonyl, an acyl group name derived from imidodicarbonic acid.

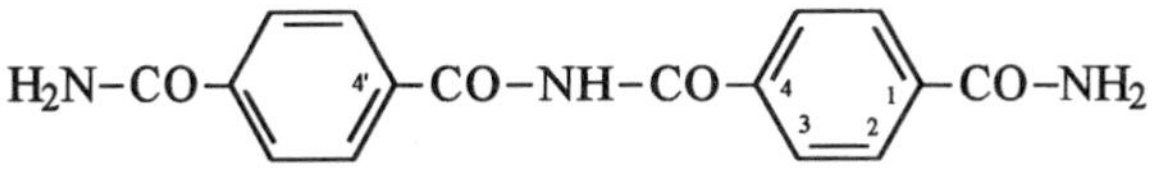

$$H_2N-CO-\underset{4'}{\bigcirc}-CO-NH-CO-\underset{3\ 2}{\overset{4\ \ \ 1}{\bigcirc}}-CO-NH_2$$

4,4′-(Iminodicarbonyl)bis(benzamide)
4,4′-(Imidodicarbonyl)bis(benzamide)

In this example, bis- is used rather than di- to avoid confusion with the name dibenzamide.

*Thioamides* and their selenium and tellurium analogs are named by the methods used for amides. For thioamides derived from monocarboxylic acids with names ending in -thioic acid or -carbothioic acid (see chapter 11), these endings become -thioamide or -carbothioamide, respectively. For thioamides derived from acids with trivial names, thio is prefixed to the name of the amide. Names such as octanethioamide, cyclohexanecarbothioamide (prior to 1972, CAS used thiocyclohexanecarboxamide), and thiobenzamide are generated. Thio analogs of polycarboxamides in which the carbonyl oxygens have been replaced by sulfur utilize suffixes such as -dithioamide and -dicarbothioamide. The prefixes thio, dithio, and so on can be used with polycarboxamides with trivial names if there is no ambiguity, but systematic names are preferred. Where an amide group without sulfur remains, it is the senior characteristic group. The $H_2N$–CS– is expressed by the prefix carbamothioyl, thiocarbamoyl (IUPAC), or aminothioxomethyl (CAS); all are acceptable.

$$\underset{4\ \ \ \ 3\ \ \ \ 2\ \ \ \ 1\ \ \ \ N}{CH_3CH_2CH_2-CS-NH_2}$$   Butanethioamide

$$\underset{N}{CH_3[CH_2]_5-CS-NH-CS-[CH_2]_5CH_3}$$   *N*-Heptanethioylheptanethioamide
Diheptanethioamide

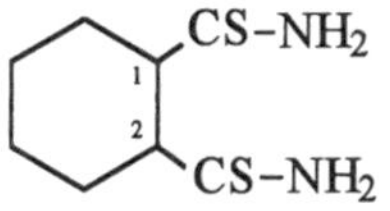

Cyclohexane-1,2-dicarbothioamide

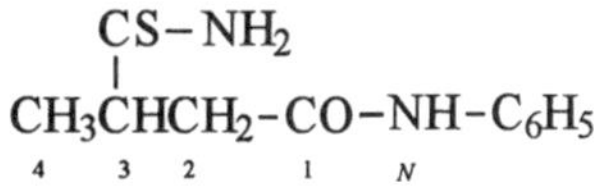

$$\underset{4\ \ \ 3\ \ \ 2\ \ \ \ 1\ \ \ \ N}{CH_3\overset{\overset{\displaystyle CS-NH_2}{|}}{C}HCH_2-CO-NH-C_6H_5}$$   3-(Carbamothioyl)-*N*-phenylbutanamide
3-(Aminocarbonothioyl)-*N*-phenylbutanamide

*Thioimides* are best named as heterocycles (see chapter 9). Suffixes such as -dicarbothioimide have been used to denote replacement of two oxygens with sulfur, but there is no corresponding suffix for the replacement of only one oxygen atom. In trivial names, the prefix thio or dithio takes care of this problem. The –CS–NH–CS– group as a prefix is denoted by iminodicarbonothioyl or imidodicarbonothioyl, which is an acyl prefix derived from imidodicarbonothioic acid; iminobis(thiocarbonyl) is ambiguous since it can also mean –CO–S–NH–S–CO–.

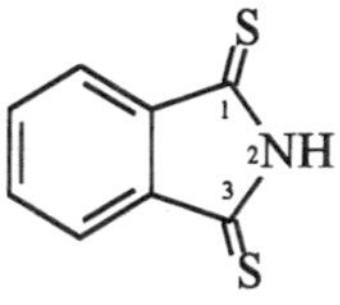

1*H*-Isoindole-1,3(2*H*)-dithione
Benzene-1,2-dicarbothioimide
Dithiophthalimide

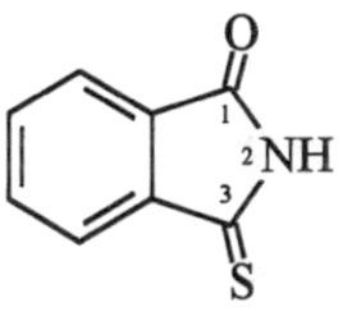

2,3-Dihydro-3-thioxo-1*H*-isoindol-1-one
Thiophthalimide

Urea, $H_2N$–CO–$NH_2$, is the diamide of carbonic acid, and its chalcogen analogs have names such as thiourea; the systematic names carbonic diamide (carbamide) and carbonothioic diamide (thiocarbamide) are also acceptable. Derivatives are named as substitution products of urea or in the same way as other diamides with the locants $N$ and $N'$ (CAS) or 1 and 3 (IUPAC) for substitution on the nitrogen atoms. Acyl derivatives may be named as such, as in $N$(or 1)-propanoylurea or on the basis of the organic amide, as in $N$-carbamoylpropanamide or N-(aminocarbonyl)propanamide. The prefixes ureido, carbamoylamino, or (aminocarbonyl)-amino for $H_2N–CO–NH–$, with locant $N$ or 1 at the nitrogen with the free valence, may be used if the parent structure carries another group to be cited as principal characteristic group. Ureylene or carbonyldiimino for $–HN–CO–NH–$ are used in names for assemblies of identical units.

$$CH_3CH_2\underset{N'}{-NH}-CO-\underset{N}{NH}-CO-CH_3$$

N-Acetyl-$N'$-ethylurea
1-Acetyl-3-ethylurea
N-(Ethylcarbamoyl)acetamide

$$CH_3-\underset{N'}{NH}-CO-NH-\text{(ring: 4,3,2,1)}-COOH$$

4-($N'$-Methylureido)benzoic acid
4-(3-Methylureido)benzoic acid
4-[(Methylcarbamoyl)amino]benzoic acid
4-[[(Methylamino)carbonyl]amino]benzoic acid

$$H_2N-CO-\text{(ring: 4',1')}-CO-NH-CO-\text{(ring: 4,3,2,1)}-CO-NH_2$$

Dimethyl 4,4'-ureylenedibenzoate
Dimethyl 4,4'-(carbonyldiimino)dibenzoate

Some trivial names are acceptable.

$$\underset{3}{HN}=\overset{\overset{2}{OH}}{\underset{1}{C}}-NH_2$$

Isourea
Carbamimidic acid

$$\underset{3}{HN}=\overset{\overset{2}{SH}}{\underset{1}{C}}-NH_2$$

Isothiourea
Carbamimidothioic acid

If the position of the double bond is unknown, the locants $N, N'$, and $O$ (or $S$) are used. Prefix names for substituting groups are 1-isoureido, [hydroxy(imino)methyl]amino, (imidocarboxy)amino, or ($C$-hydroxycarbonimidoyl)amino (the "$C$" is needed to avoid ambiguity) for $HN=C(OH)–NH–$, and 3-isoureido or [amino(hydroxy)methylidene]amino for $H_2N–C(OH)=N–$.

The structure $NH=C=NH$ is named carbodiimide. Amides of polycarbonic acids are named by the methods outlined above. However, numbering and names used by IUPAC and CAS differ for these compounds, although both are acceptable.

$$\underset{5}{H_2N}-\underset{4}{CO}-\underset{3}{NH}-\underset{2}{CO}-\underset{1}{NH_2}$$   Biuret (IUPAC)

$$\underset{N'}{H_2N}-CO-\underset{3}{NH}-\underset{2}{CO}-\underset{1}{NH_2}\ {}_{N}$$   Imidodicarbonic diamide (CAS)

Triuret (IUPAC) or diimidotricarbonic diamide (CAS), $H_2N–CO–NH–CO–NH–CO–NH_2$, is numbered analogously. For the sulfur analogs, appropriate amide names and locants are used by IUPAC. Thiuram monosulfide and disulfide are accepted by IUPAC for $(H_2N–CS)_2S$ and $(H_2N–CS)_2S_2$. For CAS, thio is prefixed to systematic names formed without sulfur replacement followed by a line or structural formula, as in thiodicarbonic diamide ($[H_2N-CO]_2S$). For general use, amide names and prefixes should be employed.

$$H_2N-CO-NH-CS-NH-CH_3$$
$$\quad\;\; 5 \quad\;\; 4 \quad\;\; 3 \quad\;\; 2 \quad\;\; 1$$

1-Methyl-2-thiobiuret
*N*-(Methylcarbamothioyl)urea
*N*-[(Methylamino)carbonothioyl]urea

$$\overset{\displaystyle CH_3}{\underset{\displaystyle |}{\phantom{x}}}$$
$$H_2N-CS-N-CS-NH-CH_3$$
$$\quad\;\; 5 \quad\;\; 4 \quad\;\; 3 \quad\;\; 2 \quad\;\; 1$$

1,3-Dimethyldithiobiuret
*N*,2-Dimethylimidodicarbonothioic diamide

*Hydrazides*, $R-CO-NHNH_2$, are named by the methods used for amides, generating names such as butanehydrazide and cyclohexanecarbohydrazide. Alternatively, functional class nomenclature can be used by replacing the word acid with hydrazide. In IUPAC recommendations, locants for substitution on the hydrazide atoms are *N* or 1′ for the imino nitrogen atom and *N*′ or 2′ for the amino nitrogen atom; for a functional class name the locants are simply 1 and 2, respectively.

$$CH_3CH_2-CO \qquad CH_3$$
$$\qquad\qquad \diagdown NN \diagup$$
$$CH_3 \diagup {}^{N\;N'} \diagdown CH_2CH_3$$

*N*′-Ethyl-*N*,*N*′-dimethylpropanehydrazide
2′-Ethyl-1′,2′-dimethylpropanehydrazide
Propanoic 2-ethyl-1,2-dimethylhydrazide

$$C_6H_5-O-CH_2CH_2-CO-NHNH-CH_3$$
$$\qquad\qquad\quad\; 3 \quad\;\; 2 \quad\;\; 1 \quad\;\; N \quad N'$$

*N*′-Methyl-3-phenoxypropanehydrazide
2′-Methyl-3-phenoxypropanehydrazide
3-Phenoxypropanoic 2-methylhydrazide

Hydrazides of carbamic and carbonic acids are also numbered and named differently by IUPAC and CAS:

$$NH_2-CO-NHNH_2$$
$$\;\; 4 \quad\;\; 3 \quad\; 2 \quad 1$$

Semicarbazide (IUPAC)

$$NH_2-CO-NHNH_2$$
$$\;\; N \qquad\qquad 1 \quad 2$$

Hydrazinecarboxamide (CAS)

$$H_2NNH-CO-NHNH_2$$
$$\;\; 5\,4 \quad\;\; 3 \quad\;\; 2 \quad 1$$

Carbonohydrazide (IUPAC)

$$H_2NNH-CO-NHNH_2$$
$$\;\; 2'\,1' \qquad\quad 1 \quad 2$$

Carbonic dihydrazide (CAS)

For *semicarbazones* and related structures, see chapter 22. The monohydrazide of carbonic acid is named hydrazinecarboxylic acid by CAS; its trivial name is carbazic acid, and it is named carbonohydrazidic acid by functional replacement nomenclature.

## Discussion

While the term "primary amide" can only mean $R-CO-NH_2$, both $R-CO-NHR'$ and $R-CO-NH-CO-R'$ have been called "secondary amides", and both $R-CO-NR'R''$ and $(R-CO)_3N$ have been called "tertiary amides". To avoid confusion, it is best to avoid the use of primary, secondary, and tertiary in referring to amides. Although $R-CO-NH-CO-R'$ structures have sometimes been called "imides", especially in the British literature, for nomenclature purposes they are amides.

In the push toward systematization, IUPAC in 1993 recommended that di- and triacyl derivatives of ammonia be named as derivatives of the parent azane, rather than as acyl derivatives of amines or amides, as IUPAC had recommended in 1979. However, an amide name (i.e., an *N*-substituted derivative of R–CO–NH$_2$), is preferred, since it clearly recognizes that amides are much higher in the order of precedence of compound classes (see chapter 3, table 3.1) than amine (or azane). For nomenclature purposes acyclic *N*-acylamides should not be viewed as imides or nitrogen analogs of anhydrides ("azaanhydride", although perhaps analogous to thioanhydride, is not an acceptable class name); they are best named as derivatives of amides, limiting "imide" to cyclic compounds containing the –CO–NH–CO– group within the ring system.

The difference between IUPAC and CAS regarding locants in amides and hydrazides is regrettable. Although *N* and *N'* are usually used as locants for nitrogen atoms in characteristic groups, the IUPAC suggestion of primed arabic numbers can cause confusion with the unprimed arabic numbers used for hydrazides in functional class nomenclature where the locants are firmly affixed to the hydrazide name, as in propanoic 2,2-dimethylhydrazide. In the case of carbonic acid derivatives, numbering from one end of a chain to the other, as IUPAC recommends, is reasonable and easy to remember. The CAS method is also reasonable to the extent that it uses *N,N'*, and so on, to denote substitution on amide nitrogen atoms but, unfortunately, the usage is not consistent with locants used for hydrazides of polybasic mononuclear acids, as illustrated by

$$\text{H}_2\text{NNH-CO-NHNH}_2 \qquad \text{carbonic dihydrazide}$$
$$\text{2' 1'} \qquad \text{1 2}$$

This does, however, reduce the need for highly primed *N* locants. Both methods are acceptable, with a preference for the IUPAC names.

Prefixes derived from the names of imides and imidic acids are easily confused. Names such as benzimidoyl, C$_6$H$_5$–C(NH)–, are often incorrectly shortened to "benzimido". Prefixes such as succinimido or cyclohexane-1,2-dicarboximido are unambiguous and acceptable, but use of heterocyclic names in forming prefixes for cyclic imides will avoid any problem.

"Toluidide", "anisidide", and "xylidide" are names found in the older literature. They should not be used because of difficulties in the interpretation of locants for substituents. For the same reason, further substitution involving names such as diacetamide is avoided. It is noted that diacetamide for HN(CO–CH$_3$)$_2$, and similar names are not derived from the names of acids, which would imply the existence of an unacceptable name such as "diacetic acid", at one time used for acetoacetic acid.

## Additional Examples

1.  HCO–NH$_2$ <br>     *N*

    Formamide

2.  CH$_3$–CO–NH–CH$_2$CH$_3$ <br>    2   1   *N*

    *N*-Ethylacetamide

3.  CH$_3$CH$_2$–CO–NH–C$_6$H$_5$ <br>   3   2   1   *N*

    *N*-Phenylpropanamide <br> Propananilide

4.

    3-Chloro-*N*$^2$-methyl-*N*$^1$-phenyl-1,2-cyclohexanedicarboxamide

5.

    *N*-(4-Bromophenyl)-4-chlorobenzamide <br> 4'-Bromo-4-chlorobenzanilide

6.

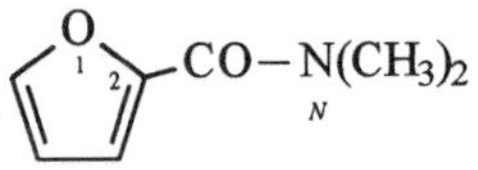

$N,N$-Dimethyl-2-furamide
$N,N$-Dimethylfuran-2-carboxamide

7.

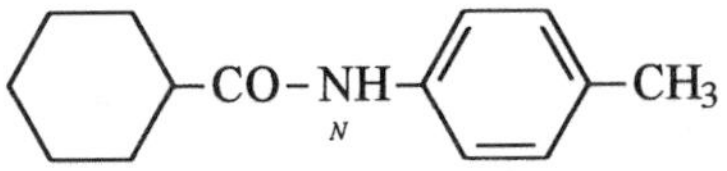

$N$-(4-Methylphenyl)cyclohexanecarboxamide
$4'$-Methylcyclohexanecarboxanilide
   (*not* Cyclohexane-*p*-toluidide)

8.

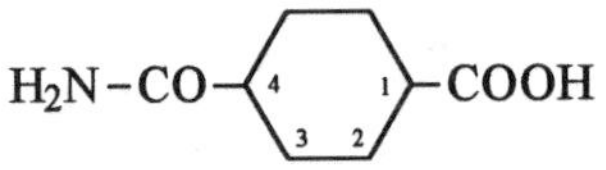

$H_2N-CO-\langle\!\langle\,^4\,\,^1\rangle\!\rangle-COOH$

4-Carbamoylcyclohexanecarboxylic acid
4-(Aminocarbonyl)cyclohexanecarboxylic
   acid

9.

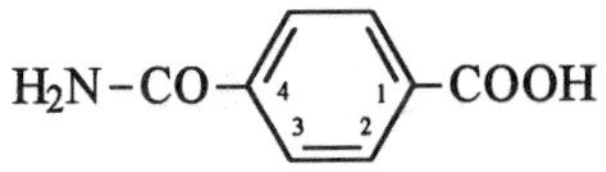

$H_2N-CO-\langle\!\langle\,^4\,\,^1\rangle\!\rangle-COOH$

4-Carbamoylbenzoic acid
Terephthalamic acid
4-(Aminocarbonyl)benzoic acid

10.    $CH_3-CO-NH-CH_2CH_2-COOH$
                       3    2    1

3-(Acetamido)propanoic acid
3-(Acetylamino)propanoic acid

11.    $C_6H_5-CO-NH-CH_2CH_2-CO-NH-CH_3$
                *N*

$N$-[2-(Methylcarbamoyl)ethyl]benzamide
$N$-[3-(Methylamino)-3-oxopropyl]benzamide

12.    $C_6H_5-NH-CO-O-CH_3$

Methyl carbanilate
Methyl phenylcarbamate

13.    $C_6H_5-NH-CO-CH_2CH=CH-CO-NH-CH_3$
        $N^5$    5   4   3   2   1   $N^1$

$N^1$-Methyl-$N^5$-phenylpent-2-enediamide

14.

                         $CH_3$
                        |
$C_6H_5-NH-CO-CH_2CH-CO-NH-CH_3$
      $N^4$    4   3   2   1   $N^1$

$N^1$-Methyl-$N^4$-phenyl-2-methyl-
   butanediamide

15.

$C_6H_5-CO-N\langle\,^4\,\,^1\rangle N-CO-CH_3$

1-Acetyl-4-benzoylpiperazine

16.

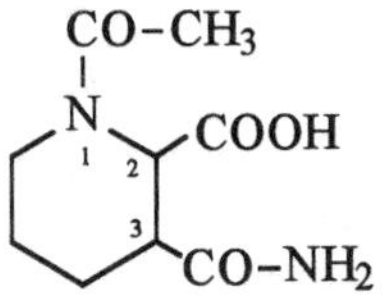

1-Acetyl-3-carbamoylpiperidine-2-
   carboxylic acid
1-Acetyl-3-(aminocarbonyl)piperidine-2-
   carboxylic acid

17.

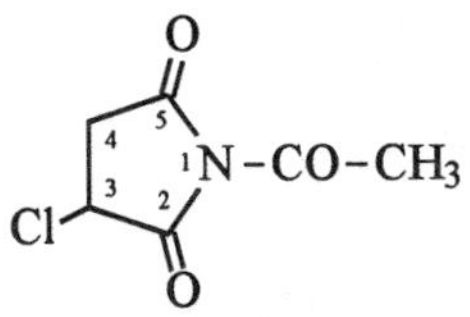

1-Acetyl-3-chloropyrrolidine-2,5-dione
$N$-Acetyl-2-chlorosuccinimide

18.

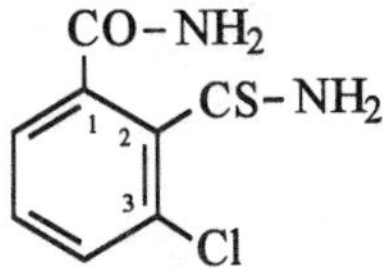

3-Chloro-2-carbamothioylbenzamide
3-Chloro-2-(aminothioxomethyl)benzamide

19.  $CH_3-CO-\underset{N}{NH}-CH_2CH_2-CS-NH-C_6H_5$

N-[2-(Phenylcarbamothioyl)ethyl]acetamide
N-[3-(Phenylamino)-3-thioxopropyl]acetamide

20. 

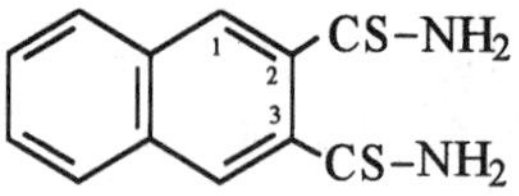

Naphthalene-2,3-dicarbothioamide

21. 

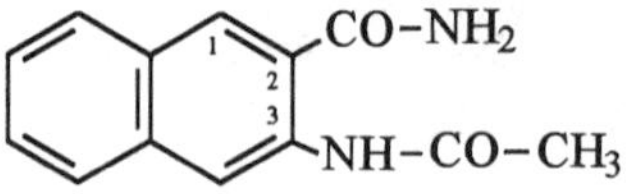

3-Acetamidonaphthalene-2-carboxamide
3-(Acetylamino)naphthalene-2-carboxamide

22.  $NH_2-CO-NH-CH_2-OH$

(Hydroxymethyl)urea
(Hydroxymethyl)carbonic diamide

23.  $CH_3CH_2-O-CO-NH-CO-O-CH_2CH_3$   Diethyl imidodicarbonate

24.  $\underset{2}{H_2N-CH_2}-\underset{1}{CO}-NH_2$

2-Aminoacetamide

25.  $CH_3-CO-\underset{N}{\overset{\overset{\displaystyle CO-CH_3}{|}}{N}}-CO-CH_3$

Triacetamide
N,N-Diacetylacetamide

26.  $CH_3-CO-\underset{N}{\overset{\overset{\displaystyle C_6H_5}{|}}{N}}-CO-CH_3$

N-Phenyldiacetamide
N-Acetyl-N-phenylacetamide

27.  $\underset{3}{Cl-CH_2}\underset{2}{CH_2}-\underset{1}{CO}-\underset{N}{\overset{\overset{\displaystyle CH_3}{|}}{N}}\underset{N'}{NH}-C_6H_5$

3-Chloro-N-methyl-N'-phenylpropanehydrazide
3-Chloro-1'-methyl-2'-phenylpropanehydrazide
3-Chloropropanoic 1-methyl-2-phenylhydrazide

28.  

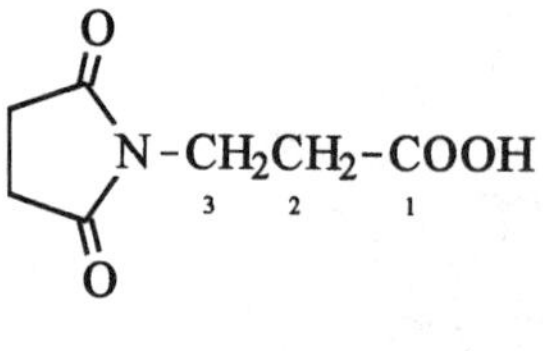

3-Succinimidopropionic acid
3-(2,5-Dioxopyrrolidin-1-yl)propanoic acid
2,5-Dioxo-1-pyrrolidinepropanoic acid

29.  

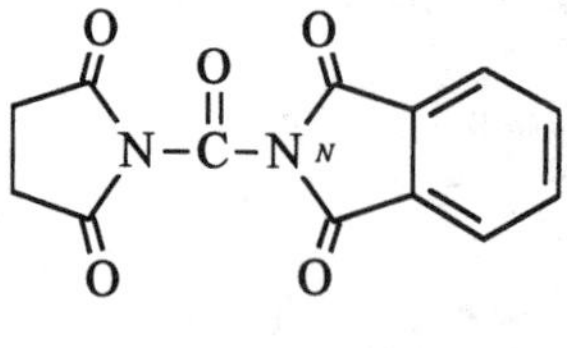

N-(Succinimidocarbonyl)phthalimide
2-[(2,5-Dioxopyrrolidin-1-yl)carbonyl]-1H-
isoindole-1,3(2H)-dione

# Amidines and Other Nitrogen Analogs of Amides

Compounds containing one or more –C(=NH)–NH$_2$ groups are called *amidines* or *carboxamidines*. The amidine group may be regarded as being derived from the carboxylic acid group, –COOH, by simultaneous replacement of –OH with –NH$_2$ and of =O with =NH; alternatively, an amidine may be considered to be the amide of the corresponding carboximidic acid, R–C(=NH)–OH (see chapter 11). Also discussed in this chapter are other nitrogen analogs of amides, including the classes called *amide oximes* (*amidoximes*), R–C(=N–OH)–NH$_2$, and their tautomers, R–C(=NH)–NH–OH, and *amidrazones*, R–C(=NNH$_2$)–NH$_2$, and their tautomers, R–C(=NH)–NHNH$_2$, as well as related compounds such as the *hydrazidines*, R–C(=NNH$_2$)–NHNH$_2$. Although the ending -amidine is used in naming analogous structures derived from sulfinic acids (see chapter 23), extension to other noncarboxylic acids has not become established. The nomenclature of carboxamides and carboximidic acids and their analogs is discussed in chapters 18 and 11, respectively.

## Acceptable Nomenclature

*Nitrogen analogs of amides* are acceptably named by two systems, neither of which is without fault. Both systems provide endings, analogous to -amide, to replace -oic acid, -ylic acid, or -ic acid in acid names. These endings are given in table 19.1; for locants on the groups represented by these endings, see the specific examples that follow in this chapter.

The compound class represented by R–C(=N–OH)–NH$_2$ is called *amide oximes* (or *amidoximes*); these structures and their tautomers, R–C(=NH)–NH–OH, and related hydroxy derivatives are best named as substitution products of the structures in table 19.1, although they can also be named as hydroxamic or hydroximic acid derivatives (see chapter 11). Prior to 1972, CAS used the suffixes -amidoxime and -carboxamidoxime.

The first column of endings in table 19.1 provides what might be called "traditional" names; substituents on the imide or hydrazone nitrogen atoms are prefixed to the class name part of the name. Examples of suffixes formed by the systematic functional replacement method (see chapter 3) are given in the second column of endings; =O and –OH in –COOH are replaced by nitrogen-containing groups. These suffixes are simply added to the name of the parent hydride. IUPAC and CAS now prefer the suffix -imidamide over -amidine although the latter can be used when the tautomer structure is unknown; -amidrazone and -hydrazidine are not used as suffixes.

$$\underset{N}{\underset{\displaystyle CH_3[CH_2]_2-\overset{\displaystyle\overset{N'}{NH}}{\overset{\displaystyle\|}{C}}-NH_2}{}}$$

Butanimidamide
Butyramidine

Table 19.1. Endings for Nitrogen Analogs of Amides and Hydrazides

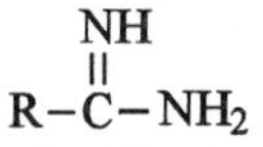

| | | |
|---|---|---|
| $\overset{\displaystyle NH}{\underset{\displaystyle (amidines)}{R-\overset{\|}{C}-NH_2}}$ | -amidine | -imidamide |
| $\overset{\displaystyle NH}{\underset{\displaystyle (amidrazones)}{R-\overset{\|}{C}-NHNH_2}}$ | -hydrazide imide | -imidohydrazide |
| $\overset{\displaystyle NNH_2}{\underset{\displaystyle (amidrazones)}{R-\overset{\|}{C}-NH_2}}$ | -amide hydrazone | -hydrazonamide |
| $\overset{\displaystyle NNH_2}{\underset{\displaystyle (hydrazidines)}{R-\overset{\|}{C}-NHNH_2}}$ | -hydrazide hydrazone | -hydrazonohydrazide |

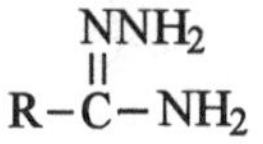

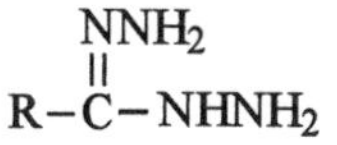

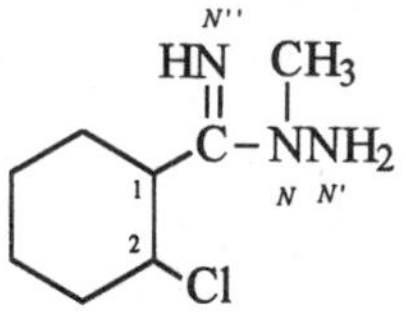

2-Chloro-*N*-methylcyclohexane-
carboximidohydrazide
2-Chloro-*N*-methylcyclohexane-
carbohydrazide imide

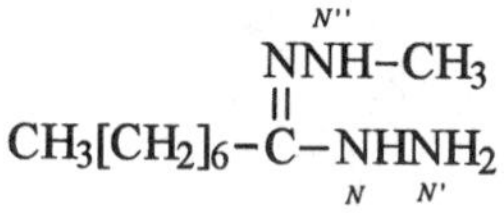

*N*″-Methyloctanehydrazonohydrazide
Octanehydrazide methylhydrazone

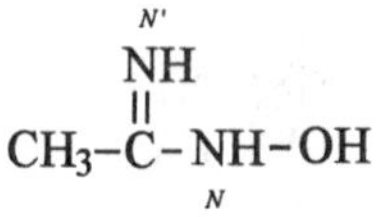

*N*-Hydroxyethanimidamide
*N*-Hydroxyacetamidine

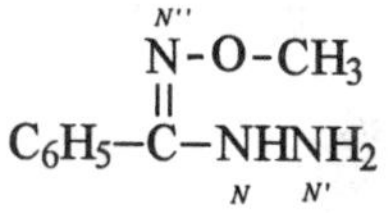

*N*″-Methoxybenzimidohydrazide
*N*-Methoxybenzenecarboximidic hydrazide

*Nitrogen analogs of amides of polybasic acids* are named by the preceding principles; see also chapter 11 on acids and chapter 18 on amides.

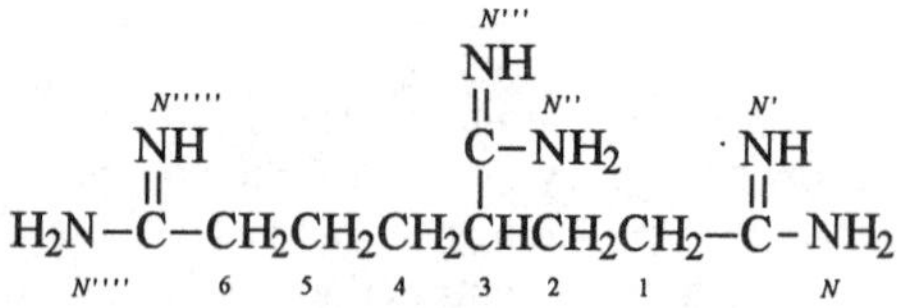

1,3,6-Hexanetricarboximidamide

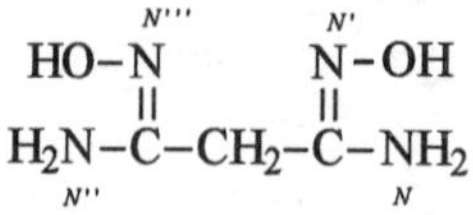

*N*′,*N*‴-Dihydroxypropane-
1,3-bis(imidamide)
*N*′,*N*‴-Dihydroxymalonamidine
Malonamidoxime

If substituents are present on one or more nitrogen atoms in polyamidines or their analogs, their locants can be confusing, as shown in the last two examples. When a chain or ring is substituted with different kinds of the groups in table 19.1, locants can become even more complicated. It is therefore preferred that one group (usually that with the most substituents) be selected as the principal characteristic group and the others named as substituents on the parent structure. Prefixes such as carbamimidoyl, amidino, or amino(imino)methyl for $H_2N-C(=NH)-$, carbamohydrazonoyl or amino(hydrazono)methyl for $H_2N-C(=NNH_2)-$, and carbonohydrazidimidoyl or hydrazinyl(imino)methyl for $H_2NNH-C(=NH)-$ are used. The individual replacement groups can also be described by amino, imino, and so on, as substituents at the end of an acyclic chain, as in 3-amino-3-iminopropanoic acid. An order of precedence has not been clearly established for the groups in table 19.1.

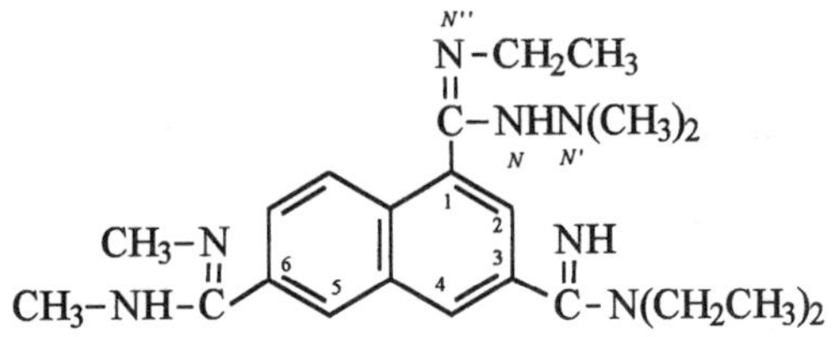

$$
\begin{array}{c}
N''\\
N-CH_2CH_3\\
\parallel\\
C-NHN(CH_3)_2\\
N\quad N'
\end{array}
$$

3-($N,N$-Diethylcarbamimidoyl)-6-($N,N'$-dimethylcarbamimidoyl)-$N''$-ethyl-$N',N'$-dimethylnaphthalene-1-carboximidohydrazide

Although guanidine is used by both IUPAC and CAS, numbering styles differ; IUPAC retains the traditional names biguanide, triguanide, and so on, while CAS employs systematic nomenclature for these compounds. Both are acceptable.

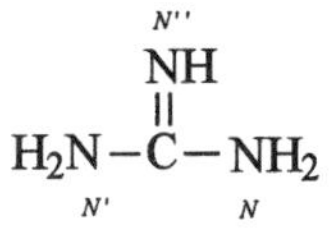

$$
\begin{array}{c}
N''\\
NH\\
\parallel\\
H_2N-C-NH_2\\
N'\qquad N
\end{array}
$$
Guanidine (CAS)
Carbonimidic diamide

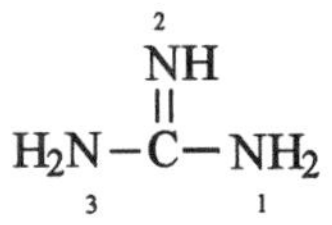

$$
\begin{array}{c}
2\\
NH\\
\parallel\\
H_2N-C-NH_2\\
3\qquad 1
\end{array}
$$
Guanidine (IUPAC)
Carbonimidic diamide

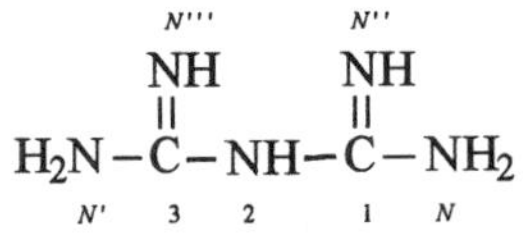

$$
\begin{array}{c}
N'''\qquad N''\\
NH\qquad NH\\
\parallel\qquad\parallel\\
H_2N-C-NH-C-NH_2\\
N'\quad 3\quad 2\quad 1\quad N
\end{array}
$$
Imidodicarbonimidic diamide (CAS)

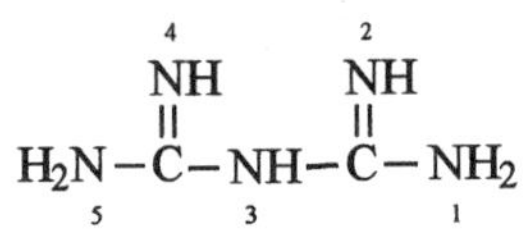

$$
\begin{array}{c}
4\qquad 2\\
NH\qquad NH\\
\parallel\qquad\parallel\\
H_2N-C-NH-C-NH_2\\
5\quad 3\quad 1
\end{array}
$$
Biguanide (IUPAC)

When one or both nitrogen atoms of an amidine (or analog) group are members of a ring system, the compound is named as a heterocycle (see chapters 9 and 18).

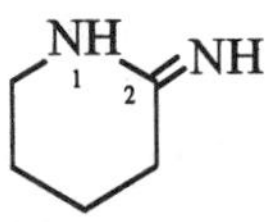

Piperidin-2-imine
2-Iminopiperidine

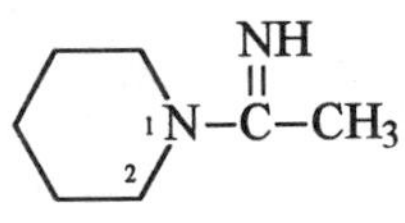

1-Acetimidoylpiperidine
1-(1-Iminoethyl)piperidine

Compounds of the types [R–C(=NH)]$_2$NH and [R–C(=NH)]$_3$N and the corresponding =NNH$_2$ derivatives are best named as *N*-substituted imidamides or amidines, hydrazonamides, and so on. The largest or most complex acyl group is expressed in the parent name as a suffix. The prefix names for R–C(=NH)– and R–C(=NNH$_2$)– are formed by changing the -ic acid ending of the corresponding acid name to -oyl (see chapter 11), giving names such as pentanimidoyl for CH$_3$CH$_2$CH$_2$CH$_2$–C(=NH)– or cyclohexanecarboximidoyl for C$_6$H$_{11}$–C(=NH)–, or by substituting an alkyl group to give 1-iminoalkyl and 1-hydrazono-alkyl, respectively.

$$\underset{N}{CH_3-\overset{\overset{NH}{\|}}{C}-NH-\overset{\overset{NH}{\|}}{C}-CH_3}$$

*N*-Acetimidoylethanimidamide
*N*-Acetimidoylacetamidine
*N*-(1-Iminoethyl)ethanimidamide

$$C_6H_5-\overset{\overset{NH}{\|}}{\underset{N}{C}}-N\Big\langle\begin{matrix}\overset{\overset{NNH_2}{\|}}{C}-[CH_2]_5CH_3\\ \\ \underset{\underset{NNH_2}{\|}}{C}-[CH_2]_5CH_3\end{matrix}$$

*N,N*-Diheptanehydrazonoylbenzimidamide
*N,N*-Diheptanehydrazonoylbenzamidine
*N,N*-Bis(1-hydrazonoheptyl)benzimidamide

Substituted amino, imidamido, or hydrazonamido prefixes, derived in the same way as carboxamido prefixes (see chapter 18), are used to name R–C(=NH)–NH– and R–C(=NNH$_2$)–NH– groups. Substituted hydrazinyl, imino, and hydrazono prefixes are used for R–C(=X)–NHNH–, R–C(=X)-N=, and R–C(=X)–NHN= groups, respectively, where X is =NH or =NNH$_2$. The R may be an acyl or alkyl group.

$$CH_3-\overset{\overset{NH}{\|}}{C}-NH-$$

Acetimidoylamino
(1-Iminoethyl)amino
Acetimidamido

$$CH_3[CH_2]_5-\overset{\overset{NNH_2}{\|}}{C}-N=$$

(1-Hydrazonoheptyl)imino
Heptanehydrazonoylimino

$$\langle\text{cyclohexyl}\rangle-\overset{\overset{NH}{\|}}{C}-NHNH-$$

2-[Cyclohexyl(imino)methyl]hydrazinyl
2-(Cyclohexanecarboximidoyl)hydrazinyl

"Guanyl" is no longer recommended, but guanidino is an approved IUPAC name for H$_2$N–C(=NH)–NH–. The groups R–C(Y)=N– and R–C(Y)=NNH–, where Y is –NH$_2$ or –NHNH$_2$, are best given names such as 1-amino(or 1-hydrazinyl)alkylideneamino and 2-[1-amino(or1-hydrazinyl)alkylidene]hydrazinyl.

## Discussion

The nomenclature of the structures considered in this chapter illustrate well the contrasts between traditional names and names that depend on a new system. It is not that the traditional names have no system whatever, but rather that so many details must be remembered and applied to contrive an accurate name. Functional replacement, if nothing else, provides simple names and lessens the need for "amidine" and its cousins in table 19.1. Supposedly, functional replacement makes life easier for the name generator, but inconsistencies, such as naming amides as such but hydrazides at the corresponding acid, do not make it easier for the user.

Prefixes such as *C*-cyclohexylcarbonimidoyl for C$_6$H$_{11}$–C(=NH)– are ambiguous without the "*C*", which is necessary to distinguish it from –C(=NC$_6$H$_{11}$)–, *N*-cyclohexylcarbonimidoyl. While the *C*-name is acceptable, it is preferred that the names of these prefixes be derived from the names of the corresponding acids, as in cyclohexanecarbonimidoyl.

Locants for substituents on the nitrogen atoms in the compounds of this chapter are a serious problem, especially where a structure contains several of the same characteristic group. Any system is cumbersome and a memory exercise. For greater specificity than primed letter locants, IUPAC recommends number superscripts on the "*N*" locants, which works with a single characteristic group but is inconsistent with the use of number superscripts to indicate position on a parent structure with amines (see chapter 21). With more than one of these characteristic groups, primes on the superscript numbers would be necessary. The recommendation in this book is to stay with primes on the *N* locants.

Complexities introduced by both history (amide + amine gives amidine) and the existence of two nitrogen atom sites for potential free bonds in amidines make analogies between amidines and amides an exercise in frustation. While there is the name acetamido for $CH_3$–CO–NH–, the analog $CH_3$–C(=NH)–NH– is usually named acetimidoylamino, although acetimidamido or ethanimidamido should be acceptable. Further, prefixes for structures with the free valence on the imino nitrogen are in a class by themselves and can only be reasonably named systematically: (1-aminoethylidene)amino. The names may be a little longer, but with prefix names in the amidine class, systematic nomenclature seems to be the way to go.

## Additional Examples

1.

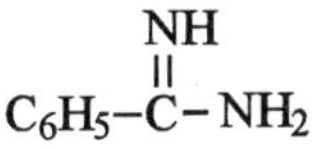

   Benzimidamide
   Benzamidine

2.

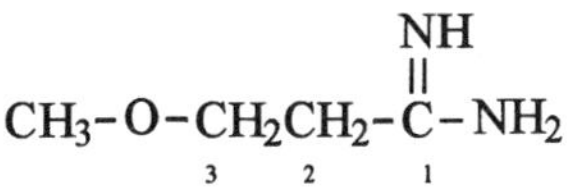

   3-Methoxypropanimidamide
   3-Methoxypropionamidine

3.

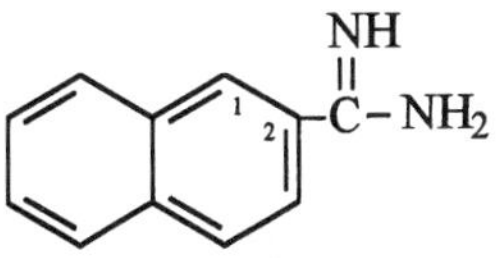

   Naphthalene-2-carboximidamide
   Naphthalene-2-carboxamidine

4.

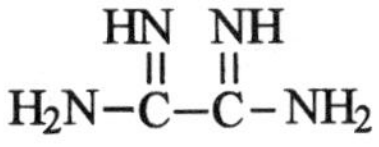

   Ethanebis(imidamide)
   Oxamidine (a contraction of Oxalamidine)

5.

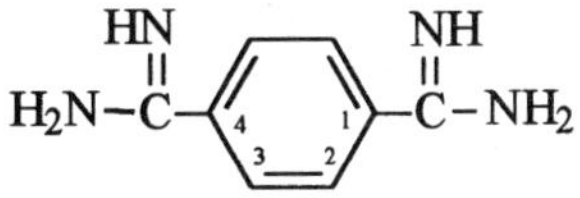

   Benzene-1,4-dicarboximidamide
   1,4-Benzenedicarboxamidine
   Terephthalamidine

6.

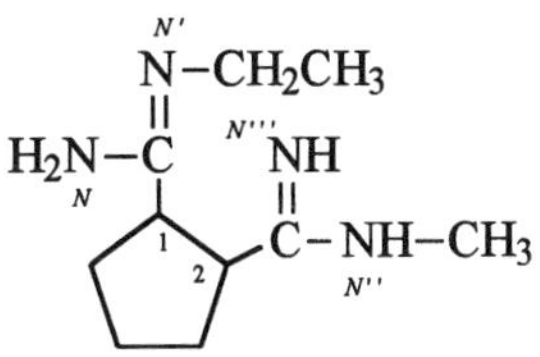

   *N'*-Ethyl-*N"*-methylcyclopentane-1,2-
      dicarboximidamide
   *N'*-Ethyl-*N"*-methylcyclopentane-1,2-
      dicarboxamidine
   *N'*-Ethyl-2-(*N*-methylamidino)cyclo-
      pentanecarboxamidine

7.

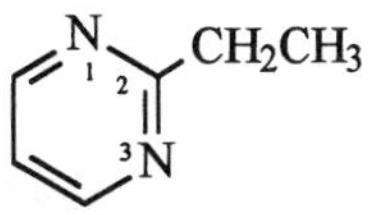

   2-Ethylpyrimidine

8.

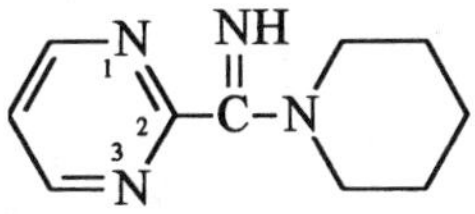

   2-(Piperidine-1-carboximidoyl)pyrimidine
   1-(Pyrimidine-2-carboximidoyl)piperidine

9.

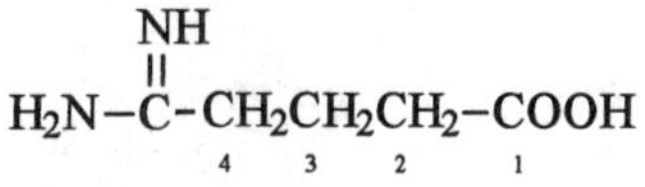

$$\underset{\underset{4\quad3\quad2\quad1}{H_2N-\overset{\displaystyle NH}{\overset{\|}{C}}-CH_2CH_2CH_2-COOH}}{}$$

4-Amidinobutanoic acid
4-Carbamimidoylbutanoic acid
5-Amino-5-iminopentanoic acid

10.

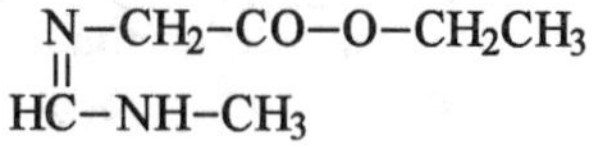

$$N-CH_2-CO-O-CH_2CH_3$$
$$\overset{\|}{HC}-NH-CH_3$$

Ethyl [[(methylamino)methylidene]amino]acetate
N-[(Methylamino)methylidene]glycine ethyl ester

11.

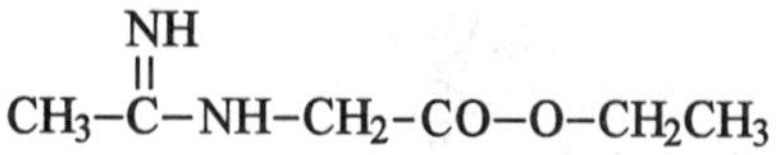

$$CH_3-\overset{\displaystyle NH}{\overset{\|}{C}}-NH-CH_2-CO-O-CH_2CH_3$$

Ethyl (acetimidoylamino)acetate
Ethyl acetimidamidoacetate
N-Acetamidoylglycine ethyl ester

12.

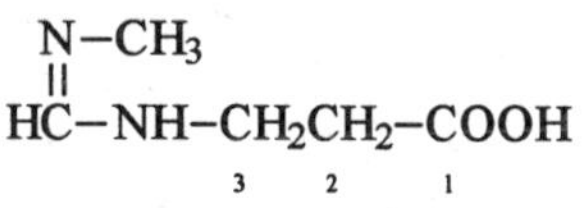

$$N-CH_3$$
$$\underset{3\quad\quad2\quad\quad1}{\overset{\|}{HC}-NH-CH_2CH_2-COOH}$$

3-[[(Methylimino)methyl]amino]propanoic acid
3-[(N-Methylformimidoyl)amino]propanoic acid
3-(N'-Methylmethanimidamido)propanoic acid
3-(N'-Methylformimidamido)propanoic acid

13.

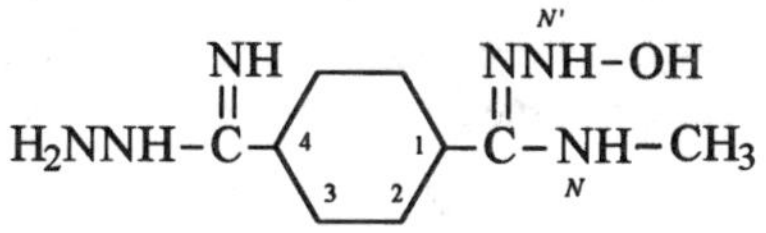

$$H_2NNH-\overset{\displaystyle NH}{\overset{\|}{C}}\!\!\left\langle\!\!\begin{array}{c}4\\3\quad2\end{array}\!\!\right\rangle\!\!\overset{\displaystyle \overset{N'}{NNH-OH}}{\underset{N}{\overset{\|}{C}}-NH-CH_3}$$

4-[Hydrazinyl(imino)methyl]-N'-hydroxy-N-methylcyclohexanecarbohydrazonamide
4-[Hydrazinyl(imino)methyl]-N-methylcyclohexanecarboxamide hydroxyhydrazone
4-[(Hydroxyhydrazono)(methylamino)methyl]cyclohexanecarboximidohydrazide
4-Hydrazinecarboximidoyl-N'-hydroxy-N-methylcyclohexanecarbohydrazonamide

14.

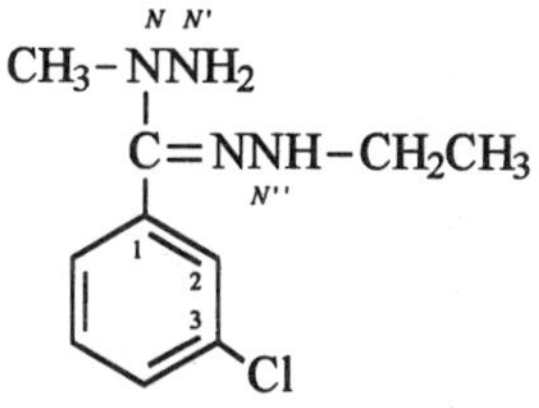

$$\overset{N\;N'}{CH_3-NNH_2}$$
$$\underset{N''}{C=NNH-CH_2CH_3}$$

3-Chloro-N''-ethyl-N-methylbenzohydrazono-
    hydrazide
3-Chloro-N-methylbenzohydrazide ethylhydrazone

15.

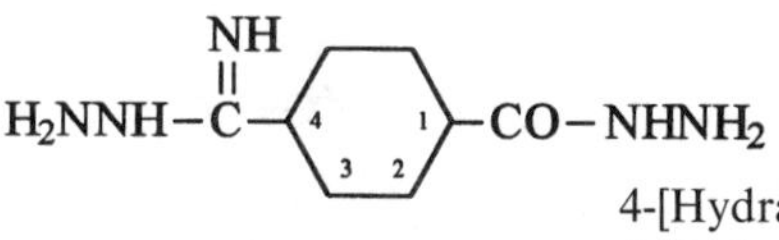

$$H_2NNH-\overset{\displaystyle NH}{\overset{\|}{C}}\!\!\left\langle\!\!\begin{array}{c}4\\3\quad2\end{array}\!\!\right\rangle\!\!-CO-NHNH_2$$

4-[Hydrazino(imino)methyl]cyclohexanecarbohydrazide
4-Carbonohydrazidimidoylcyclohexanecarbohydrazide

# 20

# Nitriles

*Nitriles* as a class are characterized by the presence of the $-C\equiv N$ group; the class is sometimes referred to as *carbonitriles*. Because of their close relationship to carboxylic acids, nitriles may be viewed as structures in which a $-COOH$ group has been replaced by a $-CN$ group. In functional class nomenclature, the class is called *cyanides*, and in this sense, the $-CN$ group is considered to be a pseudohalogen (see chapter 10).

*Nitrile oxides*, $R-C\equiv NO$, and *cyanohydrins*, $R_2C(OH)-CN$, are also covered in this chapter. Isomers of nitriles represented by $R-NC$, are often called *isocyanides*; the $-NC$ characteristic group is considered to be a pseudohalogen in chapter 10.

## Acceptable Nomenclature

The term -nitrile, when used as a suffix in names of individual compounds, denotes only $\equiv N$ triply bound to a terminal carbon atom of a hydrocarbon. The term -carbonitrile describes $-C\equiv N$; it is used as a suffix to replace -carboxylic acid in names ending in -carboxylic acid. Names of *acyclic mononitriles* and *dinitriles* are based on the names of the corresponding hydrocarbon (see chapter 5) by adding the suffix -nitrile or -dinitrile. The chain and its numbering includes the carbon atom(s) of the nitrile group(s).

$$\underset{6\quad 5\text{-}2\qquad 1}{CH_3[CH_2]_4-CN}$$
Hexanenitrile

$$\overset{\displaystyle CH_3}{\underset{7\quad 6\quad 5\quad 4\quad 3\ 2\quad 1}{NC-CH_2CH_2CH_2CHCH_2-CN}}$$
3-Methylheptanedinitrile

$$\underset{5\quad 4\quad 3\quad 2\quad 1}{CH_3CH=CHCH_2-CN}$$
Pent-3-enenitrile

Trivial names of acids (see chapter 11) may be used to derive names for nitriles where all the $-COOH$ groups have been replaced by $-CN$ groups. These names are formed by changing the ending -ic acid or -oic acid to -onitrile as in acetonitrile for $CH_3-CN$, benzonitrile for $C_6H_5-CN$, or succinonitrile for $NC-CH_2CH_2-CN$; the latter name implies that all $-COOH$ groups have been changed to $-CN$ groups. Oxalonitrile, $NC-CN$, has the trivial name cyanogen and the systematic name ethanedinitrile.

Nitriles corresponding to acids with names ending in -carboxylic acid are named by replacing that ending with -carbonitrile. Such structures include acyclic polynitriles with more than two $-CN$ groups attached to the same straight chain. They are named with the suffix -tricarbonitrile, and so on, in which case the chain numbering does *not* include the $-CN$ groups.

$$\overset{\displaystyle CN}{\underset{4\quad 3\quad 2\quad 1}{NC-CH_2CH_2CHCH_2-CN}}$$
Butane-1,2,4-tricarbonitrile

The -carbonitrile suffix(es) are also used where the –CN group(s) are attached to a ring system or to a hetero atom of a linear chain.

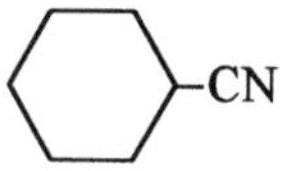

Cyclohexanecarbonitrile

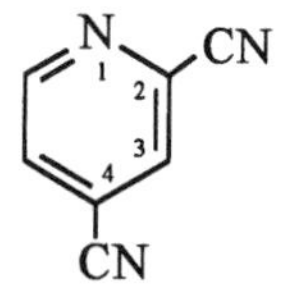

Pyridine-2,4-dicarbonitrile

$H_3Si-O-SiH_2-CN$
    3    2    1
Disiloxanecarbonitrile

$CH_3-NHNH-CN$
        2   1
2-Methylhydrazine-1-carbonitrile
2-Methylcarbazonitrile

Nitriles derived from aldehydic acids may be named with the suffix -aldehydonitrile, as in

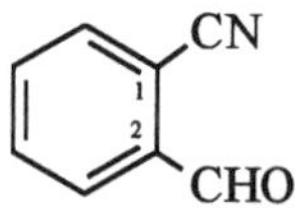

Phthalaldehydonitrile
2-Formylbenzonitrile
2-Cyanobenzaldehyde

When the nitrile function is treated as a substituting group, the prefix cyano (not nitrilo), NC–, is used.

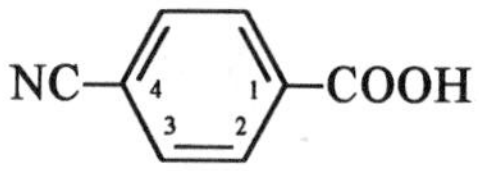

4-Cyanobenzoic acid

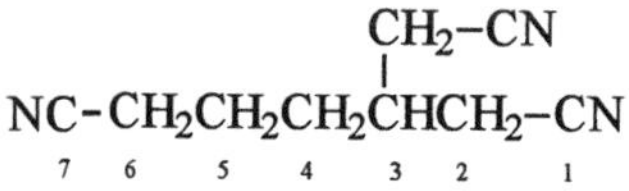

$CH_2-CN$
      |
$NC-CH_2CH_2CH_2CHCH_2-CN$
 7  6   5   4   3  2   1
3-(Cyanomethyl)heptanedinitrile

The nitrile group itself is often viewed as a pseudohalogen (see chapter 10). This leads to functional class names such as ethyl cyanide. *Acyl cyanides*, R–CO–CN, may advantageously be named similarly as analogs of *acyl halides* (see chapter 11), giving names such as hexanoyl cyanide for $CH_3[CH_2]_4-CO-CN$, but the substitutive name 2-oxoheptanenitrile is usually used.

α-Hydroxy nitriles should be named as such rather than as *cyanohydrins*, as in hydroxy-(phenyl)acetonitrile or α-hydroxybenzeneacetonitrile for $C_6H_5-CH(OH)CN$.

*Nitrile oxides*, R–C≡NO, are named by adding oxide as a separate word after the corresponding nitrile name (also see chapter 33 for the ylide form).

$CH_3CH_2CH_2-C≡NO$
 4   3   2   1
Butanenitrile oxide

Oxidocyano, for O←N≡C–, is used by CAS, but it is not recommended since oxido is defined as ¯O–.

## Discussion

Conjunctive nomenclature (see chapter 3), used primarily by CAS, generates acceptable names such as benzeneacetonitrile.

The simplest nitrile, HCN, might be named "formonitrile"; it is usually called hydrogen cyanide rather than hydrocyanic acid. Cyanamide, $H_2N-CN$, and cyanogen for NC–CN are acceptable.

Although acceptable in biochemical nomenclature, trivial names for $\alpha$-amino nitriles should not be formed by combining the suffix nitrile with the trivial name of an amino acid, as in "alaninenitrile" or "alaninonitrile"; these compounds should be named systematically as nitriles substituted by an amino group.

The prefix nitrilo is recognized as a multivalent group in which nitrogen is bound to two or three identical groups having a substituent with higher priority than amine. This prefix, a close cousin of imino ($HN=$), is seldom used for N triply bound to another atom; azanylidyne is preferred for that purpose. The term nitrido describes N replacing $=O$ and $-OH$ in functional replacement names of mononuclear oxo acids (see chapter 3).

## Additional Examples

1. $CH_2{=}CH{-}CN$
   $\phantom{xx}$ 3 $\phantom{x}$ 2 $\phantom{x}$ 1

   Acrylonitrile
   Propenenitrile

2. $C_6H_5{-}CH_2{-}CN$

   Phenylacetonitrile
   Phenylethanenitrile
   Benzyl cyanide
   Benzeneacetonitrile

3. $C_6H_5{-}CH(CN)_2$

   Phenylmalononitrile
   Phenylpropanedinitrile

4. $C_6H_5{-}CO{-}CN$

   Benzoyl cyanide
   Oxo(phenyl)acetonitrile
   $\alpha$-Oxobenzeneacetonitrile

5. $\begin{array}{cc} CH_3 & O \\ | & || \\ CH_3CH{-}C{-}CN \end{array}$
   $\phantom{x}$ 4 $\phantom{xx}$ 3 $\phantom{xx}$ 2 $\phantom{x}$ 1

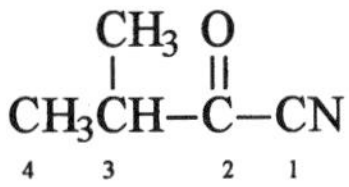

   3-Methyl-2-oxobutanenitrile
   2-Methylpropanoyl cyanide

6. $NC{-}CH_2CH_2{-}COOH$
   $\phantom{xxxx}$ 3 $\phantom{xx}$ 2 $\phantom{xx}$ 1

   3-Cyanopropanoic acid

7. $\begin{array}{c} OH \\ | \\ CH_3CH{-}CN \end{array}$
   $\phantom{x}$ 3 $\phantom{xx}$ 2 $\phantom{x}$ 1

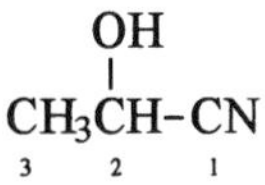

   2-Hydroxypropanenitrile

8. $\begin{array}{c} NH_2 \\ | \\ NC{-}CH_2CH{-}CN \end{array}$

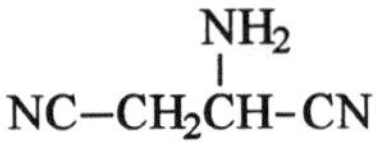

   Aminosuccinonitrile
   Aminobutanedinitrile

9. $\begin{array}{c} NC \\ | \\ CH_3CHCH_2CH_2{-}CN \end{array}$
   $\phantom{x}$ 5 $\phantom{xx}$ 4 $\phantom{x}$ 3 $\phantom{xx}$ 2 $\phantom{x}$ 1

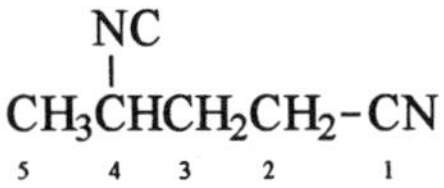

   4-Isocyanopentanenitrile

10. $=CH{-}CN$

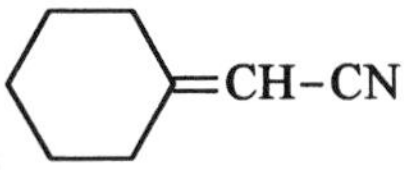

   Cyclohexylideneacetonitrile

11.

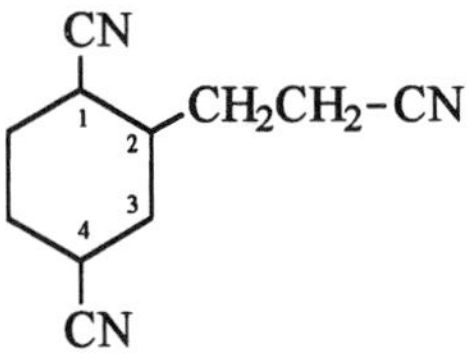

   2-(2-Cyanoethyl)cyclohexane-1,4-dicarbonitrile

# 21

# Amines and Imines

*Amines* are most simply defined as derivatives of ammonia, $NH_3$, in which one or more hydrogen atoms have been replaced by nonacyl groups; these groups are linked to the nitrogen atom through single bonds. *Primary*, *secondary*, and *tertiary amines* have the generic structures $R-NH_2$, $R_2NH$, and $R_3N$, respectively. *Imines* include compounds of the type $RR'C=N-R''$ in which either or both $R'$ and $R''$ may be hydrogen. Hydrazine (diazane), $H_2NNH_2$, hydroxylamine, $H_2N-OH$, and related compounds will be found in chapter 22. Replacement of hydrogen atoms of ammonia by acyl groups provides amides (see chapter 18). The combination of imine and amine groups attached to the same carbon atom results in an amidine (see chapter 19). Amino acids are covered in chapter 31. Ions and radicals derived from nitrogen compounds are discussed in chapter 33.

When the amino or imino nitrogen atom is part of a ring system, the structure is named as a heterocyclic compound (see chapter 9) rather than as an amine or imine.

## Acceptable Nomenclature

Generally, amines are named: (a) as derivatives of a parent hydride, such as a hydrocarbon, by means of the suffix -amine appended to the name of the parent; (b) substitutively as derivatives of ammonia, with the name(s) of substituents prefixed to "amine", considered in this case to be a parent hydride equivalent to ammonia; and (c) through use of the prefix amino denoting substitution into a parent structure. Under method (b), a primary monoamine may itself be treated as a parent structure that can be further substituted, as in 2-chloroethylamine, viewed as the 2-chloro derivative of ethylamine; alternatively, (2-chloroethyl)amine is the 2-chloro-ethyl derivative of "amine".

*Primary monoamines* are named by either methods (a) or (b), with (b) usually preferred for simple structures.

$$CH_3CH_2-NH_2$$
$$\phantom{CH_3}{}_2\phantom{CH}{}_1\phantom{-NH}{}_N$$

(a) Ethanamine
(b) Ethylamine

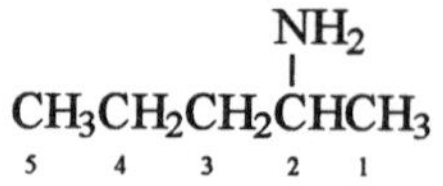

$$\overset{\displaystyle NH_2}{\underset{\phantom{CH_3CH_2CH_2}5\phantom{CH}4\phantom{CH_2}3\phantom{CH}2\phantom{CH_3}1}{CH_3CH_2CH_2CHCH_3}}$$

(a) Pentan-2-amine
(b) (1-Methylbutyl)amine
    Pentan-2-ylamine

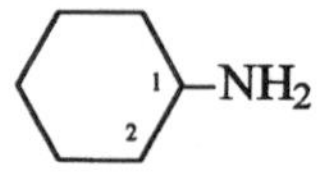

(a) Cyclohexanamine
(b) Cyclohexylamine

$$C_6H_5-CH_2-NH_2$$
$$\quad\ \ _1\qquad\ \ _N$$

(a) 1-Phenylmethanamine
Benzenemethanamine
(*not* α-Toluenamine)
(b) Benzylamine
(Phenylmethyl)amine

Method (c) can be used to emphasize a parent structure, such as a heterocyclic ring, substituted with an amino group.

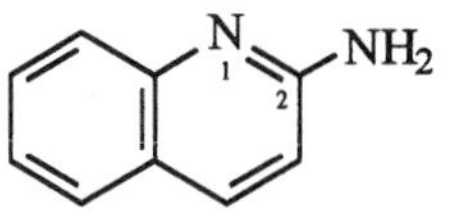

(a) Quinolin-2-amine
(b) 2-Quinolylamine
(c) 2-Aminoquinoline

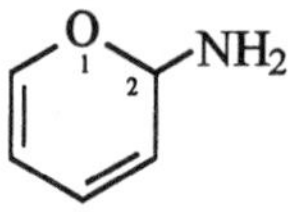

(a) 2*H*-Pyran-2-amine
(b) 2*H*-Pyran-2-ylamine
(c) 2-Amino-2*H*-pyran

The suffix -amine is also used where the amino group is attached to a heteroatom, as in silanamine, pyrrolidin-1-amine, or 2*H*-1,5,2-dithiazin-2-amine.

The trivial name aniline (phenylamine or benzenamine) is acceptable, along with anisidine (methoxybenzenamine or (methoxyphenyl)amine, 3 isomers), toluidine (methylbenzenamine or (methylphenyl)amine, 3 isomers), and benzidine ([1,1′-biphenyl]-4,4′-diamine). Other common names, such as phenetidine (ethoxybenzenamine or (ethoxyphenyl)amine, 3 isomers) and xylidine (dimethylbenzenamine or (dimethylphenyl)amine, 6 isomers) are discouraged. Greek letters are used as locants for substitution on the side chain in such compounds.

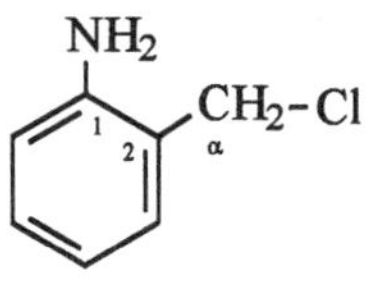

α-Chloro-*o*-toluidine
2-(Chloromethyl)benzenamine
2-(Chloromethyl)aniline

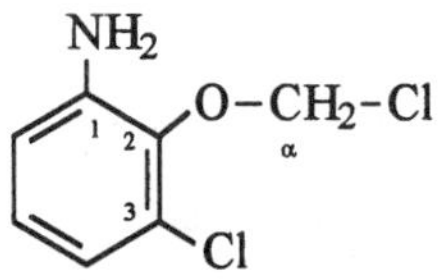

α,3-Dichloro-*o*-anisidine
3-Chloro-2-(chloromethoxy)benzenamine
3-Chloro-2-(chloromethoxy)aniline

Amines are low in the seniority list of compound classes (see chapter 3, table 3.1). In the presence of a more senior characteristic group the amine function is cited by the prefix amino.

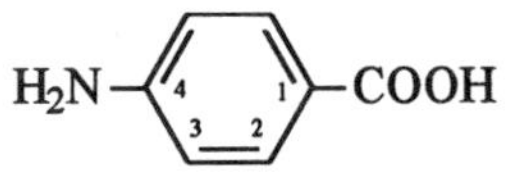

$$H_2N-\underset{\ \ 3\quad 2}{\overset{4\quad 1}{\bigcirc}}-COOH$$    4-Aminobenzoic acid

*Primary polyamines* in which the amino groups are attached to an acyclic chain or directly to a cyclic structure are named by adding the suffix diamine, triamine, and so on, with appropriate locants, to the name of the parent compound or to the name of a multivalent substituent group.

$$H_2N-CH_2[CH_2]_4CH_2-NH_2$$
$$\qquad\quad _6\quad\ _{5\text{-}2}\qquad _1$$

Hexane-1,6-diamine
Hexamethylenediamine
Hexane-1,6-diyldiamine

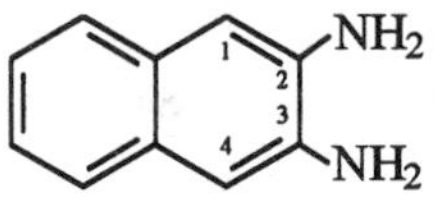

Naphthalene-2,3-diamine
Naphthalene-2,3-diyldiamine

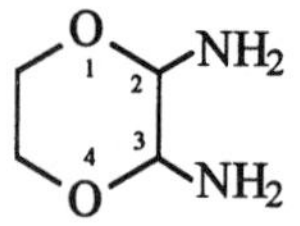

1,4-Dioxane-2,3-diamine
1,4-Dioxane-2,3-diyldiamine

Where an amino group is attached through a side chain to a cyclic structure, conjunctive nomenclature (see chapter 3) can be used; the name is formed from the name of the cyclic parent structure and the name of one (or more) of the side chain(s) with the remaining side chains denoted by prefixes. The name also may be based on substitution of the senior amine, or by citing all side chains by prefixes.

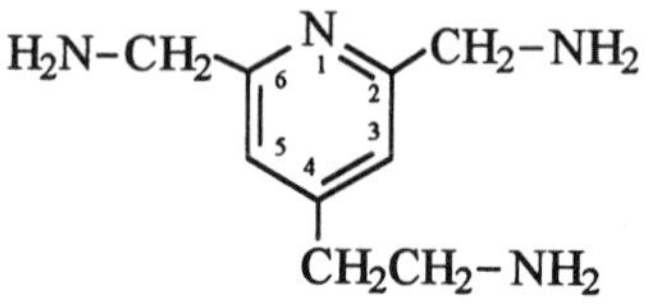

4-(2-Aminoethyl)pyridine-2,6-bis(methanamine)
[[4-(2-Aminoethyl)pyridine-2,6-diyl]-
    bis(methylene)]diamine
4-(2-Aminoethyl)-2,6-bis(aminomethyl)pyridine

*Symmetric secondary and tertiary amines* can be named: (a) as substitution products of $NH_3$; or (b) as *N*-substituted primary amines.

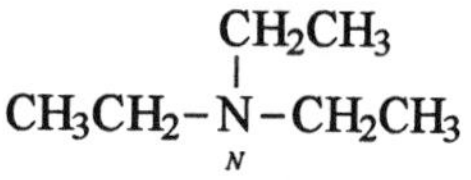

(a) Triethylamine
(b) *N,N*-Diethylethanamine

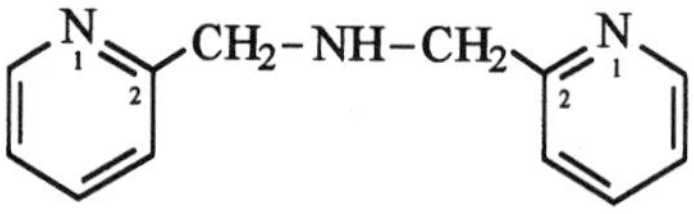

(a) Bis(2-pyridylmethyl)amine
(b) *N*-(Pyridin-2-ylmethyl)-2-
    pyridinemethanamine
    1-(2-Pyridyl)-*N*-(2-pyridylmethyl)-
    methylamine

Derivatives of these secondary and tertiary amines may be named: (a) with "amine" (that is, $NH_3$), as the parent compound; (b) as derivatives of secondary or tertiary amines with primed locants as needed; or (c) as derivatives of a primary amine.

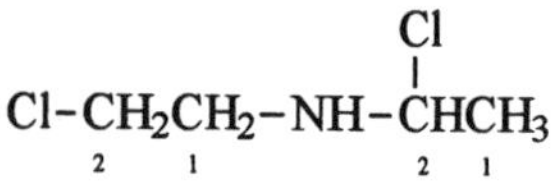

(a) (1-Chloroethyl)(2-chloroethyl)amine
(b) 1,2′-Dichlorodiethylamine
(c) 1-Chloro-*N*-(2-chloroethyl)ethylamine
    1-Chloro-*N*-(2-chloroethyl)ethanamine

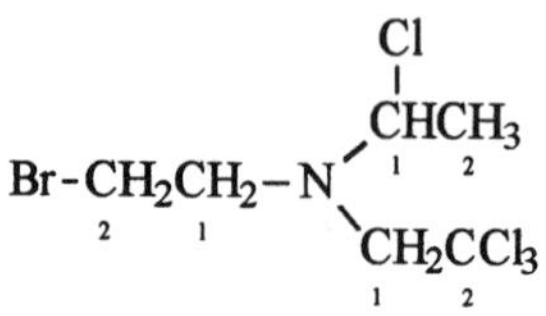

(a) (2-Bromoethyl)(1-chloroethyl)-
    (2,2,2-trichloroethyl)amine
(b) 2″-Bromo-1,2′,2′,2′-tetrachloro-
    triethylamine
(c) *N*-(2-Bromoethyl)-2,2,2-trichloro-*N*-
    (1-chloroethyl)ethylamine
    *N*-(2-Bromoethyl)-2,2,2-trichloro-*N*-
    (1-chloroethyl)ethanamine

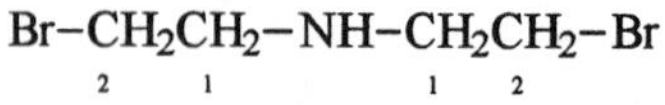

(a) Bis(2-bromoethyl)amine
(b) 2,2′-Dibromodiethylamine
(c) 2-Bromo-*N*-(2-bromoethyl)ethanamine
    2-Bromo-*N*-(2-bromoethyl)ethylamine

*Unsymmetric secondary and tertiary amines* and their derivatives are named: (a) on the basis of the parent name "amine"; or (b) as derivatives of the most senior primary or secondary amine. For this purpose, any cyclic group is usually considered to be senior to an acyclic group.

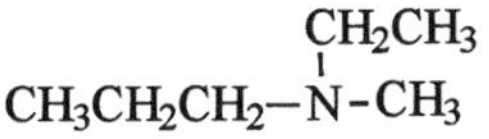

CH₃CH₂CH₂—N—CH₃ with CH₂CH₃

(a) Ethyl(methyl)(propyl)amine
(b) *N*-Ethyl-*N*-methylpropylamine
   *N*-Ethyl-*N*-methylpropan-1-amine

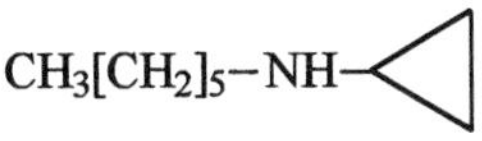

CH₃[CH₂]₅—NH—◁

(a) Cyclopropyl(hexyl)amine
(b) *N*-Hexylcyclopropylamine
   *N*-Cyclopropylhexan-1-amine
   *N*-Hexylcyclopropanamine

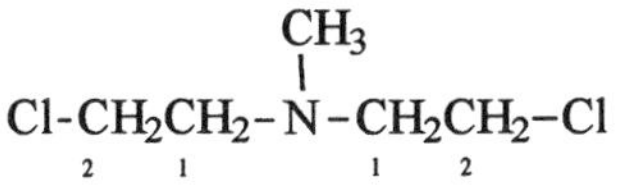

Cl-CH₂CH₂-N-CH₂CH₂—Cl with CH₃

(a) Bis(2-chloroethyl)(methyl)amine
(b) 2,2′-Dichloro-*N*-methyldiethylamine
   2-Chloro-*N*-(2-chloroethyl)-*N*-methylethanamine
   2-Chloro-*N*-(2-chloroethyl)-*N*-methylethylamine

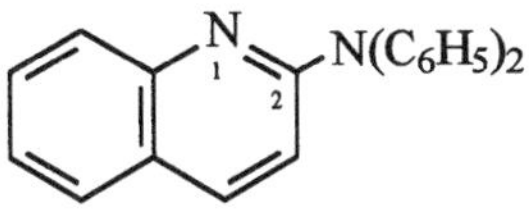

quinoline-N(C₆H₅)₂

(a) Diphenyl(quinolin-2-yl)amine
(b) *N*,*N*-Diphenylquinolin-2-amine
   2-(Diphenylamino)quinoline
   *N*-Quinolin-2-yldiphenylamine

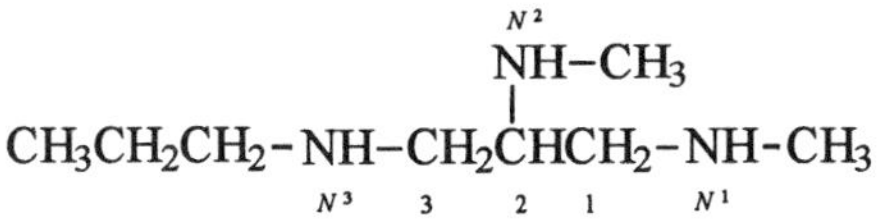

CH₃CH₂CH₂-NH-CH₂CHCH₂-NH-CH₃ with NH-CH₃

(a) *N*¹,*N*²-Dimethyl-*N*³-propyl(propane-1,2,3-triyltriamine)
(b) *N*¹,*N*²-Dimethyl-*N*³-propyl-1,2,3-propanetriamine

"*Alternating*" polyamines of the type CH₃–(NH–[CH₂]ₘ)ₙ–NH–CH₃ may be named: (a) by substitution principles; (b) by an adaptation of structure-based polymer nomenclature (see chapter 29); or (c) where *m* is one, by the (ab)ₙa method.

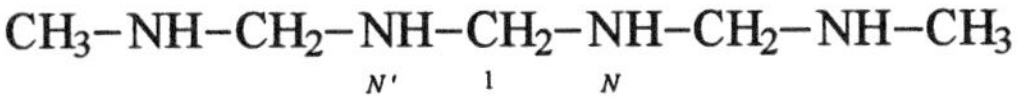

CH₃−NH−CH₂−NH−CH₂−NH−CH₂−NH−CH₃

(a) *N*,*N*′-Bis[(methylamino)methyl]methanediamine
(b) α-Methyl-ω-(methylamino)tri(iminomethylene)
(c) Pentacarbazane
   *N*,*N*′-Dimethyltricarbazane-1,5-diamine

Acyclic chains containing several nitrogen atoms can also be named by replacement nomenclature (see chapter 3). This is especially advantageous with unsymmetric arrangements and where other hetero atoms are present in the chain, even when the amine group is expressed as a suffix.

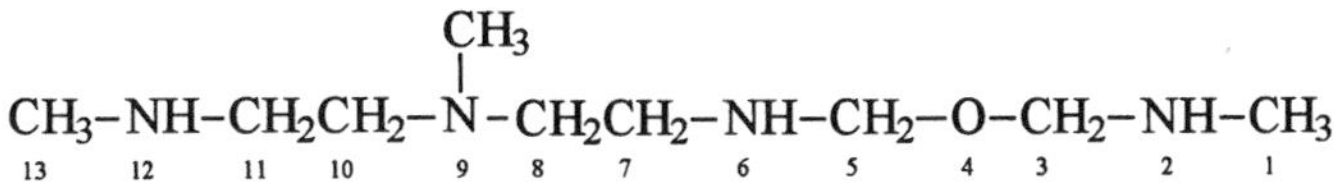

CH₃−NH−CH₂CH₂−N-CH₂CH₂−NH−CH₂−O−CH₂−NH−CH₃ with CH₃

9-Methyl-4-oxa-2,6,9,12-tetraazatridecane
*N*,*N*′,7-Trimethyl-2-oxa-4,7-diazanonane-1,9-diamine

By substitutive nomenclature, the example above could be named *N*,*N*′-dimethyl-*N*-[2-[[[(methylamino)methoxy]methyl]amino]ethyl]ethylenediamine or *N*,*N*′-dimethyl-*N*-[2-[[[(methylamino)methoxy]methyl]amino]ethyl]ethane-1,2-diamine. A substitutive name would be used by CAS, since this example does not meet CAS criteria for a replacement name.

*Imines*, R–CH=N–R″ and RR′C=N–R″ in which R″ may be hydrogen, have been called *aldimines* and *ketimines* as subclass names; the latter subclass has also been called *azomethines* and "Schiff bases". Acyclic and cyclic imines are named by methods analogous to those used for amines by adding the suffix imine to the name of a hydrocarbon, R–CH₃, corresponding to R–CH= in R–CH=NH, as in ethanimine, or through use of the -ylidene ending, as in ethylidenimine. *N*-Substituted imines can also be named as derivatives of the appropriate imine. The prefix imino may be employed where it is desired to place emphasis on another part of the structure such as a nitrogenous heterocycle or a more senior characteristic group, including the amine characteristic group itself.

$$CH_3CH{=}N{-}CH_2CH_2CH_3$$
$$\phantom{xx}{}_{2}\phantom{xxx}{}_{1}\phantom{xxx}{}_{N}\phantom{xxx}{}_{1}\phantom{xx}{}_{2}\phantom{xx}{}_{3}$$

*N*-Propylethanimine
*N*-Propylethylidenimine

$$\overset{NH}{\overset{\|}{CH_3CH_2CH_2CCH_3}}$$
$$\phantom{xx}{}_{5}\phantom{xx}{}_{4}\phantom{xx}{}_{3}\phantom{xx}{}_{2}\phantom{x}{}_{1}$$

Pentan-2-imine
Pentan-2-ylidenimine
(1-Methylbutyliden)imine

$$CH_3{-}N{=}\langle\text{ring}\rangle{=}O$$

4-(Methylimino)cyclohexa-2,5-dien-1-one

The trivial name carbodiimide is acceptable for the hypothetical structure HN=C=NH, systematically named methanediimine or methanediylidenediimine.

The prefixes imino (–NH–) and nitrilo (–N< or –N=) are used when these groups are linked to identical groups in a structure having characteristic groups higher than amine in seniority (see chapter 3, table 3.1).

$$HOOC{-}\langle\text{ring}\rangle{-}NH{-}\langle\text{ring}\rangle{-}COOH$$

4,4′-Iminodibenzoic acid

$$\overset{3''}{CH_2CH_2{-}COOH}$$
$$HOOC{-}CH_2CH_2{-}\overset{|}{N}{-}CH_2CH_2{-}COOH$$
$$\phantom{xxxx}{}_{3'}\phantom{xxxxx}{}_{3}\phantom{xx}{}_{2}\phantom{xx}{}_{1}$$

3,3′,3″-Nitrilotripropanoic acid

## Discussion

Nomenclature recommendations for amines allow a variety of names, each of which is readily understandable for the simpler compounds. This is consonant with the IUPAC rules. The main nomenclature problem lies with the determination of the parent rather than with the derivatives: with amine as the parent, ethylamine is both a derivative and a parent. Ethylamine is used as a parent primarily because it provides a larger structure for naming more complex compounds. With a hydrocarbon as the parent, the parent name ethanamine results. For well over a century, the parent name amine has stood for NH₃ when it is substituted (IUPAC now recommends azane as an alternative to amine, but this has yet to see much usage), and names like propylamine and diethylamine have stood the test of time. However, amine is also the suffix name for a characteristic group, –NH₂, that can be substituted for hydrogen in a parent structure such as a hydrocarbon; propan-2-amine is the kind of name that results. In both systems, unsymmetric secondary and tertiary amines are named as derivatives of the senior primary amine: *N*-methylpropylamine and *N*-methylpropan-1-amine are good names for CH₃CH₂CH₂–NH–CH₃, which can also be named methyl(propyl)amine. But unsubstituted simple symmetric secondary and tertiary amines are clearly better named as derivatives of NH₃, which are also used as parents. Diethylamine is widely preferred over *N*-ethylethylamine or *N*-ethylethanamine. With substitution, symmetry may be lost, and primed locants become necessary, yielding names such as 1,2′-dichlorodiethylamine. As complexity of substitution increases, names based on a senior primary amine are usually preferred.

The IUPAC rules reasonably place the name of the principal characteristic group as a suffix at the end of a chemical name; in the absence of a more senior characteristic group, amines ought to be named as amines. In a pragmatic sense, however, it may be desirable to emphasize the rest of the molecule at the expense of a single characteristic group. Heterocyclic compounds with amino substituents are a case in point. At present, IUPAC allows use of the prefix amino with nitrogenous heterocycles, presumably because of the "functional" basicity of the latter; however, this privilege can be applied to all heterocycles, at least as an alternative.

The past practice of naming $N$-substituted imines as amines has led to a perplexing variety of names. This method was allowed in the 1993 IUPAC recommendations, but IUPAC has since redefined amine to permit only groups attached to the nitrogen atom by single bonds. In CAS, imines are named as imines unless they have an $N$-alkyl or $N$-aryl substituent, in which case they are named as amines. In this book, it is recommended that *all* imines be named as imines in the absence of a group higher in the order of precedence.

Semisystematic names for $H_2N–[CH_2CH_2–NH]_2–CH_2CH_2–NH_2$ such as triethylene-tetramine are acceptable but systematic substitutive names such as $N,N'$-bis(2-aminoethyl)-ethane-1,2-diamine or, if desired, replacement names such as 3,6-diazaoctane-1,8-diamine are preferred.

The use of "imine" to denote conversion of =O to =NH in quinone-like structures, as in "benzoquinone monoimine", should be discontinued in favor of a name like 4-iminocyclo-hexa-2,5-dien-1-one.

## Additional Examples

1.

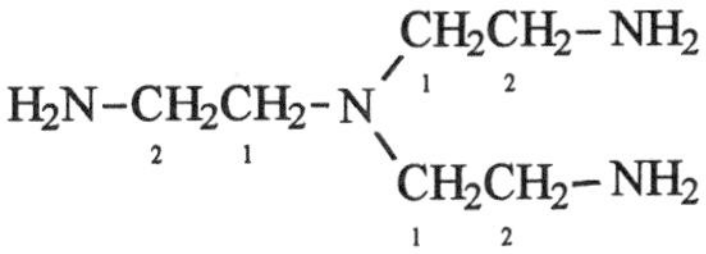

    Tris(2-aminoethyl)amine
    $N,N$-Bis(2-aminoethyl)ethane-1,2-diamine
    2,2′,2″-Nitrilotris(ethylamine)

2.   $H_2N–CH_2CH_2–OH$

    2-Aminoethanol
    (*not* Ethanolamine)

3.

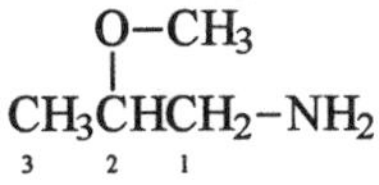

    (2-Methoxypropyl)amine
    2-Methoxypropan-1-amine
    2-Methoxypropylamine

4.

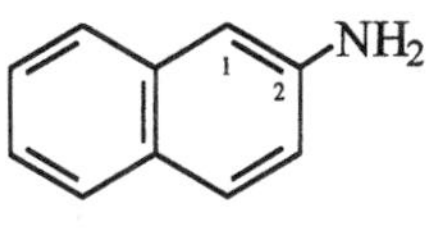

    2-Naphthylamine
    Naphthalen-2-amine

5.

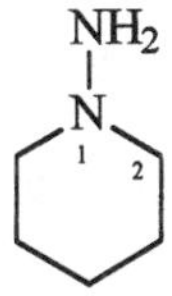

    Piperidin-1-ylamine
    Piperidin-1-amine
    1-Aminopiperidine

6.

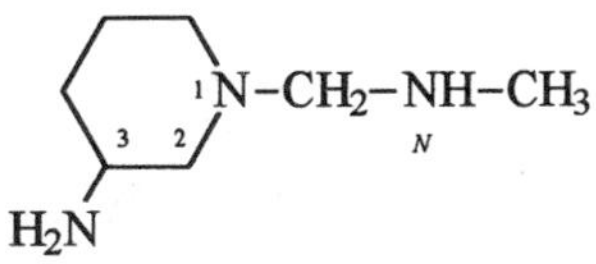

    3-Amino-$N$-methyl-1-piperidinemethanamine
    3-Amino-1-[(methylamino)methyl]piperidine
    1-[(Methylamino)methyl]piperidin-3-amine
    $N$-[(3-Aminopiperidin-1-yl)methyl]methylamine

7.

$H_2N-CH_2CH_2-NH-CH_2CH_2-NH_2$
      2      1                1      2

Bis(2-aminoethyl)amine
$N$-(2-Aminoethyl)ethane-1,2-diamine
2,2′-Diaminodiethylamine
Diethylenetriamine
2,2′-Iminodiethanamine

8.

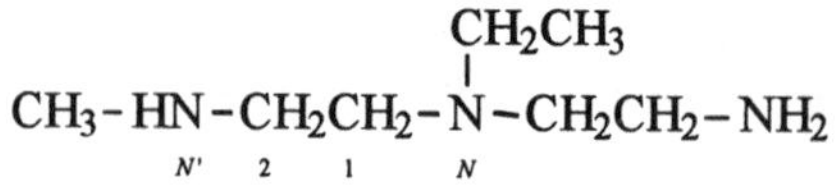

                                $CH_2CH_3$
                                    |
$CH_3-HN-CH_2CH_2-N-CH_2CH_2-NH_2$
       $N'$    2      1    $N$

$N$-(2-Aminoethyl)-$N$-ethyl-$N'$-methyl-
    ethane-1,2-diamine
2-Amino-2′-(methylamino)triethylamine

9.   $H_2N-CH_2CH_2-NH-CH_2CH_2-NH-CH_2CH_2-NH-CH_2CH_2-NH_2$
                              $N'$    2      1    $N$

$N$-(2-Aminoethyl)-$N'$-[2-[(2-aminoethyl)amino]ethyl]ethane-1,2-diamine
3,6,9-Triazaundecane-1,11-diamine
Tetraethylenepentamine

10.

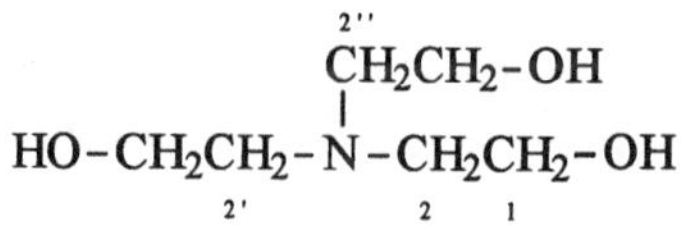

                 2″
            $CH_2CH_2-OH$
                 |
$HO-CH_2CH_2-N-CH_2CH_2-OH$
     2′           2      1

2,2′,2″-Nitrilotriethanol

11.

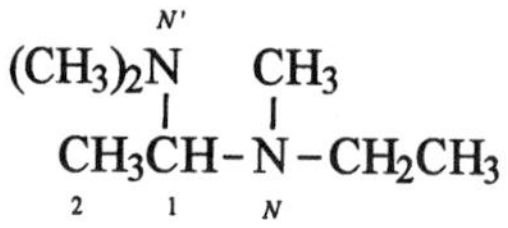

        $N'$
$(CH_3)_2N$      $CH_3$
         |         |
    $CH_3CH-N-CH_2CH_3$
         2    1   $N$

$N$-Ethyl-$N,N',N'$-trimethylethane-1,1-
    diamine
$N$-Ethyl-$N,N',N'$-trimethyl(ethane-1,1-
    diyldiamine)
$N$-[1-(Ethylmethylamino)ethyl]dimethylamine

12.

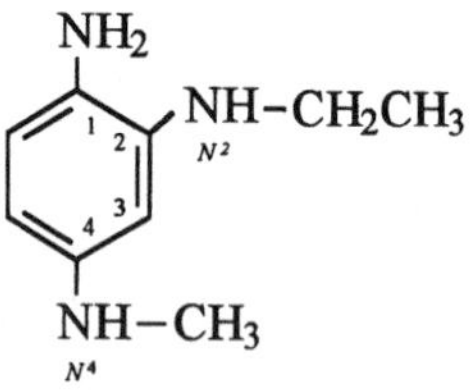

$NH_2$
|
(benzene ring)
$NH-CH_2CH_3$  ($N^2$)
    1  2
    4  3
$NH-CH_3$  ($N^4$)

$N^2$-Ethyl-$N^4$-methyl-1,2,4-benzenetriamine

13.

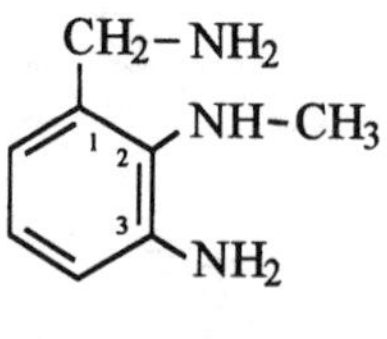

$CH_2-NH_2$
(benzene ring)
$NH-CH_3$
    1  2
       3
$NH_2$

3-Amino-2-(methylamino)benzylamine
3-(Aminomethyl)-$N^2$-methyl-1,2-
    benzenediamine
    (*not* 6-(Aminomethyl)-$N^1$-methyl-1,2-
    benzenediamine)
3-Amino-2-(methylamino)benzenemethanamine

14.

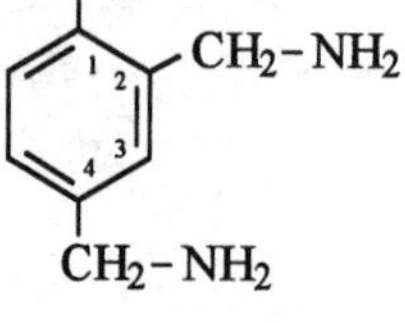

$CH_2-NH_2$
(benzene ring)
$CH_2-NH_2$
    1  2
       3
$CH_2-NH_2$

1,2,4-Benzenetrimethanamine
1,2,4-Benzenetris(methylamine)
[Benzene-1,2,4-triyltris(methylene)]triamine

15.

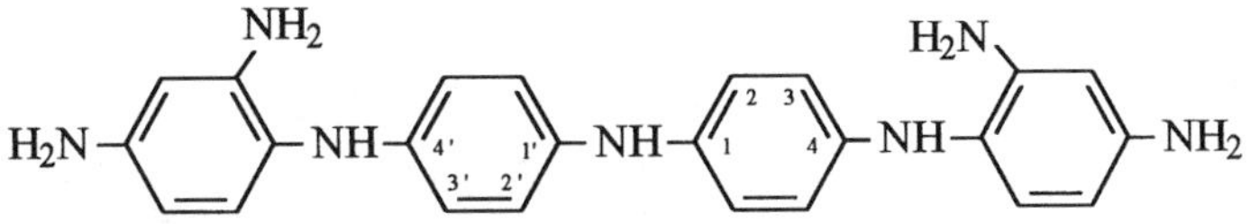

4,4'-[Bis[(2,4-diaminophenyl)amino]diphenylamine
$N^1,N^{1'}$-(Iminodi-4,1-phenylene)bis[1,2,4-benzenetriamine]
Bis[4-[(2,4-diaminophenyl)amino]phenyl]amine

16.

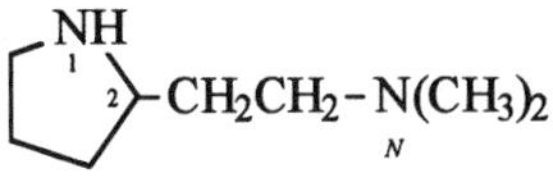

N,N-Dimethyl-2-pyrrolidineethanamine
N,N-Dimethyl-2-(pyrrolidin-2-yl)ethylamine
2-[2-(Dimethylamino)ethyl]pyrrolidine
Dimethyl[2-(pyrrolidin-2-ylethyl)]amine

17.

$C_6H_5-N=CHCH_3$

N-Phenylethanimine
N-Phenylethylidenimine
(Ethylideneamino)benzene

18.

$CH_3-N=CH-C_6H_5$

N-Methyl-1-phenylmethanimine
N-Methyl-1-phenylmethylidenimine
N-Methylbenzenemethanimine

19. $C_6H_5-CH=NH$

1-Phenylmethanimine
Phenylmethylidenimine

20. $C_6H_5-CH_2-N=CHCH_2CH_3$

N-Benzylpropan-1-ylidenimine
N-Benzylpropan-1-imine

21.

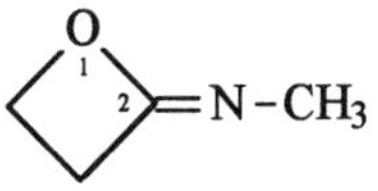

2-(Methylimino)oxetane
N-Methyloxetan-2-imine
N-Methyloxetan-2-ylidenimine

22.

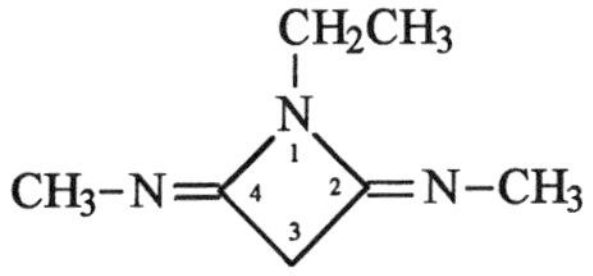

1-Ethyl-2,4-bis(methylimino)azetidine
1-Ethyl-N,N'-dimethylazetidine-2,4-diimine
1-Ethyl-N,N'-dimethylazetidine-2,4-diylidenediimine

23.

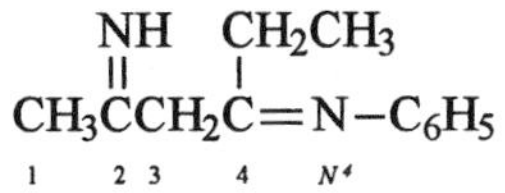

$N^4$-Phenyl-2,4-hexanediimine
$N^4$-Phenylhexane-2,4-diylidenediimine

24.

$CH_3-N=C=N-CH_2CH_3$

N-Ethyl-N'-methylcarbodiimide
N-Ethyl-N'-methylmethanediimine
N-Ethyl-N'-methylmethanediylidenediimine

25.

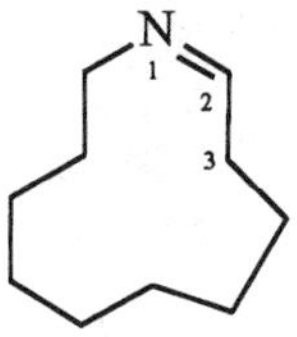

1-Azacycloundec-1-ene

26.

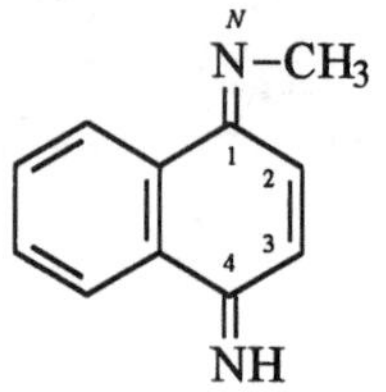

*N*-Methylnaphthalene-1,4-diimine
*N*-Methylnaphthalene-1,4-diylidenediimine
1-Imino-4-(methylimino)naphthalene

27.  $HN = CHCH_2 - NH_2$
       2    1

2-Aminoethanimine
(2-Iminoethyl)amine
2-Iminoethylamine

# 22

# Other Nitrogen Compounds

In many ways, the chemistry of nitrogen compounds is as complex as that of carbon. It will be no surprise to learn that the nomenclature of these compounds is just as complicated, not only in traditional description, but also in the systematic nomenclature methods used.

A variety of less familiar nitrogen compounds that do not fit conveniently elsewhere in this book are gathered together in this chapter. They include hydroxylamine derivatives, hydrazo, azo, and azoxy compounds, hydrazine (diazane) derivatives, polyazanes (and other compounds having more than one nitrogen atom, such as formazans), $N$-oxides, and esters and anhydrides of cyanic acid and its chalcogen analogs. Some overlap with other chapters is inevitable, particularly in cases of mixed functions.

The nomenclature of the more frequently encountered classes of nitrogen compounds has been discussed in a number of chapters. These classes include amines and imines (chapter 21), ammonium salts (chapter 33), amides, hydrazides, and imides (chapter 18), amidines (chapter 19), nitriles (chapter 20), and nitro, nitroso, azido, isocyano, and isocyanato compounds and their chalcogen analogs (chapter 10).

## Acceptable Nomenclature

*Hydroxylamine derivatives* can often be named as substitution products of the parent hydroxylamine, $H_2N–OH$, or its chalcogen analogs, such as thiohydroxylamine, $H_2N–SH$; $N, O, S, Se$, and $Te$ are used to indicate position of substituents. However, $S, Se$, and $Te$ derivatives are usually named as sulfenamides and so on (see chapter 23), and $O$ derivatives are commonly named as alkyloxy or aryloxy derivatives of amines. It is not unreasonable to extend substitutive nomenclature to -ylidene substituents of these parent structures to give names such as ethylidenehydroxylamine, but traditionally, functional class "oxime" names (see below) have been preferred because they emphasize the carbonyl function from which they are derived.

$$CH_3–NH–O–CH_2CH_3$$

$O$-Ethyl-$N$-methylhydroxylamine
$N$-Ethoxymethanamine
Ethoxy(methyl)amine
$N$-Ethoxymethylamine

$$H_2N–S–CH_3$$

$S$-Methyl(thiohydroxylamine)
Methanesulfenamide

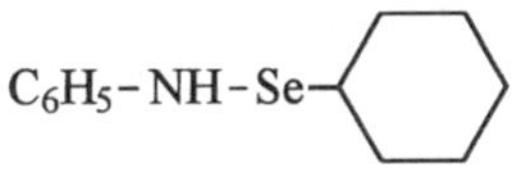

$$C_6H_5–NH–Se–\langle \rangle$$

$Se$-Cyclohexyl-$N$-phenyl(selenohydroxylamine)
$N$-Phenylcyclohexaneselenenamide

*N*-acyl derivatives, R–CO–NH–OH, are *hydroxamic acids* but best named as *N*-hydroxy amides (see chapter 18); *O*-acyl derivatives, $H_2N$–O–CO–R, are named as *O*-acylhydroxyl-amines. *Hydroximic acids*, R–C(OH)=N–OH, are preferably named as *N*-hydroxy imidic acids.

$CH_3$-CO-NH-OH

N-Hydroxyacetamide
N-Acetylhydroxylamine
Acetohydroxamic acid

$H_2N$-O-CO-$C_6H_5$

O-Benzoylhydroxylamine

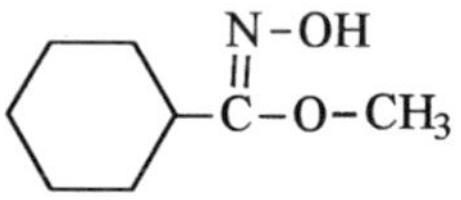

Methyl cyclohexanecarbohydroximate
Methyl *N*-hydroxycyclohexanecarboximidate

In the presence of characteristic groups of higher precedence, the prefixes hydroxyamino, HO–NH–; mercaptoamino (sulfanylamino), HS–NH–; hydroxyimino, HO–N=; aminooxy, $H_2N$–O–; aminothio (aminosulfanyl), $H_2N$–S–; and aminoseleno (aminoselanyl), $H_2N$–Se–, are used.

$CH_3$-O-NH-$CH_2CH_2$-OH
                  2      1

2-(Methoxyamino)ethanol

HO-N=CHCH$_2$-CO-N(CH$_3$)$_2$
        3    2      1      N

3-(Hydroxyimino)-*N*,*N*-dimethylpropanamide

*Oximes*, R–CH=N–OH and $R_2C$=N–OH, are alkylidene derivatives of hydroxylamine. Since they are formed from aldehydes or ketones, they are also called *aldoximes* and *ketox-imes*, respectively. Their names consist of the name of the corresponding aldehyde or ketone followed by the class term "oxime". In the presence of a higher function, the prefix hydroxy-imino is used. Substitution at the oxygen atom is denoted by *O*-alkyl and so on prefixed to the term "oxime" or by prefixes such as ethoxyimino.

$CH_3CH_2CH_2$CH=N-OH
 4     3    2     1

Butanal oxime
Butyraldehyde oxime

HO-N          O
     ‖          ‖
$CH_3$C$CH_2$C$CH_3$
 5    4    3   2 1

4-(Hydroxyimino)pentan-2-one
2,4-Pentanedione monooxime
2,4-Pentanedione 2-oxime

N-O-$CH_3$
‖
$C_6H_5$-C$CH_3$

Acetophenone *O*-methyloxime
1-Phenylethan-1-one *O*-methyloxime

*Nitrosolic acids*, R–C(NO)=N–OH, and *nitrolic acids*, R–C($NO_2$)=N–OH, have been named on the basis of the class name, as in acetonitrosolic acid or butyronitrolic acid. However, they are best named as oximes of 1-substituted aldoximes, giving names such as 1-nitrosoacetaldehyde oxime or 1-nitropentanal oxime.

*Amine oxides*, $R_3N$(O), and *nitrones*, $R_2C$=N(O)H, are named by appending, as a separate word, the term "oxide" to the name of the corresponding amine or imine. Nitrone, $CH_2$=N(O)H, with the locant α for the methylidene group, is not recommended as a name for a parent compound.

$C_6H_5$-CH=N(O)-$CH_3$
                N

N-Methylbenzenemethanimine oxide
N-Methyl-1-phenylmethanimine oxide
(*not N*-Methyl-α-phenylnitrone)

$(CH_3CH_2CH_2)_3NO$        Tripropylamine oxide

$CH_3CH=N(O)H$        Ethanimine oxide
    2     1              Ethylidenimine oxide

Oxides of oximes (*nitronic acids*), $R_2C=N(O)–OH$, have been named with the prefix *aci*-nitro for $HO–N(O)=$. The 1993 IUPAC recommendations brought the nomenclature of oxides (or acids) of nitrogen into consonance with that of phosphorus and introduced the concept of additive prefixes derived from nitrogen and phosphorus acids (see chapter 25). Examples are azonic acid, $HN(O)(OH)_2$ (prefixes are azono or dihydroxynitroryl for $(HO)_2N(O)–$ and azonoyl or hydronitroryl for $HN(O)=$ or $HN(O)<$) and azinic acid, $H_2N(O)–OH$ (prefixes are hydrohydroxynitroryl for $HO–NH(O)–$; azinoyl or dihydronitroryl for $H_2N(O)–$; hydroxy-nitroryl for $HO–N(O)=$ or $HO–N(O)<$; and nitroryl for $ON≡$ or $>(O)N–$) (see also chapter 3, table 3.4). Other phosphorus analog prefixes, such as azinylidene for $HN(O)=$ or $HN(O)<$ and azinylidyne for $ON≡$ or $>(O)N–$ are acceptable.

    *Hydrazine compounds* other than acyl derivatives (which are hydrazides, see chapter 18) are readily named substitutively on the basis of the parent hydride $H_2NNH_2$, which is tradition-ally (and by CAS) named hydrazine; the corresponding prefix is hydrazinyl (or hydrazino) for $H_2NNH–$. The numerical locants 1 and 2 are preferred to $N$ and $N'$, and in the prefix, 1 is at the nitrogen atom with the free valence. Just as acceptable, although not cited with examples in this chapter, are the IUPAC-approved names diazane and diazanyl, which correspond to hydrazine and hydrazinyl.

$C_6H_5–NHN(CH_3)_2$        1,1-Dimethyl-2-phenylhydrazine
       2  1

$CH_3–NHNH$—⟨4 1⟩—COOH        4-(2-Methylhydrazinyl)benzoic acid
   2  1       3  2

Acceptable, but not encouraged, is the substitutive prefix hydrazo for the $–NHNH–$ group where the free valences are attached to identical parent compounds; the preferred prefix name is hydrazine-1,2-diyl or diazane-1,2-diyl. Other prefix names are azino or diazanediylidene for $=NN=$; hydrazono or diazanylidene for $H_2NN=$; and hydrazin-1-yl-2-ylidene or diazan-1-yl-2-ylidene for $–NHN=$. The use of hydrazo as an analog of azo (see below) is rarely seen. When the free valences of the $–NHNH–$ group are attached to the same atom, it is named hydrazi, but the compound resulting from such substitution is usually better named as a heterocycle.

                                   1-(4-Chlorophenyl)-2-(3,5-dichlorophenyl)-
                                   1-methylhydrazine

                                     3-Ethyl-3-methyldiaziridine
                                     2-Hydrazibutane

    The unsaturated analog of hydrazine, $HN=NH$, is systematically named diazene; diimide has been used in the past. Examples of related group prefixes are: diazenyl for $HN=N–$ and azo or diazenediyl for $–N=N–$. The group $–N=N(O)–$ is azoxy prefixed by *NNO-* or *ONN-* to denote the location of the oxygen atom, as in phenyl-*ONN*-azoxy for $C_6H_5–N(O)=N–$. When the two groups affixed to $–N=N–$ are identical, a name such as azobenzene is still acceptable, but diphenyldiazene is preferred. If the groups are different, an "azo" name can be used with the largest group cited first, as in naphthalene-2-azobenzene. Alternatively, a diazene name such as 2-(phenyldiazenyl)naphthalene can be used; this kind of name is preferable when substituents are present. Azo dyes (usually known by trivial names) are traditionally named with azo as above or, where a principal characteristic group is present, with azo as a substitutive prefix, as in phenylazo.

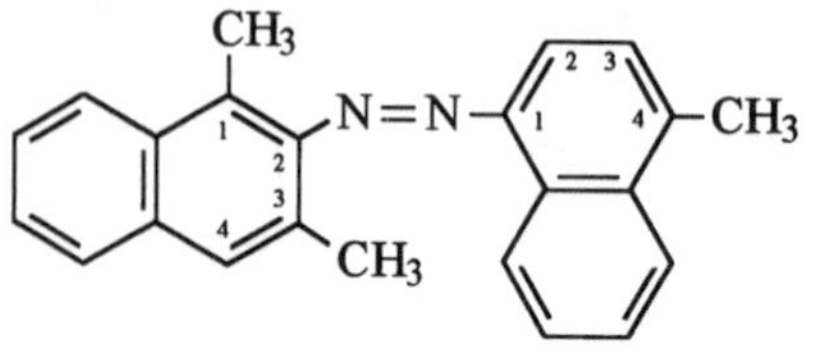

<table>
<tr><td>

$CH_3-N=N-$⬡

</td><td>

Cyclohexyl(methyl)diazene

</td></tr>
</table>

(1,3-Dimethylnaphthalen-2-yl)-(4-methylnaphthalen-1-yl)-diazene

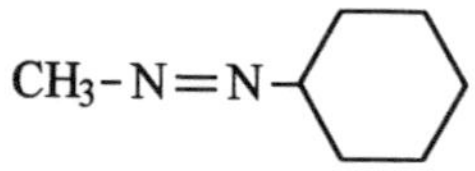

Cyclohexyl(methyl)diazene

$C_6H_5-N=N-COOH$

Phenyldiazenecarboxylic acid

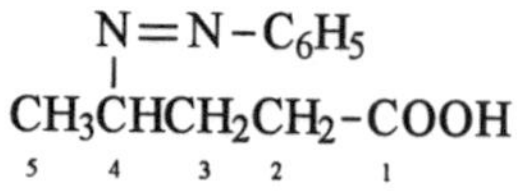

$\underset{5\ \ 4\ \ \ \ 3\ \ \ \ 2\ \ \ \ \ 1}{CH_3\overset{\displaystyle N=N-C_6H_5}{\underset{|}{CH}}CH_2CH_2-COOH}$

4-(Phenylazo)pentanoic acid
4-(Phenyldiazenyl)pentanoic acid

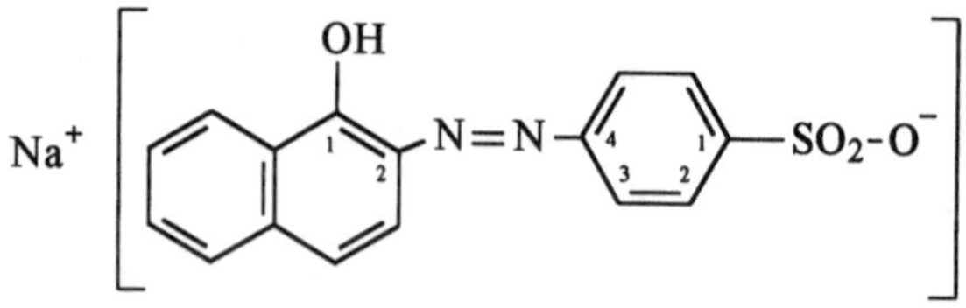

Sodium 4-(1-hydroxy-2-naphthylazo)benzene-sulfonate

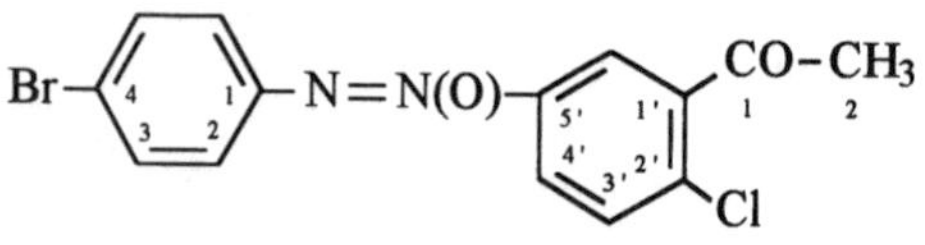

5′-[(4-Bromophenyl)-*NNO*-azoxy]-2′-chloroacetophenone
1-[5-[(4-Bromophenyl)-*NNO*-azoxy]-2-chlorophenyl]-ethan-1-one

*Formazans* are named differently by IUPAC and CAS:

$\underset{5\ \ 4\ \ \ \ \ 3\ \ \ \ \ \ 2\ \ \ \ \ 1}{H_2NN=CH-N=NH}$     Formazan (IUPAC)
Diazenecarboxaldehyde hydrazone (CAS)

Either method is acceptable.

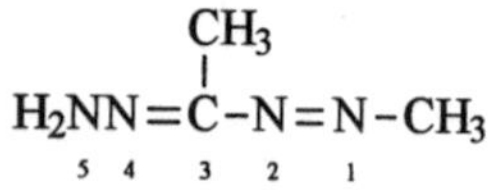

$\underset{5\ \ 4\ \ \ \ 3\ \ \ \ \ 2\ \ \ \ 1}{H_2NN=\overset{\displaystyle CH_3}{\underset{|}{C}}-N=N-CH_3}$

1,3-Dimethylformazan (IUPAC)
(1-Hydrazonoethyl)methyldiazene (CAS)

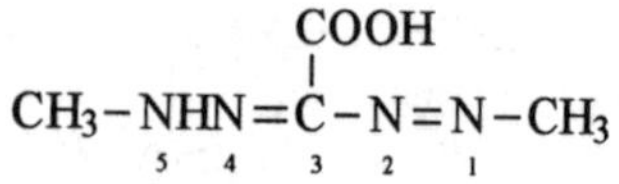

$\underset{5\ \ \ \ \ 4\ \ \ \ \ 3\ \ \ \ 2\ \ \ \ 1}{CH_3-NHN=\overset{\displaystyle COOH}{\underset{|}{C}}-N=N-CH_3}$

1,5-Dimethylformazan-3-carboxylic acid (IUPAC)
(Methylazo)(methylhydrazono)acetic acid (CAS)

Substituent names are 1-formazano for $H_2NN=CH-N=N-$, 5-formazano for $HN=N-CH=NNH-$, and formazanyl for $HN=N-C(=NNH_2)-$.

$HOOC-\langle\text{ring}\rangle-\underset{5}{N}H\underset{4}{N}=\underset{3}{C}H-\underset{2}{N}=\underset{1}{N}-\langle\text{ring, Cl}\rangle$

4-[1-[(2-Chlorophenyl)-5-
formazano]benzoic acid
(IUPAC)
4-[[[(2-Chlorophenyl)azo]-
methylene]hydrazino]benzoic
acid (CAS)

$CH_3-\underset{5}{N}H\underset{4}{N}=\underset{3}{\overset{\overset{\displaystyle CH_2CH_2-OH}{|}}{C}}-\underset{2}{N}=\underset{1}{N}-CH_3$

2-(1,5-Dimethylformazanyl)-
ethanol (IUPAC)
3-(Methylazo)-3-methyl-
hydrazono-1-propanol (CAS)

*Hydrazides of carbonic acid* were briefly mentioned in chapter 18. Related structures have a variety of seemingly unrelated traditional (but approved by IUPAC) names, some of which CAS has supplanted with more systematic functional names:

$\underset{3}{H_2}\underset{2}{N}NH-\underset{1}{COOH}$

Carbazic acid (IUPAC)
Hydrazinecarboxylic acid (CAS)
Carbonohydrazidic acid

$\underset{4}{H_2}N-\underset{3}{CO}-\underset{2}{NH}\underset{1}{NH_2}$

Semicarbazide (IUPAC)
Hydrazinecarboxamide (CAS)
Carbamohydrazide

$\underset{5}{H_2}N\underset{4}{NH}-\underset{3}{CO}-\underset{2}{NH}\underset{1}{NH_2}$

Carbonohydrazide (IUPAC)
 (*not* Carbohydrazide or Carbazide)
Carbonic dihydrazide (CAS)
Hydrazinecarbohydrazide

$\underset{5}{H}\underset{4}{N}=N-\underset{3}{CO}-\underset{2}{NH}\underset{1}{NH_2}$

Carbazone (IUPAC)
Diazenecarbohydrazide
Diazenecarboxylic hydrazide (CAS)

$\underset{5}{H}\underset{4}{N}=N-\underset{3}{CO}-\underset{2}{N}=\underset{1}{NH}$

Carbodiazone (IUPAC)
1,1′-Carbonylbis(diazene) (CAS)

$\underset{4}{H}\underset{}{N}=\overset{\overset{\displaystyle \overset{3}{OH}}{|}}{\underset{}{C}}-\underset{2}{NH}\underset{1}{NH_2}$

Isosemicarbazide (IUPAC)
Hydrazinecarboximidic acid (CAS)
Carbonohydrazidimidic acid

$\underset{5}{H_2}\underset{4}{N}N=\overset{\overset{\displaystyle \overset{3}{OH}}{|}}{\underset{}{C}}-\underset{2}{NH}\underset{1}{NH_2}$

Isocarbonohydrazide (IUPAC)
Hydrazinecarbohydrazonic acid (CAS)
Carbonohydrazidohydrazonic acid

In the last two examples, the CAS names shown are those used when "O" substituents are present; the unsubstituted forms are named as the tautomers hydrazinecarboxamide and carbonic dihydrazide, respectively. Sulfur analogs of these two examples are given names such as isothiosemicarbazide (IUPAC) or hydrazinecarboximidothioic acid (CAS) and iso-thiocarbonohydrazide (IUPAC) or hydrazinecarbohydrazonothioic acid (CAS), respectively. Substituting groups formed by the removal of one hydrogen atom from position 1 have traditional names obtained by converting the endings -ide to -ido and -one to -ono to give names such as semicarbazido and carbazono. Carbazoyl is the name for $H_2NNH-CO-$. Organic derivatives of the series above are best named substitutively, but 1-alkylidene derivatives of semicarbazide (the sulfur analog is thiosemicarbazide) are acceptably named with the functional class name *semicarbazone*.

$$NNH-CO-NH_2$$
$$\overset{\parallel}{CH_3\overset{}{C}CH_3}$$
$$\quad_{3}\quad_{2}\quad_{1}$$

2-Propanone semicarbazone
2-(1-Methylethylidene)-
   hydrazinecarboxamide
2-(Propan-2-ylidene)hydrazine-
   carboxamide
1-Isopropylidenesemicarbazide

$$\langle\text{cyclohexyl}\rangle-CH=NNH-CO-\overset{CH_3}{\overset{|}{N}}N=CH-C_6H_5$$
$$\qquad\qquad\qquad_{5}\qquad_{4}\quad_{3}\quad_{2}\quad_{1}$$

1-Benzylidene-5-(cyclohexylmethylidene)-2-methyl-
   carbonohydrazide (IUPAC)
2'-(Cyclohexylmethylidene)-1-
   methyl-2-(phenylmethylidene)-
   carbonic dihydrazide (CAS)

$$\overset{CH_3}{\overset{|}{H_2N\overset{}{N}}}-CO-O-CH_2CH_3$$
$$\quad_{3}\quad_{2}\qquad_{1}$$

Ethyl 2-methylcarbazate (IUPAC)
Ethyl 1-methylhydrazine-
   carboxylate (CAS)

Compounds such as $C_6H_5-N=N-OH$ were traditionally named benzenediazohydroxide. IUPAC now recommends phenyldiazenol and CAS names them as hydroxy derivatives of diazene. The term diazo is reserved for $N_2=$ where both nitrogen atoms are bound to the same atom.

$$N_2CHCH_3$$

Diazoethane

$$Na^+ \left[ \langle\text{naphthalene}\rangle-N=N-O \right]^-$$

Sodium 2-naphthyldiazenolate
   (IUPAC, 1993)
Sodium 2-naphthalenediazoate
   (IUPAC, 1979)
Hydroxy(naphthalen-2-yl)diazene
   sodium salt (CAS)

Higher homologs of hydrazine, $H_2N[NH]_nNH_2$, are named in a manner similar to hydrocarbons except that the final "a" of a numerical prefix is not elided, as in triazane, pentaazane (CAS uses pentazane), and so on. Locants are placed from one end of the nitrogen atom chain to the other; double bonds, if present, are given lowest locants. Corresponding prefix names are formed with the endings -yl and -ylidene as in tetraaz-3-en-1-ylidene for $HN=NNHN=$; traditional prefix names ending in -eno or -ano are now recommended by IUPAC only for use as bridge prefixes.

$$\overset{CH_3}{\overset{|}{H_2NN\overset{}{N}}}=N-CH_2-CN$$
$$\quad_{4}\quad_{3}\quad_{2}\qquad_{1}$$

(3-Methyltetraaz-1-en-1-yl)acetonitrile

When one or more nitrogen atoms of any of the above groups forms part of a ring system, the compound is usually named as a heterocycle.

*Derivatives of nitrogen-containing mononuclear inorganic acids* not treated elsewhere (see chapters 11 and 25) in this book are in most cases named conventionally as esters or by substitutive nomenclature. Some of the prefixes and class names are isocyano, isocyanide for CN–; cyanato, cyanate for NCO–; isocyanato, isocyanate for OCN–; fulminato, fulminate for CNO–; thiocyanato, thiocyanate for NCS–; and isothiocyanato, isothiocyanate for SCN–. Alkyl or aryl and acyl derivatives of HO–CN, HS–CN, and HO–NC are acceptably named as esters or as mixed anhydrides (see chapter 12); acyl derivatives of HNC, HNCO, and HNCS are named in the same way as acid halides (see chapter 11).

$CH_3CH_2-OCN$                  Ethyl cyanate

$SCN-CH_2-CO-O-CH_3$         Methyl isothiocyanatoacetate

$C_6H_5-CO-NCO$                Benzoyl isocyanate

4-Isocyanatobenzonitrile
OCN—⟨4 1⟩—CN                  4-Cyanophenyl isocyanate
                                     4-Isocyanatophenyl cyanide

OCN—⟨4 1⟩—NCO                 1,4-Phenylene diisocyanate
                                     1,4-Diisocyanatobenzene

*Replacement nomenclature* (see chapter 3) can be useful in naming complex nitrogenous structures (see also chapter 21) and their derivatives.

$$\underset{9\quad 8\quad\;\; 7\quad\;\; 6\quad 5\;\; 4\quad\; 3\quad 2\quad 1}{CH_3CH_2-NH-\overset{\overset{\displaystyle C_6H_5}{|}}{C}=NNH-CO-NH-CH_3}$$

3-Oxo-6-phenyl-2,4,5,7-tetraazanon-5-ene
2-[α-(Ethylamino)benzylidene]-*N*-methylhydrazinecarboxamide
1-[α-(Ethylamino)benzylidene]-4-methylsemicarbazide

## Discussion

For derivatives of hydrazine, names based on both the traditional name hydrazine and the IUPAC name diazane have been presented as acceptable, even though diazane has yet to be established in the literature. There are undoubted advantages to names based on "azane" for the higher homologs, and they are well-established and used by CAS. The use of diazane and diazene provides much needed systematic alternatives to hydrazo, hydrazono, azo, diazo, azino, and so forth.

Amines, imines, and the hydrazines, in that order, rank low in the order of seniority for compound classes (see chapter 3, table 3.1). For that reason, prefixes have been stressed throughout this chapter, since these groups will generally be of low rank in structures of mixed function. However, among the groups discussed here, no order of seniority has been prescribed, and seniority is a matter of choice for emphasis.

It is unusual in systematic organic nomenclature to break a bond between like hetero atoms in forming a name. For example, $H_2NNH-$ is never called "aminoamino" or HOO– "hydroxyoxy". However, with a few characteristic groups such as isocyanato, –NCO, or isocyano, –NC, it is convenient to retain the identity of the group within a name even when it is attached to the same kind of atom in a parent structure, as in 1-aminoquinoline or *N*-isocyanatoaniline.

Many amine names are possible (see chapter 21). In the following additional examples, only one such name will be cited with appropriate examples.

## Additional Examples

1.
$CH_3-NH-O-CO-$⟨1 2 3 4⟩$-CH_3$

*N*-Methyl-*O*-(4-methyl-1-naphthoyl)-hydroxylamine
*N*-[(4-Methyl-1-naphthoyl)oxy]-methanamine

2. $CH_3CH_2-CO-S-NH-CH_3$

N-Methyl-S-propanoyl-
   (thiohydroxylamine)
N-(Propanoylthio)methanamine

3. $CH_3-NH-S-CH_2CH_2-OH$
              2      1

2-[(Methylamino)thio]ethanol
2-Hydroxy-N-methylethanesulfenamide

4.

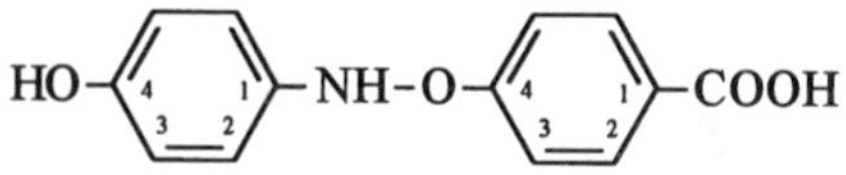

$HO-\langle\underset{3\;\;\;2}{\overset{4\;\;\;1}{\bigcirc}}\rangle-NH-O-\langle\underset{3\;\;\;2}{\overset{4\;\;\;1}{\bigcirc}}\rangle-COOH$

4-[[(4-Hydroxyphenyl)amino]oxy]-
   benzoic acid

5. $HO-N=CHCH_2CH_2-NH-SH$

N-[3-(Hydroximino)propyl]-
   (thiohydroxylamine)
3-(Mercaptoamino)propanal oxime
N-[3-(Mercaptoamino)propylidene]-
   hydroxylamine

6. $C_6H_5-NH-O-CH_2CH_3$

O-Ethyl-N-phenylhydroxylamine
N-Ethoxybenzenamine

7.  $\overset{N-OH}{\underset{\|}{HC-NH-OH}}$

N-[(Hydroxyimino)methyl]-
   hydroxylamine
(Hydroxyamino)methanal oxime
N,N'-Dihydroxyformamidine

8.  $\overset{N-OH}{\underset{\|}{CH_3-C-NO_2}}$

N-(1-Nitroethylidene)hydroxylamine
1-Nitroethanal oxime
1-Nitroacetaldehyde oxime

9.

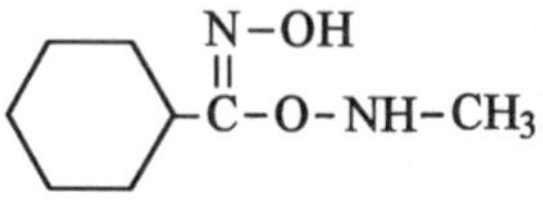

$\overset{N-OH}{\underset{\|}{C-O-NH-CH_3}}$

O-Cyclohexanecarbohydroximoyl-N-
   methylhydroxylamine
N-(Cyclohexanecarbohydroximoyloxy)-
   methanamine
O-(N-Hydroxycyclohexane-
   carboximidoyl)-N-methyl-
   hydroxylamine

10.
              $C_6H_5$
$CH_3-N\overset{|}{H}N-CO-O-CH_3$
          2      1

Methyl 2-methyl-1-phenyl-
   hydrazinecarboxylate
Methyl 3-methyl-2-phenylcarbazate
Methyl 2-methyl-1-phenylcarbono-
   hydrazidoate

11. $CH_3-O-CO-NHNH-CO-O-CH_3$
                          2      1

Dimethyl hydrazine-1,2-dicarboxylate

12.

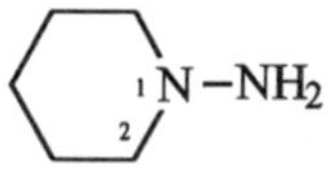

Piperidin-1-amine
   (not 1,1-Pentamethylenehydrazine)
1-Aminopiperidine

13.

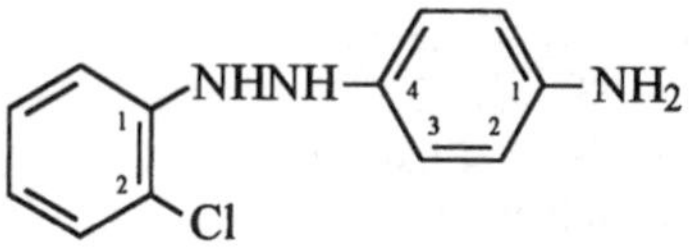

1-(4-Aminophenyl)-2-(2-chlorophenyl)-
   hydrazine
4-[2-(2-Chlorophenyl)hydrazinyl]-
   benzeneamine

14.

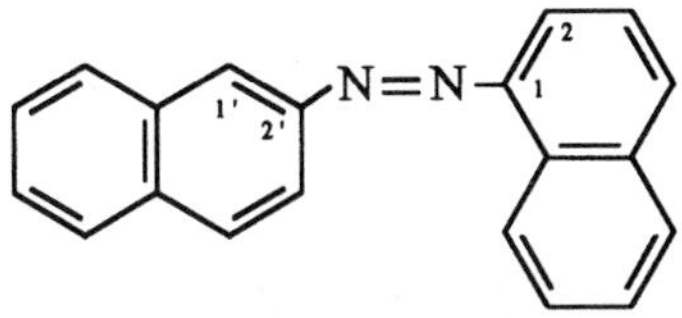

1,2'-Azonaphthalene
(1-Naphthyl)(2-naphthyl)diazene

15. 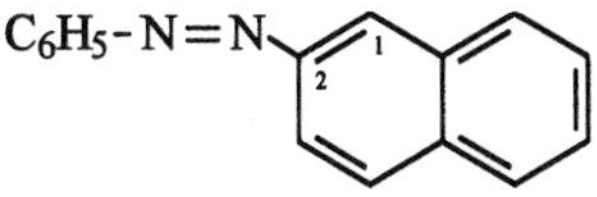 $C_6H_5-N=N$

Naphthalene-2-azobenzene
2-(Phenyldiazenyl)naphthalene
(2-Naphthyl)(phenyl)diazene

16.

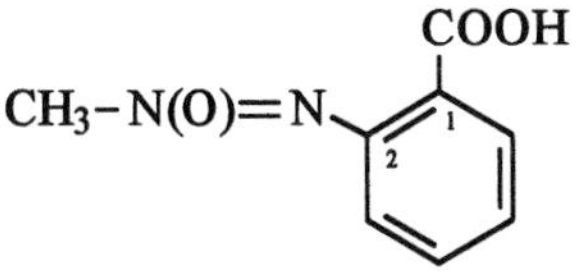

$CH_3-N(O)=N$

COOH

2-(Methyl-*ONN*-azoxy)benzoic acid

17. $C_6H_5-O-CO-CH_2-N=N(O)-CH_2-CO-O-CH_3$

Methyl [(2-oxo-2-phenoxyethyl)-*NNO*-azoxy]acetate
Phenyl [(2-methoxy-2-oxoethyl)-*ONN*-azoxy]acetate

18. $C_6H_5-N=N-CN$

Phenyldiazenecarbonitrile
Cyano(phenyl)diazene

19.

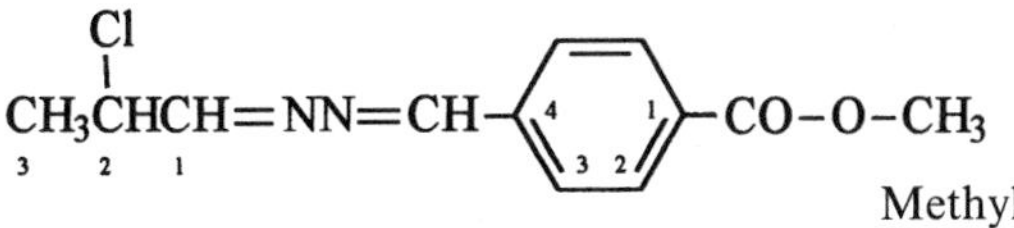

Cl
|
$CH_3CHCH=NN=CH-$⟨4 1 3 2⟩$-CO-O-CH_3$
3  2  1

Methyl 4-[[(2-chloropropylidene)-
hydrazono]methyl]benzoate

20.

NNH$-CH_2CH_3$
‖
$CH_3CH_2C-N=N-CH_3$
3     2    1

[1-(Ethylhydrazono)propyl](methyl)diazene
3,5-Diethyl-1-methylformazan

21.

CH$_3$
|
$C_6H_5-CH=NN=C-N=N-CH_2CH_3$
5   4    3    2    1

5-Benzylidene-1-ethyl-3-methylformazan
Benzaldehyde [1-(ethylazo)ethylidene]-
hydrazone

22.

CS$-CH_3$
|
$CH_3-NHN-CS-CH_3$
2      1

2-Methyl-1,1-bis(thioacetyl)hydrazine
1'-Ethanethioyl-2'-methylethane-
thiohydrazide
Ethanethioic 2-methyl-1-(1-thioxoethyl)-
hydrazide

23. $(CH_3)_2N-NCO$

*N*-Isocyanatodimethylamine
1,1-Dimethyl-2-(oxomethylidene)hydrazine
*N*-Isocyanato-*N*-methylmethanamine

24.

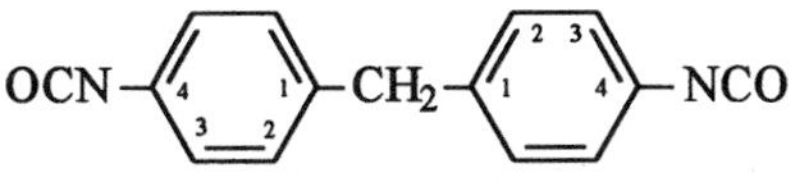

OCN$-$⟨4 1 3 2⟩$-CH_2-$⟨2 3 1 4⟩$-NCO$

1,1'-Methylenebis(4-isocyanatobenzene)
Bis(4-isocyanatophenyl)methane
4,4'-Methylenebis(phenyl isocyanate)
(*not* Diphenylmethane 4,4'-diisocyanate)

25.

NNH$_2$
‖
$H_2NN=N-C-NHNH_2$
3   2   1

Triaz-1-ene-1-carbohydrazonohydrazide

# 23

# Sulfur, Selenium, and Tellurium Acids and Their Derivatives

The nomenclature of sulfur acids (and their selenium and tellurium analogs) in which the sulfur atom is directly attached to an organic group is considered in this chapter (sulfur analogs of carboxylic acids are discussed in chapter 11). Typical derivatives of sulfur acids, such as esters, anhydrides, and amides will also be found here. Structural analogs of these classes with S or N replacing O are included, as are organic derivatives of inorganic sulfur acids such as sulfuric acid and sulfamic acid, and related selenium and tellurium acids.

Selenium and tellurium analogs of the sulfur compounds discussed in this chapter may be named by the same procedures as for sulfur compounds, with the forms selen(o) and tellur(o) replacing sulf(o) or thio. Tellurium compounds may also be named as organometallic compounds (see chapter 28).

In seniority for nomenclature purposes (see chapter 3, table 3.1), organosulfur acids rank just below the carboxylic acids in the decreasing order $R-SO_2-OH$ (sulfonic) > $R-SO-OH$ (sulfinic) > $R-S-OH$ (sulfenic).

## Acceptable Nomenclature

*Organosulfur acids* are named substitutively by appending one of the suffixes -sulfonic acid, $-SO_2-OH$, -sulfinic acid, $-SO-OH$, or -sulfenic acid, $-S-OH$, to the name of a parent hydride. This kind of nomenclature corresponds to that used in names such as cyclohexanecarboxylic acid; there is no counterpart to a suffix like -oic acid (see chapter 11). Sulfur acid suffixes can be attached at any point on a chain.

| | |
|---|---|
| $\underset{4\ \ \ 3\ \ \ 2\ \ \ 1}{CH_3CH_2\overset{\displaystyle SO_2-OH}{\overset{\displaystyle \vert}{C}HCH_3}}$ | Butane-2-sulfonic acid |
| $\underset{2\ \ 1}{H_2NNH-SO_2-OH}$ | Hydrazinesulfonic acid |
| 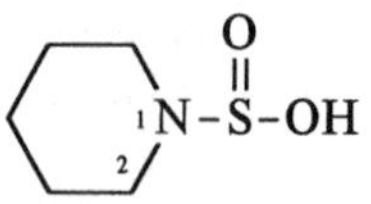 Piperidine-1-sulfinic acid | Piperidine-1-sulfinic acid |
| $CH_3-S-OH$ | Methanesulfenic acid |

The sulfenic acid suffix was recognized in the 1979 IUPAC rules, but not in the 1993 IUPAC recommendations; it continues to be used by CAS. An acceptable alternative name for $CH_3-S-OH$ is *S*-methyl thiohydroperoxide (see chapter 17). Names like "phenol-4-sulfonic acid" are discouraged; names with two characteristic group suffixes are not used in systematic organic nomenclature.

*Polybasic organosulfur acids* are named as above with appropriate locants and numerical prefixes. Where more than one acid type is present, the senior acid group is denoted by a suffix and the remaining acid groups by prefixes such as sulfo, $HO-SO_2-$, sulfino, $HO-SO-$, or sulfeno, $HO-S-$.

$$HO-SO_2-CH_2CH_2-SO_2-OH$$
$$\phantom{HO-SO_2-CH_2CH}{}_2\phantom{CH}{}_1$$

Ethane-1,2-disulfonic acid

$$HO-S-\langle\ \rangle-\overset{\overset{O}{\|}}{S}-OH$$

4-Sulfenobenzenesulfinic acid

Replacement of oxygen by sulfur in an organosulfur acid is denoted by prefixing thio, dithio, or trithio to the appropriate sulfur acid suffix or, preferably, by inserting the infix thio and so on preceded by an "o" for euphony, just ahead of the -ic ending of the suffix; further specificity is made by the use of *S*- or *O*-acid.

$$C_6H_5-SO_2-SH$$

Benzenesulfonothioic *S*-acid
Benzenethiosulfonic *S*-acid

$$CH_3CH_2CH_2-\overset{\overset{O}{\|}}{\underset{\underset{S}{\|}}{S}}-OH$$
$$\phantom{CH}{}_3\phantom{CH}{}_2\phantom{CH}{}_1$$

Propane-1-sulfonothioic *O*-acid
Propane-1-thiosulfonic *O*-acid

$$CH_3CH_2-S(S_2)-OH$$

Ethanesulfonodithioic *O*-acid
Ethanedithiosulfonic *O*-acid

A compound such as $CH_3-SSH$ can also be named as a hydrodisulfide or a disulfane; Se and Te analogs can be named as a hydrodiselenide or a diselane and as a hydroditelluride or ditellane (see chapter 17), or as a thiosulfenic acid or sulfenothioic acid.

Structures in which =O is replaced by =NH or =NNH$_2$ are named with the infixes imid(o) or hydrazon(o), respectively, placed just ahead of -ic acid in the suffix; these infixes have never been placed ahead of the sulfur acid suffix in the way that thio has been (see above). Replacement of –OH in an organosulfur acid by –OOH, –O–SH, –S–OH, or –SSH is similarly denoted by use of the appropriate peroxo infix (see chapter 3, table 3.9), as in ethanesulfonoperoxoic acid. The prefix peroxy can also be used, as in benzeneperoxysulfonic acid. Again, compounds of the type R–S–SSH can be named as hydrotrisulfides, dithioperoxysulfenic acids, or as sulfeno(dithioperoxoic) acids. Where a number of different replacements have occurred, the infixes are cited alphabetically.

$$C_6H_5-\overset{\overset{N-CH_3}{\|}}{S}-OH$$

*N*-Methylbenzenesulfinimidic acid

$$CH_3CH_2-\overset{\overset{NNH_2}{\|}}{\underset{\underset{S}{\|}}{S}}-OH$$

Ethanesulfonohydrazonothioic *O*-acid
Ethanethiosulfonohydrazonic *O*-acid

$$\langle\ \rangle-\overset{\overset{\overset{N'}{N-O-CH_3}}{\|}}{\underset{\underset{\underset{N}{N-CH_2CH_3}}{\|}}{S}}-SH$$

*N*-Ethyl-*N'*-methoxycyclohexanesulfonodiimidothioic acid
*N*-Ethyl-*N'*-methoxycyclohexanethiosulfonodiimidic acid

$$\overset{NH}{\langle\ \rangle}-S-SOH$$

1*H*-Pyrrole-2-sulfeno(thioperoxoic) *SO*-acid

SS$_2$–SSH

SS$_2$–SSH

Benzene-1,2-bis[sulfono(dithioperoxo)dithioic] acid
Benzene-1,2-bis[dithiosulfono(dithioperoxoic)] acid

Removal of an –OH group from an organosulfur acid creates an *organosulfur acyl group*, for example R–SO$_2$–, R–SO–, and R–S–, with the generic names alkanesulfonyl, alkanesulfinyl, and alkanesulfenyl, respectively, analogous in both form and nomenclature to a cyclohexanecarbonyl group generated from a cyclohexanecarboxylic acid (see chapter 11). The primary use of such acyl group names has been in functional class nomenclature, as in the naming of acid chlorides, but they are also be used for substituents. The corresponding additive prefixes are alkylsulfonyl, alkylsulfinyl, and alkylthio (not alkylmercapto), used in naming substituents.

$C_6H_5$–$SO_2$–Cl                                      Benzenesulfonyl chloride

$$C_6H_5-\overset{O}{\underset{\phantom{O}}{\overset{\|}{S}}}-\underset{4}{CH_2}\underset{3}{CH_2}\underset{2}{CH_2}-\underset{1}{COOH}$$

4-(Benzenesulfinyl)butanoic acid
4-(Phenylsulfinyl)butanoic acid

$$CH_3-\overset{N-CH_2CH_3}{\underset{\phantom{N}}{\overset{\|}{S}}}-N_3$$

*N*-Ethylmethanesulfinimidoyl azide

S–F                                      Cyclohexanesulfenyl fluoride
Cyclohexylfluorosulfane

S—⟨4 1 3 2⟩—COOH                 4-(Cyclohexylsulfanyl)benzoic acid
4-(Cyclohexylthio)benzoic acid

The abbreviated names mesyl, for methanesulfonyl or methylsulfonyl, and tosyl, for toluene-4-sulfonyl or 4-tolylsulfonyl, are acceptable, but should not be used in substituted form.

For compounds of the type R–SX$_n$, where $n$ is 3 or 5, substitutive names may be based on $\lambda^4$-sulfane, SH$_4$, or $\lambda^6$-sulfane, SH$_6$, leading to names such as pentafluoro(phenyl)-$\lambda^6$-sulfane for $C_6H_5$–SF$_5$.

*Anhydrides and thioanhydrides of organosulfur acids* are named in the same way as carboxylic anhydrides (see chapter 12). Cyclic anhydrides from symmetrical dibasic organosulfur acids can also be named in the usual way, but most cyclic anhydrides and imides of organosulfur acids are named as heterocyclic compounds.

$C_6H_5$–$SO_2$–O–$SO_2$–$C_6H_5$            Benzenesulfonic anhydride

$$C_6H_5-S-S-\overset{O}{\underset{\phantom{O}}{\overset{\|}{S}}}-CH_3$$

Benzenesulfenic methanesulfinic thioanhydride

SO$_2$
O
SO$_2$

Butane-1,4-disulfonic cyclic anhydride
2,2,7,7-Tetraoxo-1,2$\lambda^6$,7$\lambda^6$-oxadithiepane
1,2$\lambda^6$,7$\lambda^6$-Oxadithiepane 2,2,7,7-tetraoxide

$$\underset{4-1}{HO-SO_2-[CH_2]_4}-SO_2-O-SO_2-\underset{1-4}{[CH_2]_4-SO_2-OH}$$

4-Sulfobutanesulfonic 1,1′-monoanhydride
1,4-Butanedisulfonic bimol. monoanhydride

*Esters of sulfur acids*, including sulfuric, sulfurous, and sulfoxylic acids and their chalcogen analogs, in common with esters of other acids, are named by replacing the ending -ic (-ous) acid with -ate (-ite) and placing the name of the esterifying group first as a separate word with *O*, *S* or *N* specifiers as needed. Names of esterifying groups in polybasic acids are placed in alphabetical order, with number locants if required; hydrogen is similarly cited in the name of an acid ester of a polybasic acid.

$CH_3CH_2-O-SO_2-OH$          Ethyl hydrogen sulfate

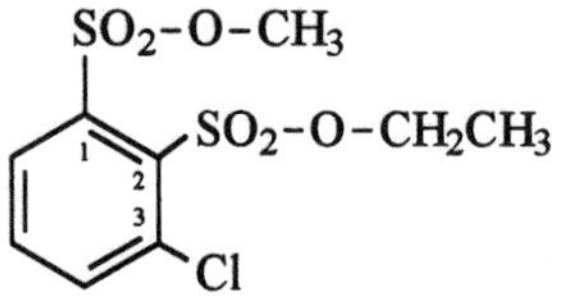

2-Ethyl 1-methyl 3-chlorobenzene-1,2-disulfonate

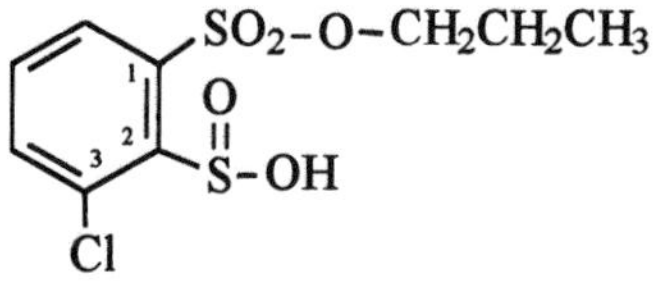

Propyl 3-chloro-2-sulfinobenzenesulfonate

*Intramolecular esters involving hydroxy sulfonic and hydroxy sulfinic acids* are best named as heterocyclic compounds. Alternatively, intramolecular esters of hydroxy sulfonic acids are known as *sultones* as a class, and they may be named in the same way as lactones (see chapter 12) with sultone describing the $-O-SO_2-$ group as part of a ring; this term is preceded by locants corresponding to the positions of the sulfonic acid group (lowest locant) and the hydroxy group in the parent hydride. Thus, 5-hydroxypentane-2-sulfonic acid, $HO-[CH_2]_3CH(CH_3)-SO_2-OH$, gives

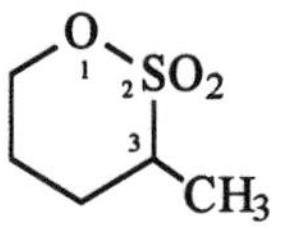

3-Methyl-1,2-oxathiane 2,2-dioxide
Pentane-2,5-sultone
3-Methyl-2,2-dioxo-1,2$\lambda^6$-oxathiane

*Salts* (see chapter 12) are named in the same way as esters, with the cation(s) cited first. Ionic substituents are denoted by prefixes such as sulfonato, $^-O-SO_2-$, sulfinato, $^-O-SO-$, and so on.

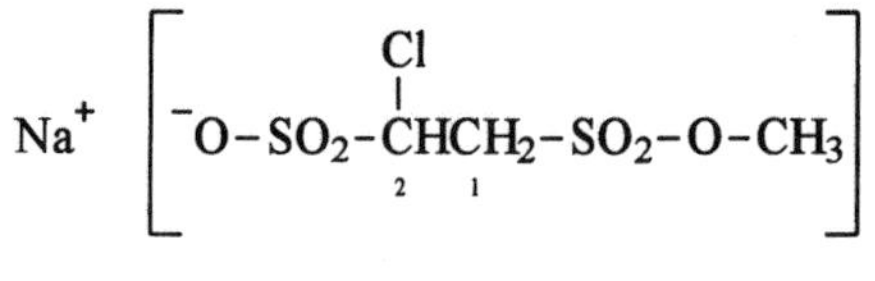

Sodium methyl 2-chloro-2-sulfonatoethanesulfonate
Sodium 2-methyl 1-chloroethane-1,2-disulfonate
Sodium 1-chloro-2-(methoxysulfonyl)-ethanesulfonate

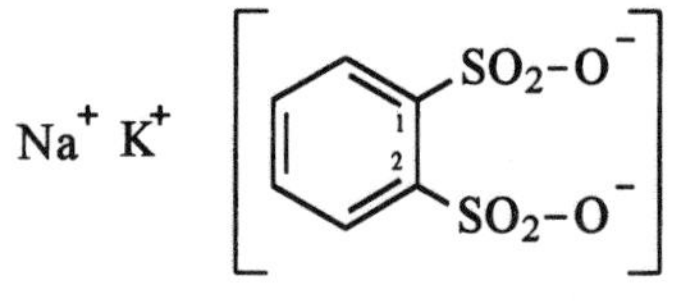

Potassium sodium benzene-1,2-disulfonate

*Amides and hydrazides of organosulfur acids* are compounds in which an $-OH$ group of the acid has been replaced by an $-NH_2$ or $-NHNH_2$ group. They are named by changing the -ic acid suffix to -amide, as in benzenesulfonamide, $C_6H_5-SO_2-NH_2$, or to -ohydrazide, as in ethanesulfinohydrazide, $CH_3CH_2-SO-NHNH_2$. This kind of name is also used where O has been replaced by S or an N group in the acid; the amide or hydrazide group is given lowest locants. Anilide, but not toluide, names may be formed (see chapter 18); amidine names are formed only from sulfinic acids.

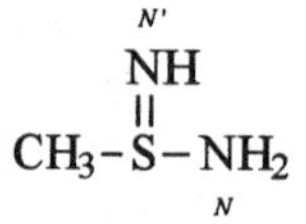

$$\underset{N}{\underset{\|}{CH_3-S}}\overset{\overset{N'}{NH}}{-NH_2}$$

Methanesulfinimidamide
Methanesulfinamidine

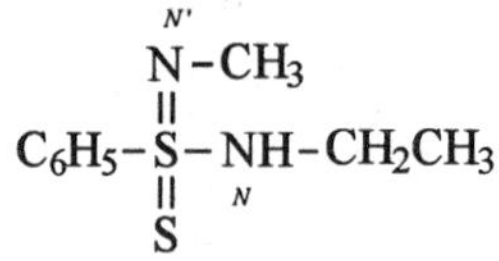

$$\underset{\underset{S}{\|}}{\underset{\|}{C_6H_5-S}}\overset{\overset{N'}{N-CH_3}}{\underset{N}{-NH-CH_2CH_3}}$$

N-Ethyl-N'-methylbenzenesulfonimidothioamide
N-Ethyl-N'-methylbenzenethiosulfonimidamide

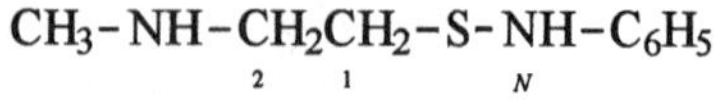

$$CH_3-NH-\underset{2}{CH_2}\underset{1}{CH_2}-S-\underset{N}{NH}-C_6H_5$$

2-(Methylamino)-N-phenylethanesulfenamide
2-(Methylamino)ethanesulfenanilide
S-[2-(Methylamino)ethyl]-N-phenyl-
  (thiohydroxylamine)

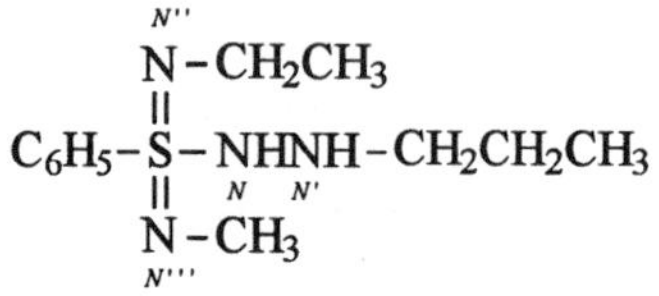

$$\underset{\underset{N'''}{\overset{\|}{N-CH_3}}}{\underset{\|}{C_6H_5-S}}\overset{\overset{N''}{N-CH_2CH_3}}{\underset{N\ N'}{-NHNH-CH_2CH_2CH_3}}$$

N''-Ethyl-N'''-methyl-N'-propyl-
  benzenesulfonodiimidohydrazide
N-Ethyl-N'-methylbenzenesulfonodiimidic
  2-propylhydrazide

Structures such as [R–SO₂]₂NH or [R–SO₂]₃N can be named as N-derivatives of a primary amide; substituting group names such as benzenesulfonyl or phenylsulfonyl are used.

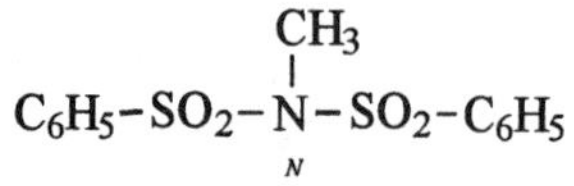

$$C_6H_5-SO_2-\underset{N}{\overset{\overset{CH_3}{|}}{N}}-SO_2-C_6H_5$$

N-(Benzenesulfonyl)-N-methylbenzenesulfonamide
N-Methyl-N-(phenylsulfonyl)benzenesulfonamide
N-Methyldibenzenesulfonamide

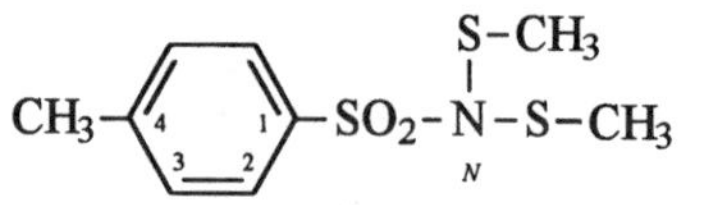

$$CH_3-\underset{3}{\overset{4}{\underset{2}{\overset{1}{\bigcirc}}}}-SO_2-\underset{N}{\overset{\overset{S-CH_3}{|}}{N}}-S-CH_3$$

4-Methyl-N,N-bis(methylthio)benzene-
  sulfonamide
(4-Methylbenzenesulfonyl)dimethane-
  sulfenamide

*Intramolecular amides of amino sulfonic acids*, like intramolecular esters, are best named as heterocyclic compounds. They can, however, also be named as *sultams*, with the principles used for lactams (see chapter 18).

The inorganic sulfur acids HO–SO₂–OH (sulfuric), HO–SO–OH (sulfurous), HO–S–OH (sulfoxylic), and their chalcogen analogs can serve as functional parents for organic esters and anhydrides and may be used with functional replacement nomenclature (see chapter 3). *Amides and imides* have been acceptably named with functional replacement prefixes and infixes.

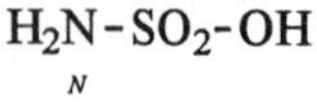

$$\underset{N}{H_2N-SO_2-OH}$$

Sulfuramidic acid
Sulfamic acid
Amidosulfuric acid

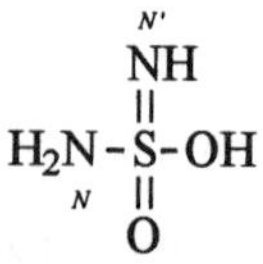

$$\underset{N}{H_2N}-\underset{\overset{\|}{O}}{\overset{\overset{N'}{NH}}{\underset{\|}{S}}}-OH$$

Sulfuramidimidic acid
Imidosulfamic acid
Amidoimidosulfuric acid

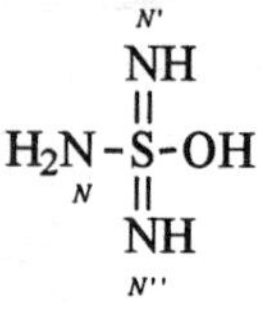

$$\underset{N}{H_2N}-\underset{\overset{\|}{\underset{N''}{NH}}}{\overset{\overset{N'}{NH}}{\underset{\|}{S}}}-OH$$

Sulfuramidodiimidic acid
Diimidosulfamic acid
Amidodiimidosulfuric acid

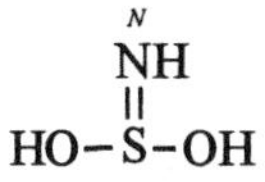

$$\underset{N}{\overset{\overset{N}{NH}}{\underset{\|}{HO-S-OH}}}$$

Sulfurimidous acid
Imidosulfurous acid

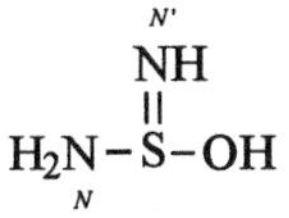

$$\underset{N}{\overset{\overset{N'}{NH}}{\underset{\|}{H_2N-S-OH}}}$$

Sulfuramidimidous acid
Sulfamimidous acid
Amidoimidosulfurous acid

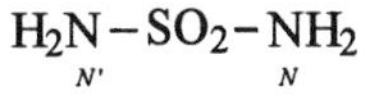

$$\underset{N' \qquad N}{H_2N-SO_2-NH_2}$$

Sulfuric diamide
Sulfuryl diamide
Sulfonyl diamide

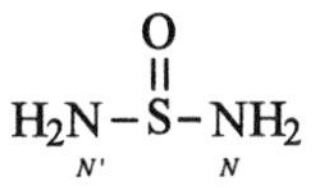

$$\underset{N' \qquad N}{\overset{\overset{O}{\|}}{H_2N-S-NH_2}}$$

Sulfurous diamide
Sulfinyl diamide
Thionyl diamide

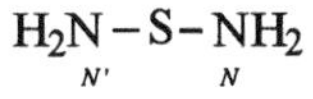

$$\underset{N' \qquad N}{H_2N-S-NH_2}$$

Sulfoxylic diamide

$$\underset{N \quad \| \quad N'}{\overset{\overset{N''}{\overset{NH}{\|}}}{\underset{NH}{\underset{N'''}{H_2N-S-NH_2}}}}$$

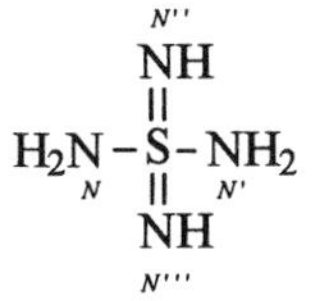

Sulfurodiimidic diamide
Diimidosulfuric diamide

Derivatives are named by substitutive nomenclature. If the parent is named by the prefix replacement method, substituents are included with the replacement prefix. In the infix method, substituents are cited in alphabetical order in front of the name of the parent with italic letter locants as needed.

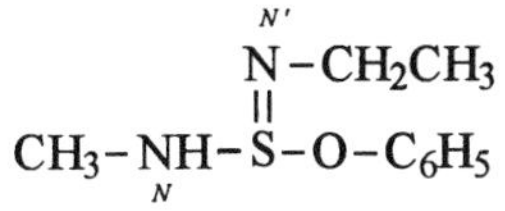

$$\underset{N}{\overset{\overset{N'}{N-CH_2CH_3}}{\underset{\|}{CH_3-NH-S-O-C_6H_5}}}$$

Phenyl $N'$-ethyl-$N$-methylsulfuramidimidite
Phenyl $N'$-ethyl-$N$-methylsulfamidimidite
Phenyl (ethylimido)(methylamido)sulfite

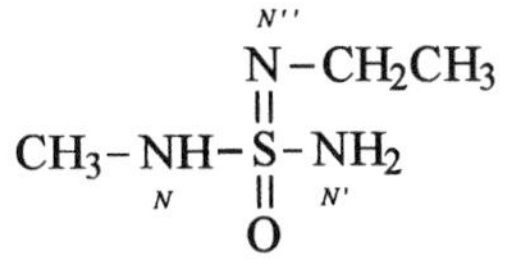

$$\underset{N \quad \| \quad N'}{\overset{\overset{N''}{N-CH_2CH_3}}{\underset{O}{CH_3-NH-S-NH_2}}}$$

$N''$-Ethyl-$N$-methylsulfurimidic diamide
(Ethylimido)sulfuric methyldiamide

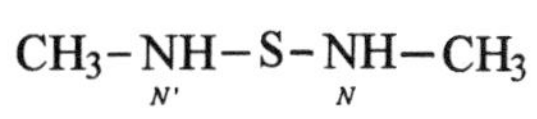

$$\underset{N' \qquad\qquad N}{CH_3-NH-S-NH-CH_3}$$

$N,N'$-Dimethylsulfoxylic diamide
Sulfoxylic bis(methylamide)

Regular substitutive prefix names for the groups R–SO$_2$–, R–SO–, and R–S– are formed from the names of the bivalent groups sulfonyl, –SO$_2$–, sulfinyl, –SO–, and thio (not "sulfenyl"), –S–. Their chalcogen analogs, thiosulfonyl or sulfonothioyl, –S(O) (S)–, dithiosulfonyl or sulfonodithioyl, –S(S)$_2$–, and thiosulfinyl or sulfinothioyl, –S (S)–, are also used. These prefixes, as well as the acyl group names such as benzenesulfonyl described earlier, are used in the presence of more senior functions. A selection is shown in table 23.1.

Prefixes for acid groups are based on bivalent group names or the names sulfo, HO–SO$_2$–, sulfino, HO–SO–, and sulfeno, HO–S–. The corresponding chalcogen replacement names formed from HO–SO$_2$– are given in table 23.2. Ester prefixes utilize the bivalent group names, as in ethoxysulfonyl or (methylthio)sulfinyl or (methylsulfanyl)sulfinyl.

## Discussion

Two of the groups in table 23.1, H$_2$N–SO– and H$_2$N–S–, were named "sulfinamoyl" and "sulfenamoyl", respectively, in the 1979 IUPAC rules. These are examples of improperly

Table 23.1. Examples of Prefixes for Sulfur-Containing Groups

| | |
|---|---|
| $C_6H_5-SO_2-$ | benzenesulfonyl<br>phenylsulfonyl |
| $\underset{\displaystyle CH_3CH_2-\overset{\displaystyle\overset{O}{\|}}{S}-}{}$ | ethanesulfinyl<br>ethylsulfinyl |
| $CH_3-S-$ | methanesulfenyl<br>methylthio<br>  (*not* methylsulfenyl or methylmercapto)<br>methylsulfanyl |
| $C_6H_5-\overset{\displaystyle\overset{O}{\|}}{S}-S-$ | benzenesulfinylthio<br>(phenylsulfinyl)thio |
| $C_6H_5-SO_2-NH-$ | benzenesulfonamido<br>(phenylsulfonyl)amino |
| $CH_3-\overset{\displaystyle\overset{N-C_6H_5}{\|}}{\underset{\displaystyle\underset{O}{\|}}{S}}-$ | *N*-phenylmethanesulfonimidoyl<br>*N*-phenyl-*S*-methylsulfonimidoyl |
| $H_2N-SO_2-$ | sulfamoyl<br>  (*not* sulfamyl)<br>sulfuramidoyl<br>aminosulfonyl |
| $H_2N-\overset{\displaystyle\overset{O}{\|}}{S}-$ | aminosulfinyl<br>  (*not* sulfinamoyl) |
| $H_2N-S-$ | aminothio<br>  (*not* sulfenamoyl) |
| $H_2NNH-\overset{\displaystyle\overset{O}{\|}}{S}-$ | hydrazinesulfinyl<br>hydrazinosulfinyl |
| $CH_3-S-NH-$ | methanesulfenamido<br>(methylthio)amino |

Table 23.2. Thio Replacement Names for Sulfo Prefixes

| | |
|---|---|
| $HO-\overset{\displaystyle\overset{O}{\|}}{\underset{\displaystyle\underset{S}{\|}}{S}}-$ | hydroxysulfonothioyl<br>hydroxy(thiosulfonyl) |
| $HS-SO_2-$ | mercaptosulfonyl<br>sulfanylsulfonyl |
| $HS_2O_2-$ (unspecified) | thiosulfo |
| $HO-S(S_2)-$ | hydroxysulfonodithioyl<br>hydroxy(dithiosulfonyl) |
| $HS-\overset{\displaystyle\overset{O}{\|}}{\underset{\displaystyle\underset{S}{\|}}{S}}-$ | mercaptosulfonothioyl<br>mercapto(thiosulfonyl) |
| $HS_3O-$ (unspecified) | dithiosulfo |
| $HS-S(S_2)-$ | trithiosulfo |

derived prefix names since there is no corresponding "sulfinamic acid" or "sulfenamic acid". Accordingly, they are not recommended here.

Nomenclature of sulfur compounds is replete with seemingly contradictory rules. The name of the group $C_6H_5–SO_2–$ is an example: should it be benzenesulfonyl or phenylsulfonyl? The former uses benzene as a parent hydride with H replaced by $–SO_2–$ (or $–SO_2–OH$ to give benzenesulfonic acid) in the same way that cyclohexane as a parent hydride can have H replaced by $–CO–$ (or $–COOH$ to give cyclohexanecarboxylic acid); "carboxylic acid", HCOOH, is really formic acid. Phenylsulfonyl appears to use the hypothetical sulfonic acid, $HSO_2–OH$, as parent with H replaced by a $C_6H_5–$ group, just as phosphonic acid, $HPO(OH)_2$ (see chapter 25), leads to phenylphosphonic acid and phenylphosphonyl. An advantage in using names such as benzenesulfonyl is that there is never a question concerning the group attached to the sulfur atom; such a question does arise when =O is replaced by =NH in sulfur-containing groups, since =NH can itself be substituted. The use of acyl group names for substituents is now allowed by IUPAC; benzenesulfonyl should be the preferred form.

Compounds of the type R–S–OH, R–O–SH, and R–SSH, as well as their alkyl derivatives can fall into a variety of classes, each providing an acceptable name. In this chapter, sulfenic acid names are stressed, but sulfide (see chapter 24) or peroxide and hydroperoxide (see chapter 17) names may serve equally well:

$$C_6H_5–S–S–CH_3$$

Methyl benzenesulfenothioate
Methyl phenyl disulfide
Methyl phenyl dithioperoxide
Methyl(phenyl)disulfane

Judicious use of hyphens or parentheses is advised in writing structures. For example, R–SS–R might also be seen as R–S(=S)–R′. This admonition also applies to other sulfur acids: compare R–SO–OH and R–S–OOH, and name accordingly.

Seniority among the variety of suffixes discussed in this chapter is often a problem. Generally, an acid group is senior to a nonacid group. One rule of thumb is that the group with sulfur in the highest oxidation state has the highest seniority, followed by the group with the largest number of oxygen atoms; S is senior to N when these atoms replace O.

A few trivial names are still acceptable for the structures covered here. Mesyl and tosyl (but not "tosylate" and so on) may be used in unsubstituted form. Sulfanilic acid (4-amino-benzenesulfonic acid), naphthionic acid (4-aminonaphthalene-1-sulfonic acid), and taurine (2-aminoethanesulfonic acid) are in use, but such names do little in terms of structure for the uninitiated. The same is true for triflic acid, $F_3C–SO_2–OH$. The best advice is not to use them.

Inorganic names such as sulfuryl and thionyl should be used only with inorganic compounds; wherever organic moieties are involved, the organic sulfonyl and sulfinyl are recommended.

## Additional Examples

1.

4-Chloropentane-2-sulfonic acid

2. $HO–S–CH_2CH_2CH_2–SO–OH$

3-Sulfenopropane-1-sulfinic acid
3-(Hydroxysulfanyl)-1-propanesulfinic acid

3.

5-(Hydroxysulfinothioyl)-3-sulfino-
benzenesulfonic acid

4.

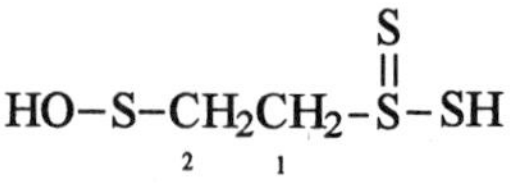

$$\underset{2}{HO-S}-\underset{1}{CH_2CH_2-\overset{\overset{\displaystyle S}{\|}}{S}-SH}$$

2-Sulfenoethanesulfinodithioic acid
2-(Hydroxysulfanyl)ethanesulfinodithioic acid

5.

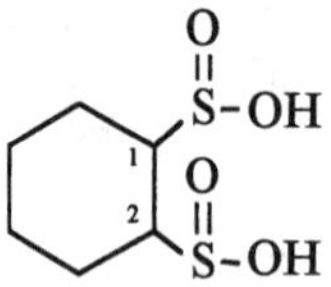

Cyclohexane-1,2-disulfinic acid

6.

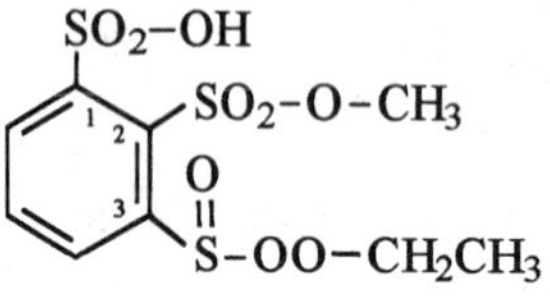

2-Methyl hydrogen 3-[(ethylperoxy)sulfinyl]-
  benzene-1,2-disulfonate (IUPAC)
Methyl 2-[(ethylperoxy)sulfinyl]-6-sulfo-
  benzenesulfonate
3-[(Ethylperoxy)sulfinyl]-2-(methoxysulfonyl)-
  benzenesulfonic acid
Ethyl 2-(methoxysulfonyl)-3-sulfo-
  benzenesulfinoperoxoate (CAS)

7.

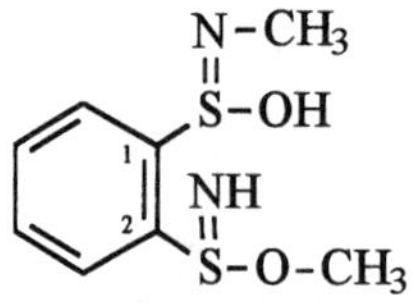

Methyl 2-(S-hydroxy-N-methylsulfinimidoyl)-
  benzenesulfinimidate
2-(S-Methoxysulfinimidoyl)-N-methylbenzene-
  sulfinimidic acid
2-Methyl hydrogen $N^1$-methyl-1,2-benzene-
  disulfinimidate

8.

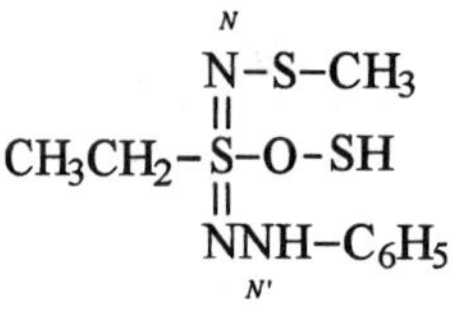

N-(Methylthio)-N'-phenylethanesulfono-
  hydrazonimido(thioperoxoic) OS-acid

9.

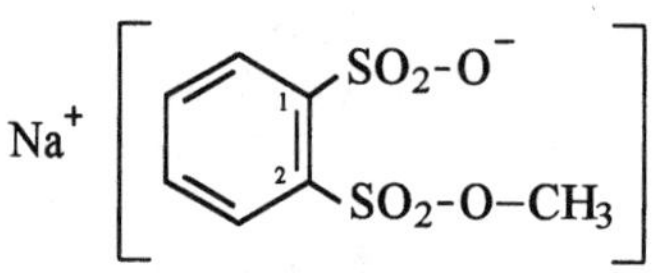

Sodium 2-(methoxysulfonyl)benzenesulfonate
Sodium methyl benzene-1,2-disulfonate

10.

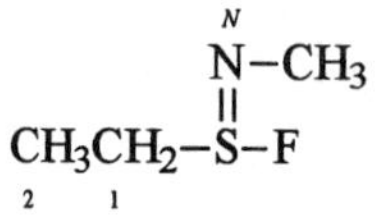

N-Methylethanesulfinimidoyl fluoride

11.

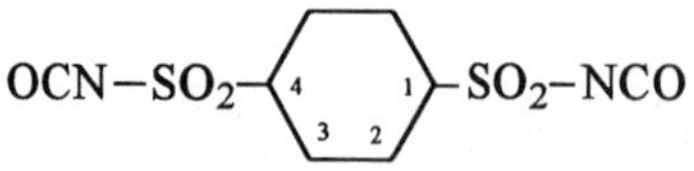

$$OCN-SO_2-\text{<ring>}-SO_2-NCO$$

Cyclohexane-1,4-disulfonyl diisocyanate

12.

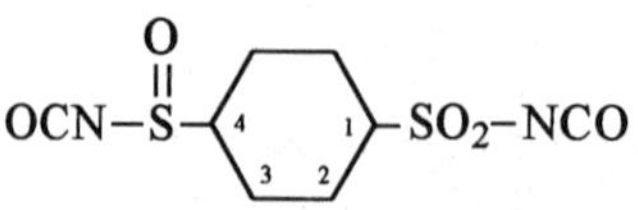

4-(Isocyanatosulfinyl)cyclohexanesulfonyl
  isocyanate

13.

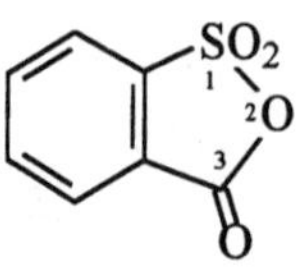

3H-2,1-Benzoxathiol-3-one 1,1-dioxide
2-Sulfobenzoic cyclic anhydride
1,1-Dioxo-3H-2,1$\lambda^6$-benzoxathiol-3-one

14.

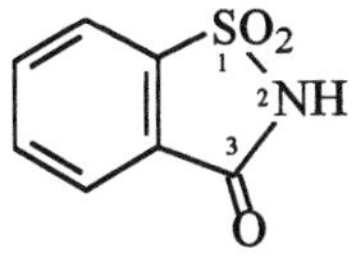

1,2-Benzoisothiazol-3(2$H$)-one 1,1-dioxide
2-Sulfobenzoic imide
1,1-Dioxo-2,1$\lambda^6$-benzoisothiazol-3(2$H$)-one

15.

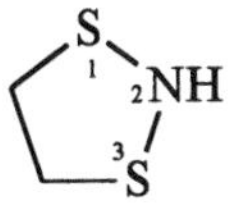

1,3,2-Dithiazolidine
1,2-Ethanedisulfenimide

16.

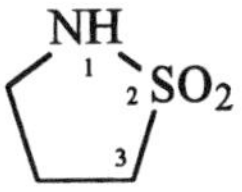

1,2-Thiazolidine 1,1-dioxide
Propane-1,3-sultam
1,1-Dioxo-1$\lambda^6$,2-thiazolidine

17.

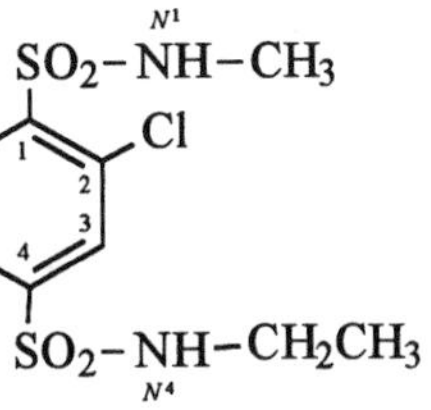

2-Chloro-$N^4$-ethyl-$N^1$-methylbenzene-1,4-disulfonamide

18.

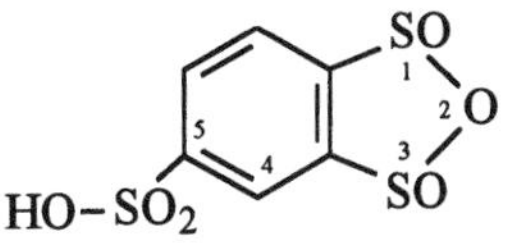

2,1,3-Benzoxadithiole-5-sulfonic acid 1,3-dioxide
4-Sulfobenzene-1,2-disulfinic cyclic 1,2-anhydride
1,3-Dioxo-5-sulfobenzo-2,1$\lambda^4$,3$\lambda^4$-oxadithiolane

19.

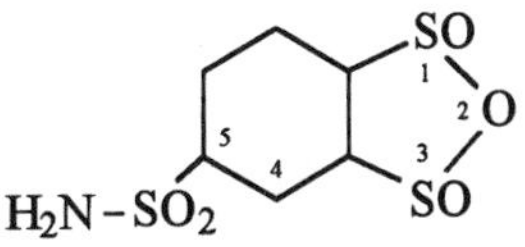

4-Sulfamoylcyclohexane-1,2-disulfinic cyclic anhydride
1,3,3a,4,5,6,7,7a-Octahydro-1,3-dioxo-5-sulfamoylbenzo-2,1$\lambda^4$,3$\lambda^4$-oxadithiolane

20.

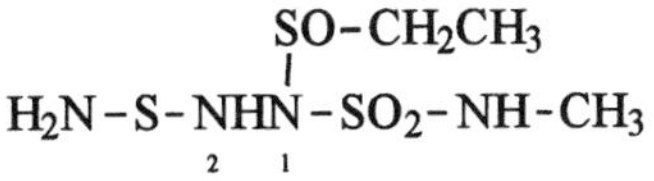

2-(Aminothio)-1-(ethylsulfinyl)-1-(methylsulfamoyl)hydrazine
$N'$-(Aminosulfanyl)-$N$-(methylsulfamoyl)-ethanesulfinohydrazide
$N$-Methylsulfamic 2-(aminothio)-1-(ethylsulfinyl)hydrazide
2-(Aminosulfanyl)-1-(ethylsulfinyl)-$N$-methylhydrazinesulfonamide

21.

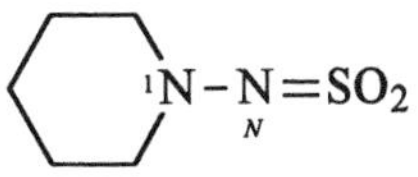

$N$-Sulfonylpiperidin-1-amine
1,1-Dioxo-$N$-piperidin-1-yl-$\lambda^6$-sulfanimine
$N$-(Dioxo-$\lambda^6$-sulfanylidene)piperidin-1-amine

# 24

# Thiols, Sulfides, Sulfoxides, Sulfones, and Their Chalcogen Analogs

The nomenclature of organosulfur compounds having structures such as R–SH, R–S–R′, R–SO–R′, and R–SO$_2$–R′ is covered in this chapter. Related structure classes, such as *thioacetals*, R$_2$C(SR′)$_2$; *disulfides*, R–SS–R′, and *polysulfides*, R–S$_n$-R′; *sulfimides*, R$_2$S=NH, and *sulfoximides*, R$_2$S(=O)(=NH); and certain cyclic systems containing sulfur are also included. Selenium and tellurium analogs of these compounds are generally named in the same way as the sulfur derivatives.

## Acceptable Nomenclature

In seniority (see chapter 3, table 3.1), thiols rank just below alcohols, but the remaining classes of compounds discussed in this chapter are near the bottom of the list. As a consequence, except for thiols, the names of most of these sulfur-containing groups are encountered in the form of prefixes in substitutive names. Functional class names (see chapter 3), such as alkyl hydrosulfide, dialkyl sulfide, dialkyl sulfoxide, and dialkyl sulfone, may also be used.

*Thiols* are generically named as *alkanethiols* or *arenethiols*. The term "mercaptan" is no longer acceptable. Certain trivial names for simple compounds such as thiophenol are often used, but not recommended. Salts of thiols are named in the same way as salts of alcohols and phenols (see chapter 15); sulfido is the prefix for ⁻S–.

$C_6H_5-CH_2-SH$

Phenylmethanethiol
$\alpha$-Toluenethiol
Benzenemethanethiol

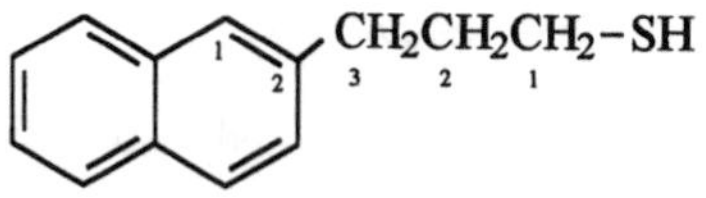

$CH_2CH_2CH_2-SH$

3-(2-Naphthyl)propane-1-thiol
2-Naphthalenepropanethiol

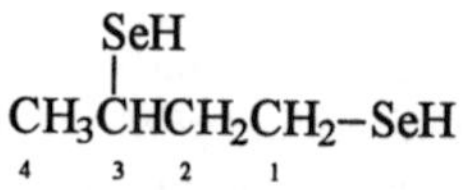

$\underset{CH_3CHCH_2CH_2-SeH}{\overset{SeH}{|}}$

Butane-1,3-diselenol

$Na^+ \left[ CH_3CH_2-S^- \right]$

Sodium ethanethiolate
Sodium ethyl sulfide

The functional class ending for thiols is hydrosulfide. Two or more sulfur atoms are denoted by inserting between hydro and sulfide a number prefix such as di-, tri-, and so on; if the number of sulfur atoms is unknown, poly- is used. An alternative nomenclature proposed by IUPAC for linear chains of sulfur atoms terminated with hydrogen atoms is based on the name sulfane, giving names such as methyldisulfane for $CH_3$–S–SH. The structure R–S–OH is best named as a sulfenic acid (see chapter 23), and R–O–SH is named as an *O*-alkyl thiohydroperoxide (see chapter 17). Prefix names for some of these groups are dithiohydroperoxy, disulfanyl (IUPAC) or thiosulfeno (CAS) for HS–S–; sulfeno for HO–S–; and mercaptooxy (CAS) or sulfanyloxy (IUPAC) for HS–O–; the name "thiohydroperoxy" alone does not specify the atom with the free valence and should not be used. The corresponding substituted forms are alkyldithio (CAS) and alkyldisulfanyl (IUPAC) for R–S–S–; alkoxythio (CAS) or alkoxysulfanyl (IUPAC) for R–O–S–; and (alkylthio)oxy (CAS) or (alkylsulfanyl)-oxy (IUPAC) for R–S–O–.

$$CH_3$$
$$|$$
$$CH_3\overset{3}{C}H\overset{2}{C}H_2-\overset{1}{S}SH$$

Isobutyl hydrodisulfide
Isobutyldisulfane
2-Methylpropane-1-sulfenothioic acid

*Sulfides*, R–S–R′, are named in the same way as ethers (see chapter 16): (a) through the use of the class name sulfide in place of ether ("thioether" is a class name but is not used in names themselves); (b) in assemblies of identical units joined through an S atom with thio or sulfanediyl for –S– in place of oxy; or (c) substitutively, by employing the generic prefix alkylthio (CAS) or alkylsulfanyl (IUPAC) for RS– ("alkylmercapto" is not used) in place of alkoxy. Numerical prefixes such as di-, tri-, and so on, preceding "sulf" or "thio" are employed in naming *polysulfides*, R–S$_n$–R′; if the sulfur chain is linear, names based on "sulfane" can be used, as in dimethyltetrasulfane for $CH_3$–SSSS–$CH_3$. Replacement nomenclature (see chapter 3) may be appropriate for hydrocarbon chains in which many hetero atoms (including S, denoted by thia) have taken the place of carbon atoms.

$$CH_3CH_2-S-C_6H_5$$

Ethyl phenyl sulfide
(Ethylthio)benzene

$$HOOC-\overset{3'}{C}H_2\overset{3}{C}H_2-S-\overset{2}{C}H_2\overset{1}{C}H_2-COOH$$

3,3′-Thiodipropanoic acid

$$CH_3-S-\langle\rangle-SSS-\langle\rangle-S-CH_3$$

Bis[4-(methylsulfanyl)phenyltrisulfane
Bis[4-(methylthio)phenyl] trisulfide

$$CH_3-SSS-\langle\rangle-SH$$

4-(Methyltrisulfanyl)benzenethiol
4-(Methyltrithio)benzenethiol
4-Mercaptophenyl methyl trisulfide

$$CH_3\overset{15}{C}H_2-S-\overset{13}{C}H_2\overset{12}{C}H_2-S-\overset{10}{C}H_2\overset{9}{C}H_2-S-\overset{7}{C}H_2-S-\overset{5}{C}H_2\overset{4}{C}H_2-S-\overset{1}{C}H_3$$

2,5,7,10,13-Pentathiapentadecane

*Thioacetals* and *thiohemiacetals* are class names for R–CH(SR)$_2$, R–CH(SR)–OR, R$_2$C(SR)$_2$, R$_2$C(SR)–OR, R–CH(SR)–OH, R–CH(OR)–SH, R$_2$C(SR)–OH, R$_2$C(SR)–SH, and so on. They are named substitutively or by using monothioacetal, dithioketal, monothiohemiacetal, dithiohemiketal, and so on, as class names. The class names "mercaptal" and "mercaptole" are no longer recommended.

$$CH_3-S \quad S-CH_2CH_3$$
$$\overset{3}{C}H_3\overset{2}{C}\overset{1}{C}H_3$$

2-(Ethylthio)-2-(methylthio)propane
Propan-2-one ethyl methyl dithioketal

$$O-CH_2CH_3$$
$$|$$
$$CH_3CH_2CH_2CH_2CH-SH$$
$$\overset{5}{\phantom{C}}\overset{4}{\phantom{C}}\overset{3}{\phantom{C}}\overset{2}{\phantom{C}}\overset{1}{\phantom{C}}$$

1-Ethoxypentane-1-thiol
Pentanal *O*-ethyl monothiohemiacetal

*Cyclic sulfides* should be named as heterocyclic compounds (see chapter 9). A sulfur atom that occurs as a bridge between adjacent carbon atoms in a chain or as a bridge in a ring may be denoted by epithio in the same way as epoxy (see chapter 16).

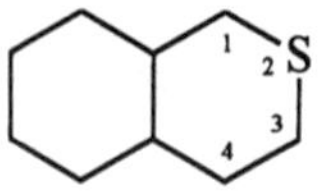

Octahydro-1*H*-2-benzothiopyran

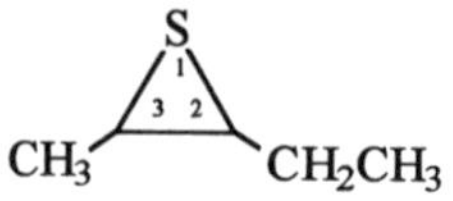

2-Ethyl-3-methylthiirane
2,3-Epithiopentane

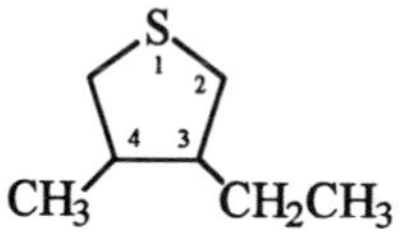

3-Ethyltetrahydro-4-methylthiophene
3-Ethyl-4-methylthiolane

7-Thiabicyclo[2.2.1]heptane

*Sulfoxides*, R–SO–R′, and *sulfones*, R–SO$_2$–R′, may be named through the corresponding class name or substitutively by use of the prefixes sulfinyl, –SO–, and sulfonyl, –SO$_2$–. Chains of –SO– or –SO$_2$– groups take terms such as disulfoxide, trisulfone, and so forth. Such compounds may also be named as oxides or as oxo derivatives of the appropriate polysulfane by means of the λ-convention (see chapter 2), in which tetravalent and hexavalent sulfur atoms are designated by λ$^4$ and λ$^6$, respectively.

$$CH_3CH_2CH_2-SO-CH_3$$
$$\phantom{CH_3CH_2CH_2}{}_3\phantom{-SO-C}{}_2\phantom{-S}{}_1$$

1-(Methylsulfinyl)propane
Methyl propyl sulfoxide

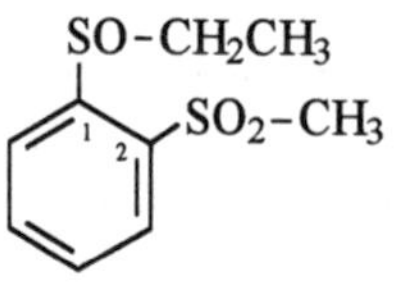

1-(Ethylsulfinyl)-2-(methylsulfonyl)benzene
2-(Ethanesulfinyl)phenyl methyl sulfone
Ethyl 2-(methanesulfonyl)phenyl sulfoxide

$$C_6H_5-SO_2-C_6H_5$$

Diphenyl sulfone
Sulfonylbis[benzene]
Benzenesulfonylbenzene

$$C_6H_5-SO-SO_2-CH_3$$
$$\phantom{C_6H_5-SO}{}_2\phantom{-S}{}_1$$

1-Methyl-2-phenyldisulfane 1,1,2-trioxide
1-Methyl-1,1,2-trioxo-2-phenyl-1λ$^6$,2λ$^4$-disulfane
[(Methanesulfonyl)sulfinyl]benzene

In ring systems, –SO–, –SO$_2$–, –S(S)–, –S(S)(O)–, and –S(S)$_2$– are denoted by adding oxide or dioxide (or sulfide, disulfide), with locants, to the name of the heterocycle or by using the prefix oxo, for O=, and the name of the heterocycle.

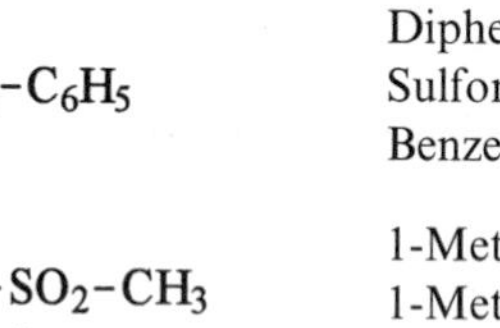

Thianthrene 5,5-dioxide
5,5-Dioxo-5λ$^6$-thianthrene

Replacement of =O by =S or =NH is shown by the prefixes thioxo or imino in place of oxo with the appropriate λ name. The hypothetical sulfimide, H$_2$S=NH, and sulfoximide, H$_2$S(=O)(=NH) (CAS calls these sulfilimine and sulfoximine, respectively, and, unfortunately, uses sulfimide for $SO_2(=NH)$), can be employed as substitutable parent structures (see Discussion).

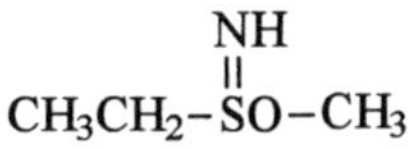

$$\overset{\displaystyle NH}{\underset{\displaystyle \phantom{x}}{\overset{\displaystyle \|}{CH_3CH_2-SO-CH_3}}}$$

S-Ethyl-S-methylsulfoximide (IUPAC)
S-Ethyl-S-methylsulfoximine (CAS)

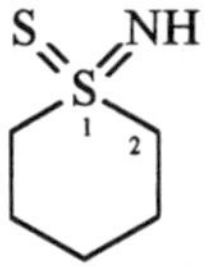

Hexahydro-1-imino-1-thioxo-$1\lambda^6$-thiopyran
Hexahydro-1-thioxo-$1\lambda^6$-thiopyran-1-imine
Tetrahydro-$1\lambda^4$-thiopyran-1(2H)-imine 1-sulfide

Structures such as R–CH=SO (*sulfines*), $R_2C=SO_2$ (*sulfenes*), and so on, are named as oxides or dioxides of thioaldehydes or thioketones (see chapters 13 and 14); as derivatives of $\lambda^4$- or $\lambda^6$-sulfanes or substitutively by using the prefixes sulfinyl, for OS= and sulfonyl, for $O_2S=$.

$CH_3CH=SO$

Ethanethial S-oxide
Ethylideneoxo-$\lambda^4$-sulfane
Sulfinylethane
(Oxo-$\lambda^4$-sulfanylidene)ethane

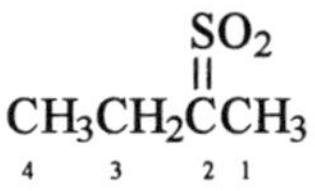

$$\overset{\displaystyle SO_2}{\overset{\displaystyle \|}{\underset{\displaystyle 4 \quad 3 \quad 2 \ 1}{CH_3CH_2CCH_3}}}$$

2-Butanethione S,S-dioxide
Butan-2-ylidenedioxo-$\lambda^6$-sulfane
2-Sulfonylbutane
2-(Dioxo-$\lambda^6$-sulfanylidene)butane

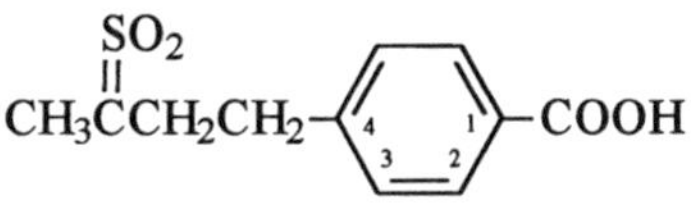

$$\overset{\displaystyle SO_2}{\overset{\displaystyle \|}{CH_3CCH_2CH_2-}}\!\!\!\text{⬡}\!\!\!-COOH$$

4-(3-Thioxobutyl)benzoic acid S,S-dioxide
4-(3-Sulfonylbutyl)benzoic acid
4-[3-(Dioxo-$\lambda^6$-sulfanylidene)butyl]benzoic acid

*Selenium and tellurium analogs* of the compounds in this chapter are generally named by replacing "sulf" or "thi" with "selen" or "tellur". However, in some cases, abbreviated forms are used, including selane (tellane), polyselane (polytellane), selanyl (tellanyl), and -selone (-tellone).

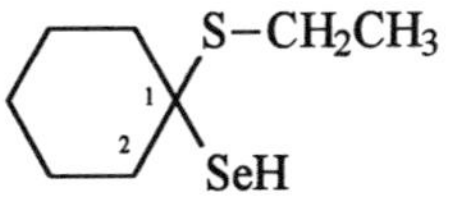

1-(Ethylsulfanyl)cyclohexane-1-selenol
Cyclohexanone S-ethyl selenothiohemiketal

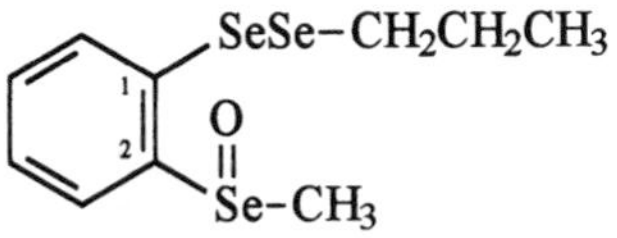

[2-(Methylseleninyl)phenyl](propyl)diselane
Methyl 2-(propyldiselanyl)phenyl selenoxide

## Discussion

In some areas of sulfur nomenclature, there is confusion and sometimes conflict among the various methods. One especially confused area is that where =NH replaces =O attached to sulfur; IUPAC and CAS use different names for both $H_2S=NH$ and $H_2S(=O)(=NH)$ and in one instance use the same name (sulfimide) for different compounds. The $\lambda$-convention (see chapter 2) may provide a reasonable alternative.

$CH_3-SH=NH$

S-Methylsulfimide (IUPAC)
S-Methylsulfilimine (CAS)
1-Methyl-$\lambda^4$-sulfanimine
Methyl(imino)-$\lambda^4$-sulfane

$$(CH_3)_2\overset{\overset{O}{\|}}{S}=N-CH_2CH_3$$

N-Ethyl-S,S-dimethylsulfoximide (IUPAC)
N-Ethyl-S,S-dimethylsulfoximine (CAS)
N-Ethyl-1,1-dimethyl-$\lambda^4$-sulfanimine S-oxide
N-Ethyl-1,1-dimethyl-1-oxo-$\lambda^6$-sulfanimine
(Ethylimino)dimethyloxo-$\lambda^6$-sulfane

$C_6H_5-N=SO_2$

1,1-Dioxo-N-phenyl-$\lambda^6$-sulfanimine
Dioxo(phenylimino)-$\lambda^6$-sulfane
N-Phenylsulfanimine S,S-dioxide

$CH_3-N=SO$

N-Methyl-1-oxo-$\lambda^4$-sulfanimine
(Methylimino)oxo-$\lambda^4$-sulfane
N-Phenylsulfanimine S-oxide

Dithiohydroperoxides (hydrodisulfides) and hydrotrisulfides are named as sulfenic acids by CAS: $CH_3CH_2-S-SH$ is ethanesulfenothioic acid and $C_6H_5-S-SSH$ is benzenesulfeno(dithioperoxoic) acid; but esters are named as di- and trisulfides, which seems inconsistent. IUPAC does not recognize "sulfenic acid" in its 1993 recommendations.

The 1993 IUPAC recommendations have placed greater emphasis on sulfane and related terms for substitutable parent structures. Several examples are included in the following section. This approach affords systematic but unfamiliar names; further development is likely before acceptance is achieved. The names can sometimes fill a nomenclatural vacuum, particularly when used with the $\lambda$-convention; for example, $(CH_3)_2SH_2$ would be named dimethyl-$\lambda^4$-sulfane. Such compounds are named by CAS as derivatives of sulfur by coordination nomenclature principles.

### Additional Examples

1.

   OH
   [structure: benzene ring with OH at position 1 and SH at position 2]
   SH

   2-Mercaptophenol
   2-Sulfanylphenol

2.

   SSH
   [structure: benzene ring with SSH at position 1, OSH at position 2, Cl at position 3]
   OSH
   Cl

   3-Chloro-2-(mercaptooxy)phenyl hydrodisulfide
   [3-Chloro-2-(sulfanyloxy)phenyl]disulfane
   3-Chloro-2-(mercaptooxy)benzenesulfenothioic acid
   O-(2-Chloro-6-dithiohydroperoxyphenyl)
      thiohydroperoxide

3.

   $$\underset{2}{\underset{\|}{CH_3\overset{OH}{\overset{|}{CH}}}}-S-\underset{}{\overset{CH_3}{\overset{|}{CH}}}-S-\underset{1}{\overset{SH}{\overset{|}{CHCH_3}}}$$

   1-[[1-[(1-Mercaptoethyl)thio]ethyl]thio]ethanol
   Ethanal 1-hydroxyethyl 1-mercaptoethyl dithioacetal
   1-[[1-[(1-Sulfanylethyl)sulfanyl]ethyl]-
      sulfanyl]ethanol

4.

   $Na^+ \left[ CH_3-O-CO-CH_2-S^- \right]$

   Sodium 2-methoxy-2-oxoethanethiolate
   Sodium methyl sulfidoacetate

5.

   $Na^+ \left[ C_6H_5-SSS^- \right]$

   Sodium benzenesulfeno(dithioperoxoate)
   Phenyl sodium trisulfide
   Sodium phenyltrisulfanide

6. $CH_3-SS-Cl$

Chloro methyl disulfide
Chloro(methyl)disulfane

7. $Cl-CH_2-S-CH_2CH_3$

Chloromethyl ethyl sulfide
[(Chloromethyl)thio]ethane

8.

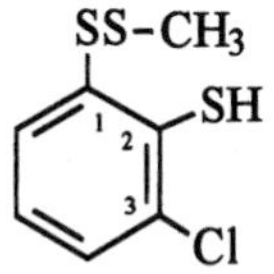

Methyl 2-mercapto-3-chlorobenzenesulfenothioate
2-Chloro-6-(methyldisulfanyl)benzenethiol
Methyl 3-chloro-2-mercaptophenyl disulfide

9. $HSS-CH_2CH_2CH_2CH_2-SSH$

Butane-1,4-diyl bis(hydrodisulfide)
Butane-1,4-disulfenothioic acid
Butane-1,4-diylbis(disulfane)

10.

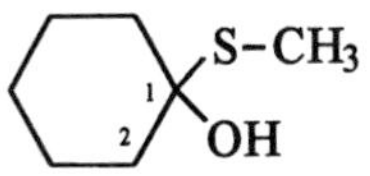

1-Hydroxycyclohexyl methyl sulfide
1-(Methylthio)cyclohexanol
Cyclohexanone $S$-methyl monothiohemiketal

11. 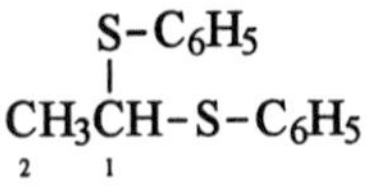 $CH_3CH-S-C_6H_5$ with $S-C_6H_5$

1,1-Bis(phenylthio)ethane
Acetaldehyde diphenyl dithioacetal
[Ethane-1,1-diylbis(thio)]dibenzene

12. $CH_3-SO-S-C_6H_5$

[(Methylsulfinyl)thio]benzene
$S$-Phenyl methanesulfinothioate
1-Methyl-2-phenyldisulfane 1-oxide
1-Methyl-1-oxo-2-phenyl-1$\lambda^4$-disulfane

13.

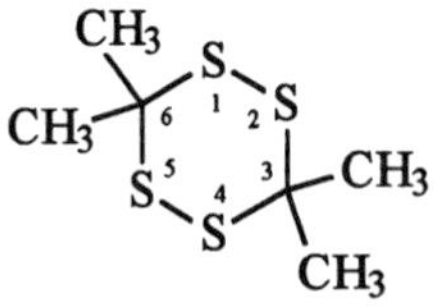

3,3,6,6-Tetramethyl-1,2,4,5-tetrathiane

14.

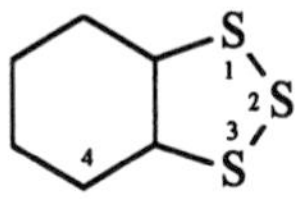

Hexahydrobenzotrithiole

15.

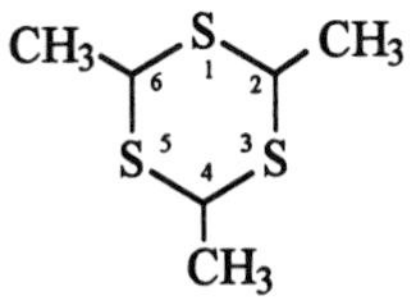

2,4,6-Trimethyl-1,3,5-trithiane

16.

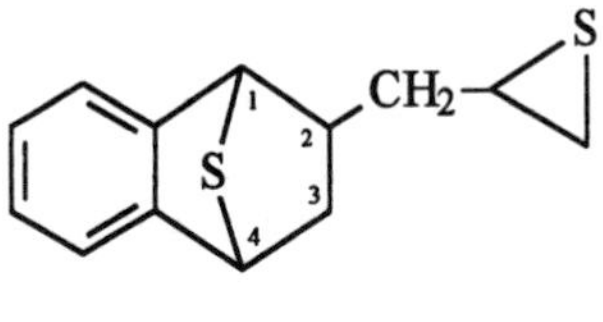

1,2,3,4-Tetrahydro-2-(thiiranylmethyl)-1,4-
    epithionaphthalene
2-(2,3-Epithiopropyl)-1,2,3,4-tetrahydro-1,4-
    epithionaphthalene

17.

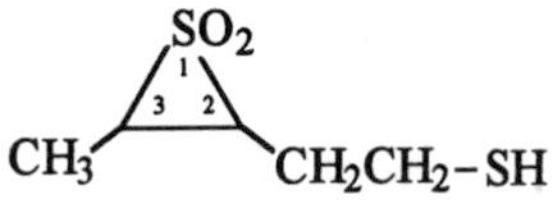

3-Methyl-2-thiiraneethanethiol 1,1-dioxide
2-(3-Methylthiiran-2-yl)ethanethiol 1,1-dioxide
3-Methyl-1,1-dioxo-1$\lambda^6$-thiirane-2-ethanethiol

18. 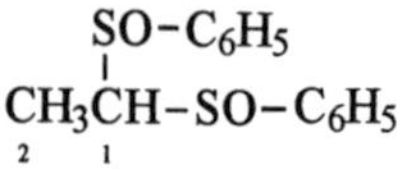    $SO-C_6H_5$
       |
       $CH_3CH-SO-C_6H_5$
        2    1

1,1-Bis(phenylsulfinyl)ethane
Ethane-1,1-diylbis(phenyl sulfoxide)
[Ethane-1,1-diylbis(sulfinyl)]benzene

19.

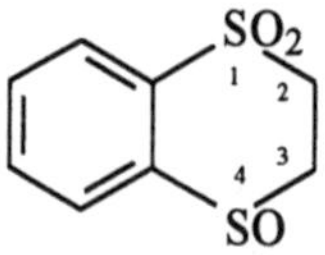

2,3-Dihydro-1,4-benzodithiin 1,1,4-trioxide
1,1,4-Trioxo-1,2,3,4-tetrahydro-1$\lambda^6$,4$\lambda^4$-benzodithiin

20.

$C_6H_5-SO_2-SO_2-CH_3$

Methyl phenyl disulfone
(Methyldisulfonyl)benzene
1-Methyl-1,1,2,2-tetraoxo-2- phenyl-1$\lambda^6$,2$\lambda^6$-
   disulfane

21.

$CH_3-SO-SO_2-S-CH_2-Br$

Bromo[[(methylsulfinyl)sulfonyl]thio]methane
1-(Bromomethyl)-3-methyltrisulfane 2,2,3-trioxide
3-(Bromomethyl)-1-methyl-1,2,2-trioxo-1$\lambda^4$,2$\lambda^6$-
   trisulfane

22. 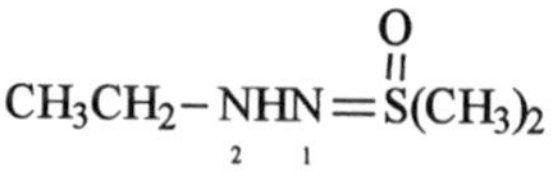              O
                ||
$CH_3CH_2-NHN=S(CH_3)_2$
        2    1

1-(Dimethyloxo-$\lambda^6$-sulfanylidene)-2-ethylhydrazine
1-(Dimethyl-$\lambda^4$-sulfanylidene)-2-ethylhydrazine
   $S$-oxide

23. 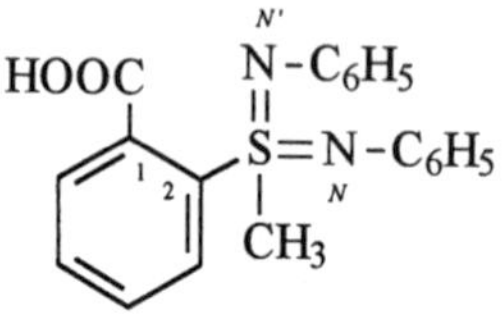
              $N'$
        $N-C_6H_5$
HOOC    ||
        $S=N-C_6H_5$
        |      $N$
        $CH_3$

2-[Methylbis(phenylimino)-$\lambda^6$-sulfanyl]benzoic acid
1-(2-Carboxyphenyl)-1-methyl-$N,N'$-diphenyl-$\lambda^6$-
   sulfanediimine

24.

$CH_3-SO_2-N=S(CH_2CH_3)_2$

$N$-(Diethyl-$\lambda^4$-sulfanylidene)methanesulfonamide
1,1-Diethyl-$N$-(methylsulfonyl)-$\lambda^4$-sulfanimine
$S,S$-Diethyl-$N$-(methylsulfonyl)sulfimide (IUPAC)
$S,S$-Diethyl-$N$-(methylsulfonyl)sulfilimine (CAS)

25.

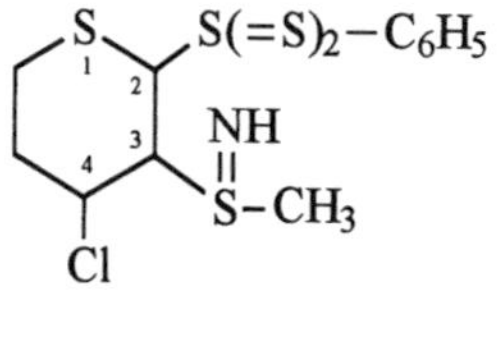

2-Benzenesulfonodithioyl-4-chlorotetrahydro-3-
   (methanesulfinimidoyl)-2$H$-thiopyran
$S$-[2-(Benzenesulfonodithioyl)chloro-
   tetrahydro-2$H$-thiopyran-3-yl]methyl-
   sulfimide (IUPAC)
4-Chlorotetrahydro-$S$-[2-(phenylsulfonodithioyl)-
   2$H$-thiopyran-3-yl]-$S$-methylsulfilimine (CAS)

26. 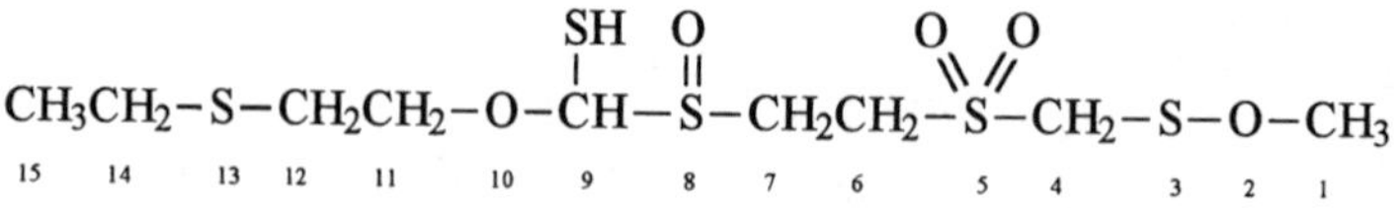                SH   O          O  O
                   |    ||         \\ //
$CH_3CH_2-S-CH_2CH_2-O-CH-S-CH_2CH_2-S-CH_2-S-O-CH_3$
 15   14    13   12   11  10  9    8   7   6   5   4   3   2   1

5,5,8-Trioxo-2,10-dioxa-3,5$\lambda^6$,8$\lambda^4$,13-tetrathiapentadecane-9-thiol (IUPAC)
Methyl 6-mercapto-2,2,5-trioxo-7-oxa-2$\lambda^6$,5$\lambda^4$,10-trithiadodecanesulfenate
Methyl 6-mercapto-7-oxa-2,5,10-trithiadodecanesulfenate 2,2,5-trioxide (CAS)

# 25

# Phosphorus and Arsenic Compounds

Phosphorus and arsenic compounds are sufficiently similar that, in general, the nomenclature of each applies to the other. Both the nomenclature and the chemistry of phosphorus compounds developed independently in the fields of organic, inorganic, and biochemistry. Today, at the crossroads of those fields, a phosphorus compound may await its fate under many names. The IUPAC organic rules list some seven ways to name phosphorus compounds, and other methods are being considered.

Inorganic chemists tend to favor coordination nomenclature[1] for phosphorus and arsenic compounds. Indeed, this nomenclature seems best suited for many of the organic compounds of antimony and bismuth (see chapter 28), since those elements are more "metallic" than phosphorus and arsenic. Organic derivatives of phosphorus and arsenic, on the other hand, look and act more like "organic" compounds, and a nomenclature that emphasizes functionality seems to be appropriate. The nomenclature recommended here for monophosphorus compounds is based on slight modifications by IUPAC of rules approved in 1952.[2] Much of it is utilized by CAS, and it is consistent with most IUPAC organic recommendations. For polyphosphorus compounds, the IUPAC inorganic rules[1] and CAS nomenclature provide the only guidance.

## Acceptable Nomenclature

Throughout this chapter, for the nomenclature of arsenic compounds the stem "ars" can replace "phos" unless otherwise noted.

### *Parent Phosphorus Hydrides*

Parent hydrides that form the basis for the substitutive nomenclature of *acyclic monophosphorus compounds* are named phosphine ($PH_3$) and phosphorane ($PH_5$). These names are interchangeable with the most recent IUPAC recommendations of phosphane for $PH_3$ and $\lambda^5$-phosphane for $PH_5$; the $\lambda$-convention indicates that the phosphorus atom has a bonding number of five (see chapter 2). All of these names are acceptable, although often only names based on phosphane and phosphorane will be used with examples in this chapter. The corresponding oxides and their analogs are named phosphine oxide ($H_3PO$); phosphine sulfide ($H_3PS$); phosphine imide, phosphoranimine, or $\lambda^5$-phosphanimine [$H_3P(=NH)$]; oxophosphine or oxophosphane (HPO); oxophosphine oxide, dioxophosphorane, or dioxo-$\lambda^5$-phosphane ($HPO_2$); phosphanimine, HP=NH, and so on.

| | |
|---|---|
| $CH_3-PH_2$ | Methylphosphane |
| $(C_6H_5)_3PH_2$ | Triphenylphosphorane<br>Triphenyl-$\lambda^5$-phosphane |

$(CH_3)_3P{=}CHCH_3$          Ethylidene(trimethyl)phosphorane

$(CH_3CH_2)_3PO$          Triethylphosphine oxide
Triethyloxo-$\lambda^5$-phosphane

$P$-Cyclohexyl-$N$-methylphosphine imide
Cyclohexylphosphine methylimide
$P$-Cyclohexyl-$N$-methyl-$\lambda^5$-phosphanimine

$CH_3CH{=}P{-}CH_3$          Ethylidene(methyl)phosphane

*Acyclic polyphosphorus hydrides* with P–P bonds take names such as diphosphine or diphosphane for $H_2P{-}PH_2$, diphosphene for $HP{=}PH$, diphosphorane or $1\lambda^5,2\lambda^5$-diphosphane for $H_4P{-}PH_4$, and so on. To avoid confusion with these names, the multiplying prefixes bis-, tris- and so on are used to denote the presence of more than one substituent $H_2P{-}$ group in the same structure.

$\underset{3\ \ \ 2\ \ \ 1}{C_6H_5{-}PHPHPH{-}CH_3}$          1-Methyl-3-phenyltriphosphane

1,2-Phenylenebis(phosphane)

[2-Methyl-6-(phenylphosphanyl)cyclohexyl]phosphine oxide
[2-Methyl-6-(phenylphosphanyl)cyclohexyl]oxo-$\lambda^5$-phosphane
$P^1$-Phenyl(3-methylcyclohexane-1,2-diyl)bis-(phosphane) $P^2$-oxide

$\overset{\overset{\textstyle O}{\textstyle \|}}{\underset{2\ \ \ \ 1}{CH_3{-}PHPH{-}CH_2CH_3}}$          1-Ethyl-2-methyldiphosphine 2-oxide
2-Ethyl-1-methyl-1-oxo-$1\lambda^5$-diphosphane

Under the IUPAC rules virtually any substituting group prefix may be used with these parent hydride names. In some instances, this method is a must; phosphorus analogs of carboxylic amides, for example, are named as acylphosphines, acylphosphanes, acylphosphine oxides, and so on, giving names such as propanoylphosphane or (1-oxopropyl)phosphane for $CH_3CH_2{-}CO{-}PH_2$. Names like dichloro(phenyl)phosphine for $C_6H_5{-}PCl_2$ are commonly used and acceptable, although phenylphosphonous dichloride is preferred on the basis of compound class seniority (see chapter 3, table 3.1). However, where an –OH group is bound to a phosphorus atom that is not part of a ring, the compound is considered to be a phosphorus acid, and it and its functional replacement derivatives preferably have names based on the parent acid names (see below) to emphasize functionality.

## Mononuclear Acids and their Derivatives

Hydroxy-substituted phosphines, phosphine oxides, sulfides, and so on, are ordinarily viewed as acids and named accordingly (compare sulfur acids, chapter 23). If the acid has a hydrogen atom bound to phosphorus, that atom may be substituted in the same way as in parent hydrides. Hydroxy-substituted derivatives of $PH_5$ not falling into one of the classes above are named substitutively, as in hydroxy(methyl)phosphorane or hydroxy(methyl)-$\lambda^5$-phosphane for $CH_3{-}PH_3{-}OH$; methylphosphoranoic acid, although recommended in the 1952 ACS rules,[2] never gained acceptance.

Parent monophosphorus acid names are given in table 25.1 (see also chapter 3, tables 3.4 and 3.5). Where tautomers are possible and the structure is unknown, the name of the penta-valent form is usually used.

Table 25.1. Mononuclear Phosphorus and Arsenic Acids

| | | | |
|---|---|---|---|
| $P(OH)_3$ | phosphorous acid | $PO(OH)_3$ | phosphoric acid |
| $As(OH)_3$ | arsenous acid | $AsO(OH)_3$ | arsoric acid |
| | | | arsenic acid |
| $HP(OH)_2$ | phosphonous acid | $HPO(OH)_2$ | phosphonic acid |
| $HAs(OH)_2$ | arsonous acid | $HAsO(OH)_2$ | arsonic acid |
| $H_2P-OH$ | phosphinous acid | $H_2PO-OH$ | phosphinic acid |
| $H_2As-OH$ | arsinous acid | $H_2AsO-OH$ | arsinic acid |
| $PO-OH$ | phosphenous acid[a] | $PO_2-OH$ | phosphenic acid[a] |
| $AsO-OH$ | arsenous acid | $AsO_2-OH$ | arsenic acid |

[a] Without regard to structure, $HPO_2$ and $HPO_3$ are named metaphosphorous acid and metaphosphoric acid, respectively.

*Esters and salts* are named in the usual way (see chapter 12) with the endings -ite and -ate replacing -ous acid and -ic acid, respectively; with phosphorous and phosphoric acids, "or" is elided to give phosphite and phosphate.

$CH_3-PO(OH)_2$        Methylphosphonic acid

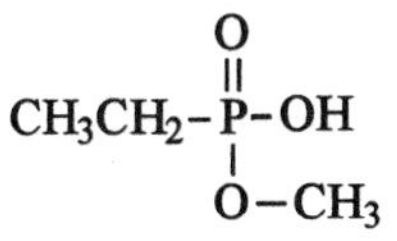

$$CH_3CH_2-\overset{\overset{\displaystyle O}{\|}}{\underset{\underset{\displaystyle O-CH_3}{|}}{P}}-OH$$        Methyl hydrogen ethylphosphonate

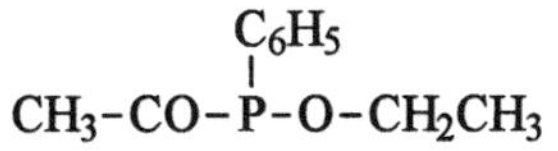

$$CH_3-CO-\overset{\overset{\displaystyle C_6H_5}{|}}{P}-O-CH_2CH_3$$        Ethyl acetyl(phenyl)phosphinite
Ethyl (1-oxoethyl)phenylphosphinite

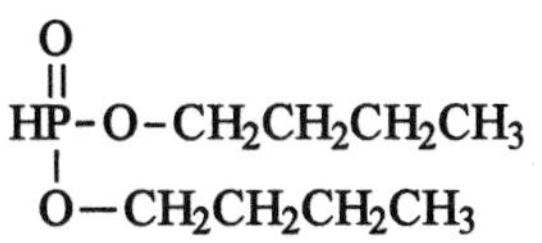

$$H\overset{\overset{\displaystyle O}{\|}}{\underset{\underset{\displaystyle O-CH_2CH_2CH_2CH_3}{|}}{P}}-O-CH_2CH_2CH_2CH_3$$        Dibutyl phosphonate

$$CH_3CH_2-O-\overset{\overset{\displaystyle O-CH_2CH_3}{|}}{P}-O-CH_2CH_3$$        Triethyl phosphite

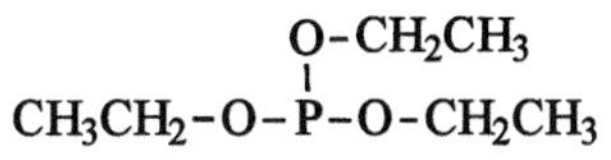

$(HO)_2P-O-CH_2CH_3$        Ethyl dihydrogen phosphite

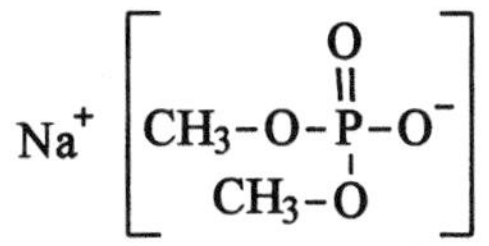

$$Na^+\left[CH_3-O-\overset{\overset{\displaystyle O}{\|}}{\underset{\underset{\displaystyle CH_3-O}{|}}{P}}-O^-\right]$$        Sodium dimethyl phosphate

Metal radical names (see chapter 28) may be used when a metal replaces hydrogen bound to a phosphorus atom; coordination nomenclature is also acceptable for this purpose.

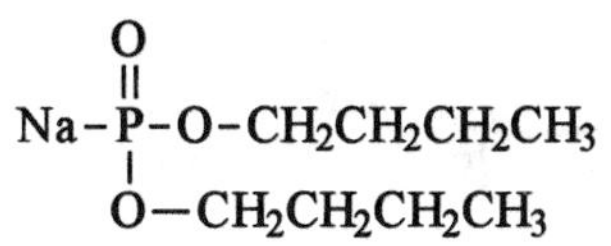

$$Na-\overset{\overset{\displaystyle O}{\|}}{\underset{\underset{\displaystyle O-CH_2CH_2CH_2CH_3}{|}}{P}}-O-CH_2CH_2CH_2CH_3$$        Dibutyl sodiophosphonate
(Dibutoxyphosphinyl)sodium
(Dibutyl phosphonato-*P*)sodium

Derivatives having the composition R–P(O)$_2$, R–O–P(O)$_2$, and so on, are polymeric. Where their structures are unknown, they may be named as metaphosphites, metaphosphates, and so on, as in trisodium trimetaphosphate, Na$_3$[P$_3$O$_4$].

*Functional derivatives of parent phosphorus acids,* in which =O or –OH or both are replaced by atoms or groups containing nitrogen, sulfur, selenium, tallurium, halogen, pseudohalogens, or peroxy groups, are best named by the infix method of functional replacement nomenclature (see chapter 3), which is used by CAS. Also acceptable is the method in which the replacing group name precedes the name of the acid. A prefix such as chloro or amido, however, should be used with care, since in functional replacement names, –Cl replaces –OH, not –H. Chlorophosphonic acid describes HP(O)Cl(OH), not Cl–PO(OH)$_2$, which is named chlorophosphoric acid or phosphorochloridic acid. On the other hand, groups such as methyl or phenyl, derived from parent hydrides, substitute H bound to P and do not replace –OH groups. In both of these methods, the senior characteristic group is expressed by a suffix such as -ate or -ite for esters and salts or a class name such as chloride or amide. The order of seniority of compound classes is the same as that used in substitutive nomenclature (see chapter 3, table 3.1). Letter locants such as *N, S, P,* and *O* are frequently required to specify the positions of substituents or derivative groups.

$$\underset{\underset{\displaystyle Cl}{|}}{\overset{\overset{\displaystyle O}{\|}}{C_6H_5-P-OH}}$$

Phenylphosphonochloridic acid
Phenylchlorophosphonic acid
    (*not* Phenyl(chloro)phosphinic acid)

CH$_3$–POCl$_2$

Methylphosphonic dichloride

$$\underset{\underset{\displaystyle Br}{|}}{\overset{\overset{\displaystyle N-C_6H_5}{\|}}{CH_3CH_2-P-SH}}$$

*P*-Ethyl-*N*-phenylphosphonobromidimidothioic acid
Ethylbromo(phenylimido)thiophosphonic acid

$$\underset{\displaystyle HP-NH-C_6H_5}{\overset{\overset{\displaystyle O-CH_3}{|}}{}}$$

Methyl *N*-phenylphosphonamidite
Methyl (phenylamido)phosphonite

$$\underset{\underset{\displaystyle \underset{N}{NH-CH_2CH_3}}{|}}{\overset{\overset{\overset{N'}{N-C_6H_5}}{\|}}{CH_3-S-P-Cl}}$$

Methyl *N*-ethyl-*N'*-phenylphosphoramidochloridimidothioate
Methyl chloro(ethylamido)(phenylimido)thiophosphate

$$\underset{\displaystyle (C_6H_5)_2P-OOH}{\overset{\overset{\displaystyle O}{\|}}{}}$$

Diphenylphosphinoperoxoic acid
Diphenylperoxyphosphinic acid

Because thio denotes S replacing O in either =O or –OH, there is a problem with isomers. This is usually solved with letter locants attached to the ester groups or the device *S*-acid or *O*-acid (see chapters 11 and 12). An alternative is to use thion(o) for =S and thiol(o) for –SH, proposed in the 1952 report[2] but never adopted officially.

$$\underset{\underset{\displaystyle O-CH_3}{|}}{\overset{\overset{\displaystyle S}{\|}}{C_6H_5-S-P-OH}}$$

*O*-Methyl *S*-phenyl *O*-hydrogen phosphorodithioate
*O*-Methyl *S*-phenyl phosphorothiolothionate

The parent compound H$_2$P≡N is systematically named phosphinic nitride, with derivatives named substitutively, such as dimethylphosphinic nitride. However, derivatives are also called *phosphonitriles* with names such as phosphonitrile chloride for Cl$_2$P≡N and phosphonitrile amide for $(H_2N)_2P{\equiv}N$. The $H_2P{\equiv}N$ structure is commonly seen in $(H_2P{=}N)_n$, which can be named polyphosphazene (see chapter 29).

## Substituting Group Prefixes

Table 25.2 shows some of the most common prefixes for substituting groups related to parent hydrides and mononuclear phosphorus acids.

While any of these prefixes are acceptable, the ones used by CAS are generally recommended. Phosphanyl and diphosphanyl are used with examples in this chapter; the corresponding phosphino names are considered to be equivalent and equally acceptable. Additional prefixes may be formed from those in table 25.2 by substitution or functional replacement methods.

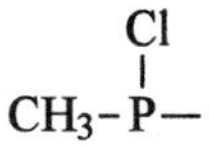

$$\overset{\displaystyle Cl}{\underset{\displaystyle CH_3-P-}{|}}$$  Chloro(methyl)phosphanyl

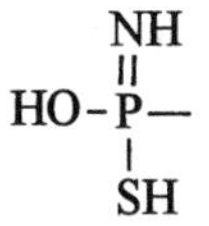

$$\overset{\displaystyle NH}{\underset{\displaystyle SH}{\overset{||}{\underset{|}{HO-P-}}}}$$  Hydroxy(mercapto)phosphinimyl
   (phosphinimyl is a contraction of phosphinimidoyl)

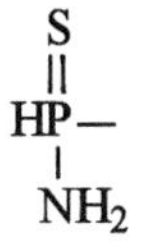

$$\overset{\displaystyle S}{\underset{\displaystyle NH_2}{\overset{||}{\underset{|}{HP-}}}}$$  Phosphonamidothioyl

$$\overset{\displaystyle S}{\underset{\displaystyle C_6H_5-S}{\overset{||}{\underset{|}{CH_3-O-P-}}}}$$  Methoxy(phenylthio)phosphinothioyl

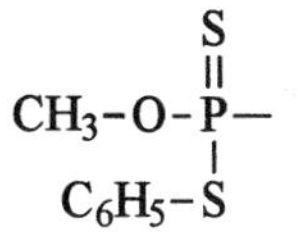

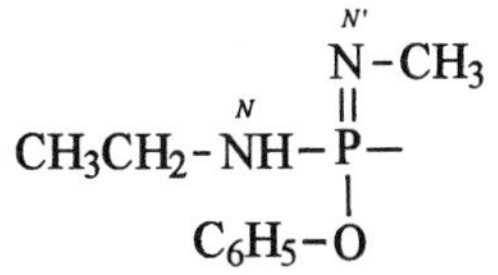

$$\overset{\displaystyle \overset{N'}{N-CH_3}}{\underset{\displaystyle C_6H_5-O}{\overset{||}{\underset{|}{CH_3CH_2-\overset{N}{NH}-P-}}}}$$  N-Ethyl-N'-methyl-P-phenoxyphosphonamidimidoyl

Examples of the use of these prefixes are as follows:

$CH_3-PH-CH_2-CO-O-C_6H_5$   Phenyl (methylphosphanyl)acetate

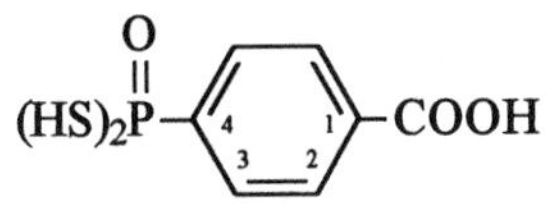

$(HS)_2\overset{O}{\overset{||}{P}}-\underset{3\quad 2}{\overset{4\quad 1}{\bigcirc}}-COOH$   4-(Dimercaptophosphinyl)benzoic acid

$\underset{2\quad 1}{H_2P-CH_2CH_2-OH}$   2-Phosphanylethanol

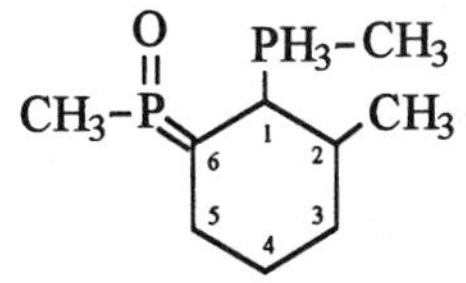

Methyl[2-methyl-6-(methylphosphinylidene)-
   cyclohexyl]phosphorane
Methyl[3-methyl-2-(methylphosphoranyl)-
   cyclohexylidene]phosphine oxide
Methyl[3-methyl-2-(methylphosphoranyl)-
   cyclohexylidene]oxophosphorane

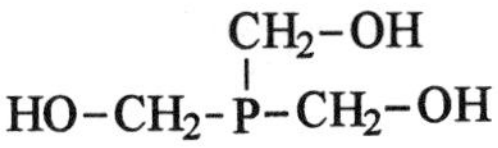

$$\underset{\displaystyle HO-CH_2-\overset{CH_2-OH}{\underset{|}{P}}-CH_2-OH}{}$$  Phosphinidynetrimethanol
Phosphanetriyltrimethanol

Table 25.2. Prefixes for Phosphorus-Containing Groups

| Group | CAS | IUPAC (1979) | IUPAC (1993) |
|---|---|---|---|
| $H_2P-$ | phosphino | phosphino | phosphanyl |
| $HP=$ | phosphinidene | phosphinediyl | phosphanylidene |
| $HP\diagdown$ | phosphinidene | phosphinediyl | phosphanediyl |
| $-P=$ | phosphinidyne | phosphinetriyl | phosphanylylidene |
| $P\equiv$ | phosphinidyne | phosphinetriyl | phosphanylidyne |
| $-P-$ (with vertical bond) | phosphinidyne | phosphinetriyl | phosphanetriyl |
| $H_4P-$ | phosphoranyl | phosphoranyl | $\lambda^5$-phosphanyl |
| $H_3P=$ | phosphoranylidene | phosphoranediyl | $\lambda^5$-phosphanylidene |
| $H_3P\diagdown$ | phosphoranylidene | phosphoranediyl | $\lambda^5$-phosphanediyl |
| $H_2P\equiv$ | phosphoranylidyne | phosphoranetriyl | $\lambda^5$-phosphanylidyne |
| $H_2P-$ (with vertical bond) | phosphoranylidyne | phosphoranetriyl | $\lambda^5$-phosphanetriyl |
| $H_2P=$ (with vertical bond) | phosphoranylidyne | phosphoranetriyl | $\lambda^5$-phosphanylylidene |
| $H_2PPH-$ | diphosphinyl | | diphosphanyl |
| $\underset{H_2P-}{\overset{O}{\overset{\|}{}}}$ | phosphinyl | phosphinoyl | phosphinoyl<br>dihydrophosphoryl |
| $\underset{HP=}{\overset{O}{\overset{\|}{}}}$ or $\underset{HP\diagdown}{\overset{O}{\overset{\|}{}}}$ | phosphinylidene | phosphonoyl | phosphonoyl<br>hydrophosphoryl |
| $\underset{H_2P-}{\overset{S}{\overset{\|}{}}}$ | phosphinothioyl | phosphinothioyl<br>thiophosphinoyl | phosphinothioyl |
| $\underset{H_2P-}{\overset{NH}{\overset{\|}{}}}$ | phosphinimyl | phosphinimidoyl | phosphinimidoyl |
| $OP\equiv$ or $OP-$ or $OP=$ (with vertical bonds) | phosphinylidyne | phosphoryl | phosphoryl |
| $(HO)_2P-$ | dihydroxyphosphino | | dihydroxyphosphanyl |
| $HO-PH-$ | hydroxyphosphino | | hydroxyphosphanyl |
| $HO-P\diagdown$ | hydroxyphosphinidene | | hydroxyphosphanediyl |
| $HO-P=$ | hydroxyphosphinidene | | hydroxyphosphanylidene |
| $\underset{(HO)_2P-}{\overset{O}{\overset{\|}{}}}$ | phosphono | phosphono | dihydroxyphosphoryl<br>phosphono |
| $\underset{HO-PH-}{\overset{O}{\overset{\|}{}}}$ | hydroxyphosphinyl | | hydrohydroxyphosphoryl |

*(Continued)*

Table 25.2. Continued

| Group | CAS | IUPAC (1979) | IUPAC (1993) |
|---|---|---|---|
| 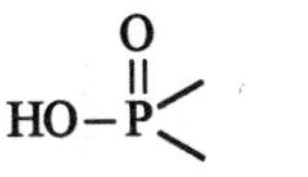 HO–P (=O) < | phosphinico | phosphinico | hydroxyphosphoryl |
| 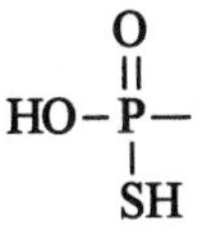 HO–P(=O)–SH | hydroxy(mercapto)-phosphinyl | | hydroxy(mercapto)-phosphoryl |
| 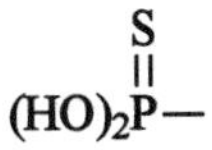 (HO)$_2$P(=S)– | dihydroxyphosphinothioyl | | dihydroxyphosphorothioyl / dihydroxythiophosphoryl |
| 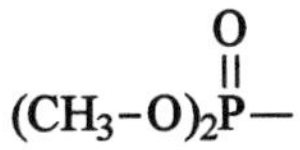 (CH$_3$–O)$_2$P(=O)– | dimethoxyphosphinyl | | dimethoxyphosphoryl |
| 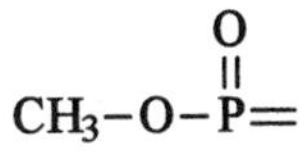 CH$_3$–O–P(=O)= or CH$_3$–O–P(=O)< | methoxyphosphinylidene | | methoxyphosphoryl |
| O$_2$P– | phospho | | |
| OP– | phosphoroso | | |

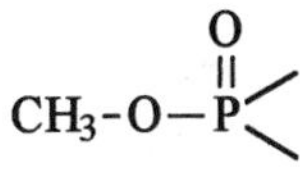

Substituents higher in seniority than phosphines (see chapter 3, table 3.1) may be expressed as suffixes, as in phosphinecarboxylic acid, phosphanecarbonitrile, or phosphoranecarboxamide.

## Polynuclear Acids and their Derivatives

As will be seen below, there are many methods but few rules governing the nomenclature of these compounds. Unsymmetrical derivatives are particularly troublesome, especially when they involve phosphorus in two valence states. Here the λ-convention (see chapter 2) can be useful. Where O is replaced by N or S, the problems increase, and often a prefix name for one portion of a structure is necessary.

Traditionally, *symmetrical phosphorus acids having a P–P bond* have been named with hypo or hypodi prefixed to the name of the corresponding monophosphorus acid, as in hypophosphoric acid, $(HO)_2PO–PO(OH)_2$ (see Discussion). Since hypophosphorous acid refers to $H_3PO_2$ (which could mean either phosphinic or phosphonous acid), the di- is necessary in hypodiphosphorous acid for $(HO)_2P–P(OH)_2$. Either prefix may be used for other acids. Solutions to these ambiguities include: (a) substitutive names based on diphosphine, diphosphane, or diphosphorane; and (b) the use of the prefix bi- to name symmetrical "P–P" acids.

| | |
|---|---|
| $(HO)_2P–P(OH)_2$ | Biphosphonous acid<br>Tetrahydroxydiphosphane<br>Hypodiphosphorous acid |
| O O<br>‖ ‖<br>HP–PH<br>│ │<br>HO  OH | Biphosphinic acid<br>1,2-Dihydroxydiphosphane dioxide<br>1,2-Dihydroxy-1,2-dioxodiphosphorane<br>Hypodiphosphonic acid |

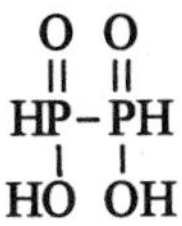

*Unsymmetrical acids* of this type may also be named through use of the prefixes in table 25.2. The senior part of the structure must be chosen. One decreasing seniority order would be: (a) that with the highest oxidation state; (b) that with the greatest number of acidic OH groups; and (c) that expressing a characteristic group class highest in the seniority order of compound classes (see chapter 3, table 3.1). Alternatively, substitutive nomenclature based on parent hydrides may be used, as in hydroxydiphosphine or hydroxydiphosphane, $H_2PPH$–OH.

$$\underset{\displaystyle (HO)_2P-\overset{\displaystyle \|}{\underset{\displaystyle }{P}}H-OH}{\overset{\displaystyle O}{}}$$

(Dihydroxyphosphanyl)phosphinic acid
1,1,2-Trihydroxydiphosphane 2-oxide
1,2,2-Trihydroxy-1-oxo-$1\lambda^5$-diphosphorane

$$CH_3-\overset{\displaystyle O}{\overset{\displaystyle \|}{P}}H-\overset{\displaystyle O}{\underset{\displaystyle O-CH_2CH_3}{\overset{\displaystyle \|}{P}}}-OH$$

Ethyl hydrogen (methylphosphinyl)phosphonate
1-Ethoxy-1-hydroxy-2-methyldiphosphane 1,2-dioxide
1-Ethoxy-1-hydroxy-2-methyl-1,2-dioxodiphosphorane

## Anhydrides of Phosphorus Acids

*Symmetric anhydrides* of phosphorus acids may be named: (a) as anhydrides (see chapter 12); (b) by placing di-, tri-, and so on, before the name of the polybasic acid from which the anhydride is derived (compare with dicarbonic acid, chapter 12); (c) by multiplicative nomenclature with names such as oxybis(ethylphosphane) or oxybis(ethylphosphine); or (d) by substitutive nomenclature based on parents such as di-, tri-, phosphoxanes, and so on. Occasionally, the phrase "anhydride with" may be useful in describing an unknown structure.

$H_2P-O-PH_2$

(a) Phosphinous anhydride
(c) Oxybis(phosphane)
      (*not* Oxydiphosphane)
(d) Diphosphoxane

$$CH_3-\overset{\displaystyle O}{\underset{\displaystyle OH}{\overset{\displaystyle \|}{P}}}-O-\overset{\displaystyle O}{\underset{\displaystyle OH}{\overset{\displaystyle \|}{P}}}-CH_3$$

(a) Methylphosphonic monoanhydride
(b) Dimethyldiphosphonic acid
(c) Oxybis[hydroxy(methyl)phosphine oxide]
      Oxybis[hydroxy(methyl)oxophosphorane]
(d) 1,3-Dihydroxy-1,3-dimethyldiphosphoxane
      1,3-dioxide
      1,3-Dihydroxy-1,3-dimethyl-1,3-dioxo-$1\lambda^5,3\lambda^5$-diphosphoxane

*Unsymmetric anhydrides* may also be named as anhydrides or as phosphoxane derivatives, where the λ-convention may be applicable. It is often convenient to use a phosphorus group prefix (see below and table 25.2) to name an anhydride as a substituted phosphorus compound, particularly when the latter is itself an acid. Where the two sides of the acid have different oxidation states, CAS uses Stock Numbers (III,V) following the name based on the higher oxidation state; this method has only limited applicability.

$$\overset{\displaystyle O}{\underset{\displaystyle OH}{\overset{\displaystyle \|}{H}P-O-P(OH)_2}}$$

Phosphonic phosphorous monoanhydride
1,1,3-Trihydroxydiphosphoxane 3-oxide
1,3,3-Trihydroxy-1-oxo-$1\lambda^5$-diphosphoxane
[(Dihydroxyphosphanyl)oxy]phosphinic acid

*Mixed anhydrides* derived from a phosphorus acid and another type of acid such as a carboxylic acid are given names such as acetic phosphoric monoanhydride for $CH_3$–CO–O–$PO(OH)_2$. Names such as "(acetyloxy)phosphonic acid" or "acetyl phosphate" are not recommended.

## Sulfur and Nitrogen Analogs and Other Anhydride Derivatives

Replacement of an anhydride oxygen atom by S may be indicated by the term thioanhydride or anhydrosulfide (see chapter 12). In names of the diphosphoric acid type, replacement by S or NH has been denoted by the prefixes thio or imido, respectively, but the resulting names, such as thiodiphosphoric acid, are often ambiguous. The structures $H_2P–E–[PH–E]_n–PH_2$, where E represents S, Se, NH, and so on, can be named phosphathiane, phosphazane, and so on, preceded by a numerical prefix corresponding to $n + 2$. Such chains with more than one kind of E atom may often be named through use of phosphorus group prefixes (table 25.2) or by functional replacement nomenclature (see chapter 3) as described in the section on mononuclear acids.

$(HO)_2PO–S–PO(OH)_2$

Phosphoric thiomonoanhydride
1,1,3,3-Tetrahydroxydiphosphathiane 1,3-dioxide
1,1,3,3-Tetrahydroxy-1,3-dioxo-$1\lambda^5,3\lambda^5$-diphosphathiane

$\underset{P \quad N}{(CH_3)_2P–\overset{\overset{\displaystyle C_6H_5}{|}}{N}–P(CH_3)_2}$

$N$-(Dimethylphosphanyl)-$P,P$-dimethyl-$N$-phenylphosphinous amide
1,1,3,3-Tetramethyl-2-phenyldiphosphazane
$N$-(Dimethylphosphanyl)-$P,P$-dimethyl-$N$-phenylphosphanamine
$N,N$-Bis(dimethylphosphanyl)aniline
(Phenylimino)bis(dimethylphosphane)

$C_6H_5–PH–S–\overset{\overset{\displaystyle O}{\|}}{\underset{\underset{\displaystyle CH_3}{|}}{P}}–OH$

Methylphosphonic phenylphosphinous thiomonoanhydride
1-Hydroxy-1-methyl-3-phenyldiphosphathiane 1-oxide
1-Hydroxy-1-methyl-1-oxo-3-phenyl-$1\lambda^5$-diphosphathiane

$\underset{5 \quad 4 \quad 3 \quad 2 \quad 1}{HO–\overset{\overset{\displaystyle O}{\|}}{P}H–N–\overset{\overset{\displaystyle C_6H_5}{|}}{P}H–NH–\overset{\overset{\displaystyle O}{\|}}{P}(CH_3)_2}$

5-Hydroxy-1,1-dimethyl-4-phenyl-triphosphazane 1,5-dioxide
5-Hydroxy-1,1-dimethyl-1,5-dioxo-4-phenyl-$1\lambda^5,5\lambda^5$-triphosphazane
$N$-[[(Dimethylphosphinyl)amino]phosphanyl]-$N$-phenylphosphonamidic acid

The last name in the example above is derived by treating the structure as a substituted phosphonamidic acid, in which –$NH_2$ has replaced an –OH of phosphonic acid.

Polyphosphorus acids and their analogs in which all the OH groups are replaced by the same function take class names such as tetraamide, diamide, dichloride, and so on as separate words. CAS uses the contracted forms diphosphoramide and diphosphonamide; this usage is inconsistent and not recommended since $(H_2N)_3PO$ is not "phosphoramide" but phosphoric triamide in CAS nomenclature.

$(H_2N)_2\overset{\overset{\displaystyle O}{\|}}{P}–O–\overset{\overset{\displaystyle O}{\|}}{P}(NH_2)_2$

Diphosphoric tetraamide
  [*not* Diphosphoramide (CAS)]
Diphosphoxanetetramine 1,3-dioxide
1,3-Dioxo-$1\lambda^5,3\lambda^5$-diphosphoxanetetramide

$\underset{\underset{\displaystyle Cl}{|}}{H\overset{\overset{\displaystyle O}{\|}}{P}}–NH–\underset{\underset{\displaystyle Cl}{|}}{\overset{\overset{\displaystyle O}{\|}}{P}H}$

$P,P'$-Imidodiphosphonic dichloride
1,3-Dichlorodiphosphazane 1,3-dioxide
1,3-Dichloro-1,3-dioxo-$1\lambda^5,3\lambda^5$-diphosphazane

$$\underset{Cl_2P-S-PCl_2}{\overset{O\quad\ O}{\overset{\|\quad\ \|}{}}}$$

*P*,*P*′-Thiodiphosphoric tetrachloride
Phosphorodichloridic thioanhydride
Tetrachlorodiphosphathiane 1,3-dioxide
Tetrachloro-1,3-dioxo-1$\lambda^5$,3$\lambda^5$-diphosphathiane

In general, unsymmetrical functional derivatives of anhydrides are named by the principles described above. The problem with S replacement (thio can mean S in either –SH or =S) can be alleviated through the group prefix method, the use of *O*- or *S*- with ester or acid terms (see chapters 11 and 12), or by the terms thiol(o) and thion(o).

$$\underset{\underset{Br}{\underset{P'}{|}}}{\overset{O}{\overset{\|}{HO-P-O-POCl_2}}}$$

*P*′-Bromo-*P*,*P*-dichlorodiphosphoric acid
Phosphorobromidic phosphorodichloridic anhydride
1-Bromo-3,3-dichloro-1-hydroxy-1,3-dioxo-1$\lambda^5$,3$\lambda^5$-
   diphosphoxane
1-Bromo-3,3-dichloro-1-hydroxydiphosphoxane
   1,3-dioxide
Diphosphoric *P*′-bromide *P*,*P*-dichloride

$$\underset{\underset{N}{}\ \ \underset{SH}{|}}{\overset{S\ \ \ \ \ \ O}{\overset{\|\ \ \ \ \ \ \|}{(H_2N)_2P-NH-P-OH}}}$$

*N*-(Phosphorodiamidothioyl)phosphoramidothioic
   *S*-acid
1,1-Diamino-3-hydroxy-3-oxo-3-sulfanyl-1-
   sulfanylidene-1$\lambda^5$,3$\lambda^5$-diphosphazane
1,1-Diamino-3-hydroxy-3-mercaptodiphosphazane
   3-oxide 1-sulfide

$$\underset{N}{(HS)_2P-NH-PO(OH)_2}$$

*N*-(Dimercaptophosphanyl)phosphoramidic acid
1,1-Dihydroxy-1-oxo-3,3-bis(sulfanyl)-1$\lambda^5$-diphosphazane
1,1-Dihydroxy-3,3-dimercaptodiphosphazane 1-oxide

$$\underset{\underset{SH}{|}\ \ \ \underset{OH}{|}}{\overset{O\ \ \ \ \ S}{\overset{\|\ \ \ \ \ \|}{HP-O-PH}}}$$

[(Mercaptophosphinyl)oxy]phosphinothioic *O*-acid
[(Mercaptophosphinyl)oxy]phosphinothionic acid
1-Hydroxy-3-oxo-3-sulfanyl-1-sulfanylidene-
   1$\lambda^5$,3$\lambda^5$-diphosphoxane
1-Hydroxy-3-mercaptodiphosphoxane 3-oxide 1-sulfide

## Cyclic Phosphorus Compounds

Generally, ring systems containing one or more phosphorus or arsenic atoms are named as heterocycles (see chapter 9). For this purpose, the order of seniority is N > P > As.

8-Phospha-3-arsabicyclo[3.2.1]octane 8-oxide
8-Oxo-8$\lambda^5$-phospha-3-arsabicyclo[3.2.1]octane

6-Oxa-5$\lambda^5$-phosphaspiro[4.5]decane

4,7-Phosphinidene-1*H*-phosphindole
4,7-Phosphano-1*H*-phosphindole

Cyclic oligomers of the composition $(RO-PO)_n$, $(RO-PO_2)_n$, and their analogs have some-times been called "cyclic metaphosphites", "cyclic metaphosphates", and so on, but they are best named as heterocyclic compounds when their structures are known. The traditional phosphazene names are acceptable.

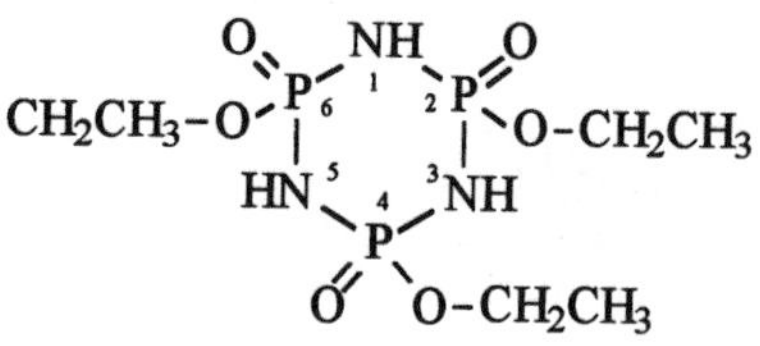

$CH_2CH_3-O-\overset{O}{\underset{6}{P}}\overset{NH}{\underset{5}{}}\overset{O}{\underset{2}{P}}-O-CH_2CH_3$

2,4,6-Triethoxy-1,3,5,2,4,6-triazatriphosphinane 2,4,6-trioxide

2,4,6-Triethoxyhexahydro-1,3,5,2,4,6-triaza-triphosphorine 2,4,6-trioxide

2,4,6-Triethoxy-2,4,6-trioxo-1,3,5,2$\lambda^5$,4$\lambda^5$,6$\lambda^5$-triphosphinane

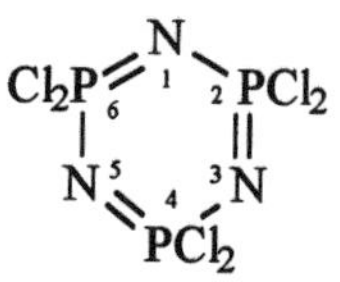

$Cl_2P\overset{N}{\underset{6}{}}\overset{}{\underset{1}{}}\overset{}{\underset{2}{}}PCl_2$

2,2,4,6,6-Hexachloro-1,3,5,2$\lambda^5$,4$\lambda^5$,6$\lambda^5$-triazatriphosphinine

2,2,4,6,6-Hexachloro-2,2,4,4,6,6-hexahydro-1,3,5,2,4,6-triazatriphosphorine

2,2,4,6,6-Hexachlorocyclotriphosphazene

The most recent IUPAC recommendations suggest that rings consisting only of alternating P and O atoms take names based on phosphoxane (see also cyclosiloxanes in chapter 26).

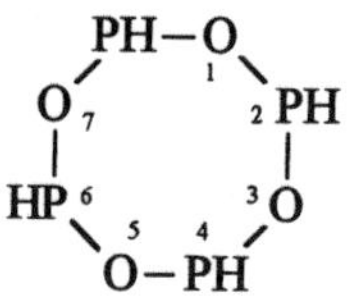

$\begin{array}{c}PH-O\\O_7 \qquad 2\,PH\\HP_6 \qquad 3\,O\\O-PH\end{array}$   Cyclotetraphosphoxane

## Replacement ("a") Nomenclature

This method (see chapter 3) is applicable to monocyclic phosphorus compounds having more than 10 ring atoms as well as to bridged and spiro systems with monocyclic components and to cyclic and acyclic compounds for which other methods are cumbersome or difficult to apply. The replacement prefix for trivalent phosphorus is phospha, and for pentavalent phosphorus it is $\lambda^5$-phospha; the corresponding terms for arsenic are arsa and $\lambda^5$-arsa. For cationic centers (see chapter 33), the prefixes are phosphonia and arsonia.

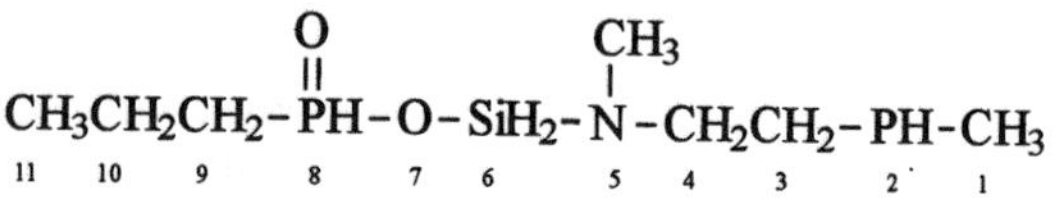

$$CH_3CH_2CH_2-\overset{O}{\underset{}{PH}}-O-SiH_2-\overset{CH_3}{\underset{}{N}}-CH_2CH_2-PH-CH_3$$

11  10  9  8  7  6  5  4  3  2'  1

5-Methyl-8-oxo-7-oxa-5-aza-2,8$\lambda^5$-diphospha-6-silaundecane
5-Methyl-7-oxa-5-aza-2,8-diphospha-6-silaundecane 8-oxide

N-Methyl-N-[2-(methylphosphino)ethyl]-1-[(propylphosphinyl)oxy]silanamine
(substitutive name)

## Discussion

As noted at the beginning of this chapter, phosphorus and arsenic compounds stand at the nomenclatural crossroads of organic and inorganic chemistry. Inorganic names are, of course, acceptable, particularly for simpler structures, the acids, and salts, and they often have the advantage of simplicity. They do not stress function, however, which through substitutive principles is the cornerstone of organic nomenclature. Additive or coordination nomenclature generates names such as dihydridomethylphosphorus for $CH_3-PH_2$ and amidochloromethoxo-oxophosphorus for $CH_3-O-P(O)Cl(NH_2)$. Indeed, this type of nomenclature is quite appropriate for compounds of the more metal-like antimony and bismuth (see chapter 28).

It must be re-emphasized that functional replacement in parent acids is considered to occur only through the formal replacement of –OH and not through substitution of –H. Thus $C_6H_5-P(O)Cl(OH)$ is phenylphosphonochloridic acid or phenylchlorophosphonic acid and not chloro(phenyl)phosphinic acid.

The IUPAC practice of using $\mu$ to denote a bridge in coordination names might be adapted to polynuclear phosphorus compounds. Names such as $\mu$-thiodiphosphoric acid for $(HO)_2P(O)-S-PO(OH)_2$ and $\mu$-imidodiphosphonic dichloride for $H(Cl)P(O)-NH-PO(Cl)H$ would be formed. For this purpose, however, the method has not been fully developed.

Many phosphorus oxo acids and esters have been named traditionally with prefixes such as meta, ortho, holo, hypo, isohypo, and pyro. Unfortunately, many of these prefixes have multiple meanings, particularly when they are used with compounds involving elements other than phosphorus and arsenic; a good memory is indispensible. Fortunately, there are better, more systematic, and equally descriptive ways to name these acids without such prefixes. For reference, in table 25.3 some traditional names of phosphorus acids are contrasted with recommended names.

Table 25.3. Traditional Names for Some Phosphorus Acids

| Structure | Traditional name | Systematic name |
|---|---|---|
| $HO-PO_2$ | metaphosphoric acid $[(HPO_3)_n, n = 1, 2, 3...]$ | phosphenic acid |
| $HO-PO$ | metaphosphorous acid $[(HPO_2)_n, n = 1, 2, 3...]$ | phosphenous acid |
| $(HO)_3PO$ | orthophosphoric acid | phosphoric acid |
| $P(OH)_5$ | holophosphoric acid | pentahydroxyphosphorane pentahydroxy-$\lambda^5$-phosphane |
| $(HO)_2PO-PO(OH)_2$ | hypophosphoric acid | phosphonophosphonic acid |
| $(HO)_2P-P(OH)_2$ | hypodiphosphorous acid | tetrahydroxydiphosphane (dihydroxyphosphanyl)phosphonous acid |
| $HO-\overset{\overset{O}{\|\|}}{P}H-\overset{\overset{O}{\|\|}}{P}H-OH$ | hypodiphosphonic acid | 1,2-dihydroxydiphosphane dioxide 1,2-dihydroxy-1,2-dioxodiphosphorane |
| $(HO)_2P-O-PO(OH)_2$ | | phosphoric phosphorous monoanhydride diphosphoric(III,V) acid |
| $(HO)_2PO-O-\overset{\overset{O}{\|\|}}{P}H-OH$ | Isohypophosphoric acid | phosphonic phosphoric monoanhydride |
| $(HO)_2PO-O-PO(OH)_2$ | pyrophosphoric acid | diphosphoric acid phosphoric monoanhydride |

In its 1993 recommendations, IUPAC introduced the use of additive nomenclature (see chapter 3) for naming pentavalent phosphorus (and nitrogen) acyl groups. With this method, phosphoryl, $P(O)\equiv$, is the parent component to which the names of atoms or groups are added to form names such as hydrophosphoryl, for $H-P(O)<$; hydrohydroxyphosphoryl, for $HO-PH(O)-$; or chloro(methoxy)phosphoryl, for $CH_3O-PCl(O)-$. These prefixes have not yet been established in the literature.

## Additional Examples

1.   $PH_2$ $\quad\quad$ $PH_3-CH_3$
$\quad\;\;|\quad\quad\quad\quad\quad\;\;|$
$CH_3CHCH_2CH_2CH_2CHCH_2CH_3$
$\quad\;\;6\;\;\;5\;\;\;4\;\;\;\;3\;\;\;\;2\;\;\;\;1$

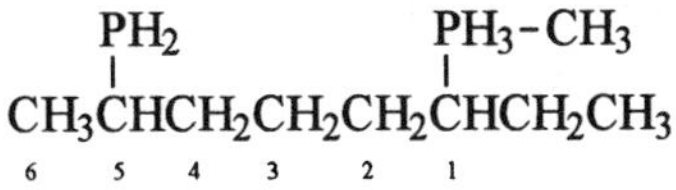

(1-Ethyl-5-phosphanylhexyl)methyl-
  phosphorane
Methyl(7-phosphanyloctan-3-yl)-$\lambda^5$-
  phosphane

2.   $H_2P-PH_3-PH_2$
$\quad\;\;3\quad\;\;2\quad\;\;1$

$2\lambda^5$-Triphosphane
Bis(phosphanyl)phosphorane
Bis(phosphanyl)-$\lambda^5$-phosphane

3. 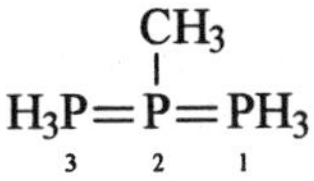   $CH_3$
$\quad\;\;|$
$H_3P=P=PH_3$
$\quad\;3\quad2\quad1$

Methylbis(phosphoranylidene)-
  phosphorane
2-Methyl-$1\lambda^5,2\lambda^5,3\lambda^5$-triphospha-1,2-diene

4.

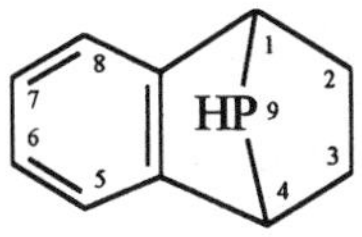

1,2,3,4-Tetrahydro-1,4-phosphinidene-
  naphthalene
1,2,3,4-Tetrahydro-1,4-phosphano-
  naphthalene

5. 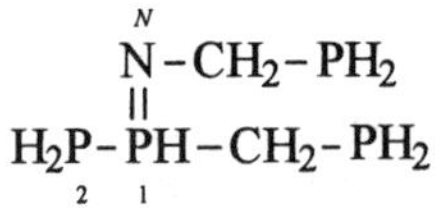   $\overset{N}{N}-CH_2-PH_2$
$\quad\;\;||$
$H_2P-PH-CH_2-PH_2$
$\quad\;\;2\quad1$

$N,1$-Bis(phosphanylmethyl)-$1\lambda^5$-
  diphosphan-1-imine
1-(Phosphanylmethyl)diphosphane
  1-(phosphanylmethyl)imide

6.   $Cl_2P-PBr_2$
$\quad\;\;2\quad1$

1,1-Dibromo-2,2-dichlorodiphosphane
(Dibromophosphanyl)phosphonous
  dichloride

7.   $H_2P-CO-NH_2$

Phosphanecarboxamide
Carbamoylphosphane

8.   $H_2P-CN$

Phosphanecarbonitrile
Cyanophosphane

9.   $H_2PPH-COOH$

Diphosphanecarboxylic acid
Carboxydiphosphane

10. 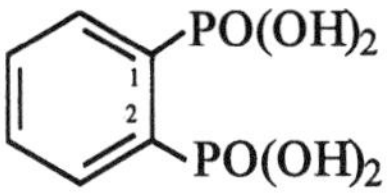   $PO(OH)_2$
$\quad\;\;PO(OH)_2$

1,2-Phenylenebis(phosphonic acid)

11. 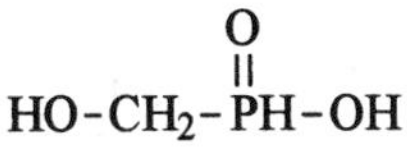   $O$
$\quad\;\;||$
$HO-CH_2-PH-OH$

(Hydroxymethyl)phosphinic acid

12.   $(HO)_2As-CH_2-COOH$

(Dihydroxyarsanyl)acetic acid

13. 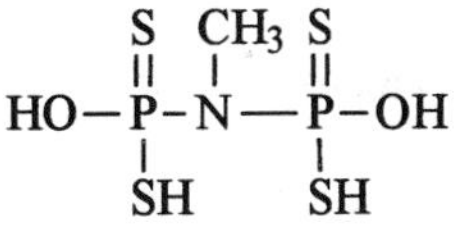   $S\;\;\;CH_3\;S$
$\quad\;\;||\quad\;\;|\quad\;||$
$HO-P-N-P-OH$
$\quad\;\;|\quad\quad\;\;|$
$\quad\;\;SH\quad\;\;SH$

(Methylimino)bis(phosphonodithioic
  $O$-acid)
$N$-Methyl-$P,P'$-imidodiphosphorodithioic
  $O,O'$-diacid
$P,P'$-(Methylimido)bis(dithiophosphoric
  $O$-acid)

14.

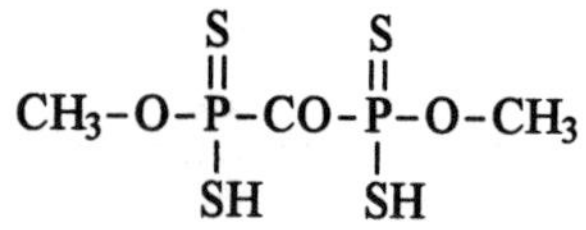

$$CH_3-O-\underset{\underset{SH}{|}}{\overset{\overset{S}{\|}}{P}}-CO-\underset{\underset{SH}{|}}{\overset{\overset{S}{\|}}{P}}-O-CH_3$$

*O,O*′-Dimethyl dihydrogen carbonyl-
bis(phosphonodithioate)

15.

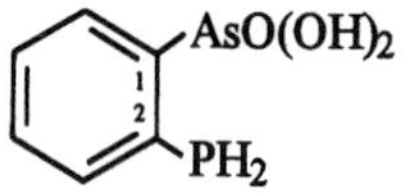

(2-Phosphanylphenyl)arsonic acid

16.

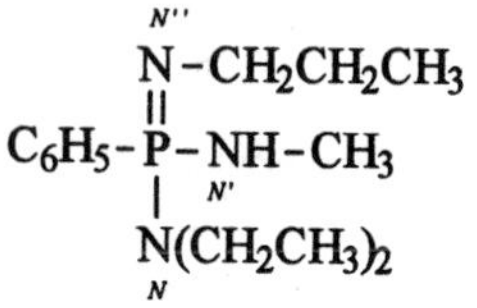

*N,N*-Diethyl-*N*′-methyl-*N*″-propyl-*P*-
phenylphosphonimidic diamide
*P*-Phenyl(propylimido)phosphonic
*N,N*-diethyl-*N*′-methyldiamide

17.

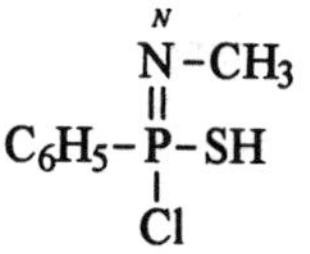

*N*-Methyl-*P*-phenylphosphono-
chloridimidothioic acid
Phenylchloro(methylimido)thiophos-
phonic acid

18.

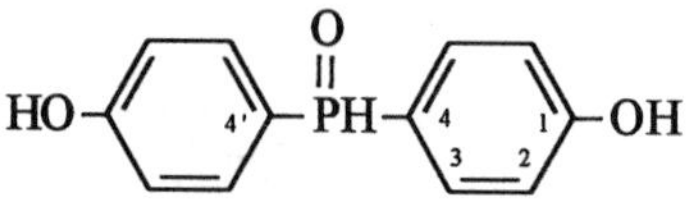

4,4′-Phosphinylidenediphenol
4,4′-(Hydrophosphoryl)diphenol

19.

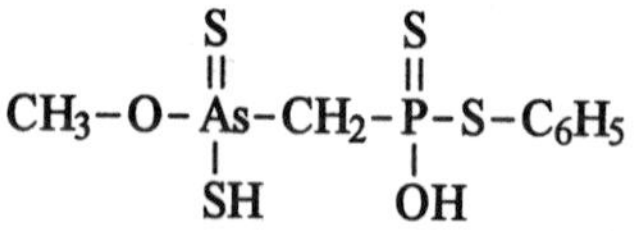

$$CH_3-O-\underset{\underset{SH}{|}}{\overset{\overset{S}{\|}}{As}}-CH_2-\underset{\underset{OH}{|}}{\overset{\overset{S}{\|}}{P}}-S-C_6H_5$$

*S*-Phenyl *O*-hydrogen[[mercapto-
(methoxy)arsinothioyl]methyl]-
phosphonodithioate

20.

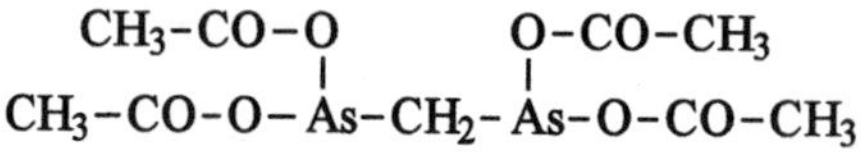

$$\begin{array}{ccc} CH_3-CO-O & & O-CO-CH_3 \\ | & & | \\ CH_3-CO-O-As-CH_2-As-O-CO-CH_3 \end{array}$$

Tetraacetic methylenebis(arsonous) tetra-
anhydride

21.

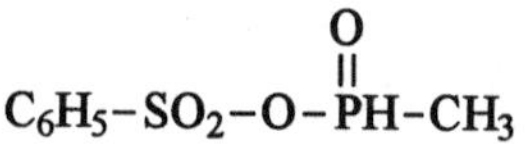

$$C_6H_5-SO_2-O-\overset{\overset{O}{\|}}{PH}-CH_3$$

Benzenesulfonic methylphosphinic
anhydride

22.

$$C_6H_5-SO_2-O-PH_4$$

[(Phenylsulfonyl)oxy]phosphorane
(Benzenesulfonyloxy]-$\lambda^5$-phosphane

23.

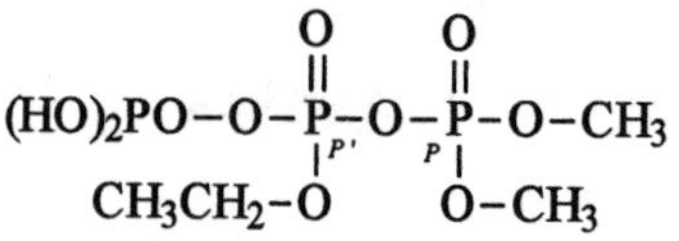

$$\begin{array}{cc} & \overset{O}{\|} \quad \overset{O}{\|} \\ (HO)_2PO-O-\underset{P'}{P}-O-\underset{P}{P}-O-CH_3 \\ CH_3CH_2-O \quad\quad O-CH_3 \end{array}$$

*P*′-Ethyl *P,P*-dimethyl dihydrogen
triphosphate

24.

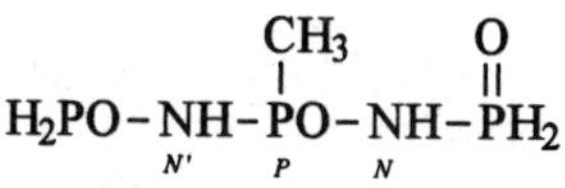

$$\begin{array}{cc} CH_3 & O \\ | & \| \\ H_2PO-NH-PO-NH-PH_2 \\ \underset{N'}{} \quad \underset{P}{} \quad \underset{N}{} \end{array}$$

*N,N*′-Diphosphinyl-*P*-methylphosphonic
diamide
3-Methyl-1,3,5-trioxo-1$\lambda^5$,3$\lambda^5$,5$\lambda^5$-
triphosphazane
3-Methyltriphosphazane 1,3,5-trioxide
Methylphosphonic bis(phosphinylamide)

25.

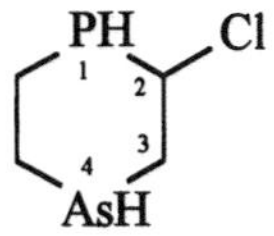

2-Chloro-1,4-phospharsinane
2-Chloro-1,4-phospharsenane

26.

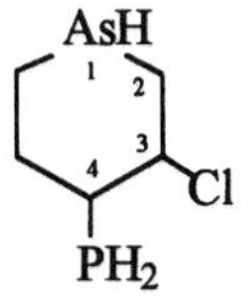

3-Chloro-4-phosphanylarsinane
(3-Chloroarsinan-4-yl)phosphane

27.

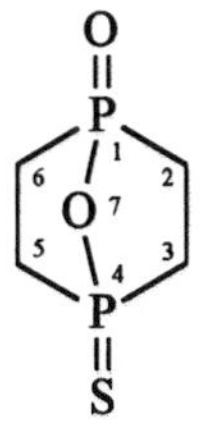

7-Oxa-1,4-diphosphabicyclo[2.2.1]heptane 1-oxide 4-sulfide
1-Oxo-4-thioxo-7-oxa-1$\lambda^5$,4$\lambda^5$-diphosphabicyclo[2.2.1]heptane

28.

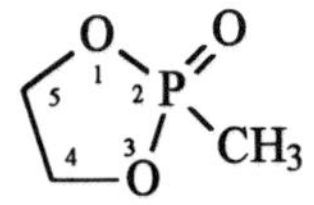

2-Methyl-1,3,2-dioxaphospholane 2-oxide
2-Methyl-2-oxo-1,3,2$\lambda^5$-dioxaphospholane

29.

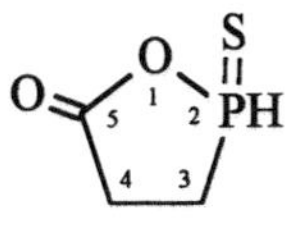

1,2-Oxaphospholan-5-one 2-sulfide
2-Sulfanylidene-1,2$\lambda^5$-oxaphospholan-5-one

30.

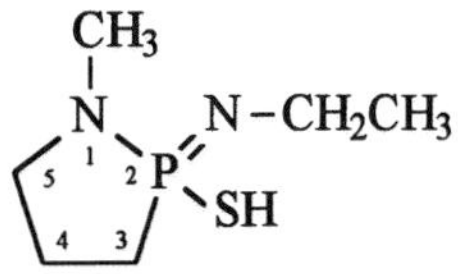

2-(Ethylimino)-1-methyl-2-sulfanyl-1,2$\lambda^5$-azaphospholane
2-(Ethylimino)-2,2-dihydro-2-mercapto-1-methyl-1,2-azaphospholane
$N$-Ethyl-1-methyl-2-sulfanyl-1,2$\lambda^5$-azaphospholan-2-imine

## REFERENCES

1. International Union of Pure and Applied Chemistry, Inorganic Chemistry Division, Commission on Nomenclature of Inorganic Chemistry. Nomenclature of Inorganic Compounds (Recommendations 1990); Blackwell, Scientific Publications: Oxford, 1990; Section I-10; pp. 143–206.
2. American Chemical Society. The Report of the ACS Nomenclature, Spelling, and Pronounciation Committee for the First Half of 1952, E. Compounds Containing Phosphorus. *Chem. Eng. News* **1952**, *30*, 4515–4522.

# Silicon, Germanium, Tin, and Lead Compounds

As might be expected from the structural resemblance of silicon and carbon compounds, the nomenclature of silicon hydrides and their derivatives tends to parallel that of the corresponding carbon structures. The names of analogous organic derivatives of germanium, tin, and lead generally follow that of silicon, but organometallic nomenclature (see chapter 28) is often acceptable for some compounds of the heavier elements. This chapter is based on Section D of the 1979 IUPAC rules, which was an adaptation of earlier IUPAC rules and a 1952 American Chemical Society report on nomenclature.[1]

## Acceptable Nomenclature

Hydrides of chains and rings of silicon atoms are called *silanes*; other Group 14 hydrides are *germanes*, *stannanes*, and *plumbanes*. The first member in the silicon series, $SiH_4$, is silane, and the acyclic homologs, $H_3Si[SiH_2]_nSiH_3$, are named disilane, trisilane, and so on, according to the total number of silicon atoms present. Alicyclic structures are given prefixes such as cyclo, bicyclo, and so on. Double and triple bonds are indicated by the suffixes -ene and -yne, as in disilene for $H_2Si=SiH_2$. The $\lambda$-convention (see chapter 2) may be used to denote atoms with other than the standard bonding number, which for Group 14 is four. However, the first members of the series $MH_2$ are usually named silylene, germylene, and so on.

| | |
|---|---|
| $H_3Si[SiH_2]_3SiH_3$ <br> 5   4-2   1 | Pentasilane |
| $H_3GeGeH_3$ <br> 2   1 | Digermane |
| $H_3SnSnH$ <br> 2   1 | $1\lambda^2$-Distannane |
| $\overset{\displaystyle SiH_2}{\underset{\displaystyle H_2Si\overset{3\ 2}{—}SiH_2}{\diagup^1\diagdown}}$ | Cyclotrisilane |
| 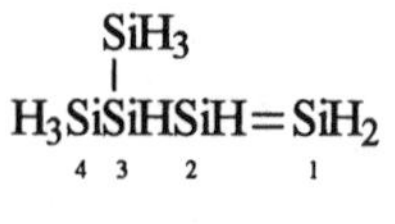 $\overset{\displaystyle SiH_3}{\underset{\displaystyle \underset{4\ 3\ 2\ 1}{H_3SiSiHSiH=SiH_2}}{\mid}}$ | 3-Silyltetrasil-1-ene |
| $\underset{4\ \ 3\ \ 2\ \ 1}{H_2Si=SiHSiH=SiH_2}$ | Tetrasila-1,3-diene |
| $\underset{2\ \ 1}{HSi\equiv SiH}$ | Disilyne |

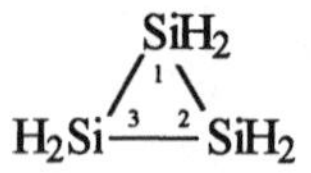

Chains terminating with silicon atoms and otherwise consisting entirely of silicon alternating with an atom other than Ge, Sn, Pb, or B, such as O, N, or S, are named siloxane, silazane, or silathiane, respectively, with a prefix giving the number of silicon atoms present. All atoms in the longest chain are numbered consecutively from one end to the other. In an attempt to broaden the recognition of functionality, CAS deviates from the IUPAC recommendations and names silazanes as amines. These names may also be used in naming cyclic structures. In polyalicycles, prefixes such as 1*Si*- or 1*N*- may be necessary to denote the bridgehead atom. Analogous germanium, tin, and lead compounds are named in the same way.

$$\underset{5\quad\;4\quad\;3\quad\;\;2\;\;\;1}{H_3Si-O-\overset{\overset{\displaystyle CH_3}{\displaystyle |}}{SiH}-O-SiH_3}$$
  3-Methyltrisiloxane

$$\underset{3\quad\;\;2\quad\;\;1}{H_3Si-NH-SiH_3}$$
  Disilazane (IUPAC)
  *N*-Silylsilanamine (CAS)
  Disilylamine

$$\underset{3\quad\;\;2\quad\;\;1}{H_3Sn-Te-SnH_3}$$
  Distannatellurane

Cyclodisilazane

Spiro[5.7]hexasiloxane

1*Si*-Bicyclo[3.3.1]tetrasilazane

Hantzsch–Widman names (see chapter 9) are recommended by IUPAC for heteromonocyclic rings containing silicon and its analogs with up to 10 members. Replacement nomenclature (see chapters 3 and 9) is used in larger monocyclic rings, polycyclic alicyclic ring systems (see chapter 6), and polycyclic fused ring systems that cannot be named by fusion principles. All silicon-containing monocyclic rings are named by replacement nomenclature by CAS. If nomenclature based on silane, siloxane, and so on, is difficult to apply, replacement nomenclature may also be applied to acyclic chains. In decreasing seniority, the terms sila, germa, stanna, and plumba are used to denote Si, Ge, Sn, and Pb replacing carbon atoms in an organic structure, with locants to indicate position. Where there is a choice in acyclic chains, the chain with the largest number of hetero atoms is selected (see chapters 3 and 9 for further criteria). Substituent chains can also be named by replacement nomenclature, but usually conventional substitutive nomenclature is used.

  Siline (IUPAC)
  Silabenzene (CAS)

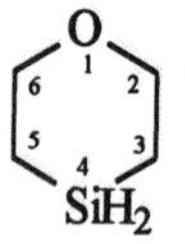

1,4-Oxasilinane (IUPAC)
1-Oxa-4-silacyclohexane (CAS)

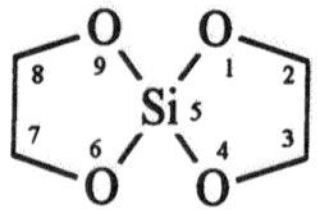

1,4,6,9-Tetraoxa-5-silaspiro[4.4]nonane

$$CH_3-SiH_2-NH-\overset{\overset{\displaystyle SiH_3}{|}}{SiHSiH_2}-NH-\overset{\overset{\displaystyle C_6H_5}{|}}{SiH}-O-CH_3$$

3-Phenyl-6-silyl-2-oxa-4,7-diaza-3,5,6,8-tetrasilanonane
$N^1$-[Methoxy(phenyl)silyl]-$N^2$-(methylsilyl)-1,2-trisilanediamine

$$\underset{5}{CH_3}-\underset{4}{SiH_2}-\underset{3}{SnH_2}-\underset{2}{GeH_2}-\underset{1}{SiH_3}$$

1,4-Disila-2-germa-3-stannapentane

Certain fused ring systems (see chapter 7) with silicon at a nonfused position are given trivial names.

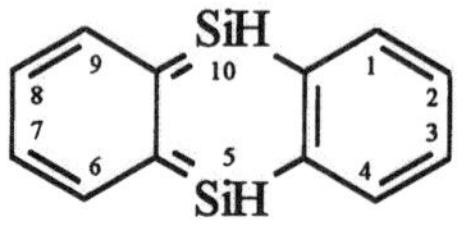

Silanthrene

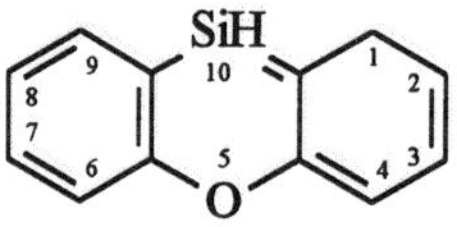

1$H$-Phenoxasilin

*Substituting groups* containing silicon and its analogs are named in the same way as carbon-containing groups (see chapter 5) by using the endings -yl, -ylidene, -ylidyne, -diyl, -triyl, and so on, with the name of the parent hydride. Usage supports silylene (corresponding to methylene) for $-SiH_2-$, although silylene is also an accepted name for the compound $SiH_2$, $\lambda^2$-silane. A double bond or a triple bond at a silicon atom is indicated by -ylidene or -ylidyne, respectively; IUPAC now uses -diyl and -triyl when each free valence is attached to a different atom. The contracted forms silyl rather than silanyl and siloxy rather than silanoxy are used to eliminate the need for multiplying prefixes such as bis- to distinguish two $-SiH_3$ groups from a disilanyl group, $-SiH_2SiH_3$. Typical examples of names for substituting groups are presented in table 26.1.

*Substitution* of other groups for hydrogen in silicon-containing parent hydrides or substituting groups is denoted in much the same way as for carbon compounds. Methylidenesilane (IUPAC) or methylenesilane (CAS) are names for $CH_2=SiH_2$. Where a characteristic group can be named by a prefix or a suffix, as in carboxy(trimethyl)silane or trimethylsilane-carboxylic acid, the suffix form is preferred when it describes a function high in the order of precedence (see chapter 3), as in silanol for $H_3Si-OH$ or silanamine for $H_3Si-NH_2$. However, $-OH$ groups attached to Ge, Sn, or Pb are expressed by the prefix hydroxy and not by the suffix -ol. Acyl derivatives are given names such as triacetylchlorosilane for $Cl-Si(CO-CH_3)_3$.

$CH_3-SiH_2-Cl$                                      Chloro(methyl)silane

$Cl-CH_2-SiH_3$                                      (Chloromethyl)silane

Table 26.1. Prefixes for Silicon-Containing Groups

| | |
|---|---|
| $H_3Si—$ | silyl |
| $H_3Ge—$ | germyl |
| $H_3Sn—$ | stannyl |
| $H_3Pb—$ | plumbyl |
| $—SiH_2—$ | silylene (CAS)<br>silanediyl (IUPAC)<br>silano (as bridging group) |
| $H_2Si=$ | silylidene |
| $HSi\equiv$ | silylidyne |
| $\overset{\mid}{\underset{\mid}{HSi}}—$ | silanetriyl (IUPAC)<br>silylidyne (CAS) |
| $—SiH=$ | silanylylidene |
| $=Si=$ | silanediylidene (IUPAC)<br>silanetetrayl (CAS) |
| $—\overset{\mid}{\underset{\mid}{Si}}—$ | silanetetrayl |
| $H_3SiSiH_2—$ | disilanyl |
| $H_3SiSiH=$ | disilanylidene |
| $\overset{\mid}{H_3Si\underset{\underset{2\ \ \ 1}{}}{SiH}}—$ | disilane-1,1-diyl (IUPAC)<br>disilanylidene (CAS) |
| $—SiH_2SiH_2—$ | disilane-1,2-diyl<br>disilano (as a bridging group) |
| $\underset{3\ \ \ \ 2\ \ \ \ 1}{—SiH_2SiH_2SiH_2—}$ | trisilane-1,3-diyl |
| $\underset{3\ \ \ 2\ \ \ 1}{H_3Si\overset{\mid}{Si}HSiH_3}$ | trisilan-2-yl (IUPAC)<br>1-silyldisilanyl (CAS) |
| 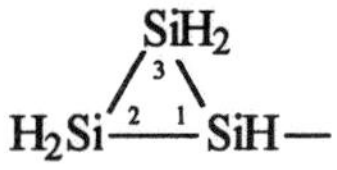 $\underset{\underset{2\ \ 1}{}}{H_2Si}\overset{\overset{SiH_2}{/^3\backslash}}{—}SiH—$ | cyclotrisilanyl |
| $\equiv SnSn\equiv$ | distannanediylidyne |
| $\overset{\mid\ \ \mid}{\underset{\mid\ \ \mid}{—SnSn—}}$ | distannanehexayl (IUPAC)<br>distannanediylidyne (CAS) |
| $H_3Si-O—$ | siloxy (contraction of silanyloxy or silyloxy) |
| $H_3Si-S—$ | silylthio |
| $H_3Si-NH—$ | silylamino |
| $\underset{3\ \ \ \ 2\ \ \ \ 1}{H_3Si-\overset{\mid}{N}-SiH_3}$ | disilylamino (CAS)<br>disilazan-2-yl (IUPAC) |

*(Continued)*

Table 26.1. Continued

| | |
|---|---|
| H$_3$Si–NH–SiH$_2$—<br>  3    2    1 | disilazan-1-yl (IUPAC)<br>disilazanyl (CAS) |
| H$_3$Si–O–SiH$_2$—<br>  3    2    1 | disiloxanyl |
| =SiH–S–SiH$_2$—<br>  3    2    1 | disilathian-1-yl-3-ylidene |
| —SiH$_2$–O–[SiH$_2$–O]$_2$–SiH$_2$—<br>  7      6      5-2          1 | tetrasiloxan-1,7-diyl |
| H$_3$Si–O–SiH–O–SiH$_3$<br>  5    4    3    2    1 | trisiloxan-3-yl (IUPAC)<br>1-(silyloxy)disiloxanyl (CAS) |

| | |
|---|---|
| C$_6$H$_5$–SiCl$_2$–CN | Dichloro(phenyl)silanecarbonitrile<br>Dichloro(cyano)phenylsilane |
| (CH$_3$)$_3$Si–NH–CH$_2$CH$_3$<br>  1    N | Ethyl(trimethylsilyl)amine<br>N-Ethyl-1,1,1-trimethylsilanamine<br>(Ethylamino)trimethylsilane<br>N-(Trimethylsilyl)ethanamine |
| H$_2$Si=NH<br>  1    N | Silanimine<br>Iminosilane |
| H$_2$Ge=O | Oxogermane<br>(*not* Germanone) |
| CH$_3$–SnH$_2$–OH | Hydroxy(methyl)stannane<br>(*not* Methylstannanol) |
| 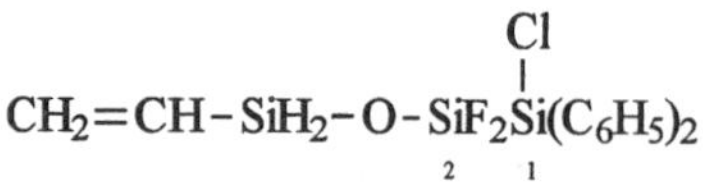 CH$_2$=CH–SiH$_2$–O–SiF$_2$Si(C$_6$H$_5$)$_2$<br>                                      Cl (above)<br>                            2          1 | 1-Chloro-2,2-difluoro-1,1-diphenyl-2-<br>(vinylsiloxy)disilane<br>1-[Chloro(diphenyl)silyl]-1,1-difluoro-3-<br>vinyldisiloxane |
| (CH$_3$)$_2$Pb(O–CO–CH$_3$)$_2$ | Diacetoxy(dimethyl)plumbane<br>Bis(acetato)dimethyllead |

Esters of silanols are named as esters, as in trimethylsilyl benzoate, or as acyloxy compounds, as in (benzoyloxy)trimethylsilane, for C$_6$H$_5$–CO–O–Si(CH$_3$)$_3$. The compound Si(OH)$_4$ is called orthosilicic acid ("ortho" is often dropped), and its esters are named as orthosilicates as in tetraethyl orthosilicate. The compound CH$_3$–Si(OH)$_3$, however, is named methylsilanetriol, not "methylsilicic acid". Analogous compounds of Ge, Sn, and Pb are named as substituted parent hydrides, such as tetraethoxystannane, (CH$_3$CH$_2$–O)$_4$Sn. Anhydrides of orthosilicates are named as derivatives of silane, disiloxane, trisiloxane, and so on, or as silyl esters.

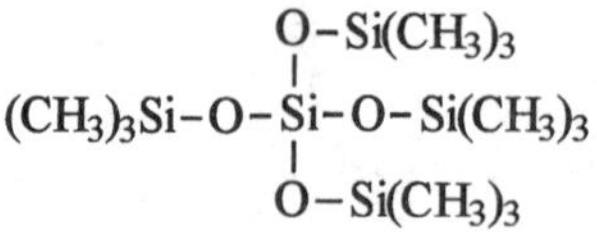

| | |
|---|---|
| (CH$_3$)$_3$Si–O–Si–O–Si(CH$_3$)$_3$<br>                    O–Si(CH$_3$)$_3$ (above)<br>                    O–Si(CH$_3$)$_3$ (below) | Tetrakis(trimethylsiloxy)silane<br>1,1,1,5,5,5-Hexamethyl-3,3-bis(trimethylsiloxy)trisiloxane<br>Tetrakis(trimethylsilyl) orthosilicate |

## Discussion

"Silene" has been often erroneously used as a name for the silicon analog of carbene and for $H_2Si=CH_2$, which is properly named substitutively as methylidenesilane. The name "silene" can only be a source of confusion; it is probably best eliminated from the lexicon.

Silicone is a class name for polysiloxanes. Names such as "dimethyl silicone", however, should never be used.

## Additional Examples

1. $(CH_3)_3Sn-CH_2-Si(CH_3)_3$ 

   Trimethyl[(trimethylstannyl)methyl]silane

2.

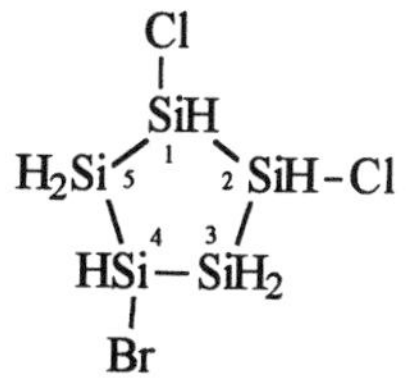

   4-Bromo-1,2-dichloropentasilolane
   4-Bromo-1,2-dichlorocyclopentasilane

3. $(CH_3)_3SiSiH_2-O-SiH_2Si(CH_3)_3$

   1,3-Bis(trimethylsilyl)disiloxane
   1,1'-Oxybis(2,2,2-trimethyldisilane)

4.

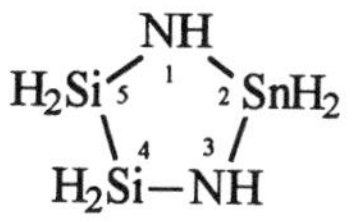

   1,3,4,5,2-Diazadisilastannolidine
   1,3-Diaza-4,5-disila-2-stannacyclopentane

5.

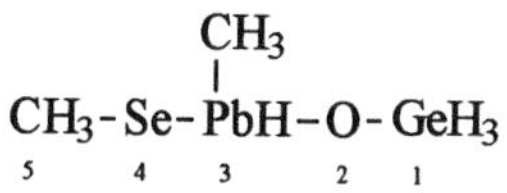

   $CH_3-Se-PbH-O-GeH_3$

   3-Methyl-2-oxa-4-selena-1-germa-3-
   plumbapentane
   [[Methyl(methylseleno)plumbyl]oxy]germane
   (Germyloxo)hydrido(methaneselenolato)-
   methyllead

6.

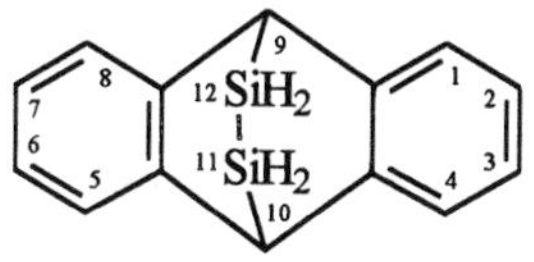

   9,10-Dihydro-9,10-disilanoanthracene

7.

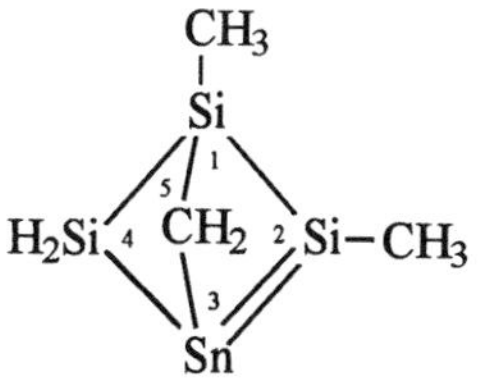

   1,2-Dimethyl-1,2,4-trisila-3-stanna-
   bicyclo[1.1.1]pent-2-ene

8.

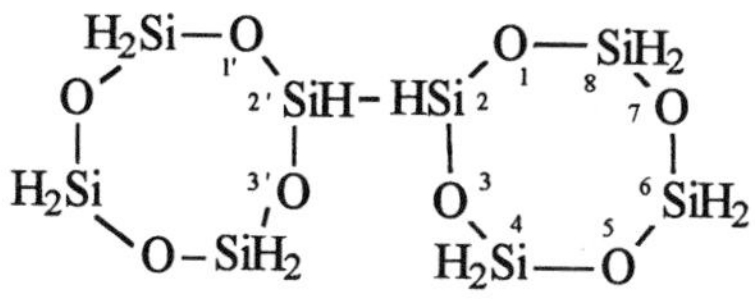

   2,2'-Bi(cyclotetrasiloxane)

9.

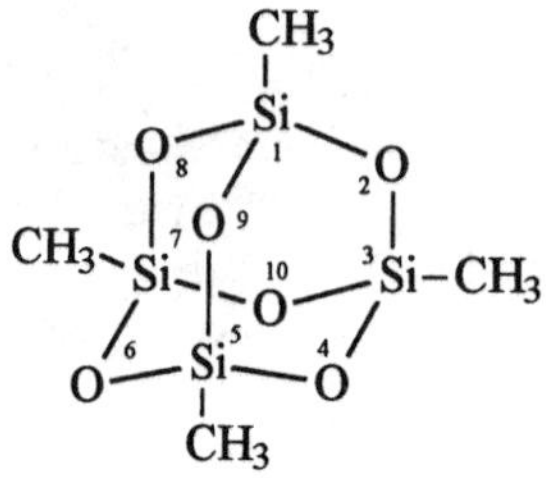

Tetramethyltricyclo[3.3.1.1$^{3,7}$]tetrasiloxane
Tetramethyl-2,4,6,8,9,10-hexaoxa-1,3,5,7-
   tetrasilatricyclo[3.3.1.1$^{3,7}$]decane

10.   $(CH_3)_3Si-CH_2-CO-Si(CH_3)_3$

Trimethyl[(trimethylsilyl)acetyl]silane
(1-Oxoethane-1,2-diyl)bis(trimethylsilane)

11.

$$CH_3-\overset{\overset{\displaystyle O}{\|}}{Si}-CH_2CH_3$$

Ethyl(methyl)oxosilane
  (*not* Ethyl(methyl)silanone)

12.   $(C_6H_5)_2\underset{1}{Si}=\underset{N}{N}-CH_3$

*N*-Methyl-1,1-diphenylsilanimine

13.   $CH_2=Si=CH_2$

Dimethylidenesilane

14.

$$CH_2=CH-\overset{\overset{\displaystyle CH_3}{|}}{Si}=CHCH_3$$

Ethylidene(methyl)(vinyl)silane

15.   $(CH_3)_2Si=SO$

Dimethyl(sulfinyl)silane
Dimethyl(oxo-$\lambda^4$-sulfanylidene)silane
Dimethylthioxosilane *S*-oxide

16.   $CH_3-SnH_2-CN$

Methylstannanecarbonitrile
Cyano(methyl)stannane

17.

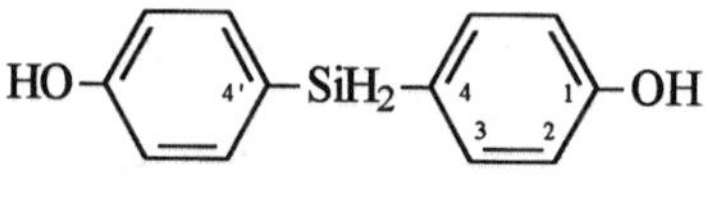

4,4′-Silanediyldiphenol

18.   $(CH_3)_2\underset{1}{Si}H-\underset{N}{N}H-C_6H_5$

1,1-Dimethyl-*N*-phenylsilanamine
(Dimethylsilyl)phenylamine
*N*-(Dimethylsilyl)aniline
*N*-(Dimethylsilyl)benzenamine

19.   $(CH_3)_3Si-\underset{N'}{N}=\underset{N}{Si}=\underset{1}{N}-Si(CH_3)_3$

*N*,*N*′-Bis(trimethylsilyl)silanediimine

20.

$$CH_3-\underset{3}{Si}H_2-\underset{2}{N}H-\overset{\overset{\displaystyle CH_3}{|}}{\underset{1}{Si}}H-NH-C_6H_5$$

1,3-Dimethyl-*N*-phenyldisilazan-1-amine
1,3-Dimethyl-3-(phenylamino)disilazane
1-Methyl-*N*-(methylsilyl)-*N*′-phenyl-
  silanediamine

21.   $\underset{N'}{H_2N}-\underset{17}{SiH_2}-\underset{16}{NH}-[\underset{15-2}{SiH_2-NH}]_7-\underset{1}{SiH_2}-\underset{N}{NH_2}$    Nonasilazane-1,17-diamine

22.   $(CH_3)_3Si-Li$

(Trimethylsilyl)lithium

23.   $Na^+\left[(CH_3)_3Si-S^-\right]$

Sodium trimethylsilanethiolate

24.  $(CH_3)_3Si-NCO$                     Isocyanatotrimethylsilane

Trimethyl(sulfinylamino)silane

25.  $(CH_3)_3Si-N=SO$               Trimethyl(thionitroso)silane $S$-oxide

1-Oxo-$N$-(trimethylsilyl)-$\lambda^4$-sulfanimine

26.  $(CH_3)_3Si-O-CO-NH-Si(CH_3)_3$     Trimethylsilyl (trimethylsilyl)carbamate

27.

$$CH_2CH_3$$
$$|$$
$$CH_3-Si(CH_3)_2-CHCH_2CH_2-SiH_2-CH_2-Si(CH_3)_2-CH_3$$
$$\quad 9 \quad\quad 8 \quad\quad\quad 7 \quad 6 \quad 5 \quad 4 \quad\quad 3 \quad 2 \quad\quad\quad 1$$

7-Ethyl-2,2,8,8-tetramethyl-2,4,8-trisilanonane

[(Trimethylsilyl)methyl][3-(trimethylsilyl)-

pentyl]silane

### Reference

1. American Chemical Society. The Report of the ACS Nomenclature, Spelling, and Pronounciation Committee for the First Half of 1952, F. Organosilicon Compounds. *Chem. Eng. News* **1952**, *30*, 4517–4522.

# 27

# Boron Compounds

Compounds containing at least one organic group attached to a boron atom through a carbon atom are called organoboron compounds. This chapter also includes other compounds containing boron, such as the carbaboranes, usually shortened to carboranes, esters and salts of boron acids, and addition compounds with amines, ethers, and other heteroatoms. Sources for the material in this chapter in addition to the general references given in the preliminary material are found at the end of this chapter.[1-4]

To name the full range of organic boron compounds requires extension and, in some cases, modification of organic nomenclature principles because of the unique bonding characteristics of boron. In addition to chains and rings formed by traditional bonds like the other nonmetals, boron also forms molecular and ionic networks of atoms characterized by triangulated clusters of atoms. Organic derivatives of such polyboron hydrides may also be considered organoboron compounds. Triangulated polyboron hydrides in which boron atoms have been replaced by carbon atoms are usually not classified as organoboron compounds, unless substituted by organic groups. Finally, boron compounds often act as Lewis acids and form "addition compounds" with compounds having atoms with non-bonding electron pairs, such as nitrogen and oxygen.

## Acceptable Nomenclature

### Borane, Polyboranes, and Heteropolyboranes

The hypothetical parent hydride $BH_3$ is called borane and the general class of neutral polyboron hydrides is known as *polyboranes*. Stoichiometric names for polyboranes containing more than one boron atom are named by adding numerical prefixes such as di-, tri-, and so on, to the name borane; a parenthetical arabic number is suffixed to the name to indicate the number of hydrogen atoms in a specific polyborane. It is necessary to designate the number of hydrogen atoms in all polyborane structures, including diborane, because the number of hydrogen atoms does not follow a regular relationship as with the hydrides of most other nonmetals. For example, the two diboranes shown below have the names diborane(4) and

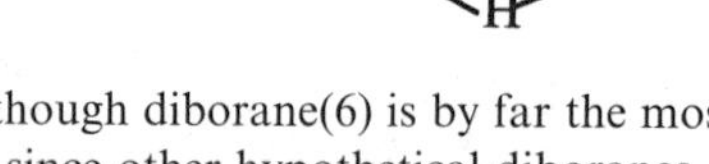

$$H_2BBH_2 \qquad H_2B\diagup\!\!\!\!\overset{\displaystyle H}{\underset{\displaystyle H}{\diagdown}}\!\!\!\!\diagdown BH_2$$

diborane(6), respectively. Although diborane(6) is by far the most common, the parenthetical (6) should never be omitted, since other hypothetical diboranes, such as $B_2H_2$ and $B_2H_3$, are possible. It is customary, however, to omit the parenthetical number for $BH_3$ even though the parent hydrides BH and $BH_2$ can be named borane(1) and borane(2), and are by CAS; the names boryl (or boranyl) and borylidene (or boranylidene) for BH and $BH_2$ are found in the recommendations by IUPAC. Other polyboranes are named in a similar manner; simple italicized prefixes, such as *catena-*, indicating a linear structure (not to be confused with the meaning of catena in the phrase "catena compounds", which refers to interlinked cyclic structures), and *cyclo-*, may be used to indicate structure (see also organoboron heterocycles later in this chapter).

$B_4H_6$          Tetraborane(6)

$\underset{4\ \ 3\ \ \ 2\ \ \ 1}{H_2BBHBHBH_2}$     *catena*-Tetraborane(6)

$$\begin{array}{c} BH_2 \\ | \\ \underset{3\ \ 2\ \ 1}{H_2BBBH_2} \end{array}$$     2-Boryl-*catena*-triborane(5)

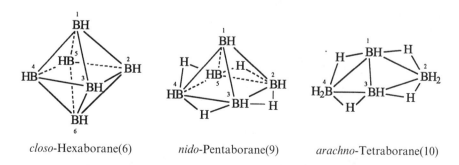

*cyclo*-Tetraborane(4)

*tetrahedro*-Tetraborane(4)

A series of italicized structural prefixes, such as *closo-*, *nido-*, and *arachno-*, describe structure and numbering of many polyboranes:

- Fully triangulated polyboranes having 4–12 boron atoms are described by the prefix *closo-*.
- A polyborane formally derived from a *closo*-polyborane by the removal of one boron atom, the one of highest connectivity if there is a choice, is described by *nido-*; to avoid ambiguity there can be only one *nido*-polyborane for each $B_n$ polyhedral polyborane structure.
- A polyborane derived from a *nido*-polyborane by the removal of one boron atom from the open face, the one of highest connectivity if there is a choice, is described by *arachno-*; to avoid ambiguity there can only be one *arachno*-polyborane for each $B_n$ polyhedral polyborane structure.

The following series of polyboranes illustrates these criteria:

*closo*-Hexaborane(6)      *nido*-Pentaborane(9)      *arachno*-Tetraborane(10)

Other descriptive prefixes are available, such as *hypho-*, *isocloso-*, and *isonido-*, but the reality is that all polyborane structures cannot be described in this manner. Those to which a prefix does not apply can only be given stoichiometric names, but see references 5–9.

*Heteropolyboranes* are polyboron hydrides in which boron atoms have been replaced by other nonmetal or semimetal (see chapter 28) atoms. They are named by an adaptation of replacement nomenclature (see chapter 3) whereby the heteroatoms, like the boron atoms of the polyborane structure, have no bonding number restrictions. The "a" prefixes of organic replacement nomenclature indicate the replacement of boron atoms by heteroatoms without regard to attached hydrogen atoms of either the boron atoms or the heteroatoms. The number of hydrogen atoms of the final structure is indicated by a parenthetical numerical suffix. *Carbaboranes* (also known as carboranes as a class) are heteropolyboranes in which the replacing atoms are carbon atoms with no bonding number restrictions; in names the prefix carba is used.

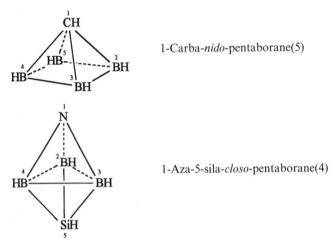

1-Carba-*nido*-pentaborane(5)

1-Aza-5-sila-*closo*-pentaborane(4)

Organic derivatives of the boranes are named substitutively with the traditional organic characteristic group prefixes and suffixes, except that the hydroxy group and its chalcogen analogs are expressed by prefixes such as hydroxy and mercapto (or sulfanyl) and not by suffixes such as -ol and -thiol; however, suffixes are used when hydroxy, mercapto, and so on, groups are attached to carbon atoms of carboranes. Hydroxy, mercapto, and so on, derivatives of $BH_3$, are named as boron acids (see below). Amino substituents that do not bridge boron atoms are usually expressed by the suffix -amine or the prefix amino; accordingly, although amino derivatives of $BH_3$ are named as amines, there is no reason why they should not be named acceptably as amides of a boron acid (see below). Similarly, halogen derivatives of $BH_3$ are usually expressed by appropriate prefixes, but again there is no reason why they should not be named acceptably as acid halides of a boron acid (see below). In names of substituted polyboranes, the parenthetical arabic number that indicates the number of hydrogen atoms in the polyborane parent hydride is used.

| | |
|---|---|
| $(CH_3)_2BH$ | Dimethylborane |
| $Cl_2B-CH_3$ | Dichloro(methyl)borane<br>Methylboronic dichloride |
| $CH_3CH_2-BH-N(CH_3)_2$ | 1-Ethyl-*N,N*-dimethylboranamine<br>1-Ethyl-*N,N*-dimethylborinic amide<br>Ethylborinic dimethylamide |
| $CH_3-HB\overset{H}{\underset{H}{\diagup\diagdown}}BH-OH$ | 1-Hydroxy-2-methyldiborane(6)<br>[*not* 2-Methyldiboran(6)-1-ol] |
| $(C_6H_5)_2B-COOH$ | Diphenylboranecarboxylic acid |

$$Cl\underset{\underset{(CH_3)_2N}{\underset{3\ 2\ 1}{}}}{\overset{\overset{Cl}{\underset{BBB}{}}}{}}\overset{Cl}{\underset{N(CH_3)_2}{}}$$

1,2,3-Trichloro-$N,N,N',N'$-tetra-methyl-*catena*-triborane(5)-1,3-diamine

$$(C_6H_5)_2B-NH-[CH_2]_4-NH-B(C_6H_5)_2$$
$$\phantom{(C_6H_5)_2B-NH-}N' \quad 4\text{-}1 \quad N$$

$N,N'$-Butane-1,4-diylbis(1,1-diphenyl-boranamine)
$N,N'$-Bis(diphenylboryl)butane-1,4-diamine

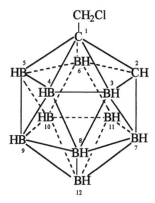

1-(Chloromethyl)-1,2-dicarba-*closo*-dodecaborane(12)

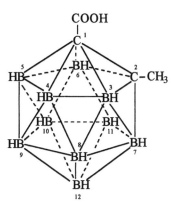

2-Methyl-1,2-dicarba-*closo*-dodeca-borane(12)-1-carboxylic acid

Substitutive prefix names for groups derived from borane are as follows: boryl, $H_2B-$; boranediyl, $-BH-$; borylidene, $HB=$ (the prefix borylene is no longer encouraged); borylidyne, $B\equiv$; boranylylidene, $-B=$; and boranetriyl, $>B-$ (CAS uses borylidyne). The contracted form boryl is used instead of boranyl to avoid the need for the prefix bis- to distinguish between two boryl, $H_2B-$, groups and one diboranyl, $H_2BH_2BH-$, group. Names for substituent prefixes derived from di- and polyboranes are formed in the usual manner; the parenthetical arabic number indicating the number of hydrogen atoms in the polyborane parent hydride is retained in the names of substituent prefixes, for example, diboran(6)yl and *nido*-penta-borane(9)-2,3-diyl.

$$(CH_3)_3Si-BF_2$$

Difluoro(trimethylsilyl)borane
(Difluoroboryl)trimethylsilane
(Trimethylsilyl)boronic difluoride

$$\underset{2'}{\overset{\overset{CH_3}{\mid}}{HO-CH_2CH}}-B(CH_3)-\underset{2}{\overset{\overset{^3CH_3}{\mid}}{CHCH_2}}-\underset{1}{OH}$$

2,2'-(Methylboranediyl)di-propan-1-ol

$$(CH_3)_2B\underset{2}{\overset{\overset{H}{}}{\underset{H}{}}}\underset{1}{BH}-\underset{3\ 2}{\underset{4}{\bigcirc}}\underset{1}{-CO-O-CH_2CH_3}$$

Ethyl 4-[2,2-dimethyldiboran(6)-1-yl]-benzoate

$$H_2N-\bigcirc_{3''}\underset{\underset{H_2N-\bigcirc_{3'}}{}}{\overset{}{B}}-\underset{2\ 3}{\overset{1}{\bigcirc}}NH_2$$

3,3',3''-Boranetriyltrianiline
3,3',3''-Boranetriyltri-benzenamine
Tris(3-aminophenyl)borane

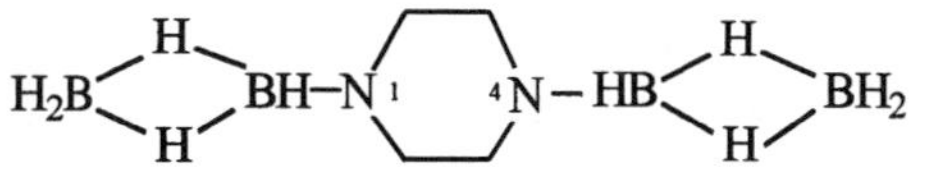

$$H_2B \overset{H}{\underset{H}{\diagup}} BH-N^1 \quad {}_4N-HB \overset{H}{\underset{H}{\diagup}} BH_2 \qquad \text{1,4-Bis[diboran(6)-1-yl]piperazine}$$

Substitution of bridging hydrogen atoms of polyboranes is indicated by the Greek letter $\mu$ prefixed to the name for the substituting group.

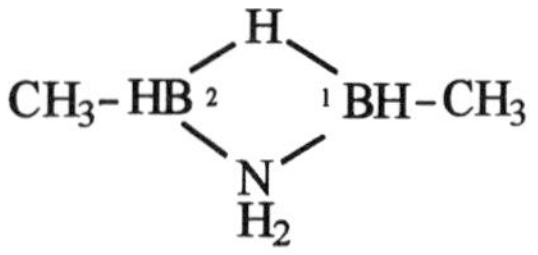

$$CH_3-HB^2 \quad {}^1BH-CH_3$$

$\mu$-Amino-1,2-dimethyldiborane(6)
[*not* 1,2-Dimethyldiborane(6)-$\mu$-amine]

Substitution of one or more bridging hydrogens can result in polyborane compounds that may also be named by replacement nomenclature.

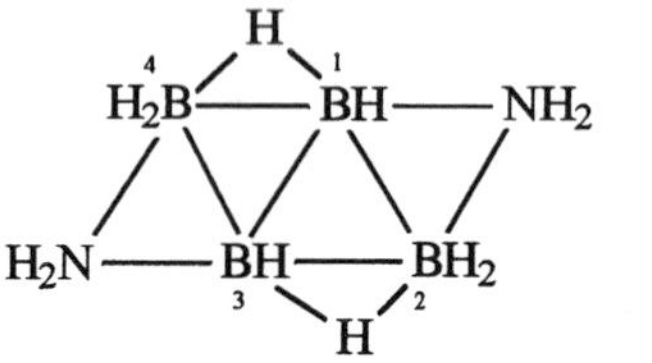

1,2:3,4-Di-$\mu$-amino-*arachno*-tetraborane(10)
1,4-Diaza-*arachno*-hexaborane(12)

Stereoisomers of disubstituted diborane(6) may be distinguished by the prefixes *cis-* and *trans-* (see chapter 30). Stereoisomers of substituted *arachno*-polyboranes may be distinguished by the prefixes *endo-* and *exo.* For an *arachno*-structure, when considered as a fragment of a dodecahedral polyborane *endo-* indicates that the substituent is directed toward the interior of the dodecahedron and *exo-* indicates that it projects away from the dodecahedron.

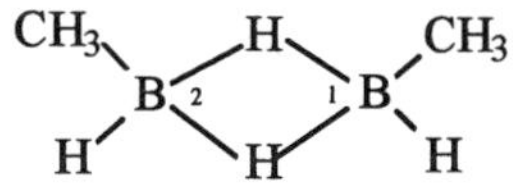

*cis*-1,2-Dimethyldiborane(6) 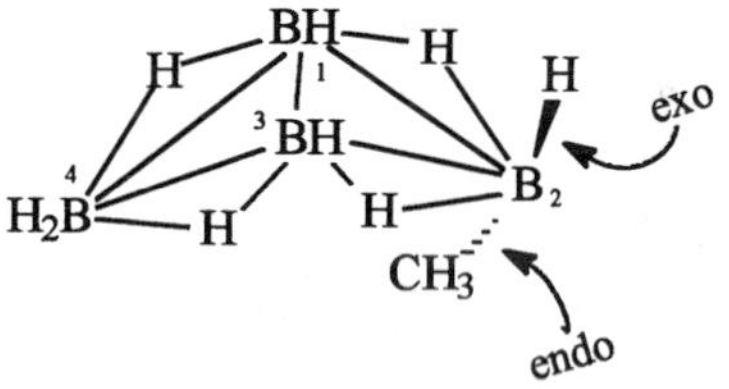          *endo*-2-Methyl-*arachno*-tetraborane(10)

## Boron Acids

Hydroxy derivatives of borane and diborane(4) are named as acids; their names are given in table 27.1. IUPAC has expressed disapproval of the names boronic acid and borinic acid, but these names continue to be widely used, are used by CAS, and are quite consistent with the names phosphonic acid and phosphinic acid (see chapter 25). CAS attaches the synonym line formula $H_3BO_3$ to the name boric acid in order to distinguish it from other boron acids named as boric acid with synonym line formulas such as $H_4B_2O_5$, $H_3B_3O_6$, and $H_2B_4O_7$. Chalcogen analogs are named by an appropriate prefix, for example, trithioboric acid, thioboronic acid, and so on. Functional replacement nomenclature in which chalcogen replacement is indicated

### Table 27.1. Names of Common Boron Acids

| | |
|---|---|
| $B(OH)_3$ | boric acid |
| $HB(OH)_2$ | boronic acid |
| $H_2B-OH$ | borinic acid |
| $(HO)_2BB(OH)_2$ | hypoboric acid |

by infixes (see chapter 3, table 3.9) should be acceptable for the mononuclear boron oxo acids, giving names like boronothioic acid. Compounds formed by replacement of all hydroxy groups of the mononuclear boron acids by $-NHNH_2$, $-NH_2$, or halo groups are officially named as derivatives of hydrazine (or diazane), as boranamines, or as halo derivatives of borane, respectively; however, again it seems reasonable to apply the principles of functional class nomenclature (see chapter 3) and functional replacement nomenclature (see chapter 3) here as well. Perhalo derivatives of borane and diborane(4) are often given names such as boron trichloride and diboron tetrafluoride.

| | |
|---|---|
| $BF_3$ | Boric trifluoride |
| | Boron trifluoride |

$$\underset{N' \quad 1 \quad N}{CH_3-NH-\overset{\overset{\displaystyle CH_2CH_3}{|}}{B}-NH-CH_3}$$

1-Ethyl-$N$,$N'$-dimethylboronic diamide
Ethylboronic $N$,$N'$-dimethyldiamide

$$\underset{1 \quad 2}{(CH_3)_2B-NHNH-CH_3}$$

1-(Dimethylboryl)-2-methylhydrazine
Dimethylborinic 2-methylhydrazide

Esters, salts, and mixed anhydrides are named in the same way as the corresponding derivatives of other mononuclear oxo acids (see chapter 12).

$$C_6H_5-O-\overset{\overset{\displaystyle O-C_6H_5}{|}}{B}-O-C_6H_5$$

Triphenyl borate

$$CH_3-S-\overset{\overset{\displaystyle O-CH_2CH_3}{|}}{B}-OH$$

$O$-Ethyl $S$-methyl hydrogen thioborate
$O$-Ethyl $S$-methyl hydrogen borothioate

$$C_6H_5-\overset{\overset{\displaystyle O-CH_2CH_3}{|}}{B}-S-CH_3$$

$O$-Ethyl $S$-methyl phenyl(thioboronate)
$O$-Ethyl $S$-methyl phenylboronothioate

$$\overset{+}{Na} \left[ (CH_3)_2B-\overset{-}{Se} \right]$$

Sodium dimethylselenoborinate
Sodium dimethylborinoselenoate

$$\overset{+}{Na} \left[ CH_3CH_2CH_2-\overset{\overset{\displaystyle OH}{|}}{B}-\overset{-}{O} \right]$$

Sodium hydrogen propylboronate

The prefix name borono for $(HO)_2B-$ is used by CAS; dihydroxyboryl is acceptable and consistent with prefix names like dihydroxyphosphino or dihydroxyphosphanyl for $(HO)_2P-$, and with names like dimethoxyboryl or hydroxy(methoxy)boryl.

$(HO)_2B-\langle\!\rangle-COOH$

4-Boronocyclohexanecarboxylic acid
4-(Dihydroxyboryl)cyclohexanecarboxylic acid

$CH_3-O-\overset{\overset{\displaystyle OH}{|}}{B}$ ... COOH

7-[Hydroxy(methoxy)boryl]-2-naphthoic acid

When one or two hydroxy groups of boric acid or one of the hydroxy groups of boronic acid has been replaced by an amino, halo or pseudohalogen group, the compound may be named by functional replacement nomenclature (see chapter 3) or as a derivative of an appropriate boron acid. As for other mononuclear oxo acids, groups expressed by functional replacement prefixes replace OH groups and do not substitute for hydrogen atoms (see also chapter 25).

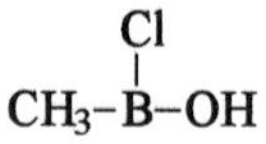

$$CH_3-\overset{\overset{\displaystyle Cl}{\displaystyle |}}{B}-OH$$

Methylboronochloridic acid
Methylchloroboronic acid
   (*not* Chloro(methyl)borinic acid)

$$(H_2N)_2B-O-CH_2CH_3$$

Ethyl borodiamidate
Ethyl diamidoborate
   (*not* Ethyl diaminoborinate)

Derivatives of hypoboric acid, $(HO)_2BB(OH)_2$, can be named in the same way as those of the phosphorus analog, hypodiphosphorus acid (see chapter 25), but they are usually named on the basis of diborane(4).

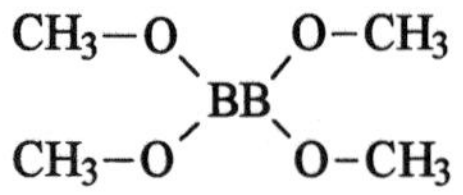

$$\begin{array}{cc} CH_3-O & O-CH_3 \\ & BB \\ CH_3-O & O-CH_3 \end{array}$$

Tetramethoxydiborane(4)
Tetramethyl hypoborate

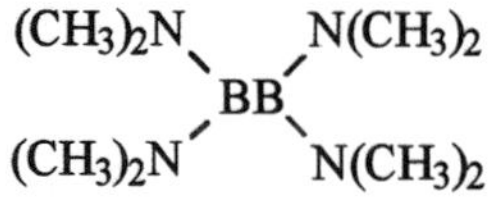

$$\begin{array}{cc} (CH_3)_2N & N(CH_3)_2 \\ & BB \\ (CH_3)_2N & N(CH_3)_2 \end{array}$$

Octamethyldiborane(4)tetramine
Octamethylhypoboric tetraamide
Hypoboric tetrakis(dimethylamide)

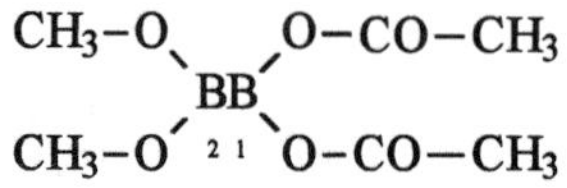

$$\begin{array}{cc} CH_3-O & O-CO-CH_3 \\ & BB \\ CH_3-O\ ^{2\,1} & O-CO-CH_3 \end{array}$$

1,1-Diacetoxy-2,2-dimethoxydiborane(4)

Symmetrical anhydrides of the mononuclear boron acids are named by attaching a numerical prefix to the name of the acid. Alternatively, they may be named as hydroxy derivatives of $(ab)_n$a parent structures. CAS names these as boric acids followed by an appropriate synonym line formula, such as $H_4B_2O_5$. Unsymmetrical or mixed anhydrides are named in the same manner as anhydrides of organic acids (see chapter 12).

$$(HO)_2B-O-B(OH)_2$$

Diboric acid
Tetrahydroxydiboroxane
Boric monoanhydride

$$CH_3-CO-O-B(OH)_2$$

Acetic boric monoanhydride
Monoacetic boric anhydride

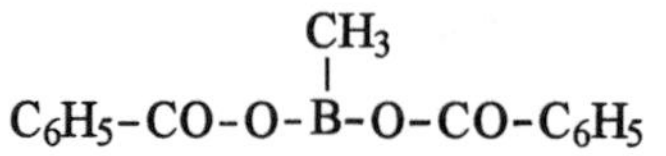

$$C_6H_5-CO-O-\overset{\overset{\displaystyle CH_3}{\displaystyle |}}{B}-O-CO-C_6H_5$$

Dibenzoic methylboronic dianhydride

$$(CH_3)_2B-O-B(CH_3)_2$$

Dimethylborinic anhydride
Tetramethyldiboroxane

Peroxy boron acids can be named using the prefix peroxy or by using replacement infixes such as peroxo and thioperoxo (see table 3.9, chapter 3). CAS describes all peroxyboric acids as perboric acid and defines the structure by means of a synonym line formula.

$$(CH_3)_2B-OO-B(CH_3)_2$$

Tetramethylperoxydiborinic acid

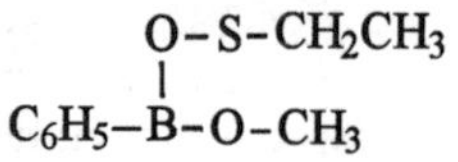

$$C_6H_5-\overset{\overset{\displaystyle O-S-CH_2CH_3}{\displaystyle |}}{B}-O-CH_3$$

*OS*-Ethyl *O*-methyl phenylborono(thioperoxoate)
*OS*-Ethyl *O*-methyl phenylmono(thioperoxy)boronate

## Acyclic Organoboron Compounds

The number of compounds having chains of boron atoms to which the principles of hydrocarbon nomenclature can be applied is relatively small because of the strong tendency for triangulation and for hydrogen atoms to bridge boron atoms. However, the principles of hydrocarbon nomenclature can be applied to compounds in which boron atoms are separated by other elements. Replacement nomenclature (see chapter 3) and the alternating atom system, $(ab)_n a$ (see also chapter 3), where applicable, are also used.

$$\underset{3'''}{\text{HOOC}-\text{CH}_2\text{CH}_2} \qquad \underset{3'}{\text{CH}_2\text{CH}_2-\text{COOH}}$$

$$\text{HOOC}-\text{CH}_2\text{CH}_2-\underset{3''}{\text{B}}-\text{CH}_2\text{CH}_2-\underset{3}{\text{B}}-\text{CH}_2\text{CH}_2-\text{COOH}$$

3,3′,3″,3‴-[Ethane-1,2-diylbis(boranetriyl)]tetrakis(propanoic acid)

$$\text{CH}_3-\text{HN}\underset{N'}{\diagdown}\underset{B}{\overset{\text{CH}_3}{|}}\underset{4}{\diagup}\underset{N}{\overset{}{}}\underset{3}{\diagdown}\underset{B'}{\overset{\text{CH}_3}{|}}\underset{2}{\diagup}\underset{N}{\overset{}{}}\underset{1}{\diagdown}\underset{B}{\overset{}{}}\underset{N}{\diagup}\text{NH}-\text{CH}_3$$

$\text{C}_6\text{H}_5 \quad \text{C}_6\text{H}_5 \quad \text{C}_6\text{H}_5$

N,N′,2,4-Tetramethyl-1,3,5-triphenyl-
    diimidotriboronic diamide
N,N′,2,4-Tetramethyl-1,3,5-triphenyl-
    triborazane-1,5-diamine
N,N′,2,4-Tetramethyl-1,3,5-triphenyl-2,4-diaza-
    1,3,5-triborapentane-1,5-diamine

$$\text{HOOC}-\underset{12}{\text{CH}_2}-\underset{11}{\text{O}}-\underset{10}{\overset{\overset{\text{Cl}}{|}}{\text{B}}}-\underset{9}{\text{O}}-\underset{8}{\text{CH}_2}\underset{7}{\text{CH}_2}-\underset{6}{\text{O}}-\underset{5}{\overset{\overset{\text{Cl}}{|}}{\text{B}}}-\underset{4}{\text{O}}-\underset{3}{\text{CH}_2}-\underset{1}{\text{COOH}}$$

4,9-Dichloro-3,5,8,10-tetraoxa-4,9-diboradodecanedioic acid

## Cyclic Boron Compounds

Cyclic organoboron compounds in which the boron atoms exhibit normal covalent bonds, are named and numbered by the methods of nomenclature for heterocyclic rings outlined in chapter 9.

Cyclohexaborane ("cyclo" hydrocarbon name)
Hexaborinane (Hantzsch–Widman name)
cyclo-Hexaborane(6) (a polyborane name)

1,5-Diborocane
1,5-Diboracyclooctane

1,3,5,2,4,6-Trioxatriborinane
    (Hantzsch–Widman name)
Cyclotriboroxane (repeating unit name)
Boroxin (traditional trivial name[a])

---

[a] The name boroxole for this compound and the names borthiole and borazole for its sulfur and nitrogen analogs should not be used because they imply five-membered rings according to the principles of the Hantzsch–Widman system.

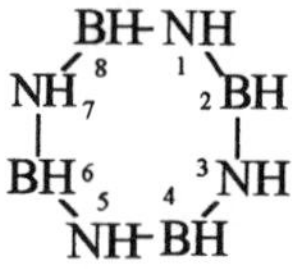

1,3,5,7,2,4,6,8-Tetrazatetraborocane
  (Hantzsch-Widman name)
Cyclotetraborazane (repeating unit name)

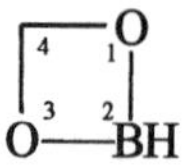

1,3,2-Dioxaboretane

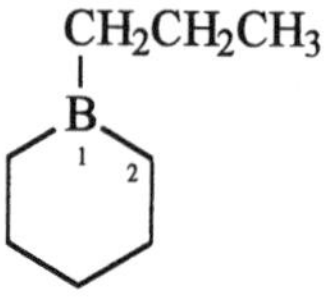

1-Propylborinane

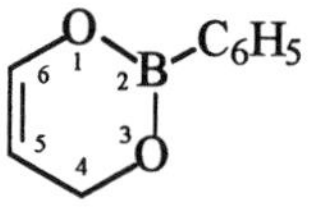

2-Phenyl-4$H$-1,3,2-dioxaborinine

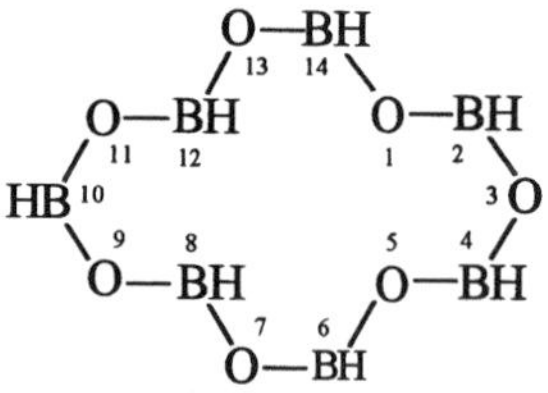

1,3,5,7,9,11,13-Heptaoxa-2,4,6,8,10,12,14-
  heptaboracyclotetradecane
  (skeletal replacement name)
Cycloheptaboroxane (repeating unit name)

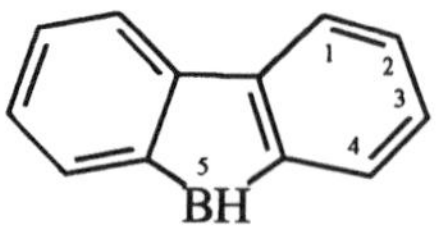

5$H$-Dibenzoborole

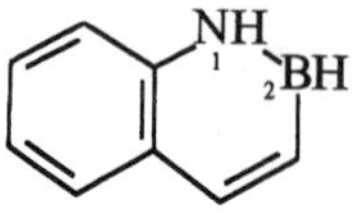

1,2-Dihydro-1,2-benzazaborine
1,2-Dihydrobenz[$e$][1,2]azaborine
  (*not* 2,1-Borazaronaphthalene)

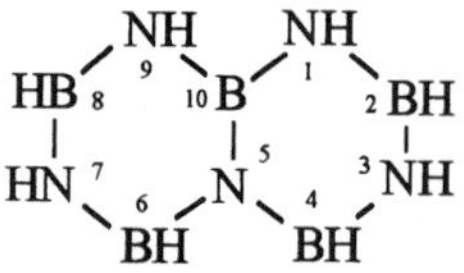

Octahydro[1,3,5,2,4,6]triazatriborino-
  [1,2-$a$][1,3,5,2,4,6]triazatriborine
Bicyclo[4.4.0]pentaborazane (repeating unit name;
  numbering is not that shown)

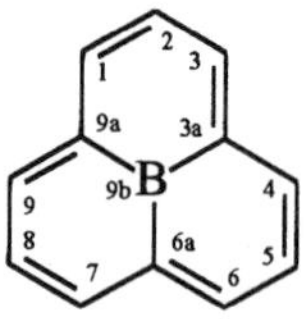

9b-boraphenylene

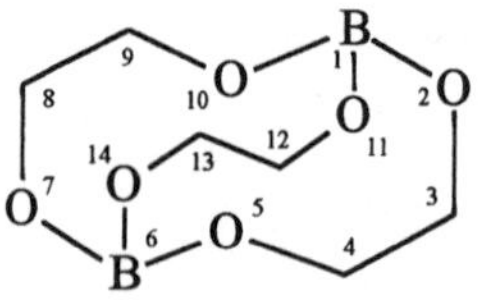

2,5,7,10,11,14-Hexaoxa-1,6-dibora-
  bicyclo[4.4.4]tetradecane

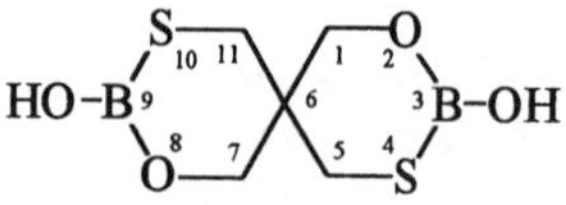

3,9-Dihydroxy-2,8-dioxa-4,10-dithia-
  3,9-diboraspiro[5.5]undecane

## Boron Radicals and Ions

Radicals and ions derived formally by the addition or removal of hydrogen atoms or hydrogen ions from a boron atom of a parent hydride are described by suffixes such as -yl, -ylidene, -ylium, -yliumyl, -uide, and -idyl; as well as replacement prefixes such as borylia, boranida, and boranuida, as described in chapter 33.

$HC\equiv C-BF^+$ — Ethynylfluoroboranylium

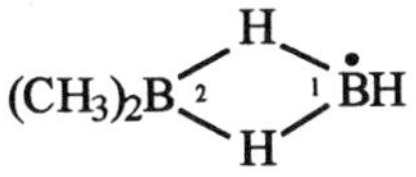

(structure) — 2,2-Dimethyldiboran(6)-1-yl

$(C_6H_4)_4B^-$ — Tetraphenylboranuide (substitutive name)[b]
Tetraphenylborate(1-) (coordination name)

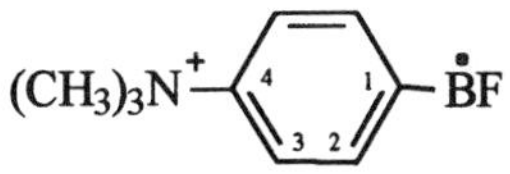

(structure) — Fluoro[4-(trimethylammoniumyl)phenyl]boryl

$(CH_3)_3P^+-CH_2CH_2-BF_3^-$ — Trifluoro[2-(trimethylphosphonio)ethyl]boranuide (substitutive name)[b]
Trifluoro[2-(trimethylphosphonio)ethyl]borate(1-) (coordination name)

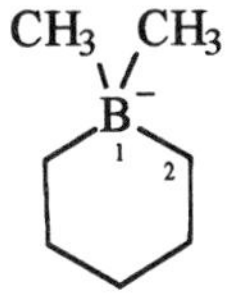

(structure) — 1,1-Dimethylborinan-1-uide
1,1-Dimethyl-1-boranuidacyclohexane

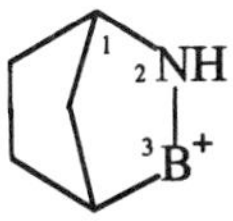

(structure) — 2-Aza-3-boryliabicyclo[2.2.1]heptane
2-Aza-3-borabicyclo[2.2.1]heptan-3-ylium

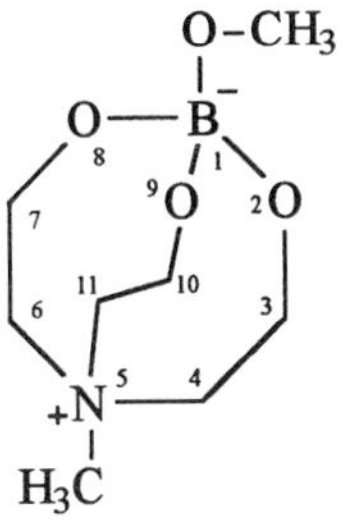

(structure) — 1-Methoxy-5-methyl-2,8,9-trioxa-5-azonia-1-boranuidabicyclo[3.3.3]undecane
1-Methoxy-5-methyl-2,8,9-trioxa-5-aza-1-bora-bicyclo[3.3.3]undecan-5-ium-1-uide

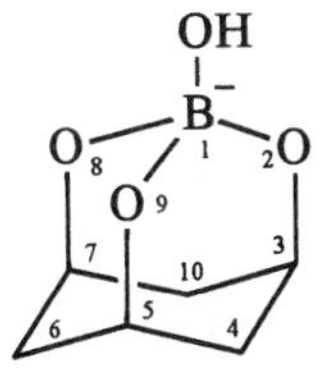

(structure) — 1-Hydroxy-2,8,9-trioxa-1-boranuidatricyclo-[3.3.1.1$^{3,7}$]cyclodecane (replacement name)[b]
1-Hydroxy-2,8,9-trioxa-1-boratricyclo-[3.3.1.1$^{3,7}$]cyclodecan-1-uide (replacement name)[b]
[1,3,5-Cyclohexanetriolato(3-)-*O,O′,O″*]hydroxo-borate(1-) (coordination name)

---

[b] Coordination nomenclature was used exclusively for such anionic boron compounds before the introduction of the suffix -uide and its corresponding replacement prefix boranuida, and it still may be used.

## Boron Addition Compounds

Organoboron compounds often act as Lewis acids (electron-pair acceptors), forming bonds between boron atoms and Lewis bases (electron-pair donors). Such bonds may be shown by means of an arrow directed from the Lewis base to the Lewis acid, as in $(CH_3)_3N{\rightarrow}BF_3$ or by a centered dot, as in $(CH_3)_2O \cdot BF_3$.

These types of compounds, called "addition compounds", may be named by the principles of coordination nomenclature, for example, (ammine)trifluoroboron, or as addition compounds (see chapter 3) through the use of a long dash (—) between the names of the components, for example, pyridine—trimethylborane. The number and position of each component may be indicated by appropriate numerical prefixes and locants or by the corresponding numbers separated by a colon (or the solidus) enclosed in parentheses following the name of the last component, which is usually the boron compound.

$OC \cdot BH_3$

Carbon monoxide—borane
(Carbon monoxide)trihydroboron
    (coordination name)

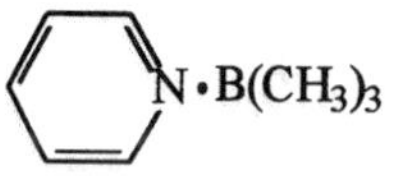

Pyridine—trimethylborane
Trimethyl(pyridine)boron
    (coordination name)

$(CH_3CH_2)_2O \cdot BF_3$

Diethyl ether—trifluoroborane
(Diethyl ether)trifluoroboron
    (coordination name)

$H_3B \cdot NH_2\text{-}CH_2CH_2\text{-}NH_2 \cdot BH_3$

Ethylenediamine—bis(borane)
Ethane-1,2-diamine—borane(1:2)
(Ethylenediamine)bis(trihydroboron)
    (coordination name)

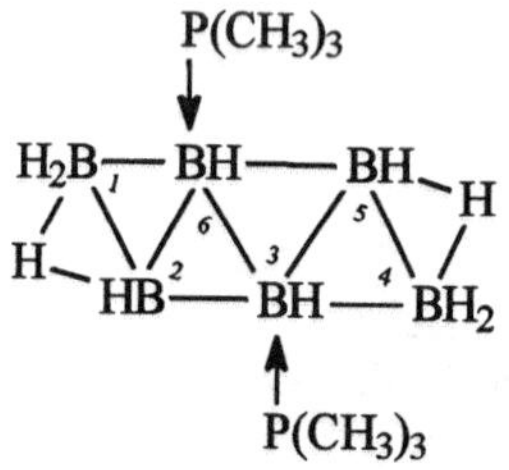

3,6-Bis(trimethylphosphane)—*arachno*-
    hexaborane(10)
3,6-Bis(trimethylphosphane)-*arachno*-
    decahydroboron (coordination name)

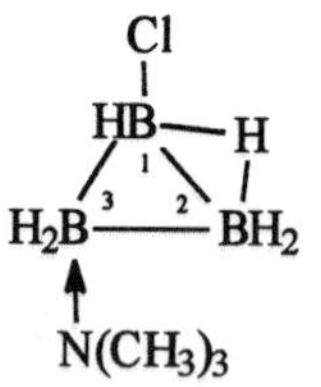

3-(Trimethylamine)—1-chlorotriborane(7)
1-Chloro-1,2-$\mu$-hydro-1,2,2,3,3-penta-
    hydro-3-(trimethylamine)-*cyclo*-
    triboron (coordination name)

The point of attachment of the Lewis base to the boron compound may be indicated by italicized atomic symbols inserted at each end of the long dash, each referring to the component nearest to it. Superscripts to indicate position are used as needed.

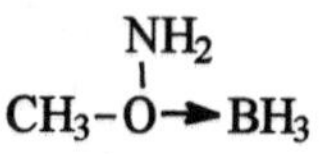

*O*-Methylhydroxylamine(*O*—*B*)borane
Trihydro(*O*-methylhydroxylamine-*O*)boron
    (coordination name)

$$H_2B-NH_2 \longrightarrow B(CH_3)_3$$

Boranamine($N$—$B$)trimethylborane
(Boranamine-$N$)trimethylboron
  (coordination name)

$$C_6H_5-NH-CO-NH_2 \longrightarrow BH_3$$
$$\quad\;\; 1 \qquad 2 \qquad 3$$

1-Phenylurea($N^3$—$B$)borane
Trihydro(1-phenylurea-$N^3$)boron
  (coordination name)

2,2′-Bipyridine($N,N'$—$B$)-10$H$-phenoxa-
  borin-10-ylium perchlorate
(2,2′-Bipyridine-$N,N'$)(oxydi-1,2-phenylene)-
  boron(1 +) perchlorate
  (coordination name)

$[ClO_4]^-$

An intramolecular Lewis base adduct of an organoboron compound may be indicated by an element symbol cited in front of the name of the complete compound.

($O^2$—$B$)-4-[(Difluoroboryl)oxy]pent-3-en-2-one
Difluoro(2,4-pentanedionato-$O,O'$)boron
  (coordination name)

($N$—$B$)-[(Dimethylboryl)oxy]quinoline
Dimethyl(8-quinolinolato-$O,N$)boron
  (coordination name)

($N$—$B$)-(1,3,2-Benzodioxaborol-2-ylmethyl)-
  aniline
(2-Aminobenzyl)[benzene-1,2-diolato(2-)-
  $O,O'$]boron (coordination name)

($N$—$B$)-2-[(Dimethylboryl)oxy]ethylamine
(2-Aminoethanolato-$O$)dimethylboron
  (coordination name)

Bis($N$—$B$)-6,6′-oxybis(8$H$-dibenzo[$d,h$][1,3,7,2]-
  dioxazaborecin-8-one)
Bis[[2-(2-hydroxybenzylidene)amino]-
  benzoato(2-)-$O,O',N$]-$\mu$-oxo-diboron
  (coordination name)

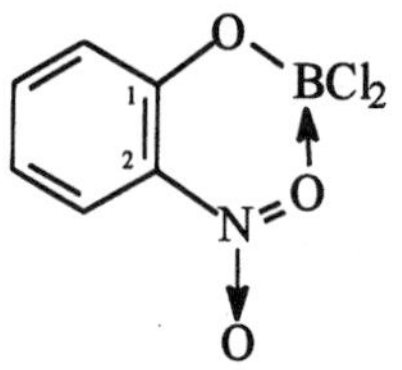

($O$—$B$)-Dichloro(2-nitrophenoxy)borane
Dichloro(2-nitrophenolato-$O^1,O^2$)boron
   (coordination name)

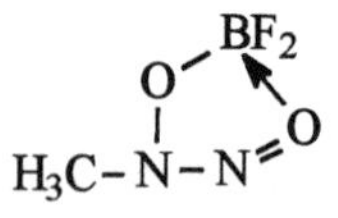

($O$—$B$)-$O$-(Difluoroboryl)-$N$-methyl-$N$-
   nitrosohydroxylamine
($N$-Nitroso-$N$-methylhydroxylaminato-$O,O'$)-
   boron (coordination name)

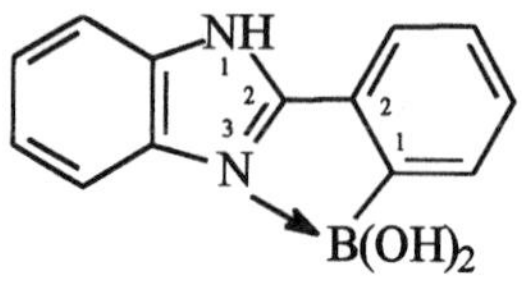

($N^3$—$B$)-[2-(1$H$-Benzimidazol-2-yl)phenyl]-
   boronic acid
[2-(1$H$-Benzimidazol-2-yl)phenyl-$N^3$]-
   dihydroxyboron (coordination name)

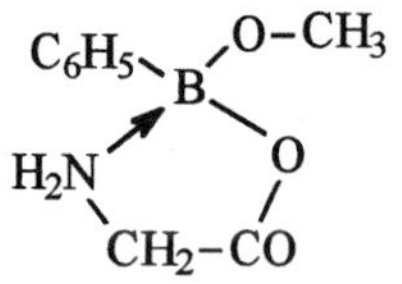

($N$—$B$)-Glycine methoxy(phenyl)borinic
   anhydride
(Glycinato-$O,N$)methoxophenylboron
   (coordination name)

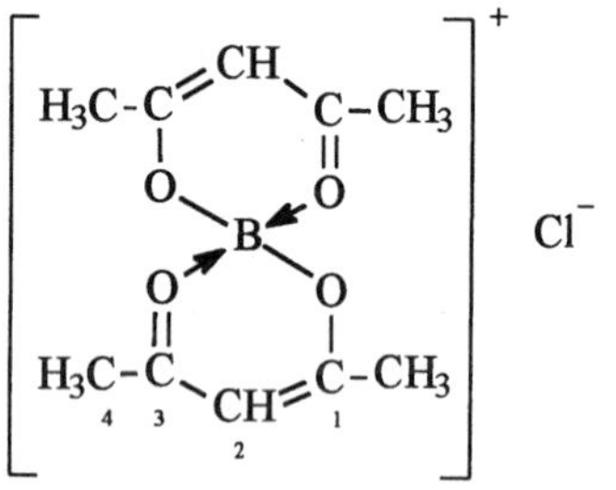

[2($O$—$B$)]-Bis[(1-methyl-3-oxobut-1-en-1-yl)oxy]-
   boranylium chloride
Bis(2,4-pentanedionato-$O,O'$)boron(1+)
   chloride

## Discussion

Compounds containing the element boron, with only six electrons in its outer shell, present a number of situations unique to organic nomenclature. To name the full range of organoboron compounds requires not only traditional organic nomenclature methods, but also adaptations of some of these methods as well as techniques borrowed from inorganic nomenclature.

Even so, there are a significant number of boron compounds whose names are not satisfactory to members of either the organic or inorganic community. For example, many neutral boron compounds can be conceptually viewed as formed by the replacement of a hydride ion in a parent compound by a neutral substituent. Organic nomenclature does not provide for such a situation, and must therefore rely on an additive approach. Existing coordination nomenclature does not easily deal with heteropolyboron hydrides and must rely on an adaptation of organic nomenclature. A combination of the substitutive techniques of organic nomenclature with the use of ligand names from coordination nomenclature to express neutral "substituents" might provide a solution to this unusual problem.

Although the structural prefixes *closo*-, *nido*-, and *arachno*- are adequate for many of the simpler polyboranes, carboranes, and other heteropolyboranes (including metallopolyboranes), they are not adequate for the full range of polyboranes. Other descriptors have been proposed to describe various structures of polyboranes and related compounds. The prefixes *hypho*- and *klado*- were introduced to indicate structures even more "open" than *arachno*-. The prefixes *iso*- and *neo*- have been used to distinguish between structural isomers of the polyboranes, for example between the two $B_{18}H_{22}$ isomers, but, once full structural detail is known, a fully structural name is recommended.. Other prefixes include *isocloso-*, *isonido-*, *isoarachno-*, *canasto-*, *anello-*, *precloso-*, *hypercloso-*, *pileo-* and *conjuncto-*. To simplify this nonsystematic approach, a systematic descriptor system has been proposed[5-9], but has yet to receive recognition.

Inorganic chemists do not like the names boronic and borinic acids and have recommended that they not be used. However, these names relate well to phosphorus and arsenic acid names, such as phosphonic and arsinic acids; they continue to be used in the literature and are used by CAS. In the hierarchy of hydrides used by CAS, boranes outrank only the hydrides of the Group 14 elements, silane, germane, stannane, plumbane, the chalcogen elements, and the hydrocarbons. The names boronic and borinic acid allow many boron compounds to be parent compounds rather than being expressed as substituents. Suggestions to name boron acids with the suffix -ol have not been taken seriously. Acid halide, hydrazide, and amide names have not been recognized officially, but are included in this book.

Names for linear polyacids derived formally by condensation of two or more molecules of the same boron acid (with elimination of water) have not, in general, been formed by traditional practices for naming linear homopolyacids of the other nonmetals; names such as diboric acid for $(HO)_2B-O-B(OH)_2$ would result. These compounds may also be named as hydroxy derivatives of $(ab)_n$ a parent structures (see chapter 3).

Although peroxy boron acids are named by the prefix per in CAS index nomenclature, use of prefixes such as peroxy and diperoxy or infixes such as peroxo would be consistent with the way peroxy acids of some of the other nonmetals are named and would produce much more specific names. The CAS name requires the use of a synonym line formula to differentiate among different perboric acids; such line formulas are usually nonspecific and should be used with care.

The capital italic letter *B* is occasionally used to indicate substitution of hydrogen atoms on boron atoms of an organoboron heterocycle. Such usage is discouraged; numerical locants are much preferred.

The prefix borazaro derives from a now obsolete method introduced in an attempt to indicate the presence of a $-NH \overset{+}{=} \overset{-}{BH}-$ linkage in a ring system. Locants preceding the prefix described the location of the linkage in the ring system; the "aro" syllable, presumably from aromatic, was meant to indicate the presence of a double bond which results in a zwitterion. This prefix has not been endorsed by nomenclature bodies; the dihydro heterocycle is isoelectronic and is recommended as the basis for naming such linkages.

In naming boron addition compounds, there have been no published recommendations for designating two or more of the same atoms. Numerical prefixes could be used, for example bis(*N—B*)-, or each italicized element symbol could be preceded by an arabic number, as (2*O—2B*)-. In this chapter the format [2(*O—B*)]- has been adopted.

## Additional Examples

1. $CH_3CH_2-B(CH_3)_2$                          Ethyldimethylborane

2. $Cl_2B-CH_2CH_2-BCl_2$                        Ethylenebis(dichloroborane)
                                                  Ethylenebis(boronic dichloride)

3.                           2-Ethyl-3-methyl-*nido*-pentaborane(9)

4. $(CH_3CH_2)_2B-CN$                             Diethylboranecarbonitrile
                                                  Diethylborinic cyanide

5.

$$CH_3$$
$$|$$
$$CH_3CH_2-B-NH-C_6H_5$$
$$\quad\quad\quad 1\quad N$$

1-Ethyl-1-methyl-*N*-phenylboranamine
*N*-(Ethylmethylboryl)aniline
1-Ethyl-1-methyl-*N*-phenylborinic amide
Ethylmethylborinic phenylamide

6.  $C_6H_5-CH_2CH_2-O-BCl_2$
$\quad\quad\quad\quad 2\quad 1$

2-Phenylethyl dichloroborinate
2-Phenylethyl borodichloridate

7.

$$O-CH_2CH_3$$
$$|$$
$$CH_3CH_2-O-B-O-CH_2CH_3$$

Triethyl borate
(*not* Ethyl borate)

8.  $CH_2CH=CHCH_2-O-B(OH)_2$
$\quad 4\quad 3\quad\quad 2\quad 1$

But-2-en-1-yl dihydrogen borate

9.

$$O-CH_3$$
$$|$$
$$CH_3CH_2CH_2-B-O-CH_3$$

Dimethyl propylboronate

10.

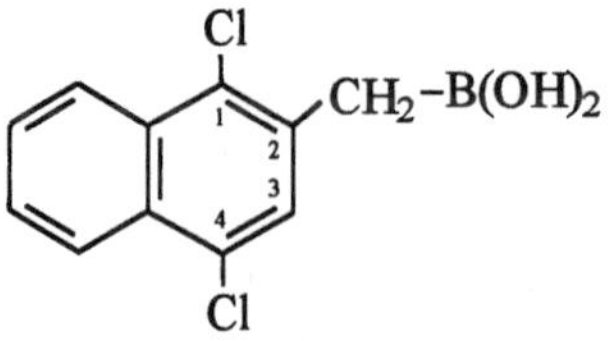

[(1,4-Dichloro-2-naphthyl)methyl]-
boronic acid

11.

$$CH_3$$
$$|$$
$$C_6H_5-CH_2CH_2-B-O-CH_2CH_3$$
$$\quad\quad\quad\quad 2\quad 1$$

Ethyl methyl(2-phenylethyl)borinate

12.

$$CH_3-CO-O$$
$$|$$
$$CH_3-CO-O-B-O-CO-CH_2CH_2CH_3$$

Boric diacetic propanoic trianhydride

13.

$$CH_3-O\quad\quad O-CH_2CH_3$$
$$\diagdown\quad\quad\diagup$$
$$BB$$
$$\diagup\quad 2\ 1\ \diagdown$$
$$CH_3-O\quad\quad\quad O-CH_2CH_3$$

1,1-Diethoxy-2,2-dimethoxydiborane(4)

14.

$$C_6H_5\ \ C_6H_5\ \ C_6H_5$$
$$|\quad\quad |\quad\quad |$$
$$HO-B-O-B-O-B-OH$$
$$\ \ 5\quad\ 4\ \ 3\quad 2\ \ 1$$

1,5-Dihydroxy-1,3,5-triphenyl-
triboroxane
Triphenyltriboronic acid

15.  $(C_6H_5)_2B-OOH$

Diphenylperoxyborinic acid
Diphenylborinoperoxoic acid

16.

$$CH_3\quad\ \ CH_3$$
$$|\quad\quad\ \ |$$
$$HO-B-OO-B-OH$$

Dimethylperoxydiboronic acid
Dioxybis(methylborinic acid)

17.

$$CH_3\ \ CH_2CH_3$$
$$|\quad\quad\ |$$
$$HOO-B-O-B-OH$$

Ethylboronic methylboronoperoxoic
monoanhydride
1-Ethyl-2-hydroperoxy-1-hydroxy-2-
methyldiboroxane

18.

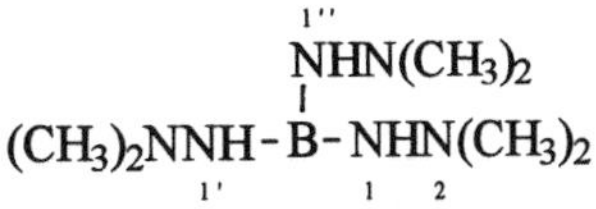

$$\underset{1'}{(CH_3)_2NNH} - \underset{1}{\overset{\overset{1''}{\overset{\displaystyle NHN(CH_3)_2}{|}}}{B}} - \underset{2}{NHN(CH_3)_2}$$

1,1',1''-Boranetriyltris(2,2-dimethyl-hydrazine)

19.

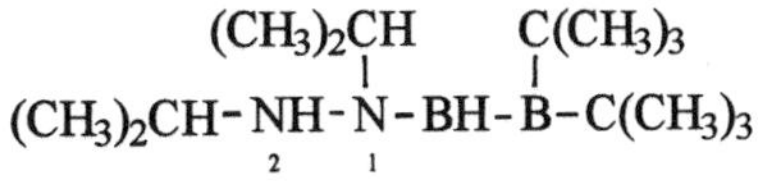

$$\underset{2}{(CH_3)_2CH} - \underset{1}{\overset{\overset{(CH_3)_2CH}{\phantom{x}}}{NH}} - \overset{\overset{C(CH_3)_3}{|}}{N} - BH - \overset{}{B} - C(CH_3)_3$$

1-(2,2-Di-*tert*-butyldiboran(4)-1-yl)-1,2-diisopropylhydrazine

20.

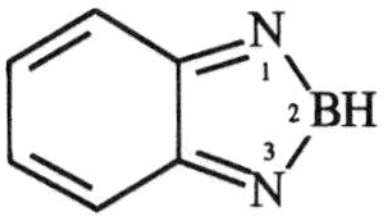

2*H*-1,3,2-Benzodiazaborole
2*H*-Benzo[*d*]-1,3,2-diazaborole

21.

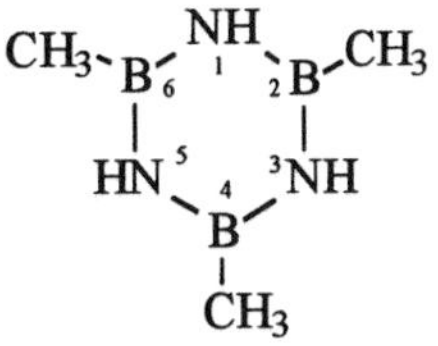

2,4,6-Trimethylborazine
Hexahydro-2,4,6-trimethyl-1,3,5,2,4,6-triazatriborine
2,4,6-Trimethylcyclotriborazane

22.

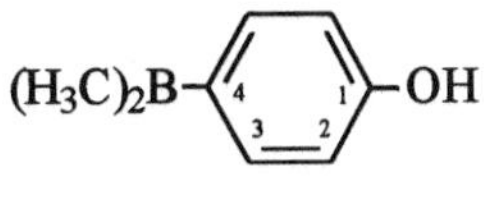

$(H_3C)_2B$——OH

4-(Dimethylboryl)phenol

23.

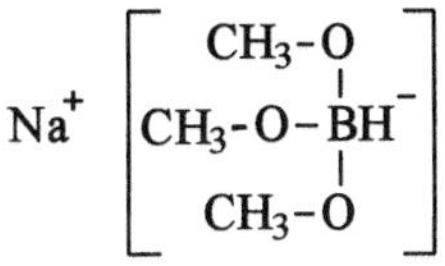

$$Na^+ \begin{bmatrix} CH_3-O \\ CH_3-O-\overset{\overset{\displaystyle |}{\phantom{}}}{\underset{\displaystyle |}{B}}H \\ CH_3-O \end{bmatrix}^-$$

Sodium trimethoxyboranuide
Sodium hydrotrimethoxoborate(1-)
 (coordination name)

24.  $(CH_3)_2S \cdot BH_2 - CH_3$

Dimethyl sulfide—methylborane
(Dimethylsulfide)methylboron
 (coordination name)

25.

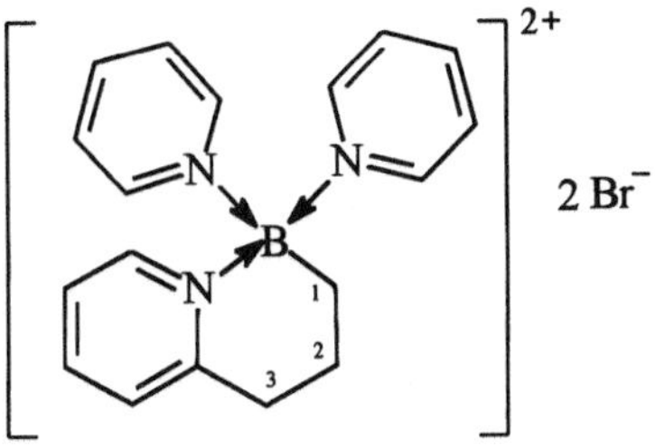

2 Br⁻

[2(*N*—*B*)]-bis(pyridine)(*N*—*B*)-[3-(2-pyridyl)propyl]borane-bis(ylium) dibromide
Bis(pyridine)[3-(2-pyridyl)propyl-boron](2+) dibromide
 (coordination name)

26.

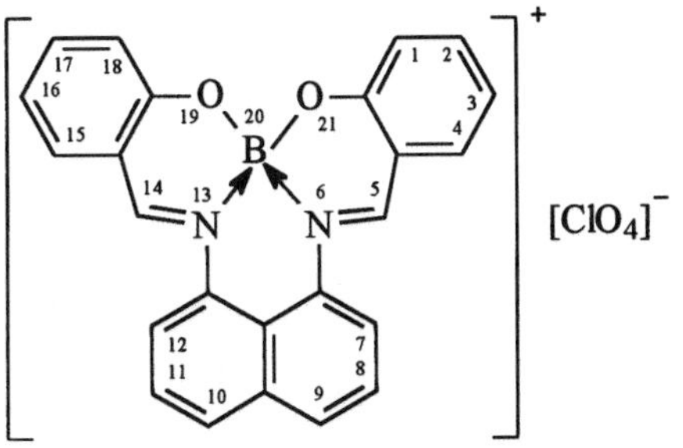

[ClO₄]⁻

[2(*N*—*B*)]-Dibenzo[*d,m*]naphtho-[1,8-*hi*]1,3,7,11,2]dioxadiaza-boracyclotetradecin-20-ylium
 perchlorate
[(2(*N*—*B*)]-Dibenzo[*d,m*]naphtho-[1,8-*hi*][1,3,7,11,2]dioxadiaza-boranyliacyclotetradecine
 perchlorate
[2,2'-[Naphthalene-1,8-diylbis-(nitrilomethylidyne)]diphenolato(2-)-*O*,*O*',*N*,*N*']boron(1+) perchlorate
 (coordination name)

**References**

1. American Chemical Society. The Nomenclature of Boron Compounds. *Inorg. Chem.* **1968**, 7, 1945–1964.
2. International Union of Pure and Applied Chemistry, Inorganic Chemistry Division, Commission on Nomenclature of Inorganic Chemistry. Nomenclature of Inorganic Boron Compounds. *Pure Appl. Chem.* **1972**, *30*, 681–710.
3. International Union of Pure and Applied Chemistry, Inorganic Chemistry Division, Commission on Nomenclature of Inorganic Chemistry. *Nomenclature of Inorganic Chemistry, Recommendations 1990*; Leigh, G. J., Ed.; Blackwell Scientific: Oxford, England, 1990; Chapter I-11, pp 207–237.
4. Block, B. P.; Fernelius, W. C.; Powell, W. H. *Inorganic Chemical Nomenclature: Principles and Practices*; American Chemical Society: Washington, DC, 1990; Chapter 11, pp 97–116.
5. Casey, J. B.; Evans, W. J.; Powell, W. H. A Descriptor System and Principles for Numbering Closed Polyboron Polyhedra with at Least One Rotational Symmetry Axis and One Symmetry Plane. *Inorg. Chem.* **1981**, *20*, 1333–1341.
6. Casey, J. B.; Evans, W. J.; Powell, W. H. A Descriptor System and Suggested Numbering Procedures for Closed Polyboron Polyhedra Belonging to $D_n$, $T$, and $C_s$ Symmetry Point Groups. *Inorg. Chem.* **1981**, *20*, 3556–3561.
7. Casey, J. B.; Evans, W. J.; Powell, W. H. Structural Nomenclature for Polyboron Hydrides and Related Compounds. 1. Closed and Capped Polyhedral Structures. *Inorg. Chem.* **1983**, *22*, 2228–2235.
8. Casey, J. B.; Evans, W. J.; Powell, W. H. Structural Nomenclature for Polyboron Hydrides and Related Compounds. 2. Nonclosed Structures. *Inorg. Chem.* 1983, *22*, 2236–2244.
9. Casey, J. B.; Evans, W. J.; Powell, W. H. Structural Nomenclature for Polyboron Hydrides and Related Compounds. 3. Linear *conjuncto*-Structures. *Inorg. Chem.* **1983**, *22*, 4132–4143.

## 28

# Organometallic Compounds

This chapter is concerned with compounds that contain at least one carbon atom of an organic structure attached directly to a metal or semimetal (metalloid) by a covalent bond, or compounds having two or more carbon atoms associated with a metal or semimetal by means of a delocalized bonding system. Complexes of large organic structures with metals, such as the chlorophylls, corrinoids, and hemes are outside the scope of this chapter. For the present purpose, the elements germanium, tin, lead, antimony, bismuth, and tellurium are considered semimetals, and the term "metal" includes these semimetals. Substitutive nomenclature as it applies to compounds of germanium, tin, and lead is discussed in chapter 26; to compounds of antimony and bismuth in chapter 25; and organic tellurium compounds are often considered along with the organosulfur and organoselenium compounds discussed in chapters 23 and 24. Organic derivatives of the semimetals usually can be named in more than one way and often the choice is one of personal preference.

Most organometallic compounds contain both "organic" and "inorganic" portions and therefore organometallic nomenclature often involves principles from both organic and inorganic nomenclature. In many instances, an organometallic compound, as defined above, can be described by two, or even more, equally acceptable names, some based on organic nomenclature and others based on inorganic nomenclature. Most of the discussion in this chapter is based on Section D in the 1979 edition of the IUPAC Organic Nomenclature Rules, a joint publication of IUPAC's Commissions on Nomenclature of Inorganic Chemistry and Nomenclature of Organic Chemistry. However, since this book is concerned with organic nomenclature, the emphasis will be on organic names for use by organic chemists; where appropriate, inorganic names will be given as acceptable alternatives. Furthermore, organic nomenclature may not be convenient for semimetallic compounds with one or more attached groups not bound by a covalent bond, or having more groups classically bound to the semimetal than can be accommodated by its bonding number, implied or stated by means of the λ-convention (see chapter 2). In such cases, inorganic nomenclature may be preferred; for detailed information inorganic nomenclature sources[1,2] should be consulted. These methods will be described briefly, with examples, throughout this chapter.

## Acceptable Nomenclature

Covalently bonded organometallic compounds of tri- and pentavalent antimony and bismuth, and of di- and tetravalent germanium, tin, and lead, may be named substitutively on the basis of the parent hydride names stibane ($SbH_3$) and $\lambda^5$-stibane ($SbH_5$); bismuthane ($BiH_3$) and $\lambda^5$-bismuthane ($BiH_5$) (see chapter 25); $\lambda^2$-germane ($GeH_2$) and germane ($GeH_4$); $\lambda^2$-stannane, ($SnH_2$) and stannane ($SnH_4$); $\lambda^2$-plumbane ($PbH_2$) and plumbane ($PbH_4$) (see chapter 26). Stibane and bismuthane are now preferred by IUPAC to the traditional, but still quite acceptable names, stibine and bismuthine. The name stiborane for $SbH_5$, analogous to arsorane and phosphorane, is still recognized by IUPAC but can be named $\lambda^5$-stibane by the λ-convention (see chapter 2). The names germylene or germylidene and stannylene or stannylidene can, and should, be reserved for use when it is desirable to reflect radical properties; names based on the $\lambda$-convention do not imply radical properties.

| | |
|---|---|
| $(C_6H_5)_3Sb$ | Triphenylstibane <br> Triphenylstibine |
| $(CH_3)_3BiCl_2$ | Dichlorotrimethyl-$\lambda^5$-bismuthane <br> Dichlorotrimethylbismuth |
| $(C_6H_5)_2Ge$ | Diphenyl-$\lambda^2$-germane <br>   (preferred to diphenygermylene or <br>   diphenylgermylidene) |
| $(CH_3CH_2)_3SnSn(CH_2CH_3)_3$ | Hexaethyldistannane |
| $(CH_3)_2PbCl_2$ | Dichlorodimethylplumbane |

Although organic derivatives of antimony and germanium with structures analogous to organic derivatives of arsenic and phosphorus acids have been named on the basis of parent acids, such as stibonic acid, $HSbO(OH)_2$, and germanonic acid, $HGeO(OH)$, this is not recommended. Such compounds should be named substitutively on the basis of the appropriate parent hydride or by means of additive, functional class, or coordination principles.

| | |
|---|---|
| $\overset{\displaystyle O}{\overset{\displaystyle \|}{(C_6H_5)_2Sb\text{-}OH}}$ | Hydroxyoxodiphenyl-$\lambda^5$-stibane (substitutive name) <br> Diphenylantimony hydroxide oxide (additive name) <br> Hydroxooxodiphenylantimony (coordination name) |
| $K^+\left[CH_3\text{-}GeO_2^-\right]$ | Potassium methyldioxogermanuide <br>   (substitutive name; see chapter 33) <br> Potassium methyloxidooxogermane <br> Potassium methyldioxogermanate (coordination name) |

Organometallic derivatives of tellurium corresponding to thiols, sulfides, sulfoxides, and sulfones may be named in the same way as these sulfur compounds (see chapters 23 and 24).

| | |
|---|---|
| $C_6H_5\text{-}TeH$ | Benzenetellurol <br> Tellurophenol |
| $Na^+\left[CH_3\text{-}Te^-\right]$ | Sodium methanetellurolate |
| $(CH_3CH_2)_2Te$ | Diethyltellane <br> Diethyl telluride (functional class name) |
| $CH_3\text{-}TeO_2\text{-}CH_2CH_3$ | Ethyl methyl tellurone (functional class name) <br> Ethyl(methyl)dioxo-$\lambda^6$-tellane <br> Ethyl(methyl)tellane dioxide (additive name) |

Organometallic compounds in which only organic groups are covalently bonded to a metal other than germanium, tin, lead, antimony, bismuth, or tellurium, are named by coordination nomenclature; that is, substituent prefix names for the organic groups are arranged alphabetically in front of the name of the metal.

$CH_3CH_2CH_2CH_2-Li$     Butyllithium

$Br-CH_2-Zn-C_6H_5$      (Bromomethyl)phenylzinc

$(C_6H_5)_3Ga$          Triphenylgallium

If ionic character is to be emphasized, binary names (see chapter 3), such as disodium acetylide ($2\,Na^+\,C_2^{2-}$), sodium methanide ($Na^+\,CH_3^-$), lithium aluminum hydride ($Li^+\,[AlH_4]^-$), and even lead tetraethyl , are often used. Cationic compounds consisting of four groups, at least one of which is an organic group, covalently bonded to one antimony or bismuth atom, are usually named as "onium" cations, for example tetraphenylstibonium (see chapter 33).

Atoms or groups other than covalently bonded organic groups can also be attached to the metal. Such atoms or groups are called "ligands" and are designated by prefix names or cited as separate words (i.e., class names), following the rest of the name. Anionic ligands generally have names ending in "o", for example, $Cl^-$, chloro, $O^{2-}$, oxo, $H^-$, hydrido, and $CH_3-CO-O^-$, acetato.

$CH_3-Mg-I$      Iodo(methyl)magnesium
Methylmagnesium iodide

$Cl-CH_2-Co(CO)_4$      Tetracarbonyl(chloromethyl)cobalt
(Chloromethyl)cobalt tetracarbonyl

Organometallic compounds with unsaturated organic groups that are not attached to the metal by simple covalent bonds, or with neutral groups, such as carbonyl and ammonia, attached to a metal are best named by coordination nomenclature. Accordingly, such names consist of the names of the attached groups (ligands) in alphabetical order, the name of the metal, the ending "ate" if the complex is anionic, and a Roman numeral (the Stock number) designating the oxidation number of the metal, or the charge on the complex enclosed in parentheses. Attachment of two or more atoms of an organic ligand to the same metal atom is denoted by the Greek letter $\eta$. A superscript number may be used to indicate the number of atoms involved in such a delocalized attachment; specific atoms may be indicated by locants prefixed to the $\eta$ symbol.

$[(CH_3)_2In][Co(CO)_4]$      Dimethylindium($1+$) tetracarbonylcobaltate($1-$)

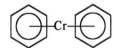

Potassium trichloro($\eta$-ethylene)platinate(II)
Potassium trichloro($\eta$-ethylene)platinate(1-)

Bis($\eta^6$-benzene)chromium
(*not* Chromocene)

Bis($\eta^8$-cycloocta-1,3,5,7-tetraene)uranium
(*not* Uranocene)

[(1,2-$\eta$)-Buta-1,3-diene]($\eta^3$-allyl)nickel($1+$)

Although attachment of a single carbon atom of an unsaturated organic ligand may be implied from the locant for the free valence of the organic substituent prefix, it may be emphasized by using the Greek letter $\kappa$ prefixed to the atomic symbol $C$ with a superscript locant; this symbol may be used in combination with $\eta$. The $\kappa$ symbol has largely replaced the $\sigma$ symbol.

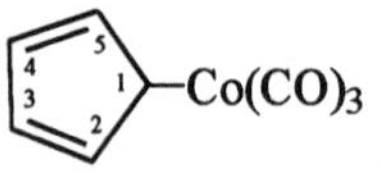

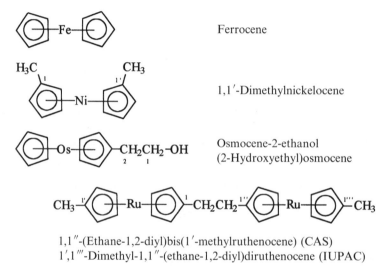

Tricarbonyl(cyclopenta-2,4-dien-1-yl-$\kappa C^1$)cobalt

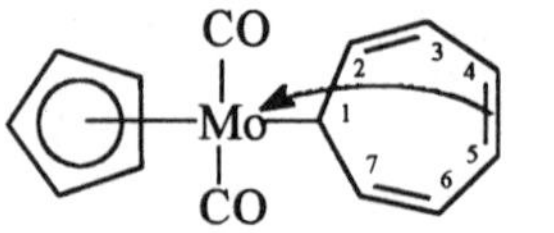

Dicarbonyl[(4,5-$\eta$, $\kappa C^1$)-cyclohepta-2,4,6-trien-1-yl]-($\eta^5$-cyclopentadienyl)molybdenum

Bis($\eta^5$-cyclopentadienyl)iron has long been called ferrocene, and "ocene" nomenclature is acceptable for bis($\eta^5$-cyclopentadienyl) complexes of other metals. Substituents of the cyclopentadiene rings are expressed as prefixes or suffixes in the usual substitutive manner.

Ferrocene

1,1'-Dimethylnickelocene

Osmocene-2-ethanol
(2-Hydroxyethyl)osmocene

1,1''-(Ethane-1,2-diyl)bis(1'-methylruthenocene) (CAS)
1',1'''-Dimethyl-1,1''-(ethane-1,2-diyl)diruthenocene (IUPAC)

Bridging and fusion principles (see chapter 7) have not been developed for naming "ocenes", although such names could be quite useful to an organic chemist. Appropriate and consistent locants for bridge atoms and fusion sites are lacking.

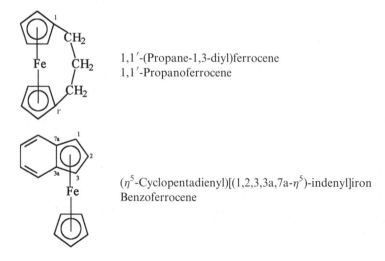

1,1'-(Propane-1,3-diyl)ferrocene
1,1'-Propanoferrocene

($\eta^5$-Cyclopentadienyl)[(1,2,3,3a,7a-$\eta^5$)-indenyl]iron
Benzoferrocene

Metallocycles are organic heterocycles in which the heteroatom is a metal other than the semimetals normally included in the nomenclature for organic heterocycles (see chapter 9). When the atoms of the ring attached to the metal are carbon atoms, derivation of names for the organic ligand according to existing rules of coordination nomenclature can be quite awkward. It is, however, possible to name metallocycles by Hantzsch–Widman and replacement nomenclature (see chapter 9). Replacement terms such as "platina" and "osma" are used; a complete list of replacement prefixes is found in IUPAC nomenclature recommendations.[1,2] Unlike the semimetals, metal atoms do not have a "standard valency" (see chapter 2), which is required in order to know the number of hydrogen atoms available for substitution, and therefore substitutive nomenclature cannot be used. Hence, atoms or groups attached to metal atoms in a metallocycle must be described as ligands according to the principles of coordination nomenclature. Although no official recommendations for naming metallocycles have been adopted, the following examples illustrate the general approach. Ligand names are used for atoms or groups attached to the metal atoms; otherwise, substitutive nomenclature is used.

1,1-Dichloro-2,3,4,5-tetramethyl-1-platina-cyclopenta-2,4-diene (skeletal replacement metallocycle name)
1,1-Dichloro-2,3,4,5-tetramethylplatinole (Hantzsch–Widman metallocycle name)
Dichloro(1,2,3,4-tetramethylbuta-1,3-diene-1,4-diyl)platinum (coordination name)
Dichloro(3,4-dimethylhexa-2,4-diene-2,5-diyl)platinum (coordination name)

Methyl 3-[1-iodo-1-methyl-1,1-bis(triethylphosphine-$P$)]platinetan-3-yl]-2-methyl-propanoate (Hantzsch–Widman metallocycle name)
Iodo[2-(3-methoxy-2-methyl-3-oxopropyl)-propane-1,3-diyl]methylbis(triethylphosphine-$P$)-platinum (coordination name)

9,9-[Methylenebis(dimethylphosphine)-$P,P'$]-10$H$-9-platinaanthracene (skeletal replacement metallocycle name)
[Methylenebis(dimethylphosphine)-$P,P'$]-(methylenedi-2,1-phenylene)platinum (coordination name)

2,5-Dimethoxy-7,7-bis(triphenylphosphine-$P$)-7-platinabicyclo[4.1.1]octane (skeletal replacement metallocycle name)
4,7-Dimethoxycycloheptane-1,3-diyl)bis-(triphenylphosphine-$P$)platinum (coordination name)

The same principles could also be used to name ocenes with atoms or groups attached to their metal atoms.

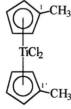

*Ti,Ti*-Dichloro-1,1′-dimethyltitanocene

Organometallic compounds can be named by citing the metal together with its ligands as prefixes. Prefix names for most metals are formed by replacing the final -ium, -um, or -y of the name of the metal by -io. For copper, silver, gold, and iron, the Latin names of the metals are modified to give cuprio, argentio, aurio, and ferrio. For tungsten (or wolfram), zinc, cobalt, and nickel, the prefix names are formed by adding the ending -io to the name of the metal; manganio is the name of the prefix derived from manganese. Because these -io prefixes do not indicate a "valence" for the metal, except for the alkali metals the number of hydrogen atoms of an organic structure replaced can often be uncertain. Hence, these prefixes must always be considered as monovalent, that is, they indicate replacement of only one hydrogen atom of an organic structure. Atoms and groups attached to metals in these prefixes in organic names are expressed by ligand names, not substitutive prefix names. The established monovalent substituent prefix names for the semimetals antimony, bismuth, germanium, tin, and lead, such as stibyl or stibanyl, germyl, and plumbyl, are much preferred to "io names", although the latter may be necessary when neutral atoms are also bonded to the semimetal.

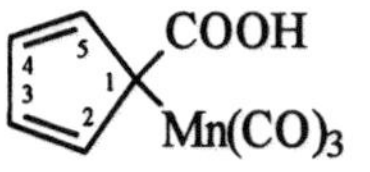

1-(Tricarbonylmanganio)cyclopenta-2,4-diene-1-carboxylic acid

Diethyl sodiomalonate

2-(Trimethylgermyl)ethanol

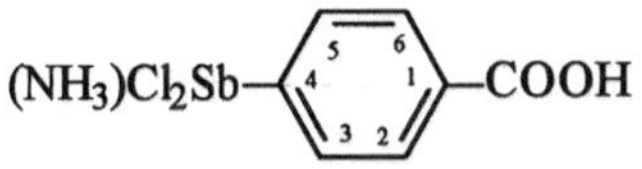

4-[(Ammine)dichloroantimonio]benzoic acid

Bivalent and trivalent substituent prefix names for the semimetals are formed in the same way as for the analogous hydrocarbon prefix names, for example, germylidyne and stibanediyl (see chapters 25 and 26). Although -ylene endings for germane, stannane, and plumbane are no longer approved by IUPAC, prefix names such as stibylene and germylene are still widely used, are used by CAS, and are acceptable. Except for stibino, stiboso, and bismuthino, there are no recognized prefix names for antimony and bismuth that correspond to those for the analogous phosphorus and arsenic compounds.

2,2'-(Methylstibanediyl)diethanol
2,2'-(Methylstibylene)diethanol

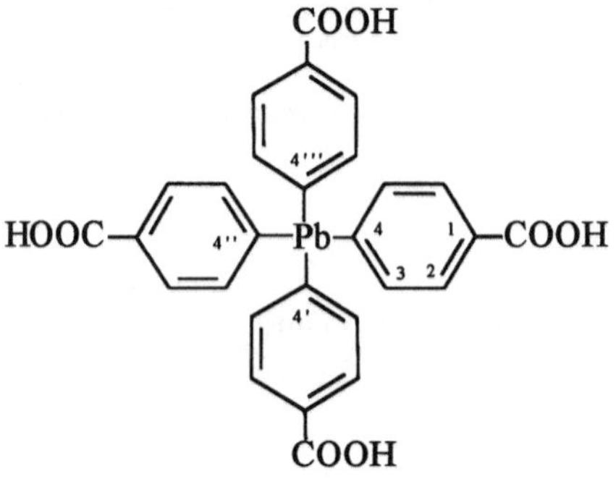

4,4',4'',4'''-plumbanetetrayltetrabenzoic acid

$(CH_3)_3C$
$(CH_3)_3C-\overset{|}{\underset{|}{Sn}}-CH_2C\equiv C-Mg-Br$
$(CH_3)_3C \quad {}_3 \quad {}_2 \quad {}_1$

Bromo[3-(tri-*tert*-butylstannyl)prop-1-yn-1-yl]magnesium
[3-(tri-*tert*-butylstannyl)prop-1-yn-1-yl]magnesium bromide
Bromo[3-[tris(1,1-dimethylethyl)-stannyl]prop-1-yn-1-yl]magnesium

$CH_3-O \qquad\qquad O$
$CH_3-O-\overset{|}{\underset{|}{Ge}}-CH_2CH_2-O-\overset{||}{Sb}(CH_3)_2$
$CH_3-O$

Dimethyloxo[2-(trimethoxygermyl)-ethoxy]stiborane
Dimethyloxo[2-(trimethoxygermyl)-ethoxy]-$\lambda^5$-stibane
Dimethyl[2-(trimethoxygermyl)ethoxy]-stibine oxide

## Discussion

Acid names corresponding to phosphonic and phosphinic acid, such as germanonic and stibinic acids, have not been encouraged by official nomenclature bodies and since 1967 have not been used by CAS.

The use of hydride names for metals, such as platinane, has never been seriously considered, although IUPAC's Organic Nomenclature Commission has been urged by IUPAC's Inorganic Nomenclature Commission to include the Group 13 metals.

"Ocene" nomenclature was developed from the need to describe bis($\eta^5$-cyclopentadienyl)-iron compounds more easily than was possible with coordination nomenclature. Extension to other rings has been specifically prohibited. Examples of potential extension through bridging and fusion principles, and the use of ligand names to describe atoms or groups attached to the metal were given in this chapter.

Combination of principles from coordination and substitutive nomenclature in order to name metallocycles in a more convenient manner was first introduced in proposals for naming polyboranes with "bare" boron atoms,[3] that is, a boron atom in a polyborane that does not have a hydrogen atom attached, as is implied by the polyborane type name (see chapter 27). This technique has not yet been given recognition by official nomenclature bodies. Examples of this combinatorial technique were given in this chapter.

## Additional Examples

1.  $CH_3-Li$

    Methyllithium

2.  $\overset{Na}{\underset{|}{\phantom{C}}} \qquad \overset{Na}{\underset{|}{\phantom{C}}}$
    $C_6H_5-\overset{|}{C}HCH_2CH_2\overset{|}{C}H-C_6H_5$
    ${}_4 \quad\; {}_3 \quad\; {}_2 \quad\; {}_1$

    (1,4-Diphenylbutane-1,4-diyl)disodium
    Disodium 1,4-diphenylbutane-1,4-diide
    1,4-Diphenyl-1,4-disodiobutane

3.  $Na-C\equiv C-Na$

    Ethynediyldisodium
    Disodium ethynediide
    Disodium acetylide
    Disodioethyne

4.  $\qquad CH_2CH_3 \qquad\qquad CH_2CH_3$
    $CH_3[CH_2]_3\overset{|}{C}HCH_2-Zn-CH_2\overset{|}{C}H[CH_2]_3CH_3$
    ${}_6 \quad {}_{5-3} \quad {}_2 \quad {}_1$

    Bis(2-ethylhexyl)zinc

5.  $\qquad CH_3$
    $CH_3CH_2-\overset{|}{Al}-CH_2CH_2CH_3$

    Ethyl(methyl)propylaluminum

6.

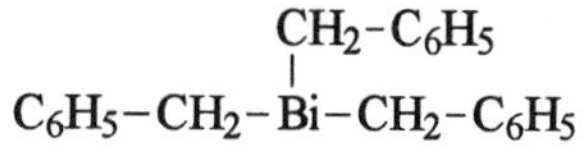

$$CH_2-C_6H_5$$
$$|$$
$$C_6H_5-CH_2-Bi-CH_2-C_6H_5$$

Tribenzylbismuthane
Tris(phenylmethyl)bismuthine
Tribenzylbismuth

7. $(CH_3CH_2)_4Pb$

Tetraethylplumbane
Tetraethyllead

8. $(C_6H_5)_2Te$

Diphenyltellane
Diphenyl telluride
Diphenyltellurium

9. $CH_3CH_2CH_2CH_2-Mg-I$

Butyliodomagnesium
Butylmagnesium iodide

10. $(CH_3)_2CHCH_2CH\!=\!CH-Hg-O-CO-CH_3$
      4    3   2     1

(Acetato)(4-methylpent-1-en-1-yl)mercury
(4-methylpent-1-en-1-yl)mercury(2+) acetate
(4-methylpent-1-en-1-yl)mercury(II) acetate

11.

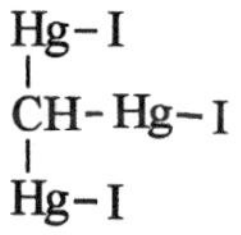

$$Hg-I$$
$$|$$
$$CH-Hg-I$$
$$|$$
$$Hg-I$$

Methylidynetris[iodomercury(2+)]
Methylidynetris[iodomercury(II)]
Methylidynetris[mercury(2+) iodide]
Methylidynetris[mercury(II) iodide]

12. $(C_6H_5)_3SbCl_2$

Dichlorotriphenyl-$\lambda^5$-stibane
Dichlorotriphenylstiborane
Dichlorotriphenylantimony

13.

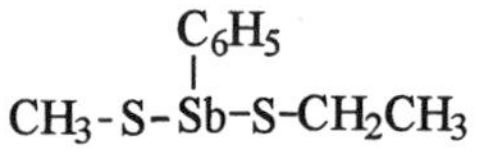

$$C_6H_5$$
$$|$$
$$CH_3-S-Sb-S-CH_2CH_3$$

(Ethylthio)(methylthio)phenylstibane
(Ethylsulfanyl)(methylsulfanyl)phenyl-
   stibane
(Ethanethiolato)(methanethiolato)
   phenylantimony

14. $(CH_3)_3Sn-O-SO_2-O—Sn(CH_3)_3$

[Sulfonylbis(dioxy)]bis(trimethylstannane)
2,2,6,6-Tetramethyl-3,5-dioxa-4-thia-2,6-
   distannaheptane 4,4-dioxide
2,2,6,6-Tetramethyl-4,4-dioxo-3,5-dioxa-
   $4\lambda^6$-thia-2,6-distannaheptane

15. $(CH_3)_2CH-AlH-CH(CH_3)_2$

Hydridodiisopropylaluminum
Hydridodipropan-2-ylaluminum
Diisopropylaluminum hydride

16.

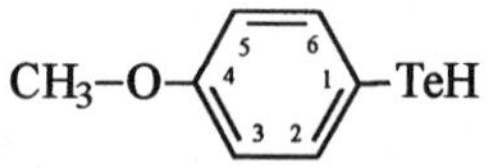

4-Methoxybenzenetellurol
Hydrido(4-methoxyphenyl)tellurium

17.

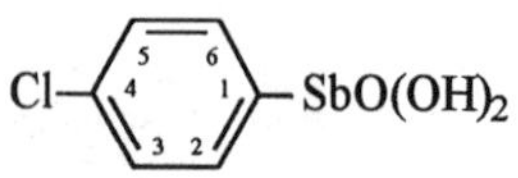

(4-Chlorophenyl)dihydroxystibane oxide
(4-Chlorophenyl)dihydroxyoxo-$\lambda^5$-stibane
(4-Chlorophenyl)dihydroxooxoantimony

18. $CH_3CH_2-Ge\!\equiv\!N$

Ethylnitrilogermane
Azanylidyne(ethyl)germane
Ethylnitridogermanium
Ethylgermanium nitride

19.

Tricarbonyl($\eta^5$-cyclopentadienyl)-
   manganese
($\eta^5$-cyclopentadienyl)manganese
   tricarbonyl

20.

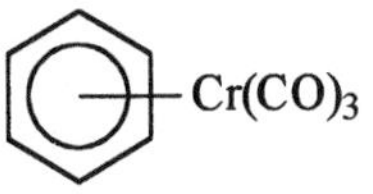

($\eta^6$-Benzene)tricarbonylchromium
($\eta^6$-Benzene)chromium tricarbonyl

21.  $(CH_3CH_2)_3Bi-Ni(CO)_2-Bi(CH_2CH_3)_3$

Dicarbonylbis(triethylbismuthane)nickel

22.  $[CH_3-O-Pb][Co(CO)_4]$

(Methanolato)lead($1+$) tetracarbonyl-
   cobaltate($1-$)
(Methanolato)lead(II) tetracarbonyl-
   cobaltate(I)
Tetracarbonyl(methoxy-$\lambda^2$-plumbyl)cobalt

23.

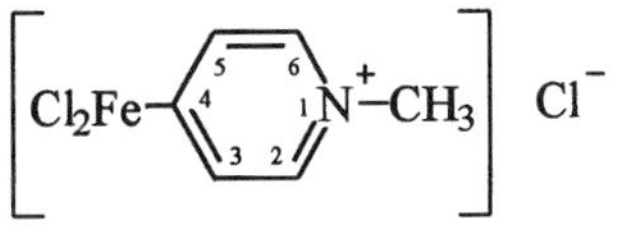

4-(Dichloroferrio)-1-methylpyridinium
   chloride

24.  $[(C_6H_5)_3Pb][Fe(CO)_4]$

[Bis(triphenyllead($1+$)] [tetracarbonyl-
   ferrate($2-$)]
Tetracarbonylbis(triphenylplumbyl)iron
Tetracarbonylbis(triphenylplumbio)iron

25.

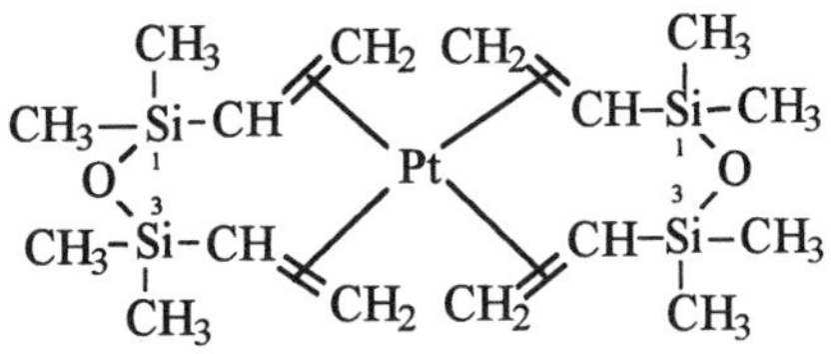

Bis[1,3-bis($\eta$-ethenyl)-1,1,3,3-tetramethyl-
   disiloxane]platinum

## REFERENCES

1. International Union of Pure and Applied Chemistry, Inorganic Chemistry Division, Commission on Nomenclature of Inorganic Chemistry. *Nomenclature of Inorganic Chemistry, Recommendations 1990*; Leigh, G. J., Ed.; Blackwell Scientific: Oxford, UK, 1990.
2. Block, B. P.; Powell, W. H.; Fernelius, W. C. *Inorganic Chemical Nomenclature, Principles and Practice*; American Chemical Society: Washington, DC, 1990.
3. Casey, J. B.; Evans, W. J.; Powell, W. H. Structural Nomenclature for Polyborane Hydrides and Related Compounds 2. Nonclosed Structures. *Inorg. Chem.* **1983**, *22*, 2236–2245.

# 29

# Polymers

In principle, it should be possible to name organic polymer molecules by the methods used for organic molecules. However, the complexity of polymer structures and often incomplete structural characterization led to a nomenclature for polymers based on the reactants from which they were made. This is called source-based nomenclature, which generates names such as polystyrene. An early attempt[1] to name polymers on the basis of structure failed because of insufficient detail in the rules. Later, enlarging on the work of the Committee on Nomenclature of the ACS Division of Polymer Chemistry,[2] the IUPAC Commission on Macromolecular Nomenclature published a self-consistent structure-based nomenclature for linear single-strand organic polymers.[3]

It is not surprising that both the source-based and structure-based nomenclature systems tend to give short and generally obvious names for simple polymers. Both types of names are acceptable, but for simple polymers, source-based names will be more familiar and are entrenched in the literature. As structural (and monomer) complexity increases, however, the structure-based system is recommended. The corresponding source-based names will often be just as cumbersome, if they can be written at all, and will carry less information. There is nothing wrong with the adjective "presumed" in front of a structure-based name for a polymer whose structure is known only by conjecture.

## Acceptable Nomenclature

### Structure-Based Nomenclature

Within the constraints of polymer structure and the concepts embodied in the system, structure-based nomenclature is consistent with organic nomenclature discussed elsewhere in this book. It has been adapted by CAS, and it is recommended by IUPAC for polymers whose structures are known or can be assumed. This nomenclature is preferred for unambiguous description of chemical structure in macromolecules.

REGULAR SINGLE-STRAND (LINEAR) POLYMERS   The chemical structure of polymers, rather than monomers, forms the basis for the nomenclature of regular single-strand polymers.[3] The 1975 IUPAC rules, currently being updated, were limited to regular single-strand polymer chains hypothetically free of branching and chain defects. To the extent that a polymer chain is fully described as a multiple of a repeating segment of the chain, it can be named "poly(repeating unit)", where the repeating unit is usually a bivalent organic group or a series of bivalent groups. However, different repeating units can often be written for a single chain. Thus, for a unique and unambiguous polymer name, it is necessary to define a starting point in the chain and a direction to move along the chain toward the point of repetition. This repeating chain segment is called the structural repeating unit or, in the IUPAC recommendations, the constitutional repeating unit (CRU). Once the CRU is defined, it can be named by the principles of organic nomenclature.

The steps to be taken, in order, are as follows: (1) write as much of the structure as necessary to include at least two repeating sequences; (2) select the smallest CRU and orient it to read left to right; (3) name the CRU. The generic name of the polymer is "poly(CRU)". For example, the polymer structure

$$\underset{\underset{\text{Cl}}{|}}{\text{—CHCH}_2}\underset{\underset{\text{Cl}}{|}}{\text{CHCH}_2}\underset{\underset{\text{Cl}}{|}}{\textbf{CHCH}_2}\underset{\underset{\text{Cl}}{|}}{\text{CHCH}_2}\underset{\underset{\text{Cl}}{|}}{\textbf{CHCH}_2}\underset{\underset{\text{Cl}}{|}}{\text{CHCH}_2}\underset{\underset{\text{Cl}}{|}}{\text{CHCH}_2}\text{—}$$

can be described by either of the units identified by boldface type. In order to give the –Cl substituent the lowest locant, reading left to right, the repeating unit –CHClCH$_2$– is chosen as the CRU; the polymer structure is written (CHClCH$_2$)$_n$, and the polymer is named poly-(1-chloroethylene) or poly(1-chloroethane-1,2-diyl).

A CRU is often composed of two or more multivalent subunits, such as a single atom (–O– or –N=), an atomic grouping (–CH$_2$CH$_2$–), or a ring system; the CRU itself does not include substituents. Identification of the preferred CRU depends upon a subunit seniority system. The CRU that is selected begins with the most senior subunit and citation moves, reading left to right, in the direction of the closest (counting only atoms in the main chain and following the shortest path in rings) subunit next in seniority. For citation as the first (most senior) subunit, the order of decreasing seniority is: (1) a heterocyclic ring; (2) a heteroatom or heteroacyclic chain; (3) a carbocyclic ring; (4) an acyclic carbon chain. This order is unaffected by the presence of substituents. In the following example, –O– is the senior subunit and the next senior subunit is the benzene ring; the preferred CRU begins with –O– and proceeds along the chain by the shortest path (a single carbon atom in this case) toward the ring.

Poly(oxymethylene-1,4-phenyleneethylene)
(*not* Poly(oxyethylene-1,4-phenylenemethylene))
Poly(oxymethylene-1,4-phenyleneethane-1,2-diyl)

Within each major seniority group, there is a suborder of seniority. Among heterocyclic rings, the descending order of seniority is given in detail in chapter 3 and discussed in chapter 9. For example, the most senior ring system is one with nitrogen in the ring, followed, in descending order, by a ring system containing at least one heteroatom other than nitrogen as high as possible in the order of heteroatoms given below, and by a ring system having the greatest number of rings. When two ring systems are identical except for the locants for the heteroatoms, the one with lower locants consistent with the fixed numbering of the rings (see chapter 9) is senior. When the systems are identical except for unsaturation, the senior subunit is that with the least hydrogenation. Further choice is based on lowest locants, the number, and the kind of substituents.

Among heteroatoms not included in rings, the descending order of seniority is

$$O > S > Se > N > P > As > Sb > Si > Ge > Sn > B > Hg$$

Other atoms, as needed, are inserted into this list according to their place in the periodic table.

Among carbocycles, the most senior carbocycle is the system with the largest number of rings, followed by the largest individual ring at the first point of difference, the largest number of atoms common to the rings, the lowest locant numbers for ring junctions, and the least hydrogenated ring. Among acyclic chains, the unit with the largest number of carbon atoms in the chain is senior. Further choice for carbocycles or chains is based on unsaturation, then the largest number of substituents, the lowest locants for substituents, and, finally, alphabetical order of substituents. These criteria differ slightly from those of chapter 3.

Frequently, there will be a choice of path to be followed from the most senior subunit to the next senior subunit. Priority is given to the shortest path in number of atoms, and paths are unidirectional. Where the most senior subunit is repeated within the CRU, the path is the shortest one between them and thence to the next senior subunit. Between equal shortest path lengths, the preferred path is the one that has the subunit of highest seniority nearest the most senior subunit. For example, if the descending order of seniority is A > B > C > D, the CRU –ABDADC– is preferred to –ADBACD– because B is closer to the first A in the former series. After all orders of seniority have been observed, the number of free valences at the ends of the CRU are minimized; accordingly, –N=CHCH$_2$– is preferred over =N–CH$_2$CH=.

To name the polymer, the CRU is written to read from left to right, with the most senior subunit at the left; the polymer name is simply the prefix poly, followed by citation of the subunits in the CRU in order as written. For example:

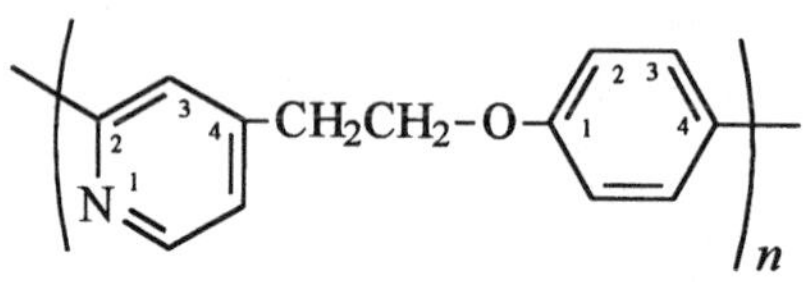

Poly(pyridine-2,4-diylethyleneoxy-1,4-phenylene)
  [*not* Poly(pyridine-4,2-diyl-1,4-phenyleneoxyethylene)]
Poly(pyridine-2,4-diylethane-1,2-diyloxy-1,4-phenylene)

Within the CRU, the largest possible subunits, exclusive of substituents, that have acceptable names according to organic nomenclature principles (see chapters 5–9 for methods of naming bivalent subunits) are chosen to form the name of the CRU. In the example above, ethylene (CAS) or ethane-1,2-diyl (IUPAC) is preferred over methylenemethylene or dimethylene. Units of three or more carbon atoms are given bivalent hydrocarbon group names, for example, pentane-1,5-diyl. The subunit –CO–[CH$_2$]$_4$–CO– can be named adipoyl, hexanedioyl, or 1,6-dioxohexane-1,6-diyl. Substituents along the main chain are named with substitutive prefixes preceding the name of the subunit to which they are bound; the combination is enclosed in parentheses or brackets. Lowest locants are used, consistent with other criteria for selecting the CRU.

In the following example the order of citation is set by the shortest path from the most senior subunit, -oxy-, to the next most senior subunit, the cyclohexene ring, in which lowest locants are assigned to the free valences; note that citation from left to right requires, in the latter subunit, that locants 3,1 rather than 1,3 be used in the polymer name.

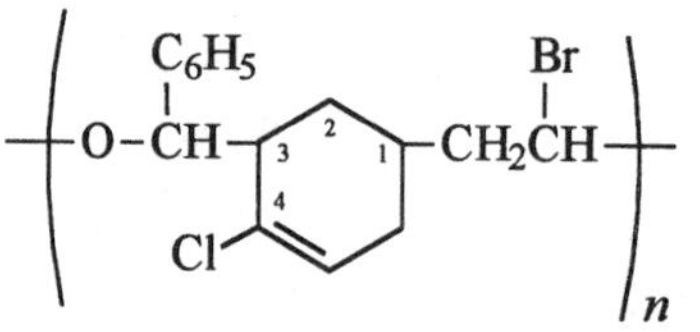

Poly[oxy(phenylmethylene)(4-chlorocyclohex-4-ene-3,1-diyl)(2-bromoethylene)]
  (*not* Poly[oxybenzylidene(6-chlorocyclohex-5-ene-1,3-diyl)(2-bromoethylene)])
Poly[oxy(phenylmethylene)(4-chlorocyclohex-4-ene-3,1-diyl)(2-bromoethane-1,2-diyl)]

Numbering of the subunits in the main chain is unaffected by the presence of substituents, which are themselves named substitutively with the appropriate subunit. In the final polymer name, acceptable names of subunits may include substituents such as the phenyl group in benzylidene.

Replacement nomenclature (see chapter 3) can also be used to form all or part of the name of a CRU. Acyclic units with heteroatoms are identified as above, and therefore begin with the most senior atom, as in –O–CH$_2$CH$_2$–NH–CH$_2$CH$_2$–, 1-oxa-4-azahexane-1,6-diyl. In units containing only heteroatoms, acceptable semisystematic names are well-established. Notable examples are the polysiloxanes and polyphosphazenes. Poly(dimethylsiloxane), +O–Si(CH$_3$)$_2$+$_n$, has the structure-based name poly[oxy(dimethylsilanediyl)], and poly-(dimethylphosphazene), +N=P(CH$_3$)$_2$+$_n$, has the more forbidding structure-based names poly[nitrilo(dimethylphosphoranylidyne)] or poly[nitrilo(dimethyl-$\lambda^5$-phosphanylylidene)].

The names of end groups can be prefixed to the name of the polymer. They are denoted by α-, for the group attached to the left end of the oriented CRU, and ω- for the group at the other end of the chain.

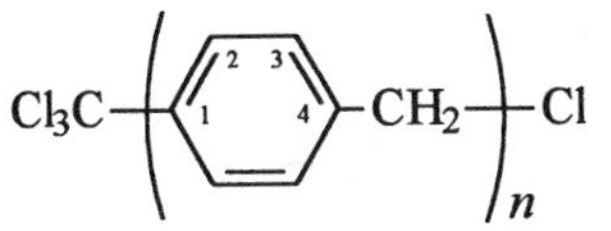

$Cl_3C$—⟨phenylene ring with positions 1,2,3,4⟩—$CH_2$—$\big]_n$—$Cl$     α-(Trichloromethyl)-ω-chloropoly(1,4-phenylenemethylene)

REGULAR DOUBLE-STRAND (LADDER AND SPIRO) POLYMERS  Polymers whose chains consist of an uninterrupted sequence of rings with adjacent rings having one atom in common (spiro polymers) or two or more atoms in common (ladder polymers) can be named on the basis of structure.[4] The repeating units are tetravalent groups that may include two or more subunits. Two pairs of locants, separated by a colon, are used to show the points of attachment. Concepts of seniority and direction are applied in the same way as with single-strand polymers.

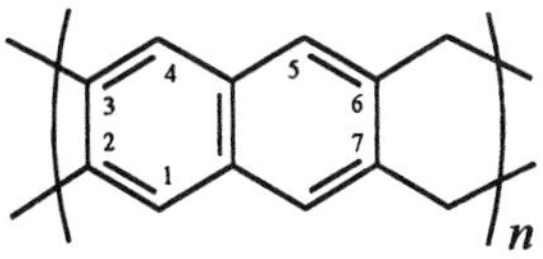

⟨naphthalene ladder structure with positions 1-7⟩$_n$     Poly[naphthalene-2,3:6,7-tetrayl-6,7-bis(methylene)]

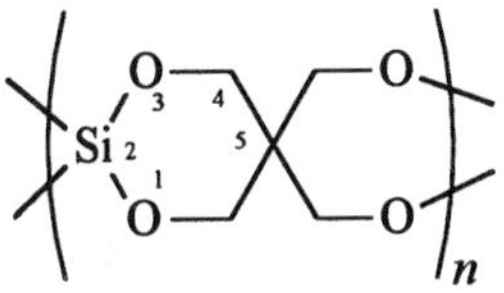

⟨dioxasilinane ring structure with Si and O atoms, positions 1-5⟩$_n$     Poly[1,3,2-dioxasilinane-2,2:5,5-tetrayl-5,5-bis-
(methyleneoxy)]
Poly[1,3-dioxa-2-silacyclohexane-2,2:5,5-tetrayl-5,5-bis(methyleneoxy)]

IRREGULAR SINGLE-STRAND POLYMERS  Structure-based polymer nomenclature can be applied to polymers whose chains consist of units in irregular sequence.[5] Basically, the structure-based units by which the polymer chain can be described are identified and named as above. Where the sequential arrangement of the units is unknown, the polymer is named poly(A/B/...), in which A, B, ... represent the structure-based names of the units, the number of which is kept to a minimum. For example, a chlorinated poly(ethylene) may have the structure

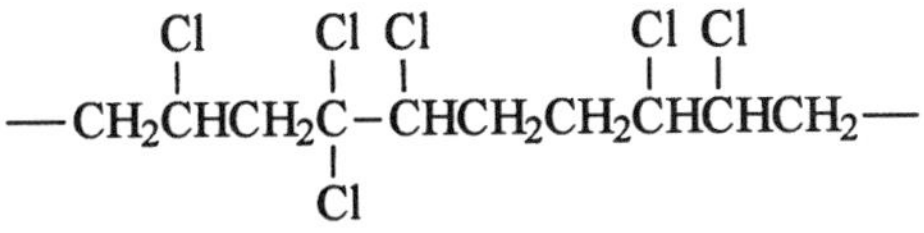

$$—CH_2CHCH_2\underset{\overset{|}{Cl}}{C}-CHCH_2CH_2CHCHCH_2—$$

(with Cl substituents shown above the chain)

Its units are –CHCl–, –CCl$_2$–, and –CH$_2$–, and its name is poly(chloromethylene/dichloromethylene/methylene).

A portion, comprising many units, of a polymer chain that is different from adjacent portions is called a "block". The name of a block copolymer is composed of the structure-based names, formed as above, of blocks separated by long dashes, as in poly(A)—poly(B). A repeating block copolymer would be named poly[poly(A)—poly(B)]. If the blocks are connected through other units, a generic name is poly(A)—$X_A$—poly(B)—$X_B$—poly(C)..., in which $X_A$, $X_B$, ... are the structure-based names of the units joining the blocks; the junction units are kept as small as possible in constructing the block copolymer name.

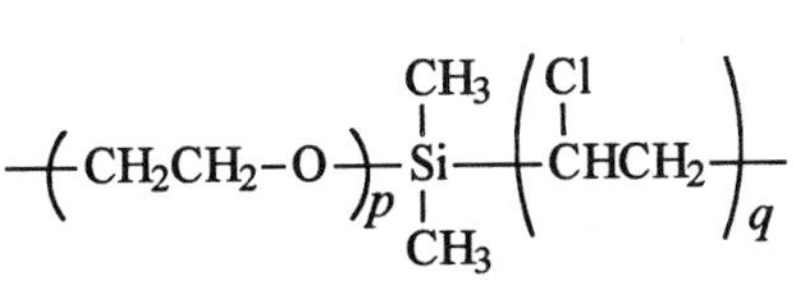

$$-\Big(CH_2CH_2-O\Big)_p-\underset{\overset{|}{CH_3}}{\overset{CH_3}{Si}}-\Big(\underset{}{\overset{Cl}{|}}CHCH_2\Big)_q-$$

Poly(ethyleneoxy)—dimethylsilanediyl—poly(1-chloroethylene)
Poly(ethane-1,2-diyloxy)—dimethylsilanediyl—poly(1-chloroethane-1,2-diyl)

In this example, the blocks are $\big(CH_2CH_2\text{-}O\big)_p$ and $\big(CHClCH_2\big)_q$, and the junction unit is –Si(CH$_3$)$_2$–.

Blocks attached to the main polymer chain as substituents are called "grafts", and the polymer itself is a "graft copolymer" if the constitution of the main chain and the grafted block differ. Such polymers are named as above by treating the grafts as substituents with structure-based names.

This example is composed of a poly(1-phenylethylene) block and a poly(methylene) block; the latter has, in addition to methylene units, random poly(2-chloroethylene) grafts that in the structure as written are named as [poly(2-chloroethylene)]methylene units. The copolymer is named poly(1-phenylethylene)—poly[methylene/[poly(2-chloroethylene)]methylene].

## Source-Based Nomenclature

POLYMERS FROM KNOWN MONOMERS    The most common way of naming homopolymers is to call them "poly(monomer)", where (monomer) is the name of the compound from which the polymer was made. This means that a single polymer can have more than one name if it can be made from more than one monomer; the virtue is that the source is highlighted, even if the exact structure of the polymer is in doubt. The name of the monomer should be enclosed in parentheses to avoid confusion between names like poly(4-chlorostyrene) and, for chlorinated styrene, polychlorostyrene. Poly(ethylene) is the traditional name for the polymer made from $CH_2{=}CH_2$, which is now named ethene, not ethylene.

Care must be used with this system. Some monomers can form more than one polymer, depending on the method of polymerization. Examples are 4-vinylbenzaldehyde and buta-1,3-diene. The latter monomer is especially troublesome, since it can provide (on a structure basis) poly(but-1-ene-1,4-diyl) or poly(1-vinylethylene), as well as head-to-tail and head-to-head isomers. The literature abounds with the names 1,4-polybutadiene and 1,2-polybutadiene, which more or less separates the main possibilities on the basis of polymerization mechanism.

POLYMER NAMES BASED ON HYPOTHETICAL MONOMERS    Polymers such as polyesters, polyamides, and so on, that are formed by condensation reactions between two different monomers, are usually named on the basis of a monomer that on cleavage and addition would give an assumed structure. An example is poly(ethylene terephthalate).

COPOLYMERS, INCLUDING BLOCK AND GRAFT COPOLYMERS[6]    Source-based nomenclature is also applicable to polymers derived from more than one species of monomer. For copolymers with monomeric units derived from monomers A, B, ..., the generic name is "poly(A-*arr*-B)" where *arr* is an italicized connective that denotes the sequential arrangement of A and B in the chain. The arrangements and the corresponding type names are:

| | | |
|---|---|---|
| unspecified | poly(A-*co*-B) | |
| statistical | poly(A-*stat*-B) | |
| random | poly(A-*ran*-B) | |
| alternating | poly(A-*alt*-B) | ABABABABABAB |
| periodic | poly(A-*per*-B-*per*-C) | ABCABCABCABC |
| block | polyA-*block*-polyB | $A_n$–$B_n$ |
| | or polyA—polyB | |
| graft | polyA-*graft*-polyB | AAAAAAAAAAAA |
| | |        &#124; |
| | | BBBBBBBB |

Arrangements that follow known statistical laws are described by *-stat-*; *-ran-* denotes a Bernoullian distribution, a special statistical arrangement. Regular arrangements of monomeric units are designated by *-alt-* and *-per-*. Except for the *-graft-* arrangement, in which the monomer name first cited is that from which the main chain is derived, no order of seniority is used for the monomer names, and therefore more than one name is possible for a given copolymer. The $\alpha$-, $\omega$- device (see above) can be used to specify end groups. In simple cases, mass or mole fractions, molar masses, or degrees of polymerization can be cited in parentheses following the copolymer name. Examples of these types of names follow:

- An unspecified copolymer of styrene and methyl methacrylate:

Poly[styrene–*co*–(methylmethacrylate)]

- A copolymer consisting of a block of poly(vinyl chloride) joined by a dimethylsilylene junction unit to a block of a statistical copolymer from buta-1,3-diene and styrene:

Poly(vinylchloride)dimethylsilylenepoly(buta-1, 3-diene-*stat*-styrene)

- A copolymer consisting of a block of polystyrene and a block of polybuta-1,3-diene to which polyacrylonitrile is grafted:

Poly[polystyrene-*block*-(polybutadiene-*graft*-polyacrylonitrile)]

## Stereoregularity

Stereochemical definitions and notation for polymers have been comprehensively described in IUPAC publications.[7,8] For example, terms such as "isotactic" and "syndiotactic" are used as unitalicized adjectives placed ahead of the name of a regular polymer to denote the sequential arrangement of chiral centers (see chapter 30) within the polymer chain.

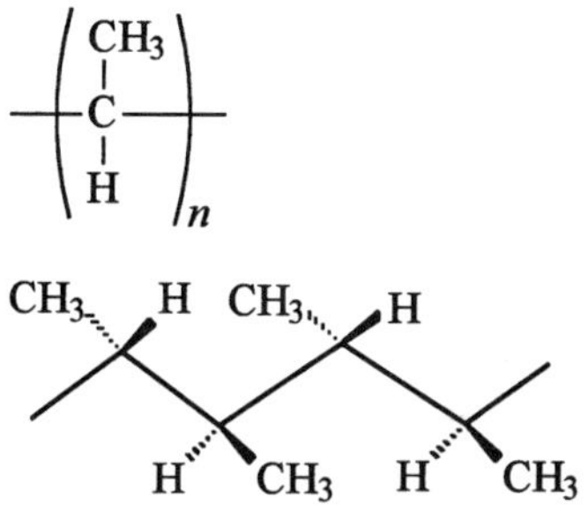

Isotactic poly(ethane-1,1-diyl)

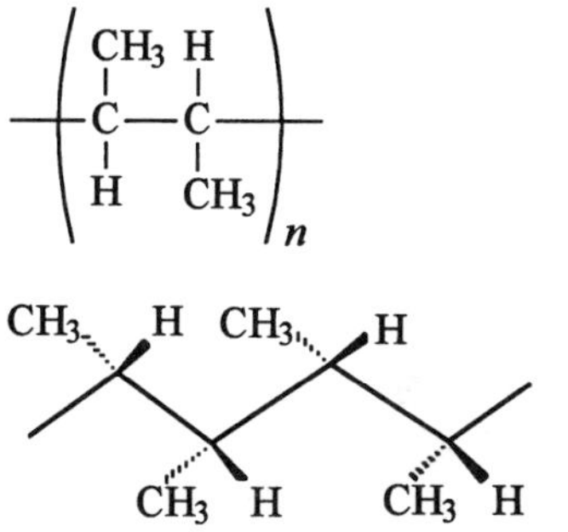

Syndiotactic poly(ethane-1,1-diyl)

Additional stereodescriptors such as *cis-* or *trans-* may be used to modify these adjectives or they may be made part of the polymer name.

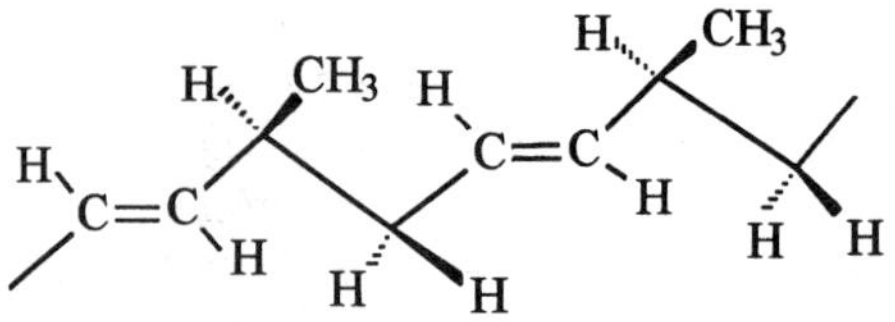

Isotactic poly(3-methyl-*trans*-but-1-en-1,4-diyl)

*trans*-Isotactic poly(3-methylbut-1-ene-1,4-diyl)

The reader should consult the original IUPAC documents for further details in this complex area of nomenclature.

## Additional Examples

Structures are in CRU form, with the structure-based name(s) given first. In each example, the final name is a source-based name. Only one source-based name is given, even when more than one monomer is possible.

1. 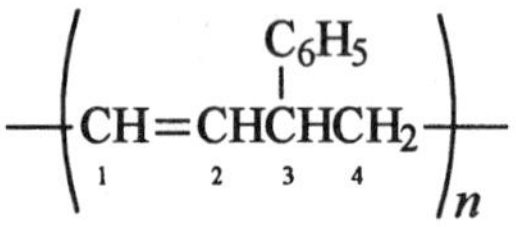   $-\left[ CH=CHCHCH_2 \right]_n$  (with $C_6H_5$ on C-3; positions 1 2 3 4)

Poly(3-phenylbut-1-en-1,4-diyl)
Poly(3-phenylbuta-1,3-diene)

2. 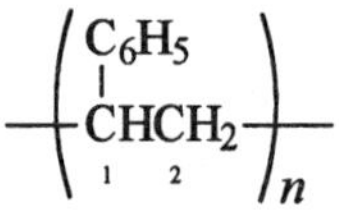   $-\left[ CHCH_2 \right]_n$  (with $C_6H_5$ on C-1; positions 1 2)

Poly(1-phenylethylene)
Poly(1-phenylethane-1,2-diyl)
Polystyrene

3. 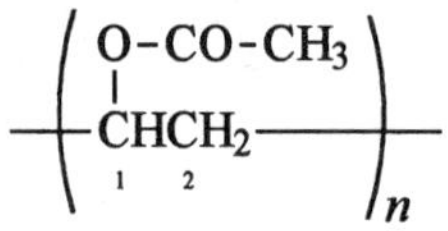   $-\left[ CHCH_2 \right]_n$  (with $O-CO-CH_3$ on C-1; positions 1 2)

Poly(1-acetoxyethylene)
Poly(1-acetoxyethane-1,2-diyl)
Poly(vinyl acetate)

4.    $-\left[ O-CH_2CH_2 \right]_n$  (positions 1 2)

Poly(oxyethylene)
Poly(oxyethane-1,2-diyl)
Poly(ethylene oxide)

5. 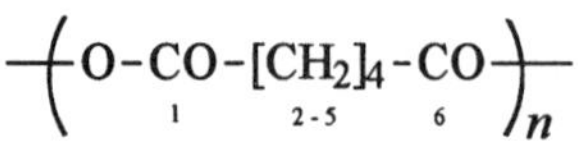   $-\left[ O-CO-[CH_2]_4-CO \right]_n$  (positions 1 2-5 6)

Poly(oxyadipoyl)
Poly[oxy(1,6-dioxohexane-1,6-diyl)]
Poly(adipic acid)

6. 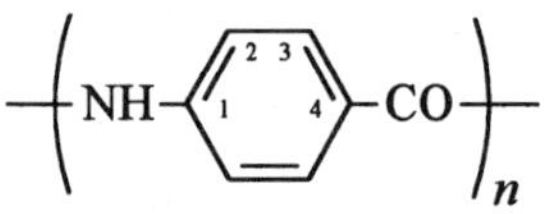   $-\left[ NH-C_6H_4-CO \right]_n$  (ring positions 1 2 3 4)

Poly(imino-1,4-phenylenecarbonyl)
Poly(terephthalamide)

7. 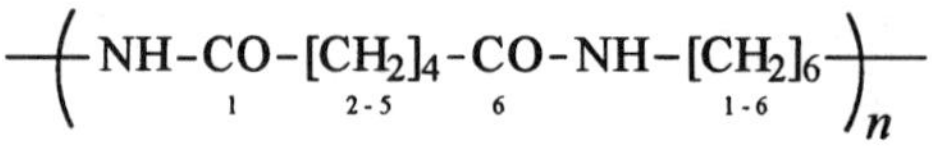   $-\left[ NH-CO-[CH_2]_4-CO-NH-[CH_2]_6 \right]_n$  (positions 1 2-5 6 1-6)

Poly(iminoadipoyliminohexane-1,6-diyl)
Poly(2,7-dioxo-1,8-diazatetradecane-1,14-diyl)
Poly(hexamethylene adipamide)

8. 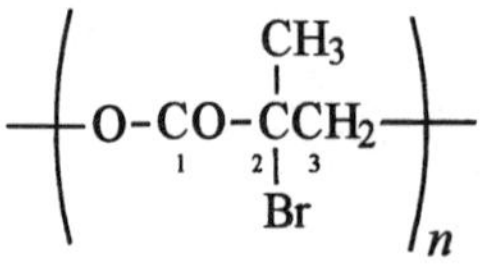   $-\left[ O-CO-CCH_2 \right]_n$  (with $CH_3$ and $Br$ on C-2; positions 1 2 3)

Poly[oxy(2-bromo-2-methyl-1-oxopropane-1,3-diyl)
Poly(2-bromo-3-hydroxy-2-methylpropanoic acid)

9.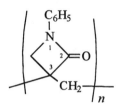

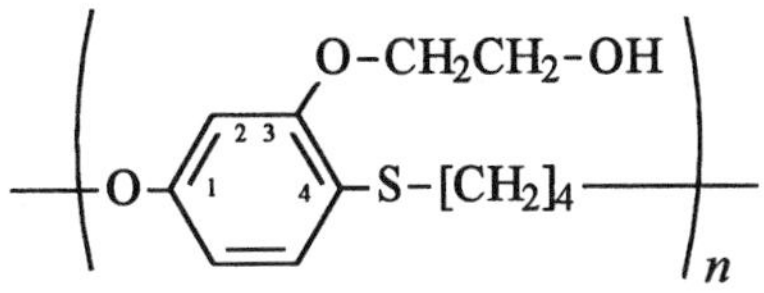

Poly[oxy-[3-(2-hydroxyethoxy)-1,4-phenylene]thiobutane-1,4-diyl]

10.

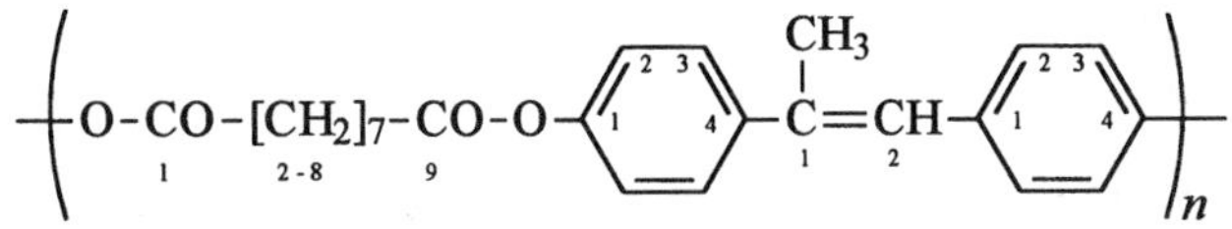

Poly[oxynonanedioyloxy-1,4-phenylene-(1-methylethene-1,2-diyl)-1,4-phenylene]

11.

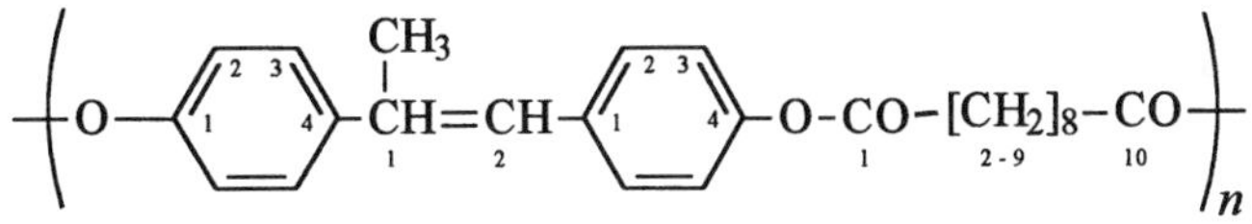

Poly[oxy-1,4-phenylene-(1-methylethene-1,2-diyl)-1,4-phenyleneoxydecanedioyl]

12.

Poly[(1-phenyl-2-oxoazetidine-3,3-diyl)methylene]

13.

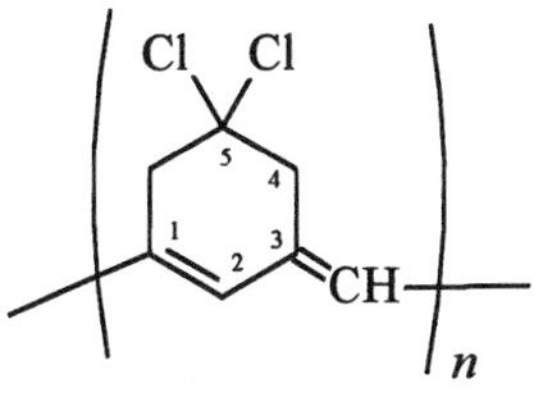

Poly[(5,5-dichlorocyclohex-1-en-1-yl-3-ylidene)-methanylylidene]

14.

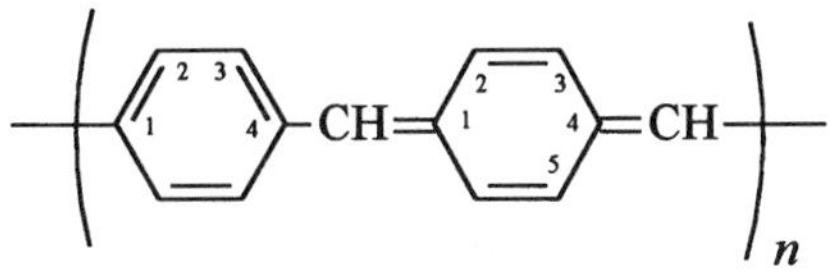

Poly(1,4-phenylenemethanylylidenecyclohexa-2,5-diene-1,4-diylidenemethanylylidene)

15. 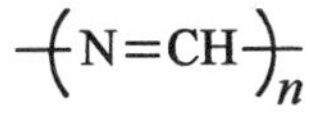 $\left(N=CH\right)_n$

Poly(nitrilomethanylylidene)

16.

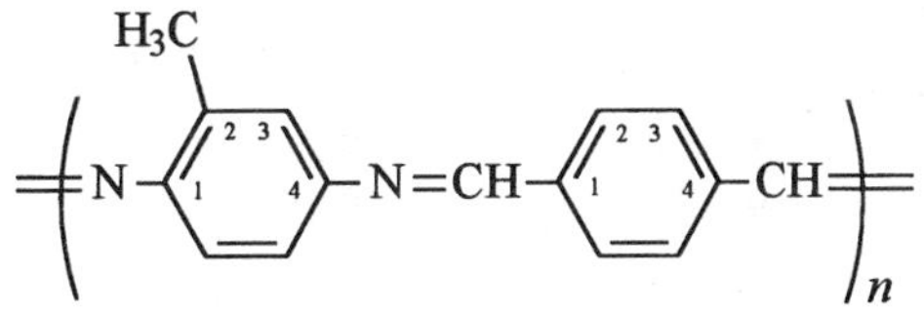

Poly[nitrilo(2-methyl-1,4-phenylene)-nitrilomethanylylidene-1,4-phenylenemethanylylidene]

17.

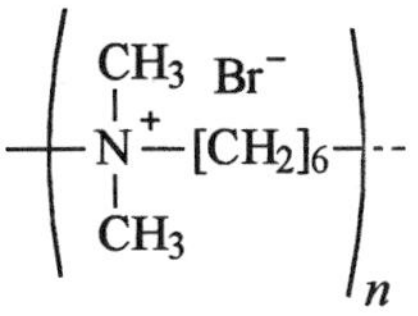

Poly[(dimethyliminio)hexane-1,6-diyl bromide]

18.

$$HO-\left(CH_2CH-O\atop \underset{2}{} \underset{1}{} \overset{CH_2-Cl}{|}\right)_n \underset{1}{CO}-\underset{2\text{-}5}{[CH_2]_4}-\underset{6}{CO}\left(O-\underset{1}{CHCH_2}\overset{CH_2-Cl}{|}_{2}\right)_n OH$$

$\alpha, \alpha'$-Adipoylbis[$\omega$-hydroxypoly[oxy-[1-(chloromethyl)ethylene]]]
$\alpha, \alpha'$-(1,6-dioxohexane-1,6-diyl)bis[$\omega$-hydroxypoly[oxy-1-(chloromethyl)ethane-1,2-diyl]

19.

$$\left(O-\underset{1}{CH_2}\underset{2}{CH}\underset{3}{CH_2}\underset{4}{CH}\underset{5}{CH_2}\underset{6}{CH_2}\right)_n$$ 

with $C_6H_5$ and $Cl$ substituents

Poly[oxy(4-chloro-2-phenylhexane-1,6-diyl)]

20.

$$\left(O-CH_2CH_2\right)_n per \left(CHCH_2\atop \overset{Cl}{|}\right)_p per \left(CHCH_2\atop \overset{C_6H_5}{|}\right)_q$$

Poly[(ethylene oxide)-*per*-(vinyl chloride)-*per*-styrene]

21.

$$\left(\left(O-CH_2CH_2\right)_p\left(\underset{1}{CH}\underset{2}{CH_2}\atop \overset{C_6H_5}{|}\right)_q\right)_n$$

Poly[poly(oxyethylene)—poly-
  (1-phenylethylene)]
Poly[poly(oxyethane-1,2-diyl)—poly-
  (1-phenylethane-1,2-diyl)]
Poly[poly(ethylene oxide)-*block*-
  polystyrene]

22.

$$Si\left(\underset{2}{CH}\underset{1}{CH_2}\atop \overset{C_6H_5}{|}\right)_p \left(\underset{2}{CH_2}\underset{1}{CH}\atop \overset{C_6H_5}{|}\right)_q \left(\underset{1}{CH_2}\underset{2}{CH}=\underset{3}{CH}\underset{4}{CH_2}\right)_r \left(\underset{1}{CH_2}\underset{2}{CH}=\underset{3}{CH}\underset{4}{CH_2}\right)_n$$

Bis[poly(but-2-ene-1,4-diyl)]-
  [poly(1-phenylethylene)]-
  [poly(2-phenylethylene)]silane
Bis[poly(but-2-ene-1,4-diyl)]-
  [poly(1-phenylethane-1,2-diyl)]-
  [poly(2-phenylethane-1,2-diyl)]-
  silane
Polystyrene-*block*-[silanetetrayl-
  bis(*graft*-polybutadiene)]-*block*-
  polystyrene

## References

1. International Union of Pure and Applied Chemistry, Physical Chemistry Division, Commission on Macromolecules, Subcommittee on Nomenclature. Report on Nomenclature in the Field of Macromolecules (1951) *J. Polymer Sci.* **1952**, *8*, 257–277.
2. American Chemical Society, Division of Polymer Chemistry, Committee on Nomenclature. A Structure-Based Nomenclature for Linear Polymers. *Macromolecules,* **1968**, *1*, 193–198.
3. International Union of Pure and Applied Chemistry, Macromolecular Division, Commission on Macromolecular Nomenclature. Nomenclature of Regular Single-Strand Organic Polymers (Rules Approved 1975). *Pure Appl. Chem.,* **1976**, *48*, 373–385; *Compendium of Macromolecular Nomenclature*; W. V. Metanomski, ed., Blackwell Scientific: Oxford, UK, 1991, Chapter 5.
4. International Union of Pure and Applied Chemistry, Macromolecular Division, Commission on Macromolecular Nomenclature. Nomenclature of Regular Double-Strand (Ladder and Spiro) Organic Polymers (Recommendations 1993). *Pure Appl. Chem.* **1993**, *65*, 1561–1580.
5. International Union of Pure and Applied Chemistry, Macromolecular Division, Commission on Macromolecular Nomenclature. Structure-Based Nomenclature for Irregular Single-Strand Organic Polymers (Recommendations 1994). *Pure Appl. Chem.* **1994**, *66*, 873–889.

6. International Union of Pure and Applied Chemistry, Macromolecular Division, Commission on Macromolecular Nomenclature. Source-Based Nomenclature for Copolymers (Recommendations 1985). *Pure Appl. Chem.* **1985**, *57*, 1427–1440; *Compendium of Macromolecular Nomenclature*: W. V. Metanomski, ed., Blackwell Scientific: Oxford, UK, 1991, Chapter 7.

7. International Union of Pure and Applied Chemistry, Physical Chemistry Division, Commission on Macromolecules, Subcommittee on Nomenclature. Report on Nomenclature Dealing with Steric Regularity in High Polymers. *Pure Appl. Chem.* **1966**, *12*, 645–656.

8. International Union of Pure and Applied Chemistry, Macromolecular Division, Commission on Macromolecular Nomenclature. Stereochemical Definitions and Notations Relating to Polymers (Recommendations 1980). *Pure Appl. Chem.* 1981, *53*, 733–52; *Compendium of Macromolecular Nomenclature*; W. V. Metanomski, ed., Blackwell Scientific: Oxford, UK, 1991, Chapter 2.

# 30

# Stereoisomers

Stereoisomers are constitutionally identical isomers that differ in the arrangement of their atoms in space. Two broad classes, configurational and conformational stereoisomers, are generally recognized. Configurational stereoisomers cannot be interconverted except by the making and breaking of bonds, while conformational stereoisomers are interconvertible by rotation about formal single bonds. This chapter will emphasize methods by which individual configurational isomers are distinguished from one another in their names. Definitions of stereochemical terms that are not a part of the names of individual compounds, such as diastereoisomer, enantiomer, and quasi-racemate, will be found in a glossary published by IUPAC[1]. For convenience, two terms used throughout this chapter are defined here.

- *Stereogenic* describes a feature of a molecule capable of producing stereoisomers. An atom or group having such a property is said to be stereogenic.
- *Chiral* describes the property of nonsuperposability of an object on its mirror image. An atom is said to be chiral if its attached atoms and/or groups are arranged in space in such a way that the structure cannot be superposed on its mirror image. Similarly, a substituted atom that can be superposed on its mirror image is said to be achiral.

Systematic names, and most semisystematic and trivial names, do not imply stereochemistry. Specific configurations are denoted by means of descriptors, usually prefixed to the part of the name to which they apply. Accordingly, enantiomers and diastereoisomers usually have the same name and numbering; however, stereogenic sites may determine the parent structure and/or direction of numbering. Methods used by Beilstein,[2] CAS, and IUPAC are included in this chapter. A few trivial names that imply configuration, such as maleic and fumaric acids (see chapter 11), have a long tradition of usage.

For any discussion of stereochemistry, conventions are indispensable for unambiguously representing three-dimensional structures in two dimensions. Several kinds of projections are used for this purpose. In the Fischer projection, an asymmetric carbon atom is implied at the intersection of two lines. Horizontal lines represent bonds to the two groups lying above the plane of the paper, and vertical lines represent bonds to the two groups lying behind the plane of the paper. In this book, bonds projecting in front of the plane of the paper are shown as solid wedges; bonds projecting behind the plane of the paper are shown as dashed wedges; and bonds in the plane of the paper are normal lines. There is no easy way to indicate relative configuration in a single structure.

This chapter is limited to the stereochemistry of carbon atoms; heteroatoms are included only if they have the same stereogenic configuration found with carbon atoms. Stereochemistry of heteroatoms in other configurations, such as trigonal, bipyramidal, and octahedral, although becoming increasingly important in organic compounds, is not considered here. The chapter is divided into several subsections. The method for determining priorities among atoms and groups, called the "sequence rule", developed by Cahn, Ingold, and Prelog[3-7] and codified in IUPAC rules and recommendations, will be described first because it is used in several ways in stereochemical nomenclature. The use of the CIP (Cahn, Ingold, Prelog) sequence rule to describe absolute configurations will then be presented, followed by methods for relative configurations of achiral and chiral stereoisomers. Achiral stereoisomers are usually identified with "geometric isomers" and chiral stereoisomers with "optical isomers", but the newer terms "cis-trans isomers" and "chiral isomers" (or "enantiomers"), respectively, are now preferred. Chiral molecules exhibit the phenomenon of optical activity, and contain at least one element of chirality, that is, a center, an axis, or a plane. Finally, a collection of more complex examples will illustrate a combination of stereochemical descriptors and provide an explanation of the principles used by CAS in its index names for *Chemical Abstracts* Volumes 76–128. Prior to Volume 76 of *Chemical Abstracts*, the stereochemistry given in index names was usually that used by the author.

Stereochemical nomenclature in specialized fields, such as amino acids, carbohydrates, cyclitols, and stereoparents, the names of which imply stereochemistry at a number of positions, is discussed in chapter 31.

## The Priority Rules for Specifying Configuration (The Sequence Rule)

The priority rules for ranking atoms or groups attached to a tetrahedral carbon atom in a decreasing sequence were developed originally to differentiate enantiomeric configurations according to the procedure developed by Cahn, Ingold, and Prelog[3-7]. These rules have been found to be useful elsewhere; for example, in this chapter they are used in several ways in describing both chiral and achiral stereochemical configurations. It is only possible here to briefly discuss the rules for determining priority among atoms and groups; for more detailed information the original papers should be consulted.

1. Atoms are prioritized in order of decreasing atomic number; for instance, $I > Br > Cl > H$. Groups are prioritized in order of decreasing atomic number of the atom directly attached to the stereogenic atom and therefore $-S-CH_3 > -O-CH_3 > -N(CH_3)_2 > -CH_3$.
2. When two identical atoms are attached to the same stereogenic atom, the one that is also attached to an atom of higher atomic number takes precedence.

$$CH_3 - \overset{\overset{\displaystyle H}{|}}{\underset{\underset{\displaystyle Cl}{|}}{C}} - CH_2 - OH$$

In this example, the chiral carbon atom identified by the arrow is attached to two carbon atoms. Clearly, $Cl > (C\text{-}1 \text{ or } C\text{-}3) > H$. However, C-1 is also attached to an oxygen atom in addition to two hydrogen atoms, whereas C-3 is attached only to three hydrogen atoms, and since $O > H$, the complete priority order is $-Cl > -CH_2-OH > -CH_3 > -H$.

When two (or three) identical atoms attached to the same stereogenic atom are attached to more than one atom of higher atomic number, these atoms must first be arranged in decreasing atomic number order and compared in that order.

$$CH_3 - CHCl - \overset{\overset{\displaystyle H}{|}}{\underset{\underset{\displaystyle Cl}{|}}{C}} - CHCl - OH$$

For the atoms attached to the chiral carbon atom identified by the arrow, again $Cl > (C\text{-}1 \text{ or } C\text{-}3) > H$. To distinguish between C-1 and C-3, the atoms attached to C-1 are arranged in order of decreasing atomic number (that is, Cl, O, H), and compared with the atoms attached to C-3, also arranged in order of decreasing atomic number, (that is, Cl,C,H). Since the first atom in each set, Cl, is the same, the second atoms are compared, and since O > C, it follows that C-1 is preferred to C-3. Thus, the complete priority order is –Cl > –CH(Cl)(OH) > –CH(Cl)CH$_3$ > –H.

Branched chains are explored in the same way with the added stipulation that once a branch has established precedence over another branch, it has priority and its atoms are always compared before atoms in less preferred branches.

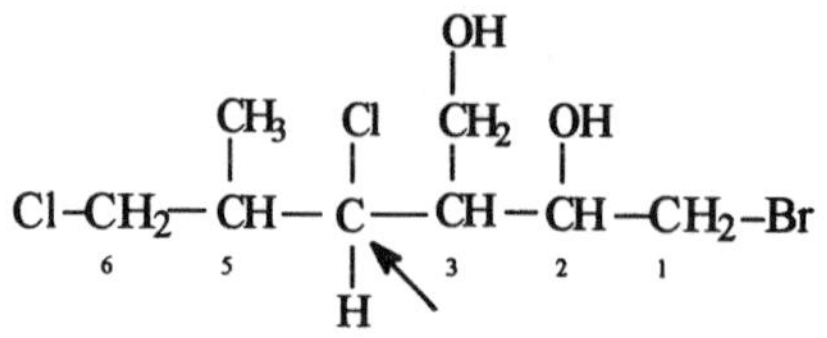

$$\text{Cl–CH}_2\text{–CH–}\underset{\underset{\text{H}}{|}}{\overset{\overset{\displaystyle \text{CH}_3}{|}}{\text{C}}}\text{–}\overset{\overset{\displaystyle \text{Cl}}{|}}{\text{CH}}\text{–}\overset{\overset{\displaystyle \overset{\text{OH}}{|}}{\text{CH}_2}}{\text{CH}}\text{–CH}_2\text{–Br}$$

For the atoms attached to the chiral carbon atom identified by the arrow, again, Cl > (C-3 or C-5) > H. The carbon atoms of C-3 and C-5 are both attached to two carbon atoms and one hydrogen atom (that is, C,C,H) and the exploration must therefore continue to the next level. For the groups attached to C-5, the next level consists of the groups –CH$_2$–Cl and –CH$_3$ in which the atoms attached to each carbon atom in order of decreasing atomic number are Cl,H,H and H,H,H, respectively; of these Cl,H,H is senior because Cl > H. For the groups attached to C-3, the next level consists of the groups –CH(OH)CH$_2$–Br and –CH$_2$-OH, in which the atoms attached to each carbon atom in order of decreasing atomic number are O,C,H and O,H,H, respectively; of these O,C,H is senior because C > H. Comparison of the two senior series Cl,H,H and O,H,H shows that the C-5 branch is preferred to the C-3 branch. The junior series of atoms O,H,H and H,H,H (or any other atoms, such as the Br atom attached to C-1) do not become involved in the comparison because these atoms are never reached in the exploration; the first point of difference determines the priority order, and in this example this point is reached in the senior series. The complete priority order around the chiral carbon atom identified by the arrow in the above structure is:

$$\text{–Cl} > \text{–CH(CH}_3\text{)CH}_2\text{–Cl} > \text{–CH(CH}_2\text{OH)CH(OH)CH}_2\text{–Br} > \text{–H.}$$

3. If an atom is attached to another atom by a multiple bond, each atom is converted to single-bond tetracovalency by replication of the atom at the other end of the multiple bond; such atoms are enclosed in parentheses in the structural formulas below. Each real atom, except for hydrogen, and each replicated atom is converted to single-bond tetracovalency by adding "phantom atoms", that is, imaginary atoms having an atomic number of zero and which are never expanded to single-bond tetracovalency; "phantom atoms" are represented by the large dots in the structural formulas below.

$$\text{–CH=CH–} \quad \equiv \quad \overset{(\cdots\text{C})}{\underset{\text{H}}{|}}\overset{(\text{C}\cdots)}{\underset{\text{H}}{|}} \qquad \overset{}{\underset{}{\diagdown}}\text{C=O} \quad \equiv \quad \overset{(\cdots\text{O})}{|}\overset{(\text{C}\cdots)}{|}\text{–C–O}\cdots$$

$$\text{–C≡C–} \quad \equiv \quad \overset{(\cdots\text{C})(\text{C}\cdots)}{|}\text{–C–C–}\underset{(\cdots\text{C})(\text{C}\cdots)}{|} \qquad \text{–C≡N} \quad \equiv \quad \overset{(\cdots\text{N})(\text{C}\cdots)}{|}\text{–C–N}\cdot\underset{(\cdots\text{N})(\text{C}\cdots)}{|}$$

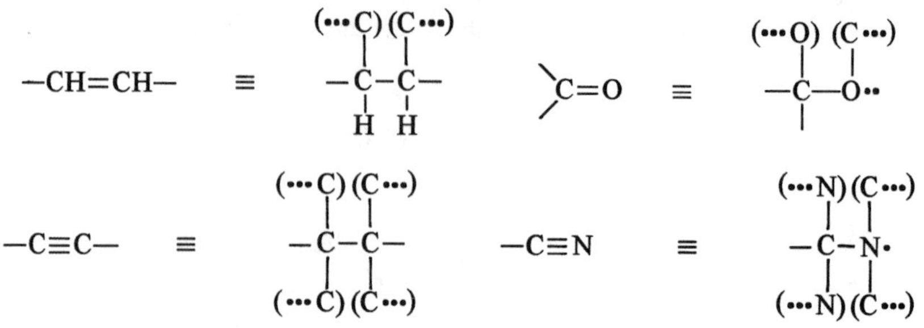

Accordingly, although an isopropyl group, $-CH(CH_3)_2$, is preferred to an allyl group, $-CH_2CH=CH_2$, because C,C,H > C,H,H, a prop-1-en-1-yl group, $CH_3-CH-CH_3$, is preferred to an isopropyl group. The reason for this is that while no difference is found on exploration of carbon atom 1 of each (both are C,C,H, because of the replicated carbon atom attached to carbon atom 1 of the prop-1-en-1-yl group), exploring carbon atom 2 gives C,C,H for the prop-1-en-1-yl group, which is preferred to the H,H,H of the isopropyl group. However, a 1,1,2-trimethylpropyl group is preferred to a 1-methylpent-1-en-1-yl group. The exploration of these two groups can be summarized as shown below each tetracovalent structure.

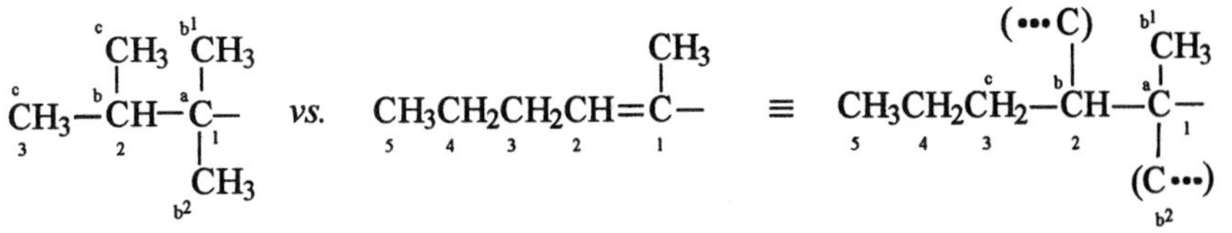

| I | | | II |
|---|---|---|---|
| 1,2,2-Trimethylpropyl | | | 1-Methylpent-1-en-1-yl |
| a  C, C, C | | = | C, C, (C) |
| b  C, C, H | | = | C, (C), H |
| b¹  H, H, H | | = | H, H, H |
| b²  H, H, H | | > | •, •, •, |

The atom marked "a" in each structure is attached to three carbon atoms, the third carbon atom in **II** being a replicated atom (enclosed in parentheses). Since no decision is reached here, each of the carbon atoms attached to C-1, marked "b", "$b^1$", and "$b^2$", are explored in turn, beginning with the "b" atom of each group, since in each case "b" is the most preferred carbon atom of the three, being attached to two further carbon atoms (in **II**, one of which is a replicated carbon atom), whereas the carbon atoms marked "$b^1$" and "$b^2$" are attached only to hydrogen atoms or phantom atoms. At the atom marked "b" the comparison is C,C,H and C,(C),H, respectively. Since no decision is reached at this step the atoms attached to the atoms marked "$b^1$" are compared. Again, no decision is attained, since the attached atoms for both are H,H,H. For these groups, the final step is to compare the atoms marked "$b^2$". Since phantom atoms (the large "dots") have by definition an atomic number of zero, the hydrogen atoms attached to "$b^2$" of group **I** are preferred and the 1,1,2-trimethylpropyl group (**I**) is preferred to the 1-methylpent-1-en-1-yl group (**II**).

Rings are treated in the same way as branched chains except that in each branch the process terminates when an atom that is the entry point to the cyclic path is reached. Such an atom is replaced by a replicate atom attached only to "phantom atoms". The exploration pathway for the two branches of a cyclopropyl group can be shown as follows:

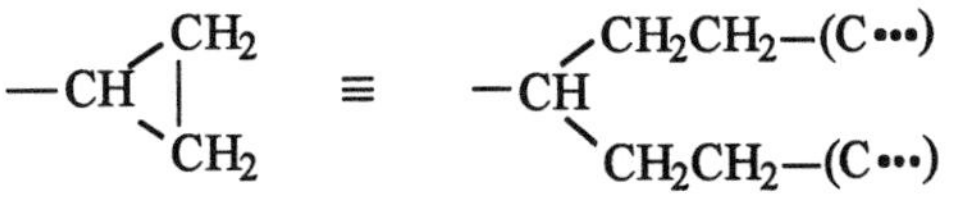

Unsaturated bonds in hydrocarbon rings are treated in the same way as unsaturated bonds in acyclic structures, replicating the atoms at each carbon-carbon double bond as described above.

Unsaturated bonds at heteroatoms in heterocyclic rings are replicated with fictitious atoms having an atomic number that is the mean of what each would have if the double bond were located at each of the possible positions. The nitrogen atom in pyridine is replicated by attaching at positions 2 and 6 fictitious atoms with the atomic number $6\frac{1}{2}$. In quinoline, the nitrogen atom is replicated by fictitious atoms with atomic numbers $6\frac{1}{2}$ and $6\frac{1}{3}$ at positions 2 and 8a, respectively. In both cases, the replicated atom attached to the nitrogen atom is carbon. In 1,8-naphthyridine, the fictitious atoms attached to positions 2, 7, and 8a have the atomic numbers $6\frac{1}{2}$, $6\frac{1}{2}$, and $6\frac{2}{3}$, respectively.

$6\frac{1}{2}$ ... N ... $6\frac{1}{2}$ ... 6

$6\frac{1}{2}$ ... N ... $6\frac{1}{2}$ ; $6\frac{1}{3}$ ; 6 ; 6

$6\frac{1}{2}$ ... N ... N ... $6\frac{1}{2}$ ; 6 ; $6\frac{2}{3}$ ; 6

4. With isotopes of the same atom, a higher mass number is preferred, for example, $^{2}H > {}^{1}H$ (or H).

5. For choice between two groups that are structurally identical, but are stereoisomeric, Z (*cis*) is preferred to E (*trans*) (see below under relative configuration), and then *R* is preferred to *S* (see below under absolute configuration).

Table 30.1, adapted from Table B of the Appendix to Section E of the IUPAC 1979 Organic Rules, gives an illustrative list of common groups in organic nomenclature in increasing seniority order according to the sequence rule.

## Table 30-1. Common Groups in Increasing Sequence Rule Seniority

*This table is to be read newspaper style.*

| | | |
|---|---|---|
| hydrogen | 2-methylphenyl (*o*-tolyl) | methoxy |
| methyl | 2,6-dimethylphenyl (2,6-xylyl) | ethoxy |
| ethyl | triphenylmethyl (trityl) | benzyloxy (phenylmethoxy) |
| propyl | 2-nitrophenyl | phenoxy |
| butyl | 2,4-dinitrophenyl | glycosyloxy |
| pentyl | formyl | formyloxy |
| hexyl | acetyl | acetoxy (acetyloxy) |
| 3-methylbutyl (isopentyl) | benzoyl | benzoyloxy |
| 2-methylpropyl (isobutyl) | carboxy | (methylsulfinyl)oxy |
| allyl (prop-2-en-1-yl) | methoxycarbonyl | (methylsulfonyl)oxy |
| 2,2-dimethylpropyl (neopentyl) | ethoxycarbonyl | fluoro |
| prop-2-yn-1-yl | (benzyloxy)carbonyl | silyl |
| benzyl |     [(phenylmethoxy)carbonyl] | phosphanyl |
| isopropyl (1-methylethyl) | [(2-methylpropan-2-yl)oxy]- | mercapto (sulfanyl) |
| vinyl (ethenyl) |     carbonyl; (tert-butoxycarbonyl) | methylthio (methylsulfanyl) |
| butan-2-yl (1-methylpropyl; | amino | methylsulfinyl |
|   sec-butyl) | ammonio (ammoniumyl) | methylsulfonyl |
| cyclohexyl | methylamino | sulfo |
| prop-1-en-1-yl | ethylamino | chloro |
| 2-methylpropan-2-yl (*tert*-butyl) | phenylamino (anilino) | germyl |
| prop-1-en-2-yl (isopropenyl) | acetylamino | arsanyl |
| ethynyl | benzoylamino | selanyl |
| phenyl | [(benzyloxy)carbonyl]amino | bromo |
| 4-methylphenyl (*p*-tolyl) | [(phenylmethoxy)carbonyl]amino | stannyl |
| 4-nitrophenyl | dimethylamino | stibanyl |
| 3-methylphenyl (*m*-tolyl) | diethylamino | tellanyl |
| 3,5-dimethylphenyl (3,5-xylyl) | nitroso | iodo |
| 3-nitrophenyl | nitro | plumbyl |
| 3,5-dinitrophenyl | hydroxy | bismuthanyl |
| prop-1-yn-1-yl | | |

## Absolute Configuration

Absolute configuration is the three-dimensional arrangement of atoms or groups around a chiral element, which may be a center, usually an atom, an axis, or a plane. The discussion here will deal only with the first of these, the chiral center.

The first method for differentiating stereoisomers was based only on the direction of rotation of polarized light, "*d*" (for dextro) if the rotation was to the right and "*l*" (for levo) if to the left. This was quite adequate for distinguishing enantiomers in compounds with one chiral center, but inadequate for compounds with more than one chiral center. A more general method was developed, based on a prescribed orientation of a Fischer projection. When the main chain of an acyclic compound is drawn vertically and the atom identified by the locant 1 is at the top of the structure, "D" is used to represent the absolute configuration of the enantiomer with the reference group on the right side of the highest numbered chiral atom, and "L" for the enantiomer with the reference group on the left. This method could be easily applied when there was a main chain and one hydrogen atom was attached to the highest numbered chiral atom. It was essentially useless when neither of the attached groups was hydrogen; until the advent of the sequence rule, there was no rational basis for choosing the reference group. However, this method was adapted to the description of absolute configuration of a carbohydrate chain by combining these descriptors with italicized word prefixes, such as *gluco-*, which indicated the relative configuration of the hydroxy groups on the chain. This is discussed more fully in chapter 31. The D/L symbols were also adopted for α-amino acids (see chapter 31).

In the specialized nomenclature for steroids, the symbols α and β are used to describe absolute configuration of both implied and stated substituents, *but only when the steroid ring structure is oriented in the prescribed manner*. This is also discussed in chapter 31.

By far the most general method for describing absolute configuration is the $R/S$ system introduced by R. S. Cahn, C. K. Ingold, and V. Prelog beginning in the early 1950s[3-7].

### *Absolute Configuration at Chiral Centers (Atoms)*

The absolute configuration of a chiral center (atom) is best expressed by the $R/S$ system, often called the CIP system. The symbol $R$ comes from the Latin word *rectus*, meaning right, and $S$ from the Latin *sinister*, meaning left. In this method, the four different groups attached to a tetrahedral chiral atom are considered in descending priority order according to the sequence rule, discussed earlier in this chapter. In the schematic below, a > b > c > d (the symbol > means "is preferred to"). The structure is oriented so that the groups "a", "b", and "c" define a plane, which can be seen as a steering wheel with three spokes, and the "d" group projects behind this plane, away from the viewer like a steering column. From the viewpoint of a driver on the side opposite the "d" group, the configuration of the chiral center is $R$ if the path traced from "a" to "b" to "c" is clockwise; if the path is counterclockwise, the configuration is $S$.

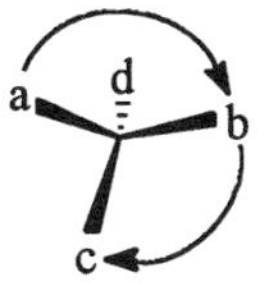

(*R*)-configuration 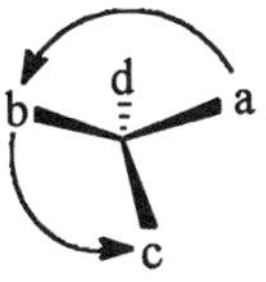          (*S*)-configuration

It is crucial that the least preferred atom or group be directed away from the viewer; if the least preferred atom or group projects toward the viewer, the resulting tetrahedral structure must be observed from behind the plane of the paper. Although the configuration of the chiral center in the structure on the right below might appear to be $S$, it is, in fact, identical to the structure on the left, which is clearly $R$.

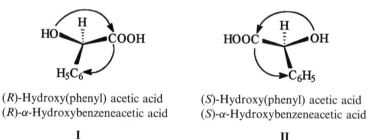

(R)-configuration                        (R)-configuration

In the two enantiomers of the compound $C_6H_5$–CH(OH)–COOH, shown in perspective in **I** and **II** below, "a" is –OH, "b" is –COOH, "c" is –$C_6H_5$, and "d" is –H. Visualization of the structure from in front of the plane of the paper reveals that in **I** the direction from "a" to "b" to "c" is clockwise, or *R*, and in **II**, the direction is counterclockwise, or *S*.

(R)-Hydroxy(phenyl) acetic acid          (S)-Hydroxy(phenyl) acetic acid
(R)-α-Hydroxybenzeneacetic acid          (S)-α-Hydroxybenzeneacetic acid

**I**                                     **II**

Assignment of *R* and *S* to each chiral center in compounds with multiple centers proceeds in the same way.

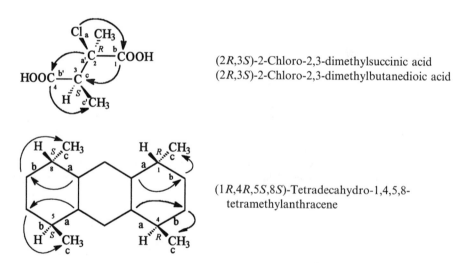

(2R,3S)-2-Chloro-2,3-dimethylsuccinic acid
(2R,3S)-2-Chloro-2,3-dimethylbutanedioic acid

(1R,4R,5S,8S)-Tetradecahydro-1,4,5,8-
tetramethylanthracene

Note that the configuration at positions 1 and 8 are *R* and *S*, respectively, even through the direction of the arrows might be taken as an indication of the opposite configuration. This is because the "d" group (that is, hydrogen) is projecting above the plane of the ring, and the direction of the arrows must be viewed from the side opposite to the "d" group. Therefore the configurations are opposite to the direction that the arrows indicate.

In the above names all the chiral centers are in the parent structure and their descriptors are therefore cited in front of the name. Descriptors for chiral centers not in the parent structure are cited in front of the part of the name to which they refer.

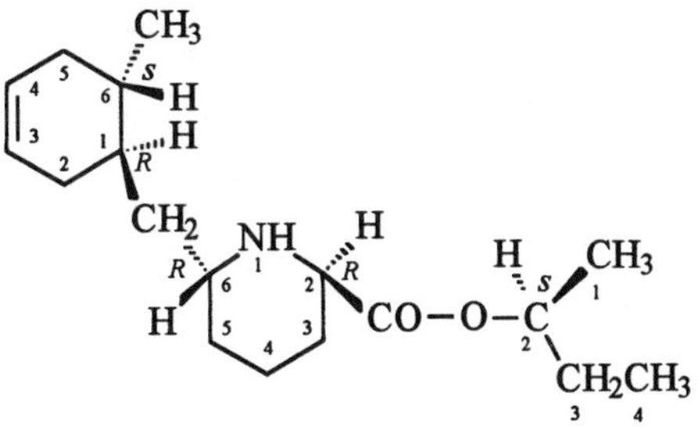

(2S)-Butan-2-yl (2R,6R)-6-[[(1R,6S)-
(6-methylcyclohex-3-en-1-yl)]methyl]-
piperidine-2-carboxylate

In this name, (2S) refers to the chiral center in the alcoholic part of the ester name, butan-2-yl; (2R,6R) refers to the chiral centers in the parent structure, piperidine-2-carboxylate; and (1R,6S) refers to the chiral centers in the (6-methylcyclohex-3-en-1-yl)methyl substituent.

## Relative Configuration

Configuration of any stereogenic element with respect to any other stereogenic element in the same molecule is termed relative configuration. These relative spatial relationships are indicated by descriptor pairs. For a plane as the reference, the descriptor pairs cis/trans, Z/E, exo/endo, and syn/anti are used under specified conditions to avoid ambiguity. For chiral centers (atoms) the descriptor pairs $R^*/S^*$ or rel-(R/S) are employed.

### Relative Configurations at Chiral Centers (Atoms)

Relative configuration among two or more chiral stereogenic centers may be described with the symbols $R^*$ and $S^*$, derived in the same way as the symbols $R$ and $S$ for absolute configuration (see above). In this method, one of the chiral centers, the reference center, is arbitrarily assigned the $R$ configuration, regardless of how it is shown in the structure, and labeled as $R^*$. The reference center is usually the chiral center in a parent structure with the lowest locant. Although there are no official rules for choosing a reference center when the parent structure does not have a chiral center, it is reasonable that the reference center be the chiral center first appearing in the name. The remaining centers are indicated as $R^*$ or $S^*$ according to whether they have the same or different chiralities, respectively, as the reference center. Hence, two centers with the same chirality are described as $(R^*,R^*)$, and two centers with different chiralities as $(R^*,S^*)$. If necessary, locants identify the chiral centers. Alternatively, the absolute descriptors $R$ and $S$ that correspond to the $R^*$ and $S^*$ of the structure as described above are cited in the name, and the whole name prefixed by the italic prefix rel- (for relative). This is the alternative now used by CAS.

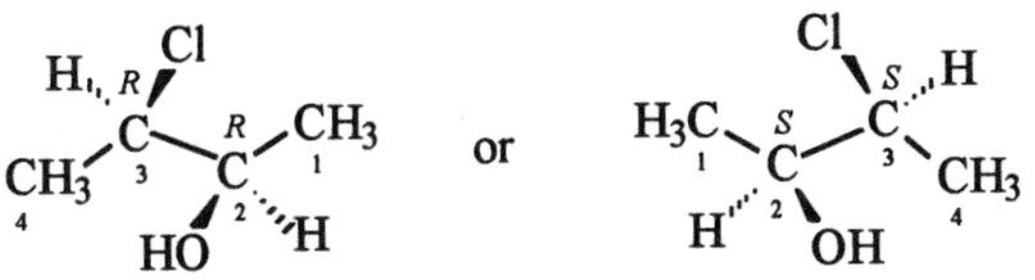

$(R^*,R^*)$-3-Chlorobutan-2-ol
(the two centers have the
same chirality)
rel-(R,R)-3-Chlorobutan-2-ol

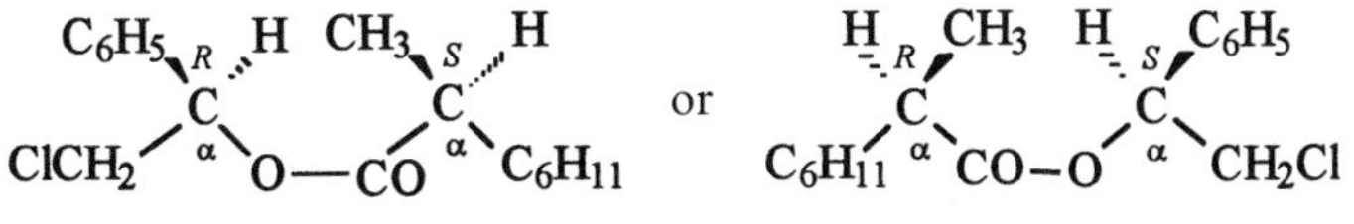

$(R^*,S^*)$-α-(Chloromethyl)benzyl 2-cyclohexylpropanoate
$(R^*,S^*)$-2-Chloro-1-phenylethyl α-methylcyclohexaneacetate
(the two centers have different chiralities)
rel-(R)-α-(Chloromethyl)benzyl (S)-2-cyclohexylpropanoate
rel-(R)-2-Chloro-1-phenylethyl (S)-α-methylcyclohexaneacetate

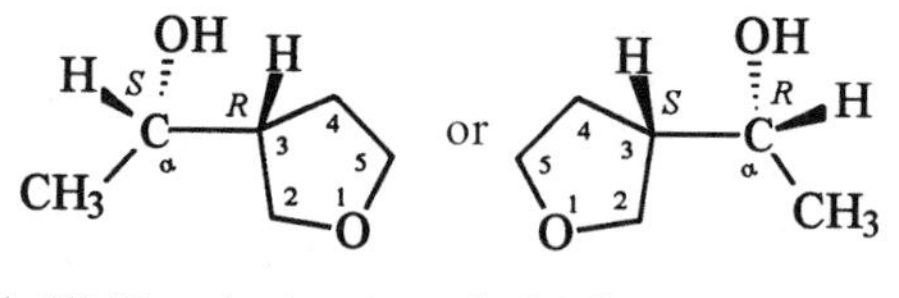

($R^*,S^*$)-Tetrahydro-$\alpha$-methyl-3-furanmethanol
($R^*,S^*$)-1-(Tetrahydro-3-furyl)ethanol
(the two centers have different chiralities)
*rel*-($\alpha R,3S$)-Tetrahydro-$\alpha$-methyl-3-furanmethanol
*rel*-($R$)-1-[($S$)-Tetrahydro-3-furyl]ethanol

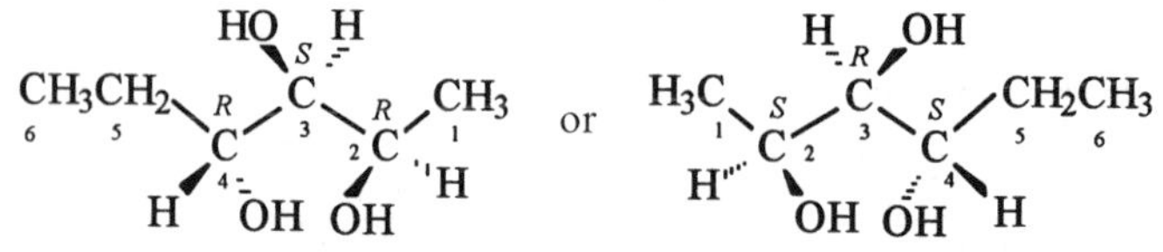

($2R^*,3S^*,4R^*$)-2,3,4-Hexanetriol
(the stereogenic center with the lowest locant, 2, is labeled $R^*$; the chirality at
position 3 is different and that at position 4 is the same as the reference center)
*rel*-($2R,3S,4R$)-2,3,4-Hexanetriol

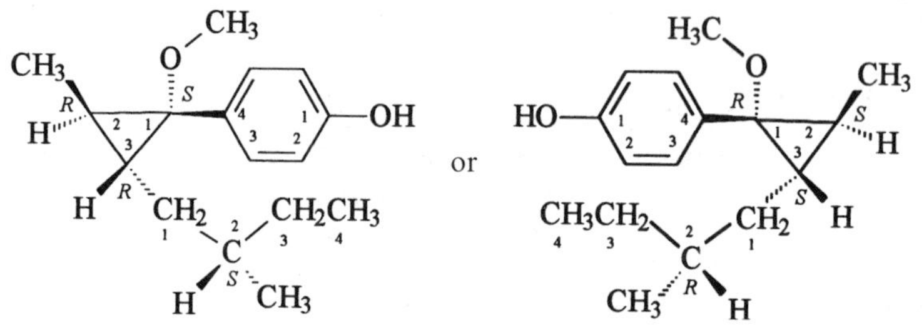

4-[($1R^*,2S^*,3S^*$)-1-Methoxy-2-methyl-3-[($2R^*$)-2-methylbutyl]cyclopropyl]phenol
(there are no chiral centers in the parent structure; the first chiral center in the
name, the 1 position of the cyclopropyl substituent, is labeled $R^*$, the chiral
centers at positions 2 and 3 of the cyclopropyl substituent have chiralities that
are different from the reference center; and the chirality at position 2 of the butyl
substituent has the same chirality as the reference center)
*rel*-4-[($1R,2S,3S$)-1-Methoxy-2-methyl-3-[($2R$)-2-methylbutyl]cyclopropyl]phenol

### Stereoisomerism at Double Bonds

The reference plane for describing stereoisomers at a double bond is the plane that is perpendicular to the molecular plane (see figure 30.1). One of the atoms or groups attached to one
atom of the double bond must be described as being on the same side, or on the opposite side,
of the reference plane from one of the atoms or groups attached to the other atom of the
double bond. By far, the most general method to do this was developed at CAS[8–10] and
codified in the IUPAC rules and recommendations. In this method the sequence rule given
earlier is applied to determine the more senior of the atoms or groups or atoms of the pair at each
end of the double bond. The configuration in which the atom or group with the higher priority at
each end of the double bond are on the same side of the reference plane is designated by $Z$ (from
the German *zusammen*, "together"), and the configuration in which the atom or group with the
higher priority at each end of the double bond are on opposite sides is designated $E$ (from the
German *entgegen*, "opposite"). These italicized symbols are enclosed in parentheses and cited in
front of the part of the name to which they refer; double bonds connecting a substituent group are
considered to be a part of the parent structure or parent substituent. Configuration of two or

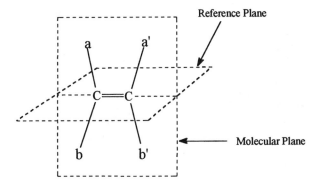

**Figure 30.1.** Stereoisomerism at double bonds.

more double bonds in the same parent, substituent, or functional derivative can be distinguished by appropriate locants; where there is a choice, Z double bonds have precedence over E double bonds for choice of parent structure and for low locants.

In the indexes for *Chemical Abstracts* Volumes 76–128, CAS cited all descriptors in front of the complete uninverted name; multiple descriptors were cited in decreasing order of the senior atom or group of the four attached to each double bond as determined by the sequence rule. This technique allowed all descriptors to be cited in front of the complete name, but it can result in extra work to determine the structure. Although all CAS names that used this technique in the CAS registry files have been updated, they still appear in the printed indexes for Volumes 76–128 of *Chemical Abstracts*. Therefore, these "old" CAS names are also mentioned in the examples below. When identical atoms or groups are attached to each of the double bond atoms in the parent chain of the compound, the descriptors *cis-* and *trans*[11] may be used instead of Z and E to indicate the relationship of these identical groups. In this way, *cis-* and *trans-* are now applied only to simple compounds such as *trans*-2,3-dichlorobut-2-ene.

The descriptors *syn-* and *anti-* are no longer recommended for configuration at double bonds.

(E)-But-2-ene
*trans*-But-2-ene

(Z)-5-Chloropent-4-enoic acid
*cis*-5-Chloropent-4-enoic acid

(E)-Diphenyldiazene
*trans*-Diphenyldiazene
(E)-Azobenzene

(Z)-(4-Chlorophenyl)phenylmethanone oxime
(Z)-N-[(4-Chlorophenyl)phenylmethylidene]-
    hydroxylamine

(1Z,4E)-1,2,4,5-Tetrachloropenta-1,4-diene
    (the Z double bond is numbered first)
(1-*cis*-4-*trans*-1,2,4,5-Tetrachloropenta-1,4-diene
    (the *cis* double bond is numbered first)
(E,Z)-1,2,4,5-Tetrachloro-1,4-pentadiene
    (*Chemical Abstracts* Volumes 76–128; the
    E double bond has a group with a Z double
    bond attached, and is therefore cited first)

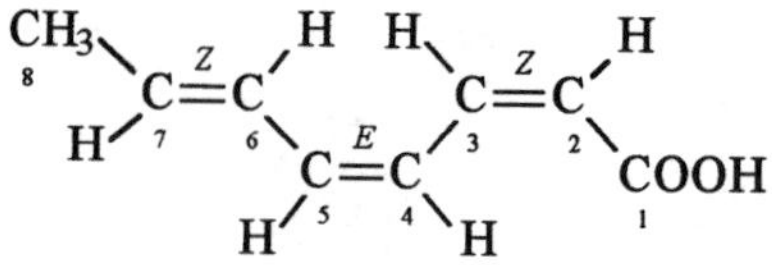

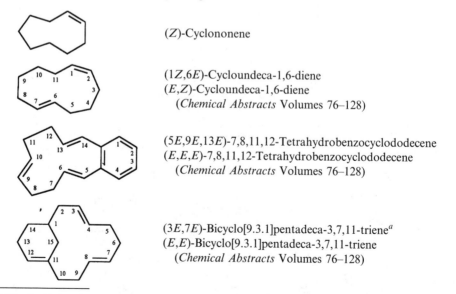

(2*Z*,4*Z*,6*E*)-Octa-2,4,6-trienoic acid
(2-*cis*-4-*cis*-6-*trans*)-Octa-2,4,6-trienoic acid
(*Z*,*Z*,*E*)-Octa-2,4,6-trienoic acid
   (*Chemical Abstracts* Volumes 76–128)

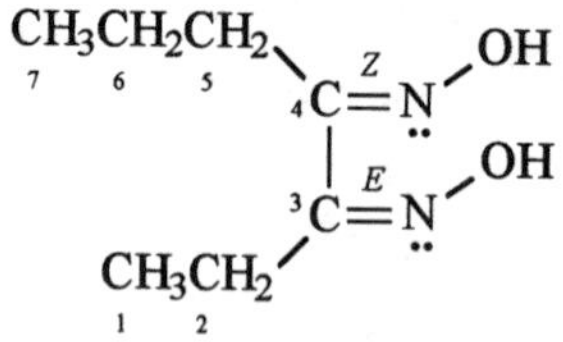

(3*Z*,4*E*)-Heptane-3,4-dione dioxime
(*E*,*Z*)-Heptane-3,4-dione dioxime
   (*Chemical Abstracts* Volumes 76–128)

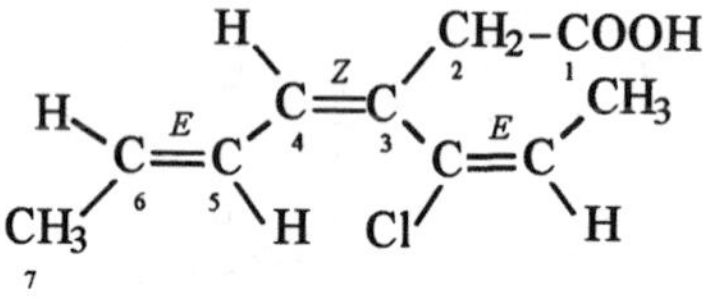

(3*Z*,5*E*)-[(1*E*)-1-Chloroprop-1-en-1-yl]-
   hepta-3,5-dienoic acid
(*E*,*Z*,*E*)-1-Chloroprop-1-en-1-yl]hepta-3,5-
   dienoic acid
    (*Chemical Abstracts* Volumes 76–128)

   *Z* and *E* or *cis*- and *trans*- are also used to describe configurations of compounds with an extended structure such that the four atoms or groups at the ends of the system are in the same plane. In the following examples "a" and "a¹" represent identical atoms or groups, or the sequence rule preferred atoms or groups.

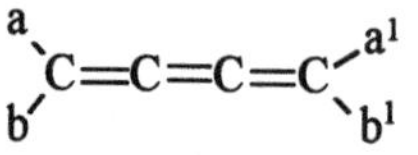

*Z* or *cis* 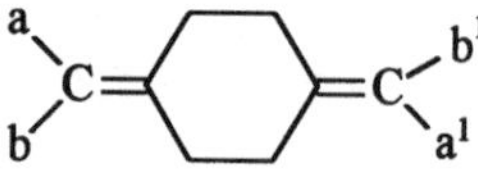       *E* or *trans*        *Z* or *cis*

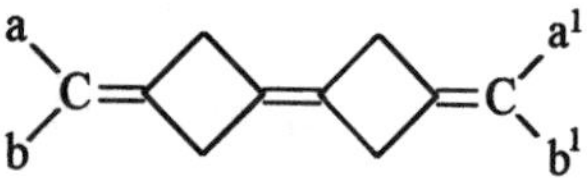

*Z* or *cis*

   Isolated double bonds in large rings may also be described by *Z* or *E*. For CAS, a large ring is defined as having more than eight ring members.

(*Z*)-Cyclononene

(1*Z*,6*E*)-Cycloundeca-1,6-diene
(*E*,*Z*)-Cycloundeca-1,6-diene
   (*Chemical Abstracts* Volumes 76–128)

(5*E*,9*E*,13*E*)-7,8,11,12-Tetrahydrobenzocyclododecene
(*E*,*E*,*E*)-7,8,11,12-Tetrahydrobenzocyclododecene
   (*Chemical Abstracts* Volumes 76–128)

(3*E*,7*E*)-Bicyclo[9.3.1]pentadeca-3,7,11-triene[a]
(*E*,*E*)-Bicyclo[9.3.1]pentadeca-3,7,11-triene
   (*Chemical Abstracts* Volumes 76–128)

---

[a] The double bond at position 11 is in a six-membered ring and not usually considered to be stereogenic.

## Stereoisomerism at Cyclic Tetrahedral Stereogenic Centers

Relative configuration in substituted monocyclic rings is indicated by reference to a planar representation of the ring. Conformations such as "chair" and "boat" forms of a ring are ignored. This method is most useful for relatively small rings; with large rings, the interpretation of the descriptors can be difficult. For rings with more than eight members CAS uses $R*/S*$ or rel-(R/S).

In achiral monocyclic compounds substituted at only two positions, relative configuration is described by cis- or trans- according to whether the senior sequence rule substituents (including hydrogen) at each position are on the same or on opposite sides, respectively, of the reference ring plane. Z and E are not used to indicate relative configuration in ring systems; their use is limited to nonchiral configurations at double bonds.

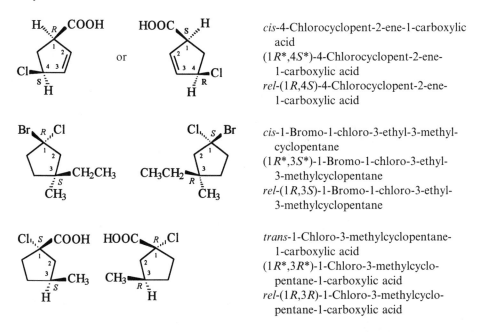

cis-1,4-Dimethylcyclohexane

trans-4-Ethyl-4-methylcyclohexan-1-ol

If the compound is chiral, $R*/S*$ or rel-(R/S) as described above can be used. The latter is the alternative now employed in CAS names. The chiral center with the lower locant is always $R*$.

cis-4-Chlorocyclopent-2-ene-1-carboxylic acid
($1R*,4S*$)-4-Chlorocyclopent-2-ene-1-carboxylic acid
rel-($1R,4S$)-4-Chlorocyclopent-2-ene-1-carboxylic acid

cis-1-Bromo-1-chloro-3-ethyl-3-methyl-cyclopentane
($1R*,3S*$)-1-Bromo-1-chloro-3-ethyl-3-methylcyclopentane
rel-($1R,3S$)-1-Bromo-1-chloro-3-ethyl-3-methylcyclopentane

trans-1-Chloro-3-methylcyclopentane-1-carboxylic acid
($1R*,3R*$)-1-Chloro-3-methylcyclo-pentane-1-carboxylic acid
rel-($1R,3R$)-1-Chloro-3-methylcyclo-pentane-1-carboxylic acid

For large rings, the descriptors cis- and trans- are potentially ambiguous (see Discussion), and therefore the $R*/S*$ or rel-(R/S) method should be used; for CAS, a ring with more than eight ring members is considered "large".

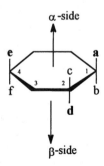

$(R^*,S^*)$-1,2-Dichlorocyclononene
*rel*-(1$R$,2$S$)-1,2-Dichlorocyclononene

For monocyclic compounds with different substituents at more than two ring atoms, three methods are available to describe relative configurational relationships between pairs of substituents (including hydrogen).

1. *The r/c/t method.* According to the stereochemical conventions given in the *Beilstein Handbook of Organic Chemistry*[2] and adapted for the IUPAC Organic Rules, the lowest numbered chiral position is selected as the reference group and its locant prefixed by an italic "*r*" (for reference). If there are two substituents at the lowest numbered chiral position, the preferred one is the suffix; if there is no suffix, the reference group is the substituent preferred by the sequence rule (see above). Other substituents are related to the group substituent by prefixing "*c*" (for *cis*) or "*t*" (for *trans*) to its locant according to whether it is on the same side of the ring as the reference group, or on the opposite side, respectively. Beilstein appends *r, c,* or *t* to the appropriate locant.

2. *The α/β method* (introduced by CAS in 1972). In this system the reference position is chosen in the same way as for (1) above. The symbols α and β are used to indicate whether a substituent is on one side of the ring plane or the other side, but they do not specify a specific side of the plane as they do in natural products (see chapter 31), and therefore can be applied no matter how the substituents are represented in the structural drawing. The α-side of the ring plane is arbitrarily defined as the side with the sequence rule preferred substituent at the lowest numbered stereogenic position. The descriptor α is appended to locants of all substituents that are on the same side of the plane as the reference substituent, and β is appended to locants for substituents that are on the other side. In figure 30.2, if the groups preferred by the sequence rule at the stereogenic centers 1, 2, and 4, are "a", "d", and "e", respectively, then the α-side is the one with the "a" substituent, the preferred substituent at the lowest numbered ring position. In a name, the locant set defining the configurations is enclosed in parentheses and prefixed to the part of the name to which it applies. This method was used for CAS index names for Volumes 76–128 of *Chemical Abstracts*. Now, it is used only for achiral compounds; the rel-(R/S) method (see below) is used for chiral compounds.

3. *The R\*/S\* or rel-(R/S) method.* As in the case of monocyclic compounds with substituents at only two positions, if the compound is chiral the symbols $R^*/S^*$ or

α-side

β-side

**Figure 30.2.** Stereochemical descriptors.

*rel*-(*R*/*S*) can be used. The latter is the alternative now used by CAS. The chiral center with the lowest locant is always *R**.

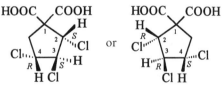

Cyclohexane-*r*-1,*c*-3,*c*-5-tricarboxylic acid
(1α,3α,5α)-Cyclohexane-1,3,5-tricarboxylic acid
(the compound is achiral)

*t*-4-Bromo-*c*-2,4-dichloro-2-methylcyclopentane-*r*-1-carboxylic acid
(1α,2α,4β)-4-Bromo-2,4-dichloro-2-methylcyclopentane-1-carboxylic acid
(1*R**,2*R**,4*S**)-4-Bromo-2,4-dichloro-2-methylcyclopentane-1-carboxylic acid
*rel*-(1*R*,2*R*,4*S*)-4-Bromo-2,4-dichloro-2-methylcyclopentane-1-carboxylic acid

*r*-1,*t*-2,*c*-4-Trichlorocyclopentane
  (*not* *r*-1,*t*-2,*t*-4-Trichlorocyclopentane; *c*-
  descriptors are preferred to *t*- descriptors)
(1α,2β,4α)-1,2,4-Trichlorocyclopentane
  [*not* (1α,2β,4β)-1,2,4-Trichlorocyclopentane; α-
  descriptors are preferred to β descriptors]
*rel*-(1*R*,2*R*)-1,2,4-Trichlorocyclopentane
  (position 4 is achiral)

*r*-2,*t*-3,*c*-4-Trichlorocyclopentane-1,1-dicarboxylic acid
(2α,3β,4α)-2,3,4-Trichlorocyclopentane-1,1-dicarboxylic acid
(2*R**,3*R**,4*S**)-2,3,4-Trichlorocyclopentane-1,1-dicarboxylic acid
*rel*-(2*R*,3*R*,4*S*)-2,3,4-Trichlorocyclopentane-1,1-dicarboxylic acid

In polycyclic ring systems other than those to which *endo/exo* and/or *syn/anti* apply (see below) relative configuration of atoms or groups at a single pair of saturated bridgeheads can be expressed by *cis*- or *trans*-. If the compound is chiral, *R**/*S** or *rel*-(*R*/*S*) can be used; again the latter is the method of choice for CAS. When additional stereogenic centers are present, α/β, *R**/*S**, or *rel*-(*R*/*S*) methods are used, as required.

*cis*-Decahydronaphthalene

trans-1,2,3,4,4a,9,9a,10-Octahydro-4a-methylanthracene
(4aR*,9aR*)-1,2,3,4,4a,9,9a,10-Octahydro-4a-methylanthracene
rel-(4aR,9aR)-1,2,3,4,4a,9,9a,10-Octahydro-4a-methylanthracene

cis-6,6-Dibromo-3-phenyl-3-azabicyclo[3.1.0]hexane
(1R*,5S*)-6,6-Dibromo-3-phenyl-3-azabicyclo[3.1.0]hexane
rel-(1R,5S)-6,6-Dibromo-3-phenyl-3-azabicyclo[3.1.0]hexane

trans-1,3,4,4a,5,6,7,8,9,10a-Decahydro-10a-hydroxybenz[a]azulen-10(2H)-one
(4aR*,10aR*)-1,3,4,4a,5,6,7,8,9,10a-Decahydro-10a-hydroxybenz[a]azulen-
    10(2H)-one
rel-(4aR,10aR)-1,3,4,4a,5,6,7,8,9,10a-Decahydro-10a-hydroxybenz[a]azulen-
    10(2H)-one

When two pairs of saturated bridgehead atoms are present, each may be *cis* or *trans*; the relationship between the pairs is expressed by *cisoid-* or *transoid-* inserted between *cis-* or *trans-*. These terms describe the relationship between the closer atoms (in terms of the smaller number of atoms) of the bridgehead pairs. If necessary, the locant of the lower numbered bridgehead atom is appended to the appropriate descriptor and separated from it by a hyphen. When two bridgehead atoms are equally close, an appropriate locant of each pair of bridgehead atoms, usually the lower locant, is appended. Since the other methods, $\alpha/\beta$, $R*/S*$, or *rel-*(R/S), are available, there is little need for the *cisoid/transoid* method beyond the most simple structures.

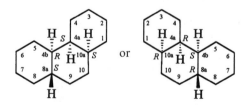

cis-cisoid-trans-Tetradecahydrophenanthrene
4aα,4bα,8aβ,10aα-Tetradecahydrophenanthrene
(4aR*,4bS*,8aR*,10aR*)-Tetradecahydrophenanthrene
rel-(4aR,4bS,8aR,10aR)-Tetradecahydrophenanthrene

H        H

(structure: tetradecahydroanthracene with labels 8, 8a R, 9, 9a S, 1, 2; 7, 6, S 10a, 10, R 4a, 3; 5, 4)

H        H

*cis-cisoid-cis*-Tetradecahydroanthracene
4aα,8aα,9aα,10aα-Tetradecahydroanthracene
(α descriptors are preferred to β descriptors)
*rel*-(4aR,8aR,9aS,10aS-Tetradecahydroanthracene
(this compound is an achiral diasteroisomer, a *meso*-compound)

(structures of octadecahydrobenz[b]acridine shown with "or" between them)

*cis*-4a-*transoid*-4a,5a-*trans*-5a-*cisoid*-5a,6a-*trans*-6a-Octadecahydrobenz[*b*]acridine
(4aα,5aβ,6aβ,10aα,11aα,12aα)-Octadecahydrobenz[b]acridine
(4aR*,5aR*,6aS*,10aS*,11aS*,12aR*)-Octadecahydrobenz[b]acridine
*rel*-(4aR,5aR,6aS,10aS,11aS,12aR)-Octadecahydrobenz[*b*]acridine

Relative steric relationships in bicyclo[x.y.z]alkane or bicyclo[x.y.z]alkene compounds, where "x" is greater than or equal to "y", both are greater than "z", and "z" cannot be "0", are indicated by the descriptive prefixes *endo-*, *exo-*, *syn-*, and *anti-*. Other restrictions applied by CAS are that (x + y) must be less than 7, there cannot be a double bond at a bridgehead, and the bridgehead atoms must be in a "normal" *cis*-configuration. Because ambiguity can occur easily in this system, CAS also applies other restrictions for the use of *syn-* and *anti-*; such refinements are beyond the scope of this discussion.

As illustrated in figure 30.3, *endo-* and *exo-* indicate that a substituent (or the senior substituent according to the sequence rule) on the "x" or "y" bridge is on the opposite side

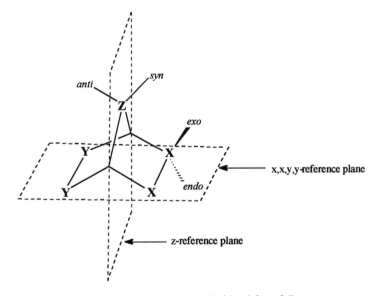

**Figure 30.3.** Stereoisomerism in bicyclo[x.y.z]alkanes.

or on the same side, respectively, as the "z" bridge with reference to the plane defined by the atoms of the "x" and "y" bridges. Locants are added as needed.

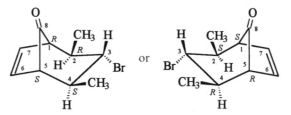

(*endo, exo*)-2,3-Dimethyllbicyclo[2.2.1]heptane
(1*R**,2*R**,3*R**,4*S**)-2,3-Dimethyllbicyclo[2.2.1]heptane
*rel*-(1*R*,2*R*,3*R*,4*S*)-2,3-Dimethyllbicyclo[2.2.1]heptane

*endo*-3-Methoxy-3-methyl-3-silabicyclo[3.2.1]octane

(2-*exo*,3-*exo*,5-*endo*)-3,5-Dimethyl-6-oxobicyclo[2.2.1]heptane-2-carboxylic acid
(1*R**,2*R**,3*R**,4*S**,5*R**)-3,5-Dimethyl-6-oxobicyclo[2.2.1]heptane-2-carboxylic acid
*rel*-(1*R*,2*R*,3*R*,4*S*,5*R*)-3,5-Dimethyl-6-oxobicyclo[2.2.1]heptane-2-carboxylic acid

(2-*exo*,3-*endo*,4-*exo*)-3-Bromo-2,4-dimethylbicyclo[3.2.1]oct-6-en-8-one
(1*R**,2*R**,4*S**,5*S**-3-*endo*)-3-Bromo-2,4-dimethylbicyclo[3.2.1]oct-6-en-8-one
*rel*-(1*R*,2*R*,4*S*,5*S*-3-*endo*)-3-Bromo-2,4-dimethylbicyclo[3.2.1]oct-6-en-8-one

Figure 30.3 also illustrates that the prefixes *syn*- and *anti*- indicate that a substituent (or the preferred substituent according to the sequence rule) on the "z" bridge projects toward or away from, respectively, the "x" or "y" bridge with the lower locant number. Locants are added as needed.

*anti*-7-Bromobicyclo[2.2.1]hept-2-ene

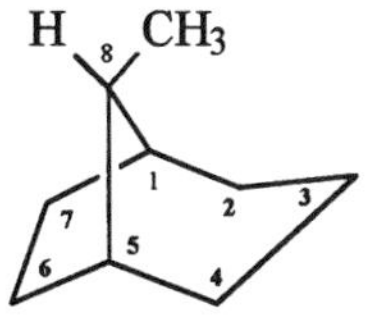

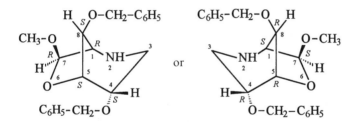

*syn*-8-Methylbicyclo[3.2.1]octane

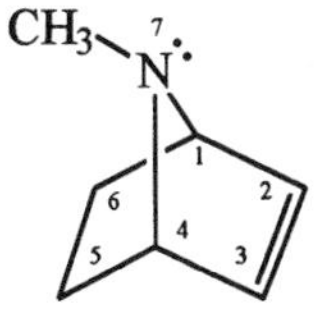

*anti*-7-Methyl-7-azabicyclo[2.2.1]hept-2-ene

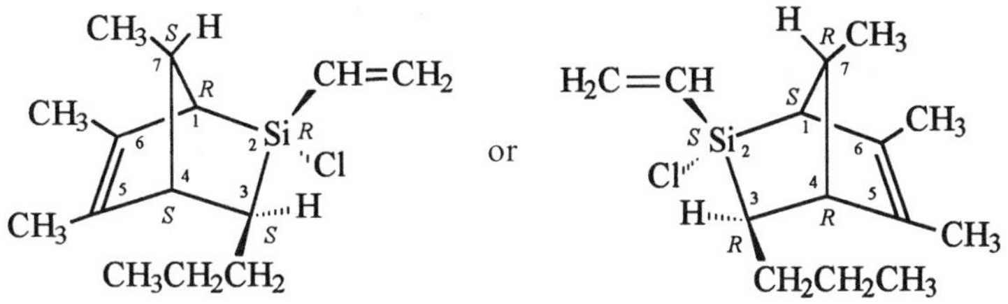

or

(2-*endo*,3-*endo*,7-*anti*)-2-Chloro-2-ethenyl-5,6,7-trimethyl-3-propyl-2-sila-
    bicyclo[2.2.1]hept-5-ene
(1*R**,2*R**,3*S**,4*S**,7*S**)-2-Chloro-2-ethenyl-5,6,7-trimethyl-3-propyl-2-
    silabicyclo[2.2.1]hept-5-ene
*rel*-(1*R*,2*R*,3*S*,4*S*,7*S*)-2-Chloro-2-ethenyl-5,6,7-trimethyl-3-propyl-2-
    silabicyclo[2.2.1]hept-5-ene

or

(4-*endo*,7-*exo*,8-*syn*)-4,8-Bis(benzyloxy)-7-methoxy-6-oxa-2-azabicyclo[3.2.1]octane
(1*R**,4*S**,5*S**,7*R**,8*S**)-4,8-Bis(benzyloxy)-7-methoxy-6-oxa-2-azabicyclo[3.2.1]octane
*rel*-(1*R*,4*S*,5*S*,7*R*,8*S*)-4,8-Bis(benzyloxy)-7-methoxy-6-oxa-2-azabicyclo[3.2.1]octane

Relative configuration among stereoisomers of hydrogenated fused and bridged fused ring systems where the ring substituents, including bridges, can be related to a reference plane defined by a planar representation of the ring system may be indicated by *cis/trans* and *α/β* as described above for monocyclic rings. The restrictions on ring size of any ring component noted above also apply. For convenience, a hydrogen atom or any single substituent at a ring fusion atom is considered to be the reference substituent and a bridge is preferred to a hydrogen atom at a bridgehead position. For chiral compounds the *R\*/S\** or *rel*-(*R/S*) methods may also be used. These methods can also apply to spiro systems in which all of the stereogenic atoms are in one ring.

Although noted earlier, it is worth repeating that when the *α/β* method is used, the symbols do not indicate a specific side of the reference plane and thus do not require a preferred orientation of the ring system. The side of the plane with the substituent preferred by the sequence rule at the lowest numbered position is arbitrarily defined as the *α*-side. As for monocyclic rings, *cis*- and *trans*- are preferred when relating substituents at only two positions.

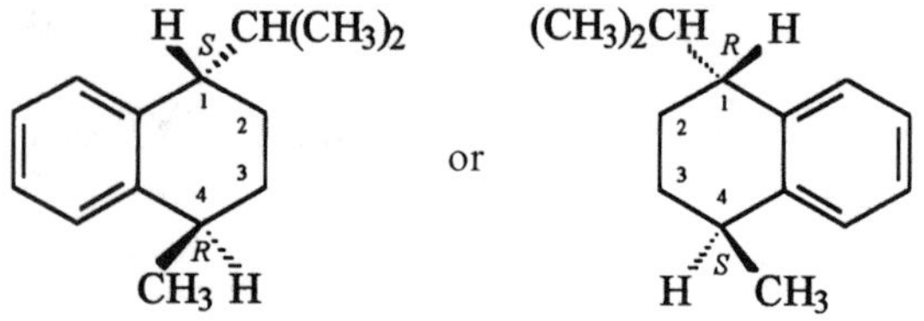

*trans*-1,2,3,4-Tetrahydro-1-isopropyl-4-methylnaphthalene
(1*R**,4*S**)-1,2,3,4-Tetrahydro-1-isopropyl-4-methylnaphthalene
*rel*-(1*R*,4*S*)-1,2,3,4-Tetrahydro-1-isopropyl-4-methylnaphthalene

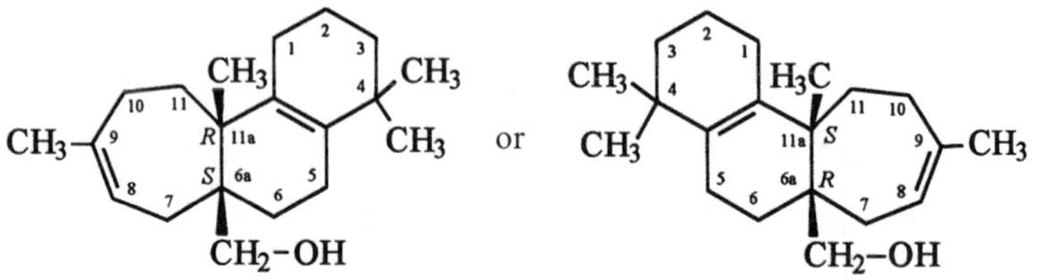

*cis*-(1,2,3,4,5,6,7,10,11,11a-Decahydro-4,4,9,11a-tetramethyl-6a*H*-
   cyclohepta[*a*]naphthalen-6a-yl)methanol
[(6a*R**,11a*S**)-1,2,3,4,5,6,7,10,11,11a-Decahydro-4,4,9,11a-tetramethyl-6a*H*-
   cyclohepta[a]naphthalen-6a-yl]methanol
*rel*-(6a*R*,11a*S*)-1,2,3,4,5,6,7,10,11,11a-Decahydro-4,4,9,11a-tetramethyl-6a*H*-
   cyclohepta[a]naphthalene-6a-methanol

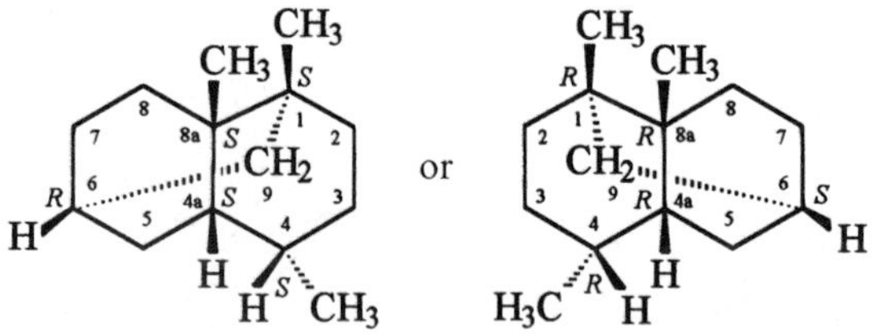

(1α,4aβ,6α,8aβ)-Decahydro-1,4,8a-trimethyl-1,6-methanonaphthalene
(1*R**,4*R**,4a*R**,6*S**,8a*R**)-Decahydro-1,4,8a-trimethyl-1,6-methanonaphthalene
*rel*-(1*R*,4*R*,4a*R*,6*S*,8a*R*)-Decahydro-1,4,8a-trimethyl-1,6-methanonaphthalene

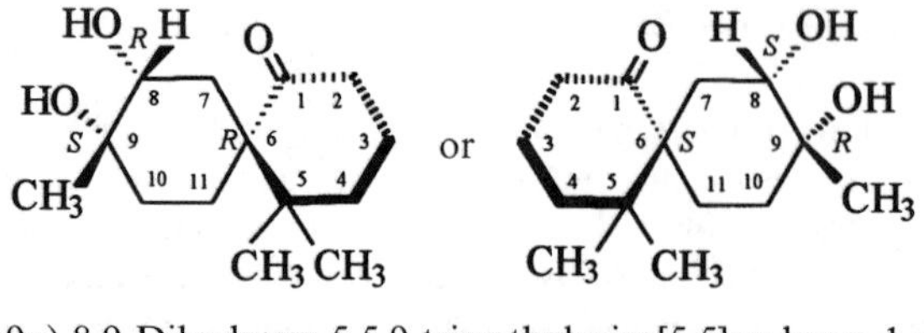

(6α,8α,9α)-8,9-Dihydroxy-5,5,9-trimethylspiro[5.5]undecan-1-one
(6*R**,8*R**,9*S**)-8,9-Dihydroxy-5,5,9-trimethylspiro[5.5]undecan-1-one
*rel*-(6*R*,8*R*,9*S*)-8,9-Dihydroxy-5,5,9-trimethylspiro[5.5]undecan-1-one

## Discussion

### Absolute configuration

A systematic nomenclature for multiple chiral centers was first developed for carbohydrates (see chapter 31). One center, the reference center, usually the highest numbered chiral center, is described by an absolute descriptor, "D" or "L", and its configuration relative to other chiral centers is given by italicized prefixes such as *glycero-* and *ribo-*. Thus, 2-deoxy-D-*ribo*-hexose and 2-dexoy-L-*ribo*-hexose are names for the two enantiomers of 2-dexoy-*ribo*-hexose.

In the nomenclature of amino acids, the D/L symbols are used for the absolute configuration of the $\alpha$-carbon atom, not the highest numbered chiral center. When ambiguity can arise, as in threonine, a subscript "g" or "s" is added ; "g" indicates that the standard is glyceraldehyde and "s", serine. Accordingly, D$_s$-threonine and L$_g$-threonine are names for the same structure. The D/L symbols also find limited use in cyclitol nomenclature (see chapter 31).

The prefix *meso-* is used for stereoisomers having two identical chiral sites of opposite chirality; this results in a total structure that is optically inactive because of the presence of a symmetry plane, center, or alternating axis. Such a prefix provides for a quicker recognition of the structure than *R* and *S* descriptors.

Clearly, the system developed by Cahn, Ingold, and Prelog with the symbols *R* and *S*, usually termed the CIP system, is the method of choice for describing configuration at tetrahedral chiral centers. Since this system is now so widely accepted, there is no need to extend the use of the D/L symbols of carbohydrate nomenclature (see chapter 31), even through the sequence rule would provide the method for choosing the reference center. It is interesting to note that the symbols "D" and "L" were used in their first CIP publication[7], but discarded in later publications.

Assignment of *R* and *S* to every center of a structure with many chiral centers can be an arduous manual task and the same is true for *R\** and *S\**; when other pairs of relative descriptors can be used, the description of stereoisomers becomes much easier. Visualization of a structure with relative descriptors seems easier because relative relationships among chiral centers should be more readily seen.

In 1972, CAS generalized the principle used in describing stereoisomers for carbohydrates to provide full descriptions of absolute stereochemistry for a very large number of organic compounds using *R* or *S* as the absolute descriptor for the reference center. This procedure is valuable for two reasons. In most cases it avoids a determination of absolute configuration for each chiral center and it directly relates enantiomers since both enantiomers have the same relative descriptor; the absolute descriptor *R* or *S* becomes the only difference, For example, the descriptors [1*R*-(1$\alpha$,3$\alpha$,4$\beta$)]- and [1*S*-(1$\alpha$,3$\alpha$,4$\beta$)]- describe two enantiomers of the compound 3,4-dimethylcyclohexanol. Furthermore, compounds that are positional isomers, some of which may be achiral, such as *trans*-1,4-dichlorocyclohexane, will have the same relative descriptor; the chiral isomers will have an additional absolute descriptor as in (1*R*-*trans*)- and (1*S*-*trans*)-1,2-dichlorocyclohexane. In 1972, the CAS system was the only method in existence that cited all configurations in front of the complete chemical name. A brief summary of the CAS system is given below; however, for details the appropriate documentation should be consulted:[10]

1. When all chiral centers are in a single parent chain, substituent, or functional derivative chain, ring or ring system, this chain, ring, or ring system is the parent for the stereodescriptor regardless of whether it is the parent structure for the name of the compound; the reference center for both the absolute and relative parts of the descriptor is the chiral center with the lowest locant.
2. When, in an acyclic structure, chiral centers are in different chains, the parent chain for the stereodescriptor is the chain with the most chiral centers; the reference center for both the absolute and relative parts of the descriptor is the chiral center with the lowest locant. When different chains have the same number of chiral centers, the chain chosen as the parent for the stereodescriptor is the one with the most preferred substituent according to the sequence rule.
3. When chiral centers are present in structures with two or more rings or ring systems, the parent ring for the stereodescriptor is the ring or ring system with the most chiral centers, even though a substituent or functional derivative chain may have more chiral centers, or the ring or ring system is not the one chosen as the parent for nomenclature. If more than one ring or ring system has the same number of chiral

centers, the ring or ring system chosen as the parent for the stereodescriptor is the preferred ring or ring system. If there is still a choice, the ring or ring system with the lowest locant set for the chiral centers is the parent for the stereodescriptor.

4. Achiral stereochemical descriptors may be included within the stereodescriptor.

However, beginning with the index to *Chemical Abstracts* Volume 129, CAS abandoned this stereodescriptor system. Chiral centers are now described relatively with *rel-(R/S)* and absolutely by the *R* and *S*. Where appropriate, *Z/E*, *cis/trans*, *syn/anti*, *α/β* are used. Stereodescriptors that apply to the parent structure, with appropriate locants, are prefixed to the complete name; descriptors that do not apply to the parent structure are cited with the portion of the name describing the part of the structure to which the descriptors apply. By means of a wide variety of examples in the Additional Examples section below, the CAS stereodescriptor used in index names for *Chemical Abstracts* Volumes 76–128 will be included in explanatory notes. Although the new CAS system will be complete for all compounds in the CAS registry file, the printed indexes for *Chemical Abstract* Volumes 76–128 will contain the previous system.

## Relative Configuration

The general class of stereoisomers differing in the relationship of atoms with respect to a reference plane is often termed "cis/trans-isomers", even though the prefixes *cis-* and *trans-* are not used in all cases. The term "geometric isomers", although still in use, is no longer recommended. *cis/trans*-Isomers are encountered in unsaturated acyclic structures, in large rings containing isolated double bonds, and in cyclic structures having substituents at saturated atoms that project above or below an idealized plane of the ring.

The prefixes *cis-* and *trans-* were introduced in 1888 by von Baeyer to differentiate stereo-isomers at double bonds and in rings. However, until the introduction of the *Z/E*-system[8,9] there was no way to unambiguously distinguish all *cis/trans* isomers for double bonds because there was no consistent general way to define the necessary reference groups. A recommenda-tion by the American Chemical Society in 1949[11] (adopted in 1953 by IUPAC), that the prefixes *cis-* or *trans-* describe the configuration at a double bond according to whether the carbon atoms of the fundamental chain adjacent to the carbon atoms of the double bond were on the same or opposite side of the double bond did not provide for double bonds not in the fundamental chain or for chains with less than four atoms. Nevertheless, *cis-* and *trans-* have long been used for configuration about carbon-carbon double bonds and are still acceptable provided they are unambiguous, as in *cis*-1,2-dichloroethene. However, the *Z/E*-system is completely general and is highly recommended, but it is only to be used for configuration at double bonds; it is not to be extended to other pairs of *cis/trans* isomers.

In large rings, *cis-* and *trans-*, as well as *α* and *β*, can become meaningless because of potential ambiguity. For instance, the structures below look like a *cis/trans* isomer pair, but are actually different conformations of the same stereoisomer:

The prefixes *syn-* (corresponding to *cis-*) and *anti-* (corresponding to *trans-*) have been used to describe configuration at the carbon–nitrogen double bond of oximes. This is no longer recommended; the *Z/E*-system is much preferred. The isomeric diazoates are still often called "normal diazoates" and "isodiazoates", without commitment as to structure; it appears that the "normal" isomer is *Z*, as in (*Z*)-benzenediazoate ion, and the "isodiazoate" is the *E*-isomer.

With the generalization of the *α/β*-system as described here and the use of *R\*/S\** and *rel-(R/S)*, the prefixes *cisoid-* and *transoid-* should no longer be needed, but their abandonment has not yet been recommended. However, the old usage of *syn-* and *anti-* for *cisoid-* and *transoid-* should be discontinued.

The prefixes *endo-* and *exo-* were proposed to describe an "inner" (more hindered) and an "outer" (less "hindered" or "open") location of a group in a bicyclo structure, respectively. Another explanation for these terms is that *endo-* describes a location "within" the dihedral angle formed by the main ring, and *exo-* describes a location away from, or outside, of this angle. It seems easier to use these prefixes with respect to a reference plane as done here, the reference plane being that defined by a planar representation of the main ring. To be the most useful, these terms must be used within the limits given above. Outside of these limits there is the possibility of ambiguity without rules for determining a "senior bridge" which would be needed to determine a reference plane. Rules for doing so have been suggested[12].

Although the *c/r/t* notation for *cis/trans*-isomers seems to be unique to the *Beilstein Handbook of Organic Chemistry*,[2] it has been codified in IUPAC recommendations.

There has been some confusion with regard to the generalization of the *α/β*-system as described in this chapter. It must be clearly understood that the *α*-side of the reference plane is chosen arbitrarily, whereas in steroid and terpene nomenclature the *α*-side is always behind the reference plane of the ring system (see chapter 31). In steroid and terpene nomenclature, the symbols *α* and *β* define absolute stereochemical configuration provided that the structure is drawn in the accepted standard orientation; to indicate a relative configuration, an additional descriptor, such as *rel-*, is needed. In the generalized approach, a preferred orientation is not required, the symbols *α* and *β* define relative configuration, and the indication of absolute configuration requires only the addition of a descriptor. This technique offers a distinct advantage for organic compounds in general because more often than not relative configuration is known before the absolute configuration, or the latter is not reported because it is not needed.

The prefixes *erythro-* and *threo-*, used to describe relative configurations in carbohydrate nomenclature (see chapter 31), have been used to describe simple organic compounds with two adjacent chiral sites in the main chain of a structure each carrying one of the same atom or group. This is of limited utility.

Although the *R\*/S\** and *rel-(R/S)* methods are quite general for relative stereochemistry they are dependent on the application of the sequence rule which requires the stereogenic centers to be chiral. The assignment of an *R* or *S* descriptor can often be difficult manually, but computer programs have made this a rather simple process. Even so, a correct interpretation of an *R* or *S* assignment may not be easy and the use of *cis/trans*, *syn/anti*, *exo/endo*, and *α/β* as far as possible should be preferable.

Optical isomers for which only relative configuration is known can be distinguished by (+) and (−) which gives only the observed direction of rotation of polarized light.

## Additional Examples

1.

(5*Z*)-5-[(2*E*, 4*E*)-5-(5,6,7,8-Tetrahydro-5,5,8,8-tetramethylnaphthalen-2-yl)-
penta-2,4-dien-1-ylidene]thiazolidine-2,4-dione

The CAS descriptor in *Chemical Abstracts* Volumes 76–128 was (*Z,E,E*); the descriptors are ranked according to the group of highest CIP priority attached to each double bond.

2.

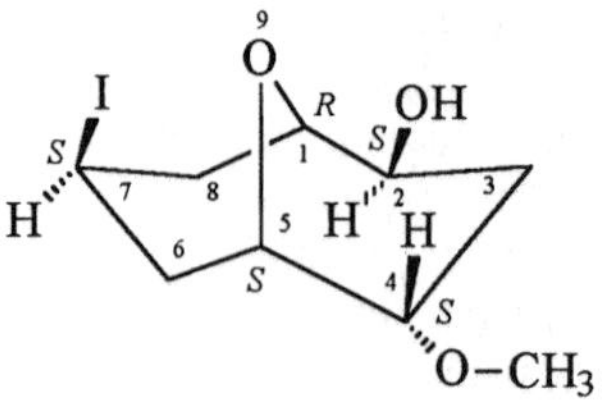

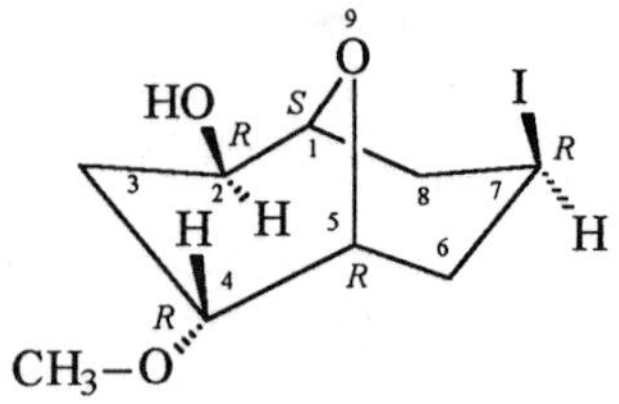

(2-*exo*,4-*endo*,7-*exo*)-7-Iodo-4-methoxy-9-oxabicyclo[3.3.1]decan-2-ol
(in ring systems where the "x" and/or the "y" bridge is larger than
in this example, only the R*/S* method should be used.)
(1R*,2S*,4S*,5S*,7S*)-7-Iodo-4-methoxy-9-oxabicyclo[3.3.1]decan-2-ol
*rel*-(1R,2S,4S,5S,7S)-7-Iodo-4-methoxy-9-oxabicyclo[3.3.1]decan-2-ol

3.

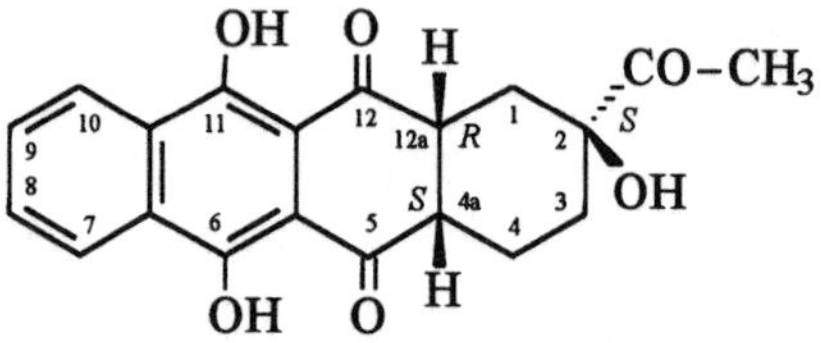

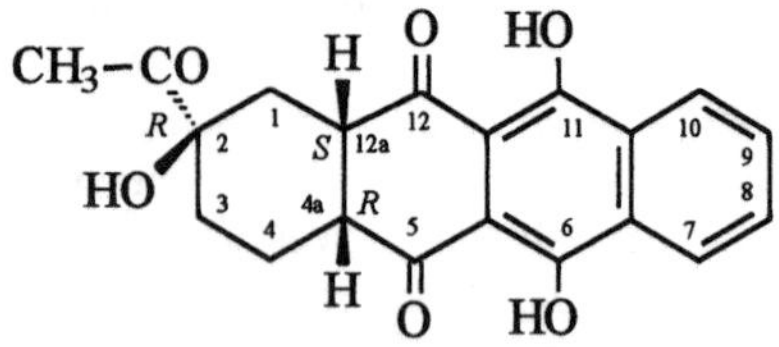

(2α,4aα,12aα)-2-Acetyl-1,2,3,4,4a,12a-hexahydro-2,6,11-trihydroxynaphthacene-
5,12-dione
(2R*,4aR*,12aS*)-2-Acetyl-1,2,3,4,4a,12a-hexahydro-2,6,11-trihydroxynaphth-
acene-5,12-dione
*rel*-(2R,4aR,12aS)-2-Acetyl-1,2,3,4,4a,12a-hexahydro-2,6,11-trihydroxynaphth-
acene-5,12-dione

4.

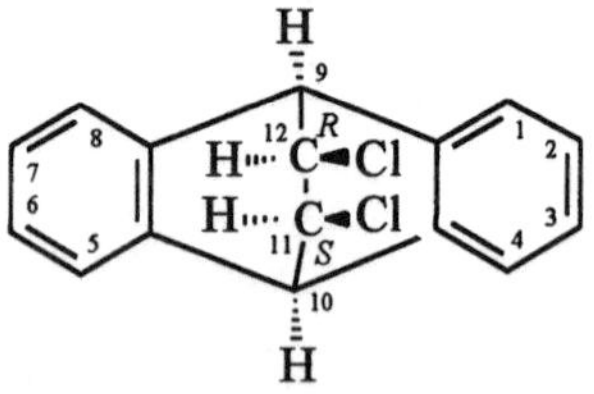

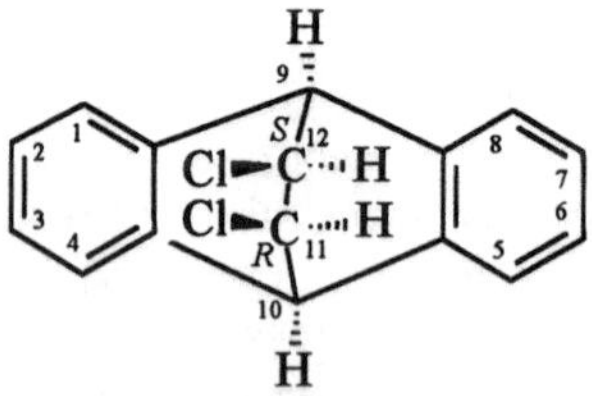

*cis*-11,12-Dichloro-9,10-dihydro-9,10-ethanoanthracene
(11R*,12S*)-11,12-Dichloro-9,10-dihydro-9,10-ethanoanthracene
*rel*-(11R,12S)-11,12-Dichloro-9,10-dihydro-9,10-ethanoanthracene

5.

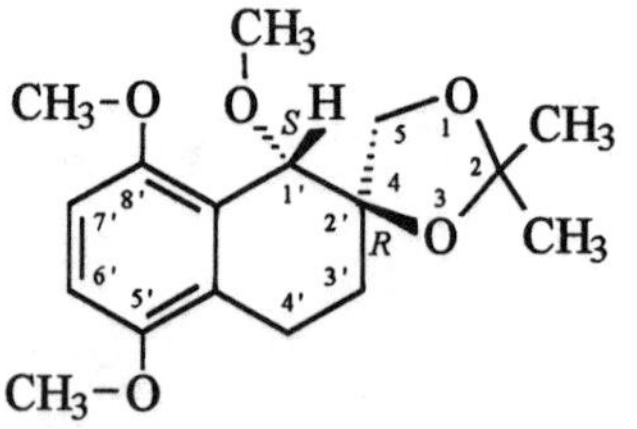

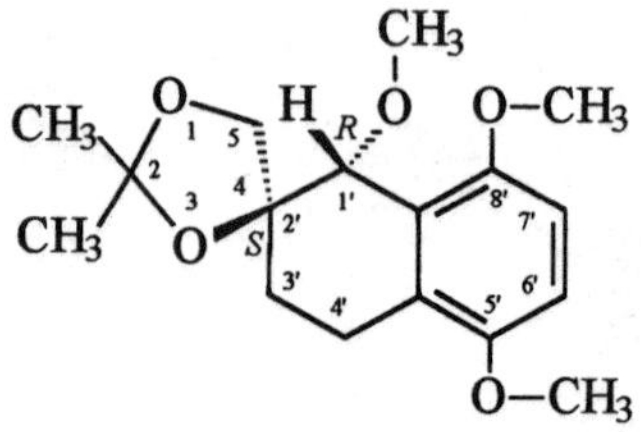

*trans*-3',4'-Dihydro-1',5',8'-trimethoxy-2,2-dimethylspiro[1,3-dioxolane-4,2'(1'H)-
naphthalene]
(1'R*,2'S*)-3',4'-Dihydro-1',5',8'-trimethoxy-2,2-dimethylspiro[1,3-dioxolane-
4,2'(1'H)-naphthalene]
*rel*-(1'R,2'S)-3',4'-Dihydro-1',5',8'-trimethoxy-2,2-dimethylspiro[1,3-dioxolane-
4,2'(1'H)-naphthalene]

6.  HOOC     COOH
    (R,R)-2,3-Dihydroxybutanedioic acid

The CAS descriptor in *Chemical Abstracts* Volumes 76–128 was [R-(R\*,R\*)]. The relative part of the descriptor, (R\*,R\*), indicates that the two centers are of like chirality. R gives the absolute configuration at either one of the centers.

7.  $CH_3CH_2CH_2$     $CH_3$
    (2R,3S)-2,3-Dichlorohexane

The CAS descriptor in *Chemical Abstracts* for Volumes 76–128 was [R-(R\*,S\*)]. The reference center for both the absolute and relative parts of the descriptor is carbon atom 2, because the preferred substituent according to the principles of the sequence rules is attached to it, that is, $-CH(Cl)CH_2CH_2CH_3$ is preferred to $-CH(Cl)CH_3$; its absolute chirality is R. The (R\*, S\*) part of the descriptor indicates that the two centers are of unlike chirality; therefore, the absolute chirality at carbon atom 3 is S.

8.
    (2S,3S,4R)-Hexane-2,3,4-triol

The CAS descriptor in *Chemical Abstracts* Volumes 76–128 was [2S-(2R\*,3R\*,4S\*)]. The reference center for both the absolute and relative parts of the descriptor is position 2, the lowest numbered chiral center in the chain; its absolute chirality is S. In the relative part of the descriptor, (2R\*,3R\*,4S\*), 2R\* marks the reference center (the reference center is always R\*), 3R\* indicates that the absolute chirality at position 3 is the same as the reference center, and 4S\* indicates that the absolute chirality at position 4 is opposite to that of the reference center. Therefore, the absolute chirality at position 3 is S and at position 4, R.

9.
    (3-*endo*,8-*anti*)-8-Methyl-8-azabicyclo-
    [3.2.1]oct-3-yl (βS)-β-hydroxy-
    benzenepropanoate
    (3-*endo*,8-*anti*)-8-Methyl-8-azabicyclo-
    [3.2.1]oct-3-yl (2S)-2-hydroxy-
    2-phenylpropanoate

10.
    [(2S,4R)-Dimethylheptyl]propanedioic acid
    [(2S,4R)-Dimethylheptyl]malonic acid

The CAS descriptor used in *Chemical Abstracts* Volumes 76–128 was [S-(R\*S\*)]. The reference center for both the absolute and relative parts of the descriptor is position 2 of the heptyl chain, the lowest numbered chiral position in the chain; its absolute chirality is S. The relative part of the descriptor, (R\*,S\*), indicates that the two centers are of unlike chirality. Therefore the absolute chirality at position 4 of the heptyl chain is R.

11.

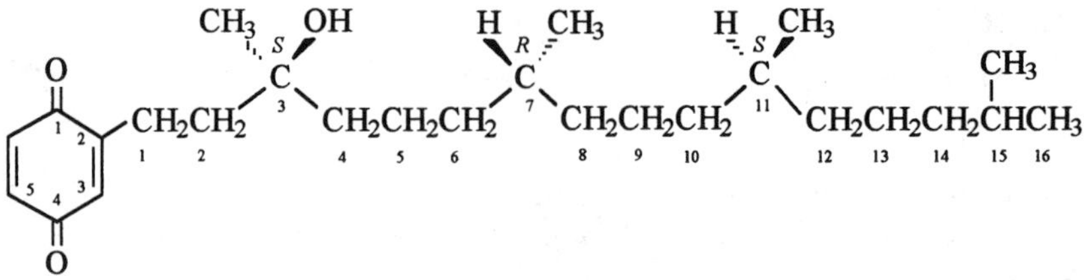

[(3S,7R,11S)-2-(3-Hydroxy-3,7,11,15-tetramethyl)]-2,5-cyclohexadiene-1,4-dione

The CAS descriptor in *Chemical Abstracts* Volumes 76–128 was [3S-(3R*,7S*,11R*)]. The reference center for both the absolute and relative parts of the descriptor is position 3, the lowest numbered chiral center in the chain; its absolute chirality is S. In the relative part of the descriptor, 3R* defines the reference center, 7S* indicates that the absolute chirality at position 7 is opposite to that of the reference center, and 11R* indicates that the absolute chirality at position 11 is the same as the reference center. Therefore, the absolute chirality at position 7 is R and at position 11, S.

12.

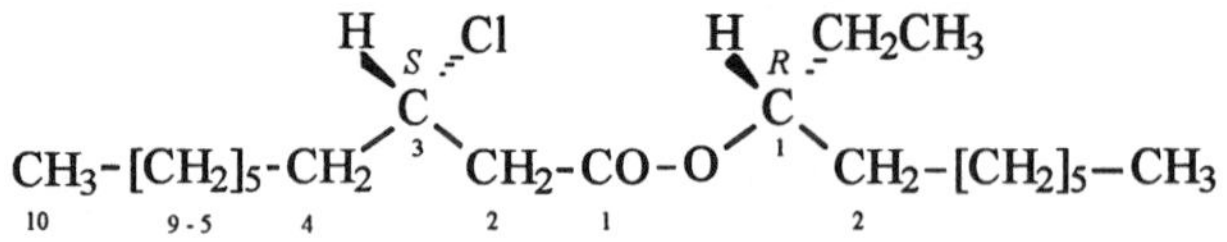

(1R)-1-Ethyloctyl (3S)-3-chlorodecanoate
(3R)-Decan-3-yl (3S)-3-chlorodecanoate

The CAS descriptor in *Chemical Abstracts* Volumes 76–128 was [S-(R*,S*)]. The reference center for both the absolute and relative portions of the descriptor is the 3 position of the acid residue because it is attached to Cl, the preferred substituent according to the principles of the sequence rule; its absolute chirality is S. The relative descriptor (R*,S*) indicates that the absolute chirality of the two centers is different and therefore the absolute chirality at the 1 position of the alcohol residue of the ester is R.

13.

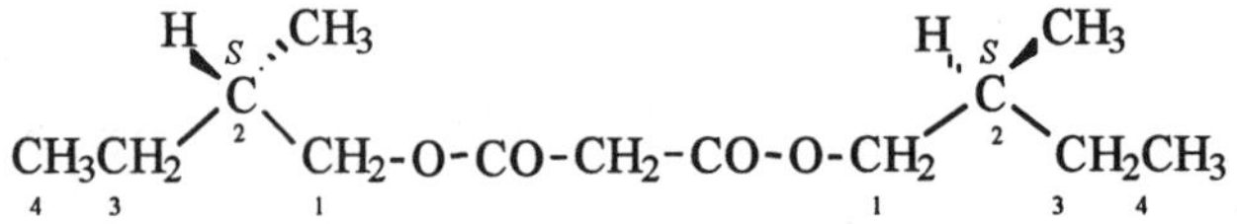

Bis[(2S)-2-methylbutyl) malonate
Bis[(2S)-2-methylbutyl)propanedioate

The CAS descriptor in *Chemical Abstracts* Volumes 76–128 was [S-(R*,R*]. The relative part of the descriptor, (R*,R*), indicates that the two centers are of like chirality and the descriptor S describes the absolute configuration at either of the centers.

14.

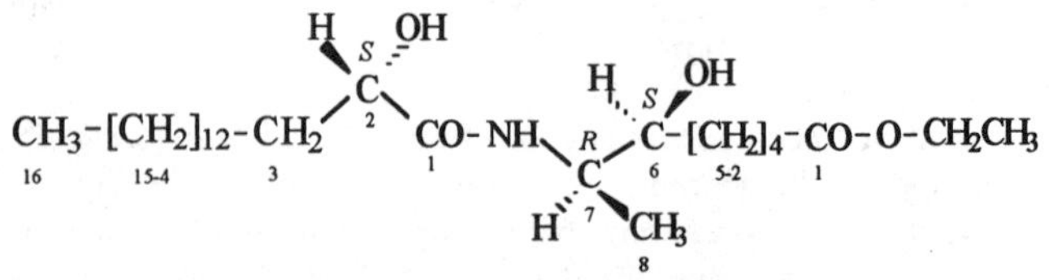

Ethyl (6S,7R)-7-[[(2S)-2-hydroxyhexadecanoyl]amino]octanoate
Ethyl (6S,7R)-7-[(2S)-2-hydroxyhexadecanamido]octanoate

The CAS descriptor in *Chemical Abstracts* Volumes 76–128 was [6S-[6R*,7S*(R*)]]. Since the octanoic acid chain has more chiral centers, position 6, the lowest numbered chiral position on this chain, is the reference center for both the absolute and relative parts of the descriptor; its absolute chirality is S. In the relative part of the descriptor, 6R* defines the reference center, the descriptor 7S* indicates that the absolute chirality at position 7 of the octanoic acid chain is different than the reference center, and the parenthetical (R*) indicates that the absolute chirality at position 2 of the hexadecanoylamino chain is the same as the reference center. Hence, the absolute chirality at position 7 of the octanoic acid chain is R and the absolute chirality at position 2 of the hexadecanoylamino chain is S.

15.

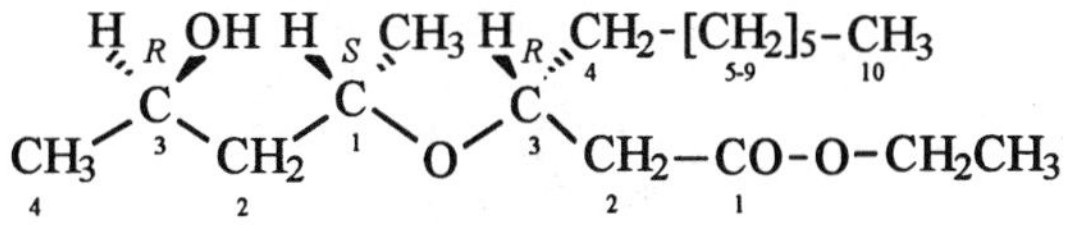

Ethyl (3R)-3-[(1S,3R)-3-hydroxy-1-methylbutoxy]decanoate
Ethyl (3R)-3-[[(1S,3R)-4-hydroxypentan-2-yl]oxy]decanoate

The CAS descriptor in *Chemical Abstracts* Volumes 76–128 was [1S-[1R*(S*),3S*]]. In this compound, the butoxy substituent to the decanoic acid chain has the most chiral centers; therefore its locants are to be used for the descriptor. Position 1, the lowest numbered chiral position in this chain is the reference center for both the absolute and relative part of the descriptor; its absolute chirality is S. In the relative portion of the descriptor, 1R* defines the reference center, the parenthetical (S*) and the 3S* indicate that the absolute chirality at position 3 of the butoxy substituent chain and position 3 of the decanoic acid chain are different than the reference substituent. Hence the absolute chirality at these positions is R.

16.

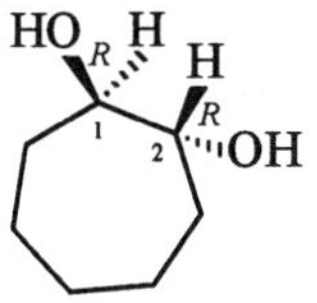

(1R,trans)-Cycloheptane-1,2-diol
(1R,2R)-Cycloheptane-1,2-diol

17.

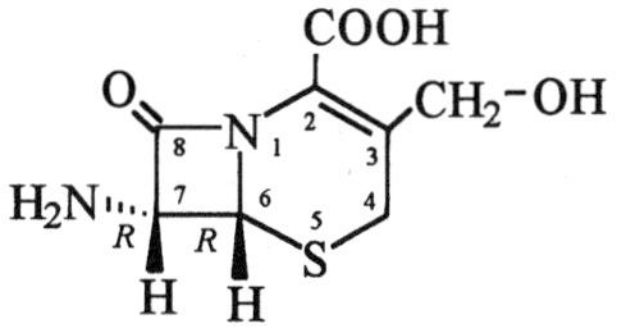

(6R-trans)-7-Amino-3-(hydroxymethyl)-8-oxo-5-thia-1-azabicyclo[4.2.0]oct-2-ene-1-carboxylic acid
(6R,7R)-7-Amino-3-(hydroxymethyl)-8-oxo-5-thia-1-azabicyclo[4.2.0]oct-2-ene-2-carboxylic acid

18.

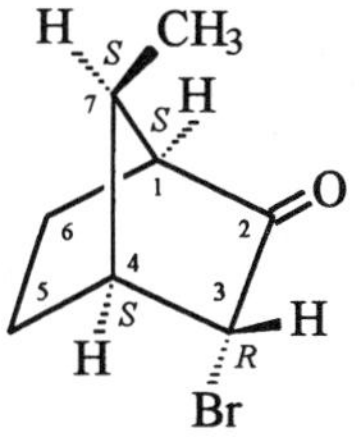

(1S-(endo,syn)]-3-Bromo-7-methylbicyclo[2.2.1]octan-2-one
(1S,3R,4S,7S)-3-Bromo-7-methylbicyclo[2.2.1]octan-2-one

19.

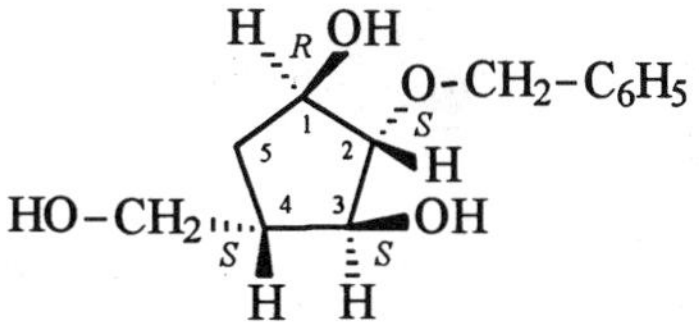

[1R-(1α,2β,3α,4β)]-2-(Benzyloxy)-4-(hydroxymethyl)cyclopentane-1,3-diol
(1R,2S,3S,4S)-2-(Benzyloxy)-4-(hydroxymethyl)cyclopentane-1,3-diol

20.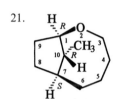

(αR,1S,4R,4aR,8aS)-1,2,3,4,4a,8a-Hexahydro-α,4-
dimethyl-α-phenyl-1-isoquinolinemethanol
(1R)-1-[(1S,4R,4aR,8aS)-1,2,3,4,4a,8a-Hexahydro-
dimethylisoquinolin-1-yl]-1-phenylethanol

The CAS descriptor in *Chemical Abstracts* Volumes 76–128 was
[1S-[1α(S*),4β,4aβ,8aα]]. The reference position for both the relative and absolute
parts of the descriptor is position 1 of the isoquinoline ring, the lowest numbered
chiral center; its absolute chirality is S. The methanol group is the reference substituent at this position and it defines the α-side of the ring reference plane. The
parenthetical (S*) indicates that the absolute chirality at the α position in the
"substituent" is different from the reference center; it is therefore R.

21.

(1R,7S,10R)-10-Methyl-2-oxabicyclo[5.2.1]decane

The CAS descriptor in *Chemical Abstracts* Volumes 76–128 was
[1R-(1R*,7S*,10R*)]. In this structure, the configuration at the bridgeheads is not
the accepted configuration for a bicyclo[x, y, z]...ane system because x + y = 7;
thus, *syn-* or *anti-* cannot be used. The reference position for both the absolute
and relative parts of the descriptor is position 1, the lowest numbered chiral center
in the ring system; its absolute chirality is R. In the relative descriptor, 1R* defines
the reference position, 7S* indicates that the absolute chirality at position 7 is
different than the reference center, and 10R* indicates that the absolute chirality
at position 10 is the same as the reference center. Therefore the absolute chirality at
position 7 is S and at position 10 it is R.

22.

[3S-(3α,6α,6aα9α,9aα)]-Octahydro-2,2,6,9-tetra-
methyl-2H-3,9a-methanocyclopent[b]oxocin
(3S,6S,6aR,9S,9aR)-Octahydro-2,2,6,9-tetramethyl-2H-
3,9a-methanocyclopent[b]oxocin

23.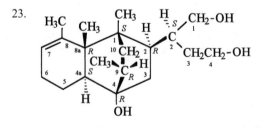

(2S)-[(1S,2R,4R,4aS,8aR,9R)-2-(1,2,3,4,4a,5,6,8a-Octahydro-4-hydroxy-1,8,8a,9-
tetra-methyl-1,4-ethanonaphthalen-2-yl)-1,4-butanediol

The CAS descriptor in *Chemical Abstracts* Volumes 76–128 was
[1S-[1α,2α(R*),4β,4aβ,8aα,9S*]]. The reference center for both the relative and

absolute parts of the descriptor is position 1 of the naphthalene ring; its absolute chirality is *S*. The bridge is the reference substituent at position 1; it defines the α-side of the ring reference plane. The parenthetical (*R**) descriptor indicates that absolute chirality at the 2 position of the parent 1,4-butanediol is the same as the reference center; it is therefore *S*. The relative configuration at position 4 is β, as determined by the hydroxy substituent. Position 4a is also β, as determined by the hydrogen atom and at 8a it is β, as determined by the methyl group. The 9*S** indicates that the absolute chirality at the 9 position is different than the reference center; it is therefore *R*.

24.

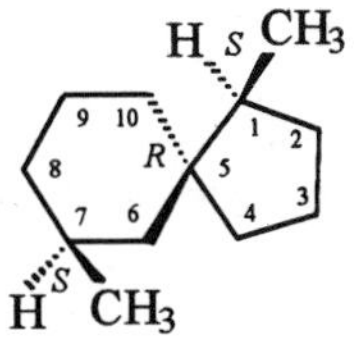

(1*S*,5*R*,7*S*)-1,7-Dimethylspiro[4.5]decane

The CAS descriptor in *Chemical Abstracts* Volumes 76–128 was [1*S*-[1α,5β(*R**)]]. The reference center for both the relative and absolute part of the descriptor is position 1, the lowest numbered chiral center in the spiro system; its absolute configuration is *S* and its methyl substituent defines the α-side of the ring reference plane, the five-membered ring. The preferred "substituent" at position 5 is the ring bond to position 6; it is also α. The parenthetical (*R**) indicates that the absolute configuration at position 7, which in such a spiro system is considered to be on a substituent chain, is the same as the reference position; therefore it is *S*.

25.

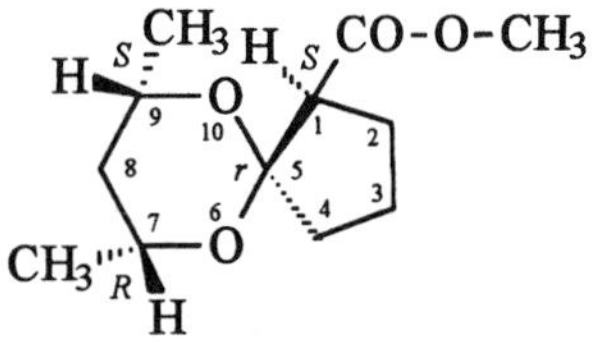

(1*S*,5*r*,7*R*,9*S*)-Methyl 7,9-dimethyl-6,10-dioxa-spiro[4.5]decane-1-carboxylate (IUPAC)
(1*S*,5α,7β,9β)-Methyl 7,9-dimethyl-6,10-dioxa-spiro[4.5]decane-1-carboxylate (CAS)

CAS does not use the pseudochirality symbols "r" and "s" and therefore configurations on the six-membered ring are described by the α/β method. The CAS descriptor in *Chemical Abstracts* Volumes 76–128 was [5(*S*)-(5α,7β,9β)]. The six-membered ring of this spiro system has more chiral sites and thus is the reference plane. The reference center for both the absolute and relative parts of the descriptor is position 5, the lowest numbered stereogenic site in the reference ring plane. The parenthetical (*S*) gives the absolute chirality at position 1. Position 5 in this structure is achiral and therefore cannot be assigned a chiral descriptor. The ring bond between positions 1 and 5 defines the α-side of the reference plane; thus the methyl groups at positions 7 and 9 are β.

26.

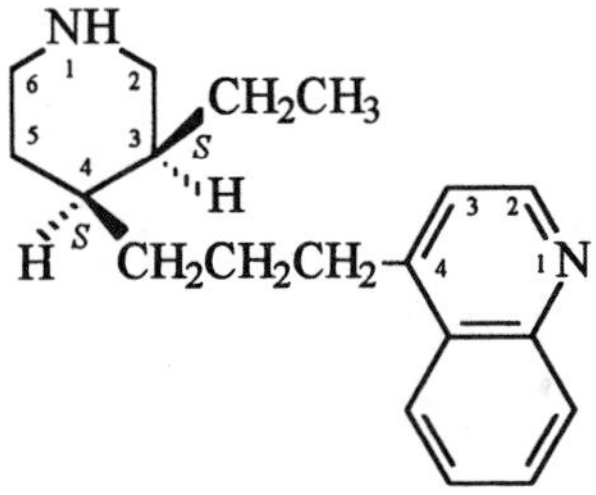

4-[[(3*S*-*cis*)-3-ethylpiperidine-4-yl]propyl]-quinoline
4-[[(3*S*,4*S*)-3-Ethylpiperidin-4-yl]propyl]-quinoline

27.

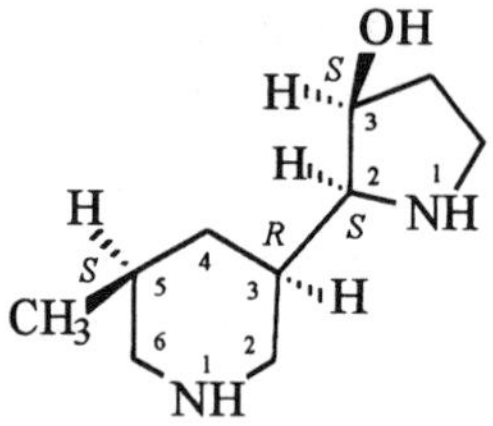

(2S,3S)-[(3R,5S)-5-Methylpiperidin-3-yl)-
pyrrolidin-3-ol

The CAS descriptor in *Chemical Abstracts* Volumes 76–128 was
[3R-[3α(2S*,3S*),5α]]. The preferred piperidine ring is the "parent" for the descriptor even though the name for the compound is based on pyrrolidine.

28.

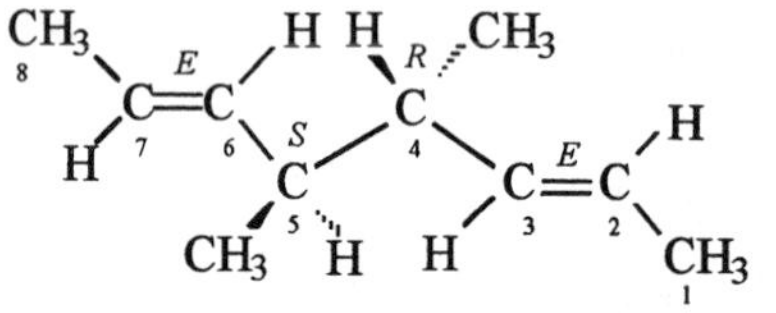

(2E,4R*,5S*,6E)-4,5-Dimethylocta-2,6-diene
rel-(2E,4R,5S,6E)-4,5-Dimethylocta-2,6-
diene

The CAS descriptor in *Chemical Abstracts* Volumes 76–238 was [R*,S*-(E,E)]. The
compound is meso (see Discussion); therefore there is no absolute chiral descriptor.

29.

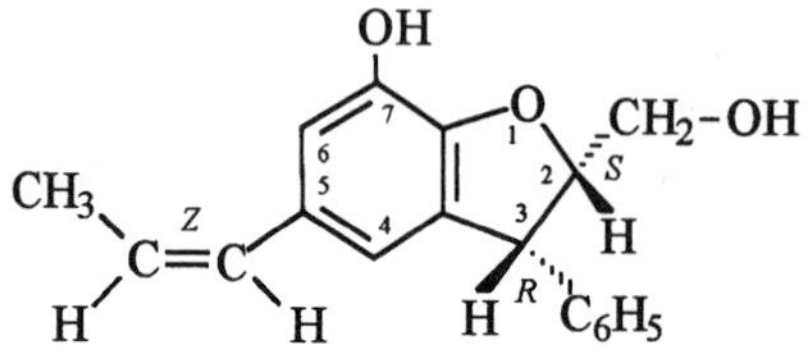

[(1R,3R)-2,3-Dihydro-7-hydroxy-3-
phenyl-5-[(1Z)-prop-1-en-1-yl]-1-
benzofuran-2-methanol

The CAS descriptor in *Chemical Abstracts* Volumes 76–128 was [2S-(2α,3α,5(Z)]]

30.

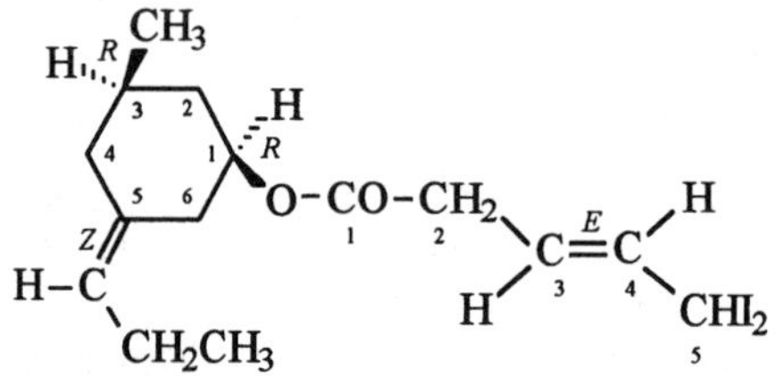

(1R,3R,5Z)-3-Methyl-5-propylidene-
cyclohexyl (2E)-5,5-diiodopent-2-
enoate

The CAS descriptor in *Chemical Abstracts* Volumes 76–128 was [1R-[1α(E),3α,5Z]].

31.

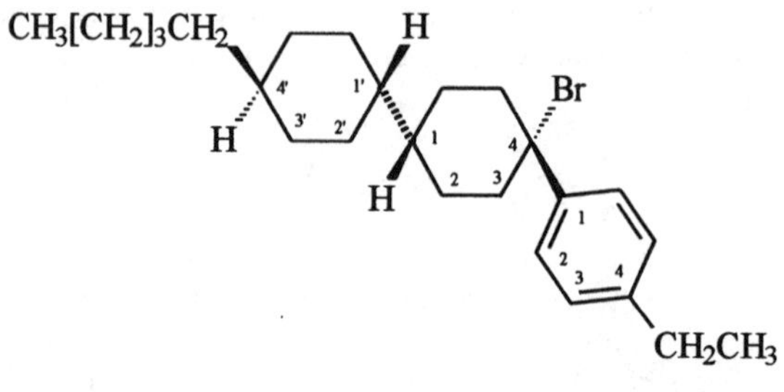

1-[(*cis,trans*)-4-Bromo-4′-pentyl[1,1′-bicyclohexyl]-4-yl]-4-ethylbenzene
(*cis,trans*)-4-(4-Ethylphenyl)-4′-pentyl-1,1′-bicyclohexyl
    (the *cis*- descriptor is associated with the unprimed cyclohexane ring.)

32.

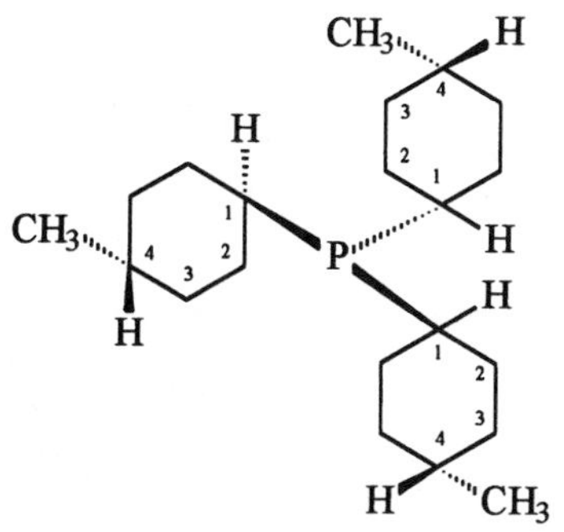

(cis-4-Methylcyclohexyl)bis(trans-4-methylcyclohexyl)phosphane

33.

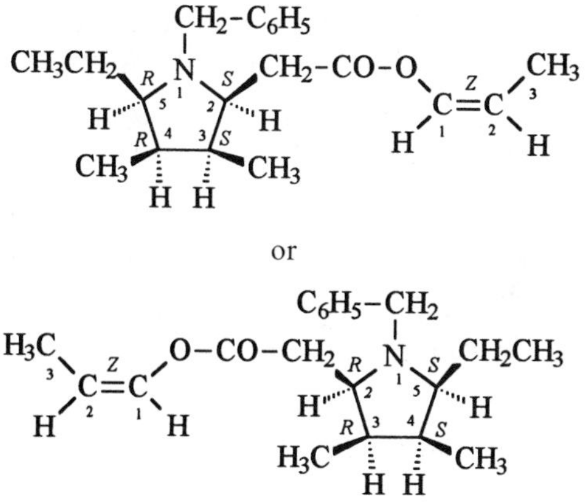

or

(1Z)-Prop-1-en-1-yl (2α,3α,4α,5α)-1-benzyl-5-ethyl-3,4-dimethyl-2-pyrrolidineacetate

(1Z)-Prop-1-en-1-yl (2R*,3R*,4S*,5S*)-1-benzyl-5-ethyl-3,4-dimethyl-2-pyrrolidineacetate

(1Z)-Prop-1-en-1-yl rel-(2R,3R,4S,5S)-1-benzyl-5-ethyl-3,4-dimethyl-2-pyrrolidineacetate

34.

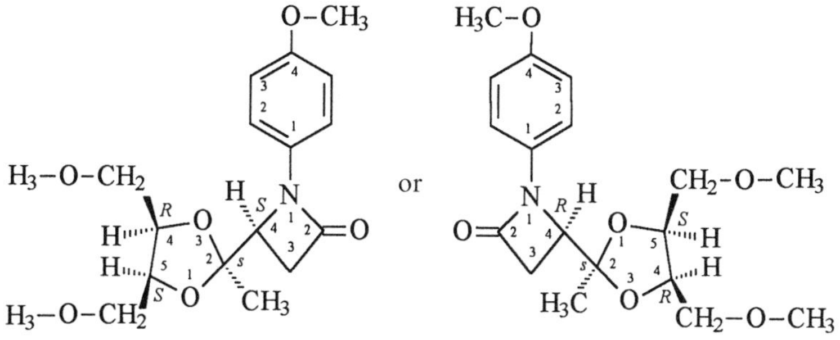

or

(4R*)-4-[(2s,4R,5S)-4,5-Bis(methoxymethyl)-2-methyl-1,3-dioxolan-2-yl]-1-4-methoxyphenyl)azetidin-2-one (IUPAC)

[the use of the asterisk in the descriptor, that is (2s*,4R*,5S*), is unnecessary since this part of the descriptor is the same for the 4S* enantiomer]

(4R*)-4-[(2α,4α,5α)-4,5-Bis(methoxymethyl)-2-methyl-1,3-dioxolan-2-yl]-1-(4-methoxyphenyl)-2-azetidinone (CAS)

(CAS does not use the "r/s" symbols for pseudoasymmetry)

**References**

1. International Union of Pure and Applied Chemistry, Organic Chemistry Division, Commission on Nomenclature of Organic Chemistry and Commission on Physical Chemistry. Basic Terminology of Stereochemistry. *Pure Appl. Chem.* **1996,** *68*, 2193–2222.
2. Stereochemical Descriptors. *Beilstein Handbook of Organic Chemistry,* 4th ed.; 5th Supp. Series, Vol. 27, Part I, 1995; pp xxiii–xxxvii.
3. Cahn, R. S.; Ingold, C. K.; Prelog, V. Specification of Molecular Chirality. *Angew. Chem., Int. Ed. Engl.* **1966,** *5*, 385–414 (errata: **1966,** *5*, 511); *Angew. Chem.* **1966,** *78*, 413-447.
4. Cahn, R. S. An Introduction to the Sequence Rule. *J. Chem. Educ.,* **1964,** *41*, 116–125 (errata: **1964,** *41*, 508).
5. Cahn, R. S.; Ingold, C. K.; Prelog, V. The Specification of Asymmetric Configuration in Organic Chemistry. *Experientia* **1956,** *12*, 81–94.
6. Cahn, R. S; Ingold, C. K. Specification of Configuration about Quadricovalent Asymmetric Atoms. *J. Chem. Soc.* **1951,** 612–622.
7. Prelog, V.; Helmchen, G. Basic Principles of the CIP-System and Proposals for a Revision. *Angew. Chem., Int. Ed. Engl.* **1982,** *21*, 567–583; *Angew. Chem.* **1982,** *94*, 614–631.
8. Blackwood, J. E.; Gladys, C. L.; Loening, K. L.; Petrarca, A. E.; Rush, J. E. Unambiguous Specification of Stereoisomerism about a Double Bond. *J. Am. Chem. Soc.* **1968,** *90*, 509–510.
9. Blackwood, J. E., Gladys, C. L.; Petrarca, A. E.; Powell, W. H.; Rush, J. E.; Unique and Unambiguous Specification of Stereoisomerism about a Double Bond in Nomenclature and Other Notations. *J. Chem. Doc.* **1968,** *8*, 30–32.
10. Blackwood, J. E.; Giles, P. M. Jr. *Chemical Abstracts* Stereochemical Nomenclature of Organic Substances in the Ninth Collective Period. *J. Chem. Inf. Comput. Sci.* **1975,** *15*, 67–72.
11. American Chemical Society. The Naming of *cis* and *trans* Isomers of Hydrocarbons Containing Olefin Double Bonds. *Chem. Eng. News* **1949,** *27*, 1303.
12. Aldler, K.; Wirtz, H.; Koppelberg, H. Several instances of "exo additions". The preparation of cyclopentanedialdehyde and its use in the Robinson–Schöpf condensation. *Ann. Chem.* **1956,** *601*, 138.

# 31

# Natural Products

The structures covered in this chapter are derived from, closely related to, or actually found as chemical compounds in nature. They run the gamut from amino acids through carbohydrates and steroids to vitamins and antibiotics. Here, the nomenclature of some major classes will be described, with no claim of completeness.

In this area of apparent nomenclatural confusion, names have often preceded characterization and determination of structure. All too frequently, proper placement of atoms and bonds has yielded a substance with a systematic name as complex as the material itself. The result, of course, is that trivial names abound. IUPAC brought some order out of chaos with Section F of its 1979 organic nomenclature rules. The general principles that follow are based largely on those provisional recommendations and revised and extended recommendations.[1]

## General Principles

Natural products and related substances, as indicated above, may be given trivial names or, where structure is established *and relatively simple*, named systematically as described elsewhere in this book. For a comparison between trivial or modified trivial names and the corresponding systematic names, the *Chemical Abstracts Index Guide*[2] should be consulted. Between these extremes there are "semisystematic" names based on accepted names for parent structures in many classes of natural products.

### Trivial Names

Trivial names for natural products are appellations of convenience, often derived from the family, genus, or species of the biological source of the compound. Since such names should not suggest chemical structure, the ending -une or, for euphony, -iune may be used as a temporary measure. As partial characterization is made, these endings may be replaced by more informative endings, such as -ane (for saturation), for example, and prefixes such as hydroxy and suffixes such as -ol can be affixed to a trivial name. With complete structural knowledge of the compound, its trivial name should be abandoned in favor of a more systematic name if a parent structure can be established. In the absence of an established parent structure, the trivial stem name may become the basis for a new parent structure. Such names usually imply a stereochemical configuration.

### Semisystematic Names

Many compounds occurring in nature can be assigned to classes having a set of parent structures, called stereoparents, that are structurally related and have common configurational detail. The names and numbering of parent structures are often derived from trivial names of compounds to which they are related. For example, lupeol, the trivial name for **I**, generates the semisystematic name lupane, the parent hydride shown in structure **II**.

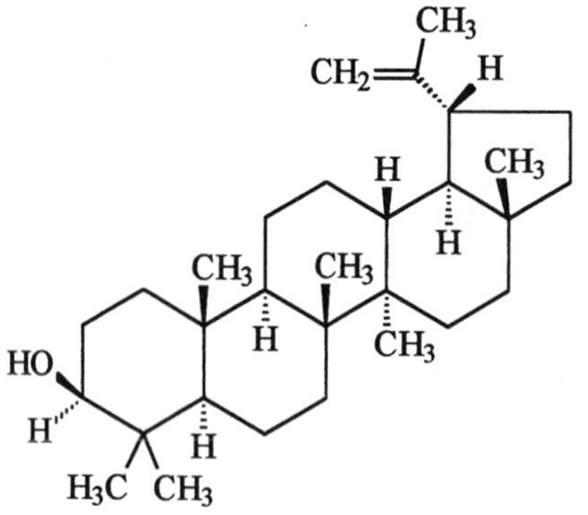

|          |          |
|:--------:|:--------:|
| **I**    | **II**   |

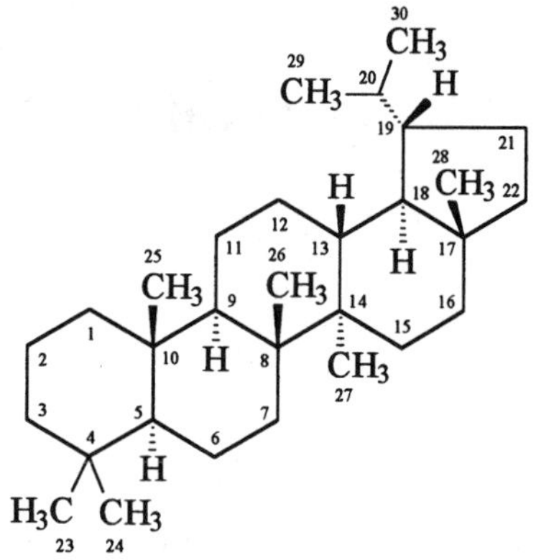

From the parent hydride **II**, compound **I** would be named, by substitutive nomenclature, 3β-hydroxylup-20(29)-ene or lup-20(29)-en-3β-ol. For the meaning of β, see the steroid section below.

Modified skeletal structures can be derived from a fundamental parent structure through one or more subtractive or additive operations denoted by nondetachable prefixes; locants indicate where, and sometimes how, the modifications occur. Generally, the stereochemical configuration of the parent is unchanged.

"Nor" (dinor, trinor, and so on) indicates removal of a methylene group from a ring or side chain to give the next lower homolog. The original locant numbers are retained.

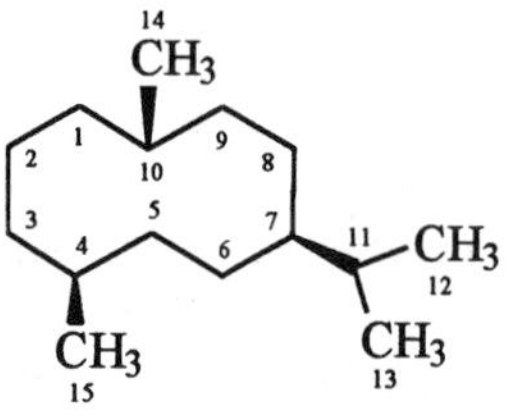

Germacrane 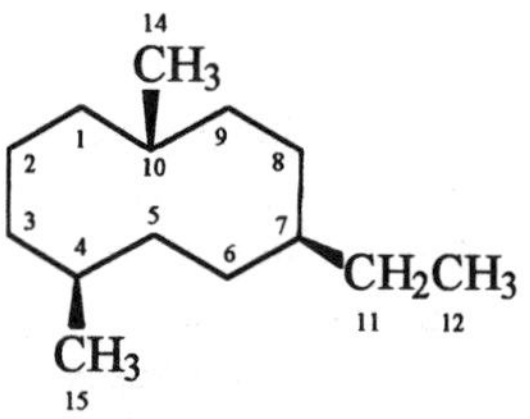            13-Norgermacrane

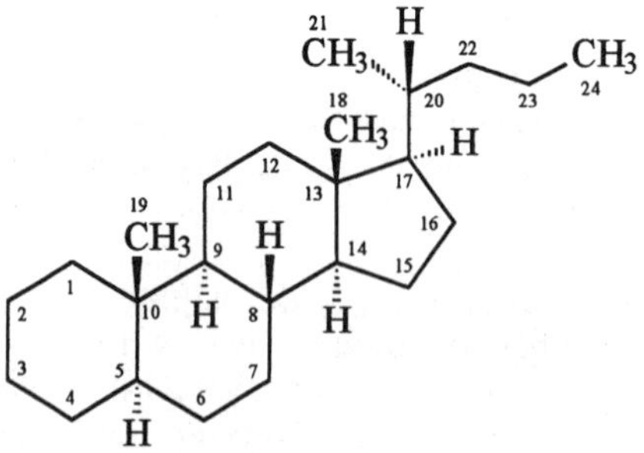

5α-Cholane 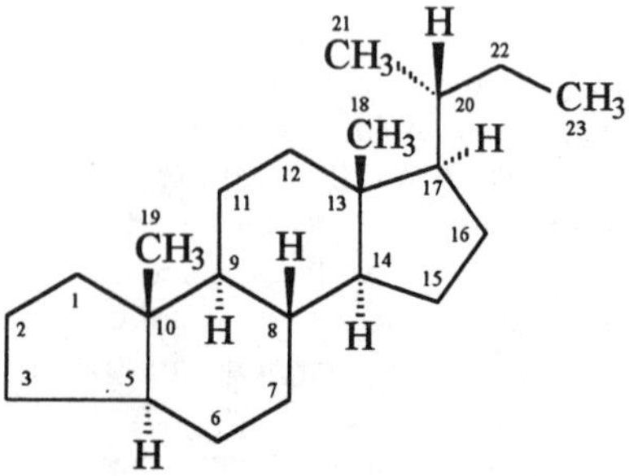            4,24-Dinor-5α-cholane
                      A,24-Dinor-5α-cholane

The name A,24-dinor-5α-cholane in the example shown above illustrates the method of labeling rings with capital letters to indicate the ring in which an operation has taken place; this method is currently used by CAS. The locant method is more exact and much preferred.

"De", as in demethyl, demethoxy, or deoxy, denotes removal of an atom or group with replacement by hydrogen where necessary.

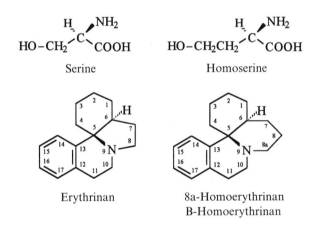

Caffeine                    1-Demethylcaffeine

2-Deoxy-D-*ribo*-hexose

1-Demethylcaffeine is sometimes called theobromine, an example of a trivial name that should *not* be used because of the implied presence of bromine.

"Homo" indicates insertion of a methylene group to give the next higher homolog:

Serine                          Homoserine

Erythrinan                  8a-Homoerythrinan
                            B-Homoerythrinan

"Cyclo" indicates formation of an additional ring by direct connection of any two atoms; new stereochemical configurations are denoted by $\alpha$, $\beta$, and $\xi$ (see the section on steroids below).

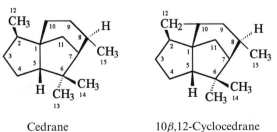

Cedrane                     10$\beta$,12-Cyclocedrane

"Seco" indicates cleavage of a ring bond:

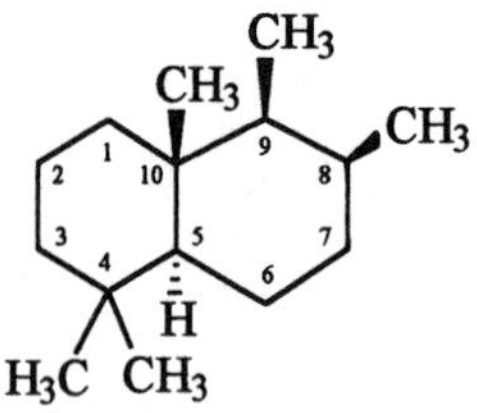

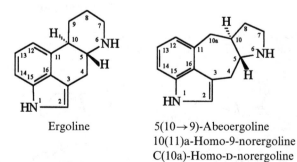

Drimane 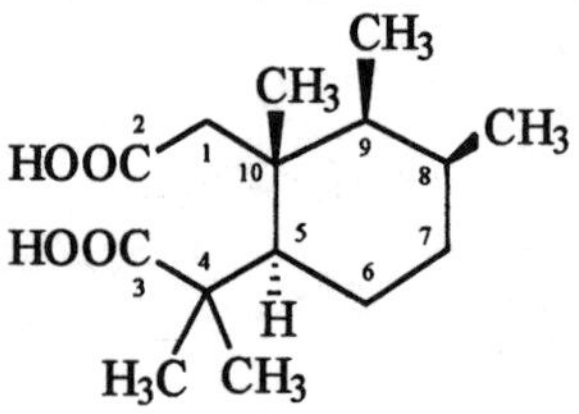                                    2,3-Secodrimane-2,3-dioic acid

"Abeo" indicates a bond migration described by x(y→z), where "x" is the locant of the atom at the stationary end of the bond and (y→z) describes the movement of the bond from the atom at locant "y" to the atom at locant "z". This rearrangement can be expressed by combinations of cyclo with seco and homo with nor.

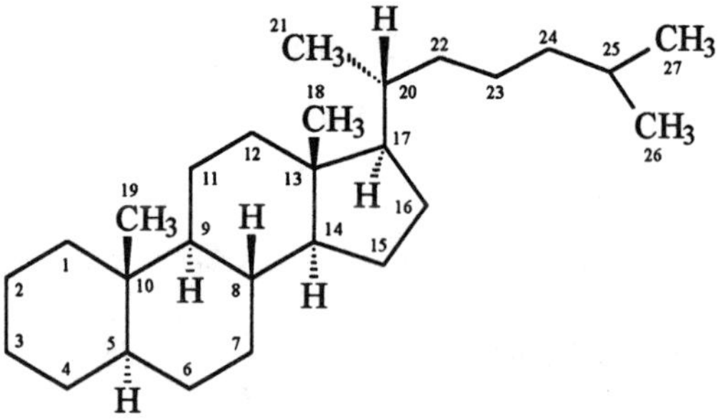

5α-Cholestane

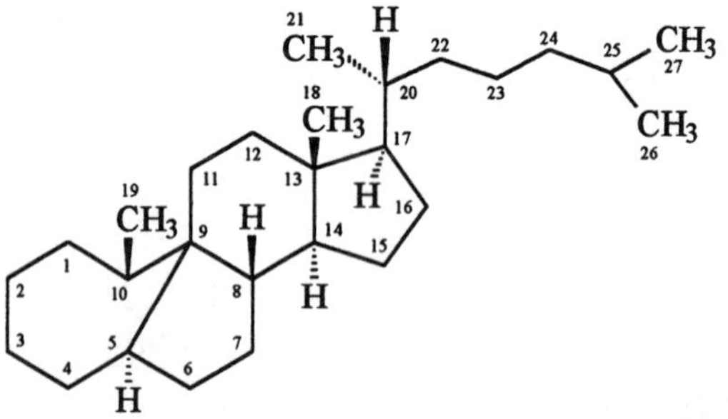

5(10→9)-Abeo-5α-cholestane
5,9-Cyclo-5,10-seco-5α-cholestane

Ergoline

5(10→9)-Abeoergoline
10(11)a-Homo-9-norergoline
C(10a)-Homo-D-norergoline

Replacement nomenclature (see chapter 3) and a slightly modified fusion nomenclature (see chapter 7) also play important roles in the nomenclature of natural products.

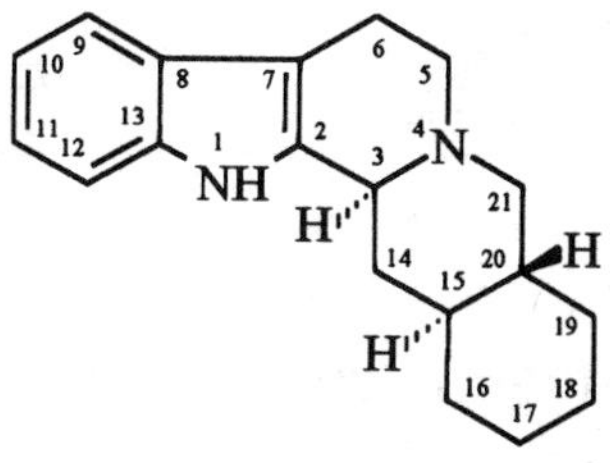

Yohimban

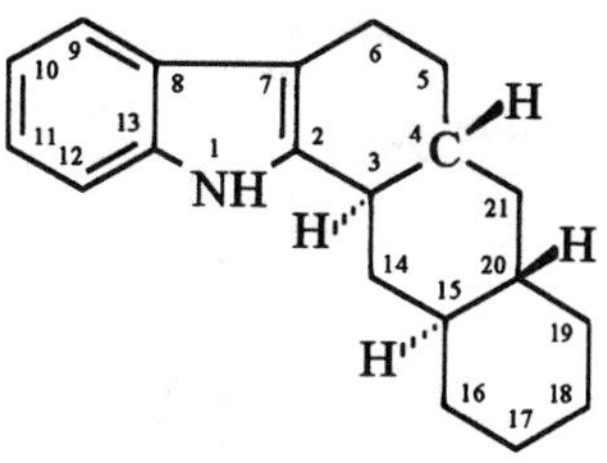

4β-4-Carbayohimban

In the example above, the replacement creates a new stereo center described by an appropriate symbol in front of the name.

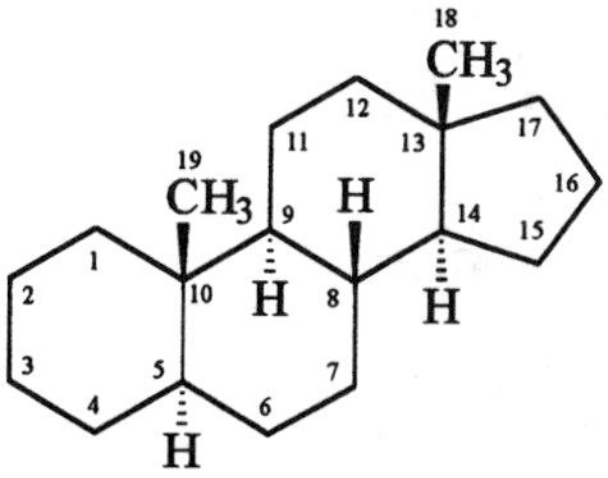

5α-Androstane

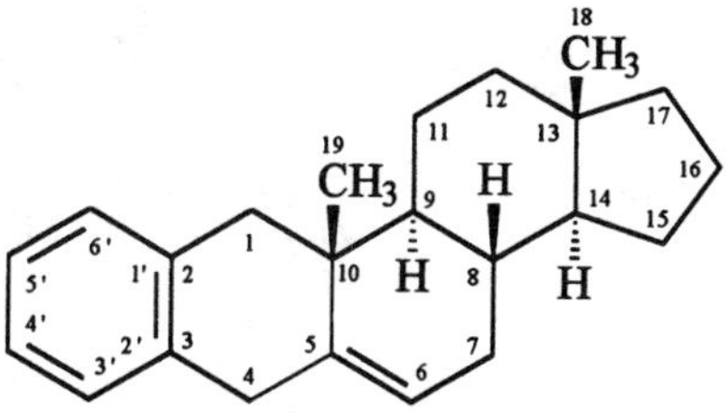

Benzo[2,3]androst-5-ene

For this example, CAS also cites the double bond in the stereoparent fusion component to give benzo[2,3]androsta-2,5-diene.

Section F of the IUPAC rules presents additional details of this semisystematic nomenclature. Its revision[1] contains an extensive compilation of fundamental parent structures with approved numbering and names in the fields of alkaloids, steroids, and terpenes.

## Specialist Nomenclature

Space does not allow a detailed discussion of the codified nomenclature for all natural product classes; a complete bibliography of published recommendations appears in section J of Appendix IV in the *Chemical Abstracts Index Guide*.[2] Here, only brief discussions of nomenclature for selected classes will be undertaken.

### Amino Acids, Peptides, and Polypeptides[3]

The names and associated symbols for some of the most common amino acids are given in table 31.1. It is recognized that under most conditions, amino acids are ionized, but for convenience the conventional unionized form is the one usually written and named. Numbering for substituents begins with the carboxy group nearest to the amino group. The corresponding acyl group is named by replacing the ending -ine with -yl except in the case of cysteine, homocysteine, asparagine, and glutamine, in which only the final "e" is replaced, and tryptophan provides tryptophyl. Acyl groups from aspartic acid are named as follows:

$$-CO-CH_2CH-CO- \quad \text{Aspartoyl}$$
(with NH$_2$ on the central CH)

$$HOOC-CH_2CH-CO- \quad \alpha\text{-Aspartyl}$$
(with NH$_2$ on the central CH)

$$HOOC-CHCH_2-CO- \quad \beta\text{-Aspartyl}$$
(with NH$_2$ on the CH)

Table 31.1. Amino Acids

| Trivial name | Symbol | Structural formula |
|---|---|---|
| alanine | Ala | $\overset{\overset{\displaystyle NH_2}{\mid}}{CH_3CH-COOH}$ |
| $\beta$-alanine | $\beta$Ala | $H_2N-CH_2CH_2-COOH$ |
| arginine | Arg | $\overset{\overset{\displaystyle NH}{\parallel}}{H_2N-C-NH-[CH_2]_3-}\overset{\overset{\displaystyle NH_2}{\mid}}{CH-COOH}$ |
| asparagine | Asn | $H_2N-CO-CH_2\overset{\overset{\displaystyle NH_2}{\mid}}{CH-COOH}$ |
| aspartic acid | Asp | $HOOC-CH_2\overset{\overset{\displaystyle NH_2}{\mid}}{CH-COOH}$ |
| cysteine | Cys | $HS-CH_2\overset{\overset{\displaystyle NH_2}{\mid}}{CH-COOH}$ |
| cystine | Cys &#124; Cys | $HOOC-\overset{\overset{\displaystyle NH_2}{\mid}}{CH}CH_2-SS-CH_2\overset{\overset{\displaystyle NH_2}{\mid}}{CH-COOH}$ |
| glutamine | Gln | $H_2N-CO-CH_2CH_2\overset{\overset{\displaystyle NH_2}{\mid}}{CH-COOH}$ |
| glutamic acid | Glu | $HOOC-CH_2CH_2\overset{\overset{\displaystyle NH_2}{\mid}}{CH-COOH}$ |
| glycine | Gly | $H_2N-CH_2-COOH$ |
| histidine | His | 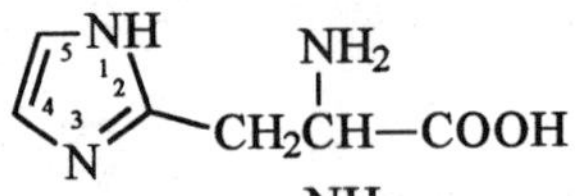 (imidazole ring) $-CH_2\overset{\overset{\displaystyle NH_2}{\mid}}{CH-COOH}$ |
| leucine | Leu | $(CH_3)_2CHCH_2\overset{\overset{\displaystyle NH_2}{\mid}}{CH-COOH}$ |
| isoleucine | Ile | $CH_3CH_2\overset{\overset{\displaystyle CH_3}{\mid}}{CH}-\overset{\overset{\displaystyle NH_2}{\mid}}{CH-COOH}$ |
| lysine | Lys | $H_2N-CH_2[CH_2]_3\overset{\overset{\displaystyle NH_2}{\mid}}{CH-COOH}$ |
| methionine | Met | $CH_3-S-CH_2CH_2\overset{\overset{\displaystyle NH_2}{\mid}}{CH-COOH}$ |
| phenylalanine | Phe | $C_6H_5-CH_2\overset{\overset{\displaystyle NH_2}{\mid}}{CH-COOH}$ |
| proline | Pro | 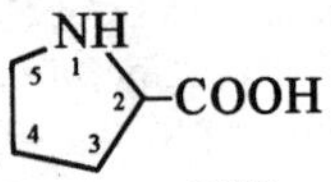 (pyrrolidine ring) $-COOH$ |
| serine | Ser | $HO-CH_2\overset{\overset{\displaystyle NH_2}{\mid}}{CH-COOH}$ |
| threonine | Thr | $CH_3\overset{\overset{\displaystyle OH}{\mid}}{CH}-\overset{\overset{\displaystyle NH_2}{\mid}}{CH-COOH}$ |

*(Continued)*

Table 31.1. Continued

| Trivial name | Symbol | Structural formula |
|---|---|---|
| tryptophan | Trp | 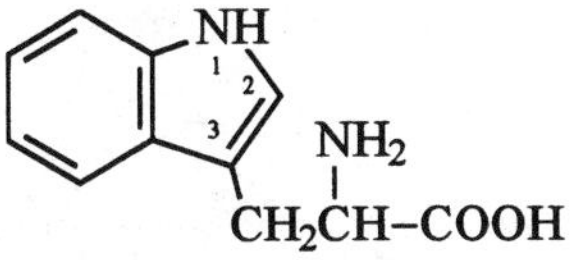 |
| tyrosine | Tyr | 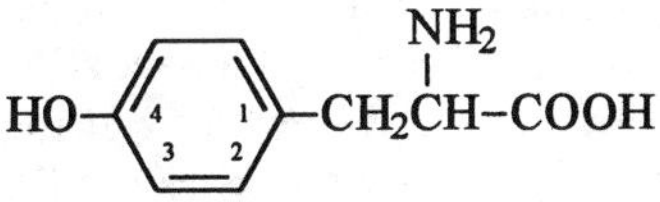 $HO-$ (ring) $-CH_2CH-COOH$ with $NH_2$ |
| valine | Val | $(CH_3)_2CHCH-COOH$ with $NH_2$ |

In addition to the above, CAS uses

| Trivial name | Symbol | Structural formula |
|---|---|---|
| α-glutamine (in peptides only) | Agn | $H_2N-CO-CHCH_2CH_2-COOH$ with $NH_2$ |
| homocysteine | Hcy | $HS-CH_2CH_2CH-COOH$ with $NH_2$ |
| homoserine | Hse | $HO-CH_2CH_2CH-COOH$ with $NH_2$ |
| isovaline | Iva | $CH_3CH_2C-COOH$ with $NH_2$ and $CH_3$ |
| norleucine | Nle | $CH_3CH_2CH_2CH_2CH-COOH$ with $NH_2$ |
| norvaline | Nva | $CH_3CH_2CH_2CH-COOH$ with $NH_2$ |
| ornithine | Orn | $H_2N-CH_2[CH_2]_2CH-COOH$ with $NH_2$ |

Analogous acyl group names are formed from glutamic acid. Generally, derivatives such as amides, esters, and so on are named in the same way as those of carboxylic acids (see chapters 12 and 18); examples are alaninamide, aspartamide, and methyl glycinate. ω-Monoamides of the dibasic amino acids aspartic and glutamic acid are named asparagine and glutamine, respectively; the corresponding acyl group names are asparaginyl and glutaminyl. Reduction of the carboxy group to the corresponding aldehyde or alcohol produces compounds with trivial names ending in -al or -ol, as in alaninal, for $CH_3CH(NH_2)-CHO$, or serinol, for $HO-CH_2CH(NH_2)-CH_2-OH$. Such names stress the amino acid portion of the compounds and correspond to biochemical usage but have not been incorporated into recommended systematic organic nomenclature.

Absolute stereochemical configuration at the chiral α-carbon atom of an α-amino acid is described by the small capital letters "D" and "L" prefixed to names of the amino acids or their analogs. These prefixes indicate a formal relationship to D-serine and thus to D-glyceraldehyde:

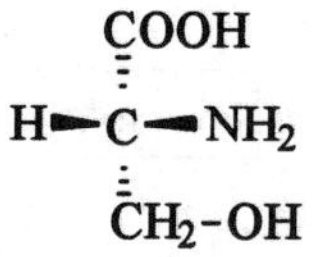

$$\begin{array}{cc}
\text{COOH} & \text{CHO} \\
\text{H}\!-\!\text{C}\!-\!\text{NH}_2 & \text{H}\!-\!\text{C}\!-\!\text{OH} \\
\text{CH}_2\text{-OH} & \text{CH}_2\text{-OH} \\
\text{D-Serine} & \text{D-Glyceraldehyde}
\end{array}$$

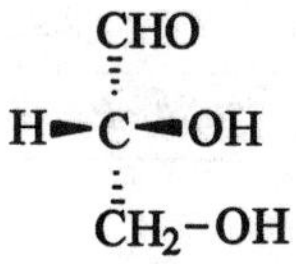

The Fischer–Rosanoff projection method is commonly used to describe the configuration of α-amino acids. All groups are connected to the chiral α-carbon atom by straight lines; the amino group and the hydrogen atom are viewed as above the plane of the paper and the main chain groups as below that plane. Oriented with the carboxyl group at the top, the D-series has the amino group to the right and the L-series has the amino group to the left of the chiral atom.

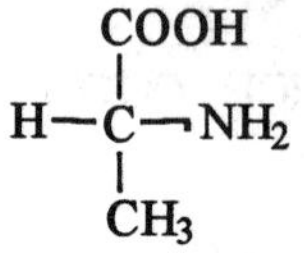

$$\begin{array}{cc}
\text{COOH} & \text{COOH} \\
\text{H}\!-\!\text{C}\!-\!\text{NH}_2 & \text{H}_2\text{N}\!-\!\text{C}\!-\!\text{H} \\
\text{CH}_3 & \text{CH}_2\text{CH(CH}_3)_2 \\
\text{D-Alanine} & \text{L-Leucine}
\end{array}$$

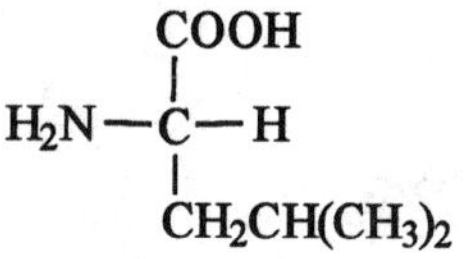

Naturally occurring α-amino acids are "L", which, except for L-cysteine and L-cystine, equates to the S-configuration in the CIP R/S system (see chapter 30); the exceptions have the R-configuration. A racemic mixture is denoted by the prefix DL.

Stereochemical configurations of α-amino acids having a second chiral center are best described by the CIP R/S method, as in (2S,3R)-threonine. This avoids the problem of configurational descriptors with different reference centers, that is, the amino acid reference center or carbohydrate reference center, in the same name. The prefix allo was used in the past to distinguish such amino acids, but it is now recommended that its use be limited to alloisoleucine and allothreonine.

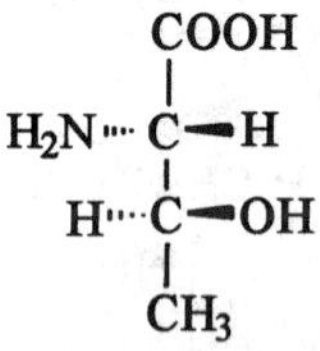

$$\begin{array}{cc}
\text{COOH} & \text{COOH} \\
\text{H}_2\text{N}\cdots\text{C}\!-\!\text{H} & \text{H}_2\text{N}\cdots\text{C}\!-\!\text{H} \\
\text{H}\cdots\text{C}\!-\!\text{OH} & \text{HO}\cdots\text{C}\!-\!\text{H} \\
\text{CH}_3 & \text{CH}_3 \\
\text{L-Threonine} & \text{L-Allothreonine}
\end{array}$$

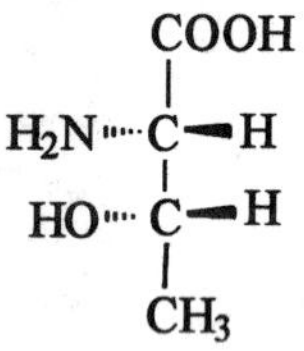

*Peptides* are substances formed by the condensation of the carboxy group from one amino acid molecule with the amino group of another. In brief, they are named by using amino acid acyl group names:

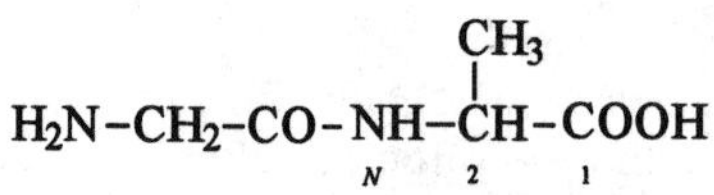

$$\underset{N\quad\;2\quad\;1}{\text{H}_2\text{N}-\text{CH}_2-\text{CO}-\text{NH}-\overset{\text{CH}_3}{\underset{|}{\text{CH}}}-\text{COOH}}$$          *N*-Glycylalanine

$$(\text{CH}_3)_2\text{CH}\underset{N}{\text{CH}}-\text{CO}-\text{NH}-\underset{N}{\text{CH}}-\overset{\text{O}}{\underset{|}{\text{C}}}-\text{NH}-\text{CH}_2-\text{COOH}$$          *N*-(*N*-Valylseryl)glycine

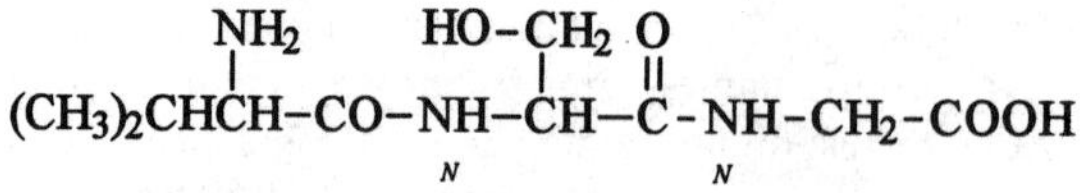

with NH₂ and HO-CH₂ substituents shown above the chain.

Peptide names always begin with the acyl group having the nonacylated amino group and end with the name of the amino acid having the free carboxy group. The symbols shown in table 31.1 are often used in place of names: *N*-(*N*-L-alanyl-L-leucyl)tryptophan may be described by Ala–Leu–Trp. While IUPAC does not use *N* locants or enclosing marks, CAS does so with chains having more than 12 amino acid residues. These compounds are examples of *dipeptides* and *tripeptides*. With very long chains, called *polypeptides* (if they are long enough, they are called *proteins*), a system of single letter symbols may be used.

Nomenclature for *peptides* is also extensively covered in the 1983 recommendations,[3] which should be consulted for methods of denoting substitution, and so on.

## Carbohydrates

Nomenclature in the ramified field of carbohydrates is well developed and documented.[4] The term carbohydrate loosely applies to the general classes of polyhydroxylated aldehydes and ketones. Here, discussion will be concentrated on the simplest carbohydrates, the *mono-saccharides*; these include structures of the type H–[CH–OH]$_n$–CHO, called *aldoses*, and H–[CH–OH]$_n$–CO–[CH–OH]$_m$–H, called *ketoses*. For oligosaccharides and polysaccharides, the IUPAC recommendations[4] should be consulted.

The three-carbon aldose is named glyceraldehyde. Trivial names that imply the relationship among the hydroxyl groups are generally used for acyclic aldoses with 4–6 carbon atoms. These and their prefix forms with corresponding abbreviations are given in table 31.2 for the "D" forms; the "L" forms are their mirror images. An alternative systematic nomenclature for aldoses involves the use of stem names such as tetrose, pentose, octose, and so on, depending on the number of carbon atoms in the chain. Numbering of the chain begins with the terminal carbonyl group.

Absolute stereochemical configuration of a monosaccharide is indicated by the small capital letters "D" or "L" prefixed to the trivial name of the monosaccharide. These prefixes describe the configuration of the highest numbered asymmetric carbon atom in the chain. The prefix "D" indicates that the hydroxy group is to the right of that atom and "L" places it at the left when the carbon atom numbered 1 is at the top of a Fischer projection. Racemic mixtures are denoted by DL.

$$
\begin{array}{cc}
& \text{CHO} \\
\text{CHO} & \text{HO–C–H} \\
\text{HO–C–H} & \text{H–C–OH} \\
\text{H–C–OH} & \text{HO–C–H} \\
\text{CH}_2\text{–OH} & \text{CH}_2\text{–OH} \\
\text{D-Threose} & \text{L-Xylose}
\end{array}
$$

For monosaccharides the relative configuration for consecutive, but not necessarily contiguous, chiral carbon atoms is denoted by the italicized prefixes given in table 31.2, which shows only the D forms. These prefixes may be used to generate systematic names for tetroses, pentoses, and hexoses, as in D-*erythro*-tetrose and L-*gluco*-hexose. They are combined to name monosaccharides with more than six carbon atoms in the main chain. Prefixes for groups of four >CH–OH units, if possible, beginning with that next to the carbon atom numbered 1, are cited in sequence starting with the group furthest from the carbon atom numbered 1; the final group of >CH–OH units often represents less than four >CH–OH units. Achiral carbon atoms in the chain are ignored in the assignment of configurational prefixes.

$$
\begin{array}{cc}
\text{CHO} & \text{CHO} \\
\text{H–C–OH} & \text{H–C–OH} \\
\text{CH}_2 \quad \}\ \text{D-}ribo & \text{HO–C–H} \\
\text{H–C–OH} & \text{H–C–OH} \quad \}\ \text{D-}gluco \\
\text{H–C–OH} & \text{H–C–OH} \\
\text{CH}_2\text{–OH} & \text{H–C–OH} \}\ \text{D-}glycero \\
& \text{CH}_2\text{–OH}
\end{array}
$$

3-Deoxy-D-*ribo*-hexose        D-*Glycero*-D-*gluco*-heptose

Table 31.2. Aldoses

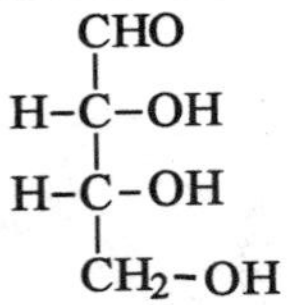

```
      CHO                     CHO
   H–C–OH                  HO–C–H
   H–C–OH                  H–C–OH
    CH2–OH                  CH2–OH
  D-Erythrose             D-Threose
   D-erythro               D-threo
```

```
      CHO            CHO            CHO            CHO
   H–C–OH         HO–C–H         H–C–OH         HO–C–H
   H–C–OH         H–C–OH         HO–C–H         HO–C–H
   H–C–OH         H–C–OH         H–C–OH         H–C–OH
    CH2–OH         CH2–OH         CH2–OH         CH2–OH
  D-Ribose      D-Arabinose    D-Xylose       D-Lyxose
   D-ribo        D-arabino      D-xylo         D-lyxo
```

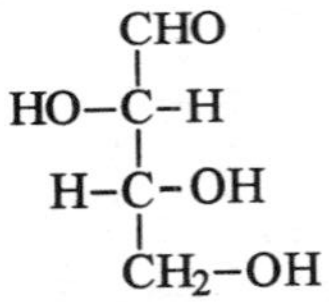

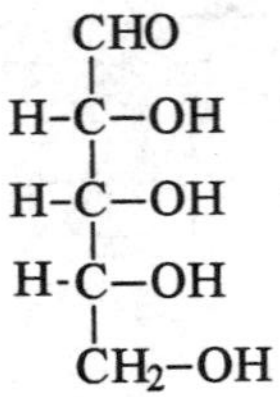

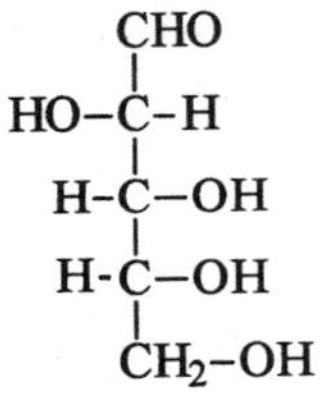

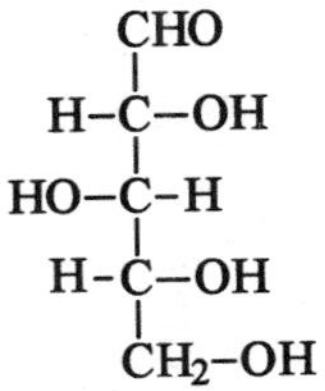

```
      CHO        CHO        CHO        CHO        CHO        CHO        CHO        CHO
   H–C–OH     HO–C–H     H–C–OH     HO–C–H     H–C–OH     HO–C–H     H–C–OH     HO–C–H
   H–C–OH     H–C–OH     HO–C–H     HO–C–H     H–C–OH     H–C–OH     HO–C–H     HO–C–H
   H–C–OH     H–C–OH     H–C–OH     H–C–OH     HO–C–H     HO–C–H     HO–C–H     HO–C–H
   H–C–OH     H–C–OH     H–C–OH     H–C–OH     H–C–OH     H–C–OH     H–C–OH     H–C–OH
    CH2–OH     CH2–OH     CH2–OH     CH2–OH     CH2–OH     CH2–OH     CH2–OH     CH2–OH
  D-Allose   D-Altrose  D-Glucose  D-Mannose  D-Gulose   D-Idose   D-Galactose D-Talose
   D-allo      D-altro    D-gluco    D-lyxo     D-gulo     D-ido     D-galacto   D-talo
```

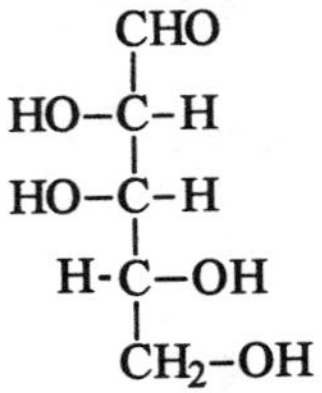

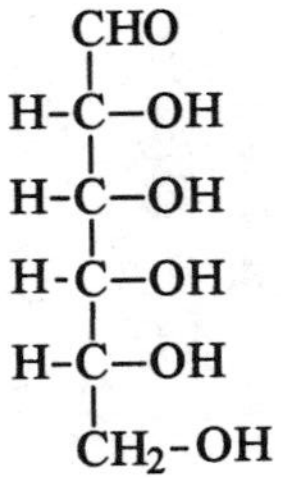

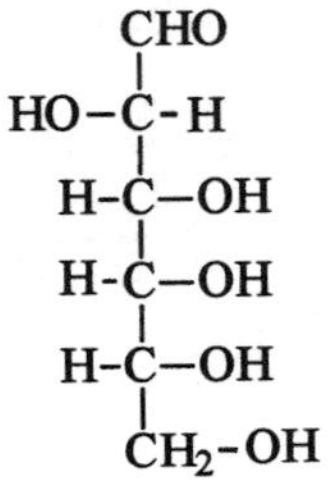

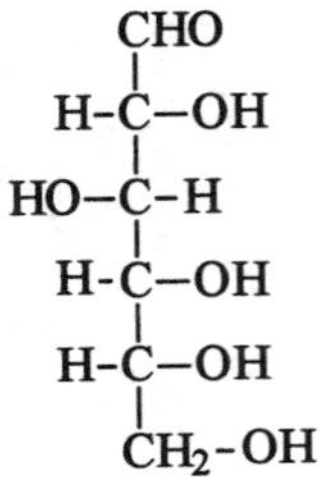

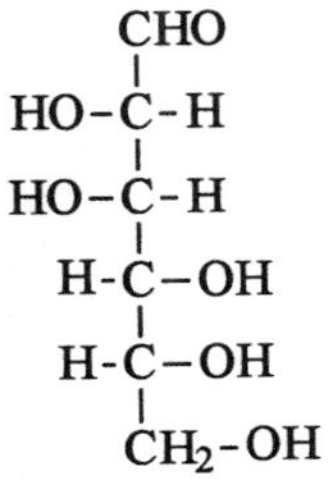

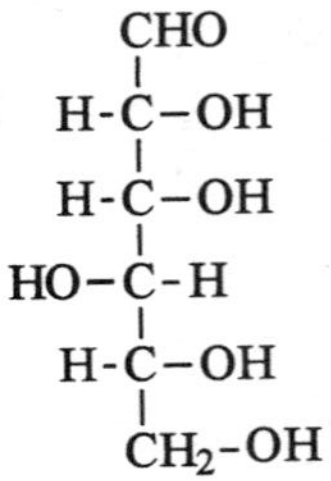

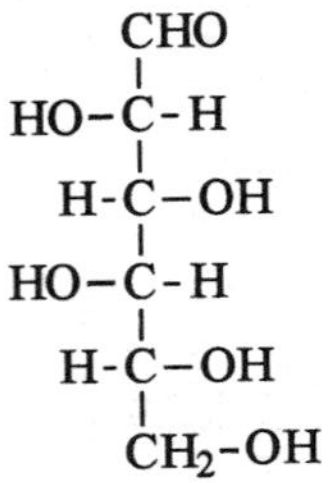

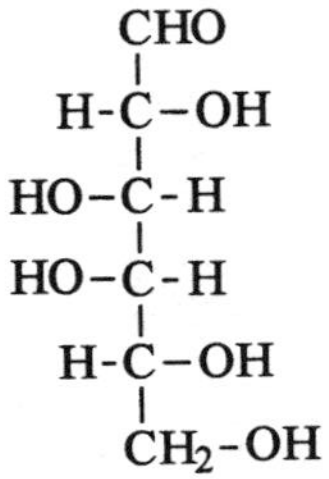

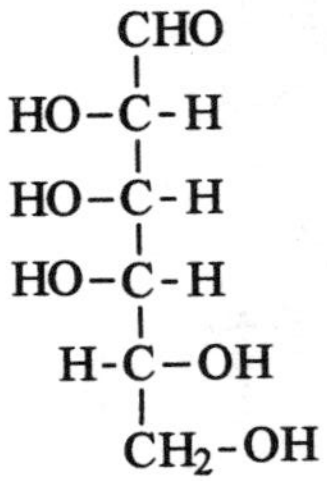

*Ketoses* take stem names such as tetrulose, hexulose, and so on. The position of the carbonyl group is given the lowest possible locant. To the stem name is appended a configurational prefix derived from the trivial name that describes the appropriate number of asymmetric carbon atoms (see table 31.2). For example, in the five-carbon 2-ketose (two >CH–OH groups) series, one set of prefixes is D-*threo*- and D-*erythro*-, and the systematic names are D-*threo*-pent-2-ulose and D-*erythro*-pent-2-ulose. As with aldoses, names are formed by combining the prefixes of table 31.2 with the stem name for carbon chains longer than six atoms. CAS assigns prefixes corresponding to the sets of chiral centers on either side of the carbonyl group when there are more than four asymmetric carbon atoms.

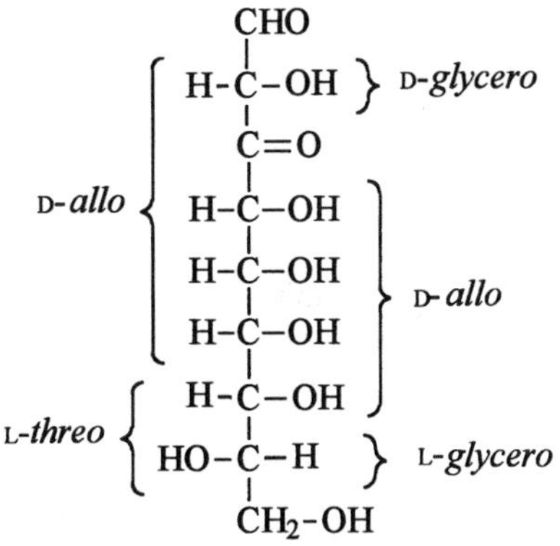

L-*glycero*-D-*allo*-D-*glycero*-3-Nonulose (IUPAC)
L-*threo*-D-allo-Non-3-ulose (CAS)

A few trivial names, such as D-ribulose (D-*erythro*-pent-2-ulose) and D-fructose (D-*arabino*-hex-2-ulose) are still acceptable.

The cyclic forms of carbohydrates are hemiacetals and hemiketals. In a Haworth representation, the plane of the ring is almost perpendicular to that of the paper. To generate a Haworth representation from a Fischer projection, it is necessary to reorient C-5. For D-glucopyranose, the process is as follows:

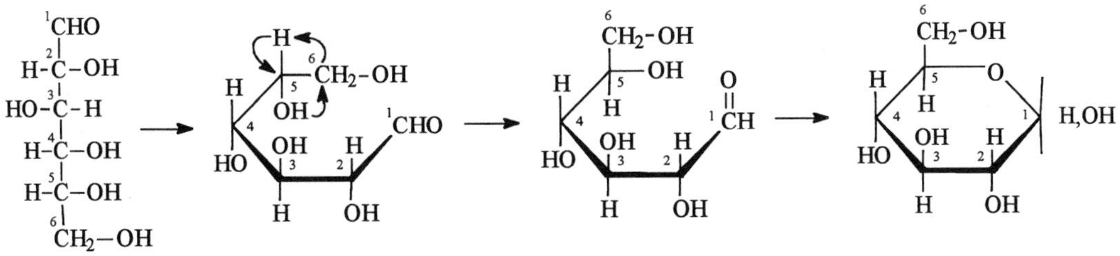

It is seen that there are two possible cyclic hemiacetals, called anomers, designated $\alpha$ and $\beta$, for each ring size from a given monosaccharide. In the D-series, the $\alpha$-anomer is the one with the 1-hydroxy group below the plane of the ring, and the $\beta$-anomer has that group above the plane. In the L-series, the $\alpha$- and $\beta$-anomers have the 1-hydroxy group above and below, respectively, the plane of the ring. Their names begin with the open-chain prefix (table 31.2) and, depending on ring size, end in furanose (5-membered ring), pyranose (6), septanose (7), and so on.

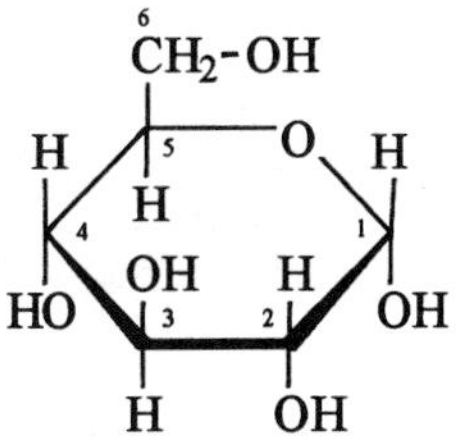

$\alpha$-D-Glucopyranose

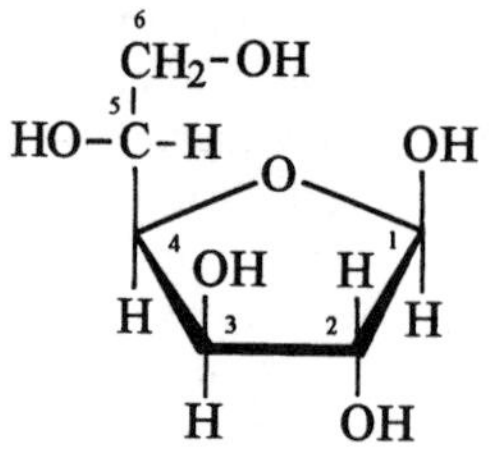

$$\overset{6}{C}H_2-OH$$
$$HO-\overset{5}{C}-H$$

β-D-Glucofuranose

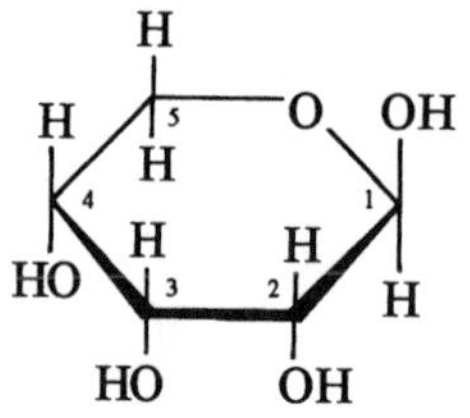

β-D-Ribopyranose

IUPAC/IUB has published specific recommendations on conformational nomenclature for 5- and 6-membered ring monosaccharides[5], unsaturated monosaccharides[6], and branched-chain monosaccharides[7]. Each of these subjects is also included in the 1996 carbohydrate document[4].

Oxidation or reduction of the terminal groups of monosaccharides gives various derivative classes that are named systematically (table 31.3) with configurational prefixes, the appropriate numerical stem, and a suffix indicating the class. Names such as D-*gluco*-hexodialdose and D-*erythro*-pent-2-ulosonic acid are formed. The names of symmetrical compounds, which are not optically active, may be preceded by the prefix *meso*- for clarity.

Many derivatives of carbohydrates are possible; most can be named by modifications of organic nomenclature principles, as in D-glucar-6-amic acid. *Deoxy sugars*, in which a hydroxyl group has been replaced by a hydrogen atom (effectively the removal of an oxygen atom from a hydroxyl group), are named with a locant and the prefix deoxy, as in 2-deoxy-D-*ribo*-hexose. Substitution of the resulting $CH_2$ group with an amino group would generate a name such as 2-amino-2-deoxy-D-glucose. The hydrogen atom of a hydroxy group might also be exchanged, producing compounds with names such as 4,6-di-*O*-methyl-α-D-galacto-pyranose or 2,4-di-*O*-acetyl-D-glucopyranose. *Glycosides* are mixed acetals or ketals derived from cyclic monosaccharides; they are best named by citing the substituting group as a separate word followed by the name of the cyclic monosaccharide with the final "e" replaced by -ide.

Table 31.3. Monosaccharide Derivative Classes

|  | R | R' | Class | Ending |
|---|---|---|---|---|
|  | CHO | $CH_2-OH$ | aldose | ose |
| R | CHO | CHO | dialdose | dialdose |
| \| | CHO | COOH | uronic acid | uronic acid |
| $[CH_2-OH]_n$ | $CH_2-OH$ | $CH_2-OH$ | alditol | itol |
| \| | COOH | $CH_2-OH$ | aldonic acid | onic acid |
| R' | COOH | COOH | aldaric acid | aric acid |
| R |  |  |  |  |
| \| |  |  |  |  |
| $[CH_2-OH]_n$ |  |  |  |  |
| \| | $CH_2-OH$ | $CH_2-OH$ | ketose | ulose |
| CO | CHO | $CH_2-OH$ | aldoketose | osulose |
| \| | COOH | $CH_2-OH$ | ketoaldonic acid | ulosonic acid |
| $[CH_2-OH]_m$ |  |  |  |  |
| \| |  |  |  |  |
| R' |  |  |  |  |

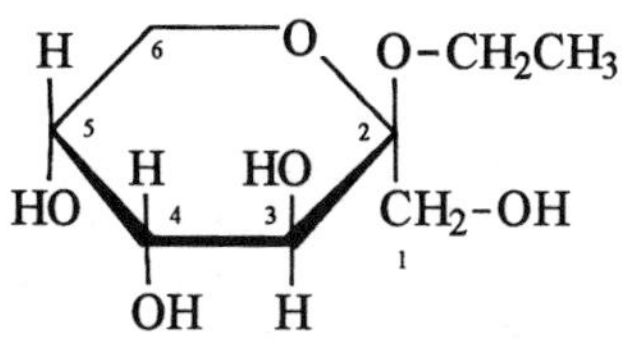

Ethyl β-D-fructopyranoside

Esters, acyl halides, and amides derived from monosaccharide acids follow the usual systematic practices (see chapters 11, 12, and 18), giving names such as D-gluconoyl chloride and methyl β-D-galactopyranuronate.

*Disaccharides* are compounds in which two monosaccharides are joined by a glycosidic linkage. Where the linkage is conceptually formed by the reaction between the two glycosidic hydroxy groups, they are named as glycosyl glycosides; these nonreducing disaccharides have no free hemiacetal group. Glycosyl prefixes, derived after removal of the anomeric hydroxy group, are named by replacing the ending -ose with -osyl.

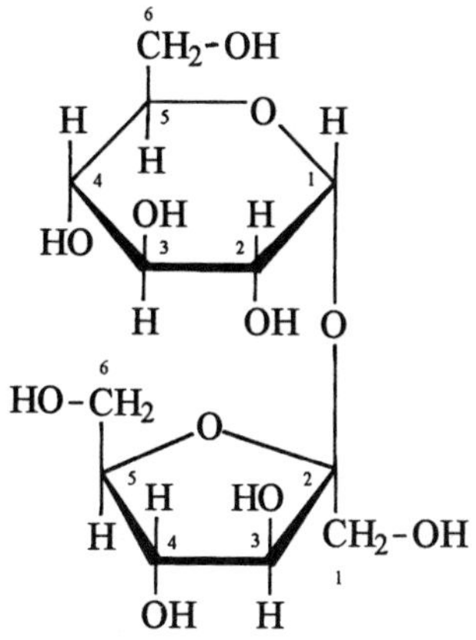

β-D-Fructofuranosyl α-D-glucopyranoside
Sucrose

Reducing disaccharides (with a free hemiacetal group) formed by replacement of an alcoholic hydrogen atom of one glycosyl unit with another glycosyl unit are named as *O*-glycosylglycoses; alternatively, the linkage between the monosaccharide units is given by the appropriate locants separated by an arrow and enclosed in parentheses.

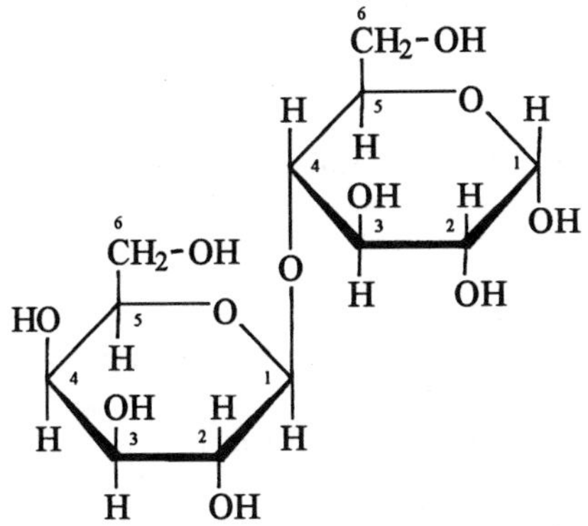

β-D-Galactopyranosyl-(1→4)-α-D-glucopyranose
4-*O*-β-D-Galactopyranosyl-α-D-glucopyranose
α-Lactose

## Cyclitols[8]

Cycloalkanes in which three or more ring atoms are substituted with hydroxyl groups are called *cyclitols*. Cyclitols other than the inositols are usually named by substitutive nomenclature, with stereochemistry depicted by $R/S$, $R^*/S^*$, *rel*-$(R/S)$, and $α/β$, as appropriate (see chapter 30).

Inositol stereoparents, that is 1,2,3,4,5,6-hexahydroxycyclohexanes, are named with itali-cized prefixes denoting the position, using lowest locants, of the –OH groups above and below the ring depicted as a plane. The prefixes and, in parentheses, the locants of –OH groups above the plane, a "/", and the locants for those below the plane are: cis-, (1,2,3,4,5,6/0), epi-, (1,2,3,4,5/6), allo- (1,2,3,4/5,6), myo-, (1,2,3,5/4,6), muco- (1,2,4,5/3,6), neo-, (1,2,3/4,5,6), chiro-, (1,2,4/3,5,6), and scyllo-, (1,3,5/2,4,6). Substituted inositols retain the numbering and configurational prefix of the parent inositol.

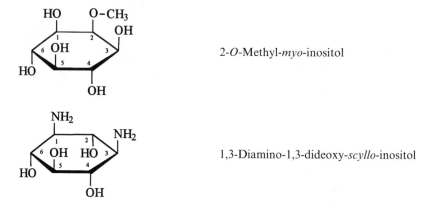

2-O-Methyl-myo-inositol

1,3-Diamino-1,3-dideoxy-scyllo-inositol

Inositols in which an –OH group has been replaced by a group higher in seniority than –OH (see chapter 3, table 3.1) may be named systematically.

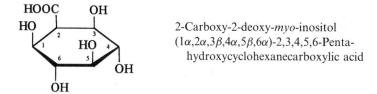

2-Carboxy-2-deoxy-myo-inositol
(1α,2α,3β,4α,5β,6α)-2,3,4,5,6-Penta-
    hydroxycyclohexanecarboxylic acid

The fractional method of representation may also be used for rings other than those with six members.

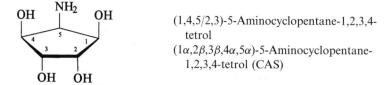

(1,4,5/2,3)-5-Aminocyclopentane-1,2,3,4-
    tetrol
(1α,2β,3β,4α,5α)-5-Aminocyclopentane-
    1,2,3,4-tetrol (CAS)

Absolute configuration of inositols is expressed by "D" and "L". For a planar ring repre-sentation, if the lowest numbered asymmetric position is above the plane of the ring, and the numbering is clockwise, the configuration is "L"; if the numbering is counterclockwise, it is "D". The position to which the configurational descriptors refer is usually placed ahead of the name. Absolute stereochemistry of other cyclitols may be indicated in the same way; CAS uses systematic stereochemical descriptors (see chapter 30).

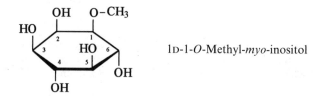

1D-1-O-Methyl-myo-inositol

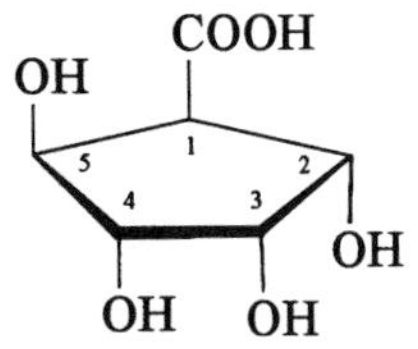

1L-(1,5/2,3,4)-2,3,4,5-Tetrahydroxy-
cyclopentanecarboxylic acid
(1R,2S,3S,4R,5S)-Tetrahydroxy-
cyclopentanecarboxylic acid
[1R-(1α,2α,3β,4β,5β)]-2,3,4,5-Tetra-
hydroxycyclopentanecarboxylic acid
(*Chemical Abstracts* Volumes 76–128)

## Steroids[9]

The *steroid* parent ring skeleton is cyclopenta[a]phenanthrene, with numbering and letters identifying the rings:

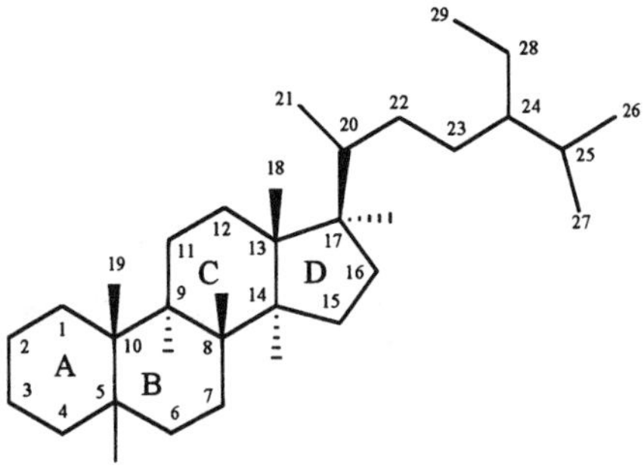

Methyl groups, numbered 19 and 18, respectively, are usually found at C-10 and C-13, and an acyclic chain may be at C-17; numbering on that chain begins with 20 on the carbon joined to the ring. Common steroids without a C-17 substituent are gonane, estrane (methyl group at C-13), and androstane (methyl groups at C-10 and C-13).

Absolute stereochemistry in a parent steroid name is defined as shown above for the asymmetric centers at positions 8, 9, 10, 13, 14, and 17. With assumed fixed orientation of the parent ring structure in the plane of the paper, atoms or groups that project behind the plane are termed α, and those projecting in front are termed β; unknown configurations are represented by a wavy line and termed ξ. Configurations at position 5, abnormal configurations at implied centers, and configurations of additional asymmetric centers are specified by α or β. Some common examples with C-17 side chains and their configurations are given in table 31.4.

## Table 31.4. Steroid Side Chains

| Side-chain structure | R' | Implied configuration | Example |
|---|---|---|---|
| $H_3C$—CH—CH$_2$—CH$_2$—R' (21, 20, 22, 23, 17) | $CH_3$ | 20R | 5α-cholane |
| | $CH_2CH(CH_3)_2$ | 20R | 5α-cholestane |
| $H_3C$—...—CH$_3$ (21, 20, 22, 23, 24, 25, 17) with R' | $CH_3$ | 20R,24S | 5α-ergostane |
| | $CH_2CH_3$ | 20R,24S | 5α-poriferastane (IUPAC) |
| | | | 5α-stigmastane (CAS) |
| $CH_2CH_3$ (20, 21, 17) | – | – | 17α-pregnane |

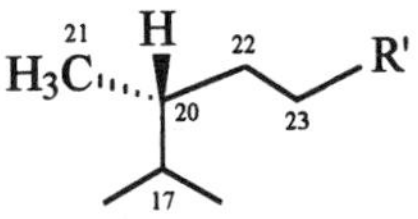

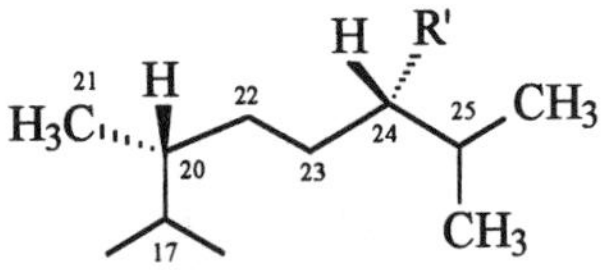

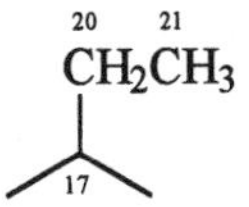

Implied configurations on C-17 side chains also occur in the following steroid analogs:

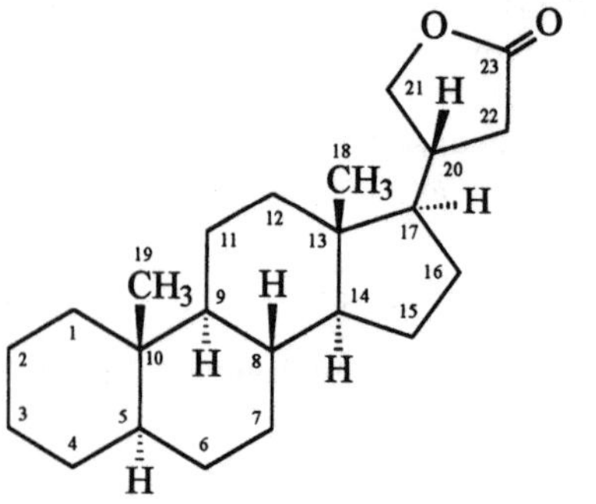

5α-Cardanolide

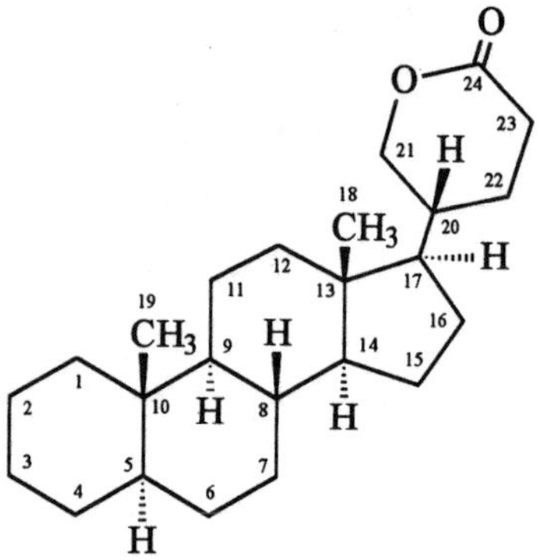

5α-Bufanolide

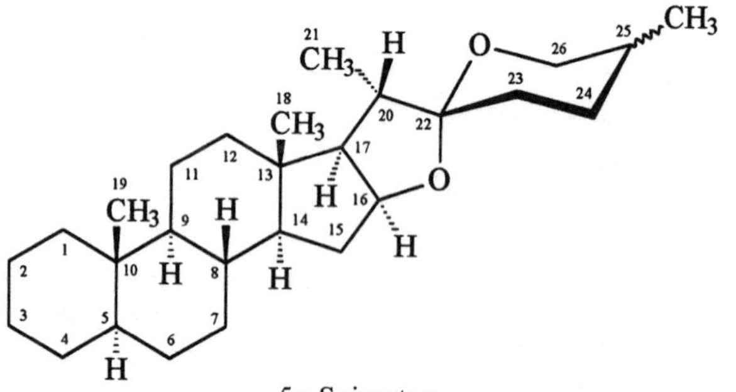

5α-Spirostan

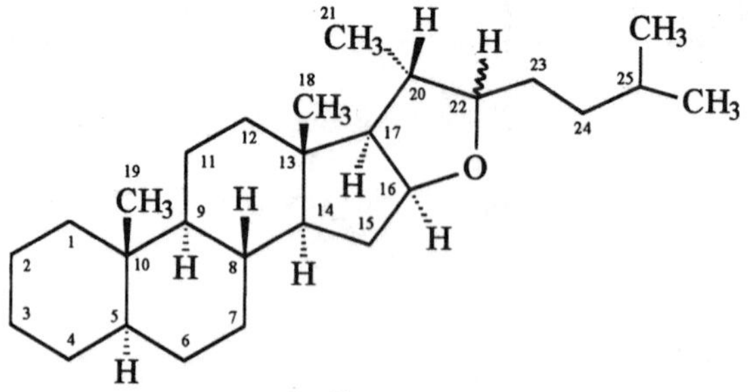

5α-Furostan

Other configurations must be specified. For the spirostans, configurations at positions 16 and 17, if different from that shown above, are given by 16β(H) and 17β(H).

Configurations at double bonds in side chains are described by an italicized *E* or *Z*.

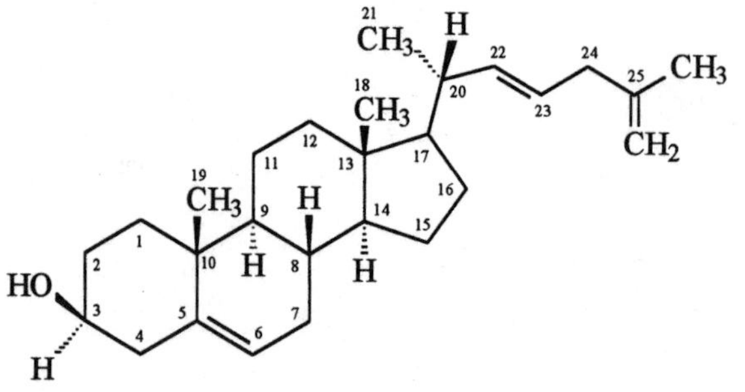

(22*E*)-Cholesta-5,22,25-
trien-3β-ol

Derivatives are named by substitutive nomenclature as in 5α-cholest-6-en-3β-ol or pregnan-21-oic acid, and the presence of additional rings may be indicated by bridge prefixes (see chapters 7 and 9) or by a modification of fusion nomenclature (see chapter 7).

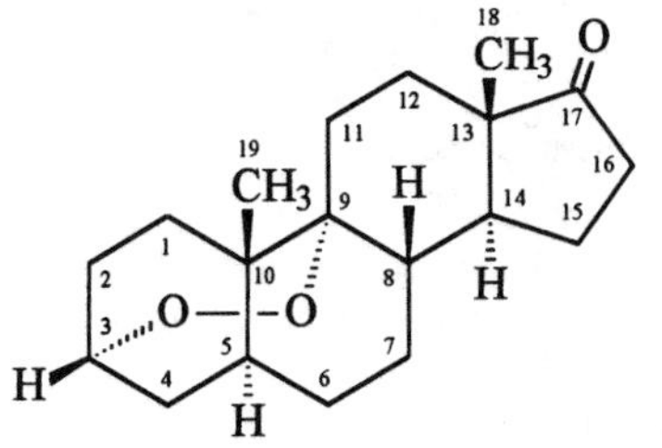

3α,9-Epidioxy-5α-androstan-17-one

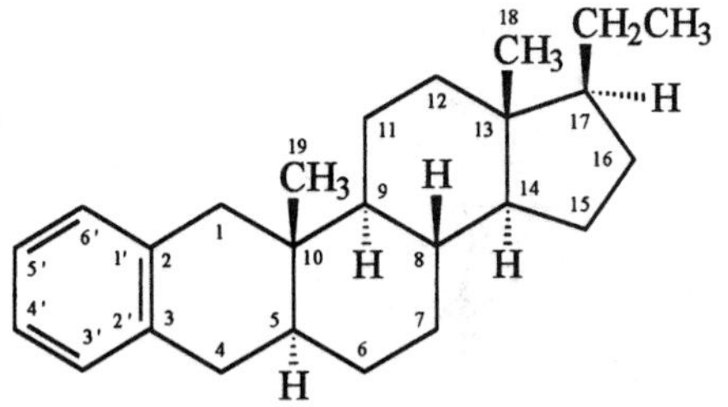

Benzo[2,3]-5α-pregn-2-ene (CAS)
Benzo[2,3]-5-α-pregnane (IUPAC)

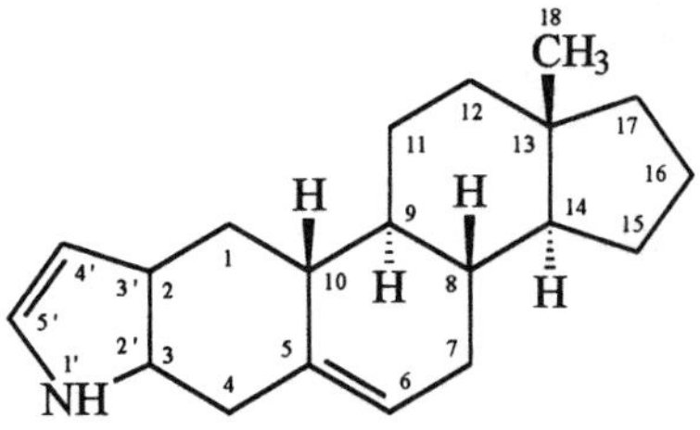

1′H-Pyrrolo[3′,2′:2,3]-5α-estr-5-ene (IUPAC)
1′H-Estra-2,5-dieno[3,2:b]pyrrole (CAS)

Steroids with a hydroxyl group at C-3 are often called *sterols*, even though another substituent, higher in seniority, may provide a final suffix to the name.

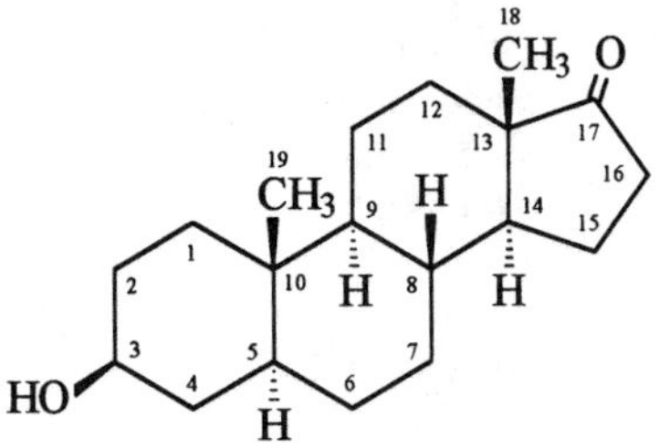

3β-Hydroxy-5α-androst-17-one
3β-Androsterone

Modifications of steroid stereoparent structures may be named as indicated under "General Principles" at the beginning of this chapter. As mentioned earlier, letters can be used to identify the rings involved and locant numbers used for side chains. For example, 7-nor (or B-nor) denotes the loss of a carbon atom at position 7 in the parent structure, the numbering of which is retained with omission of locant 7 in this case. Because of its potential for ambiguity, the letter method is not recommended. CAS no longer uses homo or nor to describe ring reduction or enlargement in steroids; such steroids are named systematically.

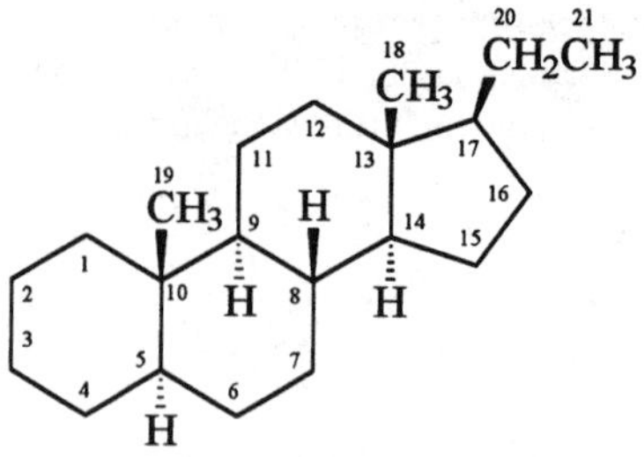

5α-Pregnane

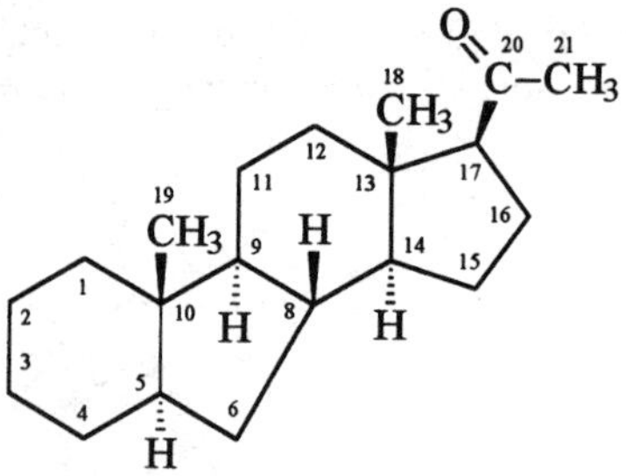

7-Nor-5α-pregnan-20-one
B-Nor-5α-pregnan-20-one

The following examples illustrate the use of locants with homo. The locant 4a indicates addition of an atom to a ring at position 4; insertion into a fusion bond is depicted by the locants of that bond followed by "a".

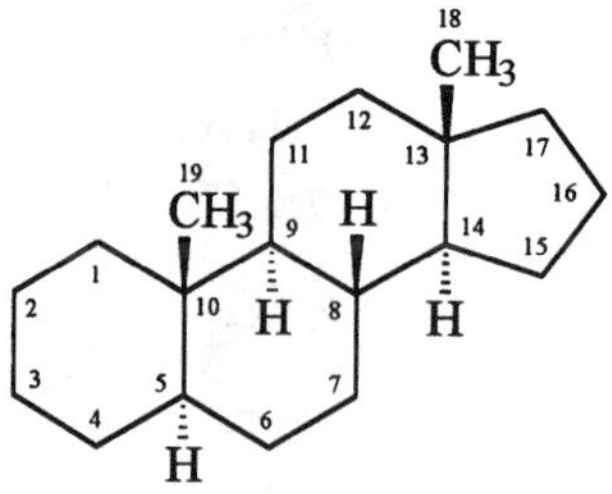

5α-Androstane

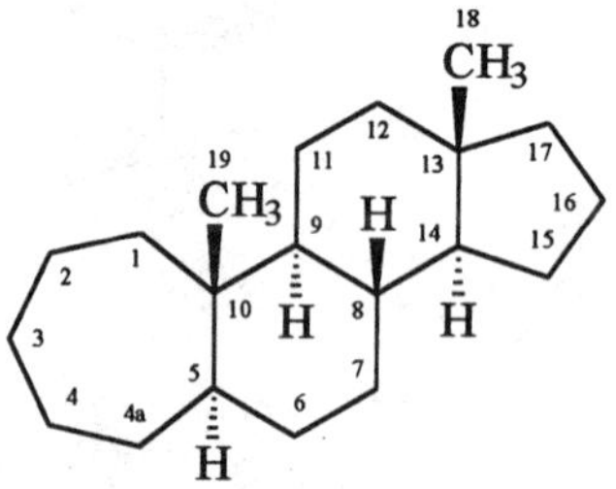

4a-Homo-5α-androstane

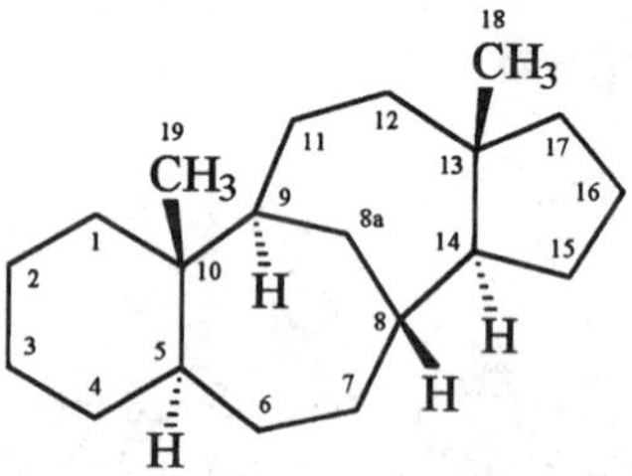

8(9)a-Homo-5α-androstane

Other prefixes include cyclo for the formation of an additional ring, as in 3α,5-cyclo-5α-androstan-17-one; and seco, for ring scission, as in 2,3-seco-5α-cholestane. The most important seco steroids are the D vitamins. However, these seco names are cumbersome and designation of stereochemistry is often a problem; hence, a set of trivial names has been developed.[10] An example is (5Z,7E)-(3S)-9,10-secocholesta-5,7,10(19)-trien-3-ol, for which the trivial names cholecalciferol or calciol are recommended. Replacement nomenclature (see chapter 3) is used to denote replacement of a carbon atom by a heteroatom, as in 4-oxa-5α-androstan-3-one.

## Terpenes and Related Compounds

*Terpenes* are cyclic and acyclic hydrocarbons or hydrocarbon derivatives whose structures are formed by joining isoprene (2-methylbuta-1,3-diene) units head-to-tail to provide the subclasses *monoterpenes* (two units), *sesquiterpenes* (three units), *diterpenes* (four units), *tetraterpenes* (eight units), and so on. Lupane (structure II, at the beginning of this chapter) is an example of a cyclic *triterpene*. The simpler terpenes should be named conventionally as alicyclic hydrocarbons (see chapter 6) or hydrocarbon derivatives with appropriate stereochemical designations, but most have stereoparent names as well. Stereoparent numbering is shown in the following examples.

p-Menthane
1-Isopropyl-4-methylcyclohexane

Thujane
1-Isopropyl-4-methylbicyclo[3.1.0]hexane

Carane
3,7,7-Trimethylbicyclo[4.1.0]heptane

Bornane
1,7,7-Trimethylbicyclo[2.2.1]heptane

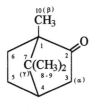

2-Bornanone
1,7,7-Trimethylbicyclo[2.2.1]heptan-2-one
Camphor

Pinane
2,6,6-Trimethylbicyclo[3.1.1]heptane

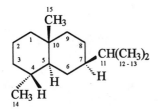

Eudesmane
(1R,4aR,7R,8aS)-Decahydro-7-isopropyl-
   1,4a-dimethylnaphthalene
[1R-(1α,4aβ,7β,8aα)]-Decahydro-7-iso-
   propyl-1,4a-dimethylnaphthalene
      (*Chemical Abstracts* Volumes 76–128)

The trivial name eudesmane implies the stereochemistry indicated in the systematic name. This implication also holds for most terpene parents having trivial names. However, as with steroids, absolute configuration is dependent on the orientation of the structure, but, unlike steroids, standard orientations have not been established. Further, numbering of stereoparents based on biochemical sources may differ from that in accepted semisystematic names. Commonly used orientations and numbering, with absolute configurations, can be found in the *Chemical Abstracts* volume and collective indexes published by CAS; abnormal configurations and additional asymmetric centers can then be denoted by α, β, R, and S. Systematic names are usually given as synonyms for the trivial names in the CAS indexes.

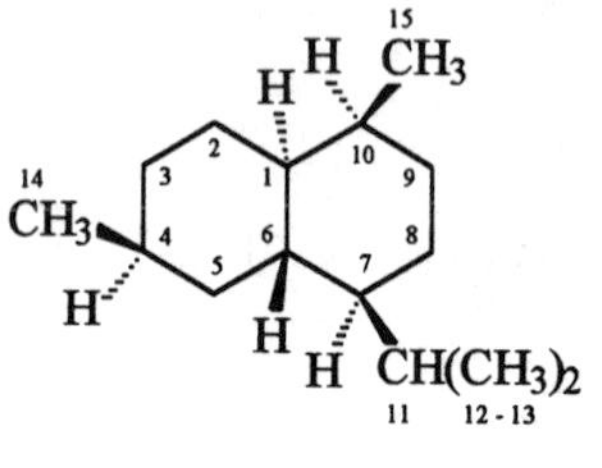

Cadinane
(1*S*,4*S*,4a*S*,6*S*,8a*S*)-Decahydro-4-isopropyl-1,6-
 dimethylnaphthalene
[1*S*-(1α,4α,4aα,6α,8aβ)-Decahydro-4-isopropyl-1,6-
 dimethylnaphthalene
 (*Chemical Abstracts* Volumes 76–128)

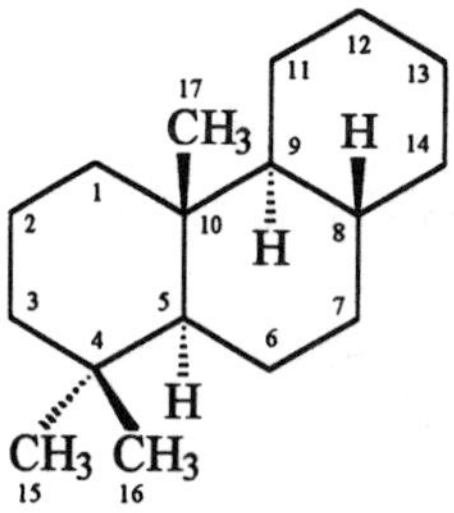

Podocarpane
(4a*R*,4b*S*,8a*R*,10a*S*)-Tetradecahydro-1,1,4a-
 trimethylphenanthrene
[4a*R*-(4aα,4bβ,8aα,10aβ)]-Tetradecahydro-1,1,4a-
 trimethylphenanthrene
 (*Chemical Abstracts* Volumes 76–128)

Modifications of terpene stereoparents, as with other stereoparent classes, are described by prefixes such as nor and homo, as defined in this chapter under "General Principles". In addition, the nondetachable prefixes friedo and neo are still acceptable, but no longer used by CAS or recommended by IUPAC. Friedo indicates that angular methyl groups have been shifted from their normal positions. The shift itself is denoted by symbols such as *D:C* (for position 14 to 13 and 8 to 14) placed ahead of the prefix.

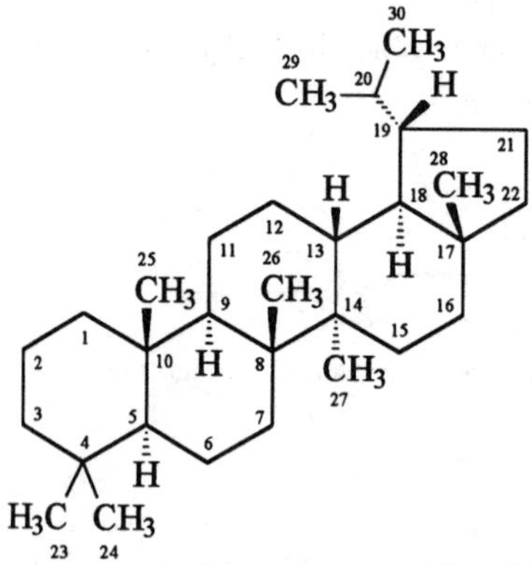

Lupane

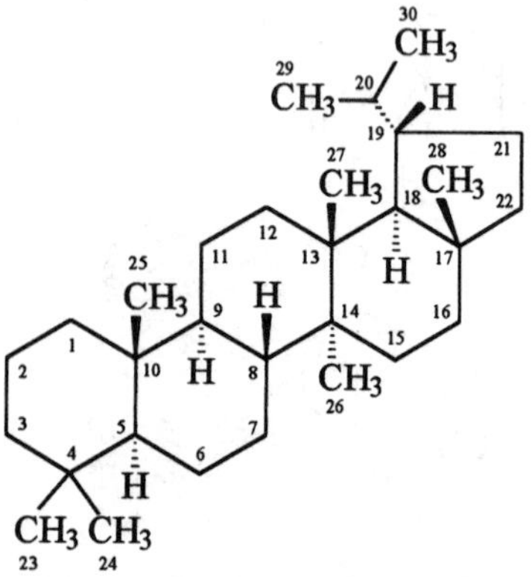

*D:C*-Friedolupane

The prefix neo denotes the rearrangement of a six-membered ring to a five-membered ring plus migration of appended methyl groups; again, symbols such as *A:B* (to indicate the nature of the rearrangement) are placed ahead of the prefix.

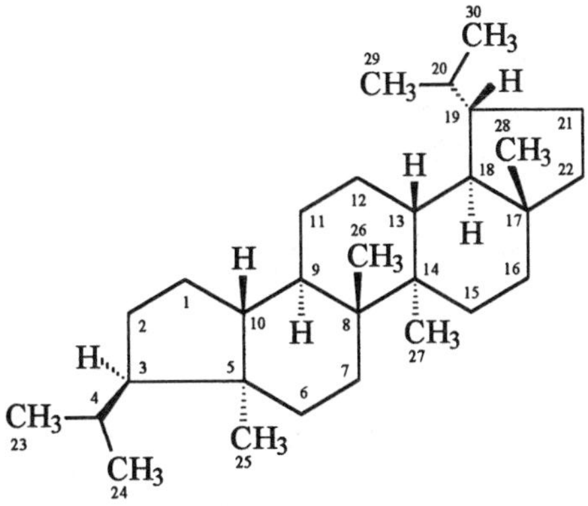

*A:B*-Neolupane

Characteristic groups and substituents are named substitutively.

*Carotenes*[11] are acyclic tetraterpenes; their oxygenated derivatives are called *xanthophylls*. Arrangement of the isoprene units in a carotene is reversed at the center of the molecule; all carotenes are therefore formally derived from the following structure:

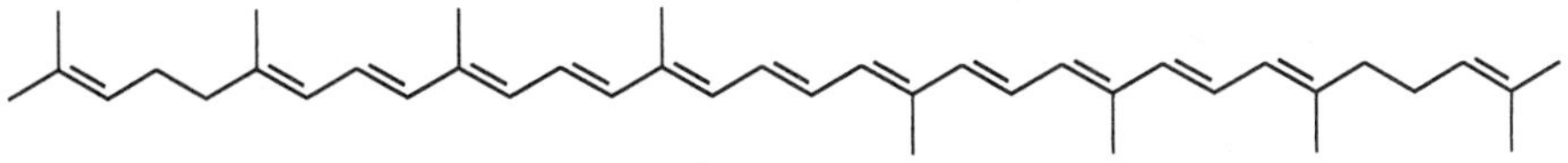

This chain may be hydrogenated, dehydrogenated, cyclized, oxidized, or any combination of these processes; it can be rearranged or degraded as long as the two central methyl groups are retained.

Where the broken circle indicates the presence of two double bonds, the compound corresponding to the stem name carotene is as follows:

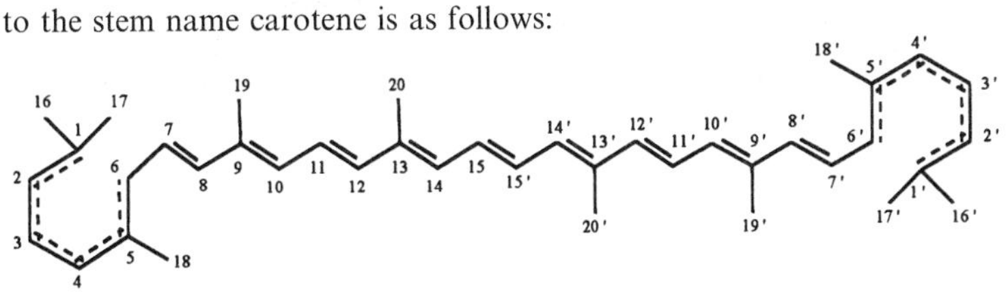

Individual carotene hydrocarbons are named with the stem name carotene preceded by two Greek letters denoting the nature of each of the end groups attached at C7 and C7′, respectively. These letters and the structures depicted are as follows:

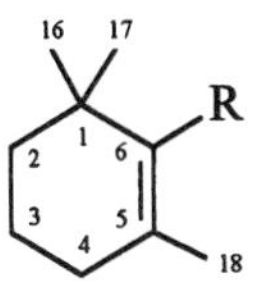

$\beta$ (beta)  $\psi$ (psi)  $\varepsilon$ (epsilon)

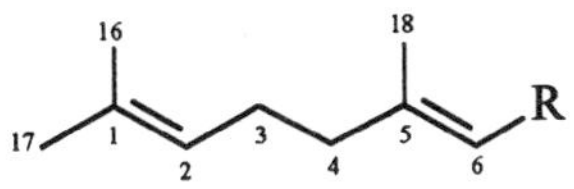

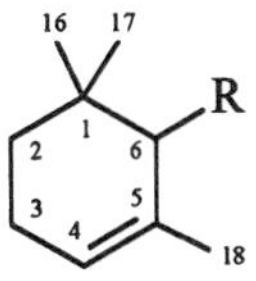

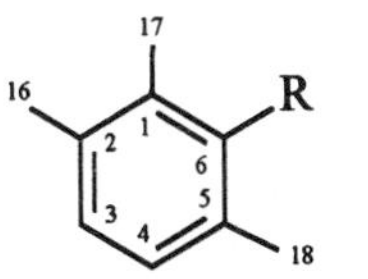

$\phi$ (phi)  $\kappa$ (kappa)  $\chi$ (chi)

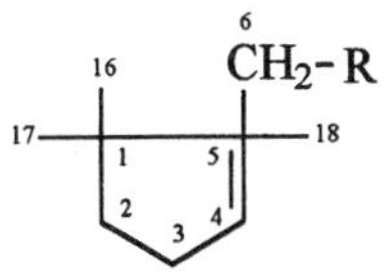

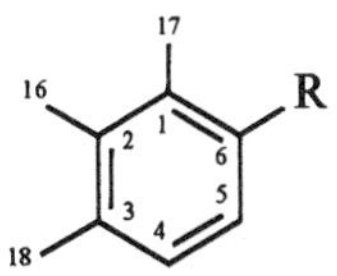

where R is

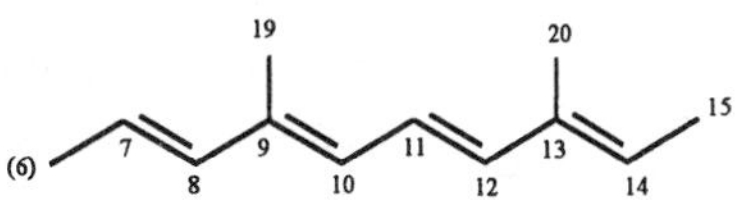

The symbols are cited in Greek alphabetical order; the part of the structure corresponding to the second symbol is given primed locants.

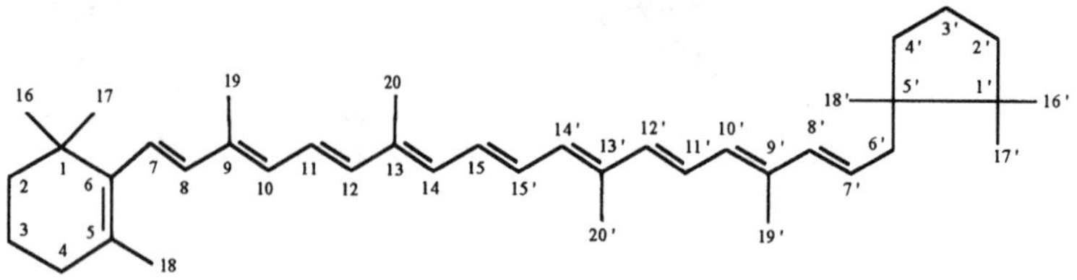

$\beta, \kappa$-Carotene

Ahead of these symbols will be cited any skeletal modifying prefixes. The stem name carotene implies *trans-* or *E* at all double bonds; *cis-* or *Z* must be cited where this is appropriate. Absolute configurations are indicated by *R* or *S*.

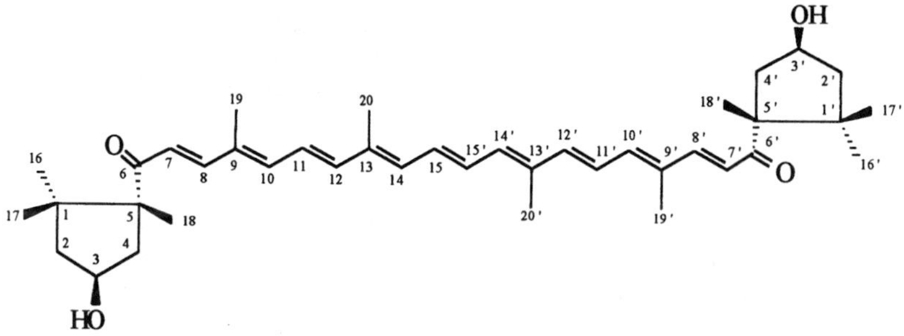

(3*S*,5*R*,3′*S*,5′*R*)-3,3′-Dihydroxy-*κ*,*κ*-carotene-6,6′-dione

The prefix *retro-* signifies a shift by one position of all double bonds in the conjugated system. The unitalicized prefix "apo" preceded by a locant indicates that all of the molecule beyond that locant has been replaced by an appropriate number of hydrogen atoms; locants on the chain are unchanged.

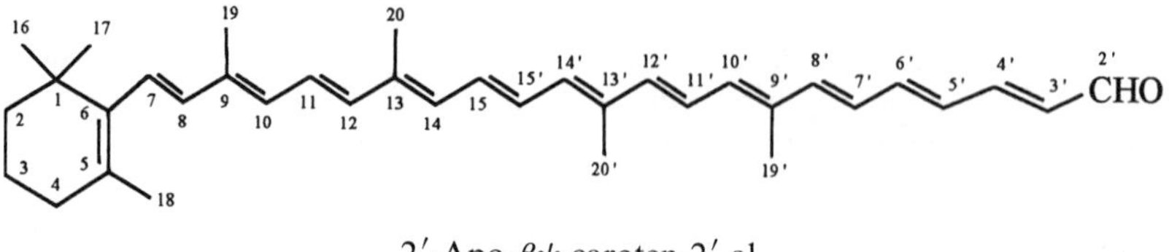

2′-Apo-*βψ*-caroten-2′-al

The carotene recommendations[11] have an extensive list of semisystematic names, trivial names, and their structures.

*Retinoids*[12] and *prenols*[13] are related to terpenes. Retinoids have four isoprene units joined head-to-tail in the stereoparent structure. Their general structure is as follows:

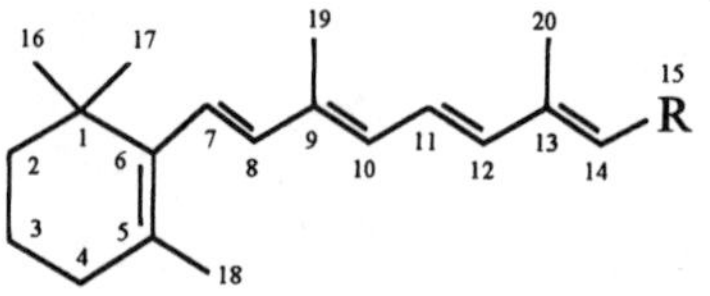

R = CH$_2$–OH    Retinol
                Vitamin A
R = CHO         Retinal
R = COOH        Retinoic acid

The systematic name for (3*R*)-13-*cis*-retinol is (2*E*,4*E*,6*E*,8*E*)-3,7-dimethyl-9-(2,6,6-trimethyl-cyclohex-1-en-1-yl)nona-2,4,6,8-tetraen-1-ol. For this kind of configuration, *all-E-* is sometimes used.

The IUPAC/IUB document[12] has many illustrative examples of trivial and semisystematic names. Structural modification prefixes, including *retro-*, but not apo-, can be used in naming retinoids.

*Prenols* have the following general structure:

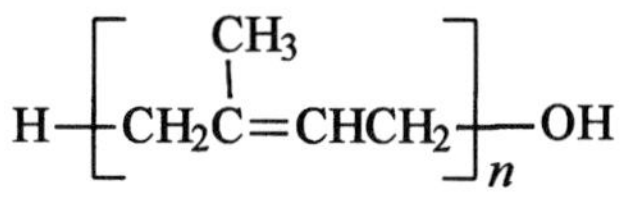

$$H-\left[CH_2\overset{\overset{\displaystyle CH_3}{|}}{C}=CHCH_2\right]_n-OH$$

Configurations at the double bonds are denoted by *cis-* and *trans-* with appropriate numerical prefixes, beginning with the unit furthest from the hydroxy group.

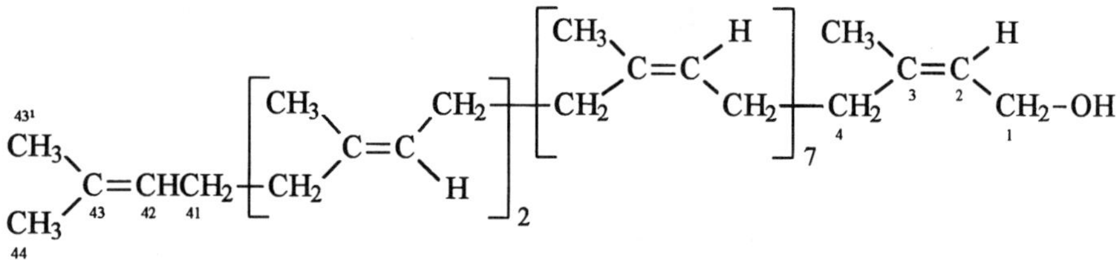

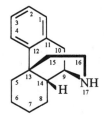

*ditrans,octacis*-Undecaprenol
Bactoprenol

(2Z,6Z,10Z,14Z,18Z,22Z,26Z,30Z,34E,38E)-3,7,11,15,19,23,27,31,35,39,43-
Undecamethyltetraconta-2,6,10,14,18,22,26,30,34,38,42-undecaen-1-ol
(substitutive name)

Both retinoids and prenols are best named systematically, despite the length and seeming complexity of such names, as illustrated above.

### Alkaloids

Rules for the systematic nomenclature of *alkaloids* have not been formulated, although the appendix to the revised recommendations for Section F[1] does contain an extensive list of parent alkaloid structures.

These nitrogenous bases derived from plant sources have structures that range from the simple to the very complex. The simpler structures are readily named by systematic nomenclature (coniine, for example, is 2-propylpiperidine), but even here, trivial names are commonly employed. More complex alkaloids with common structural features and known absolute stereochemistry are named by CAS as derivatives of stereoparents, the names of which imply absolute stereochemistry. A partial list of these stereoparent names, including common alkaloids such as morphinan and yohimban, appears in Appendix IV of the *Chemical Abstracts Index Guide*.[2] Specific derivatives are named as modifications of the stereoparent names through the use of prefixes such as nor, homo, seco, and so on, with meanings as described under "General Principles" (above).

Morphinan (a stereoparent name)

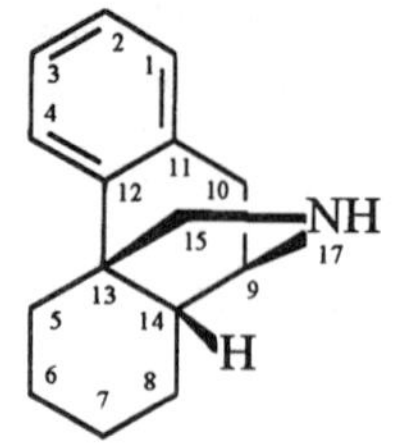

16-Normorphinan
D-Normorphinan

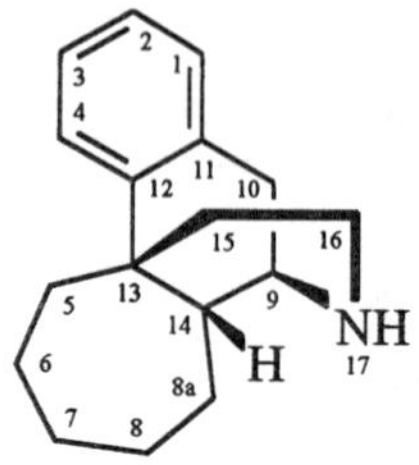

8a-Homomorphinan
C-Homomorphinan

## Tetrapyrroles

The variety of trivial names in use for the various classes of *tetrapyrroles* (*porphyrins*) has been much reduced in 1986 IUPAC document[14]. This area includes macrocyclic *porphyrins*, their metal coordination complexes, and linear *bilins*, including tri- and dipyrroles. Vitamin $B_{12}$, the chlorophylls, and hemes are cobalt, magnesium, and iron complexes, respectively, of tetra-pyrroles.

Some recommended names for fundamental parent structures are as follows:

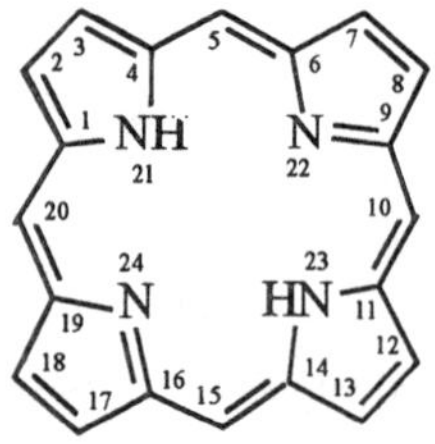

Porphyrin (IUPAC)
21*H*,23*H*-Porphine (CAS)

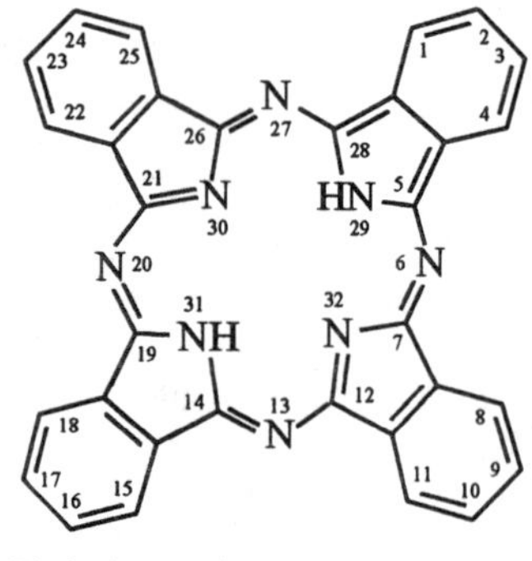

Phthalocyanine (IUPAC)
29*H*,31*H*-Phthalocyanine (CAS)
29*H*,31*H*-Tetrabenzoporphyrazine
 (21*H*,23*H*-Porphyrazine is a CAS
 name for 21*H*,23*H*-5,10,15,20-
 tetraazaporphyrin)

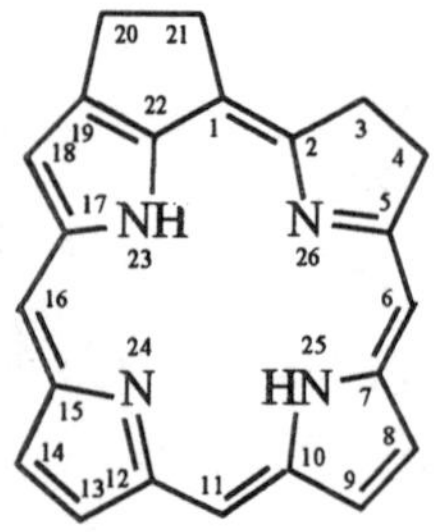

Phorbine

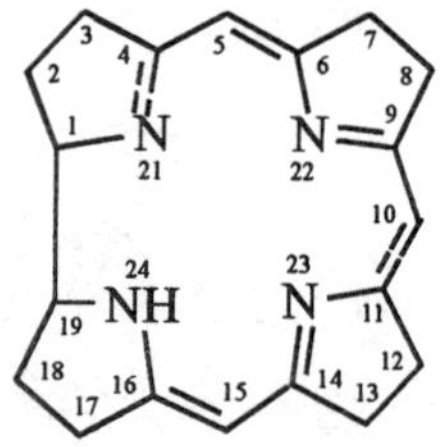

Corrin

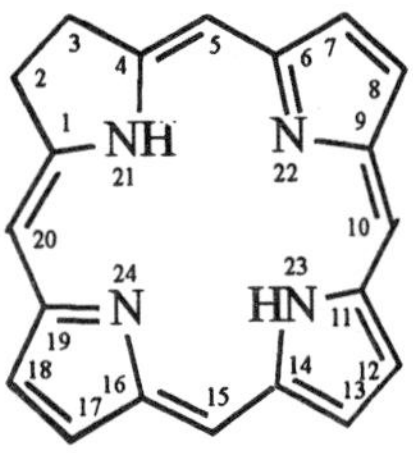

Chlorin
2,3-Dihydro-21$H$,23$H$-
porphine (CAS)

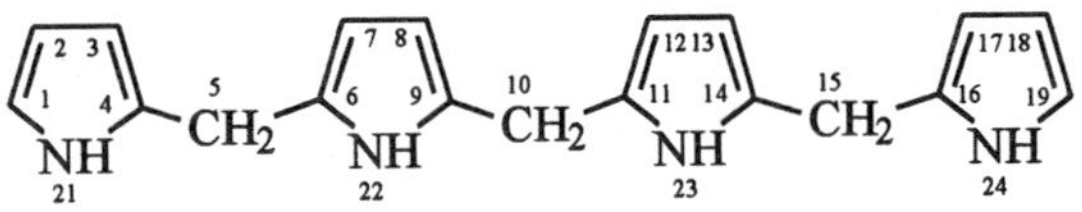

Bilane (IUPAC)
5,10,15,22,23,24-Hexahydro-21$H$-bilane (CAS)

Parent structures such as those above can be modified according to the "General Principles" described above.

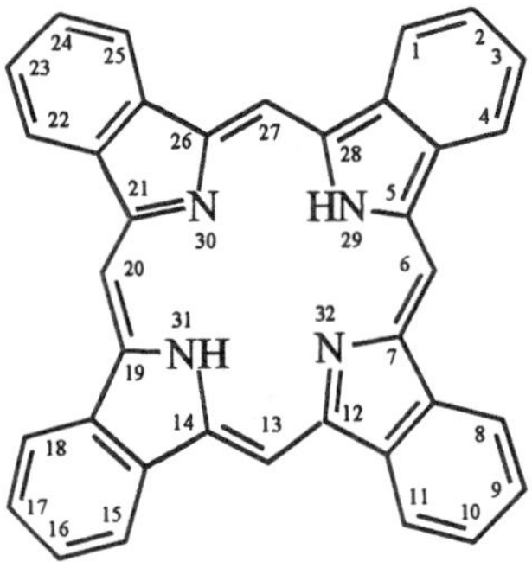

29$H$,31$H$-Tetrabenzo[$b,g,l,q$]porphine (CAS)
Tetrabenzo[$b,g,l,q$]porphyrin (IUPAC)

IUPAC/IUB uses porphyrin numbering and locants and superscript locants such as $2^2, 2^3$, and so on, to indicate positions on the "benzo" rings. Derivatives are named by substitutive nomenclature. Systematic, semisystematic, and Fischer names are compared for many types of compounds, and a number of acceptable trivial names, with structures, are given in an appendix to the IUPAC recommendations.[14]

## Prostaglandins

*Prostaglandins* and *thromboxanes* are related substances whose names are based on the following stereoparents:

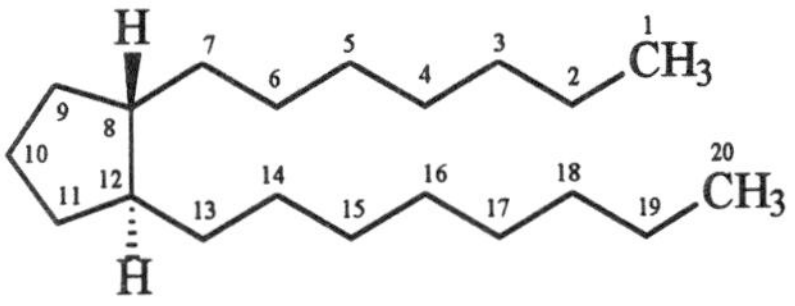

Prostane

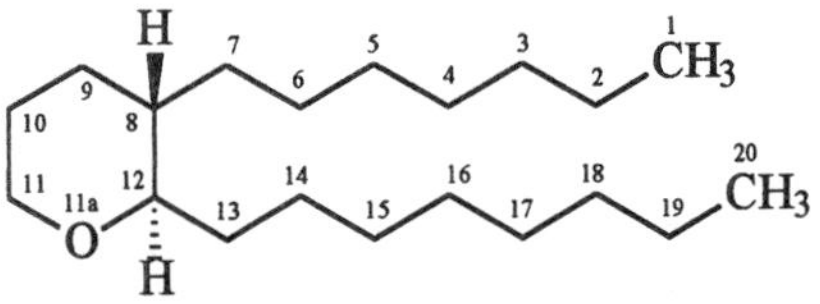

Thromboxane

The absolute configuration of prostane and thromboxane at positions 8 and 12 is as shown above. Configurations at other asymmetric carbon atoms in the ring are given by $\alpha$ and $\beta$ and in the chains by $R$ and $S$; at double bonds, configuration is denoted by $E$ and $Z$. Derivatives are named by substitutive nomenclature principles.

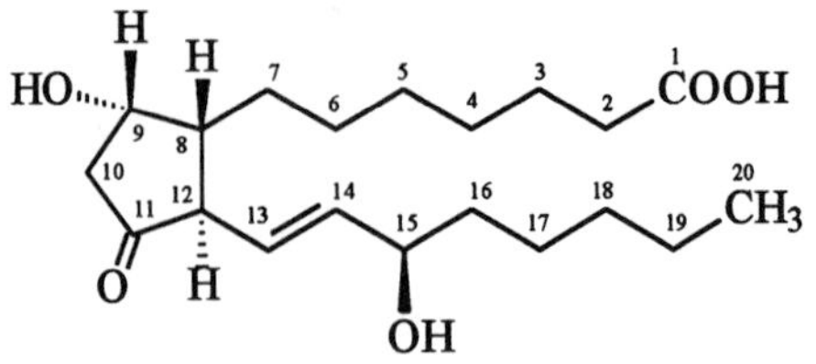

(9α,13E,15S)-9,15-Dihydroxy-11-oxoprost-13-en-1-oic acid

## Nucleosides and Nucleotides

*Nucleosides* are mainly *N*-glycosyl derivatives of substituted heterocyclic bases such as purine and pyrimidine (see chapter 9); *nucleotides* are esters of nucleosides with phosphoric acids. The bases themselves should be named systematically but have common trivial names such as adenine (1*H*-purin-6-amine) and uracil [pyrimidine-2,4(1*H*,3*H*)-dione]. These trivial names give rise to stereoparent names for nucleosides based on purine (adenosine, guanosine, inosine, and xanthosine) and pyrimidine (cytidine, thymidine, and uridine); derivatives are then named substitutively.

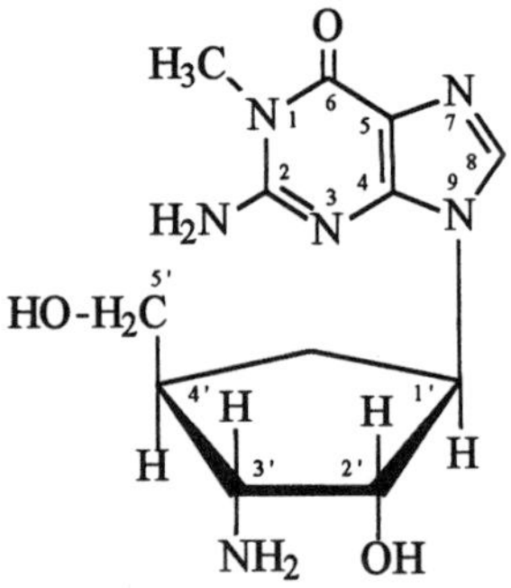

3′-Amino-3′-deoxy-1-methylguanosine

The corresponding nucleotides are generally named as phosphoric acid esters of the stereo-parents, as in adenosine 5′-(trihydrogen diphosphate). However, the monophosphates have trivial names ending in -ylic acid. The location of the phosphate group is denoted by an appropriate primed locant.

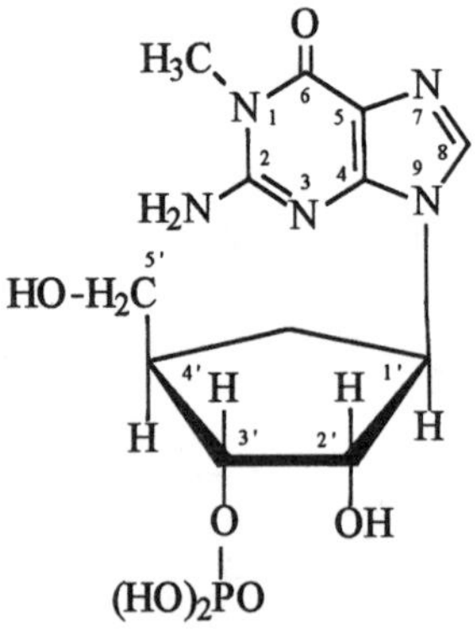

3′-Guanylic acid

## Vitamins

Vitamins belong to a variety of structural classes. Where structures are known, vitamins may be named systematically or on the basis of stereoparent names. For vitamins with well-established trivial names, the corresponding systematic name can be found in the *Chemical Abstracts Index Guide*[2].

Table 31.5. Trivial, Stereoparent, or Parent Names for B Vitamins

| | |
|---|---|
| vitamin $B_1$ | thiamin |
| vitamins $B_2$ and G | riboflavin |
| vitamin $B_3$ | niacin, niacinamide, nicotinamide |
| vitamin $B_4$ | adenine |
| vitamin $B_5$ | pantothenic acid (a $\beta$-alanine derivative) |
| vitamin $B_6$ group[15] | pyridoxol, pyridoxal, pyridoxamine |

Vitamin A structures are retinoids (see above); they are named as derivatives of stereoparents such as retinol, retinal, and retinoic acid.

Compounds with vitamin B activity include widely diverse structures. Some of these are given in table 31.5. One of the most important, vitamin $B_{12}$, is a tetrapyrrole (see above); its derivatives are considered to be corrinoids,[16] with trivial names such as cobinic acid, cobyrinic acid, and cobamic acid.

Vitamin C is L-ascorbic acid, and derivatives are named as carbohydrates. Vitamin $C_2$ is a monosaccharide derivative of $2H$-benzopyran-2-one.

Vitamin D compound nomenclature is given in an IUPAC/IUB document[10]; the members of this group are seco steroids and can be named as such (see above).

The nomenclature of *tocopherols*, including vitamin E, has been described in another IUPAC/IUB document.[17] Individual members of the vitamin K group, quinones with isoprenoid or phytyl side chains, are named systematically.

## Miscellaneous Natural Products

*Folic acid*[18] and its derivatives are based on the heterocyclic structure having the trivial name pteroic acid:

Pteroic acid
4-[[(2-Amino-1,4-dihydro-4-oxopteridin-6-yl)methyl]amino]benzoic acid

Folic acid (or "folate") itself is named $N$-[4-[[(2-amino-1,4-dihydro-4-oxopteridin-6-yl)methyl]-amino]benzoyl-L-glutamic acid. Additional glutamic acid or glutamate residues are indicated by the numerical prefixes di-, tri-, and so on, as in pteroyldiglutamate. IUPAC/IUB recommendations list a number of symbols to represent substituents and thereby shorten the names, always an objective in biochemical nomenclature.

*Lipids* are a large and diverse group of compounds, usually derived from long-chain saturated and unsaturated acyclic acids, carotenoids, or steroids. They may be in the form of esters or other reaction products with glycerol, inositol, sugars, serine, choline, or phosphorus or sulfur acids. There is no established nomenclature, but wherever possible, systematic organic nomenclature is recommended despite the wealth of trivial names. For the glycerol derivatives, stereospecific numbering, indicated by the prefix "*sn*", is used. In this method, the carbon atom above the asymmetric carbon atom with its hydroxyl group to the left in a Fischer projection is given the number 1. Alternatively, the usual "D" and "L" prefixes may be used.

*sn*-Glycerol 3-(dihydrogen phosphate)
L-Glycerol 3-(dihydrogen phosphate)

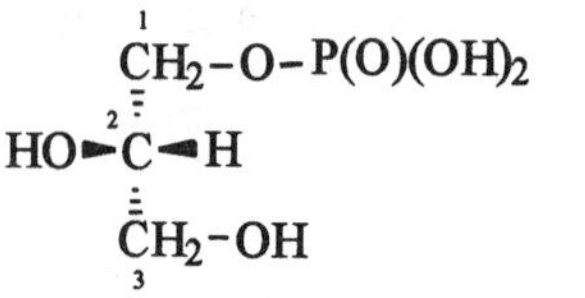

sn-Glycerol 1-(dihydrogen phosphate)
L-Glycerol 1-(dihydrogen phosphate)

*Antibiotics* and some drugs also comprise a heterogeneous group of natural products. Most have trivial names such as nicotine and lysergic acid diamide (LSD), and trade names and generic names abound. Where structures are known, systematic nomenclature can be employed. Cocaine, for example, is methyl (1R,2R,3S,5S)-3-(benzoyloxy)-8-methyl-8-azabicyclo[3.2.1]-octane-2-carboxylate.

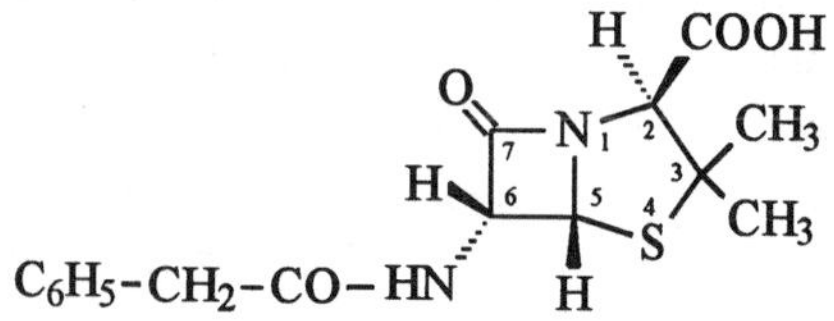

Penicillin G
(2S,6R)-3,3-Dimethyl-7-oxo-6-[(phenylacetyl)amino]penam-2-carboxylic acid (IUPAC)[1]
(2S,5R,6R)-3,3-Dimethyl-7-oxo-6-[(phenylacetyl)amino]-4-thia-1-azabicyclo[3.2.0]-
    heptane-2-carboxylic acid (CAS)

In many cases, a stereoparent can be established with a consequent reduction in the length of names of derivatives.

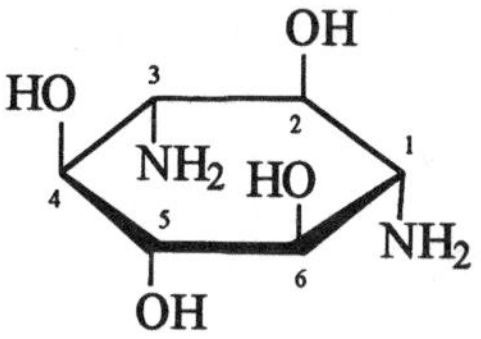

D-Streptamine
1,3-Diamino-1,3-dideoxy-*scyllo*-inositol

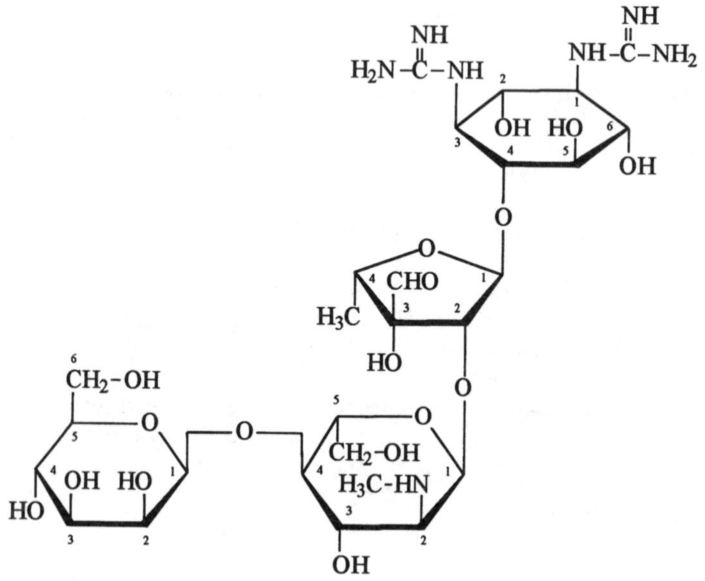

Streptomycin-B
O-β-D-Mannopyranosyl-(1→4)-O-2-deoxy-2-(methylamino)-α-L-glucopyranosyl-
    (1→2)-O-5-deoxy-3-C-formyl-α-L-lyxofuranosyl-(1→4)-N,N'-
    bis(aminoiminomethyl)-D-streptamine

It is obvious that trivial names in this field will find more usage in practice than systematic names.

## REFERENCES

1. International Union of Pure and Applied Chemistry, Division of Organic Chemistry, Commission on Nomenclature of Organic Chemistry. Revised Section F: Natural Products and Related Compounds (Recommendations 1999). *Pure Appl. Chem.*, **1999**, *71*, 587–643.
2. American Chemical Society, Chemical Abstracts Service. *Chemical Abstracts Index Guide 1999*; Chemical Abstracts Service: Columbus OH, 1999.
3. International Union of Pure and Applied Chemistry and International Union of Biochemistry, Joint Commission on Biochemical Nomenclature (JCBN). Nomenclature and Symbolism for Amino Acids and Peptides (Recommendations 1983). *Pure Appl. Chem.* **1984**, *56*, 595–624; *Eur. J. Biochem.* **1984**, *138*, 9–17; International Union of Biochemistry and Molecular Biology. *Biochemical Nomenclature and Related Documents*, 2nd. ed.; Liébecq, C. Ed.; Portland Press: London, 1992; pp 39–69 (includes additions and corrections).
4. International Union of Pure and Applied Chemistry and International Union of Biochemistry and Microbiology, Joint Commission on Biochemical Nomenclature (JCBN). Nomenclature of Carbohydrates (Recommendations 1996). *Pure Appl. Chem.* **1996**, *68*, 1919–2008.
5. International Union of Pure and Applied Chemistry and International Union of Biochemistry, Joint Commission on Biochemical Nomenclature (JCBN). Conformational Nomenclature for Five- and Six-Membered Ring Forms of Monosaccharides and their Derivatives (Provisional Recommendations). *Pure Appl. Chem.*, **1981**, *53*, 1901–1906; International Union of Biochemistry and Molecular Biology. *Biochemical Nomenclature and Related Documents*, 2nd ed.; Liébecq, C. Ed.; Portland Press: London, 1992; pp 158–161 (see also ref 4).
6. International Union of Pure and Applied Chemistry and International Union of Biochemistry, Joint Commission on Biochemical Nomenclature (JCBN). Nomenclature of Unsaturated Monosaccharides (Provisional Recommendations). *Pure Appl. Chem.* **1982**, *54*, 7–10; International Union of Biochemistry and Molecular Biology. *Biochemical Nomenclature and Related Documents*, 2nd ed.; Liébecq, C., Ed.; Portland Press: London, 1992; pp 162–164 (includes corrections, see also ref. 4).
7. International Union of Pure and Applied Chemistry and International Union of Biochemistry, Joint Commission on Biochemical Nomenclature (JCBN). Nomenclature of Branched-Chain Monosaccharides (Provisional Recommendations). *Pure Appl. Chem.* **1982**, *54*, 211–215; International Union of Biochemistry and Molecular Biology. *Biochemical Nomenclature and Related Documents*, 2nd ed.; Liébecq, C., Ed.; Portland Press: London, 1992; pp 165–168 (includes corrections, see also ref. 4).
8. International Union of Pure and Applied Chemistry, Organic Chemistry Division, Commission on Nomenclature of Organic Chemistry; International Union of Pure and Applied Chemistry and International Union of Biochemistry, Commission on Biochemical Nomenclature. Rules for Cyclitol Nomenclature (Recommendations, 1973). *Biochem. J.* **1976**, *151*, 25–31; International Union of Biochemistry, Nomenclature Committee. Numbering of Atoms in *myo*-Inositol (Recommendations, 1988). *Biochem. J.* **1989**, *258*, 1–2; International Union of Biochemistry and Molecular Biology. *Biochemical Nomenclature and Related Documents*, 2nd ed.; Liébecq, C., Ed.; Portland Press: London, 1992; pp 149–155, 156–157.
9. International Union of Pure and Applied Chemistry and International Union of Biochemistry, Joint Commission on Biochemical Nomenclature (JCBN). The Nomenclature of Steroids (Recommendations, 1989). *Pure Appl. Chem.* **1989**, *61*, 1783–1822; International Union of Biochemistry and Molecular Biology. *Biochemical Nomenclature and Related Documents*, 2nd ed., Liébecq, C., Ed.; Portland Press: London, 1992; pp 192–221.
10. International Union of Pure and Applied Chemistry and International Union of Biochemistry, Joint Commission on Biochemical Nomenclature (JCBN). Nomenclature of Vitamin D (Provisional). *Pure Appl. Chem.* **1982**, *54*, 1511–1516; International Union

of Biochemistry and Molecular Biology. Nomenclature of Vitamin D (Recommendations 1981). *Biochemical Nomenclature and Related Documents*, 2nd ed.; Liébecq, C., Ed.; Portland Press: London, 1992; pp 242–246.

11. International Union of Pure and Applied Chemistry, Commission on Nomenclature of Organic Chemistry, and International Union of Biochemistry, Commission on Biochemical Nomenclature. Nomenclature of Carotenoids (Rules Approved 1974). *Pure Appl. Chem.* **1975**, *41*, 405–431; International Union of Biochemistry and Molecular Biology. *Biochemical Nomenclature and Related Documents*, 2nd ed.; Liébecq, C., Ed.; Portland Press: London, 1992; pp 226–238.

12. International Union of Pure and Applied Chemistry and International Union of Biochemistry, Joint Commission on Biochemical Nomenclature (JCBN). Nomenclature for Retinoids (Provisional). *Pure Appl. Chem.* **1983**, *55*, 721–726; International Union of Biochemistry and Molecular Biology. *Biochemical Nomenclature and Related Documents*, 2nd ed.; Liébecq, C., Ed.; Portland Press: London, 1992; pp 247–251.

13. International Union of Pure and Applied Chemistry and International Union of Biochemistry, Joint Commission on Biochemical Nomenclature (JCBN). Nomenclature of Prenols (Recommendations 1986). *Pure Appl. Chem.* **1987**, *59*, 683–689; International Union of Biochemistry and Molecular Biology. *Biochemical Nomenclature and Related Documents*, 2nd ed.; Liébecq, C., Ed.; Portland Press: London, 1992; 252–255.

14. International Union of Pure and Applied Chemistry and International Union of Biochemistry, Joint Commission on Biochemical Nomenclature (JCBN). Nomenclature of Tetrapyrroles (Recommendations 1986). *Pure Appl. Chem.* **1987**, *59*, 779–832; International Union of Biochemistry and Molecular Biology. *Biochemical Nomenclature and Related Documents*, 2nd ed; Liébecq, C., Ed.; Portland Press: London, 1992; pp 278–329.

15. International Union of Pure and Applied Chemistry and International Union of Biochemistry, Commission on Biochemical Nomenclature (CBN). Nomenclature for Vitamins $B_6$ and Related Compounds (Recommendations 1973). *Biochemistry*, **1974**, *13*, 1056–1058; International Union of Biochemistry and Molecular Biology. *Biochemical Nomenclature and Related Documents*, 2nd ed.; Liébecq, C., Ed.; Portland Press: London, 1992; pp 269–271.

16. International Union of Pure and Applied Chemistry and International Union of Biochemistry, Commission on Biochemical Nomenclature (CBN). Nomenclature of Corrinoids (Rules Approved 1975). *Pure Appl. Chem.* **1976**, *48*, 495–502; International Union of Biochemistry and Molecular Biology. *Biochemical Nomenclature and Related Documents*, 2nd ed.; Liébecq, C., Ed.; Portland Press: London, 1992; pp. 272–277.

17. International Union of Pure and Applied Chemistry and International Union of Biochemistry, Joint Commission on Biochemical Nomenclature (JCBN). Nomenclature of Tocopherols and Related Compounds (Recommendations 1981). *Pure Appl. Chem.* **1982**, *56*, 1507–1510; International Union of Biochemistry and Molecular Biology. *Biochemical Nomenclature and Related Documents*, 2nd ed.; Liébecq, C., Ed.; Portland Press: London, 1992, pp 239–241.

18. International Union of Pure and Applied Chemistry and International Union of Biochemistry, Joint Commission on Biochemical Nomenclature (JCBN). Nomenclature and Symbols for Folic Acid and Related Compounds (Recommendations 1986). *Pure Appl. Chem.* **1987**, *59*, 833–836; International Union of Biochemistry and Molecular Biology. *Biochemical Nomenclature and Related Documents*, 2nd ed.; Liébecq, C. Ed.; Portland Press.: London, 1992, pp 266–268.

# 32

# Isotopically Modified Compounds

This chapter deals with the nomenclature of organic compounds in which one or more atoms has an isotopic composition measurably different from that occurring naturally.

Until publication of IUPAC recommendations for naming isotopically modified compounds, the extended and modified Boughton system used by CAS was the only general method for describing the presence of isotopic atoms. These systems will be covered in some detail but without implying a preference, since each has advantages and disadvantages. Both systems describe isotopic substitution, that is, structures that are essentially isotopically pure at each specified position; however, the IUPAC system also provides for isotopic labeling where only a relatively small number of molecules of a compound may actually be isotopically substituted.

## Acceptable Nomenclature

### Isotopically Substituted Compounds

A compound is isotopically substituted if essentially all of its molecules contain the indicated isotope(s) at the specified position(s); in all other respects, the isotopic composition is the natural one.

In the extended Boughton system developed by CAS, an isotopic descriptor is added after the name, or after the part of the name to which it refers, and is separated from the rest of the name by a hyphen. Deuterium and tritium in names are denoted by the italic letters $d$ and $t$, followed by a subscript number giving the number of isotopic atoms; in formulas the capital letters D and T are used. For example, $Cl_3CD$ is named trichloromethane-$d$. For atoms of other elements the isotopic descriptor used in a name consists of the italicized element symbol, a left superscript number that gives the mass number, and a right subscript that gives the number of isotopic atoms, if more than one; italicized locants for the positions of the modified atoms in the appropriate part of the structure precede the descriptor. In formulas, the element symbol is not italicized. Mass numbers are not included in the formulas of the CAS Formula Index.

In the IUPAC system, the isotopic descriptor is not italicized, is enclosed in parentheses, and is inserted into the name just ahead of the part of the name having independent numbering to which the descriptor refers with no intervening punctuation except when followed by a locant, for example, ($^{15}N$) and (1,2-$^{14}C_2$). The element symbols $^2H$ and $^3H$ are used for deuterium and tritium in both names and formulas, as in $Cl_3C^2H$ and trichloro($^2H$)methane.

In both systems, when more than one atom at the same position can be isotopically substituted, the right subscript is always cited, even when only monosubstitution has occurred; for example, $Cl_2CHD$ or $Cl_2H^2H$ is named dichloromethane-$d_1$ (CAS) or dichloro($^2H_1$)methane (IUPAC). When several isotopic symbols occur in a formula or must be inserted at the same place in a name, they are ordered first according to the alphabetical order of the symbols, and then in order of their increasing mass numbers.

$^{14}CH_4$

Methane-$^{14}C$ (CAS)
($^{14}C$)Methane (IUPAC)

$DO-CH_2CD_2-OD$
     2      1

$^2HO-CH_2C^2H_2-O^2H$
       2      1

1,2-Ethane-*1,1-d$_2$*-diol-*d$_2$* (CAS)

$(1,1-^2H_2)$Ethane-1,2-$(^2H_2)$diol (IUPAC)

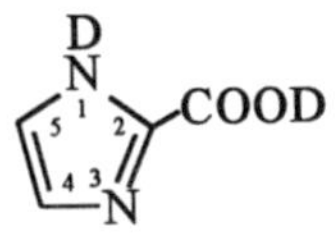

1*H*-Imidazole-*1-d*-2-carboxylic acid-*d* (CAS)

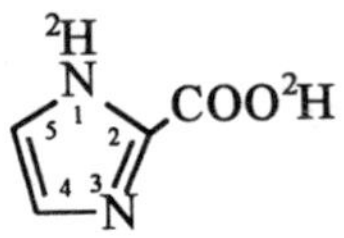

$(1-^2H_1)$-1*H*-Imidazole-2-$(^2H)$carboxylic acid (IUPAC)

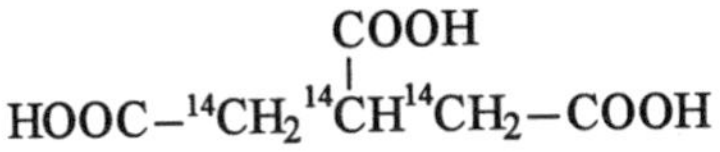

$$\overset{\text{COOH}}{\underset{}{|}}$$
$HOOC-^{14}CH_2{}^{14}CH^{14}CH_2-COOH$

1,2,3-Propanetricarboxylic-*1,2,3-$^{14}C_3$* acid (CAS)
$(1,2,3-^{14}C_3)$-Propane-1,2,3-tricarboxylic acid (IUPAC)

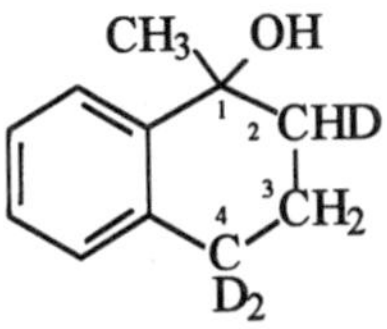

1,2,3,4-Tetrahydro-*4-d*-1-methyl-1-naphthalen-*2,4-d$_2$*-ol (CAS)
(hydro prefixes are detachable (see chapter 2); naphthalene is the parent; hence two deuterium atoms are expressed after its name and one after the hydro prefixes)

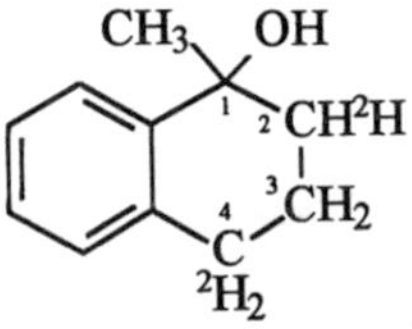

1-Methyl-$(2,4,4-^2H_3)$-1,2,3,4-tetrahydro-1-naphthol (IUPAC)
(hydro prefixes are nondetachable (see chapter 2); tetrahydronaphthalene is the parent hydride and all deuterium atoms are expressed in front of it)

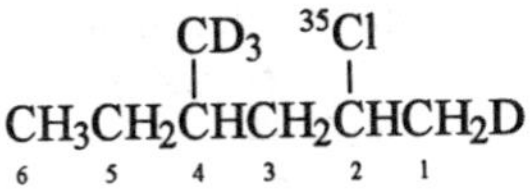

$$\overset{CD_3\ \ ^{35}Cl}{\underset{6\ \ \ 5\ \ \ \ 4\ \ \ 3\ \ \ 2\ \ \ 1}{|\ \ \ \ \ |}}$$
$CH_3CH_2CHCH_2CHCH_2D$

2-(Chloro-$^{35}Cl$)-4-(methyl-*d$_3$*)hexane-*1-d* (CAS)

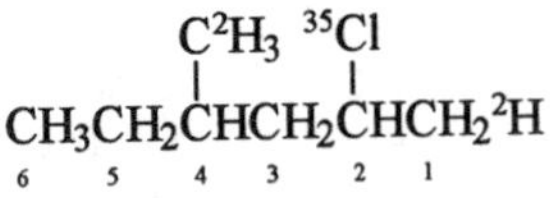

$$\overset{C^2H_3\ \ ^{35}Cl}{\underset{6\ \ \ \ 5\ \ \ \ 4\ \ \ 3\ \ \ 2\ \ \ 1}{|\ \ \ \ \ \ \ |}}$$
$CH_3CH_2CHCH_2CHCH_2{}^2H$

2-$(^{35}Cl)$Chloro-4-[$(^2H_3)$methyl]-$(1-^2H_1)$hexane(IUPAC)

$CH_3CHT-$$-CHTCH_3$

1,4-Di(ethyl-*1-t*)benzene (CAS)

$CH_3CH^3H-$$-CH^3HCH_3$

1,4-Di[$(1-^3H_1)$ethyl]benzene (IUPAC)

$DHN-^{14}CO-^{15}NH_2$
*N'*          *N*

Urea($^{14}C,N'-d,N-^{15}N$) (CAS)

$H^2HN-^{14}CO-^{15}NH_2$
3(*N*)        1(*N'*)

$(^{14}C,3-^2H_1,1-^{15}N)$Urea (IUPAC)
$(^{14}C,N'-^2H_1,N-^{15}N)$Urea (IUPAC)

In both systems, isotopic substitution at a position in a structure that does not have a locant may be indicated by an italicized word, by an italicized isotopic symbol acting as a locant, or by including the isotopic symbol in a group formula.

$C_6H_5-{}^{14}CO-NH_2$        Benzamide-*carbonyl*-$^{14}C$ (CAS)
                            (*carbonyl*-$^{14}C$)Benzamide (IUPAC)

$CH_3-O-CO-{}^{18}O-CH_2CH_3$   $^{18}O$-Ethyl $O$-methyl carbonate-$^{18}O$ (CAS[a])
                            $^{18}O$-Ethyl $O$-methyl ($^{18}O_1$)carbonate (IUPAC)

$CH_3-C^{18}O-OOH$            Ethaneperoxoic-*carbonyl*-$^{18}O$ acid (CAS)
                            (*carbonyl*-$^{18}O$)Peroxyacetic acid (IUPAC)

$CH_3-CO-O^{18}OH$            Ethaneperoxoic-*peroxy*-$^{18}O$ acid (CAS[b])
                            (O-$^{18}O$)Peroxyacetic acid (IUPAC)

An important principle in naming an isotopically substituted compound is to keep its name as close as possible to the name of the corresponding unmodified compound. Even though parentheses may be used to enclose a simple substituent (see chapter 2), the multiplicative prefix di- is not changed to bis- in names for isotopically substituted compounds; for instance, 2,6-di(ethyl-*1-t*)-2*H*-pyran not 2,6-bis(ethyl-*1-t*)-2*H*-pyran and tri(silyl-$^{29}Si$)phosphine not tris(silyl-$^{29}Si$)phosphine). However, a name for an isotopically substituted compound may differ from its corresponding unmodified compound when identical parts of an unmodified structure are not identically modified at equivalent positions. Where there is ambiguity, these groups must be expressed separately. The CAS system cites the unmodified prefix before the modified one; in the IUPAC system the reverse is the case.

$(CH_3)_3{}^{29}Si-{}^{13}CH_3$   Trimethylmethyl-$^{13}C$-silane-$^{29}Si$ (CAS)
                            ($^{13}C$)Methyltrimethyl($^{29}Si$)silane (IUPAC)

$$\begin{array}{c} {}^{37}Cl \\ | \\ {}^{35}Cl-CH \\ | \\ Cl \end{array}$$   Chlorochloro-$^{35}Cl$-chloro-$^{37}Cl$-methane (CAS)
                            ($^{35}Cl_1,{}^{37}Cl_1$)Trichloromethane (IUPAC)

$$\begin{array}{cccc} CH_3CH_2 & & CH_2CHD_2 & \\ | & & | & \\ CH_3[CH_2]_2CHCH_2CHCH_2-OH & & & \\ 7 \quad 6\text{-}5 \quad\quad 4 \quad 3 \quad\quad 2 \quad 1 & & & \end{array}$$    $$\begin{array}{cccc} CH_3CH_2 & & CH_3CH^2H_2 & \\ | & & | & \\ CH_3[CH_2]_2CHCH_2CHCH_2-OH & & & \\ 7 \quad 6\text{-}5 \quad\quad 4 \quad 3 \quad\quad 2 \quad 1 & & & \end{array}$$

4-Ethyl-2-(ethyl-*2,2-d₂*)-1-heptanol (CAS)    2-(2,2-$^2H_2$)Ethyl-4-ethylheptan-1-ol (IUPAC)

One must be aware of two differences between the systems. First, in the CAS system, isotopic substitution in a parent compound or parent substituent prefix may result in a change in the numbering of the parent structure from that of the unmodified compound.

---

[a] This name is the uninverted form of the CAS index name, which is carbonic-$^{18}O$ acid, $^{18}O$-ethyl $^{16}O$-methyl ester; the use of $^{16}O$ for the unmodified oxygen is misleading, since it should only be used for the pure oxygen-16 isotope and not the "natural" mixture. For oxygen, this is probably acceptable, but it might not be acceptable for all other elements. The inverted form for the modified Boughton name probably should be carbonic-$^{18}O$ acid, $^{18}O$-ethyl $O$-methyl ester.

[b] This name does not differentiate between the oxygen atoms of the peroxy group. The use of the isotopic symbol in a group formula would do so, as in ethaneperoxoic $O^{18}O$-acid and ethaneperoxoic $^{18}OO$-acid. CAS does not usually use italic letters, such as $O$-acid and $S$-acid, to differentiate between tautomeric thioacids or the different (thioperoxoic) acids. Any differentiation of these structures in esters is made through the use of italic letters preceding the name of the ester group as in $O$-methyl or $S$-methyl.

$$F_3CCH_2D$$
$$\overset{}{2}\ \overset{}{1}$$

2,2,2-Trifluoroethane-*1-d* (CAS)

$$CH_2{}^2HCF_3$$
$$\overset{}{2}\ \ \ \ \overset{}{1}$$

1,1,1-Trifluoro(2-$^2H_1$)ethane (IUPAC)

$$\overset{Cl}{\underset{}{|}}\ \overset{OH}{\underset{}{|}}$$
$$CH_3CHCHCD_2CH_3$$
$$\overset{}{5}\ \ \ \overset{}{4}\ \ \ \overset{}{3}\ \ \ \overset{}{2}\ \ \ \overset{}{1}$$

4-Chloro-3-pentan-*2,2-d₂*-ol (CAS)
(*not* 2-Chloro-3-pentan-*2,2-d₂*-ol)

$$\overset{HO}{\underset{}{|}}\ \overset{Cl}{\underset{}{|}}$$
$$CH_3C^2H_2CHCHCH_3$$
$$\overset{}{5}\ \ \ \ \overset{}{4}\ \ \ \ \overset{}{3}\ \ \ \overset{}{2}\ \ \ \overset{}{1}$$

2-Chloro(4,4-$^2H_2$)pentan-3-ol (IUPAC)

5-Chlorophenol-*2-d* (CAS)
(*not* 3-Chlorophenol-*6-d*)

3-Chloro(6-$^2H_1$)phenol (IUPAC)

Second, in a choice between otherwise identical parent structures, both systems prefer the parent with the maximum number of modified atoms. However, the CAS system chooses a parent with the earliest alphabetic isotope symbol, whereas the IUPAC system prefers a parent that has the isotope symbol with the higher atomic number. The CAS system prefers a chain with a deuterium isotope over an identical chain with a nitrogen isotope, and a chain with a nitrogen isotope is preferred over an identical chain with a tritium isotope. The IUPAC system prefers the opposite. If a further choice is needed, the CAS system prefers a lower mass number; the IUPAC system prefers a higher mass number.

$$\overset{CH_2T}{\underset{}{|}}$$
$$Cl-CH_2[CH_2]_2CH^{14}CH_3$$
$$\overset{}{5}\ \ \ \ \overset{}{4\text{-}3}\ \ \ \overset{}{2}\ \ \overset{}{1}$$

5-Chloro-2-(methyl-*t*)pentane-
1-$^{14}C$ (CAS)

$$\overset{CH_2{}^3H}{\underset{}{|}}$$
$$^{14}CH_3CH[CH_2]_2CH_2-Cl$$
$$\overset{}{5}\ \ \ \overset{}{4}\ \ \ \ \overset{}{3\text{-}2}\ \ \ \overset{}{1}$$

1-Chloro-4-($^3H_1$)methyl(5-$^{14}C$)-
pentane (IUPAC)

$$\overset{CH_2\text{-}{}^{81}Br}{\underset{}{|}}$$
$$CH_3CH_2CHCH_2\text{-}{}^{79}Br$$
$$\overset{}{4}\ \ \ \overset{}{3}\ \ \ \overset{}{2}\ \ \ \ \overset{}{1}$$

1-(Bromo-$^{79}Br$)-2-(bromo-$^{81}Br$-
methyl)butane (CAS)

$$\overset{CH_2\text{-}{}^{79}Br}{\underset{}{|}}$$
$$CH_3CH_2CHCH_2\text{-}{}^{81}Br$$
$$\overset{}{4}\ \ \ \overset{}{3}\ \ \ \overset{}{2}\ \ \ \ \overset{}{1}$$

1-($^{81}Br$)Bromo-2-[($^{79}Br$)bromomethyl]-
butane (IUPAC)

When stereochemical descriptors and isotopic descriptors occur at the same place in a name the former are cited first, even in names of stereoparents where the stereodescriptor refers to the parent structure (see chapter 30). This situation occurs much less frequently in the CAS system because the descriptor is placed after the part of the name to which it refers.

(*S*)-Ethan-*1-d*-ol (CAS)

(*S*)-(1-$^2H_1$)Ethanol (IUPAC)

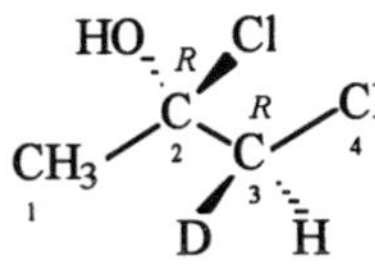

(2R,3R)-2-Chloro-2-butan-
3-d-ol (CAS)

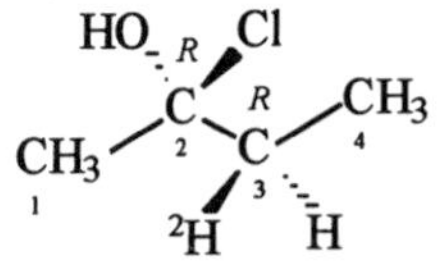

(2R,3R)-2-Chloro(3-$^2$H$_1$)butan-
2-ol (IUPAC)

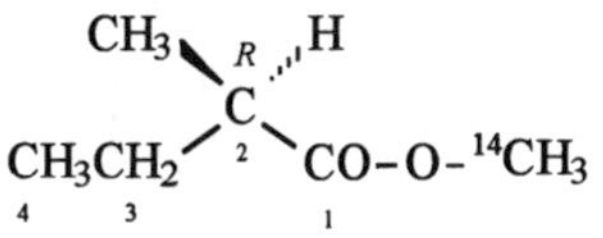

Methyl-$^{14}$C (R)-2-methylbutanoate (CAS)
($^{14}$C)Methyl (R)-2-methylbutanoate (IUPAC)

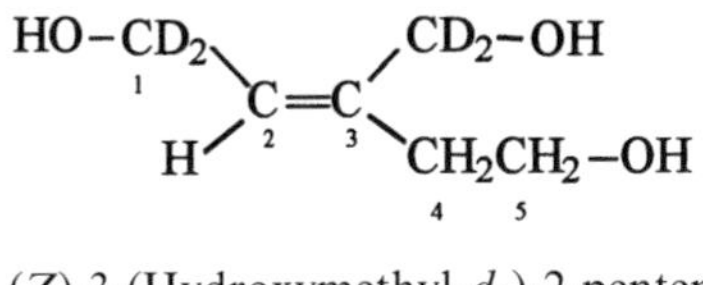

(Z)-3-(Hydroxymethyl-d$_2$)-2-pentene-
1,1-d$_2$-1,5-diol (CAS)

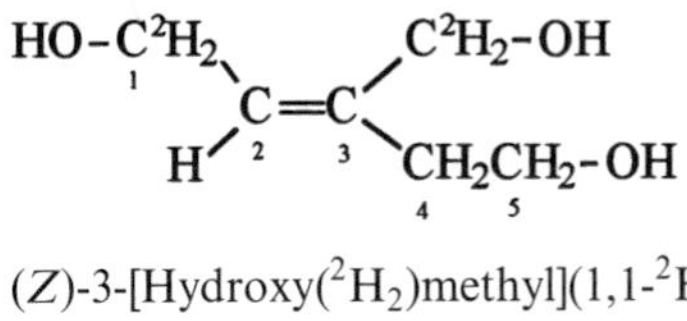

(Z)-3-[Hydroxy($^2$H$_2$)methyl](1,1-$^2$H$_2$)-
pent-2-ene-1,5-diol (IUPAC)

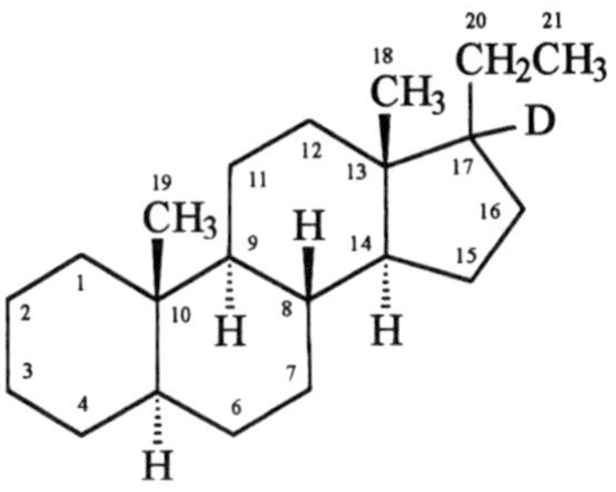

5α-Pregnane-17-d (CAS)

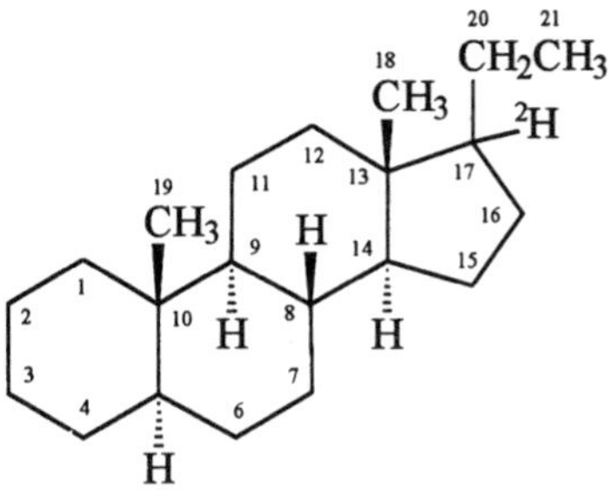

5α-(17-$^2$H$_1$)Pregnane (IUPAC)

## Isotopically Labeled "Compounds"

Although a mixture of an isotopically unmodified compound with one or more corresponding isotopically substituted compounds is really a mixture, for nomenclature purposes such mixtures are considered as if they were as compounds.

The CAS system does not differentiate between isotopically labeled compounds and isotopically substituted compounds, nor among different types of isotopically labeled compounds. Hence, a CAS name defines only an isotopically substituted compound or an isotopically substituted component of an isotopically labeled compound.

The IUPAC recommendations define several kinds of isotopically labeled compounds based on various compositions of these mixtures.

1. A *specifically labeled* isotopically modified compound is a mixture of a single isotopically substituted compound with the corresponding isotopically unmodified compound; both the position and number of each isotopic atom is known. This situation is indicated in both the formula and name in the same way as for isotopically substituted compounds (see above), except that the isotopic descriptor is enclosed in square brackets.

A mixture of the isotopically substituted methane $^{14}$CH$_4$ with isotopically unmodified methane CH$_4$ is the specifically labeled compound [$^{14}$C]H$_4$, which has the IUPAC name [$^{14}$C]methane. Schematically, this may be shown as follows:

| Isotopically substituted compound | + | Isotopically unmodified compound | = | Specifically labeled compound |
|---|---|---|---|---|
| $^{14}CH_4$ | | $CH_4$ | | $[^{14}C]H_4$ |
| $(^{14}C)$Methane | + | Methane | = | $[^{14}C]$Methane |

The same principles that apply to locants and subscripts in formulas and names for isotopically substituted compounds in the IUPAC system described above also apply to formulas and names of specifically labeled compounds.

$$CH[^2H_2]CH_2\text{-}[^{18}O]H \qquad\qquad [2,2\text{-}^2H_2]Ethan[^{18}O]ol$$
$$\;\;\;_2\qquad\;\;\;_1$$

$$CH_2[^2H_2]CH_2CH[^3H_1]\text{-}COOH \qquad [4,4\text{-}^2H_2,2\text{-}^3H_1]Butanoic\ acid$$
$$\;\;_4\qquad\quad_3\quad_2\qquad\;\;_1$$

$$[^{14}C]H_3CF_3 \qquad\qquad\qquad 1,1,1\text{-Trifluoro}[2\text{-}^{14}C_1]ethane$$
$$\;\;_2\quad\;_1$$

2. A *selectively labeled* isotopically modified compound is a mixture of more than one isotopically substituted compound with the corresponding isotopically unmodified compound in such a way that the position(s), but not necessarily the number, of each isotopically labeled atom is known; such a mixture may also be considered as a mixture of specifically labeled compounds. Selective labeling cannot occur in a compound with only one atom of an element that can be modified; only specific labeling can occur.

A unique structural formula for a selectively labeled compound cannot be written; the isotopic descriptor, preceded by necessary locants, is enclosed in square brackets and placed in front of the formula of the compound, or in front of parts of a formula that have independent numbering. A name for a selectively labeled compound is constructed in the same way as a name for a specifically labeled compound except that no multiplying subscripts follow the atomic symbols, other than in the very specific case described below, and identical locants referring to the same element are not repeated.

A mixture of the isotopically substituted methanes $CH_3{}^2H$, $CH_2{}^2H_2$, $CH^2H_3$, and $C^2H_4$, or any two or more of these, with isotopically unmodified methane $CH_4$ is the selectively labeled compound $[^2H]CH_4$, which has the IUPAC name $[^2H]$methane. Schematically, this may be shown as follows:

| Isotopically substituted compound | + | Isotopically unmodified compound | = | Specifically labeled compound |
|---|---|---|---|---|
| $CH_3{}^2H$ $(^2H_1)$Methane $CH_2{}^2H_2$ $(^2H_2)$Methane $CH^2H_3$ $(^2H_3)$Methane $C^2H_4$ $(^2H_4)$Methane (or any two or more of these) | + | $CH_4$ <br><br> Methane | = | $[^2H]CH_4$ <br><br> $[^2H]$Methane |

A mixture of the isotopically substituted propanoic acids $CH_3CH^2H_1-COOH$ $[(2\text{-}^2H_1)$propanoic acid$]$ and $CH_3C^2H_2-COOH$ $[(2,2\text{-}^2H_2)$propanoic acid$]$ with isotopically unmodified propanoic acid $CH_3CH_2-COOH$ is the selectively labeled compound $[2-^2H]CH_3CH_2-COOH$, which has the IUPAC name $[2-^2H]$propanoic acid (*not* $[2,2-^2H_2]$propanoic acid).

A mixture of the isotopically substituted methyl propanoates
$^{14}CH_3CH_2-CO-O-CH_3$ [methyl $(3-^{14}C)$propanoate],
$CH_3CH_2-CO-O-^{14}CH_3$ [$(^{14}C)$methylpropanoate], and
$^{14}CH_3CH_2-CO-O-^{14}CH_3$ [$(^{14}C)$methyl $(3-^{14}C)$propanoate],
or any two of these, with isotopically unmodified methyl propanoate
$CH_3CH-CO-O-CH_3$
is the selectively labeled compound
$[3-^{14}C]CH_3CH_2-CO-O-[^{14}C]CH_3$,
which has the IUPAC name $[^{14}C]$methyl $[3-^{14}C]$propanoate.

There is a special case of selective labeling in which the mixture with the corresponding isotopically unmodified compound is produced from two or more isotopically substituted compounds in which the position of isotopic substitution is known. In such cases, the position of the isotopic substitution in each may be indicated by subscripts to the atomic symbol; the absence of isotopic substitution at a particular position is indicated by the cipher 0. Subscripts referring to the same atomic symbol but in different isotopically substituted components are separated by a semicolon. Subscripts must be cited in the same order that they are considered for each isotopically substituted compound.

A mixture of the isotopically substituted ethanols $CH_2^2HCH_2-OH$ [$(2-^2H_1)$ethanol] and $CH^2H_2CH_2-OH$ [$(2,2-^2H_2)$ethanol], but *not* $C^2H_3CH_2-OH$ [$(2,2,2-^2H_3)$ethanol], with isotopically unmodified ethanol $CH_3CH_2OH$ is the selectively modified compound $[2-^2H_{1;2}]CH_3CH_2-OH$, which has the IUPAC name $[2-^2H_{1;2}]$ethanol.

A mixture of only the isotopically substituted ethanols $CH_3CH_2-^{18}OH$ [ethan$(^{18}O)$ol] and $CH^2H_2CH_2-OH$ [$(2,2-^2H_2)$ethanol] with isotopically unmodified ethanol $CH_3CH_2-OH$, is the selectively modified compound $[2-^2H_{0;2},^{18}O_{1;0}]CH_3CH_2-OH$, which has the IUPAC name $[2-^2H_{0;2},^{18}O_{1;0}]$ethanol.

A selectively labeled isotopically modified compound in which all atoms of an element are labeled, but not in the same isotopic amount, may be described by replacing the locants of the isotopic descriptor by the capital letter G to indicate "general" labeling. If all positions are labeled in the same isotopic amount, the capital letter U indicates "uniform" labeling. Locants may follow the U to describe labeling in the same isotopic amount at each of the indicated positions.

A mixture of the isotopically substituted propanoic acids
$CH_3CH_2-^{14}COOH$, $CH_3^{14}CH_2-COOH$, $^{14}CH_3CH_2-COOH$,
$^{14}CH_3CH_2-^{14}COOH$, $^{14}CH_3^{14}CH_2-COOH$, $CH_3^{14}CH_2-^{14}COOH$, and
$^{14}CH_3^{14}CH_2-^{14}COOH$
with isotopically unmodified propanoic acid $CH_3CH_2-COOH$ may be described by the formula $[G-^{14}C]CH_3CH_2-COOH$, which has the IUPAC name $[G-^{14}C]$propanoic acid.

A mixture of equal amounts of only the isotopically substituted propanoic acids $CH_3CH_2-^{14}COOH$, $CH_3^{14}CH_2-COOH$, and $^{14}CH_3CH_2-COOH$ with isotopically unmodified propanoic acid $CH_3CH_2-COOH$ may be described by the formula $[U-^{14}C]CH_3CH_2-COOH$ which has the IUPAC name $[U-^{14}C]$propanoic acid.

A mixture of equal amounts of only the isotopically substituted propanoic acids $CH_3CH_2-^{14}COOH$, $CH_3^{14}CH_2-COOH$, and $^{14}CH_3CH_2-COOH$ with isotopically unmodified propanoic acid $CH_3CH_2-COOH$ may be described by the formula $[U-1,3-^{14}C]CH_3CH_2-COOH$, which has the IUPAC name $[U-1,3-^{14}C]$propanonic acid.

3. A *nonselectively labeled* compound is a mixture of isotopically substituted compounds such that neither the position nor the number of the labeling isotopic atoms is known. A formula and name of a nonselectively labeled compound is derived in the same way as for selectively labeled compounds except that neither locants nor subscripts are cited.

$[^{14}C]CH_3CH_2-COOH$                    $[^{14}C]$Propanoic acid

$[^3H]C_6H_5-Cl$                                 Chloro$[^3H]$benzene

For nonselective labeling to occur, the element to be modified must be at different positions in a structure. Accordingly, nonselective labeling cannot occur when there are only atoms of an element to be modified at the same position in a structure. For example, $CH_4$ and $CCl_3CH_2-CCl_3$ cannot be nonselectively labeled with a hydrogen isotope. In such cases, only specific or selective labeling can occur.

4. An *isotopically deficient* compound is an isotopically labeled compound in which the isotopic content of one or more of the elements has been reduced to less than that which occurs naturally. This situation is indicated in both the formula and name by prefixing *def*- to the isotopic nuclide symbol without an intervening hyphen.

$[def^{13}C]CHCl_3$                              $[def^{13}C]$Chloroform

## Discussion

A nomenclature for isotopically modified compounds became necessary with the separation and study of the properties of the isotope of hydrogen with mass 2 by H. C. Urey[1,2]. This challenge was met promptly by the ACS Committee on Nomenclature, Spelling, and Pronunciation, in a report issued in 1935,[3] in which the use of D and T as symbols for the hydrogen isotopes of mass 2 and 3, respectively, was recommended, mainly because of convenience in formulas and in oral communication. For compounds, the committee recognized two methods. In the first, deuterium was to be expressed by the prefix "deuterio", thus "substituting" for normal hydrogen in the compound, giving names such as dideuteriomethane. In the other, the deuterium was to be indicated as part of a parent structure. Because of a strong desire to minimize the difference between the name of the unmodified compound and the modified compound, the committee recommended a modification of the system suggested by Boughton.[4] In 1947, this modified Boughton system was extended to include compounds in which other elements were isotopically modified.[5] CAS began using this extension in its index nomenclature in 1957 for both organic and inorganic compounds, modifying it in 1967 only to be consistent with the recommendations of the IUPAC Inorganic Nomenclature Commission[6] with regard to the citation of the mass number as a left superscript. A French Nomenclature Commission proposal to IUPAC in 1963 was very similar.[7]

An alternative system proposed jointly by the Editorial Board of the Biochemical Society and the Editors to the Chemical Society was formally recommended in the *Handbook for Chemical Society Authors* in 1960.[8] In this system, isotopic modification was indicated by the symbol of the isotope enclosed in square brackets inserted into the name directly preceding the part of the name to which it refers, for example, $[1-^2H_2]$ethanol. This system also appeared in the 1967 edition of the *ACS Handbook for Authors*[9] and was endorsed and further developed by the IUB Committee of Editors of Biochemical Journals and adopted by the Council of Biology Editors in the third edition (1972) of its style manual.[10] In 1979, the IUPAC Commission on Nomenclature of Organic Chemistry officially adopted recommendations for describing isotopically modified compounds based on this alternative system.

The existence of two methods for describing isotopic substitution should not be a problem; the choice is simply the one that better suits the purpose. For instance, putting the isotopic descriptor after the part of the name affected means that its characters have a lesser effect on alphabetization, an important consideration in an alphabetically organized index. On the other hand, the use of D and T in formulas can affect the position of the formula in a formula index. For example, $CH_2Cl_2$, might occur considerably earlier in a formula index than $CCl_2D_2$. And $CClT_3$ would be even later in the index.

Although the IUPAC recommendations for isotopically modified compounds provide an exact method for defining mixtures of isotopically substituted compounds, its usefulness at the present time appears to be limited. The composition of the mixture is usually unknown; the amount of isotopically substituted compound is usually very small, although measurable; and an exact knowledge of the composition of the the labeled compound is usually unimportant for the purpose of a publication.

Although not recommended, deuterated and tritiated compounds have been described by the prefixes deutero (or deuterio) and tritio, as in 4-deuterobenzoic acid and *N*-tritioacetamide. Deutero is preferred to deuterio mainly because of analogy to hydro. The prefix "per" has also been used to describe fully deuterated compounds, although its use is subject to the same considerations as for any other fully substituted compound (see chapter 10); perdeuterotoluene is quite acceptable, but perdeuterophenol would not be acceptable.

## Additional Examples

### *Isotopically Substituted Compounds*

1. $^{13}CHCl_3$

    Trichloromethane-$^{13}C$ (CAS)
    Chloroform-$^{13}C$
    Trichloro($^{13}C$)methane (IUPAC)
    ($^{13}C$)Chloroform (IUPAC)

2. $H^{18}O$  $^{14}CH_2-NH_2$

    1-(Aminomethyl-$^{14}C$)cyclopentanol-$^{18}O$ (CAS)
    1-[Amino($^{14}C$)methyl]cyclopentan($^{18}O$)ol (IUPAC)
    1-[Amino($^{14}C$)methyl]($^{18}O$)cyclopentanol (IUPAC)

3. H ND$_2$

H N$^2$H$_2$

2-Cyclohexen-1-amine-$d_2$ (CAS)

Cyclohex-2-en-1-($^2H_2$)amine (IUPAC)
($N,N$-$^2H_2$)Cyclohex-2-en-1-amine (IUPAC)

4.

    1-Cyclopenten-1-yl-$1$-$^{14}C$-benzene (CAS)
    (1-$^{14}C$)Cyclopent-1-en-1-ylbenzene (IUPAC)

5.

    1-Cyclopenten-1-ylbenzene-$1$-$^{14}C$ (CAS)
    Cyclopent-1-en-1-yl(1-$^{14}C$)benzene (IUPAC)

6.

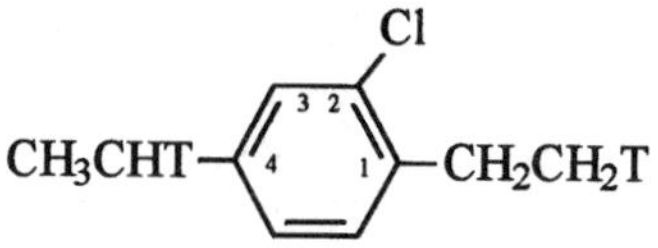

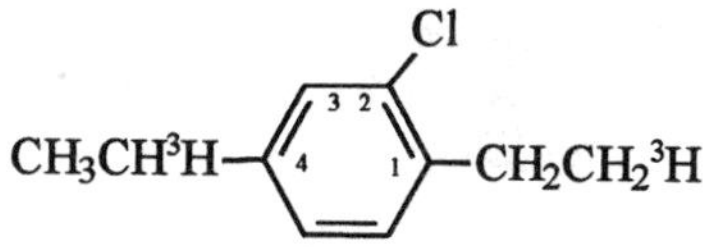

$CH_3CHT-$ [ring, Cl at top, positions 3 2 / 4 1] $-CH_2CH_2T$

2-Chloro-4-(ethyl-*1-t*)-1-
(ethyl-*2-t*)benzene (CAS)

$CH_3CH^3H-$ [ring, Cl at top, positions 3 2 / 4 1] $-CH_2CH_2^3H$

2-Chloro-4-[(1-$^3H_1$)ethyl]-1-
[(2-$^3H_1$)ethyl]benzene (IUPAC)

7.

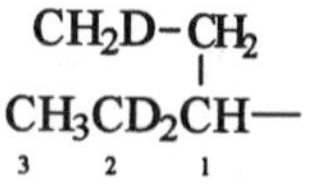

$CH_2D-CH_2$
$CH_3CD_2CH-$
   3    2    1

1-(Ethyl-*2-d*)propyl-*2,2-d$_2$* (CAS)

$CH_3C^2H_2CHCH_2CH_2{}^2H$
   5     4     3    2     1

(1,4,4-$^2H_3$)Pentan-3-yl (IUPAC)

8.    $CD_3-{}^{127}I$

Iodo-$^{127}I$-methane-*d$_3$* (CAS)

$C^2H_3-{}^{127}I$

($^{127}I$)Iodo($^2H_3$)methane (IUPAC)

9.

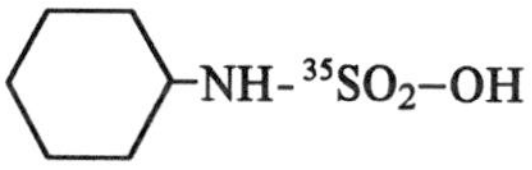

[cyclohexyl]$-NH-{}^{35}SO_2-OH$

Cyclohexylsulfamic-$^{35}S$ acid (CAS)
Cyclohexyl($^{35}S$)sulfamic acid (IUPAC)

10.

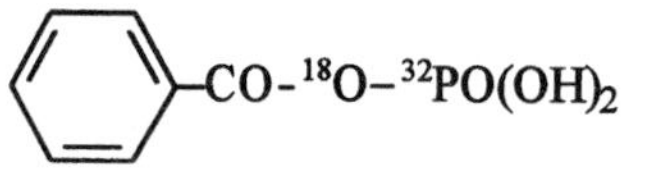

[phenyl]$-CO-{}^{18}O-{}^{32}PO(OH)_2$

Benzoic-$^{18}O$ acid $^{18}O$-monoanhydride with
   phosphoric-$^{32}P$ acid (CAS)
Benzoic ($^{32}P$)phosphoric ($^{18}O$)-mono-
   anhydride (IUPAC)

11.    $(CH_3CH_2-O)_2{}^{32}PO-CH_2-{}^{32}PO(O-CH_2CH_3)_2$

Tetraethyl methylenebis(phosphonate-$^{32}P$) (CAS)
Tetraethyl methylenebis[($^{32}P$)phosphonate] (IUPAC)

12.

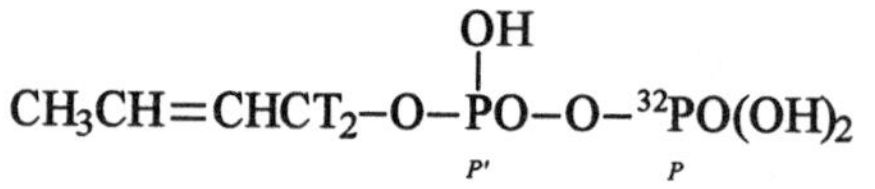

$CH_3CH=CHCT_2-O-\underset{P'}{PO}-O-\underset{P}{{}^{32}PO(OH)_2}$ with OH on the P'

$P'$-2-Butenyl-*1,1-t$_2$* trihydrogen
   diphosphate-$^{32}P$ (CAS)

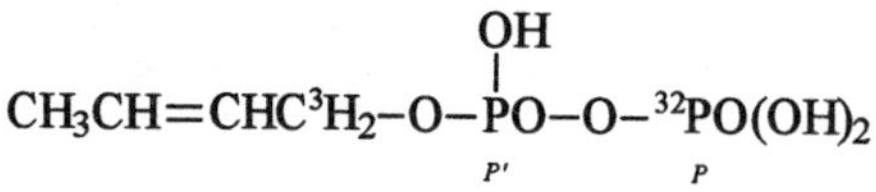

$CH_3CH=CHC^3H_2-O-\underset{P'}{PO}-O-\underset{P}{{}^{32}PO(OH)_2}$ with OH on the P'

$P'$-(1-$^3H_2$)But-2-enyl trihydrogen
   (P-$^{32}P$)diphosphate (IUPAC)

13.

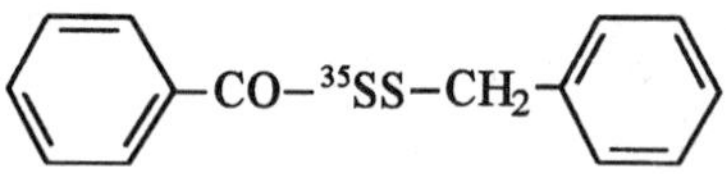

[phenyl]$-CO-{}^{35}SS-CH_2-$[phenyl]

$^{35}S{}^{32}S$-(Phenylmethyl)benzenecarbo-
   (dithioperoxoate-$^{35}S$) (CAS[a])
(S$^{35}$S)-Benzyl benzenecarbo[($^{35}$S)-dithio-
   peroxoate] (IUPAC)

---

[a] This name is the uninverted form of the CAS name, which is benzenecarbo(dithioperoxic-$^{35}S$) acid, $^{35}S{}^{32}S$-(phenylmethyl) ester; the use of $^{32}S$ for the unmodified sulfur atom is misleading, since it should only be used for the pure sulfur-32 isotope and not the "natural" mixture. The modified Boughton name probably should be benzenecarbo(dithioperoxic-$^{35}S$) acid, $^{35}SS$-(phenylmethyl) ester.

14.

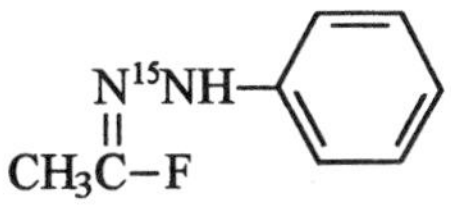

*N*-Phenylethanehydrazonoyl-*amino*-$^{15}N$ fluoride (CAS)
$N'$-Phenylethane($N'$-$^{15}$N)hydrazonoyl fluoride (IUPAC)

15.

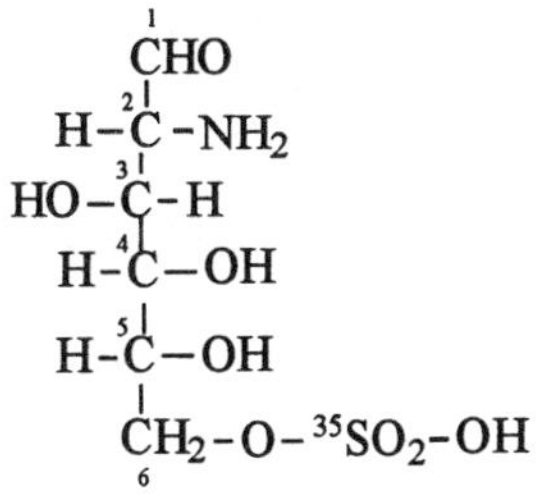

2-Amino-2-deoxy-D-glucose 6-(hydrogen
   sulfate-$^{35}S$) (CAS)
2-Amino-2-deoxy-D-glucose 6-[hydrogen
   ($^{35}$S)sulfate)] (IUPAC)

17.

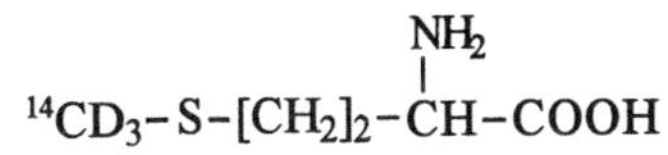

$$^{14}CD_3-S-[CH_2]_2-\overset{\overset{NH_2}{|}}{C}H-COOH$$

DL-Methionine-*methyl*-$^{14}C$-
*methyl*-$d_3$ (CAS)

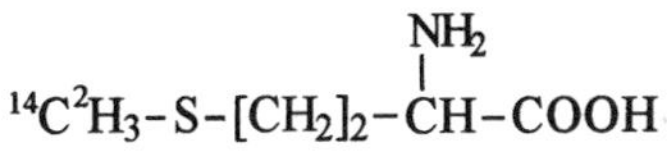

$$^{14}C^2H_3-S-[CH_2]_2-\overset{\overset{NH_2}{|}}{C}H-COOH$$

DL-[*methyl*-$^{14}$C,$^2$H$_3$]Methionine (IUPAC)

18.

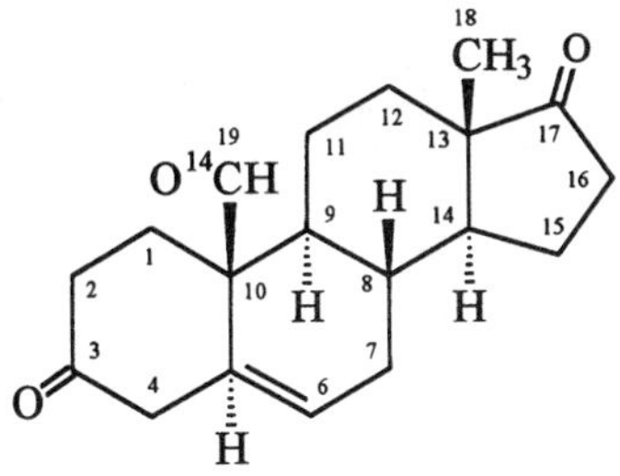

Androst-5-en-19-al-$^{14}C$ (CAS)
Androst-5-en-19-($^{14}$C)al (IUPAC)

## Specifically Labeled Isotopic Compounds

| *Isotopically substituted compound* | | *Isotopically unmodified compound* | | *Specifically labeled compound* |
|---|---|---|---|---|

19.

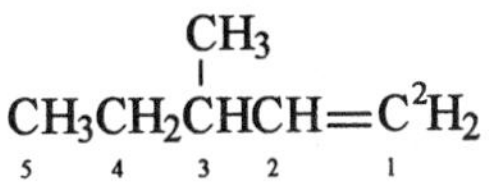

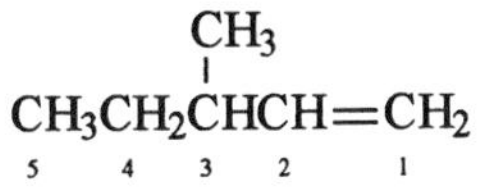

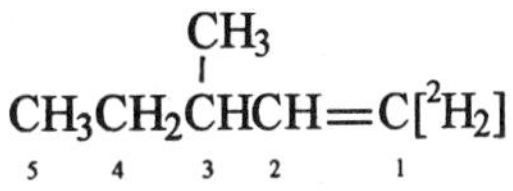

$$\overset{\overset{CH_3}{|}}{CH_3CH_2\underset{5\quad4\quad3}{CHCH}=\underset{2\quad1}{C^2H_2}}$$ + $$\overset{\overset{CH_3}{|}}{CH_3CH_2\underset{5\quad4\quad3}{CHCH}=\underset{2\quad1}{CH_2}}$$ = $$\overset{\overset{CH_3}{|}}{CH_3CH_2\underset{5\quad4\quad3}{CHCH}=\underset{2\quad1}{C[^2H_2]}}$$

3-Methyl[1,1-$^2$H$_2$]pent-1-ene

20.

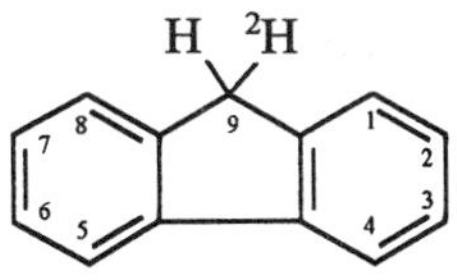

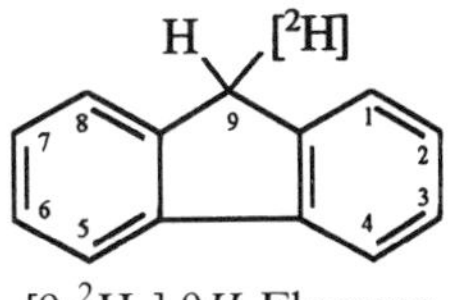

+ 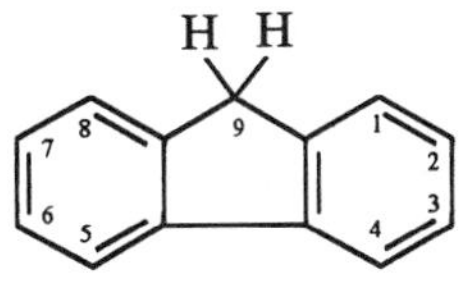 =

[9-$^2$H$_1$]-9*H*-Fluorene

21.

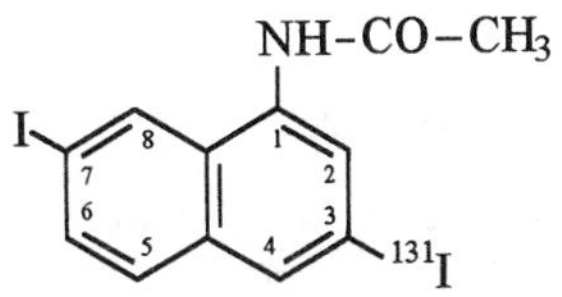

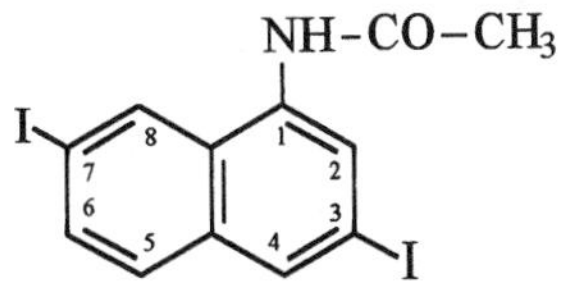

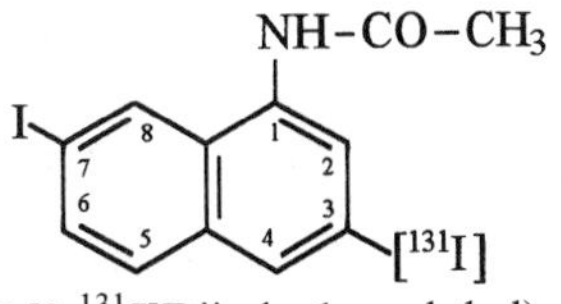

+ =

($N$-3,7-[3-$^{131}$I]Diiodo-1-naphthyl)-
acetamide

## Selectively Labeled Isotopic Compounds

| Isotopically substituted compound | + | Isotopically unmodified compound | = | Specifically labeled compound |
|---|---|---|---|---|

$CH_3{}^{14}CH_2{}^{14}CH_2-CO-NH_2$

22.  $CH_3{}^{14}CH_2CH_2-CO-NH_2$  +  $CH_3CH_2CH_2-CO-NH_2$  =  $[2,3-{}^{14}C]CH_3CH_2CH_2-CO-NH_2$

$CH_3CH_2{}^{14}CH_2-CO-NH_2$

(or any two of these)

$[2,3-{}^{14}C]$Butanamide

$CH_3{}^{14}CH_2CH_2-OH$

23.  $CH_3CH_2CH_2-{}^{18}OH$  +  $CH_3CH_2CH_2-OH$  =  $[2-{}^{14}C,{}^{18}O]CH_3CH_2CH_2-OH$

$CH_3{}^{14}CH_2CH_2-{}^{18}OH$

(or any two of these)

$[2-{}^{14}C,{}^{18}O]$Propan-1-ol

$CH_3CH{}^2H-OH$

24.  $CH{}^2H_2CH_2-OH$  +  $CH_3CH_2-OH$  =  $[1-{}^2H_{1;0},2-{}^2H_{0;2}]CH_2CH_2-OH$

(and only these two)

$[1-{}^2H_{1;0},2-{}^2H_{0;2}]$Ethanol

### REFERENCES

1. Urey, H. C.; Brickwedde, F. G.; Murphy, G. M. A Hydrogen Isotope of Mass 2. *Phys. Rev.* **1932**, [2], *39*, 164–165.
2. Urey, H. C.; Murphy, G. M.; Brichwedde, F. G. A Name and Symbol for H[2*]. *J. Chem. Phys.* **1932**, *1*, 512–513.
3. Report of the Committee on Nomenclature, Spelling, and Pronunciation, Report on Nomenclature of the Hydrogen Isotopes and their Compounds. *Ind. Eng. Chem.(News Ed.)* **1935**, *13*, 200–201.
4. Boughton, W. A. Naming Hydrogen Isotopes. *Science* **1934**, *79*, 159–160.
5. Otvos, J. W.; Wagner, C. D., A System for Isotopic Compounds. *Science* **1947**, *106*, 409–411.
6. International Union of Pure and Applied Chemistry, Inorganic Chemistry Division, Commission on Nomenclature of Inorganic Chemistry. *Nomenclature of Inorganic Chemistry, 1957 Rules;* Butterworths: London, 1958; American version with comments, *J. Am. Chem. Soc.* **1960**, *82*, 5523–5544.
7. Commission francaise de Nomenclature. Nomenclature des Molécules Organiques Marquées. *IUPAC Info. Bull.* December **1963**, No. 20, 27–30; *Bull. Soc. Chim. France,* **1967**, 3581-3582.
8. The Chemical Society. Collated Editorial Reports on Nomenclature, 1950–1959. *Handbook for Chemical Society Authors;* Special Publication No. 14; The Chemical Society: London, 1960; Chapter 10.
9. American Chemical Society. Handbook for Authors of Papers in the Journals of the Americal Chemical Society. American Chemical Society Publications: Washington, D.C., 1967, p. 42.
10. Council of Biology Editors, Committee on Form and Style. CBE Style Manual, 3rd ed. American Institute of Biological Sciences: Washington, D.C., 1972, pp. 100–101.

# 33

# Radicals, Ions, and Radical Ions

Nomenclature for organic structures that have unpaired electrons or formal charges on one or more atoms is discussed in this chapter. Also included are structures containing atoms with two or three nonbonding (free) electrons. These electrons may be paired with antiparallel spins: singlet, or unpaired with parallel spins: triplet, that are not nonbonding electron pairs ("lone pairs"). However, the nomenclature described in this chapter is not to be used to provide a differentiation between a singlet and a triplet electronic state; that distinction, if necessary, should only be made by adding "singlet" or "triplet" as a separate word to the name of the radical. Also included in this chapter are species that may be called "reactive intermediates"; however, this does not remove the need for unambiguous structurally descriptive names.

Names for salts using the names for the cations and anions given in this chapter are formed in the usual manner following the principles of binary nomenclature, as in sodium formate. Salts of acids and alcohols are discussed in the appropriate chapters. Examples of salts are also given in the additional examples section.

Until recently, nomenclature for radicals and ions in some areas has been inconsistent and poorly defined. In 1993, the IUPAC Commission on Nomenclature of Organic Chemistry published recommendations that provide a consistent description of radicals and ions in terms of classical valence structures.[1] Those recommendations did not cover delocalized or indefinite structures, an important area of radical and ion chemistry, except in so far as such species can be implied by the omission of locants and the use of descriptive phrases. Structural names for delocalized organic radicals and ions could be very useful, for instance, for describing ligands in coordination names; therefore, a brief description of the way these structures are treated by CAS is included in the discussion section.

In structural formulas, an unpaired electron is usually indicated by an enlarged dot written above or following the appropriate atom and its hydrogen atoms as in $CH_3 \cdot$ and $CH_3\overset{\bullet}{C}HCH_3$ and by an enlarged superscript dot following a molecular formula that is usually enclosed in square brackets as in $[C_6H_6]^{\bullet}$. Cations and anions are indicated in formulas by means of plus and minus signs placed above or as superscripts to the appropriate atom. Radicals that carry a net positive or negative charge are called radical cations and radical anions, respectively; in molecular formulas, the plus or minus sign follows the enlarged superscript dot as in $[CH_4]^{\bullet-}$ and $[C_6H_6]^{\bullet+}$. The opposite order, as in $[CH_4]^{-\bullet}$, is used in mass spectrometry.[2]

## Acceptable Nomenclature

For the purposes of organic nomenclature most radicals and ions are considered, at least formally, as derived by the removal of hydrogen atoms or by the addition or removal of hydrogen ions from atoms of a parent structure or from certain characteristic groups expressed as suffixes. Acyl radicals and corresponding cations are exceptions; they may be

371

considered as derived from acids by loss of OH groups in the form of hydroxyl radicals or
hydroxide ions.

## Radicals and Ions Derived from Parent Hydrides

Suffixes (and the prefix "ylo") are used to describe radicals and ions derived formally from
parent hydrides by loss of hydrogen atoms or by loss or gain of hydrogen ions. The suffixes are
summarized in table 33.1 and in chapter 3, table 3.7; each is illustrated in table 33.2 and
described in more detail in the following paragraphs.

### Table 33.1. Suffixes for Radicals and Ions from Parent Hydrides

| | Parent suffix | Suffix for substituent prefix |
|---|---|---|
| 1. Loss of $H^\bullet$ | -yl | —[a] |
| 2. Loss of $H^-$ | -ylium | -yliumyl |
| 3. Addition of $H^+$ | -onium | -oniumyl |
| | -ium | -iumyl |
| 4. Loss of $H^-$ | -ide | -idyl |
| 5. Addition of $H^-$ | -uide | -uidyl |

[a] This operation has no ending to use as a substituting group. The prefix "-ylo" is
   added in front of the name of the substituting group.

### Table 33.2. Formation of Names for Radicals and Ions and Corresponding Prefixes

| Loss of $H^\bullet$ | | |
|---|---|---|

$$CH_4 \xrightarrow{-H\bullet} CH_3\bullet \xrightarrow{-H\bullet} \bullet CH_2-$$

methane          methyl               -ylomethyl

$$PH_5 \xrightarrow{-H\bullet} PH_4\bullet \xrightarrow{-H\bullet} \bullet PH_3-$$

phosphorane       phosphoranyl           ylophosphoranyl
$\lambda^5$-phosphane   $\lambda^5$-phosphanyl    -ylo-$\lambda^5$-phosphanyl

Loss of $H^-$

$$CH_4 \xrightarrow{-H^-} [CH_3]^+ \xrightarrow{-H\bullet} [CH_2]^+-$$

methane          methylium            methyliumyl

$$PH_5 \xrightarrow{-H^-} [PH_4]^+ \xrightarrow{-H\bullet} [PH_3]^+-$$

phosphorane       phosphoranylium        phosphoranyliumyl
$\lambda^5$-phosphane   $\lambda^5$-phosphanylium   $\lambda^5$-phosphanyliumyl

Addition of $H^+$

$$SH_2 \xrightarrow{+H^+} [SH_3]^+ \xrightarrow{-H\bullet} [SH_2]^+-$$

sulfane          sulfonium            sulfoniumyl
                 sulfanium            sulfaniumyl

Loss of $H^+$

$$CH_4 \xrightarrow{-H^+} [CH_3]^- \xrightarrow{-H\bullet} [CH_2]^--$$

methane          methanide            methanidyl

Addition of $H^-$

$$BH_3 \xrightarrow{+H^-} [BH_4]^- \xrightarrow{-H\bullet} [BH_3]^--$$

borane           boranuide            boranuidyl

*Radicals and cations* derived by loss of one hydrogen atom or one hydride ion, respectively, from an atom of a parent hydride are named in one of two ways. If the loss is from the terminal atom of a saturated unbranched acyclic hydrocarbon, from an atom of a monocyclic hydrocarbon, or from the parent hydrides methane, silane, germane, stannane, or plumbane, the "ane" ending of the name of the hydride is replaced by -yl for radicals or -ylium for cations. Otherwise the final "e" of the name of the parent hydride is replaced by -yl or -ylium, as appropriate. The latter method may also be applied to all hydrocarbons, including those just mentioned above, for example, methanyl rather than methyl for $CH_3 \bullet$, but it is not applied to silane, germane, stannane, or plumbane (see Discussion). Neither method can be used with the names phosphine and arsine because of ambiguity with the acyl group names phosphinyl for $[H_2P(O)-]$, and arsinyl for $[H_2As(O)-]$ (see Discussion).

Many traditional names for radicals and cations are also acceptable.

| | | | |
|---|---|---|---|
| $CH_3CH_2\bullet$ | Ethyl | (cyclobutyl cation) | Cyclobutylium |
| $H_2NNH\bullet$ | Hydrazyl (traditional) Hydrazinyl Diazanyl | $H_3Ge^+$ | Germylium |
| $H_3Si\bullet$ | Silyl | $CH_3CH=CHCH_2^+$ 4 3 2 1 | But-2-en-1-ylium |
| $CH_3\overset{\bullet}{C}HCH_3$ 3 2 1 | Propan-2-yl 1-Methylethyl Isopropyl (traditional) | $\overset{+}{H_3SiSiHSiH_3}$ 3 2 1 | Trisilan-2-ylium |

Propan-2-yl, 2H-Pyran-6-yl, Bicyclo[2.2.1]heptan-2-ylium, Napththalen-4a(8aH)-ylium shown with structures.

| | |
|---|---|
| $H_2N\bullet$ | Aminyl[a] Azanyl[a] Amino (traditional) Amidogen (CAS) |
| $H_2P\bullet$ | Phosphanyl (*not* Phosphinyl) |

2-Methyl-2-propanylium
1,1-Dimethylethylium
*tert*-Butylium

The names carbene or methylene, nitrene or aminylene, and silylene (not silene) may be used for radicals derived by the loss of two hydrogen atoms from the parent hydrides methane, azane (ammonia), and silane, respectively; the radical derived from methane by the removal of three hydrogen atoms is named carbyne. All radicals, including the ones just described, derived by the loss of two or three hydrogen atoms from the same atom of a parent hydride are named by using -ylidene and -ylidyne in the same way as described above for -yl. The λ-convention (see chapter 2) can also be used to name such radicals (see Discussion).

| | | | |
|---|---|---|---|
| $(C_6H_5)_2C^{2\bullet}$ | Diphenylcarbene Diphenylmethylene Diphenylmethylidene Diphenyl-$\lambda^2$-methane | $C_6H_5-C^{3\bullet}$ | Phenylcarbyne Phenylmethylidyne Phenyl-$\lambda^1$-methane |

---

[a] The name "azane" is a systematic name for the parent hydride $NH_3$ (trivial name, ammonia); the name "aminyl" is derived from the name "amine", which is considered to be a parent hydride name for $NH_3$ as well as a class name.

$C_6H_5-CH_2-SiH^{2\bullet}$

Benzylsilylene
(Phenylmethyl)silylidene
Benzyl-$\lambda^2$-silane

Cyclohexylnitrene
Cyclohexylaminylene
Cyclohexylazanylidene
Cyclohexyl-$\lambda^1$-azane

$CH_3-P^{2\bullet}$

Methylphosphanylidene
Methylphosphinidine
   (traditional)
Methyl-$\lambda^1$-phosphane

$CH_3\overset{2\bullet}{\underset{3\quad 2\quad 1}{C}CH_3}$

Propan-2-ylidene
1-Methylethylidene
   (*not* Dimethylcarbene)

Cyclopenta-2,4-dien-1-ylidene
$1\lambda^2$-Cyclopenta-2,4-diene

$CH_3\overset{\overset{\displaystyle O}{\|}}{\underset{4\quad 3\quad 2\quad 1}{C}}\overset{2\bullet}{C}CH_3$

3-Oxobutan-2-ylidene
1-Methyl-2-oxopropylidene
   [*not* Acetyl(methyl)carbene or acetyl(methyl)methylene[a]]
3-Oxo-$\lambda^2$-butane

Naphthalen-1(2*H*)-ylidene
2*H*-1$\lambda^2$-Naphthalene

*Cationic heterocycles* resulting from skeletal bonding at a heteroatom rather than the addition or removal of hydrogen ions have been described by trivial names such as pyrylium, by replacement nomenclature with names such as 4a-thioniafluorene (see below), or simply by replacing the final "e" of the name of the heterocycle by -ium. The development of the $\lambda$-convention (see chapter 2) provides for a consistent systematic method for naming such cations based on the bonding number of the parent hydride as given by the $\lambda$ symbol. The well-established contracted and trivial names furylium, pyrylium, thiopyrylium, xanthylium, thioxanthylium, and quinolizinium are acceptable. IUPAC also accepts flavylium and chromenium; CAS uses indolizinium and pyrrolizinium.

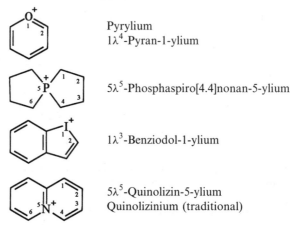

Pyrylium
$1\lambda^4$-Pyran-1-ylium

$5\lambda^5$-Phosphaspiro[4.4]nonan-5-ylium

$1\lambda^3$-Benziodol-1-ylium

$5\lambda^5$-Quinolizin-5-ylium
Quinolizinium (traditional)

*Cations* derived formally by the addition of one hydrogen cation (hydron[b]) to the mononuclear parent hydride of an element of the nitrogen, chalcogen, or halogen groups in its

---

[a]  These names are discouraged; in systematic organic nomenclature, except for radicals whose names require termination of the chain at position 1 and for acids and their analogs named by the Geneva rules (see, for example, chapter 11), breaking an unbranched chain of identical atoms in formation of a name is avoided.

[b]  Hydron is a generic name proposed by the IUPAC Commission on Physical Organic Chemistry[3] for the hydrogen cation, i.e., the naturally occurring mixture of protons, deuterons, and tritons. Accordingly, the name proton is restricted to the hydrogen cation with the mass number of 1, i.e., $^1$H.

Table 33.3. Mononuclear Parent Cations

| $H_4N^+$ | ammonium[a] | $H_3O^+$ | oxonium[b] | $H_2F^+$ | fluoronium |
|---|---|---|---|---|---|
| $H_4P^+$ | phosphonium | $H_3S^+$ | sulfonium | $H_2Cl^+$ | chloronium |
| $H_4As^+$ | arsonium | $H_3Se^+$ | selenonium | $H_2Br^+$ | bromonium |
| $H_4Sb^+$ | stibonium | $H_3Te^+$ | telluronium | $H_2I^+$ | iodonium |
| $H_4Bi^+$ | bismuthonium | | | | |

[a] The name nitronium has been used for $O_2N^+$ (which could be named nitrylium by applying recommendations in the 1979 edition of the IUPAC organic rules to the radicals given in the 1990 IUPAC inorganic recommendations[4]) and therefore cannot be used for naming $H_4N^+$.

[b] The name hydronium (hydrated $H^+$) has been used; it is not recommended.

standard valence state may be named by adding the ending -onium to a root for the name of the element as shown in table 33.3. "Carbonium" is not recommended (see Discussion). The addition of a hydrogen cation (hydron ) to any position of a neutral parent hydride, including the mononuclear parent hydrides mentioned just above and the mononuclear parent hydrides of the carbon family, gives cations that are named by adding -ium together with an appropriate locant, if necessary, to the name of the parent hydride with elision of any final "e" from the name of the parent hydride. The resulting cationic parent hydride is substituted by the usual methods of substitutive nomenclature.

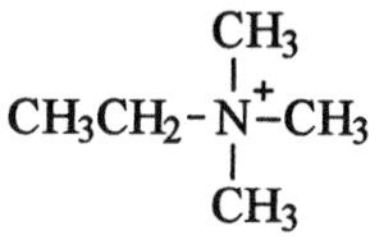

Ethyltrimethylammonium
Ethyltrimethylazanium

$CH_3CH_2C{\equiv}O^+$

Propylidyneoxonium
Propanylidyneoxonium

$(C_6H_5)_2I^+$

Diphenyliodonium
Diphenyliodanium

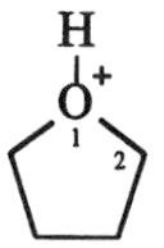

Tetrahydrofuranium
Oxolanium

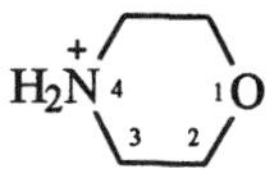

Morpholin-4-ium

$CF_3-\overset{+}{S}F_4$

Tetrafluoro(trifluoromethyl)-$\lambda^4$-sulfanium

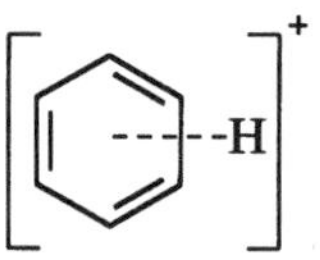

or $[C_6H_7]^+$

Benzenium

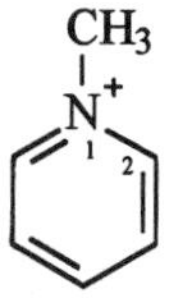

1-Methylpyridinium

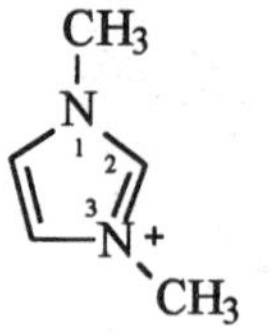

CH₃
│
N
[imidazolium ring structure]
N⁺
│
CH₃

1,3-Dimethyl-1*H*-imidazol-3-ium

*Anions* derived formally by the loss of a hydrogen cation (hydron) from, or by the addition of a hydride ion to, any position of a neutral parent hydride may be named by replacing the final "e" of the name of the parent hydride by -ide or -uide, respectively. The suffix -uide appeared for the first time in recommendations from IUPAC in 1993.[1] Previously there was no method for naming such anions by principles of substitutive nomenclature. Coordination nomenclature can also be used for simple mononuclear anions, such as $[H_4B]^-$ and $[Cl_6P]^-$.

Although the λ-convention (see chapter 2) can be used to raise the valence of a skeletal atom so that the suffix -ide can be used to generate the name of an anion with the right number of hydrogen atoms, this method may require the use of a bonding number that is unacceptably artificial. Thus, when there is a choice the standard valency state of the skeleton is preferred.

The trivial name "amide" is the recommended name for the anion $H_2N^-$; derivatives are named by substitutive principles. Although the names hydroxide for $HO^-$, hydroperoxide for $HOO^-$, hydrosulfide for $HS^-$, and hydrodisulfide for $HSS^-$ are acceptable, substitution of the hydrogen atom in these groups is not acceptable. Anions such as $CH_3-O^-$ and $CH_3-OO^-$ are considered as derived from alcohols (phenols) and hydroperoxides, for which see below. The trival name acetylide is retained for $[C_2]^{2-}$.

| | | |
|---|---|---|
| $C_6H_5-NH^-$ | Phenylamide<br>Phenylazanide | $(NC)_3C^-$  Tricyanomethanide |

$HC≡Si^-$          Methylidynesilanide

$(C_6H_5)_4B^-$          Tetraphenylboranuide (substitutive name)
Tetraphenylborate(1-) (coordination name)

$(CH_3)_4P^-$          Tetramethylphosphanuide
Tetramethyl-λ⁵-phosphanide  } (substitutive names)
Tetramethylphosphoranuide
Tetramethylphosphate(1-) (coordination name)

$CH≡CCH_2CH_2^-$          But-3-yn-1-ide

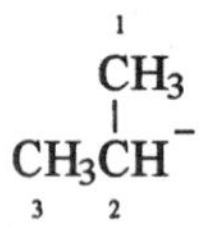

CH₃
│
CH₃CH⁻
3   2          Propan-2-ide
1-Methylethanide

[cyclopentadienide ring structure]          Cyclopenta-2,4-dien-1-ide

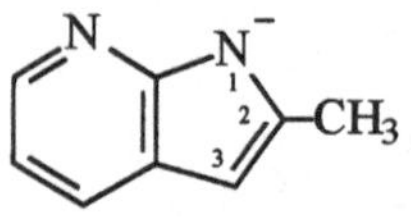

[pyrrolopyridine ring structure]—CH₃          2-Methyl-1*H*-pyrrolo[2,3-*b*]pyridin-1-ide

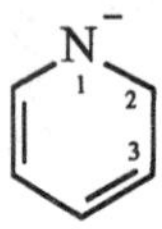

[pyridinide ring structure]          1(2*H*)-Pyridinide

*Two or more radicals* located on different skeletal atoms of a parent hydride are described by combining -yl, -ylidene, and -ylidyne, in that order; the final "e" of the name of the parent hydride, if present, is replaced by the combined ending. The -yl position is preferred for lowest locant, followed by the -ylidene site, and then the -ylidyne site. Similarly, *two identically charged ions* on different atoms of a parent hydride but derived by different methods are named with the suffixes -iumylium and -iduide, respectively. The numerical prefixes di-, tri-, and so on, describe two or more radical or ionic positions on the same parent hydride derived by the same method, except for -ylium positions for which the numerical prefixes bis-, tris-, and so on, and parentheses are used. Each position is identified by an appropriate locant, as needed.

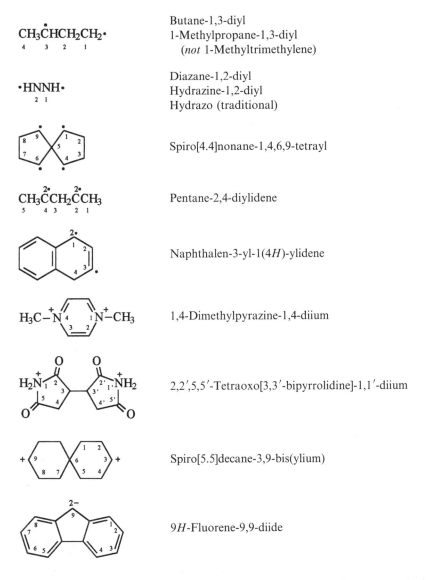

Butane-1,3-diyl
1-Methylpropane-1,3-diyl
(*not* 1-Methyltrimethylene)

Diazane-1,2-diyl
Hydrazine-1,2-diyl
Hydrazo (traditional)

Spiro[4.4]nonane-1,4,6,9-tetrayl

Pentane-2,4-diylidene

Naphthalen-3-yl-1(4*H*)-ylidene

1,4-Dimethylpyrazine-1,4-diium

2,2′,5,5′-Tetraoxo[3,3′-bipyrrolidine]-1,1′-diium

Spiro[5.5]decane-3,9-bis(ylium)

9*H*-Fluorene-9,9-diide

*Polyradicals, polycations, or polyanions,* that is, radicals or ions on different parent hydrides in the same structure, are named by means of multiplicative nomenclature (see chapter 3), where applicable, or by combinations of cationic or anionic endings (see also zwitterionic compounds, below). For numberings the endings -ylium and -uid have preference for lowest locants.

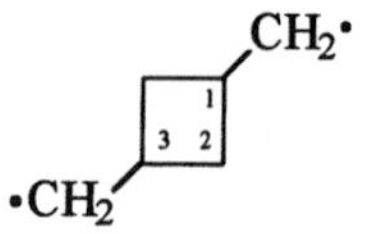

Cyclobutane-1,3-diyldimethyl

$(CH_3)_3\overset{+}{P}-CH_2CH_2-\overset{+}{P}(CH_3)_3$          Ethylenebis(trimethylphosphonium)

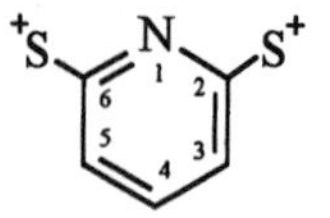

Pyridine-2,6-diylbis(sulfanylium)

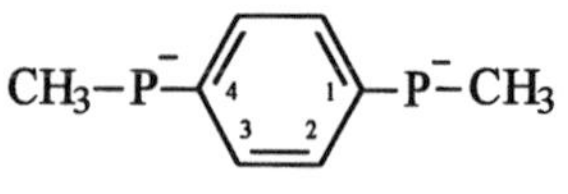

$CH_3-\overset{-}{P}-$     $-\overset{-}{P}-CH_3$          1,4-Phenylenebis(methylphosphanide)

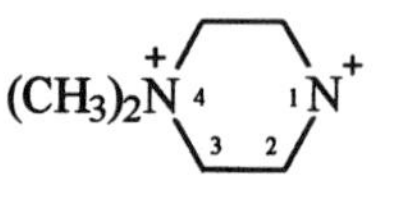

$(CH_3)_2\overset{+}{N}$     $\overset{+}{N}$          4,4-Dimethylpiperazin-4-ium-1-ylium

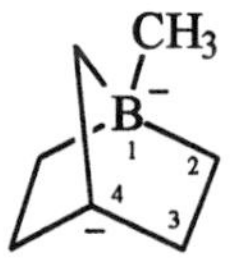

1-Methyl-1-borabicyclo[2.2.1]heptan-4-id-1-uide

Replacement nomenclature (see chapter 3) can be used to describe ionic positions of parent ions. Cationic replacement prefixes are formed by changing the final "a" of the replacement prefix for the corresponding neutral hetero atom to -onia, except for bismuth (the cationic replacement prefix for bismuth is bismuthonia), or by replacing the final "e" of the name of the corresponding mononuclear parent hydride with -ylia. These prefixes indicate cationic heteroatoms having a bonding number one greater or one less than that of the corresponding neutral heteroatom, respectively.

$-\overset{+}{\underset{|}{N}}-$   or   $=\overset{+}{N}-$   or   $=\overset{+}{N}=$   azonia          $-\overset{+}{N}-$   azanylia

$-\overset{+}{\underset{|}{S}}$   or   $=\overset{+}{S}-$   thionia          $-\overset{+}{B}-$   boranylia

$-\overset{+}{I}-$   iodonia          $-\overset{+}{\underset{|}{Si}}$   or   $=\overset{+}{Si}-$   silanylia

Anionic replacement prefixes are formed by converting the final "e" of the corresponding neutral parent hydride to -ida or -uida. Such prefixes indicate an anionic heteroatom with a bonding number one less or one greater than that of the corresponding neutral heteroatom, respectively.

$-\overset{-}{P}-$   phosphanida          $\overset{|^-}{>}\!P$   or   $=\overset{-}{\underset{|}{P}}$   or   $=\overset{-}{P}=$   phosphanuida

$-\overset{-}{\underset{|}{S}}$   or   $=\overset{-}{S}-$   sulfanuida $\lambda^4$-sulfanida          $-\overset{-}{\underset{|}{B}}-$   or   $=\overset{-}{\underset{|}{B}}$   or   $=\overset{-}{B}=$   boranuida

In names, ionic replacement prefixes follow the corresponding neutral replacement prefix. An ylia prefix follows an onia prefix, and an uida prefix follows an ida prefix. For numbering, ionic replacement prefixes are preferred to their corresponding neutral replacement prefixes and, when there is a choice, an ylia prefix is preferred to an onia prefix and an uida prefix is preferred to an ida prefix for lowest locants.

IUPAC prefers the suffixes -ium, -ylium, -ide, and -uide with neutral replacement names to the use of ionic replacement prefixes. CAS uses the aminium suffix (see cations derived from characteristic groups, below) rather than an azonia replacement prefix for acyclic nitrogen cations.

2-Methyl-2-phosphonianaphthalene
2-Methylisophosphinolinium
2-methyl-2λ⁵-isophosphinolin-2-ylium

2,3,7,8-Tetrachloro-5-arsoniaspiro[4.4]nona-2,7-diene
2,3,7,8-Tetrachloro-5λ⁵-arsaspiro[4.4]nona-2,7-dien-5-ylium

1*H*-3a,5a-Diazoniacyclohepta[*def*]phenanthrene

3,3,6,6,9,9,12,12-Octamethyl-3,6,9,12-tetrazoniatetradecane
3,3,6,6,9,9,12,12-Octamethyl-3,6,9,12-tetrazatetradecane-3,6,9,12-tetraium (IUPAC)
*N*,*N*′-Bis[2-(ethyldimethylammonio)ethyl]-*N*,*N*,*N*′,*N*′-tetramethyl-1,2-ethanediaminium (CAS)

8,13,13-Trimethyl-3,6,15-trioxa-18-thia-8-thionia-13-azoniaeicosane
8,13,13-Trimethyl-3,6,15-trioxa-8,18-dithia-13-azaicosane-8,13-diium (IUPAC)
4-[[(2-Ethoxyethoxy)methyl]methylsulfonio]-*N*-[[2-(ethylthio)ethoxy]methyl]-*N*,*N*-dimethyl-1-butanaminium (CAS)

1-Methyl-4-aza-1-azoniabicyclo[2.2.1]heptane
1-Methyl-1,4-diazabicyclo[2.2.1]heptan-1-ium

1,1-Bis(trifluoromethyl)-1-boranuidacyclohexane
1,1-Bis(trifluroromethyl)-1-boracyclohexan-1-uide

5,5-Dichloro-5λ⁵-phosphanuidaspiro[4.4]nonane
5,5-Dichloro-5λ⁵-phosphaspiro[4.4]nonan-5-uide

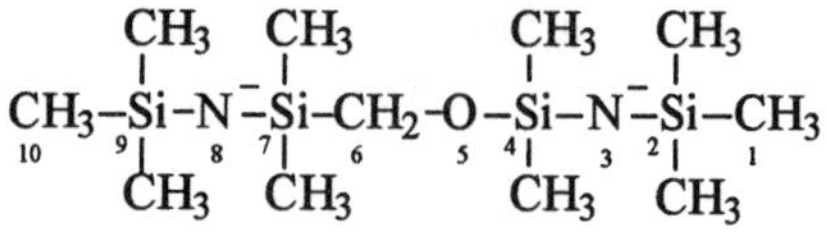

$$CH_3-\underset{\underset{\overset{|}{CH_3}}{10}}{\overset{\overset{|}{CH_3}}{\underset{9}{Si}}}-\underset{\underset{\overset{|}{CH_3}}{8}}{\overset{\overset{|}{CH_3}}{N}}-\underset{\underset{\overset{|}{7}}{Si}}{Si}-CH_2-\underset{6}{O}-\underset{\underset{\overset{|}{CH_3}}{5}}{\overset{\overset{|}{CH_3}}{Si}}-\underset{\underset{4}{N}}{N}-\underset{\underset{\overset{|}{CH_3}}{3}}{\overset{\overset{|}{CH_3}}{Si}}-CH_3$$

2,2,4,4,7,7,9,9-Octamethyl-5-oxa-3,8-diazanida-2,4,7,9-tetrasiladecane

2,2,4,4,7,7,9,9-Octamethyl-5-oxa-3,8-diaza-2,4,7,9-tetrasiladecane-3,8-diide

*Substituent prefixes* expressing radical or ionic sites must be used when all radical or all ionic sites cannot be expressed by the name of a parent radical or parent ion derived from a parent hydride.

A *radical site in a substituent* is described by the term "ylo"[1]. It is prefixed directly in front of the name of the substituent prefix together with a numerical prefix and appropriate locant(s). When the substituent is derived from a parent hydride, the prefix ylo indicates removal of a hydrogen atom from the parent substituent prefix.

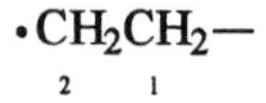

·CH₂CH₂—    2-Yloethyl

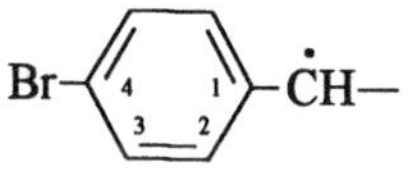

(4-Bromophenyl)ylomethyl
4-Bromo-α-ylobenzyl

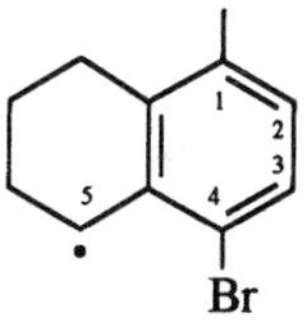

4-Bromo-5-ylo-1-naphthyl
4-Bromo-5-ylonaphthalen-1-yl

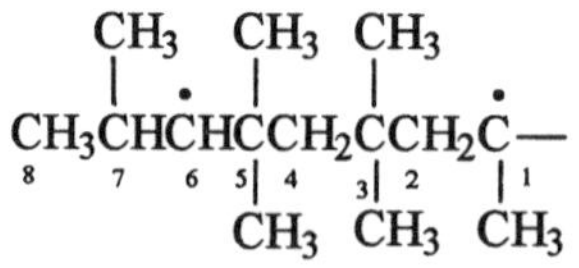

1,3,3,5,5,7-Hexamethyl-1,6-diylooctyl
4,4,6,6,8-Pentamethyl-2,7-diylononan-2-yl

IUPAC also recommends the use of ylo with prefixes that do not have hydrogen atoms that can be removed, for example, ylooxy for • O– (not ylohydroxy), and (ylooxy)carbonyl for • O–CO– (not ylocarboxy); in such cases the ylo prefix indicates only the presence of a free radical site.

*Prefixes* for substituents derived from *ionic parent hydrides* are systematically generated by adding suffixes such as -yl, -ylidene, and -diyl to the name of the parent ion. Alternatively, names for monovalent prefixes derived from the mononuclear parent cations of the Group 15, 16, and 17 elements (see table 33.3) are formed by changing the -onium ending of the parent cation to -onio. Also, names for monovalent prefixes derived from polynuclear parent cations with the free valence at a cationic hetero atom may be formed by changing the -ium suffix of the parent cation to -io. The traditional names for the chalcogen anions ⁻O –, ⁻S –, and so on, are oxido-, sulfido-, and so on.

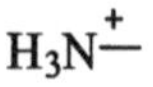

H₃N⁺⁻    Ammonio
        Ammoniumyl

C₆H₅–I⁺⁻    Phenyliodonio
           Phenyliodoniumyl

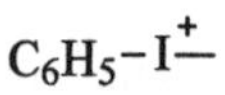

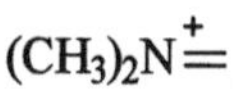

(CH₃)₂N⁺⁼    N-Methylmethaniminiumylidene
            Dimethylazaniumylidene

HN=NNH₂⁺⁻    2-Triazen-1-io
            2-Triazen-1-ium-1-yl

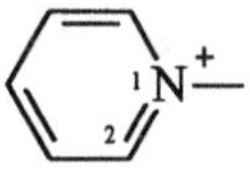

Pyridinio
Pyridin-1-ium-1-yl

Pyridin-1-ium-4-yl

$^-CH_2-$    Methanidyl                          $^-N=$    Amidylidene
                                                          Azanidylidene

$H_3B\overset{-}{-}$    Boranuidyl                 Cyclopentan-4-ide-1,2-diyl

$^{2-}N-$    Imidyl

$-N\overset{-}{-}$    Azanidediyl

Although any parent hydride can theoretically be chosen for the derivation of the parent radical, cation, or anion, the one containing the most radical, cationic, or anionic sites is generally chosen. Normally, -ylidyne radical sites are preferred to -ylidene sites which are preferred to -yl sites; -ylium sites are preferred to -ium sites; and -uide sites are preferred to -ide sites. Further choice goes to the parent radical or ion derived from the preferred parent hydride.

•$CH_2$                                            2-[3-Bromo-5-(ylomethyl)phenyl]ethylidene
           $-CH_2CH$:                                   (-ylidene site preferred to -yl site)
Br

•$CH_2CH_2-$ •                                     4-(2-Yloethyl)cyclohexyl
                                                      (ring preferred to chain)

•             $N$ •                                4-(4-Ylocyclohexyl)piperidin-1-yl
                                                      (heterocycle preferred to carbocycle)

$(CH_3)_3N^+-$ $-S(CH_3)_2^+$                       Dimethyl[[4-(trimethylammonio)phenyl]sulfonium
                                                      (sulfur cationic site preferred to nitrogen cationic site)

$(CH_3)_3N^+-$ $N^+$                               4-(Trimethylammonio)piperidin-1-ylium
                                                      ("ylium" site preferred to "ium" site)

                $CH_3$
                |
$^-C\equiv CCH_2CH_2-\overset{-}{B}-CH_3$           (3-Butyn-4-id-1-yl)trimethylboranuide
                |                                     ("uide" site preferred to "ide" site)
                $CH_3$

## Radicals and Ions Derived from Characteristic Groups

*Anions* derived by loss of a proton from acids and alcohols have long been named by suffixes, such as -oate, -carboxylate, -thioate, -sulfonate, -olate, and -thiolate, added to the name of a parent hydride. These anions were mentioned in connection with salts in chapters 12, 15, 23, and 24, and are included here only for the sake of completeness. Although radicals and ions with free valences or charges located on characteristic groups can be named on the basis of small parent hydrides as described above, just as for anions of acids and alcohols it would often be convenient to name such radicals and ions on the basis of larger parent structures.

Suffixes for radicals and ions derived from characteristic groups containing nitrogen are well established. They are summarized in table 33.4 and illustrated by examples in table 33.5.

Table 33.4. Suffixes for Radical and Ionic Sites on Nitrogen Characteristic Groups

| Characteristic group | *Radical sites* Loss of H• | *Cationic sites* Addition of $H^+$ | Loss of $H^-$ | *Anionic sites* Loss of $H^+$ |
|---|---|---|---|---|
| amine | -aminyl | -aminium | -aminylium | aminide |
| imine | -iminyl | -iminium | -iminylium | iminide |
| amide | -amidyl | -amidium | -amidylium | —[a] |
| imide | -imidyl | -imidium | -imidylium | —[a] |
| nitrile | —[b] | -nitrilium | —[b] | —[a] |

[a] The suffixes amidide and imidide are not recommended.
[b] The suffixes involving removal of a hydrogen atom or ion are not applicable.

Nitriles do not have hydrogen atoms and thus only the addition operations are feasible, and, of those, only the addition of a proton.

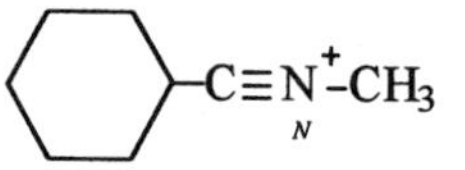

$$CH_3CH_2CH_2CH_2CH_2C\equiv \overset{+}{N}H$$

*N*-Methylcyclohexanecarbonitrilium          Hexanenitrilium

The suffix -ide is not recommended for use with the suffixes -amide or -imide. Anions derived from amides and imides are named on the basis of the parent anion amide or azanide, for $H_2N^-$, or an appropriate heterocycle.

$$CH_3-CO-\overset{-}{N}H$$

Acetylamide
Acetylazanide

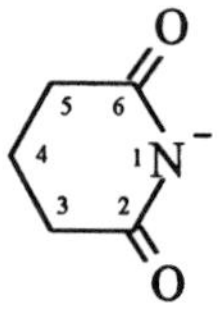

2,6-Dioxopiperidin-1-ide

Radical or ionic suffixes used with trivial names that imply the presence of two or more of a given characteristic group apply to all of them. With systematic names, except for imides, two or more identical radical or ionic suffixes are indicated by the prefixes bis-, tris-, and so on. Two radical or ionic sites derived from a diimide are named on the basis of an appropriate heterocycle.

•HN–CH₂CH₂–NH•          Ethane-1,2-bis(aminyl)

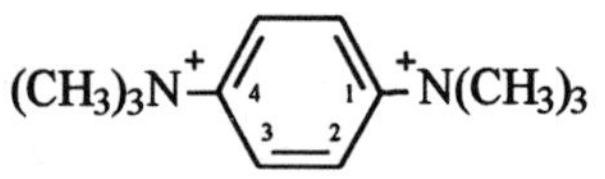

*N,N,N',N',N',N'*-Hexamethylbenzene-1,4-bis(aminium)

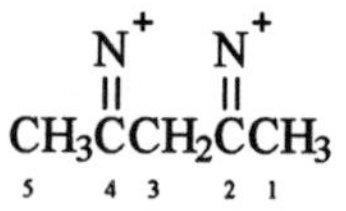

Pentane-2,4-bis(iminylium)

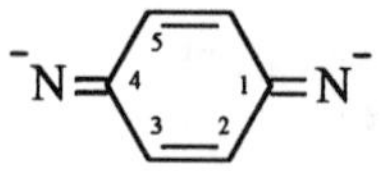

Cyclohexa-2,5-diene-1,4-bis(iminide)

•HN–CO–CH₂–CO—NH•          Malonamidyl
                            Propanebis(amidyl)

Table 33.5. Radicals and Ions on Nitrogen Suffix Groups

| | Loss of H• | Addition of H⁺ | Loss of H⁻ | Addition of H⁻ |
|---|---|---|---|---|
| $C_6H_5-NH_2$<br>benzenamine<br>aniline<br>phenylamine | $C_6H_5-NH\bullet$<br>benzenaminyl<br>anilinyl<br>phenylaminyl | $C_6H_5-NH_3^+$<br>benzenaminium<br>anilinium<br>phenylaminium<br>phenylammonium | $C_6H_5-NH^+$<br>benzenaminylium<br>anilinylium<br>phenylaminylium<br>phenylammoniumyl | $C_6H_5-NH^-$<br>benzenaminide<br>anilinide<br>phenylamide |
| $NH$<br>$\parallel$<br>$CH_3CCH_3$<br>3  2  1<br>propan-2-imine | $N\bullet$<br>$\parallel$<br>$CH_3CCH_3$<br>3  2  1<br>propan-2-iminyl | $NH_2^+$<br>$\parallel$<br>$CH_3CCH_3$<br>3  2  1<br>propan-2-iminium | $N^+$<br>$\parallel$<br>$CH_3CCH_3$<br>3  2  1<br>propan-2-iminylium | $N^-$<br>$\parallel$<br>$CH_3CCH_3$<br>3  2  1<br>propan-2-iminide |
| $CH_3-CO-NH_2$<br>acetamide | $CH_3-CO-NH\bullet$<br>acetamidyl | $CH_3-CO-NH_3^+$<br>acetamidium | $CH_3-CO-NH^+$<br>acetamidylium | — |
| 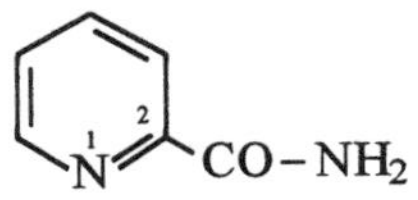 <br>pyridine-2-carboxamide | 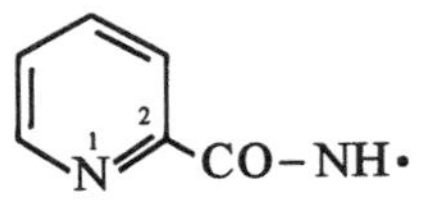 <br>pyridine-2-carboxamidyl | 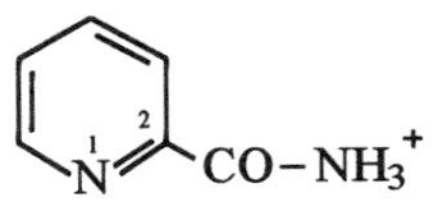 <br>pyridine-2-carboxamidium | 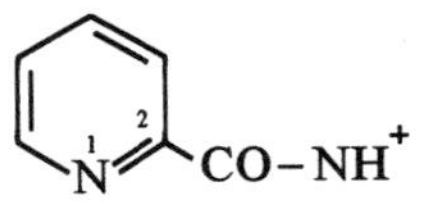 <br>pyridine-2-carboxamidylium | |
| 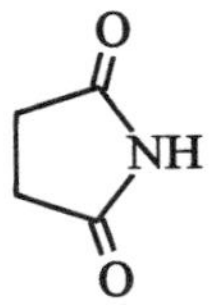 <br>butanimide<br>succinimide<br>2,5-pyrrolidinedione | 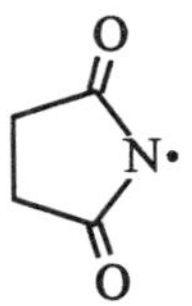 <br>butanimidyl<br>succinimidyl<br>2,5-dioxopyrrolidin-1-yl | 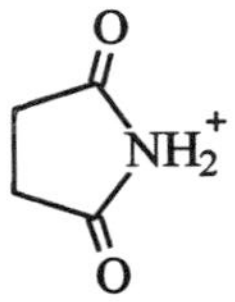 <br>butanimidium<br>succinimidium<br>2,5-dioxopyrrolidin-1-ium | 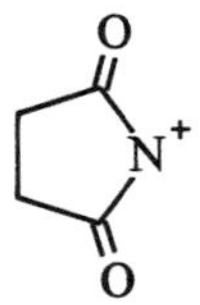 <br>butanimidylium<br>succinimidylium<br>2,5-dioxopyrrolidin-1-ylium | — |

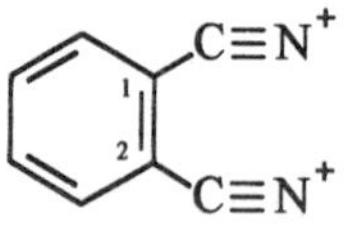

Phthalonitrilium
Benzene-1,2-bis(carbonitrilium)

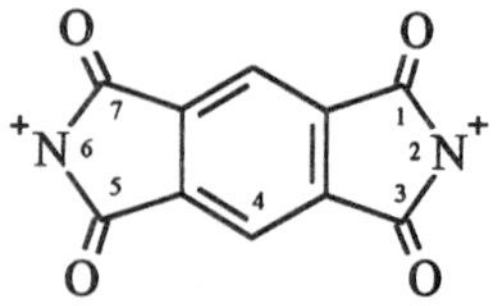

5,7-Dihydro-1,3,5,7-tetraoxobenzo[1,2-c:4,5-c']-
dipyrrole-2,6(1H,3H)-diylium

*Acyl radicals* are derived by the conceptual removal of the hydroxyl group from all acid characteristic groups expressed by the name of an organic acid. They are best named by means of an appropriate acyl substituent prefix. The corresponding cations derived by the conceptual removal of hydroxide ions are named by adding the ending -ium to the name of the acyl radical. These radicals and cations may also be named systematically as derivatives of a parent radical derived from a parent hydride.

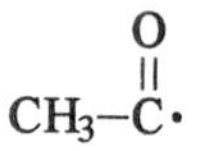

Acetyl
1-Oxoethyl

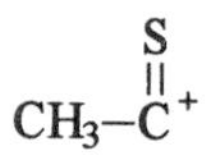

Thioacetylium
Ethanethioylium
1-Thioxoethylium

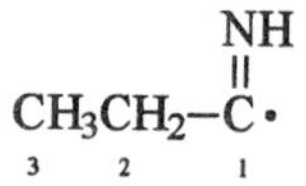

Propanimidoyl
1-Iminopropyl

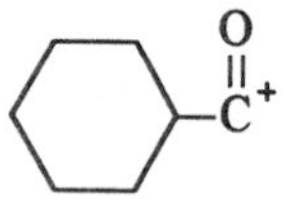

Cyclohexanecarbonylium
Cyclohexyloxomethylium

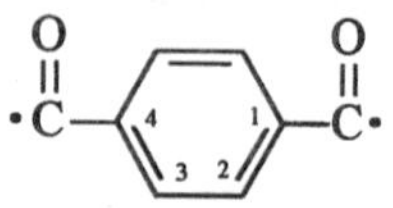

Benzene-1,4-dicarbonyl
1,4-Phenylenebis(carbonyl)
1,4-Phenylenebis(oxomethyl)
Terephthaloyl

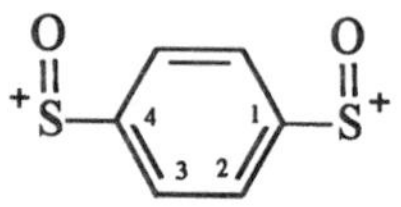

Benzene-1,4-disulfinylium
1,4-Phenylenebis(oxo-$\lambda^4$-sulfanylium)

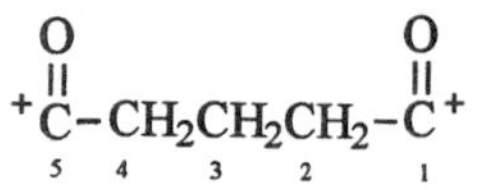

Glutarylium
Pentanedioylium
1,5-Dioxopentane-1,5-bis(ylium)

Acyl radicals and acylium ions derived from mononuclear oxo acids such as phosphinic acid and arsonic acid are named on the basis of the parent hydrides phosphane, phosphorane ($\lambda^5$-phosphane), arsane or arsorane ($\lambda^5$-arsane).

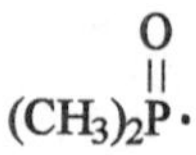

Dimethyloxophosphoranyl
Dimethyloxo-$\lambda^5$-phosphanyl
(*not* Dimethylphosphinyl or dimethylphosphinoyl)

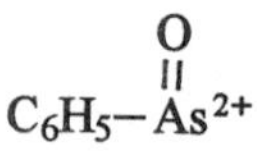

Oxophenylarsoranebis(ylium)
Oxophenyl-$\lambda^5$-arsanebis(ylium)
(*not* Phenylarsonoylium)

Radicals and cations conceptually derived by removal of hydrogen as an atom or as a hydride ion from an oxygen or sulfur atom of an acid, a peroxyacid, a hydroxy, or a hydroperoxy characteristic group, or a sulfur analog, are named by combining the name for a parent hydride or acid residue with an appropriate ending: -oxyl, -oxylium, -thiyl, -thiylium, -peroxyl (or -dioxyl), -peroxylium (or -dioxylium), -perthiyl- (or -perthiylium). For radical sites on sulfur atoms, IUPAC prefers names based on the parent hydride names sulfane and disulfane. Substituents are expressed in the usual way and, except for the contracted forms given above, the compound parent radical name is enclosed in parentheses or brackets. Names for selenium and tellurium analogs should be based on the parent hydride radical or ionic names selanyl, selanylium, tellanyl, and tellanylium.

$CH_3-O\cdot$   Methoxyl           $C_6H_5-O^+$   Phenoxylium

$CH_3CH_2-OO\cdot$   Ethylperoxyl / Ethyldioxyl

$C_6H_5-S^+$   Phenylthiylium / Phenylsulfanylium (*not* Phenylsulfenylium)

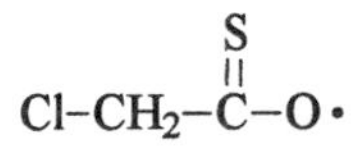

$Cl-CH_2-\overset{\overset{S}{\|}}{C}-O\cdot$   Chloro[(thioacetyl)oxyl][a]

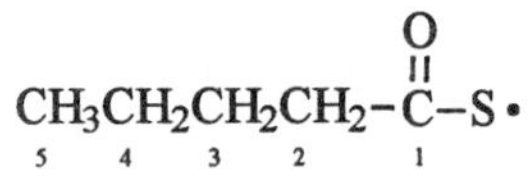

$CH_3CH_2CH_2CH_2-\overset{\overset{O}{\|}}{C}-S\cdot$   Pentanoylthiyl / (1-Oxopentyl)sulfanyl

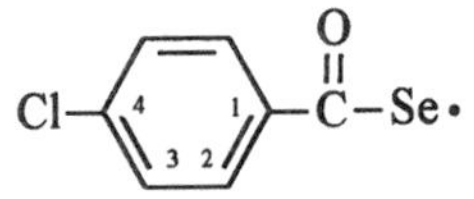

(4-Chlorobenzoyl)selanyl

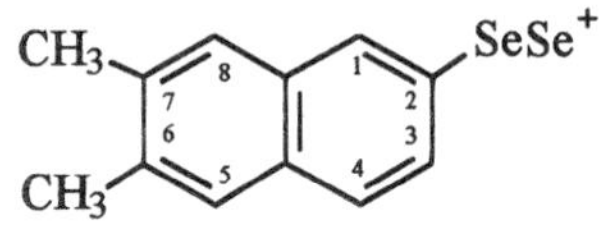

(6,7-Dimethylnaphthalen-2-yl)diselanylium

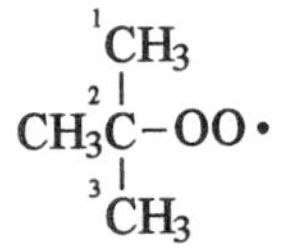

$CH_3\overset{1}{\underset{3}{\overset{2}{C}}}-OO\cdot$   2-Methyl(propan-2-ylperoxyl) / 2-Methyl(propan-2-yldioxyl) / *tert*-Butylperoxyl / *tert*-Butyldioxyl

$C_6H_5-OO^+$   Phenylperoxylium / Phenyldioxylium

Cations derived by addition of a proton to an acid oxygen atom or other chalcogen atom are named by changing the acid class name to acidium, diacidium, and so forth. They can also be named as derivatives of an appropriate mononuclear onium atom, such as oxonium (see table 33.3).

$CH_3-C(O,S)H_2^+$   Thioacetic acidium          $CH_3-\overset{\overset{O}{\|}}{C}-OH_2^+$   Acetyloxonium

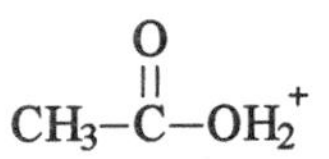

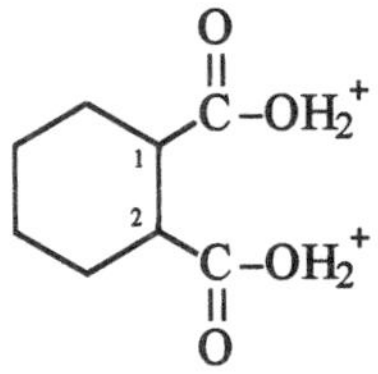

Cyclohexane-1,2-dicarboxylic diacidium / [Cyclohexane-1,2-diylbis(carbonyl)]bis(oxonium)

Some trivial names, such as urea and guanidine, imply a characteristic group; cations conceptually derived by the addition of a hydrogen cation have names such as uronium and guanidinium.

---

[a] Brackets are used to enclose the compound parent radical because parentheses are necessary to define unambiguously the acyl parent radical itself. Furthermore, the contracted name thioacetoxyl should not be used because it could describe a radical with the radical site on either the oxygen or the sulfur atom.

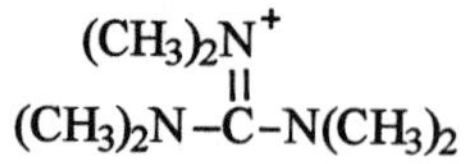

$$(CH_3)_2N^+$$
$$\|$$
$$(CH_3)_2N-C-N(CH_3)_2$$       Hexamethylguanidinium

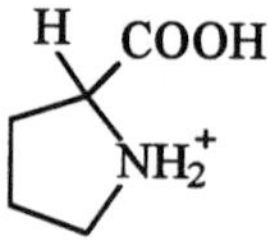

Prolinium

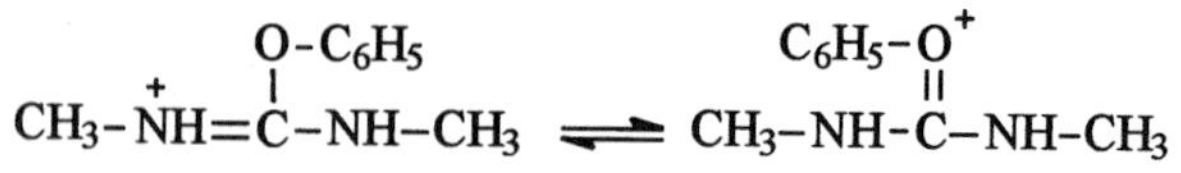

$$CH_3-\overset{+}{N}H=\overset{O-C_6H_5}{\underset{|}{C}}-NH-CH_3 \;\rightleftharpoons\; CH_3-NH-\overset{C_6H_5-\overset{+}{O}}{\underset{\|}{C}}-NH-CH_3$$

1,3-Dimethyl-2-phenyluronium

Systematic names for each of the uronium structures are $N$-methyl-1-(methylamino)-1-phenoxymethaniminium, $N$-[(methylamino)phenoxymethylidene]methanaminium, or methyl-[(methylamino)phenoxymethylidene]ammonium and [bis(methylamino)methylidene]phenyl-oxonium.

Diazonium ions are cations consisting of an $-N_2^+$ group attached to a parent hydride and are named by replacing the final "e" of the name of a parent hydride, if present, with -diazonium. The prefixes bis-, tris-, and so on, are used to describe two or more diazonium groups. Diazonium ions may also be named structurally on the basis of the parent cation diazenylium, $HN=N^+$.

$$C_6H_5-\overset{+}{N}_2$$       Phenyldiazonium
Phenyldiazenylium

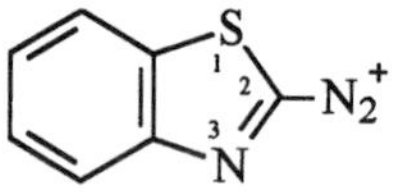

Benzothiazole-2-diazonium
Benzothiazol-2-yldiazenylium

Traditional names for radicals corresponding to some of the characteristic groups of substitutive nomenclature are given in table 33.6. The corresponding names hydroxylium for $HO^+$, hydroperoxylium for $HOO^+$, nitrylium for $O_2N^+$, and nitrosylium for $ON^+$ are well established.

Radicals and ions derived from tautomeric characteristic groups, such as -carbothioic acid and -imidamide (-amidine), for which specificity is required, and radical and ionic sites on characteristic groups other than those specifically mentioned above are named on the basis of another characteristic group or on the name of a parent hydride radical or ion.

### Table 33.6. Traditional Names for "yl" Radicals Corresponding to Some Characteristic Groups

| | | | |
|---|---|---|---|
| HOOC• | carboxyl[a] | OC$^{2\bullet}$ | carbonyl |
| HO• | hydroxyl[a] | OS$^{2\bullet}$ | sulfinyl |
| HOO• | hydroperoxyl[a] | O$_2$Se$^{2\bullet}$ | sulfonyl |
| | hydrodioxyl[a] | OSe$^{2\bullet}$ | seleninyl |
| NO• | nitrosyl | O$_2$Se$^{2\bullet}$ | selenonyl |
| O$_2$N• | nitryl | OTe$^{2\bullet}$ | tellurinyl |
| OCl• | chlorosyl[b] | O$_2$Te$^{2\bullet}$ | telluronyl |
| O$_2$Cl• | chloryl[b] | | |
| O$_3$Cl• | perchloryl[b] | | |
| NC• | cyanyl | | |
| CN• | isocyanyl | | |
| N$_3$• | azidyl | | |

[a] Substitution of hydrogen is not allowed.
[b] Similarly for other halogen analogs.

$C_6H_5-CO-NH-O\cdot$          Benzoylaminoxyl

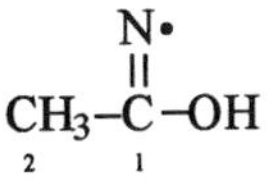

$$\underset{2 \quad 1}{CH_3 - \overset{\overset{N\cdot}{\|}}{C} - OH}$$

1-Hydroxy-1-ethaniminyl
(1-Hydroxyethanylidene)azanyl
(1-Hydroxyethylidene)aminyl

$$\underset{2 \quad 1}{CH_3CH_2CH_2-CO-NHNH\cdot}$$

2-Butanoylhydrazyl
2-Butanoylhydrazinyl

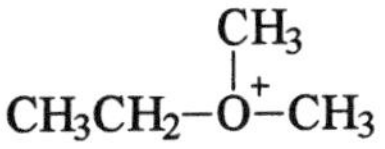

$$CH_3CH_2-\overset{\overset{CH_3}{|}}{\overset{+}{O}}-CH_3$$

Ethyldimethyloxonium

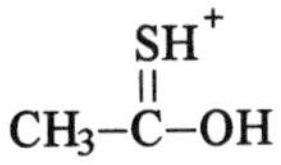

$$CH_3-\overset{\overset{SH^+}{\|}}{C}-OH$$

(1-Hydroxyethylidene)sulfonium
(1-Hydroxyethylidene)sulfanium

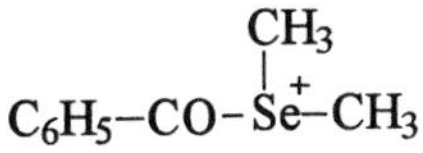

$$C_6H_5-CO-\overset{\overset{CH_3}{|}}{\overset{+}{Se}}-CH_3$$

Benzoyldimethylselenonium
Benzoyldimethylselanium

$CH_3-CO-\overset{+}{Cl}-CH_3$      Acetyl(methyl)chloronium

$$\underset{2 \quad 1}{H_2N\overset{+}{N}-CO-CH_3}$$

1-Acetyldiazan-1-ylium
1-Acetylhydrazin-1-ylium

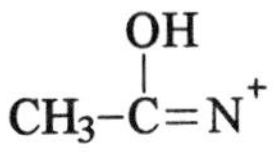

$$CH_3-\overset{\overset{OH}{|}}{C}=\overset{+}{N}$$

(1-Hydroxyethylidene)azanylium
(1-Hydroxyethylidene)aminylium
(1-Hydroxyethylidene)nitrenium
1-Hydroxyethaniminium

$CH_3-CO-N^{2+}$      Acetylazanebis(ylium)

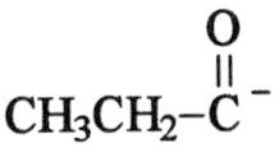

$$CH_3CH_2-\overset{\overset{O}{\|}}{C}^-$$

1-Oxopropan-1-ide

$C_6H_5-CO-NH^-$      Benzoylamide
Benzoylazanide

$C_6H_5-O-S^-$      Phenoxysulfanide

$(CH_3)_3C-OOO^-$      *tert*-Butyltrioxidanide
(2-Methylpropan-2-yl)trioxidamide

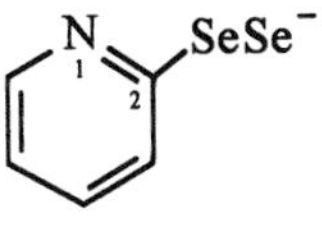

(2-Pyridyl)diselanide

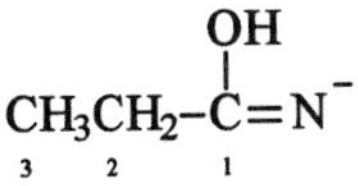

$$\underset{3 \quad 2 \quad 1}{CH_3CH_2-\overset{\overset{OH}{|}}{C}=N^-}$$

(1-Hydroxypropylidene)amide
(1-Hydroxypropan-1-ylidene)azanide

    Anionic suffixes, such as -carboxylate and -olate, are added to the names of cationic parent hydrides to name the resulting zwitterionic compounds (see below) as well as to anionic parent hydrides. Cationic suffixes are added to cationic parent hydride names, but are *not* added to anionic parent hydride names.

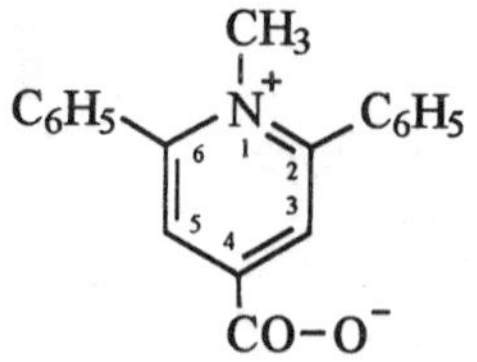

1-Methyl-2,6-diphenylpyridin-1-ium-4-carboxylate

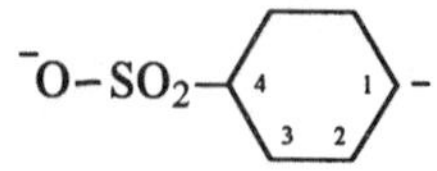

Cyclohexan-1-ide-4-sulfonate

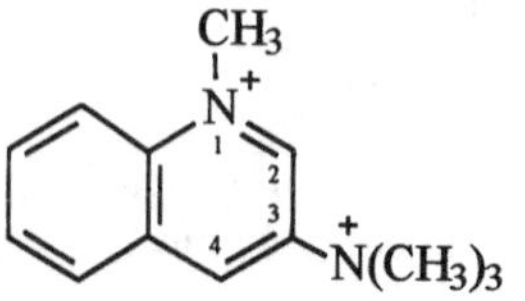

$N,N,N$-Tetramethylquinolin-1-ium-3-aminium

*Zwitterionic compounds* are neutral compounds with compensating ionic sites. They are named by adding cationic or anionic suffixes to the name of a parent hydride or by attaching ionic replacement prefixes to the name of a parent hydride. A cationic suffix or prefix is cited before an anionic suffix or prefix. The order of the suffixes from the end of the parent hydride name is -ium, -ylium, -ide, -uide; and the order of the ionic replacement prefixes in front of the parent hydride name is -onia, -ylia, -ida, -uida. The preference for low locants is the reverse of the order of citation, that is, -uide or -uida is the most preferred.

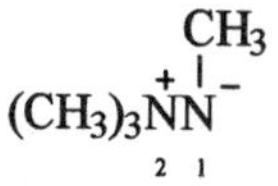

1,2,2,2-Tetramethylhydrazin-2-ium-1-ide
1,2,2,2-Tetramethyldiazan-2-ium-1-ide

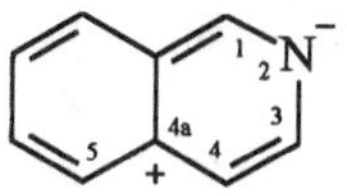

Isoquinolin-4a(2$H$)-ylium-2-ide

When all cationic and anionic sites of a zwitterion cannot be expressed by means of a single parent structure, including ionic suffixes, the zwitterion is named by expressing parts of the structure containing cationic sites as substituent prefixes to the name of an anionic parent structure.

(Trimethylammonio)acetate
(Trimethylammoniumyl)acetate

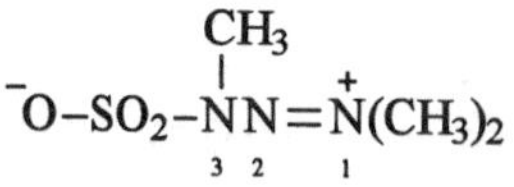

1,1,3-Trimethyltriaz-1-en-1-ium-3-sulfonate

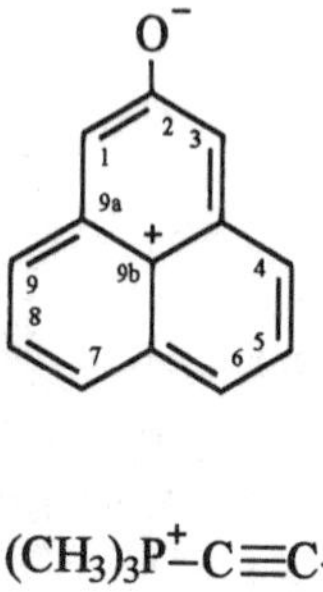

9b$H$-Phenalen-9b-ylium-2-olate

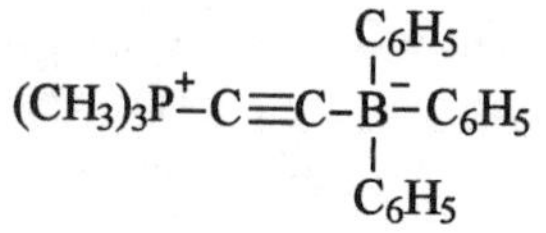

Triphenyl[(trimethylphosphonio)ethynyl]boranuide
Triphenyl[(trimethylphosphoniumyl)ethynyl]boranuide

$$CH_3-\overset{+}{N}N\overset{\overset{\displaystyle CH_3}{|}}{H}N-CH_2C\equiv C^-$$

3-(1,3-Dimethyltriazan-3-ylium-1-yl)prop-1-yn-1-ide

*Radical ions* are structures with at least one radical site and one cationic and/or anionic site, on the same or different atoms of a parent hydride or suffix. They are named by adding a radical suffix, such as -yl, -ylidene, and -diyl, to the name of the parent ion or zwitterion, eliding a final "e" of the name of the parent ion or zwitterion, if present, before "y". Radical sites are given preference for the assignment of low numbers.

$H_2C^{\bullet+}$ — Methaniumyl

$(CH_3)_3N^{\bullet+}$

*N,N*-Dimethylmethanaminiumyl
Trimethylammoniumyl
Trimethylazaniumyl

$\overset{+}{C}H_2CH=CHCH_2\bullet$ — But-2-en-4-ylium-1-yl

$CH_3-Si^{(2\bullet)+}$ — Methylsilyliumylidene

$CH_3-C\equiv N^{\bullet+}$ — Acetonitriliumyl

$$\overset{\quad Cl\quad Cl}{\overset{|\quad\ |}{{}^-CCl_2C=\overset{\bullet}{C}CCl_2}}$$

Hexachlorobut-2-en-4-id-1-yl

$(CH_3)_3B^{\bullet-}$ — Trimethylboranuidyl

## Discussion

Some special requirements are encountered in extending the principles of systematic nomenclature to radicals and ions. Unpaired electrons and ionic charges are often considered to be delocalized. Flexibility in name specificity is often desirable to allow some degree of structural ambiguity. Because organic nomenclature is based on classical structures, unambiguous names for parent structures are the basis for naming derivatives, including radicals and ions. Some ambiguity can be introduced by the deliberate omission of locants for unsaturation and/or positions of radical and/or ionic sites. If the name of a single canonical form is inadequate to represent the actual structure, descriptive words, such as "delocalized" or "hybrid" may be added to the name. Although there are no official rules for naming radicals and ions delocalized over several or all atoms of a structure, there are a few useful techniques that can be applied.

Delocalization in names of structures with a single $CH_2$ group in an otherwise conjugated double bond structure can be indicated simply by the omission of locants for double bonds, indicated hydrogen, or position of the radical or ionic site.

$C_{10}H_7\bullet$

naphthalenyl
naphthyl

(x) = • : cyclopentadienyl
(x) = + : cyclopentadienylium
(x) = − : cyclopentadienide

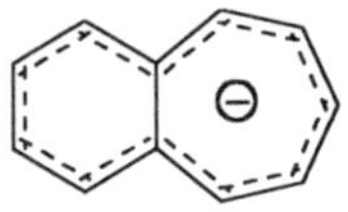

Benzocycloheptenide

$$CH_2 = CH = \overset{\bullet}{C}H = CH = CH_2 \qquad \text{Pentadienyl}$$

The position of a radical or ionic site is always 1 in a monocyclic system. Thus, in a delocalized cyclohexadienyl radical (or ion) the $CH_2$ group will have the locant 6.

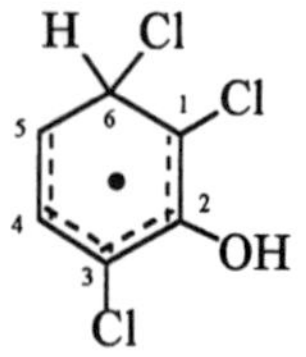

1,3,6-Trichloro-2-hydroxycyclohexadienyl

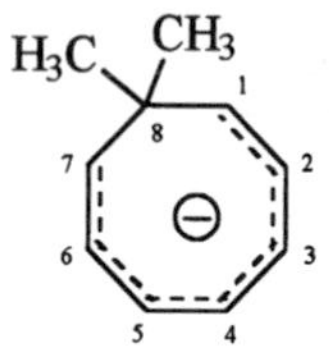

8,8-Dimethylcyclooctatrienide

Delocalization in fully conjugated ring systems can be indicated by the use of a question mark instead of a locant for hydro prefixes.

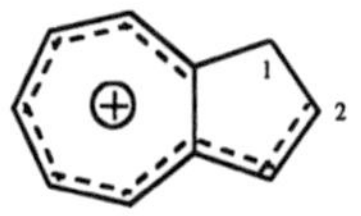

1,?-Dihydroazulenylium

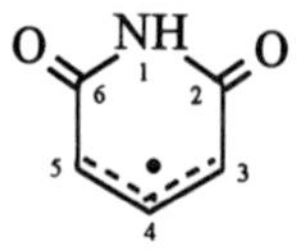

1,2,6,?-Tetrahydro-2,6-dioxopyridinyl

Partial delocalization and delocalization in nonconjugated systems, in nonplanar systems, and in bridged or "nonclassical systems" can be indicated by attaching the prefix *deloc-*, followed by appropriate locants, to the structurally definitive name of the corresponding radical or ion.

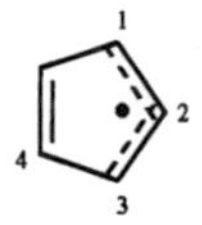

(*deloc*-1,2,3)-Cyclopenta-2,4-dien-1-yl

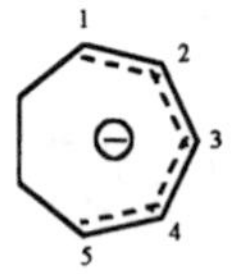

(*deloc*-1,2,3,4,5)-Cyclohepta-2,4-dien-1-ide

$$H_2C = CHCH_2CH = \overset{+}{C}H = CH_2 \qquad (deloc\text{-}1,2,3)\text{-Hexa-2,5-dien-1-ylium}$$

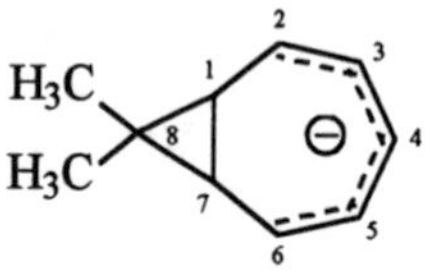

(*deloc*-2,3,4,5,6)-8,8-Dimethylbicyclo-[5.1.0]octa-3,5-dien-2-ide

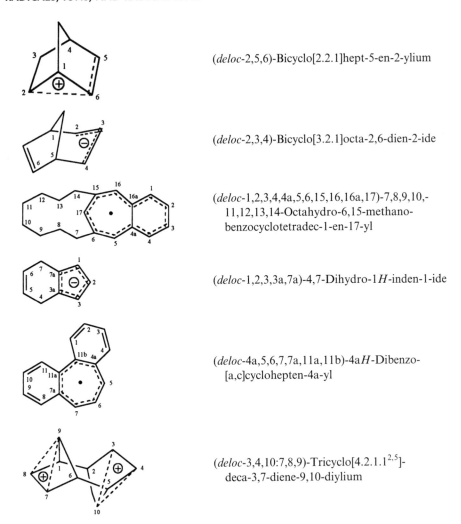

(*deloc*-2,5,6)-Bicyclo[2.2.1]hept-5-en-2-ylium

(*deloc*-2,3,4)-Bicyclo[3.2.1]octa-2,6-dien-2-ide

(*deloc*-1,2,3,4,4a,5,6,15,16,16a,17)-7,8,9,10,-
11,12,13,14-Octahydro-6,15-methano-
benzocyclotetradec-1-en-17-yl

(*deloc*-1,2,3,3a,7a)-4,7-Dihydro-1*H*-inden-1-ide

(*deloc*-4a,5,6,7,7a,11a,11b)-4a*H*-Dibenzo-
[a,c]cyclohepten-4a-yl

(*deloc*-3,4,10:7,8,9)-Tricyclo[4.2.1.1$^{2,5}$]-
deca-3,7-diene-9,10-diylium

The previous edition of this guidebook provided two ways to form names for radicals and ionic compounds systematically, one as independent entities and one when attached to, or associated with, other atoms, ions, or groups. For instance, the radical $H_3C\bullet$ would, by itself, be called methyl radical, but, as a substituent prefix, simply methyl. By itself $(CH_3)_3C^+$ would be called *tert*-butyl cation, but, in the name of an ion pair, 2-methylpropan-2-ylium or 1,1-dimethylethylium, and similarly $(C_6H_5)_3C^-$ would be triphenylmethyl anion and triphenyl-methanide, respectively. This distinction no longer seems important and, although there are times when the functional class type name is useful, there is little reason to apply it rigorously to radicals and ions as independent entities.

The development of systematic methods for naming ions and radicals has closed some of the gaps that existed in the past. Names are now possible for the cationic fragments observed in mass spectrometry, for instance, $C^+$ and $Cl-C^+$ may be named methyliumylidyne and chloromethyliumylidene.

## Radicals

The systematic method for naming monovalent radicals derived from silane, germane, stannane, and plumbane is not desirable because the multiplicative prefixes bis-, tris-, and so on, would be required to indicate two or more such radicals in multiplicative nomenclature; a name such as disilanyl describes a radical derived from the parent hydride disilane.

The traditional names phosphine and arsine for $PH_3$ and $AsH_3$ cannot be used for naming radicals derived from these parent hydrides because the resulting names, phosphinyl and arsinyl, have long been the accepted names for acyl groups derived from phosphinic and arsinic acids. Although such acyl names have not been used for the corresponding antimony and bismuth groups, consistency with phosphorus and arsenic nomenclature would dictate that names for radicals names derived from $SbH_3$ and $BiH_3$ be formed from the parent hydride names stibane and bismuthane.

Although names analogous to carbene and nitrene, such as phosphene, are sometimes used, it is not recommended that any of these be used to name specific compounds. Furthermore, the analogous silicon name cannot be used without ambiguity since the compound $CH_2{=}SiH_2$ is commonly called "silene".

Names for bivalent radicals, such as carbene, and the use of the suffix -ylidene do not reflect a particular electronic state of the free (nonbonding) electrons. It is not the purpose of systematic structural nomenclature for organic radicals to distinguish between a singlet and a triplet electronic state. Such a distinction, if needed, should be made by adding the separate word "singlet" or "triplet" following the name of the radical, for example, triplet dichloro-carbene, or dichloromethylidene triplet. The λ-convention (see chapter 2) can be used to describe bivalent and trivalent radical sites as long as the atom on which the free (nonbonding) electrons are considered to reside is in a standard valence state before the formal removal of hydrogen atoms.

$(C_6H_5)_2C^{2\bullet}$    Diphenyl-$\lambda^2$-methane          $CH_3{-}N^{2\bullet}$    Methyl-$\lambda^1$-azane

$C^{2\bullet}$    $\lambda^2$-Cyclohexane          $C_6H_5{-}C^{3\bullet}$    Phenyl-$\lambda^1$-methane

Since radicals appear at the top of almost everyone's seniority list for choosing a preferred parent structure, there has not been a great demand for a method of describing the presence of a radical site in a substituent. When this was necessary, a bivalent prefix has been used with one free valence indicating the site of the radical in the substituent and the other the position of substitution into the parent structure. This method is not free from ambiguity, however, since the name methylenecyclohexane could be interpreted as indicating either a • $CH_2{-}$ or $H_2C{=}$ substituent. The 1993 recommendations from IUPAC include the use of the name methylidene for the latter, which provides some relief from this kind of ambiguity, but there are other cases in which ambiguity can occur. The IUPAC recommendations for naming radicals and ions[1] provide an answer to this difficulty; that is, the prefix "ylo" added non-detachably in front of the name of the normal substitutive prefix for the group. Thus, the name 2-yloethyl unequivocally describes the presence of a radical position at the "2" position of an ethyl substituent.

The traditional radical names for certain characteristic groups in table 33.6 may be considered as formed either by adding the letter "l" to the name of a substitutive prefix name, if the letter "l" is not already present, in accordance with the recommendation in the 1979 edition of the IUPAC organic rules, or by replacing an "o" at the end of the name of a substituent prefix with -yl. However, care must be exercised in extending the latter because of potential ambiguity with established names with other meanings, for instance, the name chloryl means the radical $O_2Cl\bullet$ , as defined above, and not $Cl\bullet$ .

## Cations

There are two large classes of cationic compounds: those in which the cationic atom has a bonding number one higher that the corresponding neutral atom, and those in which the cationic atom has a bonding number one less than the corresponding neutral atom. Those of the first class are generally called "ium compounds" and those of the second "ylium compounds".

The ending -ium, when added to the name of a parent hydride or to certain characteristic group suffixes, is almost universally accepted as designating the addition of a proton to an atom, usually one that carries an unshared pair of electrons. There are exceptions. One is found in "ocenium" names such as ferrocenium, where the cation is the result of the loss of an electron from the metal atom. A second exception occurs in trivial names, such as quinolizinium, where the cation is the result of additional skeletal bonds formed by the sharing of a nonbonding electron pair of a heteroatom. A third exception, though perhaps less obvious, is when -ium follows -yl in names of cations derived by adding -ium to the traditional radical names given in table 33.6; here, the cation formation is by loss of an unpaired electron. The names hydroxylium, hydroperoxylium, nitrylium, and nitrosylium are well-established. Although this method can be extended to other characteristic groups with established radical names ending in -yl, giving names such as chlorosylium for $OCl^+$ and carboxylium for $HOOC^+$, it cannot be extended to bivalent group names such as carbonyl and sulfonyl because $OC^+$ and $O_2S^+$ are radical cations. Nevertheless, the names hydroxysulfonylium for $HO-O_2S^+$ and chlorosulfonylium for $ClO_2S^+$ have been proposed.[5]

There are four main types of "ium compounds":

1. ordinary salts formed by the addition of a hydrogen cation to one or more atoms having an unshared pair of electrons such as nitrogen, phosphorus, or sulfur;
2. fully substituted cationic compounds;
3. cationic compounds in which the cation is a member of a ring structure and is *not* attached to hydrogen atoms; and
4. carbon centered cations.

The second and third types of "ium compounds" are often included in a subclass called "onium compounds", even though only those in which the cationic heteroatom is not a part of a ring system have names ending in "onium". The term carbonium has long been used to describe hydrocarbon cations derived formally by the loss of a hydride ion from a neutral carbon atom. More recently, the analogous term siliconium, and even boronium, has appeared. However, "carbonium" in names for specific cations has been inconsistently used and often has been confusing.[6] Therefore, these terms should not be used either as class names or for naming specific individual cations such as $H_5C^+$ or $H_5Si^+$; the proper names for these cations are methanium and silanium. Similarly, names such as carbenium and silenium have been used to describe classes of cations derived formally by addition of a hydrogen cation (hydron) to the mononuclear divalent species $CH_2$ and $SiH_2$. However, except for the names carbenium, carbynium, and nitrenium, which are acceptable alternatives to methylium, methyliumyl and azanylium or aminylium, such names should not be used for naming specific compounds; the general systematic ending -ylium should be used. The general class name carbocation includes cationic hydrocarbons of both types. Modifying terms, such as tricoordinate and pentacoordinate, can be added, if it is necessary to differentiate between the two types.

Since -ium and -onium endings have often been used in naming ordinary salts, for example, dimethylammonium chloride rather than dimethylamine hydrochloride for $[(CH_3)_2NH_2]^+ Cl^-$, class names such as quaternary ammonium compounds and ternary sulfonium compounds have been introduced. Further, an IUPAC publication[7] includes the first and second types of "ium compounds" described above in its definition of "onium compounds".

Although the naming of ordinary salts as "ium compounds" was recommended in the earlier edition of this guidebook and seems to be favored in IUPAC recommendations, the "salt" method for naming amine salts is preferred by CAS in order to collect all salts of the same parent structure together in the formula index, and it is necessary when a cationic site cannot be explicitly expressed. Thus, it is acceptable to name $[CH_3-NH_3]^+ Cl^-$ both as methylamine (or methanamine) hydrochloride and as methylammonium chloride or methanaminium chloride. However, the "salt" method, should not be extended to quaternary compound names such as pyridinium methiodide.

Quaternary acyclic ammonium compounds are preferably named on the basis of the parent hydride ammonium or on the basis of a parent hydride to which -aminium has been attached as in dimethyl(phenyl)ammonium chloride and $N,N$-dimethylbenzenaminium chloride or $N,N$-dimethylanilinium chloride. Other fully substituted heteroatomic cations can be named as "onium compounds", as trimethyloxonium chloride and tetramethylphosphonium iodide, or as derivatives of parent cationic hydrides, such as phosphanium, arsanium, and sulfanium.

The endings -onio and -iumyl should be used for expressing "ium" cationic organic substituents as prefixes; prefixes for mononuclear cations ending in -inio, such as -aminio should be avoided.

The ending -ylium, when added as a suffix to the name of a parent hydride or to certain characteristic group suffixes, is a composite suffix designating cation formation by the loss of a hydride ion (or loss of a hydrogen atom and the residual electron); in this instance, the separate meanings of "yl" and "ium" are not involved. In order to avoid possible confusion the multiplicative prefixes bis-, tris-, and so on, are used to indicate two or more cationic "ylium sites".

Trivial names such as pyrylium, chromenylium, isochromenylium, xanthylium and their sulfur analogs, such as thiopyrylium, are commonly used. There seems to be little reason to keep flavylium or thioflavylium as they are merely 2-phenyl derivatives of chromenylium and thiochromenylium.

In cationic structures in which the positive charge is on a heteroatom in a cyclic system, the locant that describes the position of the charge is often omitted, as in quinolinium, especially when the cationic atom is fully substituted as in 1-methylquinolinium. The modern tendency is to cite such locants even though they may be redundant.

### Anions

The traditional methods for naming anions derived by loss of a hydrogen cation (hydron) from acids, alcohols, phenols, and chalcogen analogs need little discussion. An -ic acid ending is changed to -ate; an -ous acid ending becomes -ite; and an -ol ending of an alcohol, phenol, or chalcogen analog, such as -thiol, becomes -olate. An alternative method for anions derived from alcohols and phenols based on functional class nomenclature leads to names such as methyl oxide and phenyl oxide; these names are disappearing from use, except for a few common contractions such as methoxide and isopropoxide. These anions are described further in chapters 12, 15, 23, 24, and 25.

Derivation of names for anions derived by loss of a hydrogen cation (hydron) from a nitrogen atom of a nitrogenous characteristic group is much less straightforward. All except cyclic imides can be named as derivatives of the parent anion $H_2N^-$, but it may be more convenient to name such anions derived from amines and imines with -aminide and -iminide attached to an appropriate parent hydride, as in ethanaminide for $CH_3CH_2-NH^-$. Anions derived from amides and inorganic imides, such as phosphine imide, are probably best named as derivatives of the amide ion, and anions from cyclic imides are best named anions derived from heterocyclic parent hydrides.

The contracted name methide (from methanide) for $CH_3^-$ has been used but is not recommended because contracted names for other mononuclear anions such as "phosphide" cannot be used; phosphide is the name for $P^{3-}$.

There is no anion counterpart to acylium cations. It is tempting to use names ending in ylide, but this breaks down quickly because of names like acetylide and the use of this term to describe a class of zwitterions.

Until recently, substitutive nomenclature has not been able to name anions formally derived by the addition of a hydride ion to a parent hydride, the anionic analogs of cations derived by the addition of a hydrogen cation (hydron) to a parent hydride. Such anions have been named by coordination nomenclature as hexachlorophosphate(1−), a method not suitable for complex organic parent compounds. Although the -ide suffix could be combined with a parent hydride described by the λ-convention (see chapter 2), this method requires unacceptably high bonding numbers in some cases; for instance, $[PCl_6]^-$ would have to be named hexachloro-$\lambda^7$-phosphanide. However, IUPAC recommendations now include the suffix -uide to indicate that a hydride ion has been added to the parent hydride to create an anionic site. Thus, $[PCl_6]^-$ would be called hexachloro-$\lambda^5$-phosphanuide or hexachlorophosphoranuide. For simple compounds, this method does not offer a clear advantage over coordination names, but in complicated organic structures, it offers the best systematic method so far proposed to describe this kind of anionic structure specifically and unambiguously.

## Cationic and Anionic Sites in a Single Structure

Until 1993, when comprehensive IUPAC recommendations for naming radicals and ions were published,[1] there was little guidance for naming compounds with both cationic and anionic sites in the same structure. Compounds with equal numbers of cationic and anionic sites in the same structure, including ionic characteristic groups cited as suffixes, are generally called zwitterions and are best named systematically by using appropriate operational suffixes and ionic characteristic group suffixes. A method commonly used in an index environment, and which is often useful when the site of the anion is unknown, is to add the phrase "hydroxide inner salt" after the name of the cation as in (2-carboxyethyl)trimethylphosphonium hydroxide inner salt. In this method, the hydroxide ion is cited as the counterion to the cation, and the phrase "inner salt" indicates removal of a molecule of water, the hydroxide ion and a proton from some site on the cation structure. The term hydroxide is no longer cited in CAS index names. This method is not recommended for general use.

Mesoionic structures can be named by selecting one of the canonical structures to name specifically, or the hydroxide inner salt method could be used.

"Ylides" are a special class of zwitterions, usually restricted to structures in which a carbanionic site is adjacent to a cationic heteroatom, as in $(CH_3)_3N^+ - CH_2{}^-$. They have been named by citing the name of the carbanion, derived by adding the ending -ide to the name of appropriate hydrocarbon radical, as a separate word following the name of the partially substituted cation. Accordingly, names for the "ylide" above would be trimethylammonium methylide or $N,N$-dimethylmethanaminium methylide. When the cationic heteroatom and the carbanion are present in the same cyclic structure, the word "ylide" follows the name of the cation. Ylides of phosphonium and arsonium cations have been named as -ylidene derivatives of phosphorane or arsorane; now the $\lambda$-convention (see chapter 2) allows all ylides of acyclic cations to be named as -ylidene derivatives of an appropriate mononuclear parent hydride. This is very acceptable, unless it is desirable to emphasize the zwitterionic nature of the compound. In the case of nitrogen, however, most chemists do not consider $\lambda^5$-azane, $NH_5$, to be chemically reasonable. In spite of all this, it seems better to use the systematic method now recommended by IUPAC, even for nitrogen.

When it is necessary to describe ionic centers in different parts of the structure, it is recommended that the parts of a structure containing the cationic sites be cited as prefixes to the part containing the anionic sites. This is consistent with the general order of citation of cations and anions in binary nomenclature, at least in English. It also means that in an order of precedence of compound classes, anions are being preferred to cations, and this is what is recommended by IUPAC in its 1993 recommendations. CAS continues to prefer cationic compounds to anionic compounds.

## Radical Ions

Radicals usually appear at the top of an order of precedence of compound classes. Hence it is quite logical that names of radical ions should end in a radical ending. This also permits the same name, or one very similar, both for the name of specific radical ions and for substituent prefix names for expressing ionic substructures. Hence, the ending -ylide for naming radical anions given in the 1979 edition of the IUPAC recommendations has been abandoned in favor of -idyl. The latter is consistent with its cationic counterpart, -iumyl; -ylide also conflicts with the use of "ylide" for a certain class of zwitterionic structures.

The phrases "radical cation" and "radical anion" are in general use but one might observe that the correct terminology should be cation radical and anion radical, since the radical ending follows ionic endings in systematic nomenclature. Obviously, both are acceptable and the former have been used in this chapter.

Trivial radical anion names such as in benzophenone ketyl for $(C_6H_5)_2\overset{\bullet}{C}-O^-$, and $p$-benzosemiquinone anion for $\bullet O-C_6H_4-O^-$ are well established by usage, but the systematic names, oxidodiphenylmethyl or diphenylmethanone radical anion and 4-oxidocyclohexa-2,5-dienyloxyl or cyclohexa-2,5-diene-1,4-dione radical anion, are preferred.

## Additional Examples

### Radicals

1.  $C_6H_5\bullet$

    Phenyl
    Benzenyl

2.  $CH_3\overset{\bullet}{C}HCH_2\bullet$
    $\quad_3\quad_2\quad_1$

    Propane-1,2-diyl
    1-Methylethane-1,2-diyl
       (*not* Methylethylene)

3.

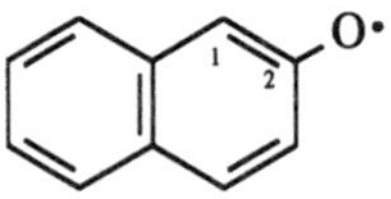

    Naphthalen-2-yloxyl
    2-Naphthyloxyl

4.  $CH_3-NHNH\bullet$
    $\qquad\quad_2\quad_1$

    2-Methyldiazanyl
    2-Methylhydrazinyl
    2-Methylhydrazyl

5.
    $$CH_3CH_2CH_2CH_2-\overset{\overset{\displaystyle O}{\|}}{C}\bullet$$
    $\quad_5\quad_4\quad_3\quad_2\qquad_1$

    Pentanoyl
    1-Oxopentyl

6.  $$\overset{\overset{\displaystyle CH_3}{|}}{\bullet HN-CHCH_2-NH\bullet}$$
    $\qquad\qquad_2\quad_1$

    Propane-1,2-bis(aminyl)
    Propane-1,2-diylbis(aminyl)
    Propane-1,2-diylbis(azanyl)

7.  $Cl_3C-N^{2\bullet}$

    (Trichloromethyl)nitrene
    (Trichloromethyl)aminylene
    (Trichloromethyl)azanylidene
    Trichloromethyl)-$\lambda^1$-azane

8.  $(CH_3)_2Si^{2\bullet}$

    Dimethylsilylene
    Dimethylsilanylidene
    Dimethyl-$\lambda^2$-silane
       (*not* Dimethylsilene)

9.  $CH_3CH_2CH^{2\bullet}$
    $\quad_3\quad_2\quad_1$

    Propan-1-ylidene
    Propylidene
    $1\lambda^2$-Propane
       (*not* Ethylcarbene or ethylmethylidene)

10. 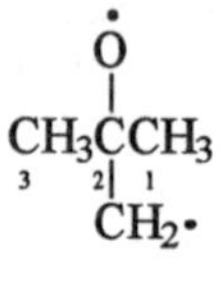
    $$\overset{\overset{\displaystyle \overset{\bullet}{O}}{|}}{\underset{\underset{\displaystyle CH_2\bullet}{|}}{CH_3\overset{}{C}CH_3}}$$
    $\quad_3\quad_2|\;_1$

    2-(Ylomethyl)(propan-2-yloxyl)
    1,1-Dimethyl-2-yloethoxy

11.

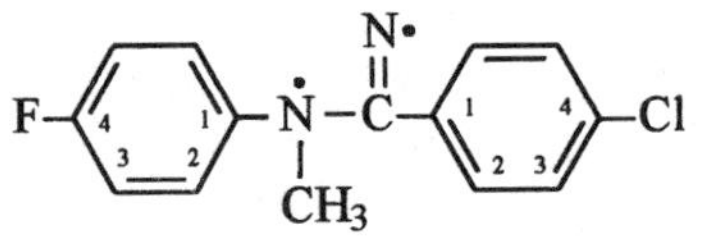

N-[4-Chloro-α-yloimino)benzyl]-4-fluoro-N-methylbenzenaminyl
[4-Chloro-α-(yloimino)benzyl](4-fluorophenyl)(methyl)aminyl
[4-Chloro-α-(yloimino)benzyl](4-fluorophenyl)(methyl)azanyl

12.

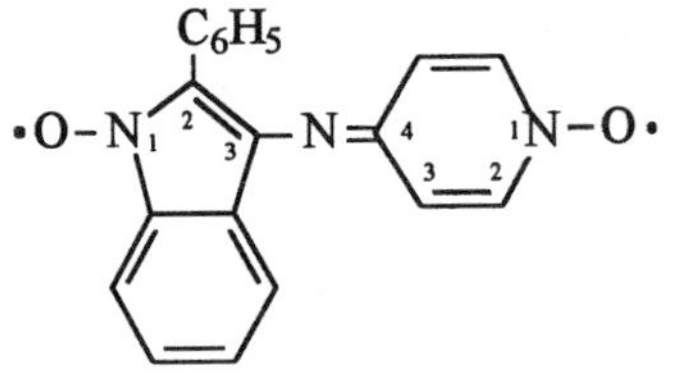

2-Phenyl-3-[[1-(ylooxy)pyridin-4(1$H$)-ylidene]-
amino](1$H$-indol-1-yloxyl)

13.

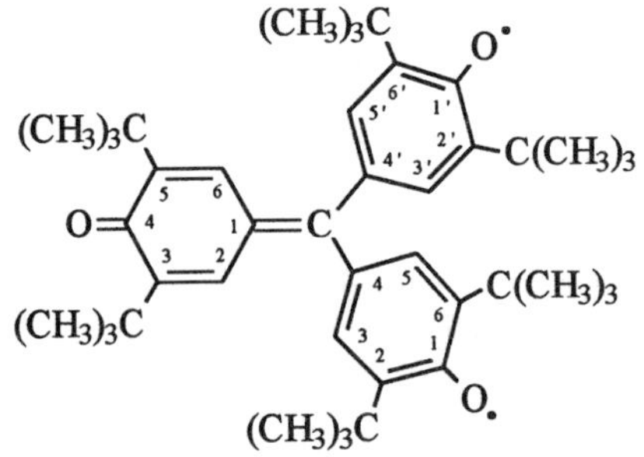

2,2',6,6'-Tetra-*tert*-butyl-4,4'-[(3,5-di-*tert*-butyl-4-oxocyclohexa-2,5-diene-1-ylidene)-
methylene]bis(phenyloxyl) (IUPAC)
4,4'-[[3,5-Bis(1,1-dimethylethyl)-4-oxo-2,5-cyclohexadien-1-ylidene]methylene]-
bis[2,6-bis(1,1-dimethylethyl)(phenyloxyl)] (CAS)

14.

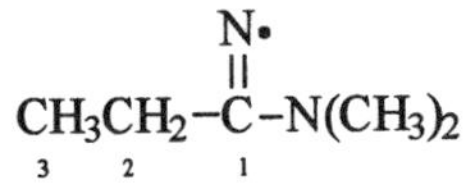

1-(Dimethylamino)propan-1-iminyl
[1-(Dimethylamino)propan-1-ylidene]azanyl
[1-(Dimethylamino)propylidene]aminyl

15. $(CH_3)_2NN^{2\bullet}$
   2  1

2,2-Dimethyldiazanylidene
2,2-Dimethylhydrazinylidene (traditional name)
2,2-Dimethylhydrazono (trivial name)
   [*not* (Dimethylamino)nitrene, (dimethylamino)aminylene
   or (dimethylamino)azanylidene]

16. $CH_3CH_2CH_2CH_2C^{3\bullet}$
   5   4   3   2   1

Pentylidyne
(*not* Butylcarbyne)

## Cations

17.

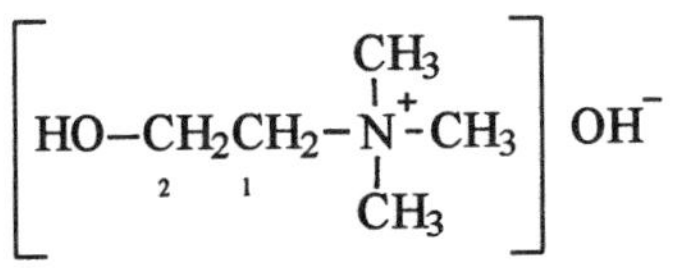

(2-Hydroxyethyl)trimethylammonium hydroxide
(2-Hydroxyethyl)trimethylazanium hydroxide
2-Hydroxy-$N,N,N$-trimethylethanaminium
   hydroxide
   (*not* Choline)

18.

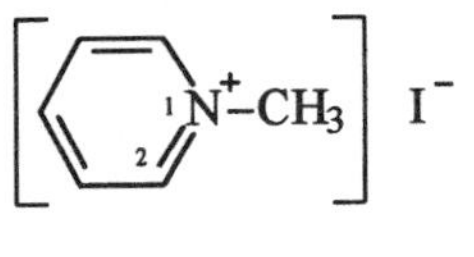

1-Methylpyridinium iodide
(*not* Pyridine methiodide)

19.

$$\left[(CH_3)_3\overset{+}{N}-CH_2CH_2-\overset{+}{N}(CH_3)_3\right]\ 2\,Cl^-$$

Ethylenebis(trimethylammonium) dichloride
Ethylenebis(trimethylazanium) dichloride
$N,N,N,N',N',N'$-Hexamethylethane-
   1,2-bis(aminium) dichloride

20.

$$2\left[(CH_3)_3\overset{+}{P}-C_6H_5\right]\ [SO_4]^{2-}$$

Bis(trimethylphenylphosphonium) sulfate
Bis(trimethylphenylphosphanium) sulfate
Trimethyl(phenyl)phosphonium sulfate (2:1)

21.

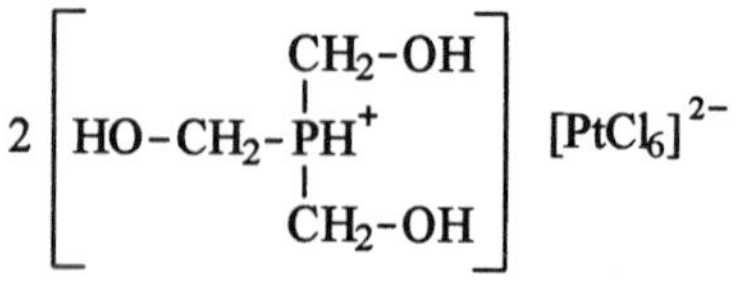

$$2\left[\begin{array}{c}CH_2-OH\\|\\HO-CH_2-\overset{+}{P}H\\|\\CH_2-OH\end{array}\right]\ [PtCl_6]^{2-}$$

Bis[tris(hydroxymethyl)phosphanium]-
   hexachloroplatinate(2-)
Bis[tris(hydroxymethyl)phosphonium]-
   hexachloroplatinate(2-)
   [*not* Bis[tris(hydroxymethyl)phosphinium
      hexachloroplatinate(2-)]

22.

$$\left[(CH_3CH_2CH_2)_2\overset{+}{I}\right]\ ClO_4^-$$

Dipropyliodonium perchlorate
Dipropyliodanium perchlorate

23.

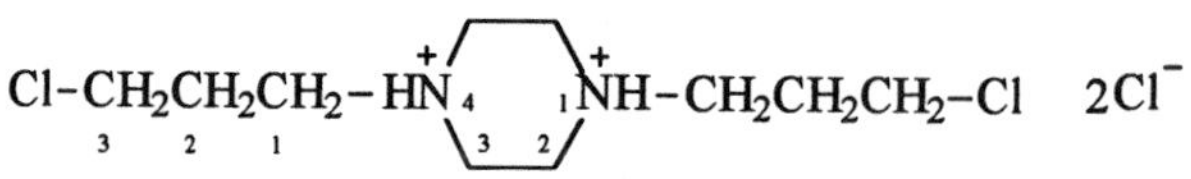

$$Cl-CH_2CH_2CH_2-\overset{+}{HN}\underset{\underset{3\ \ 2}{}}{\overset{/\diagup\diagdown}{\phantom{x}}}\overset{+}{NH}-CH_2CH_2CH_2-Cl\quad 2Cl^-$$

1,4-Bis(3-chloropropyl)piperazinium dichloride
1,4-Bis(3-chloropropyl)piperazine dihydrochloride

24.

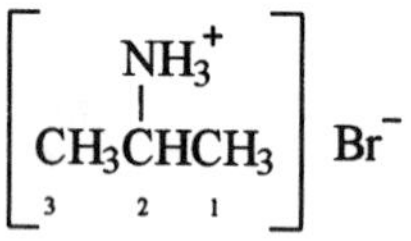

$$\left[\begin{array}{c}\overset{+}{NH_3}\\|\\CH_3CHCH_3\end{array}\right]\ Br^-$$

2-Propanaminium bromide
2-Propanamine hydrobromide
(1-Methylethyl)ammonium bromide

25.

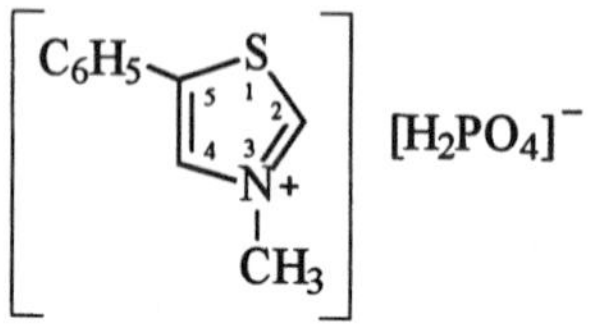

$$\left[\begin{array}{c}C_6H_5\diagup\overset{S}{\phantom{x}}\\ \overset{+}{N}\\|\\CH_3\end{array}\right]\ [H_2PO_4]^-$$

3-Methyl-5-phenyl-1,3-thiazolium (dihydrogen
   phosphate)
3-Methyl-5-phenyl-1,3-thiazolium phosphate (1:1)

26.

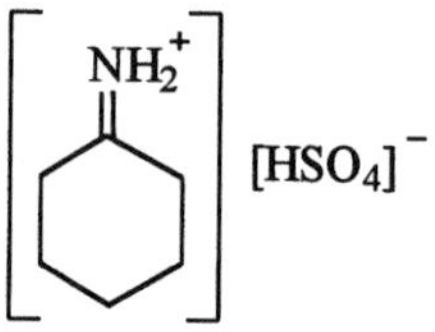

$$\left[\begin{array}{c}\overset{+}{NH_2}\\ \text{(cyclohexane ring)}\end{array}\right]\ [HSO_4]^-$$

Cyclohexaniminium (hydrogen sulfate)
Cyclohexaniminium sulfate (1:1)
Cyclohexanimine dihydrogen sulfate

27.

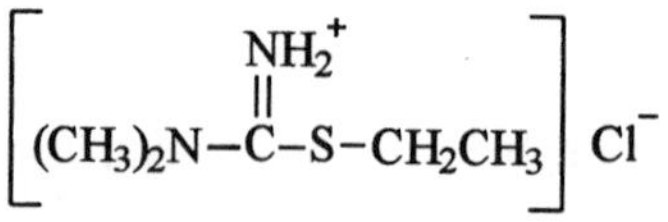

1-(Dimethylamino)-1-(ethylthio)methaniminium
   chloride

$$\left[\begin{array}{c}\overset{+}{NH_2}\\ \|\\(CH_3)_2N-C-S-CH_2CH_3\end{array}\right]\ Cl^-$$

[(Dimethylamino)(ethylthio)methylidene]-
   ammonium chloride
Ethyl $N,N$-dimethylcarbamimidothioic acid
   monohydrochloride

28.  $CH_3CH{=}\overset{+}{OH}$

Ethylideneoxonium

29.  $H_2C^{2+}$

Methanebis(ylium)

30.  $CH_3CH_2-\overset{+}{S}$

Ethylsulfanylium
Ethylthiylium

31. $\overset{+}{S}=\underset{2}{S}-\underset{1}{S}-C_6H_5$      1-Phenyl-1$\lambda^4$-disulfen-1-ylium

32. 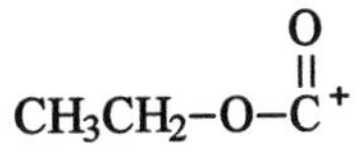 $CH_3CH_2-O-\overset{\overset{O}{\|}}{C}{}^+$      Ethoxyoxomethylium

33. $\underset{4}{C}H_3\underset{3}{C}H_2\underset{2}{C}H_2\underset{1}{C}H_4^+$      1-Butanium

34.

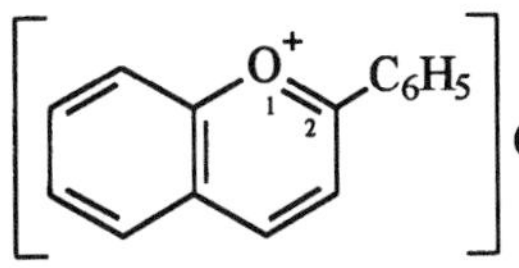

2-Phenylchromenylium chloride
2-Phenyl-1$\lambda^4$-benzopyran-1-ylium chloride
(*not* Flavylium chloride)

## Anions

35. 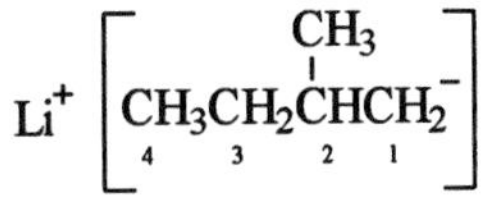 $Li^+ \left[ \underset{4}{C}H_3\underset{3}{C}H_2\underset{2}{\overset{\overset{CH_3}{|}}{C}}H\underset{1}{C}H_2^- \right]$      Lithium 2-methylbutan-1-ide

36. 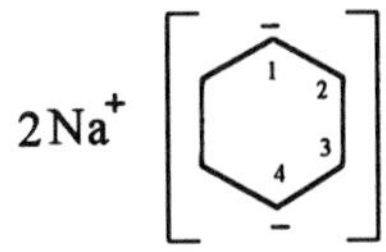 $2\,Na^+$      Disodium cyclohexane-1,4-diide

37. $Ba^{2+} \left[ CH_3CH_2-CO-O^- \right]_2$      Barium propanoate
Barium dipropanoate

38. 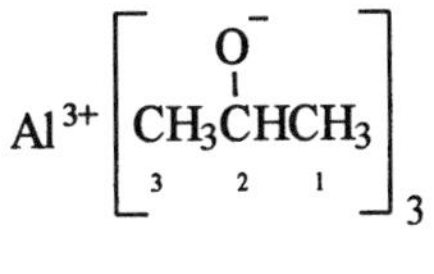 $Al^{3+} \left[ \underset{3}{C}H_3\underset{2}{\overset{\overset{O^-}{|}}{C}}H\underset{1}{C}H_3 \right]_3$      Aluminum tripropan-2-olate
Aluminum tris(1-methylethanolate)
Aluminum triisopropoxide

39. 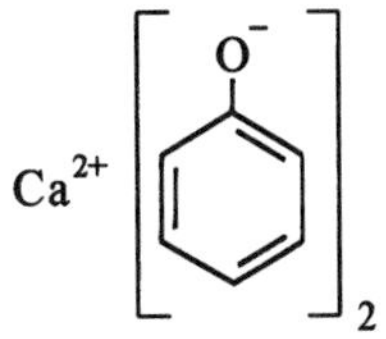 $Ca^{2+}$      Calcium diphenolate
Calcium diphenoxide

40. 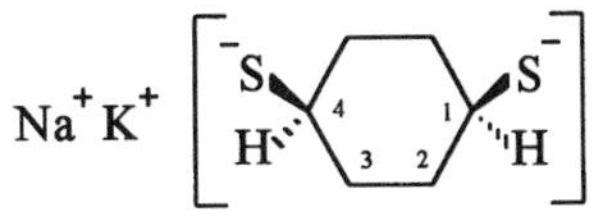 $Na^+K^+$      Potassium sodium *cis*-cyclohexane-1,4-bis(thiolate)

41. 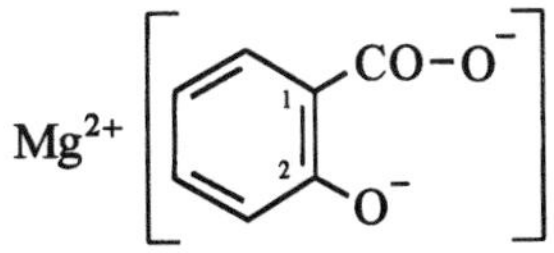 $Mg^{2+}$      Magnesium 2-oxidobenzoate

42. 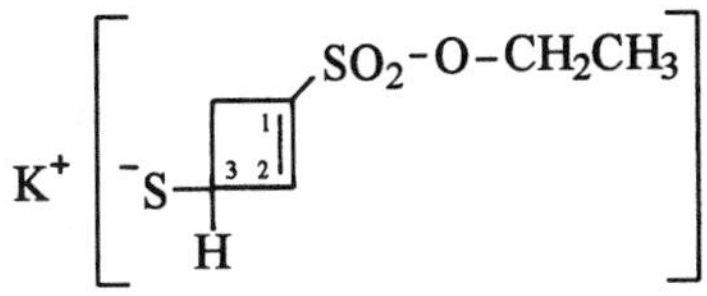 $K^+$      Potassium 1-ethyl 3-sulfidocyclobutane-1-sulfonate

43.

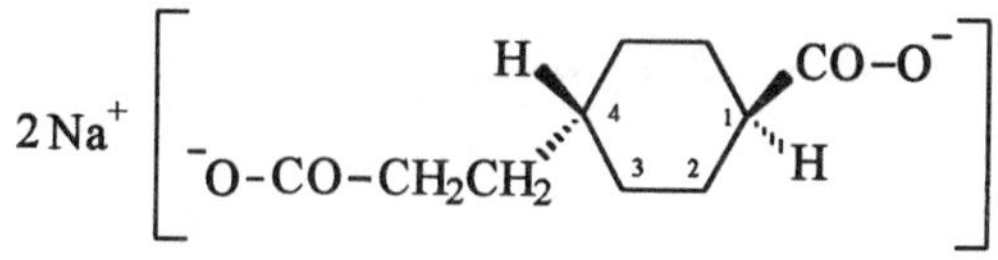

$$2\,\text{Na}^+ \left[ {}^-\text{O-CO-CH}_2\text{CH}_2 \overset{H}{\underset{3\ 2}{\diagdown}}\overset{4}{\diagup}\overset{1}{\underset{}{\diagup}}\overset{\text{CO-O}^-}{\underset{H}{}} \right]$$

Disodium *trans*-4-(2-carboxylatoethyl)cyclohexanecarboxylate
Disodium *trans*-4-carboxylatocyclohexanepropanoate

44. $\text{Li}^+ \left[ \text{CH}_3\text{-SF}_2^- \right]$     Lithium difluoro(methyl)sulfanuide
                           Lithium difluoro(methyl)-$\lambda^4$-sulfanide

45. $\text{Na}^+ \left[ (\text{CH}_3\text{CH}_2)_2\text{I}^- \right]$     Sodium diethyliodanuide
                           Sodium diethyl-$\lambda^3$-iodanide

## Radical Ions

46.

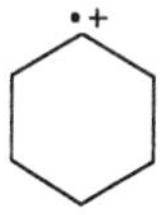

Cyclohexaniumyl[a]

47.

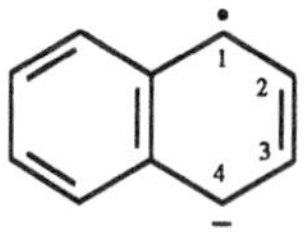

1,4-Dihydronaphthalen-4-id-1-yl

48. CH$_3$

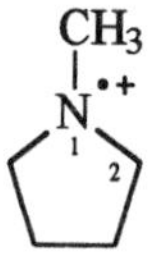

1-Methylpyrrolidin-1-ium-1-yl

49.

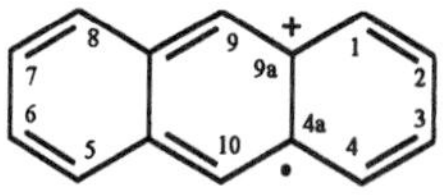

Anthracen-9a-ylium-4a(9a*H*)-yl[b]

## Zwitterions and Related Compounds

50. $(\text{CH}_3[\text{CH}_2]_{15})_2\overset{+}{\text{S}}\text{-CH}_2\text{-CO-O}^-$

Dihexadecylsulfonio)acetate
(Carboxymethyl)dihexadecylsulfonium
   hydroxide inner salt (CAS)[c]

51.

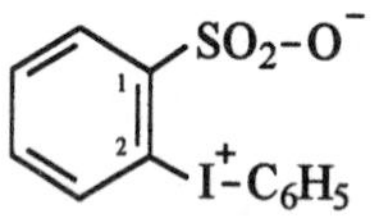

SO$_2$-O$^-$

$\overset{+}{\text{I}}$-C$_6$H$_5$

2-(Phenyliodonio)benzesulfonate
Phenyl(2-sulfophenyl)iodonium hydroxide inner salt (CAS)[c]

52.

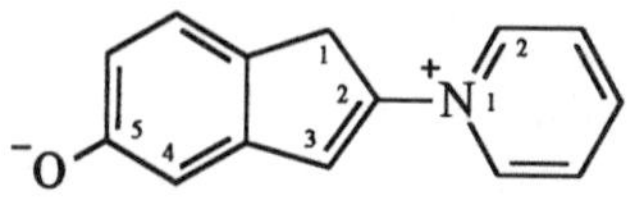

2-Pyridinio-1*H*-inden-5-olate
1-(5-Hydroxy-1*H*-inden-2-yl)pyridinium
   hydroxide inner salt (CAS)[c]

---

[a] The locant 1 is assumed for both -ium and -yl.

[b] Since the two operations described by -ium, and -yl are independent operations, the added hydrogen required by the formation of the radical site provides the hydrogen necessary for the generation of the -ylium cation.

[c] This is the format that was used by CAS until recently; the hydroxide term is no longer cited. The descriptive phrase is now simply "inner salt".

53.

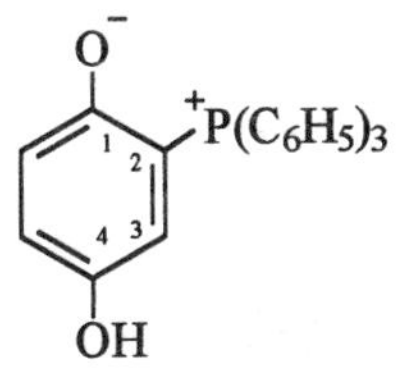

4-Hydroxy-2-(triphenylphosphonio)phenolate
(2,5-Dihydroxyphenyl)triphenylphosphonium
  hydroxide inner salt (CAS)[a]

54.

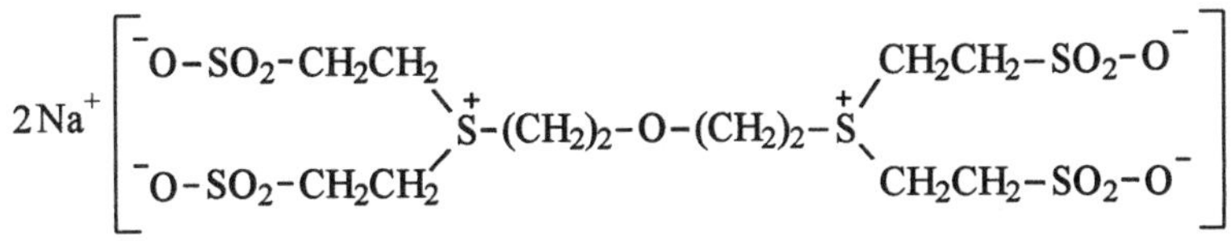

Disodium 2,2′,2″,2‴-[oxybis(ethylenesulfoniumtriyl]tetrakis(ethanesulfonate)
Disodium [oxybis(2,1-ethanediyl)]bis[[2-(sulfooxy)ethyl]-2-[(sulfonatooxy)ethyl]-
sulfonium] dihydroxide bis(inner salt) (CAS)[a]

55.

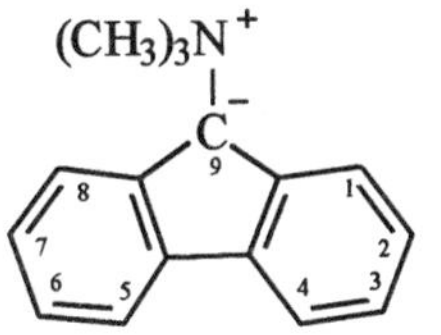

9-(Trimethylammonio)-9H-fluoren-9-ide
Trimethylammonium 9H-fluoren-9-methylide
  (an "-ylide" class name)
N,N-Dimethylmethanaminium 9H-fluoren-9-methylide (CAS)

56.

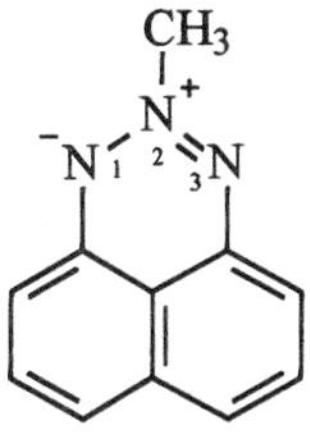

2-Methyl-1H-naphtho[1,8-de][1,2,3]triazin-2-ium-1-ide
2-Methyl-1H-naphtho[1,8-de]-1,2-3-triazin-2-ium hydroxide
  inner salt (CAS)[a]

57.

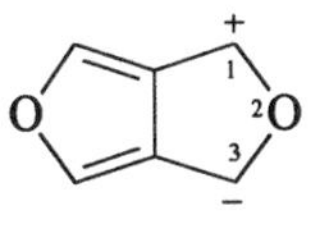

1H,3H-Furo[3,4-c]furan-1-ylium-3-ide[b]

58.

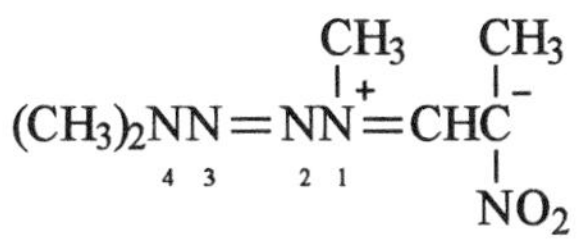

1-(1,4,4-Trimethyltetraaz-2-en-1-ium-1-ylidene)-
  2-nitropropan-2-ide
1,4,4-Trimethyl-1-(2-nitropropylidene)-2-
  tetrazen-1-ium hydroxide inner salt (CAS)[a]

59.

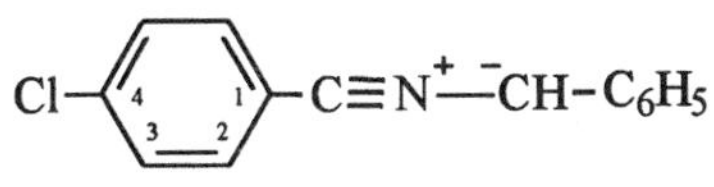

[[(4-Chlorophenyl)methylidyne]ammoniumyl]-
  (phenyl)methanide
4-Chlorobenzonitrilium phenylmethylide
  (an "ylide" class name)

---

[a] This is the format that was used by CAS until recently; the hydroxide term is no longer cited. The descriptive phrase is now simply "inner salt".

[b] This name describes one of the resonance forms of the actual mesoionic compound; such compounds have been named as "mesoionic didehydro" derivatives of the corresponding neutral parent structure.

60.

$$\left[ CH_3[CH_2]_5 \underset{24 \quad 23-19}{-} \overset{\overset{CH_3}{|}}{\underset{\underset{CH_3}{|}}{P}} \overset{+}{\underset{18}{-}} \overset{-}{-} CH \underset{17}{-} \overset{\overset{CH_3}{|}}{\underset{\underset{CH_3}{|}}{P}} \overset{+}{\underset{16}{-}} [CH_2]_6 \underset{15-10}{-} \overset{\overset{CH_3}{|}}{\underset{\underset{CH_3}{|}}{P}} \overset{+}{\underset{9}{-}} \overset{-}{-} CH \underset{8}{-} \overset{\overset{CH_3}{|}}{\underset{\underset{CH_3}{|}}{P}} \overset{+}{\underset{7}{-}} [CH_2]_5 CH_3 \right] 2 Br^-$$

7,7,9,9,16,16,18,18-Octamethyl-7,9,16,18-tetraphosphatetracosane-7,9,16,18-tetraium-8,17-diide dibromide

7,7,9,9,16,16,18,18-Octamethyl-7,9,16,18-tetraphosphoniatetracosane-8,17-diide dibromide

7,7,9,9,16,16,18,18-Octamethyl-7,9,16,18-tetraphosphoniatetracosane-8,17-diylide dibromide

61.  $(C_6H_5)_3P=CH-C_6H_5$      or      $(C_6H_5)_3\overset{+}{P}\overset{-}{-}CH-C_6H_5$

Benzylidenetriphenylphosphorane

(Triphenylphosphoniumyl)(phenyl)methanide
Triphenylphosphonium phenylmethylide

**REFERENCES**

1. International Union of Pure and Applied Chemistry, Organic Chemistry Division, Commission on Nomenclature of Organic Chemistry. Revised Nomenclature for Radicals, Ions, and Related Species. *Pure Appl. Chem.* **1993**, *65*, 1357–1455. (A revision of Subsection C-0.8, Free Radicals, Ions, and Radical Ions: in *Nomenclature of Organic Chemistry, Section C: Characteristic Groups Containing Carbon, Hydrogen, Oxygen, Nitrogen, Halogen, Sulfur, Selenium, and Tellurium;* Klesney, S. P., Rigaudy, J., Eds.; Pergamon Press: Oxford, England, 1979; pp 133–143.

2. International Union of Pure and Applied Chemistry, Physical Chemistry Division, Commission on Molecular Structure and Spectroscopy. Recommendations for Nomenclature and Symbolism for Mass Spectroscopy, Recommendations 1988. *Pure Appl. Chem.* **1991**, *63*, 1541–1566.

3. International Union of Pure and Applied Chemistry, Organic Chemistry Division, Commission on Physical Organic Chemistry. Nomenclature for Hydrogen Atoms, Ions, or Groups and for Reactions Involving Them, Recommendations 1988. *Pure Appl. Chem.* **1988**, *60*, 1115–1116.

4. International Union of Pure and Applied Chemistry, Inorganic Chemistry Division, Commission on Nomenclature of Inorganic Chemistry. *Nomenclature of Inorganic Chemistry (Recommendations 1990)*; Leigh, G. J., Ed.; Blackwell Scientific: Oxford; Recommendation I-4.4.3, pp 48–49.

5. International Union of Pure and Applied Chemistry, Organic Chemistry Division, Commission on Physical Chemistry. Nomenclature for Organic Chemical Transformations, Recommendations 1988. *Pure Appl. Chem.* **1989**, *61*, 725–768; Table I.

6. Hurd, C. D. Carbonium Nomenclature. *J. Chem. Educ.* **1971**, *48*, 490.

7. International Union of Pure and Applied Chemistry, Organic Chemistry Division. Commissions on Nomenclature of Organic Chemistry and on Physical Organic Chemistry. Glossary of Class Names of Organic Compounds and Reactive Intermediates Based on Structure. *Pure Appl. Chem.* **1995**, *67*, 1307–1375.

# Appendix A

## Prefixes

This appendix contains a number of prefixes for the purpose of identification and provides equivalent names; inclusion or exclusion does not imply a recommendation for use. An "=" refers to an acceptable alternative name. The "<" symbol in a structure indicates that the group is attached to two different atoms.

Prefixes derived from hydrazine almost always have equivalent names based on diazane or diazene. Similarly, most complex prefixes beginning with benzene or cyclohexane have equivalents beginning with phenyl or cyclohexyl.

A more comprehensive list of phosphorus prefixes is given in table 25.2, chapter 25; each has an equivalent arsenic or antimony (beginning with "stib") prefix. Additional prefixes will be found in tables 3.2–3.6, 3.9–3.11, 23.1–23.2, and 26.1.

acetamido = acetylamino  $CH_3-CO-NH-$
acetimidoyl  $CH_3-C(NH)-$
acetohydrazonoyl  $CH_3-C(NNH_2)-$
acetohydroximoyl  $CH_3-C(N-OH)-$
acetonyl = 2-oxopropyl = 2-oxopropan-1-yl
  $CH_3-CO-CH_2-$
acetonylidene = 2-oxopropylidene =
  2-oxopropan-1-ylidene = 2-oxopropane-
  1,1-diyl  $CH_3-CO-CH=$
acetoxy = acetyloxy  $CH_3-CO-O-$
acetyl  $CH_3-CO-$
acetylamino = acetamido  $CH_3-CO-NH-$
acetylimino  $CH_3-CO-N=$  or
  $CH_3-CO-N<$
adipoyl = hexanedioyl  $-CO-[CH_2]_4-CO-$
$\beta$-alanyl  $H_2N-CH_2CH_2-CO-$
allophanyl = carbamoylcarbamoyl
  $H_2N-CO-NH-CO-$
allyl = prop-2-en-1-yl  $CH_2=CHCH_2-$
allylidene = prop-2-en-1-ylidene =
  prop-2-en-1,1-diyl  $CH_2=CHCH=$
amidino = guanyl = carbamimidoyl
  $H_2N-C(NH)-$
amidinoamino = guanidino = carbamimidoylamino  $H_2N-C(NH)-NH-$
amino  $H_2N-$
aminoacetyl = glycyl  $H_2N-CH_2-CO-$
aminocarbonyl = carbamoyl  $H_2N-CO-$

aminosulfinyl  $H_2N-SO-$
  (*not* sulfinamoyl)
aminosulfonyl = sulfamoyl = sulfuramidoyl
  $H_2N-SO_2-$
aminothio = aminosulfanyl  $H_2N-S-$
ammonio = ammoniumyl  $H_3N^+-$
aminoxy = aminooxy  $H_2N-O-$
amyl: *see* pentyl
anilino = phenylamino  $C_6H_5-NH-$
aspartoyl  $-CO-CH_2CH(NH_2)-CO-$
$\alpha$-aspartyl  $HOOC-CH_2CH(NH_2)-CO-$
$\beta$-aspartyl  $HOOC-CH(NH_2)CH_2-CO-$
azanetriyl = nitrilo  $-N<$
azanylidyne = nitrilo  $N\equiv$
azanylylidene = nitrilo  $-N=$
azido  $N_3-$
azimino = triaz[1]eno (only as a bridge)
  $-N=NNH-$
azinico = hydroxyazonoyl = hydroxyazinylidene = hydroxynitroryl
  $HO-N(O)(S)<$
azino = hydrazinediylidene  $=NN=$
azinoyl = azinyl = dihydronitroryl
  $H_2N(O)-$
azinylidene = azonoyl = hydronitroryl
  $HN(O)<$  or  $HN(O)=$
azo = diazenediyl  $-N=N-$
azono  $(HO)_2N(O)-$

azonoyl = azinylidene = hydronitroryl
  HN(O) <  or  HN(O)=
azoxy –N(O)=N–  or  –N=N(O)–
benzal: *see* benzylidene
benzamido = benzoylamino = (phenylcarbonyl)amino $C_6H_5$–CO–NH–
benzenecarbohydroximoyl = benzohydroximoyl $C_6H_5$–C(N–OH)–
benzenecarboximidoyl = benzimidoyl
  $C_6H_5$–C(NH)–
benzenedicarbonyl: *see* phthaloyl,
  isophthaloyl, terephthaloyl
benzenesulfenamido = (phenylthio)amino
  $C_6H_5$–S–NH–
benzenesulfinyl = phenylsulfinyl
  $C_6H_5$–SO–
benzenesulfonamido = benzenesulfonyl-
  amino = (phenylsulfonyl)amino
  $C_6H_5$–SO$_2$–NH–
benzenesulfonyl = phenylsulfonyl
  $C_6H_5$–SO$_2$–
benzhydryl = diphenylmethyl $(C_6H_5)_2$CH–
benzhydrylidene = diphenylmethylidene =
  diphenylmethylene $(C_6H_5)_2$C=; or
  diphenylmethylene = diphenyl-
  methanediyl $(C_6H_5)_2$C<
benzimidoyl = benzenecarboximidoyl
  $C_6H_5$–C(NH)–
benzohydroximoyl = benzenecarbohydroximoyl $C_6H_5$–C(N–OH)–
benzoyl = phenylcarbonyl = benzene-
  carbonyl $C_6H_5$–CO–
benzoylamino = benzamido = benzenecarbonylamino = (phenylcarbonyl)amino $C_6H_5$–CO–NH–
benzoylhydrazinyl = 2-(phenylcarbonyl)-
  hydrazinyl = 2-benzenecarbonyl-
  hydrazinyl $C_6H_5$–CO–NHNH–
benzoylimino = (phenylcarbonyl)imino =
  benzenecarbonylimino $C_6H_5$–CO–N=
  or $C_6H_5$–CO–N<
benzoyloxy = (phenylcarbonyl)oxy =
  benzenecarbonyloxy $C_6H_5$–CO–O–
benzyl = phenylmethyl $C_6H_5$–CH$_2$–
benzylidene = phenylmethylidene =
  phenylmethylene $C_6H_5$–CH=; or
  phenylmethylene = phenylmethanediyl
  $C_6H_5$–CH<
benzylidyne = phenylmethylidyne
  $C_6H_5$–C≡  or  phenylmethanetriyl
  $C_6H_5$–C$\overset{\diagup}{\diagdown}$  or phenylmethanyl-
  ylidene $C_6H_5$–C=
benzyloxy = phenylmethoxy
  $C_6H_5$–CH$_2$–O–

biimino = diazano (only as a bridge)
  –NHNH–
biphenylyl $C_6H_5$–$C_6H_4$–
boranediyl HB<
boranetriyl –B<
boranylidene HB=
boranylidyne B≡
boranylylidene –B=
borono (HO)$_2$B–
boryl (a necessary contraction of boranyl)
  H$_2$B–
borylene = boranediyl HB<  or 
  borylidene HB=
borylidyne = boranylidyne B≡  or 
  boranetriyl –B<  or  boranylylidene –B=
bromo Br–
butan-1-yl = butyl $CH_3CH_2CH_2CH_2$–
butan-2-yl = 1-methylpropyl = *sec*-butyl
  $CH_3CH_2CH(CH_3$–
butanedioyl = succinyl
  –CO–CH$_2$CH$_2$–CO–
butane-1,1-diyl $CH_3CH_2CH_2$CH<
butane-1,4-diyl = tetramethylene
  –CH$_2$CH$_2$CH$_2$CH$_2$–
butane-2,3-diyl = 1,2-dimethylethylene

$$\overset{|\quad|}{CH_3CHCHCH_3}$$
$$\scriptstyle 4\quad3\quad2\quad1$$

butanethioyl $CH_3CH_2CH_2$–CS–
butano (only as a bridge)
  –CH$_2$CH$_2$CH$_2$CH$_2$–
butanoyl = butyryl $CH_3CH_2CH_2$–CO–
but-2-enedioyl = maleoyl or fumaroyl
  –CO-CH=CH–CO–
but-1-en-1-yl $CH_3CH_2$CH=CH–
but-2-en-1-yl $CH_3$CH=CHCH$_2$–
but-2-ene-1,4-diyl –CH$_2$CH=CHCH$_2$–
but[1]eno (only as a bridge)
  –CH=CHCH$_2$CH$_2$–
but-2-enoyl $CH_3$CH=CH–CO–
but-2-enylene: *see* but-2-ene-1,4-diyl
butoxy = butyloxy = butan-1-yloxy
  $CH_3CH_2CH_2CH_2$–O–
butyl = butan-1-yl $CH_3CH_2CH_2CH_2$–
*sec*-butyl = butan-2-yl = 1-methylpropyl
  $CH_3CH_2CH(CH_3$–
*tert*-butyl = 2-methylpropan-2-yl =
  1,1-dimethylethyl $(CH_3)_3$C–
butylidene = butan-1-ylidene
  $CH_3CH_2CH_2$CH=
butyryl = butanoyl $CH_3CH_2CH_2$–CO–
caproyl: *see* hexanoyl
capryl: *see* decanoyl
capryloyl: *see* octanoyl

carbamido: *see* ureido
carbamimidoyl = amidino = guanyl
 $H_2N-C(NH)-$
carbamimidoylamino = guanidino =
 amidinoamino $H_2N-C(NH)-NH-$
carbamothioyl = amino(thiocarbonyl) =
 aminocarbonothioyl $H_2N-CS-$
carbamoyl = aminocarbonyl $H_2N-CO-$
carbamoylamino = ureido =
 (aminocarbonyl)amino
 $H_2N-CO-NH-$
(carbamoylamino)acetyl =
 *N*-carbamoylglycyl = hydantoyl =
 [(aminocarbonyl)amino]acetyl
 $H_2N-CO-NH-CH_2-CO-$
carbamoylcarbonyl = oxamoyl =
 aminooxoacetyl $H_2N-COCO-$
*N*-carbamoylglycyl = hydantoyl =
 [(aminocarbonyl)amino]acetyl =
 (carbamoylamino)acetyl
 $H_2N-CO-NH-CH_2-CO-$
2-carbamoylhydrazinyl = semicarbazido =
 2-(aminocarbonyl)hydrazinyl
 $H_2N-CO-NHNH-$
2-carbamoylhydrazono = semicarbazono =
 2-(aminocarbonyl)hydrazinylidene
 $H_2N-CO-NHN=$
carbazimidoyl = hydrazinecarboximidoyl
 $H_2NNH-C(NH)-$
carbazono = 2-diazenecarbonylhydrazinyl =
 2-(diazenylcarbonyl)hydrazinyl
 $HN=N-CO-NHNH-$
carbazoyl = hydrazinecarbonyl =
 hydrazinylcarbonyl $H_2NNH-C(O)-$
carboethoxy = ethoxycarbonyl
 $CH_3CH_2-O-CO-$
carbomethoxy = methoxycarbonyl
 $CH_3-O-CO-$
carbonimidoyl = iminomethylene =
 iminomethylidene $HN=C=$ or
 iminomethylene = iminomethanediyl
 $HN=C<$
carbonochloridoyl = chloroformyl =
 chlorocarbonyl $Cl-CO-$
carbonothioyl = thiocarbonyl $-CS-$
carbonyl $-CO-$
carbonyldiimino = ureylene
 $-HN-CO-NH-$
carboxy $HOOC-$
carboxycarbonyl = oxalo =
 hydroxyoxoacetyl $HO-COCO-$
carboxylato $^-O-CO-$
chloro $Cl-$
chlorocarbonyl = chloroformyl =
 carbonochloridoyl $Cl-CO-$

chlorooxy $Cl-O-$
chlorosyl $OCl-$
chloryl $O_2Cl-$
cinnamoyl = 3-phenylprop-2-enoyl
 $C_6H_5-CH=CH-CO-$
cinnamyl = 2-phenylprop-2-en-1-yl
 $C_6H_5-CH=CHCH_2-$
crotonoyl: *see* but-2-enoyl
cyanato $NCO-$
cyano $N\equiv C-$
cyclohexane-1,1-diyl

[structure: cyclohexane-1,1-diyl ring]

cyclohexanecarbonyl = cyclohexylcarbonyl
 $C_6H_{11}-CO-$
cyclohexane-1,4-diyl

[structure: cyclohexane-1,4-diyl ring]

cyclohexylcarbonyl = cyclohexanecarbonyl
 $C_6H_{11}-CO-$
cyclohexyl $C_6H_{11}-$
cyclohexylene: *see* cyclohexane-1,4-diyl
cyclohexylidene

[structure: cyclohexylidene ring]

cyclopropyl

[structure: cyclopropyl ring]

cyclotrisilanyl

[structure: cyclotrisilanyl, $H_2Si$ and $SiH-$ and $H_2Si$]

decanoyl $CH_3[CH_2]_8-CO-$
decyl = decan-1-yl $CH_3[CH_2]_9-$
diazano = biimino (only as a bridge)
 $-HNNH-$
diazanyl = hydrazinyl = hydrazino
 $H_2NNH-$
(diazanylidenemethyl)diazenyl =
 1-formazano = (hydrazonomethyl)azo
 $H_2NN=CH-N=N-$
(diazenecarbonyl)hydrazinyl = carbazono
 $HN=N-CO-NHNH-$
diazenediyl = azo $-N=N-$
diazenyl $HN=N-$
(diazenylhydrazono)methyl = formazanyl =
 3-formazanyl
 $HN=N-\overset{|}{C}=NNH_2$
(diazenylmethylene)hydrazinyl =
 5-formazano = 5-formazanyl
 $HN=N-CH=N-NH-$

diazoamino = triaz-1-ene-1,3-diyl
$-N=NNH-$
dichloroboryl $Cl_2B-$
dichloroiodo = dichloro-$\lambda^3$-iodanyl $Cl_2I-$
dichlorophosphino = dichlorophosphanyl
$Cl_2P-$
dihydronitroryl = azinoyl = azinyl $H_2N(O)-$
dihydroxycarbonimidoyl = hydroxy-
(hydroxyimino)methyl
$HO-C(N-OH)-$
dihydroxyboryl: *see* borono
dihydroxynitroryl: *see* azono
dihydroxyphosphanyl = dihydroxyphosphino $(HO)_2P-$
dihydroxyphosphinothioyl = dihydroxy-
phosphorothioyl = dihydroxy-
thioxophosphoranyl $(HO)_2P(S)-$
dihydroxyphosphoryl: *see* phosphono
2,3-dihydroxypropanoyl = glyceroyl
$HO-CH_2-CH(OH)-CO-$
1,1-dimethylethyl = 2-methylpropan-
2-yl = *tert*-butyl $(CH_3)_3C-$
1,2-dimethylethylene = butane-2,3-diyl

$$\overset{|}{C}H_3\overset{|}{C}HCHCH_3$$

1,1-dimethylpropyl = *tert*-pentyl =
2-methylbutan-2-yl $CH_3CH_2C(CH_3)_2-$
dioxy = peroxy $-OO-$
$\alpha,\alpha$-diphenylbenzyl = trityl =
triphenylmethyl $(C_6H_5)_3C-$
diphenylmethyl = benzhydryl =
$\alpha$-phenylbenzyl $(C_6H_5)_2CH-$
diphosphanyl = diphosphino $H_2PPH-$
disilane-1,1-diyl $H_3SiSiH<$
disilanyl $H_3SiSiH_2-$
disilanylidene $H_3SiSiH=$
disilazan-1-yl $H_3Si-NH-SiH_2-$
disilazan-2-yl = disilylamino $(H_3Si)_2N-$
disiloxanyl $H_3Si-O-SiH_2-$
disulfanyl = thiosulfeno = dithiohydroperoxy $HSS-$
dithio = disulfanediyl $-SS-$
dithiocarboxy $HSSC-$
dithiohydroperoxy = thiosulfeno =
disulfanyl $HSS-$
dithiosulfo (unspecified) $HS_3O-$
dithiosulfonyl = sulfonodithioyl $-S(S)_2-$
dodecanoyl = lauroyl $CH_3[CH_2]_{10}-CO-$
dodecyl = dodecan-1-yl $CH_3[CH_2]_{11}-$
epidioxy (only connecting two different
atoms in the same chain or ring)
$-OO-$
epidithio (only as a bridge) $-SS-$
epimino (only as a bridge) $-NH-$

epithio (only as a bridge) $-S-$
epoxy (only connecting two separate atoms
in the same chain or ring) $-O-$
epoxyimino (only as a bridge) $-O-NH-$
epoxymethano (only as a bridge) $-O-CH_2-$
epoxythio (only as a bridge) $-O-S-$
ethanedioyl = oxalyl $-COCO-$
ethane-1,1-diyl $CH_3CH<$
ethane-1,2-diyl = ethylene $-CH_2CH_2-$
ethanethioyl = thioacetyl $CH_3-CS-$
ethano (only as a bridge) $-CH_2CH_2-$
ethanoyl: *see* acetyl
ethanylidene = ethylidene $CH_3CH=$
ethene-1,2-diyl = vinylene $-CH=CH-$
etheno (only as a bridge) $-CH=CH-$
ethenyl = vinyl $CH_2=CH-$
ethenylidene = vinylidene $CH_2=C=$
ethoxy = ethyloxy $CH_3CH_2-O-$
ethoxycarbonyl = carboethoxy
$CH_3CH_2-O-CO-$
ethyl = ethanyl $CH_3CH_2-$
ethylene = ethane-1,2-diyl $-CH_2CH_2-$
ethylenebis(oxy) $-O(CH_2CH_2-O-$
ethylenedioxy $-CH_2CH_2-O-O-$
ethylidene = ethanylidene $CH_3CH=$
ethylthio = ethylsulfanyl $CH_3CH_2-S-$
fluoro $F-$
formamido = formylamino $HCO-NH-$
1-formazano = formazano = (diazanylidenemethyl)diazenyl = (hydrazonomethyl)azo $H_2NN=CH-N=N-$
5-formazano = 5-formazanyl = (diazenylmethylene)hydrazinyl =
$HN=N-CH=N-NH-$
formazanyl = 3-formazanyl = (diazenylhydrazono)methyl

$$HN=\underset{1}{N}-\underset{2}{\overset{|}{C}}=\underset{3}{N}\underset{4}{N}\underset{5}{H_2}$$

formimidoyl $HC(NH)-$
formyl $HCO-$
formylamino = formamido $HCO-NH-$
formyloxy $HCO-O-$
fulminato $CNO-$
fumaroyl = (*E*)-but-2-enedioyl

galloyl = 3,4,5-trihydroxybenzoyl
$3,4,5-(HO)_3C_6H_2-CO-$
germyl $H_3Ge-$
germylene = germanediyl $H_2Ge<$ or
germanylidene $H_2Ge=$
glutaryl = pentanedioyl $-CO-[CH_2]_3-CO-$

glyceroyl = 2,3-dihydroxypropanoyl =
2,3-dihydroxypropionyl
$HO-CH_2CH(OH)-CO-$
glycyl = 2-aminoacetyl $H_2N-CH_2-CO-$
guanidino = amidinoamino = carbamimidoylamino $H_2N-C(NH)-$
guanyl = carbamimidoyl = amidino
$H_2N-C(NH)-$
heptyl = heptan-1-yl $CH_3[CH_2]_6-$
hexamethylene = hexane-1,6-diyl $-[CH_2]_6-$
hexanedioyl = adipoyl $-CO-[CH_2]_4-CO-$
hexanoyl $CH_3[CH_2]_4-CO-$
hexyl = hexan-1-yl $CH_3[CH_2]_5-$
hydantoyl = N-carbamoylglycyl =
(carbamoylamino)acetyl =
[(aminocarbonyl)amino]acetyl
$H_2N-CO-NH-CH_2-CO$
hydrazinecarbonyl = carbazoyl =
hydrazinylcarbonyl $H_2NNH-CO-$
hydrazinecarboximidoyl = carbazimidoyl
$H_2NNH-C(NH)-$
hydrazine-1,2-diyl = hydrazo $-NHNH-$
hydrazinediylidene = azino $=NN=$
hydrazinyl = hydrazino = diazanyl
$H_2NNH-$
hydrazinylidene = hydrazono $H_2NN=$
hydrazinylylidene $-HNN=$
hydrazo = hydrazine-1,2-diyl $-NHNH-$
hydrazono = hydrazinylidene $H_2NN=$
hydrazono(hydroxy)methyl = C-hydroxycarbonohydrazonoyl $HO-C(=NNH_2)-$
hydrohydroxynitroryl = hydroxyazinyl =
hydroxyazinoyl $HO-NH(O)-$
hydronitroryl = azonoyl = azinylidene
$HN(O)<$ or $HN(O)=$
hydroperoxy $HOO-$
hydroperoxycarbonyl $HOO-CO-$
hydroseleno = selanyl $HSe-$
hydroxy $HO-$
hydroxyamino $HO-NH-$
hydroxyazinyl = hydroxyazinoyl =
hydrohydroxynitroryl $HO-NH(O)-$
hydroxyazinylidene = aci-nitro $HO-N(O)=$
or azinico $HO-N(O)<$ or
hydroxyazonoyl = hydroxynitroryl
$HO-N(O)=$ or $HO-N(O)<$
2-hydroxybenzoyl = salicyloyl
$2-HO-C_6H_4-CO-$
2-hydroxybenzyl = salicyl
$2-HO-C_6H_4-CH_2-$
2-hydroxybutanedioyl = maloyl
$-CO-CH(OH)CH_2-CO-$
hydroxy(hydroxyimino)methyl =
dihydroxycarbonimidoyl
$HO-C(=N-OH)-$

hydroxyimino $HO-N=$ or $HO-N<$
hydroxy(imino)methyl = C-hydroxycarbonimidoyl $HO-C(NH)-$
hydroxy(mercapto)phosphinyl =
hydroxy(mercapto)phosphoryl =
hydroxymercaptooxophosphoranyl
$(HS)(HO)P(O)-$
hydroxynitroryl = hydroxyazonoyl =
hydroxyazinylidene $HO-N(O)<$
or $HO-N(O)=$; aci-nitro $HO-N(O)=$
or azinico $HO-N(O)<$
hydroxyphosphinidine = $HO-P<$ or
$HO-P=$ or hydroxyphosphanylidene
$HO-P=$
hydroxyphosphinyl = hydrohydroxyphosphoryl $HO-PH(O)-$
hydroxyphosphinylidene = phosphinico
$HO-P(O)<$ or hydroxyphosphoryl =
hydroxyphosphonoyl $HO-P(O)<$ or
$HO-P(O)=$
2-hydroxypropanoyl = lactoyl = 2-hydroxy-
propionyl $CH_3CH(OH)-CO-$
hydroxysulfanyl = sulfeno = hydroxythio
$HO-S-$
hydroxysulfonothioyl = hydroxy(thiosulfonyl) $HO-S(O)(S)-$
hydroxythio = sulfeno = hydroxysulfanyl
$HO-S-$
hydroxy(thiosulfonyl) = hydroxysulfonothioyl $HO-S(O)(S)-$
imidodicarbonyl = iminodicarbonyl
$-CO-NH-CO-$
imino = azanylidene $HN=$ or azanediyl
$HN<$
iminomethylene = iminomethanediyl
$HN=C(NH)<$ or iminomethylidene
$HN=C=$
iodo $I-$
iodoso = iodosyl $OI-$
iodyl $O_2I-$
isobutyl = 2-methylpropyl = 2-methyl-
propan-1-yl $(CH_3)_2CHCH_2-$
isocarbazido = isocarbonohydrazido =
2-(C-hydroxycarbonohydrazonoyl)-
hydrazinyl $H_2NN=C-(OH)-NHNH-$
isocyanato $OCN-$
isocyano $CN-$
isopentyl = 4-methylbutyl = 4-methyl-
butan-1-yl $(CH_3)_2CHCH_2CH_2-$
isophthaloyl = benzene-1,3-dicarbonyl

isopropenyl = propen-2-yl = 1-methyl-
  ethenyl   $CH_2=C(CH_3)-$
isopropyl = 1-methylethyl = propan-2-yl
  $(CH_3)_2CH-$
isopropylidene = 1-methylethylidene =
  propan-2-ylidene   $(CH_3)_2C=$
isosemicarbazido = 2-(C-hydroxycarbonimidoyl)hydrazinyl
  $HN=C(OH)-NHNH-$
isothiocyanato   $SCN-$
1-isoureido = (C-hydroxycarbonimidoyl)-
  amino   $HN=C(OH)-NH-$
3-isoureido = (aminohydroxymethylidene)-
  amino   $H_2N(C(OH)=N-$
keto: *see* oxo
lactoyl = 2-hydroxypropanoyl =
  2-hydroxypropionyl   $CH_3CH(OH)-CO-$
lauroyl = dodecanoyl   $CH_3[CH_2]_{10}-CO-$
maleoyl = (*Z*)-but-2-enedioyl

$$-OC_{\ 4}\diagdown C{=}C_{\diagup}{}^{CO-}_{1}$$
$$\qquad H{\diagup}{}_{3}\quad {}_{2}{\diagdown}H$$

malonyl = propanedioyl   $-CO-CH_2-CO-$
maloyl = 2-hydroxybutanedioyl
  $-CO-CH_2CH(OH)-CO-$
mercapto = sulfanyl   $HS-$
mercaptocarbonyl = sulfanylcarbonyl
  $HS-CO-$
mercapto(dithiosulfonyl) = mercaptosulfonodithioyl = trithiosulfo
  $HS-S(S)_2-$
mercaptooxy = sulfanyloxy   $HSO-$
mercaptophosphinyl = hydromercapto-
  phosphoryl = mercaptooxo-
  phosphoranyl   $HS-PH(O)-$
mercaptosulfonyl   $HS-SO_2-$
mesityl = 2,4,6-trimethylphenyl
  $2,4,6-(CH_3)_3C_6H_2-$
mesyl: *see* methanesulfonyl
methacryloyl = 2-methylprop-2-enoyl
  $CH_2=C(CH_3)-CO-$
methanediylidene   $=C=$
methanesulfinamido = (methylsulfinyl)-
  amino   $CH_3-SO-NH-$
methanesulfinyl = methylsulfinyl
  $CH_3-SO-$
methanesulfonamido = (methylsulfonyl)-
  amino   $CH_3-SO_2-NH-$
methanesulfonyl = methylsulfonyl
  $CH_3-SO_2-$
methanetetrayl   $>C<$
methanetriyl   $-CH<$
methano (only as a bridge)   $-CH_2-$
methanoyl: *see* formyl

methanylylidene   $-CH=$
methoxalyl = methoxyoxoacetyl
  $CH_3-O-COCO-$
methoxy   $CH_3-O-$
methoxycarbonyl = carbomethoxy
  $CH_3-O-CO-$
methoxythio   $CH_3-O-S-$
methyl = methanyl   $CH_3-$
4-methylbutan-1-yl = isopentyl =
  4-methylbutyl   $(CH_3)_2CHCH_2CH_2-$
2-methylbutan-2-yl = *tert*-pentyl =
  1,1-dimethylpropyl   $CH_3CH_2C(CH_3)_2-$
methyldioxy = methylperoxy   $CH_3-OO-$
methyldithio = methyldisulfanyl   $CH_3-SS-$
methylene = methanediyl   $CH_2<$ or
  methylidene   $CH_2=$
methylenebis(oxy)   $-O-CH_2-O-$
methylenedioxy   $-CH_2-O-O-$
1-methylethenyl = propen-2-yl =
  isopropenyl   $CH_2=C(CH_3)-$
1-methylethylidene = isopropylidene =
  propan-2-ylidene   $(CH_3)_2C=$
1-methylpropyl = butan-2-yl = sec-butyl
  $CH_3CH_2CH(CH_3)-$
2-methylpropyl = isobutyl = 2-methyl-
  propan-1-yl   $CH_3CH(CH_3)CH_2-$
2-methylpropan-2-yl = *tert*-butyl =
  1,1-dimethylethyl   $CH_3C(CH_3)_2-$
methylsulfanyl = methylthio   $CH_3-S-$
*S*-methylsulfinimidoyl = methanesulfinimidoyl   $CH_3S(NH)-$
methylsulfinyl = methanesulfinyl
  $CH_3-SO-$
*S*-methylsulfonimidoyl = methanesulfon-
  imidoyl   $CH_3-S(O)(NH)-$
methylsulfonyl = methanesulfonyl
  $CH_3-SO-$
methylthio = methylsulfanyl
  (*not* methylsulfenyl or methylmercapto)
  $CH_3-S-$
(methylthio)oxy   $CH_3-S-O-$
(methylthio)sulfonyl = (methylsulfanyl)-
  sulfonyl   $CH_3-S-SO_2-$
methyltrisulfanyl = methyltrithio
  $CH_3-SSS-$
morpholino = morpholin-4-yl

$$O_{\ 1}\diagdown {}^{2\ \ 3}{}_{4}N-$$

naphthoyl = naphthalenecarbonyl =
  naphthylcarbonyl   $C_{10}H_7-CO-$
naphthyl = naphthalenyl   $C_{10}H_7-$
neopentyl = 2,2-dimethylpropyl =
  2,2-dimethylpropan-1-yl   $(CH_3)_3CCH_3-$

nitrilo = azanylidyne N≡ or azanetriyl
  –N< or azanylylidene –N=
nitro O$_2$N–
*aci*-nitro = hydroxyazinylidene =
  hydroxyazonoyl HO–N(O)=
nitroryl ON≡
nitroso ON–
nonyl = nonan-1-yl CH$_3$[CH$_2$]$_8$–
octanoyl CH$_3$[CH$_2$]$_6$–CO–
octyl = octan-1-yl CH$_3$[CH$_2$]$_7$–
oleoyl = octadec-9-enoyl
  CH$_3$[CH$_2$]$_7$CH=CH[CH$_2$]$_7$–CO–
oxalo = carboxycarbonyl = hydroxyoxoacetyl HO–CO–CO–
oxalyl = ethanedioyl –CO–CO–
oxamoyl = carbamoylcarbonyl =
  aminooxoacetyl H$_2$N–CO–CO–
oxido ⁻O–
oxo O
3-oxobutanoyl = 3-oxobutyryl
  = acetoacetyl CH$_3$–CO–CH$_2$–CO–
2-oxo-2-phenylethyl = phenacyl
  (*not* benzoylmethyl) C$_6$H$_5$–CO–CH$_2$–
2-oxopropyl = acetonyl = 2-oxopropan-1-yl
  CH$_3$–CO–CH$_2$–
2-oxopropylidene = acetonylidene =
  2-oxopropan-1-ylidene CH$_3$–CO–CH=
oxy –O–
pentafluorothio = pentafluoro-λ$^6$-sulfanyl
  F$_5$S–
pentanedioyl = glutaryl
  –CO–CH$_2$CH$_2$CH$_2$–CO–
pentanoyl CH$_3$CH$_2$CH$_2$CH$_2$–CO–
pent-2-enoyl CH$_3$CH$_2$CH=CH–CO–
pentyl = pentan-1-yl
  CH$_3$CH$_2$CH$_2$CH$_2$CH$_2$–
*tert*-pentyl = 2-methylbutan-2-yl =
  1,1-dimethylpropyl CH$_3$CH$_2$C(CH$_3$)$_2$–
perchloryl O$_3$Cl–
perfluorobutyl = nonafluorobutyl =
  nonafluorobutan-1-yl
  CF$_3$CF$_2$CF$_2$CF$_2$–
peroxy = dioxy –OO–
peroxycarboxy: *see* hydroperoxycarbonyl
phenacyl = 2-oxo-2-phenylethyl
  C$_6$H$_5$–CO–CH$_2$–
phenethyl = 2-phenylethyl
  (not benzylmethyl) C$_6$H$_5$–CH$_2$CH$_2$–
phenoxy = phenyloxy C$_6$H$_5$–O–
phenyl C$_6$H$_5$–
phenylamino = anilino C$_6$H$_5$–NH–
phenylazo = phenyldiazenyl C$_6$H$_5$–N=N–
phenylcarbonyl = benzoyl = benzene-
  carbonyl C$_6$H$_5$–CO–

(phenylcarbonyl)amino = benzamido =
  benzoylamino = benzenecarbonylamino
  C$_6$H$_5$–CO–NH–
(phenylcarbonyl)hydrazinyl = benzoylhydrazinyl = benzenecarbonyl-
  hydrazinyl C$_6$H$_5$–CO–NHNH–
phenylene –C$_6$H$_4$–
2-phenylethenyl = styryl = 2-phenylvinyl
  C$_6$H$_5$–CH=CH–
3-phenylprop-2-enoyl = cinnamoyl
  C$_6$H$_5$–CH=CH–CO–
3-phenylprop-2-en-1-yl = cinnamyl
  C$_6$H$_5$–CH=CHCH$_2$–
phenylsulfinyl = benzenesulfinyl
  C$_6$H$_5$–SO–
phenylsulfonyl = benzenesulfonyl
  C$_6$H$_5$–SO$_2$–
(phenylthio)amino = benzenesulfenamido
  C$_6$H$_5$–S–NH–
phosphanyl = phosphino H$_2$P–
λ$^5$-phosphanyl = phosphoranyl H$_4$P–
phosphanylidene = phosphinidene HP=
phosphinediyl = phosphanediyl HP<
phosphinico = hydroxyphosphoryl =
  hydroxyphosphinylidene HO–P(O)<
phosphinidene = phosphanylidene HP=
phosphinimidoyl = phosphinimyl
  H$_2$P(NH)–
phosphino = phosphanyl H$_2$P–
phosphinothioyl = dihydrophosphorothioyl
  H$_2$P(S)–
phosphinyl = phosphinoyl = dihydro-
  phosphoryl H$_2$P(O)–
phosphinylidene = phosphonoyl = hydrophosphoryl HP(O)= or HP(O)<
phosphinylidyne = phosphoryl P(O)≡ or
  –P(O)< or –P(O)=
phospho O$_2$P–
phosphonato (⁻O)$_2$P(O)–
phosphono = (HO)$_2$P(O)–
phosphonoyl = phosphinylidene = hydrophosphoryl HP(O)= or HP(O)<
phosphoranyl = λ$^5$-phosphanyl H$_4$P–
phosphorochloridonitridoyl Cl–P(≡N)–
phosphorodiamidothioyl = diamino-
  phosphinothioyl = (H$_2$N)$_2$P(S)–
phosphoryl = phosphinylidyne –P(O)<
  or P(O)≡ or –P(O)=
phthaloyl = o-phthaloyl = benzene-
  1,2-dicarbonyl

plumbyl  $H_3Pb-$

propanedioyl = malonyl  $-CO-CH_2-CO-$

propane-1,3-diyl = trimethylene
$-CH_2CH_2CH_2-$

propane-1,1,1-triyl  $CH_3CH_2C\overset{3\ 2\ 1}{\underset{}{\diagdown}}$

propano (only as a bridge)
$-CH_2CH_2CH_2-$

propanoyl = propionyl
$CH_3CH_2-CO-$

propan-2-yl = isopropyl = 1-methylethyl
$(CH_3)_2CH-$

propan-2-ylidene = 1-methylethylidene =
isopropylidene  $(CH_3)_2C=$

propan-1-yl-3-ylidene  $=CHCH_2CH_2-$

propanylidyne = propylidyne  $CH_3CH_2C\equiv$

propargyl = prop-2-ynyl  $HC\equiv CCH_2-$

prop-2-enoyl = acryloyl  $CH_2=CH-CO-$

propen-2-yl = 1-methylethenyl =
isopropenyl  $CH_2=C(CH_3)-$

prop-1-en-1-yl  $CH_3CH=CH-$

prop-2-en-1-yl = allyl  $CH_2=CHCH_2-$

prop-2-en-1-ylidene = allylidene
$CH_2=CHCH=$

propionamido = propanamido
$CH_3CH_2-CO-NH-$

propionyl = propanoyl  $CH_3CH_2-CO-$

propionyloxy = propanoyloxy
$CH_3CH_2-CO-O-$

propoxy = propyloxy = propan-1-yloxy
$CH_3CH_2CH_2-O-$

propyl = propan-1-yl  $CH_3CH_2CH_2-$

propylene = propane-1,2-diyl =
1-methylethylene = 1-methylethane-
1,2-diyl  $-CH_2CH(CH_3)-$

propylidyne = propanylidyne  $CH_3CH_2C\equiv$
or propane-1,1,1-triyl  $CH_3CH_2C\overset{3\ 2\ 1}{\underset{}{\diagdown}}$

or propan-1-yl-1-ylidene  $CH_3CH_2C\overset{3\ 2\ 1}{\underset{}{=}}$

prop-2-ynyl = propargyl  $HC\equiv CCH_2-$

salicyl = 2-hydroxybenzyl
$2\text{-}HO-C_6H_4-CH_2-$

salicyloyl = 2-hydroxybenzoyl
$2\text{-}HO-C_6H_4-CO-$

selanyl = hydroseleno  $HSe-$

selenothiocarboxy  $H\{S,Se\}C-$

semicarbazido = 2-carbamoylhydrazinyl =
(aminocarbonyl)hydrazinyl
$H_2N-CO-NHNH-$

semicarbazono = 2-carbamoylhydrazono =
(aminocarbonyl)hydrazo
$H_2N-CO-NHN=$

silanediyl  $H_2Si<$

silanediylidene  $=Si=$

silanetetrayl  $>Si<$

silanylylidene  $-HSi=$

siloxy = silyloxy = (necessary contractions of
silanyloxy)  $H_3Si-O-$

silyl  $H_3Si-$

silylene = silylidene  $H_2Si=$  or
silanediyl  $H_2Si<$

silylidene  $H_2Si=$

stannyl  $H_3Sn-$

stannylene = stannanediyl  $H_2Sn<$  or
stannylidene  $H_2Sn=$

stibyl  $H_2Sb-$

stibylene = stibinediyl = stibanediyl
$HSb<$  or  stibanylidene  $HSb=$

styryl = 2-phenylethenyl = 2-phenylvinyl
$C_6H_5-CH=CH-$

succinyl = butanedioyl
$-CO-CH_2CH_2-CO-$

sulfamoyl = aminosulfonyl = sulfuramidoyl
$H_2N-SO_2-$

sulfanyl = mercapto  $HS-$

sulfanylcarbonyl = mercaptocarbonyl
$HS-C(O)-$

sulfanylidene = thioxo  $S=$

sulfanyloxy = mercaptooxy  $HS-O-$

sulfeno = hydroxythio = hydroxysulfanyl
$HO-S-$

sulfido  $^-S-$

sulfino  $HO-SO-$

sulfinothioyl = thiosulfinyl = thiosulfino
$-S(S)-$

sulfinyl  $-SO-$

sulfo  $HO-SO_2-$

sulfonato  $^-O_3S-$

sulfonodithioyl = dithiosulfonyl  $-S(S)_2-$

sulfonothioyl = thiosulfonyl  $-S(O)(S)-$

sulfonyl  $-SO_2-$

sulfonylbis(oxy)  $-O-SO_2-O-$

sulfonyldioxy  $-SO_2-O-O-$

sulfuramidoyl = sulfamoyl = aminosulfonyl
$H_2N-SO_2-$

terephthaloyl = benzene-1,4-dicarbonyl

tetramethylene = butane-1,4-diyl
$-CH_2CH_2CH_2CH_2-$

thio = sulfanediyl (*not* sulfenyl)  $-S-$

thioacetyl = ethanethioyl  $CH_3-CS-$

thiocarbamoyl = carbamothioyl
$H_2N-C(S)-$

thiocarbonyl = carbonothioyl  $-CS-$

thiocarboxy  $H\{O,S\}C-$

thiocyanato  $NCS-$

thioformyl   HCS–
thiohydroperoxy   H{S, O}– or H{O, S}–
thiophosphinoyl = phosphinothioyl
   H₂P(S)–
thiosulfeno = disulfanyl = dithiohydroperoxy   HS–S–
thiosulfinyl = sulfinothioyl   –S(S)–
thiosulfo (unspecified)   HS₂O₂–
thiosulfonyl = sulfonothioyl   –S(O)(S)–
thioxo = sulfanylidene   S=
tolyl = methylphenyl   CH₃–C₆H₄–
tosyl = (4-methylphenyl)sulfonyl =
   4-methylbenzenesulfonyl
   CH₃–C₆H₄–SO₂–
triazano (only as a bridge)   –HNNHNH–
triaz[1]eno = azimino (only as a bridge)
   –HNN=N–

traz-l-en-1-yl   H₂NN=N–
trimethylene = propane-1,3-diyl
   –CH₂CH₂CH₂–
trithio = trisulfanediyl   –SSS–
trithiosulfo = mercaptosulfonodithioyl
   HS–S(S)₂–
trityl = triphenylmethyl = α,α-diphenyl-
   benzyl   (C₆H₅)₃C–
ureido = carbamoylamino = (aminocarbonyl)amino   H₂N–CO–NH–
ureylene = carbonyldiimino
   –NH–CO–NH–
vinyl = ethenyl   CH₂=CH–
vinylene = ethene-1,2-diyl   –CH=CH–
vinylidene = ethenylidene   CH₂=C=   or
   ethene-1,1-diyl   CH₂=C<

# Appendix B

## Common Endings

From most of these endings, generic names for compound classes can be formed by prefixing terms such as alkan(e), aren(e), and so on. For example, acyclic amides have the class name alkanamide.

-adiene: two double bonds
-al  –(C)HO
-amide  –(C)O–NH$_2$
-amidine = -imidamide  –(C)(NH)–NH$_2$
-amine  –NH$_2$
-ate: anion, salt, or ester of an "ic" acid
-carbaldehyde = -carboxaldehyde  –CHO
-carbo(dithioperoxo)thioic acid  –CS–SSH
-carbo(thioperoxoic) *SO*-acid  –CO–SOH
-carbo(thioperoxoic) *OS*-acid  –CO–OSH
-carbodithioic acid  –CS–SH
-carbohydrazonamide  –C(NNH$_2$)–NH$_2$
-carbohydrazonic acid  –C(NNH$_2$)–OH
-carbohydroxamic acid  –CO–NH–OH
-carbohydroximic acid  –C(N–OH)OH
-carbolactone  –CO–O–  (as part of a ring)
-carbonitrile  –C≡N
-carbonyl chloride  –CO–Cl
-carboperoxoic acid = -peroxycarboxylic
  acid  –CO–OOH
-carboselenaldehyde  –CHSe
-carbothialdehyde = -carbothioaldehyde
  –CHS
-carbothioamide  –CS–NH$_2$
-carbothioic *S*-acid  –CO–SH
-carbothioic *O*-acid  –CS–OH
-carboxamide  –CO–NH$_2$
-carboxamidine = -carboximidamide
  –C(NH)–NH$_2$
-carboximidic acid  –C(NH)–OH
-carboximidothioic acid  –C(NH)–SH
-carboxylate  –COO$^-$  or  –COO$^-$ M$^+$
  or  –COOR$^a$
-carboxylic acid  –COOH

-dicarboximide  –CO–NH–CO–
  (as part of a ring)
-dicarboxylic anhydride  (–CO)$_2$O
-dioic anhydride  [–(C)O]$_2$O
-dithioic acid  –(C)S–SH
-dithiosulfinic acid = -sulfinodithioic acid
  –S(S)–SH
-dithiosulfonic *S*-acid = -sulfonodithioic
  *S*-acid  –S(O)(S)–SH
-dithiosulfonic *O*-acid = -sulfonodithioic
  *O*-acid  –SS$_2$–OH
-diyl: two single bonds to different atoms
-ene: one double bond in acyclic and
  alicyclic parent hydrides
-hydrazonic acid  –(C)(NNH$_2$)–OH
-hydroxamic acid  –(C)O–NH–OH
-hydroximic acid  –(C)(N–OH)–OH
-ide: anion derived by loss of H$^+$ from a
  parent hydride
-imidamide = -amidine  –(C)(NH)–NH$_2$
-imide  –(C)O–NH–(C)O–
  (as part of a ring)
-imidic acid  –(C)(NH)–OH
-imine  =NH
-ite: anion, salt, or ester of an "ous" acid
-ium: cation usually derived by addition
  of H$^+$
-lactam  –CO–NH–  (as part of a ring)
-lactone  –CO–O–  (as part of a ring)
-nitrile  –(C)≡N
-oate  –(C)OO$^-$  or  –(C)OO$^-$ M$^+$ or
  –(C)OOR$^a$
-oic anhydride  [–(C)O]$_2$O
-oic acid  –(C)OOH

---

$^a$ M = a metal; R = an organic moiety.

-ol   –OH
-olactone   –(C)O–O–   (as part of a ring)
-olate   –O⁻   (loss of H⁺ from –OH)
-one   >(C)=O
-onium: mononuclear cation by addition of H⁺
-oyl chloride   –(C)O–Cl
-peroxoic acid   –(C)O–OOH
-peroxothioic *OO*-acid   –(C)S–OOH
-peroxy...oic acid   –(C)O–OOH
-peroxycarboxylic acid = -carboperoxoic acid   –CO–OOH
-selen...acid: *see* -sulf...acid analog
-selenal   –(C)HSe
-selenol   –SeH
-selenolate   –Se⁻   (loss of H⁺ from –SeH)
-sulfenic acid   –S–OH
-sulfenothioic acid   –S–SH
-sulfinic acid   –SO–OH
-sulfinimidic acid   –S(NH)–OH
-sulfinohydrazonic acid   –S(NNH₂)–OH
-sulfinohydroximic acid   –S(N–OH)–OH
-sulfonamide   –SO₂–NH₂
-sulfonate   –SO₂–O⁻ or –SO₂–O⁻ M⁺ or   –SO₂–OR*ᵃ*
-sulfonic acid   –SO₂–OH
-sulfonimidic acid   –S(O)(NH)–OH
-sulfono(dithioperoxoic) acid   –SO₂–SSH
-sulfono(thioperoxoic) *OS*-acid   –SO₂–O–SH
-sulfonodiimidic acid   –S(NH)₂–OH
-sulfonohydrazonic acid   –S(O)(NHNH₂)–OH
-sulfonohydrazonimidic acid   –S(NH)(NHNH₂)–OH
-sulfonohydroximic acid   –S(O)(N–OH)–OH
-sultam   –SO₂–NH–   (as part of a ring)

-sultone   –SO₂–O–   (as part of a ring)
-tellur...: *see* sulfur... analog
-tetrathiosulfonoperoxoic acid = -sulfono-(dithioperoxo)dithioic acid   –S(S)₂–SSH
-thioic *S*-acid   –(C)O–SH
-thioic *O*-acid   –(C)S–OH
-thiol   –SH
-thiolate   –S⁻   (loss of H⁺ from -SH)
-thiosulfinic *O*-acid = -sulfinothioic *O*-acid   –S(S)–OH
-thiosulfinic *S*-acid = -sulfinothioic *S*-acid   –SO–SH
-thiosulfonic *O*-acid = -sulfonothioic *O*-acid   –S(O)(S)–OH
-thiosulfonic *S*-acid = -sulfonothioic *S*-acid   –SO₂–SH
-thiosulfonimidic *S*-acid = -sulfonimido-thioic *S*-acid   –S(O)(NH)–SH
-thiosulfonimidic *O*-acid = sulfonimido-thioic *O*-acid   –S(S)(NH)–OH
-thiosulfonimidic *S*-acid = -sulfonimido-thioic *S*-acid   –S(O)(NH)–SH
-thio(thioperoxoic) *OS*-acid   –(C)S–OSH
-trithiosulfonic acid = -sulfonotrithioic acid   –S(S)₂–SH
-triyl: three single bonds to different atoms
-uide: anion derived by addition of H⁺ to a parent hydride
-yl: single bond or free radical by loss of H·
-ylidene: two single bonds to the same atom
-ylidyne: three single bonds to the same atom
-ylium: cation derived by loss of H⁻ from a parent hydride
-yne: a triple bond in an acyclic or alicyclic parent hydride

---

*ᵃ* M = a metal; R = an organic moiety.

# Appendix C

## Glossary

**Affix**  A syllable or name component added as a prefix, inserted as an infix, or appended as a suffix to the name of a parent hydride, parent compound, suffix, or prefix. For example, chloro, thio, and aza are prefixes in the names chlorocyclohexane, thiocarbonic acid, thiosulfonyl, and 9b-azaphenalene; peroxo, amido, and thio are infixes in the names hexaneperoxoic acid, phosphoramidic acid, and carbonothioyl; and -ol and -sulfonic acid are suffixes in the names propan-2-ol and benzenesulfonic acid.

**Bonding number**  The bonding number, $n$, of an atom of a parent hydride is the sum of the number of bonding equivalents, that is classicial valence bonds, of that atom to any adjacent atom. For example, the bonding number of the sulfur atom in $SH_4$ is four and that of the phosphorus atom in $PH_5$ is five.

**Characteristic group**  An atom or heteroatomic group attached to a parent hydride or a parent compound, but not derived from a parent hydride other than mononuclear nitrogen, chalcogen, or halogen parent hydrides, that falls into one of the following classes:

1. a single heteroatom, such as –Br (bromo), =S (thioxo), and ≡N (nitrilo);
2. an acyclic heteroatomic group consisting of a single heteroatom attached only to hydrogen atoms, as –OH (hydroxy) and $-NH_2$ (amino), or to one or more heteroatoms that differ from it, as –ClO (chlorosyl), $-NO_2$ (nitro), and $-BrO_3$ (perbromo), which may themselves be attached to hydrogen atoms or to still other heteroatoms, as in $-P(O)(OH)_2$ (phosphono) and $-SO_2-OOH$ (hydroperoxysulfonyl);
3. an acyclic heteroatomic group consisting of a heteroatom or a heteroatomic group-attached to a single monovalent carbon atom, such as –CHO (formyl), $-CO-NH_2$ (carbamoyl), and –CN (cyano), or that contains an isolated carbon atom, such as –NCO (isocyanato) and –NC (isocyano).
4. an acyclic heteroatomic group consisting of like heteroatoms and any attached hydrogen atoms that may be described by functional class nomenclature (see chapter 3) and are not formed by removal of hydrogen from a parent hydride, such as –OOH(hydroperoxy), –OO– (peroxy or dioxy), $=N_2$ (diazo), $-N_3$ (azido), and $-S_3H$ (trithio).

Historically, the term characteristic group was introduced during the codification of hydrocarbon nomenclature as an alternative to the term functional group because agreement on the meaning of the latter term could not be achieved. Today, heteroatomic groups, such as $H_2NNH-$ (hydrazinyl) and $C_5H_{10}N-$ (piperidin-1-yl), with names derived from heteroacyclic or heterocyclic parent hydrides are not considered as characteristic groups. Likewise, atomic groupings that form part of a cyclic structure derived from a cyclic parent hydride are not viewed as characteristic groups. Curiously, unsaturation between two carbon atoms in an acyclic parent hydride is considered by IUPAC as a special kind of functionality.

**Class (generic) name**   A name applicable to all members of a specific structural group, such as alkane, amine, and acyl halide. Some class names serve as the name for the first member of the series, such as phosphane and silane.

**Descriptor**   A symbol or series of symbols or words that indicates a modification or refinement of the structure described by the name that follows. Three kinds of descriptors very important in organic chemical nomenclature are as follows: structural descriptors, such as *tert-*, *as-*, and *retro-*; stereochemical descriptors, as *cis-*, *trans-*, *E/Z*, and *R/S*; and isotopic descriptors, such as *d* and $^{15}N$ or $^{15}N$.

**Detachable prefixes**   Substituent prefixes that do not modify the structure of a parent hydride or parent compound. These prefixes are cited in alphabetical order in the name in front of any nondetachable prefixes (see below). In the volume indexes to *Chemical Abstracts*, they appear after the parent hydride or parent compound name (heading parent) following the inversion comma (see the Introduction to this book). For example, "chloro" in the name chloromethane is a detachable prefix; its name in a *Chemical Abstracts* volume index is Methane, chloro-.

**Functional class name**   A name in which the chosen principal characteristic group is expressed as a class term written as a separate word following the name of a parent structure or a name derived from the name of a parent structure, for example, phosphane oxide, methyl chloride, ethyl acetate, and acetyl chloride.

**Functional derivative**   A modification of an acid, an acid related group or a carbonyl characteristic group expressed by a separate word. These include anhydrides, esters, hydrazides, salts, oximes, hydrazones, semicarbazones, azines, and acetals.

**Functional parent compound**   A mononuclear or polynuclear structure whose name expresses or implies one or more characteristic groups. The compound must have one or more hydrogen atoms attached to at least one of its skeletal atoms or to at least one of its characteristic groups, or has one or more characteristic groups that can form at least one kind of functional derivative, such as ester, salt, and oxime. Examples of functional parent compounds are guanidine, acetone, carbonic acid, acetic acid, phosphoramidic dichloride, sulfamide, cholesterol, and urea.

**Heterogeneous parent hydride**   A parent hydride having different kinds of skeletal atoms, such as disiloxane, 1*H*-indole, 3,6,9,12-tetraoxatetradecane, bicyclo[3.3.1]tetrasiloxane and 9b-boraphenalene.

**Homogeneous parent hydride**   A parent hydride in which all skeletal atoms are of the same kind, such as pentane, diazene, cyclopentagermane or pentagerminane, and anthracene.

**Infix**   A letter or syllable or word fragment that is inserted into the name of a parent hydride, suffix, or prefix, for example, "thio" in the names carbonothioic acid and carbonothioyl, and "or" and "in" in names such as phosphorane and phosphinane.

**Locants**   Symbols that specify positions of various features in a structure or partial structure that are not implied by its name. Locants may be arabic numbers, italicized atomic symbols, Greek or Roman letters, or italicized words or their abbreviations.

**Nondetachable prefixes**   Prefixes that denote a modification of the skeletal structure of a parent hydride; they are always placed immediately in front of the name of the parent hydride. They are not separated from the name of the parent hydride and are generally not subject to inversion in an alphabetical index such as a *Chemical Abstracts* volume index. There are two main groups of nondetachable prefixes; those that signify addition, removal, or rearrangement, such as nor, homo, seco, iso, *sec*; and those that indicate replacement of a skeletal atom of a parent hydride by another atom, such as aza or thia. Hydro and dehydro prefixes have been considered both as nondetachable and detachable.

**Parent compound** A structure whose name includes or implies a suffix and/or nondetachable prefixes (see also functional parent compound). Examples are benzoic acid, cyclohexane-carboxylic acid, and thiourea.

**Parent hydride** A system of one or more atoms to which *only* hydrogen atoms are attached and whose name defines an unbranched acyclic, cyclic, or acyclic/cyclic structure that is acceptable for substitution by other atoms or groups. Examples are methane, phosphine, sulfane, pentane, pyridine, phenanthrene, bicyclo[2.2.2]octasilane, biphenyl, toluene, disiloxane, and 2,4,6,8-tetraoxanonane.

**Prefix** A name component appearing in front of the name of a parent hydride, parent compound, prefix or suffix, such as methyl in the name methylcyclohexane, di in the name dimethyl, and thio in the name thiocarbonyl. Prefixes indicating a numerical values such as di- and bis- are called multiplying prefixes.

**Principal characteristic group** The characteristic group cited as a suffix to the name of a parent hydride, included in the name of a functional parent compound, or expressed as a class name in functional class nomenclature.

**Radical** A molecular group, such as $CH_3\bullet$, $NH_2\bullet$, and $Cl\bullet$, which has one or more unpaired electrons. In the past this term was used to designate a substituent group in contrast to "free radical". The use of this term to designate a substituent group is no longer recommended.

**Radicofunctional name** A functional class name in which the name derived from a parent structure is that of a substituent group (formerly called a radical), such as ethyl methyl ketone and acetyl chloride.

**Senior, seniority** Terms referring to priorities given in a prescribed hierarchical order. A senior feature is usually preferred in the formation of a name.

**Skeletal atom** An atom in the framework of the structure of a parent hydride or parent compound, or substitutent group derived from a parent hydride or parent compound.

**Substituent** An atom or group that substitutes (replaces, exchanges) for one or more hydrogen atoms of a parent hydride or functional parent compound and is expressed in a substitutive name by either a prefix or a suffix. Substituents may be characteristic groups or groups derived from parent hydrides, such as $-CH_3$, $-NHNH_2$, $-P(CH_3)_2$. Amino and carboxylic acid are both substituents in the name 4-aminopyridine-2-carboxylic acid.

**Suffix** A name component that is cited following the name of a parent hydride; sulfonic acid is the suffix in the name methanesulfonic acid. Suffixes are not the same as endings, such as -ene in the name cyclohexene and -ole in the name pyrrole.

# List of Tables and Figures

## Tables

## Figures

# Index